The Physics of Submicron Lithography

MICRODEVICES
Physics and Fabrication Technologies

Series Editors: Ivor Brodie and Julius J. Muray†

SRI International
Menlo Park, California

ELECTRON AND ION OPTICS
Miklos Szilagyi

GaAs DEVICES AND CIRCUITS
Michael Shur

ORIENTED CRYSTALLIZATION ON AMORPHOUS SUBSTRATES
E. I. Givargizov

THE PHYSICS OF MICROFABRICATION
Second Edition
Ivor Brodie and Julius J. Muray

PHYSICS OF SUBMICRON DEVICES
David Ferry and R. O. Grondin

THE PHYSICS OF SUBMICRON LITHOGRAPHY
Kamil A. Valiev

SEMICONDUCTOR LITHOGRAPHY
Principles, Practices, and Materials
Wayne M. Moreau

SEMICONDUCTOR PHYSICAL ELECTRONICS
Sheng S. Li

† *Deceased.*

A Continuation Order Plan is available for this series. A continuation order will bring delivery of each new volume immediately upon publication. Volumes are billed only upon actual shipment. For further information please contact the publisher.

The Physics of Submicron Lithography

Kamil A. Valiev
Academy of Sciences of the USSR
Moscow, USSR

Plenum Press • New York and London

Library of Congress Cataloging-in-Publication Data

Valiev, Kamil′ Akhmetovich.
The physics of submicron lithography / Kamil A. Valiev.
p. cm. -- (Microdevices)
Includes bibliographical references and index.
ISBN 0-306-43578-0
1. Lithography, Electron beam. 2. X-ray lithography. 3. Ion beam lithography. 4. Physics. I. Title. II. Series.
TK7874.V336 1992
621.381'531--dc20 91-41122
CIP

ISBN 0-306-43578-0

A Division of Plenum Publishing Corporation
233 Spring Street, New York, N.Y. 10013

Printed in the United States of America

Foreword

This book is devoted to the physics of electron-beam, ion-beam, optical, and x-ray lithography. The need for this book results from the following considerations.

The astonishing achievements in microelectronics are in large part connected with successfully applying the relatively new technology of processing (changing the properties of) a material into a device whose component dimensions are submicron, called photolithography. In this method the device is imaged as a pattern on a metal film that has been deposited onto a transparent substrate and by means of a broad stream of light is transferred to a semiconductor wafer within which the physical structure of the devices and the integrated circuit connections are formed layer by layer. The smallest dimensions of the device components are limited by the diffraction of the light when the pattern is transferred and are approximately the same as the wavelength of the light. Photolithography by light having a wavelength of $\lambda \approx 0.4\ \mu$m has made it possible to serially produce integrated circuits having devices whose minimal size is 2–3 μm in the pattern and having 10^4–10^5 transistors per circuit.

A new stage in the development of microelectronics is connected with converting to devices of submicron size. Processing materials into devices whose components are submicron in size is possible with electron, ion, and x-ray lithographic methods that respectively use beams of corpuscular radiation having a cross section of less than a micron and x-ray radiation in the 1–10 nm wavelengths and ultraviolet radiation in the λ = 100–300 nm wavelengths. That which today is the ultimate in materials diagnostics, electron and ion beams of submicron cross section used to ascertain the composition and other properties of submicron amounts of a material, becomes the technology for mass producing devices whose elements are submicron in size. The creation of submicron technology will unconditionally be the most important achievement in science and technology for the last quarter of the 20th century.

Converting to submicron lithography and submicron device sizes demands that an expert in the field of microelectronics be "scientifically reequipped." He must study the physics of electron and ion beams and the physics of x-ray and ultraviolet radiation. The physics of each effect must include the topics of forming the beams and their interactions with matter. It is precisely these considerations that have guided the authors in selecting the material for this book.

A great deal of literature has been devoted to the physics of corpuscular beams and x-ray radiation. As a rule, the topics of forming beams of particles and high energy photons are studied in the literature from the viewpoint of accelerator techniques, high energy physics, and nuclear energy. Material diagnostics is another area of application for corpuscular beam physics and the physics of x-ray radiation, which pertains to methods of electron beam and ion beam microscopy. Naturally, in working on this book we made extensive use of the literature in the fields indicated.

Our book is intended for specialists who are developing or using the methods of processing materials into a device whose components are submicron in size. It will be useful to experienced workers striving to attain the level of requirements imposed by the new "submicron" stage in the development of microelectronics. Finally, it will be essential for undergraduate and graduate students specializing in microelectronics.

Contents

Introduction

The lithographic systems used in microelectronics are designed to perform two major operations: 1) create the topological pattern for an IC layer by printing the different pattern elements serially in time, and 2) simultaneously transfer a mask pattern in parallel to a sensitive layer deposited onto the wafer being processed. Creating the topological pattern is most often done to prepare the mask used in the parallel transfer operation; however, in recent years a serial pattern is more often created right on the semiconductor wafer for the purpose of fabricating prototype ICs or in the technology of custom circuits. Parallel pattern transfer can be used to set the number of copies of the masks and to form resist masks on semiconductor wafers.

Optical, x-ray, electron beam, and ion beam systems designed to perform the operations indicated above are shown in Fig. I-1. The nomenclature of the relatively "old" methods widely used in industry is shown as solid rectangles; systems in development are shown as dashed rectangles. Optical lithography systems have a 25-year history; nonetheless, they are currently undergoing rapid development in connection with the use of ultraviolet light in a vacuum (e.g., a laser wavelength of $\lambda = 248$ nm). For this reason, new developments in optical lithography systems are primarily associated with creating laser image generators and laser steppers.

X-ray lithographic sets are a variation of optical sets for aligning and exposing with a gap. They have been developed to use soft x-ray radiation in the $\lambda = 0.5$–4 nm wavelengths. Decreasing the radiation wavelength to less than $\sim$1 nm reduces the diffraction fuzziness at the edge of the mask image to $\sim$0.1 nm. Going into the x-ray region of electromagnetic radiation (in comparison with the optical region) significantly changes the radiation sources (synchrotrons or other x-ray sources) and the nature of the mask constructed on the difference between the absorption coefficients of the mask substrate material and the material that forms the mask pattern. The lack of materials that are transparent in the x-ray region denies us the opportunity to construct x-ray lens optical systems in the same sense of the word as it is used in the optical region. Therefore, only a shadow printing method with a controllable gap of the type used in align-and-expose optical sets can be used to transfer the mask pattern. On the other hand, the gap deprives us of the control needed to obtain the necessary resolution over the entire area of a 150-mm-diameter wafer; therefore, the wafer is exposed by the stepwise align and expose method used in optical systems. Thus, an x-ray lithographic set is a hybrid of an align and expose set and a stepper. Inherent to the method of x-ray lithography, a diffraction resolution of 0.1 μm requires an alignment system that will guarantee an accuracy of roughly 0.01 μm. X-ray lithographic systems are making gradual progress toward the parameters that would make them better than optical steppers. It is to be hoped that

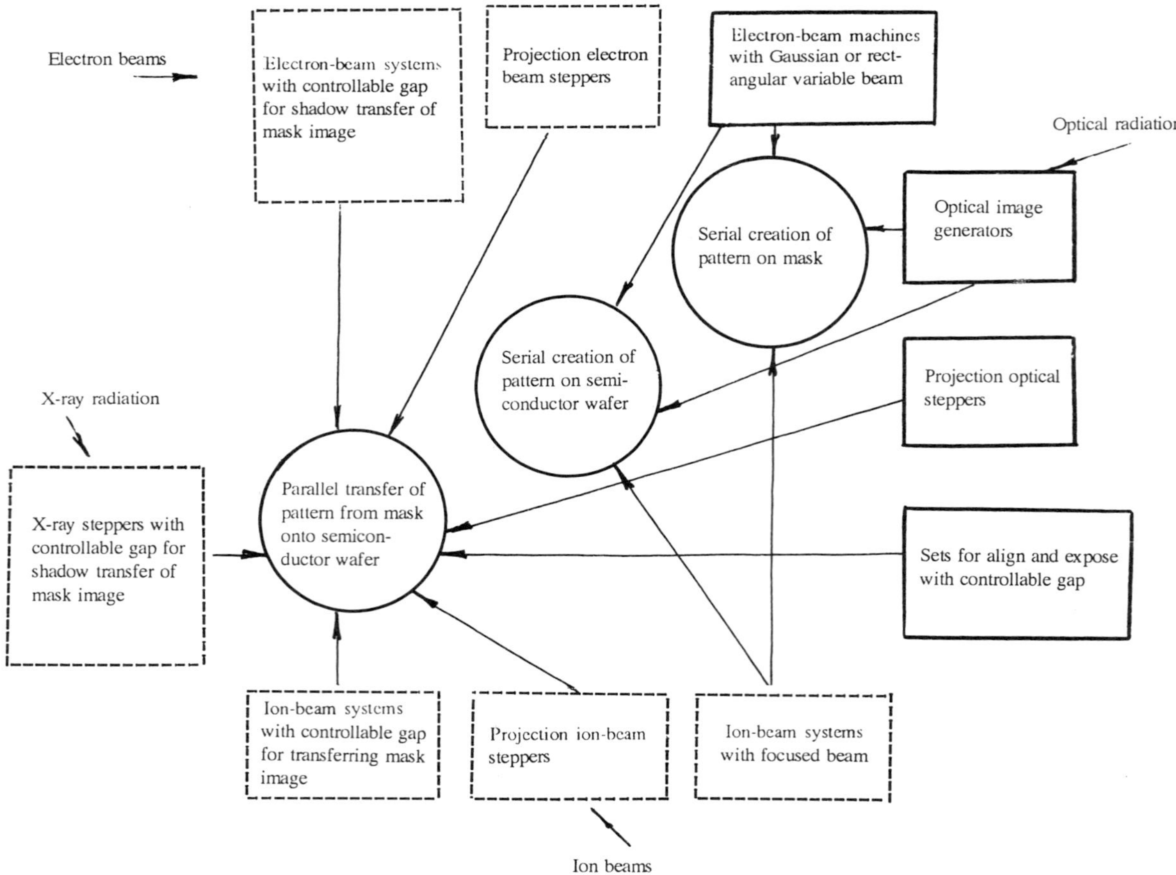

FIGURE I-1. Fundamental types of lithography machines that use electron and ion beams, visible and x-ray radiation and the means for employing them in microelectronics.

x-ray steppers will find wide application once they have reached this level of parameters. The construction of special lithographic sources of synchrotron radiation (COSI-compact synchrotron sources) has begun in several countries. The problem of creating thin but durable membranes to be used as the substrates for x-ray masks must also be solved.

In the last decade electron-beam systems having variable size beams (most often rectangular) have found widespread application in industry. The dimensions of an electron-beam die are varied in increments (e.g., 0.1 μm) over the range of from 0.2 to 5 μm. These electron-beam sets have been widely used to create masks for fabricating VLSI as well as to expose an electron resist right on the semiconductor wafer. In the latter case the electron-beam set must have a system for seeking and detecting the alignment marks on the semiconductor wafer.

In investigating the processes of electron-beam lithography the beam diameter may reach magnitudes of about 0.01 μm.

Ion-beam systems having a thin focused probe are most often thought of as systems with nanometer (0.1–0.01μm) resolution.

The beam cross section is characterized by a Gaussian-distributed ion current; in systems having micron resolution it is not difficult to form a rectangular cross section in the beam. Light ions (helium ions, protons) are used when exposing polymer ion-beam resists. By using other ions, new (in comparison with electron beams) processing methods such as doping and forming thin layers of a new chemical compound can be obtained. The formation of point defects in insulators during ion bombardment allows them to be thought of as inorganic ion resists when the presence of defects produces a change in their dissolution rate in specific solvents. These processing methods however, such as doping and forming new chemical compounds (here we have in mind reactions between the atoms in the composition of the substrate and the implanted ions) require a dosage on the order of 10^{18} ions/cm^2, which reduces the productivity of serially acting systems to unacceptably low levels. These processing methods may be acceptable in practice in systems with parallel processing in which wafer processing is done by a wide ion beam through stencil masks.

The use of liquid metal sources of ion current whose brightness is close to that of electron sources having field emission is a characteristic of ion-beam sources. It must also be noted that electrostatic lenses must be used exclusively in ion-beam machines, because their focal length is independent of the particle mass, whereas it is proportional to the particle mass for magnetic lenses (at a fixed particles energy).

In recent years a great deal of attention has been given to developing electron-beam systems that use a wide beam of electrons and are designed for transferring the mask pattern. The familiar projection-type electron beam systems are analogs of optical projection steppers. The masks in these systems are foils with apertures constructed with a 10$\times$ magnification. Projection-type electron-beam systems have not been widely accepted. A reason for this may be aberrations of the electron optics that introduce distortions into the reduced image, as well as the difficulty of fabricating and using metal foil stencil masks. A second type of electron-beam system is analogous to shadow printing systems having a controlled gap. Stencil masks in the 1$\times$ scale are used in these systems. Diffraction effects in shadow printing systems have no practical significance. Distortions in the mask pattern are determined by the initial angular divergence in the wide beam of electrons that illuminate the stencil mask. Finally, developments of electron-beam sets for transferring a mask image in the 1$\times$ scale, in which the mask serves as the photocathode are familiar: illumination by ultraviolet radiation causes photoelectrons to be emitted in the transparent portion of a mask from a CsI layer covering the mask.

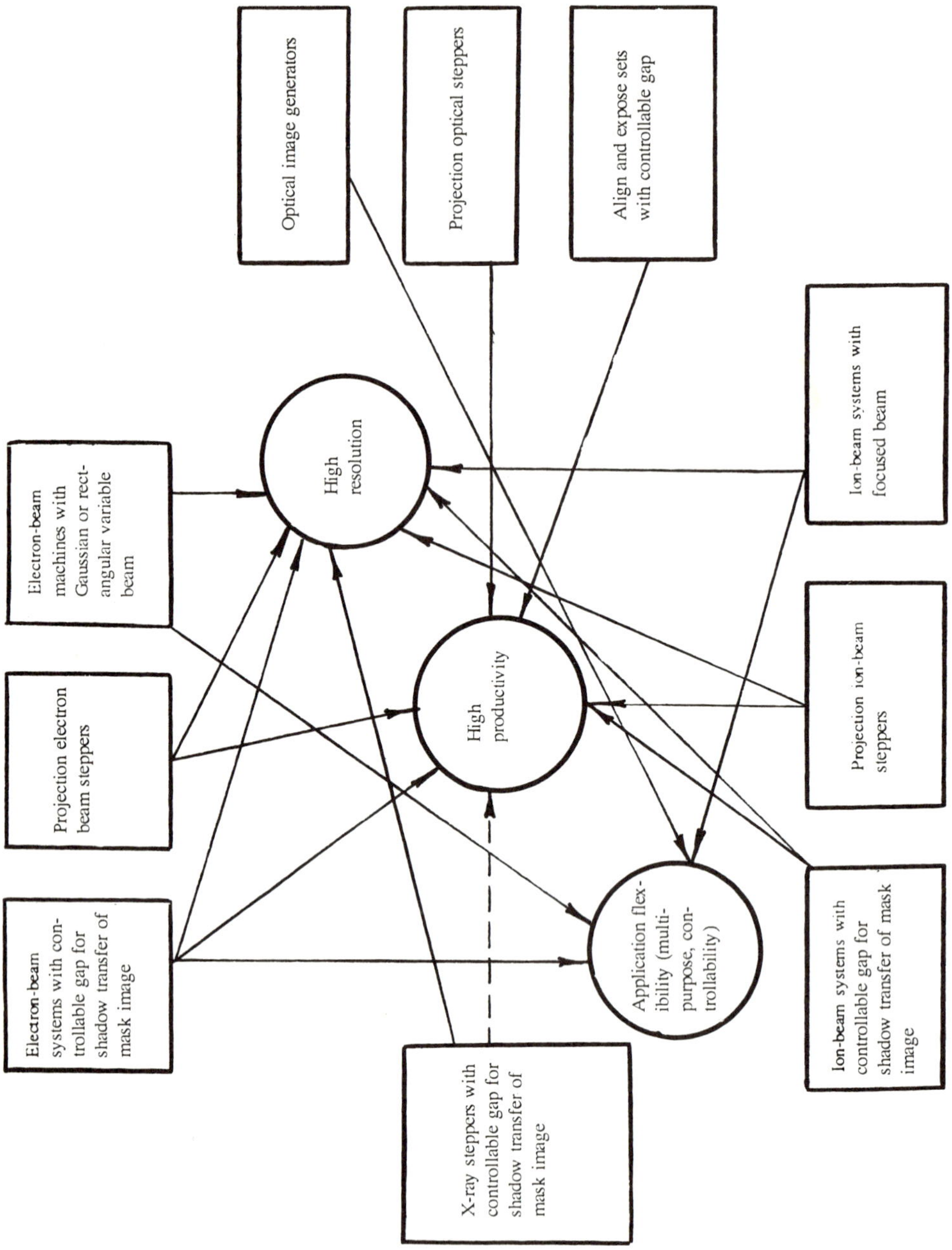

FIGURE I-2. Classification of lithography systems according to the level of resolving power achievable, productivity, and flexibility of application. Only a few types of systems have all of the necessary qualities.

A longitudinal magnetic field focuses accelerated photoelectrons onto the surface of a semiconductor wafer. To date, not one of the different types of electron-beam sets that use a wide electron beam has come beyond the development stage. The same is true of ion-beam projection sets.

Attention must now be turned to the scale of reduction with which the systems that transfer the mask pattern in parallel operate. Systems with 10× and 5× reduction must have complicated, expensive lens systems; but, in return, the mask fabricating machines that go along with them are relatively simple. Conversely, machines for fabricating masks in the 1× scale must have stringent parameters, and systems for making a direct shadow mask image transfer in the 1× scale are much simpler in comparison with projection systems.

Lithographic systems can be characterized by three groups of parameters: 1) geometrical resolution (in the tenths of a micron); 2) application flexibility (convenience of creating different patterns, ease of making an accurate alignment, etc., and 3) productivity. The classification of lithographic systems according to the indicated groups of parameters is shown in Fig. I-2. Not every image generator has high productivity, because they are serial acting machines. High resolution and application flexibility are characteristic of electron-beam image generators. High productivity is inherent to all systems that use parallel image transfer; however, some of these do not have application flexibility or high resolution.

In the operation of each lithographic system we must deal with three "subsystems": 1) the radiation source; 2) the optical system that shapes the radiation from the source into a stream having the required parameters, and 3) the process by which the radiation stream interacts with the target. To properly understand how the system operates, each "subsystem" must be analyzed. We will first examine electron-beam and ion-beam systems. Among these systems the most important in a practical sense are the image generators used to fabricate the masks and expose the semiconductor wafers. Their designation for fabricating masks compels us to examine them prior to optical or x-ray lithographic systems. The latter serve primarily to transfer the mask image onto a semiconductor wafer.

The task of this book—to describe radiation sources and optical systems for shaping a radiation stream and the processes by which a radiation stream interacts with a resist layer or other target—has not in every case been fulfilled in the necessary measure. For example, lasers as a source of ultraviolet radiation have not been examined in detail other than to describe the properties of laser radiation. The reason for this is the limited size of the book as well as the impossibility of providing a brief description of laser processes that would be comprehensive enough to establish the connection between the laser processes and the properties of their radiation.

It must also be noted that optical, electron, and ion systems having a focused beam of light or particles (a probe) are widely used for local studies into the properties of materials. For this reason the information we will present is contained in part in the literature devoted to probe methods of material diagnostics. We will also point out along the way the possibilities of using the information presented in probe diagnostic methods.

1

Forming Electron Beams of Submicron Cross Section

1-1. THERMIONIC EMISSION. THE RICHARDSON–DUSHMAN EQUATION FOR THE THERMIONIC CURRENT

In this chapter we will examine the issues in obtaining bright sources of electrons and forming electron beams of submicron cross section. Thermionic emission and field emission are the usual methods of obtaining electron beams having a brightness on the order of 10^6 A/cm^2·steradian. The first sections of this chapter discuss the physics of these processes (Sections 1–5). Subsequent sections discuss electron optics as a method of forming a beam having a submicron cross section; in this presentation we will in part follow the sources.[1–8]

A thermionic current is emitted from the surface of a heated solid by "evaporating" that fraction of the metal's conduction electrons whose kinetic energy is sufficient to do the so-called work function, φ.[9] When this happens, we will understand the kinetic energy to be the energy of motion perpendicular to the surface of the solid, because as an electron passes through the surface the braking forces responsible for the work function are normal to the surface and do not affect the velocity component of the electron that lies in the plane of the surface.

A current of density $j_x = n_0 v_x e$ from within the metal impinges on its surface, where n_0 is the electron density, 10^{22}–10^{23} cm^{-3}; v_x is the velocity of the electrons, about 10^8 cm/sec for electrons in metal and whose energy is near the Fermi energy E_{F}, and e is the electron charge 1.6×10^{-19}. Therefore, the total electron current impinging on the surface of the metal from within is an astronomical $j_x \approx 10^{12}$ A/cm^2. However, only a small fraction of these electrons escape from the metal. This is because the work function is 2–5 eV; the number of electrons having this much energy in excess of the Fermi energy is negligibly small.

We will go about calculating the thermionic current, beginning from the statements presented. We choose the x axis to be normal to the metal surface and place the y and

z axes on the plane of the surface. The density of electrons having momenta in the interval p, $p + dp$ is

$$dN = 2\frac{d^3p}{h^3}f(p), \qquad f(p) = \left(1 + \exp\frac{E - E_F}{kT}\right)^{-1} \tag{1-1}$$

Then for the density of the thermionic current we will have

$$j_x = \frac{2e}{h^3}\int_{-\infty}^{+\infty} dp_y \int_{-\infty}^{+\infty} dp_z \int_{v_{x\min}}^{\infty} v_x f(p)\, dp_x \tag{1-2}$$

The energy diagram for the conduction electrons in the metal is shown in Fig. 1-1. Considering the fact that

$$E(p) = (p_x^2 + p_y^2 + p_z^2)/2m, \qquad v_x = \partial E/\partial p_x \tag{1-3}$$

we rewrite Eq. (1-2) as

$$j_x = \frac{2e}{h^3}\int_{-\infty}^{+\infty} dp_y \int_{-\infty}^{+\infty} dp_z \int_{E_{\min}}^{+\infty} f(E)\, dE \tag{1-4}$$

We find the minimal electron energy that will guarantee electron emission using the condition

$$p_x^2/2m \geq E_F + \varphi \tag{1-5}$$

which yields

$$E_{\min} = E_F + \varphi + (p_y^2 + p_z^2)/2m \tag{1-6}$$

Now we can perform the integration without difficulty:

$$I = \int_{E_{\min}}^{\infty} \frac{dE}{1 + \exp\left[\dfrac{E - E_F}{kT}\right]} = kT \ln\left[1 + \exp\left(-\frac{E_{\min} - E_F}{kT}\right)\right] \tag{1-7}$$

Because the inequality

$$\frac{E_{\min} - E_F}{kT} = \frac{\varphi + (p_y^2 + p_z^2)/2m}{kT} \gg 1 \tag{1-8}$$

is obvious, we can set

$$\ln\left[1 + \exp\left(-\frac{E_{\min} - E_F}{kT}\right)\right] \approx \exp\left(-\frac{\varphi}{kT} - \frac{p_y^2 + p_z^2}{2mkT}\right) \tag{1-9}$$

and then the integration with respect to dp_y and dp_z is easily done:

$$\begin{aligned} j_x &= 2ekT/h^3 \exp(-\varphi/kT) \int_{-\infty}^{+\infty} \exp(-p_y^2/2mkT)\, dp_y \int_{-\infty}^{+\infty} \exp(-p_z^2/2mkT)\, dp_z \\ &= (4\pi mek^2/h^3)T^2 \exp(-\varphi/kT) \\ &= A_0 T^2 \exp(-\varphi/kT) \end{aligned} \tag{1-10}$$

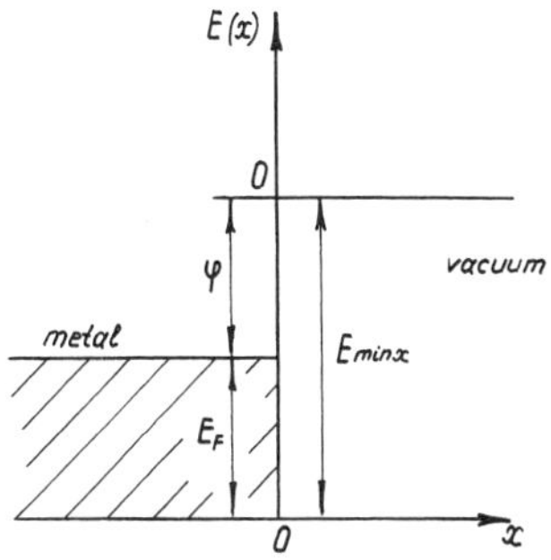

FIGURE 1-1. Energy diagram for the electrons being emitted from a metal in a vacuum; φ is the work done in escaping the metal; E_F is the Fermi energy; $E_{\min\,x}$ is the minimal energy of electron motion in the x direction that will ensure its being emitted.

TABLE 1-1

Metal	A_0, A/cm^2 · grad^2	φ, eV
W	75	4.5
Ta	55	4.2
Ni	30	4.6
Ag	–	4.8
Cs	160	1.8
Pt	32	5.3
Ba (on W)	1.5	1.56
Cs (on W)	3.2	1.36
Cr	48	4.60
LaB_6	40	2.4–2.7

The formula obtained for a thermionic current has been named the Richardson–Dushman equation. A theoretical value of A_0 in Eq. (1-10) is 120 A/cm^2 · grad^2. Experimental values for A_0 and the work required to escape the metal are given in Table 1-1.

The field-emission and thermionic-emission currents (see below) are exponential functions of the cathode work function. Therefore, the reproducibility and stability of the emission current are in large part determined by the reproducibility and stability of the cathode work function. However, holding the cathode work function to a constant value is a complex technological problem. Changes in the work function are, as a rule, associated with changes in the chemical composition of the monatomic layer at the emitter surface. For example, the work required to escape LaB_n ($n = 4, 6, 9$) varies from 2.52 to 3.35 eV; i.e., it increases as the boron content increases.[11] Hence it follows that during cathode operation any processes that lead to enriching the surface monolayer with boron will increase the work function and reduce the thermionic current. Similar effects are seen in cathodes having a complex composition. For example, separating molybdenum from the surface of a rhenium cathode with 1% Mo leads to a reduction in the work function.[12] The work function changes during plastic and elastic deformation of monocrystalline metals, as was established in experiments with monocrystalline copper.[13] Finally, data are available on the temperature dependence of the work function.[14]

The thinnest possible electron beam (probe) having a high current density is required for electron lithography in electron-beam sets. The minimal cross section of a beam in the focal plane of the set is, from the point of view of electron optics, the image of the hot cathode. Hence we have the requirement that the area of the hot cathode be as small as possible, but provide the largest possible current. The most common hot cathodes are those made from tungsten and lanthanum hexaboride. They provide thermionic emission currents of 1–10 and 10–100 A/cm^2, respectively.

Hot cathodes of arbitrary shape can be obtained by electrophoretically depositing LaB_6 on the surface of refractory metals (rhenium, tantalum). The thickness of the coating is about 150 μm. After deposition the coating is annealed at a temperature of 1650°C and a pressure of 10^{-8} torr. The refractory metals actively interact with the boron that diffuses from the LaB_6 coating by forming compounds that deteriorate. The formation of these compounds prevents the preliminary carburization (carbonization) of the metal surface.(15) Monocrystalline hot cathodes fabricated from LaB_6 are characterized by greater stability in the emission current than are cathodes made of LaB_6 and a ceramic.(16)

1-2. THE ENERGY SPECTRUM OF THERMAL ELECTRONS

The focusing properties of an electron optical system depend on the energy of the electrons. Prior to passing through the electron lenses the electrons are accelerated to an energy on the order of 10^4 eV. The intrinsic energy of the electrons with which they were emitted from the cathode is added to this energy. Because, as we have seen, emission is a random process, the primary energy of the electrons will be a random variable which can be described by some (e.g., Gaussian) distribution. The magnitude of the total mean-squared scatter in the energy of the electrons causes so-called chromatic aberrations in an electron optic system.

We can point to several causes of scatter in the energy of the electrons in electron-beam systems.(17,18)

If thermionic emission is thought of as the process of evaporating electrons off of a solid, the electron gas above the solid will be in equilibrium with, and at the same temperature as the solid. The energy distribution of the electrons will be Maxwellian and have a mean-squared scatter of $\sqrt{\overline{\Delta E}^2} = \sqrt{\bar{E^2} - (\bar{E})^2} = \sqrt{3/2}\,kT \approx 0.1$–$0.2$ eV. The bar over a quantity, as always, means average value.

A second source of nonuniformity in the energy of the electrons is the roughness of the hot cathode surface. These nonuniformities cause the cathode work function to vary from point to point: $\varphi = \varphi(y, z)$. According to experiments conducted on monocrystals, each crystallographic plane has its own work function. The surface of a polycrystalline specimen (hot cathode) is itself characterized by a random shift in the crystallographic planes that belong to different crystal grains (crystallites). In addition to this, the work function depends on the purity of the surface—on the presence of adsorbed atoms, etc. The nonuniformity in the work function creates subsurface electric fields $E_s \sim \Delta\varphi/\Delta l$ that either brake or accelerate the thermionic electron current.

A third important source of nonmonochromaticity in the electrons are fluctuations in the accelerating voltages: $U = U_0 + \delta U(t)$. Stabilizing an accelerating voltage of 1–3×10^4 eV to within 10^{-4} provides a scatter of 1–3 eV in the electrons' energy.

A fourth source of nonuniformity in the energy of the electrons is that the emitting surface of a hot cathode being heated by an electric current (a filament cathode) is not equipotential. For example, a tungsten filament cathode can operate at temperatures of 2500–2700°K. At these temperatures the specific resistance of the cathode material increases and the voltage drop across the resistance makes the emitting filament substantially nonequipotential. Therefore, the accelerating voltage—the potential difference between the cathode and anode—has a distribution that causes scatter in the energy of the electrons. The function for this distribution will depend on the cathode geometry, the material from which it is made, and its temperature; all of which determine its resistance. The magnitude of the respective energy scatter can easily be estimated by estimating the voltage drop across the emitting portion of a wire filament cathode: $\Delta U \approx \rho(T(i_{\text{h}}))(\Delta l/\pi r^2)i_{\text{h}}$. Here, $\rho(T(i_{\text{h}}))$ is the electrical resistivity of the cathode material at a given temperature T (and heating current i_{h}), Δl is the length of the emitting portion of the cathode, and πr^2 is the cross-sectional area of the cathode. V-shaped tungsten cathodes whose emitting surface is the apex of the V-shaped bend are used in electron-beam lithography equipment. Apparently, the voltage drop and the nonequipotentiality of the emitting surface can be neglected in this case, owing to the smallness of ΔU.

Finally, we must consider the coulomb interactions among electrons in the beam, which may produce an exchange of energy between the electrons as they travel from the cathode to the target and lead to an additional increase in the spread of the energy distribution in comparison with the spread of the temperature distribution ΔE_T. This increase in the spread of the energy distribution is called the Boersch effect,[19] or anomalous spreading.[17] It is easy to see that anomalous spreading must be an increasing function of the current density j_p because when this happens the probability of electrons colliding with one another increases, and decreases as the accelerating voltage increases because U_p reduces the time during which the electrons travel the path from cathode to target. According to reference [20], e.g., $\Delta E = \Delta E_T C j_p^{1/3} U_p^{-1/6}$, where C is a constant. In the general case when the electron beam passes through an optical system the current density in different sections will be different, and in sections where the beam is relatively wide, electron collisions with one another will be less frequent. The Boersch effect was calculated in reference [21] from a kinetic equation for the system of electrons: when this was done, the following result was obtained for the spread in the energy distribution as a result of the beam passing through a region of minimal cross section of radius r_0 (crossover):

$$\Delta E^2 = \frac{e^3 L r_0 \sqrt{m}\,\beta}{\sqrt{E}(4\pi\varepsilon_0)^2}\pi^{5/2}\left\{\ln\left(\frac{T_\perp}{T_\parallel}\right) + 4\ln 2 - 6\right\} \tag{1-11}$$

Here, $L = \ln(\lambda_D m v^2)/(2e^2)$ is a Coulomb logarithm, λ_D is the Debye shielding radius, m is the particle mass, β is the brightness of the source in A/cm^2 · steradian, and $T_\perp$ and $T_\parallel$ are the temperatures of the Maxwellian distribution of the transverse and lengthwise energy of the electrons, ignoring anomalous spreading. If the electron optical system comprises several lenses that make several reductions (crossovers) of the beam, the spreads introduced by each crossover must be added, computing the spread in each of them with due regard for the spread accumulated in the remaining crossovers.[21]

It is essential to note that Boersch spreading according to Eq. (1-11) increases as the beam particle mass increases: $\Delta E \sim m^{1/4}$, i.e., all other conditions being equal, it will be substantially greater for ions.

Acting together, all factors cause scattering within 3–4 eV in the energy of the electrons.[3,17] This scatter leads to electron-beam spreading in the focal plane due to chromatic aberration.

1-3. THERMIONIC CURRENT AND THE SPACE-CHARGE FIELD. LANGMUIR'S LAW

The current as a function of the voltage (difference of potentials) in a two-electrode system is determined by the difference in applied potential as well as by the space-charge field and is therefore nonlinear. We will examine a one-dimensional problem concerning the current between a plane-parallel cathode and anode. In a system of electrons in a vacuum the collisions between the electrons do not produce a loss in the total momentum, and therefore there are no ohmic losses in the electron current. Neglecting the initial velocities of the electrons, we write the current equations as

$$\partial^2\varphi(x)/\partial x^2 = -4\pi\rho(x), \qquad \rho(x) = -en(x) \tag{1-12}$$

$$mv^2(x)/2 = -e\cdot -\varphi(x) = e\varphi(x) \tag{1-13}$$

$$j_x = -en(x)v(x) = \text{const} \tag{1-14}$$

Excluding the concentration and velocity of the electrons, we find from Eqs. (1-12) and (1-13) that

$$\frac{\partial^2\varphi(x)}{\partial x^2} = -4\pi\sqrt{\frac{m}{2e}}\, j_x\varphi^{-1/2} = kj_x\varphi^{-1/2} = a\varphi^{-1/2} \tag{1-15}$$

We must formulate the boundary conditions for the problem. A rather general form of the boundary conditions will be

$$\varphi(0) = 0, \qquad \varphi(d) = U_a, \qquad \left.\frac{\partial\varphi}{\partial x}\right|_{x=0} = E_0 \tag{1-16}$$

The magnitude of E_0 depends on the specific conditions. When the cathode is completely shielded by the electron space charge from the anode-generated field (strong emission from the cathode, small difference of potentials), $E_0 = 0$. When shielding is completely absent (a strong field at a large difference of potentials, weak emission), the field at the cathode approaches the value $E_0 = U_a/d$. With partial shielding, $0 < E_0 < U_a/d$.

We will integrate Eq. (1-15) for the initial conditions given by Eq. (1-16). Equation (1-15) can be rewritten as

$$0.5d(d\varphi/dx)^2 = 2ad(\varphi^{1/2}) \tag{1-17}$$

Integrating Eq. (1-17) and giving due regard to the initial conditions we obtain

$$d\varphi/dx = (4a\varphi^{1/2} + E_0^2)^{1/2} \tag{1-18}$$

Separating variables and introducing a new variable $y = \sqrt{\varphi}$, we write

$$\int\frac{d\varphi}{[4a\varphi^{1/2} + E_0^2]^{1/2}} = 2\int\frac{y\,dy}{[4ay + E_0^2]^{1/2}} = \int dx \tag{1-19}$$

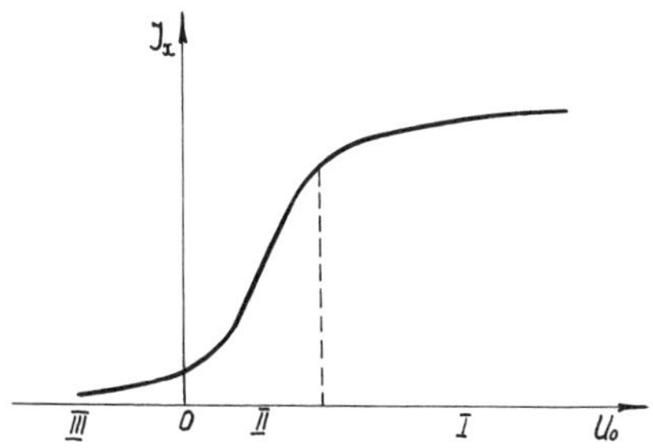

FIGURE 1-2. Volt–ampere characteristics of a planar diode.

The integration produces

$$\frac{1}{4a^2}\left[\frac{1}{3}(E_0^2 + 4ay) - E_0^2\right][E_0^2 + 4ay]^{1/2} = x + C \tag{1-20}$$

Returning to the variable φ and using the boundary condition $\varphi(0) = 0$, we find

$$(2kj_x\sqrt{\varphi} - E_0^2)(4kj_x\sqrt{\varphi} + E_0^2)^{1/2} = 6k^2j_x^2x - E_0^3, \qquad k = 4\pi\sqrt{\frac{m}{2e}} \tag{1-21}$$

The nonlinear relation obtained relates the current j_x, the difference of potentials φ, and the field E_0 at the cathode. From this relation we will obtain special formulas for when total cathode shielding is present and for when shielding is absent.

As we have seen, when complete shielding is present, we must assume that $E_0 = 0$; which, in keeping with Eq. (1-21), corresponds to the conditions $j_x \gg E_0^2/(4k\varphi^{1/2})$. Then, instead of Eq. (1-21), we find

$$j_x(x) = \frac{4}{9}\frac{1}{kx^2}\varphi^{3/2}(x) \tag{1-22}$$

The relation obtained is named Langmuir's law.

It is not hard to find the magnitude of the field at the cathode when the emission currents are small. In this case we have

$$j_x \ll E_0^2/4k\varphi^{1/2}, \qquad 4kj_x\varphi^{1/2}/E_0^2 \ll 1 \tag{1-23}$$

and expand the root factor on the left-hand side of Eq. (1-21) with respect to the indicated small parameter, preserving the quadratic term in the expansion. Reducing the relation obtained leads directly to the result

$$E_0 = U_a/d \tag{1-24}$$

In these conditions the system current is equal to the thermionic current.

It must be noted that the degree of shielding changes with the accelerating field if the density of the thermionic current is held constant. The cathode is shielded at small differences of potential and unshielded at large potential differences. These conditions correspond to sections II and I on the volt–ampere characteristics of a vacuum diode (Fig. 1-2).

When we take the initial velocities of the electrons being emitted into consideration, there is a minimum potential at some plane between the cathode and anode, which means that there is no electrical field present in this section: $E_x = 0$ at some $x = \delta$ near the cathode. Electrons pass through the section $x = \delta$ by diffusion and only in the region $x > \delta$ the electrons experience drift because of the anode field. On the contrary, in the $x < \delta$ region the electrons are decelerated by the space-charge field from the emitted electrons. The section $x = \delta$ is called a virtual cathode.

The electron guns in electron-beam systems operate in the saturation mode because a maximal current is required from the smallest area.

1-4. THE SCHOTTKY EFFECT IN THERMIONIC EMISSION

The picture of thermionic emission painted in the preceding sections is simplified at certain times. We have seen that there is a field $E_0 \neq 0$ at the cathode surface when the cathode is not completely shielded from the anode. An electron being accelerated by this field increases its energy, and at a point x its potential energy decreases by an amount $-eE_0x$. In addition, an emitted electron near the surface of the metal experiences an image force—the attraction of the effective positive charge it induced in the metal. For this reason the potential energy of an electron a distance x from the cathode will be

$$\varepsilon(x) = E_F + \varphi - eE_0x - e^2/4x \tag{1-25}$$

As can easily be verified, the potential energy of an electron, given by Eq. (1-25), is mapped as a curve having a maximum at $x_0 = (e/4E_0)^{1/2}$; the energy at this maximum is

$$\varepsilon(x_0) = E_F + \varphi - e^{3/2}E_0^{1/2} \tag{1-26}$$

As is apparent from Eq. (1-26) and Fig. 1-3, the anode field and the image forces acting together lower the work function: the effective work function is

$$\varphi_{\text{eff}} = \varphi - e^{3/2}E_0^{1/2} \tag{1-27}$$

The reduction in the work function that we have described is called the Schottky effect. This reduction produces an increase in thermionic current

$$j_x = AT^2 \exp\left[-\frac{\varphi - e^{3/2}E_0^{1/2}}{kT}\right] = j_x(E_0 = 0)\exp\left[\frac{e^{3/2}E_0^{1/2}}{kT}\right] \tag{1-28}$$

For a field $E_0 = 10^6$ V/cm at the cathode the potential barrier is lowered by 0.36 V; the increase in thermionic current when $kT = 0.18$ eV (the cathode temperature is 2000°) is $\exp(\Delta\varphi/kT) = e^2 \approx 8$ times.

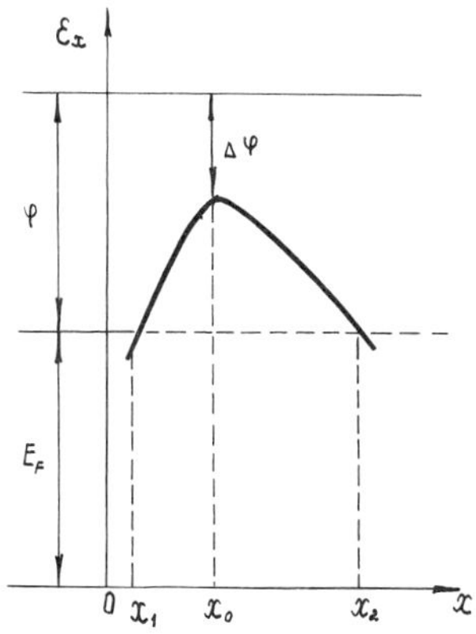

FIGURE 1-3. Energy diagram for the Schottky effect. An electric field $E_x = E_0$ perpendicular to the surface of the metal and imaging forces reduce the height of the potential barrier to $\Delta\varphi = e\sqrt{eE_0}$.

TABLE 1-2. The Parameters of the Energy Barrier for Electrons at the Fermi Level in a Metal (Tungsten) as a Function of the Voltage E_0 of the External Electric Field

E_0	10^5	10^6	10^7	10^8
x_0, Å	60	20	7	2
Δx, Å	1790	179	17.9	1.79
$\Delta\varphi$, eV	0.12	0.38	1.2	3.8

1-5. FIELD EMISSION OF ELECTRONS

The thermionic current as a function of field voltage and temperature given by Eq. (1-28), with due regard for the Schottky effect at high field voltages at the cathode, has not been confirmed experimentally.[22] It was shown afterwards that the increase in thermionic current due to the Schottky effect in fields of 10^6–10^8 V/cm is overshadowed by the more dominant current increase due to electrons tunneling through the energy barrier. The tunneling current was given the name field emission because there was no need to supply energy to the cathode in order to ensure emission (if we ignore the strong static field generated at the cathode). Experiments in, and a theory of field emission were presented in Elinson's monograph.[22]

Table 1-2 gives values for the extent of potential barrier lowering and barrier width for several values E_0 (x_0 is the distance from the surface of the metal to the crest of the barrier, Δx is barrier width, and $\Delta\varphi$ is the reduction in barrier height). The cathode work function has been taken to be 4.4 eV (tungsten). From the formula

$$\Delta x = x_2 - x_1 = [(\varphi/eE_0)^2 - e/E_0]^{1/2} \tag{1-29}$$

the barrier width at a height $E = E_F$ has been found, which is easily obtained from Eq. (1-25).

From Table 1-2 it can be seen that the barrier width remains large in fields of 10^5–10^6 V/cm and is practically nontransparent to electrons; noticeable tunneling through the barrier occurs in fields larger than 10^7–10^8 V/cm. It is interesting to note the significantly steeper drop in the barrier on the left near the emitter surface.

The following terminology has been adopted for emission current: the tunneling current of electrons having energy $E < E_F$ is called field-emission current, the tunneling current of electrons having energy $E_F < E < E(x_0)$ is, appropriately, called the thermal field-emission current because there is talk of the tunneling emission of thermally excited electrons, and, finally, the emission of electrons having energy greater than the barrier height $E > E(x_0)$ will be pure thermionic emission. Obviously, the answer to the question which current dominates depends on the specific conditions of the experiment.

If we consider the fact that the thermionic current density can reach values on the order of 10^2 A/cm^2 (cathodes with special coatings to reduce the work function to values on the order of 1.5–2 eV), then field emission can provide current densities near 10^6 A/cm^2. Earlier we saw that the current due to electrons impinging on the metal surface from inside is on the order of 10^{12} A/cm^2 and, therefore, to obtain a current density of 10^6 A/cm^2 it is necessary to choose a mode of field emission in which the transparency of the barrier is approximately 10^{-6}.

According to quantum theory, the transparency of a barrier described by the potential energy $U(x)$, for an electron with energy E, is given by[23]

$$D(E) = \exp\left[-2\int_{x_1}^{x_2}\left(\frac{2m}{\hbar^2}\right)^{1/2}(U(x)-E)^{1/2}\,dx\right] = \exp(-2Q(E)) \qquad (1\text{-}30)$$

Here x_1 and x_2 are the points of intersection on a graph of the potential energy $U(x)$ by a horizontal line at a height E (the beginning and end of a barrier for an electron with energy E). For a given function $D(E)$ of the barrier transparency the tunnel current is given by

$$j_t = \frac{2e}{h^3}\int dp_y\int dp_z\int\frac{p_x}{m}D[E(p_x,p_y,p_z)]f(p_x,p_y,p_z)\,dp_x \qquad (1\text{-}31)$$

The integration is done with respect to electron energies below the barrier height.

The integral in the transparency formula, written with regard for the image forces

$$Q(E) = \int_{x_1}^{x_2}\left(\frac{2m}{\hbar^2}\right)^{1/2}\left[E_F+\varphi-eE_x x-\frac{e^2}{4x}-E\right]^{1/2}dx \qquad (1\text{-}32)$$

is not expressed as elementary functions and reduces to elliptic tabulated integrals.[22] Accordingly, the tunneling current can be found only by numerical integration.

The integration can be performed to the end if we neglect the image forces (an acute-angled barrier). Then

$$Q(E) = \left(\frac{2m}{\hbar^2}\right)^{1/2}\int_{x_1}^{x_2}\left[E_F+\varphi-eE_x x-E\right]^{1/2}dx \qquad (1\text{-}33)$$

In this case $x_1 = 0$ and x_2 is found from the equation $E_F+\varphi-eE_x x = E$:

$$x_2 = (E_F+\varphi-E)/eE_x \qquad (1\text{-}34)$$

By making the substitution of variables $x' = -x + x_2$ the integral, Eq. (1-33), is easily evaluated:

$$Q(E) = \left(\frac{2m}{\hbar^2}\right)^{1/2} \int_0^{x_2} \sqrt{eE_x x'}\, dx' = \frac{2}{3}\left(\frac{2m}{\hbar^2}\right)^{1/2} \frac{(E_F + \varphi - E)^{3/2}}{eE_x} \tag{1-35}$$

In order to continue calculating the tunneling current it is important to linearize the expression obtained with respect to the electron energy E. We can rewrite Eq. (1-35) as

$$Q(E) = \frac{2}{3}\left(\frac{2m}{\hbar^2}\right)^{1/2} \frac{\varphi^{3/2}}{eE_x}\left(1 - \frac{E - E_F}{\varphi}\right)^{3/2} \tag{1-36}$$

Considering the fact that the electrons tunnel primarily in the narrow range of energies near the Fermi energies we can say that

$$(E - E_F)/\varphi \ll 1 \tag{1-37}$$

Then

$$Q(E) = \frac{2}{3}\left(\frac{2m}{\hbar^2}\right)^{1/2} \frac{\varphi^{3/2}}{eE_x}\left(1 - \frac{3}{2}\frac{\Delta E}{\varphi}\right), \qquad \Delta E = E - E_F \tag{1-38}$$

$$D(E) = D_0 \exp\left[2\left(\frac{2m}{\hbar^2}\right)^{1/2} \frac{\varphi^{1/2}\Delta E}{eE_x}\right]$$

$$D_0 = \exp\left[-\frac{4}{3}\left(\frac{2m}{\hbar^2}\right)^{1/2} \frac{\varphi^{3/2}}{eE_x}\right] \tag{1-39}$$

Here D_0 is the barrier transparency for electrons at the Fermi level. As Nordheim has shown,[24] we can introduce a Nordheim correcting function $\theta(\sqrt{e^3 E_x}/\varphi)$ in the formula obtained, which monotonically decreases from 1 to 0 as the argument varies over the range from 0 to 1, which provides a description of the barrier transparency that takes the image forces into account:

$$D_0(E) = \exp\left[-\frac{4}{3}\left(\frac{2m}{\hbar^2}\right)^{1/2} \frac{\varphi^{3/2}}{eE_x}\theta\left(\frac{\sqrt{e^3 E_x}}{\varphi}\right)\right] \tag{1-40}$$

A table of Nordheim functions is given in reference [22]. Numerical estimates can be made from Eq. (1-39). They show that transparency values $D_0 \geq 10^{-6}$ when $\varphi = 2$ eV lie in the range of fields $E_x = 10^7$ V/cm, when $\varphi = 3$ eV they lie in the range of fields $E_x = 2.2 \times 10^7$ V/cm, and when $\varphi = 5$ eV the transparency indicated is achieved in fields greater than 5×10^7 V/cm.[22]

We will now turn to calculating the field emission current. We need to calculate the integral

$$j_t = 2e\left(\frac{m}{\hbar}\right)^3 \iint_{-\infty}^{+\infty} dv_y\, dv_z \int_0^{\infty} dv_x \frac{D(E)v_x}{1 + \exp[(m/2kT)(v_x^2 + v_y^2 + v_z^2) - E_F/kT]} \tag{1-41}$$

In doing this we must allow for the fact that

$$\Delta E = E - E_F = mv_x^2/2 - E_F \tag{1-42}$$

in the transparency function given by Eq. (1-39) and the transparency coefficient depends only on v_x. Integration with respect to the transverse velocity components is easily accomplished when we convert to polar coordinates:

$$\rho^2 = v_y^2 + v_z^2, \qquad dv_y\, dv_z = \rho\, d\rho\, d\chi$$

$$\begin{aligned} I &= \iint_{-\infty}^{+\infty} \frac{dv_y\, dv_z}{1 + \exp[\alpha(v_y^2 + v_z^2) + \beta]} \\ &= \int_0^{2\pi} d\chi \int_0^{\infty} \frac{\rho\, d\rho}{1 + \exp(\alpha\rho^2 + \beta)} = \frac{\pi}{\alpha} \ln(1 + e^{-\beta}) \end{aligned} \tag{1-43}$$

(the meaning of the notations α and β is obvious). As a result, we find

$$j_t = \frac{4\pi kTe}{m}\left(\frac{m}{h}\right)^3 \int_0^{\infty} v_x D(v_x) \ln\left[1 + \exp\left(-\frac{mv_x^2 - 2E_F}{2kT}\right)\right] dv_x \tag{1-44}$$

Further calculations are performed on the assumption of low temperatures. Considering the fact that $mv_x\, dv_x = d\Delta E$, we write

$$j_t = \frac{4\pi kTe}{m^2}\left(\frac{m}{h}\right)^3 \int_{-E_F}^{\infty} D(\Delta E) \ln(1 + e^{-\Delta E/kT})\, d\Delta E \tag{1-45}$$

At low temperatures the tunneling electrons have energies $E < E_F$ and $\Delta E < 0$; therefore,

$$\ln(1 + e^{-\Delta E/kT}) \simeq \Delta E/kT \tag{1-46}$$

Also, the upper limit of integration in Eq. (1-45) can be set to 0. Substituting the $D(\Delta E)$ from Eqs. (1-39) and (1-40) into the integral, we find

$$j_t = \frac{e^3 E_x^2}{8\pi\hbar\varphi} \exp\left[-\frac{4}{3}\left(\frac{2m}{\hbar^2}\right)^{1/2} \frac{\varphi^{3/2}}{eE_x}\theta\left(\frac{\sqrt{e^3 E_x}}{\varphi}\right)\right] \tag{1-47}$$

Numerical values of j_t can be found from[22]

$$j_t = 1.55 \times 10^{-6} \frac{E^2}{\varphi} \exp[-6.85 \times 10^7\, E_x^{-1} \varphi^{3/2} \theta(3.62 \times 10^{-4}\, E_x^{1/2} \varphi^{-1})]\ \text{A/cm}^2 \tag{1-48}$$

φ is in eV; E_x is in V/cm. At limited temperatures Eq. (1-45) can be written as a series.[22]

Equation (1-47) was subjected to comprehensive experimental verification through observations of the field emission current as a function of the electric field voltage E_x and by changes in the current as a function of the work function φ. In every case good agreement was established between theory and experiment.[22]

We now pause to discuss the means for obtaining a high-voltage field at the emitter. These means are based on the fact that the field voltage near the surface of a charged conductor is inversely proportional to the radius of curvature of its surface. If the sphere has been charged to a potential Φ, the charge on it is $Q = r\Phi$. The voltage of the field near the surface of the conducting sphere will be

$$E_r = Q/r^2 = \Phi/r \tag{1-49}$$

i.e., inversely proportional to the sphere's radius of curvature. It is easy to see that when the radius of curvature is 1 μm and the potential is 10^4 V, a field 10^8 V/cm arises. Emitter surfaces of such small radius are obtained by point etching methods.(22) A precise determination of the curvature and shape of a point surface and, accordingly, the field at the emitter surface is a rather complex affair.(22)

Considering the dependence of the field voltage on the curvature of the point surface, we can construct a Fowler–Nordheim graph for the field-emission current at points I/V^2 from $1/V$. These graphs are straight lines with a slope of $S = -2.97 \times 10^3 \varphi^{3/2}/\beta$; φ is in eV, β in cm^{-1}. Ordinarily, the value β, where r is the radius of the point, is taken as $\beta = 1/5r$.(25)

At the present time only thermionic cathodes are being used in electron-beam lithography systems. However, increasing the brightness of the emitters in order to increase the resolving power and the throughput of electron-beam devices for submicron lithography is a current problem.(26) A fundamental roadblock to using the field-emission operating mode in emitters is fluctuations in the field-emission current from the emitter during extended operation. Apparently, the main reasons for current fluctuation are changes in the condition of the emitter surface due to restructuring of the surface layers by diffusing atoms, change in the composition of adsorbed atoms (due to ionic sputtering of ions from the residual gases), and absorbents having an uncontrollable composition precipitating from the residual gases. We can assume that the indicated reasons for fluctuation are not insurmountable.(22) Maintaining and operating a field-emission cathode in a high vacuum is a necessary condition for stability in the field-emission current.

Stable field emission of electrons has been observed in cathodes fabricated from monocrystals of TiC.(25) A point having a radius less than 0.1 μm was fabricated by etching and subjected to annealing at a temperature of 1900–2100°C for 15 sec. This type of cathode produced a stable field-emission current of 10^{-5} A in a 10^{-8} torr vacuum. Substantial changes were observed in the work function when oxygen was present: 0.15 eV for a plane of a 100 monocrystal and 0.6 eV for a plane of a 111.(25)

The use of heated field-emission cathodes is preferred.(27) Heating can also be used to purify the cathode after the system has been evacuated. Careful design and fabrication of the geometry control and power supply systems for field-emission or thermionic-emission guns can ensure that the current is stable to within 100 ppm.(27) The experiments conducted in reference [14] using lasers to heat field-emission cathodes are of interest. Lasers have the advantage of not requiring auxiliary electrical and mechanical connections to the cathode, and there are no magnetic fields from a heating current. The heating laser can be focused at some distance from the emitting point, which eliminates the possibility of electrons being emitted due to photoemission. The point temperature as a function of the laser spot radius ρ at a constant beam power is estimated as an inverse proportion: $T \sim 1/\rho$.(14) In taking currents on the order of 5×10^7 A/cm^2 a vacuum arc appears at the emitter, which melts and reshapes the emitter, leading to catastrophic changes in its properties. Apparently, this phenomenon puts an upper limit on the currents that can be taken from the emitter. We can point to the widespread use of field-emission cathodes to obtain powerful current pulses in different areas of science and technology.[27,28–31] Specifically, field-emission currents of 1.1 kA and brightness 74 kA/cm · steradian have been obtained.[32] Graphitic materials have been used for field-emission cathodes. The maximum field voltage used in these experiments did not exceed 650 kV/cm, which is much lower than needed for field emission from pointed cathodes. It has been assumed in this case that electron emission is associated with nonmetallic inclusions in the surface layer of the cathode.[32,33] Such large electric fields can be generated in these inclusions as well as in points having small radii, and electron heating effects will appear. The assumption regarding this emission mechanism has been supported by direct observations of the emitting spots on the cathode surface. Large-area thermionic-emission cathodes are being developed as LaB_6 or LaMo disks up to 30 cm in diameter. (This material is obtained by sintering 98% Mo and 2% La_2O_3). In doing this emission current densities of 20 A/cm^2 for LaB_6 disks and 11 A/cm^2 for LaMo disks have been achieved.[11,34] These wide electron beams can be used in lithography systems with shadow masks (see below).

1-6. ELECTRON OPTICS. THE ELECTROOPTIC ANALOG

We now turn to discussing the theory of generating the electron and ion beams that are the fundamental tools for creating a submicron pattern for devices. The equation for the motion of charged particles having charge q and mass m in electric and magnetic fields **E** and **H**

$$m\frac{d\bar{v}}{dt} = q\mathbf{E} + \frac{q}{c}\bar{v} \times \mathbf{H} \tag{1-50}$$

is the initial equation for analyzing the processes involved in generating a beam. Multiplying Eq. (1-50) by $\bar{v}$ and integrating between points 1 and 2 on the trajectory gives us

$$\left(\frac{mv^2}{2}\right)_2 - \left(\frac{mv^2}{2}\right)_1 = \int_1^2 q\mathbf{E}\, d\mathbf{S} \tag{1-51}$$

Because $\mathbf{E} = -(d\varphi/dn)\mathbf{n}$, where $\mathbf{n}$ is the normal to the equipotential surface $\varphi(r) =$ const, we can write

$$\Delta\left(\frac{mv^2}{2}\right) = -q\int_1^2 \frac{\partial\varphi}{\partial n}\, dn = q(\varphi_1 - \varphi_2) \tag{1-52}$$

The electron optical properties of an electrostatic and a magnetic field are significantly different. We will examine the properties of an electrostatic field. Let a surface separate two spaces in which the potentials are φ_1 and φ_2 (this surface actually models a layer in which the difference of potentials $\varphi_2 - \varphi_1$ increases). Because the voltage of the electric field is normal to the surface $\mathbf{E} = -(d\varphi/dn)\bar{n}$, only the normal component of the electron's velocity changes as it passes through a thin layer. This determines the law of refraction for an electron according to Eq. (1-52) as it passes through a thin layer (see

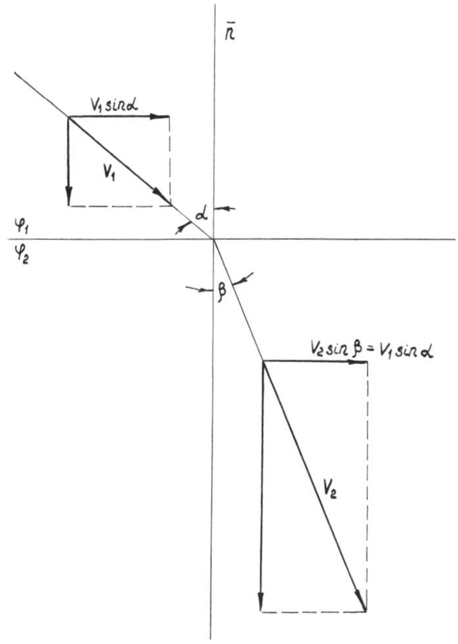

FIGURE 1-4. Refraction of an electron beam at a thin layer for which the potential changes from φ_1 to φ_2. The law of refraction is $\sin d / \sin \beta = \sqrt{\varphi_2/\varphi_1}$.

Fig. 1-4):

$$v_1 \sin\alpha = v_2 \sin\beta; \qquad \frac{mv_1^2}{2} = q\varphi_1, \qquad \frac{mv_2^2}{2} = q\varphi_2 \tag{1-53}$$

$$\frac{\sin\alpha}{\sin\beta} = \frac{v_2}{v_1} = \frac{\sqrt{\varphi_2}}{\sqrt{\varphi_1}} = \frac{n_2}{n_1}; \qquad n_1 = \sqrt{\varphi_1}, \qquad n_2 = \sqrt{\varphi_2} \tag{1-54}$$

The potential function $\varphi(r)$ is a continuous, monotonic function (discontinuities in it occur only on the charged surfaces of the metallic electrodes that produce the same field $\varphi(r)$ in the vacuum; inside the metal the field is $\bar{E} = 0$). We can construct a picture of the refractive properties of the field if we introduce a system of equipotential surfaces having the approximate values of the potential $\varphi_n(r)$, $\varphi_n + \delta\varphi_n$ and consider the potential in the space between the equipotentials to be constant but changing abruptly by an amount $\delta\varphi_n$ at the equipotential itself. Then an electron between the equipotentials moves at constant velocity and is refracted, intersecting the equipotential. (This procedure for drawing the trajectory of an electron corresponds roughly to a machine computation of the properties of electron optics).

We will compare the properties of ordinary optical lenses and electrostatic electron optical systems. The coefficients of refraction for glass optical lenses are limited to the range of 1–2.5; inside the lenses $n(r)$ is constant (if we overlook details, e.g., such as multilayer interferometer filters deposited on the lens surfaces) and $n(r)$ changes abruptly at the edge of the lenses; n is a constant in time (exceptions are, e.g., the electrooptic elements for deflecting beams of light). For an electrostatic field $n = \sqrt{\varphi}$ is almost unbounded from above, varies continuously in space, and can easily be varied in time. Further, the function $n(r)$ for optical lenses need not be subordinated to an equation since the function $n^2 = \varphi(r)$ for an electrostatic field must be subordinated to the Laplace equation, $\Delta\varphi = 0$. The latter restriction makes it difficult to eliminate a number of types of aberrations in electron optical systems.

The mass and charge of a charged particle are not found in the law of refraction $\sin\alpha/\sin\beta = (\varphi_2/\varphi_1)^{1/2}$ [according to Eqs. (1-54) and (1-53), when the algebraic sign of the charge is changed, the polarity of the fields, i.e., the sign of the potential, must also be changed]. This means that the particle trajectories are independent of the particles' charge and mass in a given electrostatic field; particles having different charge and mass traverse the same trajectories for different periods of time. This property of electrostatic systems bears the name of scaling law. The scaling law makes it possible in principle to use the same electrostatic optical systems for focusing electron and ion currents (see the comment above regarding the sign of the charge and potential).

1-7. THE FUNDAMENTAL EQUATION OF ELECTROSTATIC OPTICS[2,3]

We need an equation in the form $f(x, y, z) = 0$ for the trajectory of a charged particle in an electrostatic field, eliminating the variable t. The scaling theorem states that this is possible. We will do this for the practically important special case of axially symmetric fields (having cylindrical symmetry). In this system the electrodes that create the fields are solids of rotation whose axes coincide with common axis of symmetry (the optical axis). Electrodes of this type are cylinders of finite length and different radii, rings, and circular diaphragms. In this system the field potential is conveniently given as a function in cylindrical coordinates z, r, χ, and the potential will be independent of the azimuthal angle χ: $\varphi = \varphi(z, r)$. The z axis coincides with the optical axis of the system. The particles move near the optical axis z; in doing this the particle is deflected from the axis a distance $r(z)$. Thus, the unknown function for the trajectory is the dependence $r = f(z)$.

The second assumption which greatly simplifies the trajectory equation is that the electron is deflected only a small amount from the optical axis. Another way of saying this is to say that the electron beams make small angles θ with the optical axis. These beams are called paraxial beams. A beam that emerges from a single point (e.g., electron emission from one point on the body) is called homocentric. When there are aberrations in an optical system it is impossible for a homocentric beam passing through the optical system to again be converged at a single point, i.e., it is impossible to obtain a sharp image of the point on the body. Image blurring is caused by aberrations that are inherent in the system. It turns out that a paraxial beam having an infinitesimal beam angle and extracted from a homocentric beam via a small diaphragm will give infinitesimal blurring of the point image. The beam angle attainable in electron-beam systems is governed by practical considerations. By decreasing the beam angle we lower the beam current. Specifying the beam current places a lower limit on the beam angle and, consequently, on the resolving power of the electron-beam system.

Magnetic lenses are used in modern electron-beam systems rather than electrostatic lenses. Electrostatic focusing is done only in the electron guns of these systems. For this reason we must discuss the theory of electrostatic and magnetic lenses in order to understand how these systems operate.

We will return to deriving the equation for the trajectory of an electron in an electrostatic field. We write the equation of motion in the variables r and z:

$$d^2r/dt^2 = -|e|E_r \tag{1-55}$$

$$d^2z/dt^2 = -|e|E_z \tag{1-56}$$

The second equation, which describes the motion of an electron along the optical axis, does not play a major role in analyzing the trajectory. For electrons in paraxial beams we can assume (to within second-order infinitesimals) that

$$v_z = v\cos\theta(z) \sim v(1-\theta^2/2) \simeq v = \sqrt{2e\varphi/m} \tag{1-57}$$

To solve the radial motion equation, we must find E_r. We will isolate a small cylinder of radius r and length dz along the optical axis and apply Gauss' theorem: the voltage of an electric field passing through the surface of the cylinder is 0 (there is no charge within the cylinder):

$$-E_z(z)\pi r^2 + E_z(z+dz)\pi r^2 + E_r(r,z)2\pi r\,dz = 0 \tag{1-58}$$

From here we find the radial component E_r of the electric field:

$$E_r = -\frac{r}{2}\frac{\partial E_z}{\partial z} = \frac{r}{2}\varphi''_{zz} \tag{1-59}$$

As we can see, the radial field component E_r is proportional to r. Those components of a paraxial beam that have a larger deflection r experience a strong force $E_r = -eE_r$ that directs them back to the axis; it is mainly this phenomenon that governs the focusing ability of the E_r field. Using the identity

$$\frac{d}{dt}\frac{dr}{dt} = \frac{dz}{dt}\frac{d}{dz}\left(\frac{dz}{dt}\frac{dr}{dz}\right) = v_z\frac{d}{dz}\left(v_z\frac{dr}{dz}\right) \tag{1-60}$$

and substituting v_z and E_r from Eq. (1-57) and (1-59), we obtain

$$2\sqrt{\varphi}\frac{d}{dz}\left(\sqrt{\varphi}\frac{dr}{dz}\right) = -\frac{r}{2}\varphi''_{zz} \tag{1-61}$$

or

$$\frac{d^2r}{dz^2} + \frac{\varphi'}{2\varphi}\frac{dr}{dz} + \frac{\varphi''}{4\varphi}r = 0 \tag{1-62}$$

To solve Eq. (1-62) in a paraxial approximation of the coefficients φ, φ', φ'' we use

$$\varphi(r,z) \approx \varphi(0,z), \qquad \varphi'(r,z) \approx \varphi'(0,z), \qquad \varphi''(r,z) \approx \varphi''(0,z) \tag{1-63}$$

It turns out that this approximation is inadequate for calculating the aberrations. In order to do this we must first expand the functions $\varphi(r,z)$, $\varphi'(r,z)$, $\varphi''(r,z)$ as a series in r and keep all terms up to r^3 inclusive (see below).

For an arbitrary function $\varphi(r,z)$ the solution of Eq. (1-62) cannot be written in analytic form. However, a number of important results can be obtained directly from analyzing this very equation.

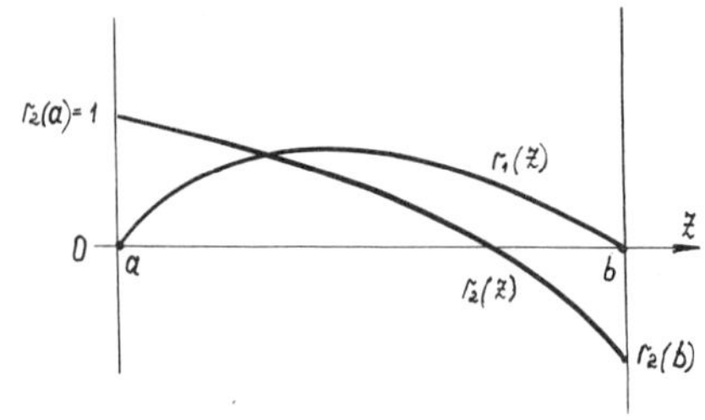

FIGURE 1-5. Two special solutions of the equation for the trajectory of electrons emerging from points on the plane of the "object" when $z = a$ parallel to the z axis and at an angle 45° to it. The plane is $z = b$. There is an "image" plane. The ratio $r_2(b)/r_2(a)$ is equal to the linear magnification M (or "reduction" if $M < 1$). The quantity $r_1'(b)/r_1'(a) = \gamma_b/\gamma_a$ is equal to the angular magnification of the system; $\gamma_a = \tan\theta_1$, $\gamma_b = \tan\theta_2$.

1-8. THE HELMHOLTZ–LAGRANGE EQUATION

We will take two points $r_1(a) = 0$ and $r_2(a) = 1$ in the plane $z = a$ (the object plane) from which the trajectories emerge respectively at angles $\pi/4$ and 0 ($r_1'(a) = 1$ and $r_2'(a) = 0$, see Fig. 1-5). We choose an image plane $z = b$ such that in this plane a trajectory emerging from point $r_1(a) = 0$ again intersects the optical axis: $r_1(b) = 0$. The trajectory emerging from point $r_2(a) = 1$ intersects the plane $z = b$ at point $r_2(b)$. The two trajectories described, $r_1(z)$ and $r_2(z)$, are, naturally, particular solutions of Eq. (1-62). The general solution of Eq. (1-62) for the beams trajectory $r(z)$ can then be given as the superposition of two independent particular solutions $r_1(z)$ and $r_2(z)$:

$$r(z) = c_1 r_1(z) + c_2 r_2(z) \tag{1-64}$$

In fact, by changing c_2 we obtain trajectories that emerge from any point $r(a) = c_2$ in an object; by setting c_1 we obtain trajectories that emerge at an arbitrary angle $r'(a) = c_1 = \tan\theta$ to the optical axis of the system.

The ratio

$$r(b)/r(a) = c_2 r_2(b)/c_2 = r_2(b) = h_b/h_a \tag{1-65}$$

is a linear magnification of the optical system located between the planes $z = a$ and $z = b$.

The ratio of the slopes of the trajectory in the $z = b$ and $z = a$ planes is an angular magnification of the system:

$$\frac{r_1'(b)}{r_1'(a)} = r_1'(b) = \gamma_b/\gamma_a \tag{1-66}$$

A beam of trajectories having spread $\tan\theta \le c_1$ and emerging from a point in the object plane $z = a$ enters a point in the image plane $z = b$ within a core having a spread angle $\tan\theta \le c_1 r_1'(b)$.

We will establish the relationship between angular and linear magnification in the system. The particular solutions $r_1(z)$ and $r_2(z)$ satisfy Eq. (1-61):

$$\frac{d}{dz}\left(\sqrt{\varphi}\frac{dr_1}{dz}\right) = -\frac{r_1}{4}\frac{\varphi''}{\sqrt{\varphi}} \tag{1-67}$$

$$\frac{d}{dz}\left(\sqrt{\varphi}\frac{dr_2}{dz}\right) = -\frac{r_2}{4}\frac{\varphi''}{\sqrt{\varphi}} \tag{1-68}$$

Multiplying the first equation by r_2, the second by r_1, and subtracting, we find

$$d/dz(\sqrt{\varphi}\, r_2 r_1' - \sqrt{\varphi}\, r_1 r_2') = 0 \tag{1-69}$$

or

$$\sqrt{\varphi}(r_2 r_1' - r_1 r_2') = \text{const} \tag{1-70}$$

Equating the left-hand side of Eq. (1-70) in the planes $z = a$ and $z = b$, we find

$$\sqrt{\varphi(a)} = \sqrt{\varphi(b)}\, r_2(b) r_1'(b) \tag{1-71}$$

or, with the help of Eqs. (1-65) and (1-66),

$$h_a \gamma_a \sqrt{\varphi(a)} = h_b \gamma_b \sqrt{\varphi(b)} \tag{1-72}$$

If angles γ_a and γ_b are not small, we must write[2,3]

$$h_a \sin\gamma_a \sqrt{\varphi(a)} = h_b \sin\gamma_b \sqrt{\varphi(b)} \tag{1-73}$$

Equation (1-73) relates the medium's coefficients of refraction $\sqrt{\varphi}$, the angular aperture of the beams $\sin\gamma$, and the sizes of an object and its image in two conjugate planes. This equation also makes it possible for us to analyze specific properties of the elements of electron-beam systems. We have already said that electrons are preliminarily focused in an electron gun having an accelerating potential γ_b. An example that applies to an electron gun is the size h_a being equal to the size of the emitting region of the cathode. Then h_b is the image size in the emitting region of the cathode and produced by the system of electrostatic lenses in an electron gun:

$$h_b = h_a \frac{\sin\gamma_a}{\sin\gamma_b} \frac{\sqrt{\varphi(a)}}{\sqrt{\varphi(b)}} \tag{1-74}$$

This type of image is called electron gun crossover. It is quite natural to think of the $\varphi(b)$ in Eq. (1-74) as being equal to the potential in the crossover plane. But what is the cathode potential $\varphi(a)$?

The ratio of the potentials in two planes appears in many equations in electron optics, an example of which is Eq. (1-74). Because the potential is determined to within an additive constant, the ratio of potentials becomes, it would seem, uncertain. Therefore, we must clearly keep in mind which electrode has zero potential. We have adopted the convention that the cathode (emitter) is at zero potential. As can be seen from deriving the trajectory equation, Eq. (1-62), the quantity $\varphi = (1/e)(mv^2/2)$ yields the electrons' kinetic energy, which in the plane of the cathode is equal to the initial energy of the electrons (see Section 1-2).

The conditions for minimizing the crossover size are obvious from Eq. (1-74): the emitter size, the initial spread in the velocities of the electrons being emitted, and the input beam aperture γ_a must all be reduced (by choosing an appropriately sized diaphragm), and the accelerating voltage $\varphi(b)$ and the gun's output aperture must be made as large as possible.

The cathode image in the crossover plane serves as a virtual cathode which is projected by successive (as a rule, magnetic) lenses. In order to get a feeling for the size of the crossover we will assume that $\varphi(a) = 1$ V, $\varphi(b) = 10^4$ V, and $\sin\gamma_a / \sin\gamma_b = 1.0$; the size of the cathode emitting area is $h_a = 1$ mm. Then the crossover size is 10 μm. Understandably, magnification of the crossover size due to aberrations has not been considered. In actual systems the crossover size is 100 μm. If the beam cross section in the exposure plane of an electron resist is to be 0.1 μm, the magnetic lenses of the system must produce a crossover image having a reduction of 1000.

1-9. THE EQUATION FOR A THIN ELECTROSTATIC LENS

We have seen that the refracting properties of an electrostatic lens are governed by the E_r component of the field. Let E_r be nonzero in some region along the optical axis z and the width of the region be small in comparison with the focal lengths of the lenses. Such a lens is called thin. If, in addition, after passing through a lens (the region $E_r \neq 0$) the r coordinate of the electron changes slightly (the direction of electron motion changes significantly), the lens is called weak. It is also important to distinguish symmetric lenses that have an identical potential on both sides of the refracting region $E_r \neq 0$, and immersion lenses for which this condition is not satisfied. A symmetric lens has one focal length; an immersion lens has two: a forward and rear focus. We will find these focal lengths and the lens formulas.

We will integrate Eq. (1-61) along the trajectory of an electron that emerges from a point A in the area of the object to a point B in the area of the image (Fig. 1-6):

$$\int_A^B d\left(\sqrt{\varphi}\frac{dr}{dz}\right) = -\frac{1}{4}\int_A^B \frac{r\varphi''}{\sqrt{\varphi}}\,dz \tag{1-75}$$

$$\sqrt{\varphi}\frac{dr}{dz}\bigg|_B - \sqrt{\varphi}\frac{dr}{dz}\bigg|_A = -\frac{1}{4}\int_A^B \frac{r\varphi''}{\sqrt{\varphi}}\,dz \tag{1-76}$$

If in the area of the object $dr/dz|_A = 0$ the electron moves parallel to the optical axis and after being refracted in the lens intersects the optical axis at the second focal point with a deflection $dr/dz|_B = -r_0/f_2$, then, assuming that $r \approx r_0$ for a thin lens and taking r_0 outside the integral, we find

$$-\sqrt{\varphi(b)}\,\frac{r_0}{f_2} = -\frac{r_0}{4}\int_A^B \frac{\varphi''}{\sqrt{\varphi}}\,dz \tag{1-77}$$

and

$$\frac{1}{f_2} = \frac{1}{4\sqrt{\varphi(b)}}\int_A^B \frac{\varphi''}{\sqrt{\varphi}}\,dz \tag{1-78}$$

Similarly, a beam moving parallel to the z axis after passing through the lens emerges from the first focal point in the object space:

$$dr/dz|_A = \frac{r_0}{f_1}, \qquad dr/dz|_B = 0$$

$$\frac{1}{f_1} = \frac{1}{4\sqrt{\varphi(a)}}\int_A^B \varphi''\varphi^{-1/2}\,dz \tag{1-79}$$

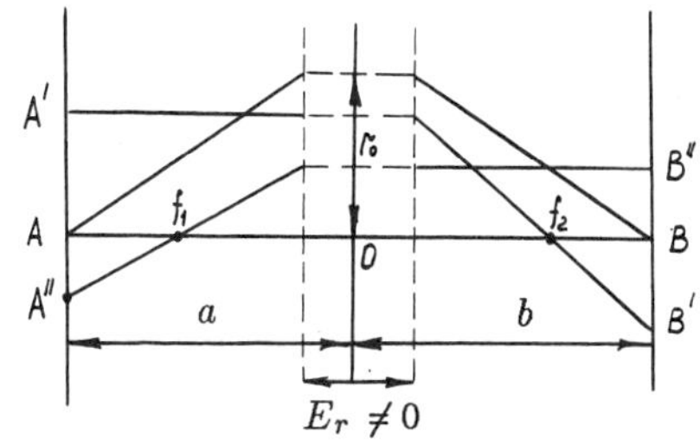

FIGURE 1-6. The path taken by beams in a "thin" and "weak" electrostatic lens. The beams $A'B'$ and $A''B''$ define the backward (f_2) and forward (f_1) foci of the lens, the projections a and b of beam AB are supported by the lens formula $f_1/a + f_2/b = 1$.

Thus, the ratio of the focal lengths depends on the ratio of potentials in the areas of the object and the image

$$f_1/f_2 = -(\varphi_A/\varphi_B)^{1/2} \tag{1-80}$$

If the points A and B are chosen as in Fig. 1-6, we can write

$$-\sqrt{\varphi_B}\,\frac{r_0}{b} - \sqrt{\varphi_A}\,\frac{r_0}{a} = -\frac{r_0}{4}\int_A^B \varphi''\varphi^{-1/2}\,dz \tag{1-81}$$

which can be rewritten as

$$f_1/a + f_2/b = 1 \tag{1-82}$$

The relationship obtained is the formula for a thin immersion lens.

For a symmetric lens, a special case of an immersion lens ($\varphi_A = \varphi_B = \varphi$), we obtain

$$\frac{1}{f} = \frac{1}{4\sqrt{\varphi}}\int_A^B \varphi''\varphi^{-1/2}\,dz \tag{1-83}$$

$$1/a + 1/b = 1/f \tag{1-84}$$

Calculating focal lengths according to the formulas obtained requires that the functions $\varphi(z)$ and $\varphi''(z)$ be determined beforehand. For cylindrical electrodes a direct solution of the electrostatic problem of finding $\varphi(z)$ and $\varphi''(z)$ can only be obtained by numerical methods. Occasionally, the unknown fields are represented by empirical formulas of the type (for a symmetric lens)

$$\varphi(z) = \varphi_0 + Ae^{-bz^2}, \qquad \varphi_0\left(1 - \frac{1}{2}e^{-\frac{1}{2}bz^2}\right), \qquad \varphi_0 + Ath(bz), \qquad \text{etc.} \tag{1-85}$$

However, the final results obtained by means of these formulas make it possible for us to analyze the properties of a given electrode system in only a qualitative sense. Fundamental approaches to calculating $\varphi(z)$ fields will be discussed later.

In complex situations the dependence of f on the electrode potentials (for a given electrode geometry) is determined experimentally because it is mainly the ability to change the system's optical properties by changing the potentials that is the major advantage of electron optical systems which makes it possible to make system adjustments not by changing the geometry (the mutual placement) of the lenses, as in ordinary optics,

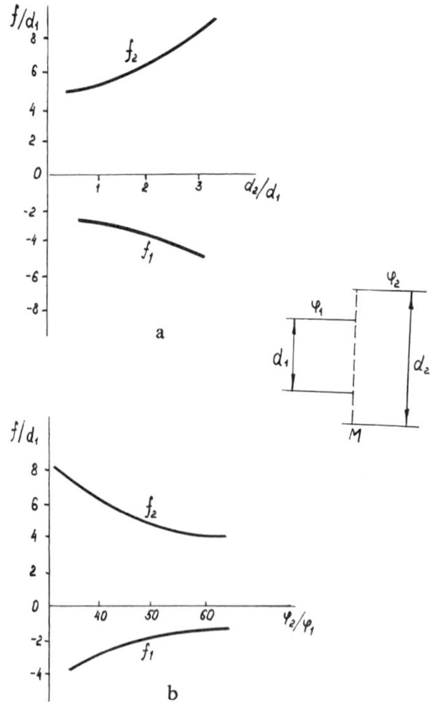

FIGURE 1-7. Change in the focal lengths and cylindrical lens as a function of the ratio of potentials φ_2/φ_1 and the ratio of cylinder diameters d_2/d_1 (when $\varphi_2/\varphi_1 = 4$).

but by changing the potentials.* Naturally, the systems designer must take the system geometry into account in order to determine how the system's optical properties are governed by the geometry. We will present some results obtained from these investigations.

Two-Cylinder Lenses. The focal lengths as a function of the potentials ratio and the ratio of the cylinder diameters are shown in Figs. 1-7a,b. The scaling theorem, which says that changing the electrode potentials while preserving their ratio does not change the focal lengths, is valid here. On the other hand, changing the radii of the cylinders while preserving their ratio changes the foci in proportion to the radius of the cylinders. The focal lengths are markedly changed when the ratio φ_2/φ_1 is varied in the range $\varphi_2/\varphi_1 < 5$ and comparatively weakly when $\varphi_2/\varphi_1 > 5$. This means that lenses having short focal lengths at a large potentials ratio φ_2/φ_1 will be highly stable to small changes in the potentials, but at the same time, have a low degree of adjustability.

Symmetric Cylindrical Lenses. A symmetric lens is characterized by two potentials: the external electrodes have the same potential φ_1; the potential of the internal electrode is φ_2 (see Fig. 1-8). The dependence $f[(\varphi_2 - \varphi_1)/\varphi_1]$ is shown in Fig. 1-9. We can see that when $\varphi_1 = \varphi_2$, $f = \infty$; the lens ceases to exist. When $(\varphi_2 - \varphi_1)/\varphi_1$ is varied between the limits $-1 \leq (\varphi_2 - \varphi_1)/\varphi_1 \leq \infty$, the lens is convergent, $f > 0$.

*It is interesting to note that Eq. (1-78), which gives the focal length of a lens via a linear integral from the second derivative of the potential, does not contain a direct indication that focal length is merely a function of the electrode potentials. However, when we consider the fact that the simultaneous specification of the potentials and the geometry of the system electrodes determine the potential $\varphi(z)$ and all its derivatives, the existence of the $f(\varphi_1, \ldots, \varphi_i)$ functions becomes understandable.

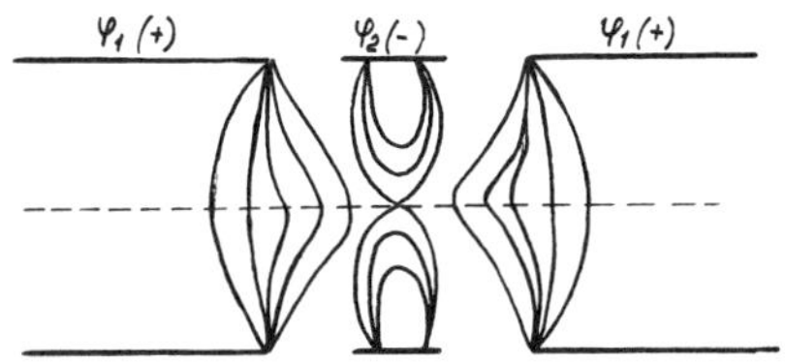

FIGURE 1-8. A three-electrode symmetric lens. The shape of the equipotential surfaces is shown.

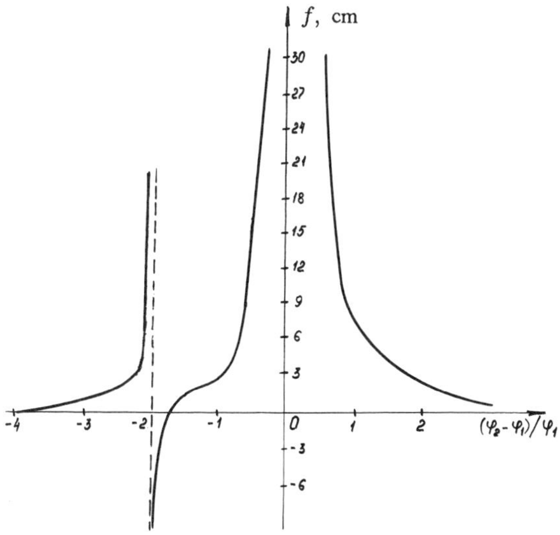

FIGURE 1-9. The focal length of the lens shown in Fig. 1-8 as a function of $(\varphi_2 - \varphi_1)/\varphi_1$.

When $\varphi_2 = 0$, the sole accelerating potential is φ_1 (the cathode potential is 0): the focal length of this type of lens is very small; the lens has little chromatic aberration. In the region $(\varphi_2 - \varphi_1)/\varphi_1 < -1$ the negative potential of the center electrode is so high that electrons leaving the cathode with zero velocity and accelerated by the potential φ_1 cannot overcome the barrier created by the negative potential φ_2 and are reflected; the lens acts like a mirror.

Symmetric lenses are used to produce images in electrostatic electron microscopes and in other devices.

A Three-Electrode Immersion Lens. Three-electrode systems are frequently used in electron guns and for that reason are quite interesting to us. A typical arrangement of a cathode at zero potential $\varphi_c = 0$, a net (Wehnelt) at potential φ_1 that can be positive or negative with respect to the cathode and an anode having a large accelerating potential φ_2 is shown in Fig. 1-10. The equipotential surfaces and electron trajectories are also shown.

The lens power $1/f$ of a three-electrode immersion lens as a function of the ratio of potentials φ_1/φ_2 is shown in Fig. 1-11. As can be seen, the optical power of the lens is high when the grid potential is negative. In order for the anode field to penetrate to the cathode the point in the grid must be substantially larger than the point in the anode and the distance between the grid and anode must be small; in addition, the ratio φ_1/φ_2 must also be small: $\varphi_1/\varphi_2 \approx 0.1$–$0.2$. It is required of electron guns in electron-beam lithography units that the lens magnification be small and then the crossover will be smaller than the emitting region of the cathode. The sizes of the points in the electrodes

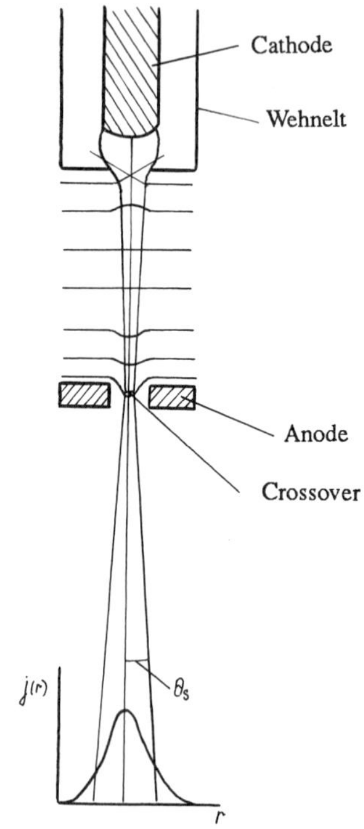

FIGURE 1-10. Diagram of a three-electrode electron gun. The Wehnelt electrode potential determines the shape of the field near the cathode surface. The beam emerging from the gun has a minimal diameter (crossover) d_c and spread angle θ_s.

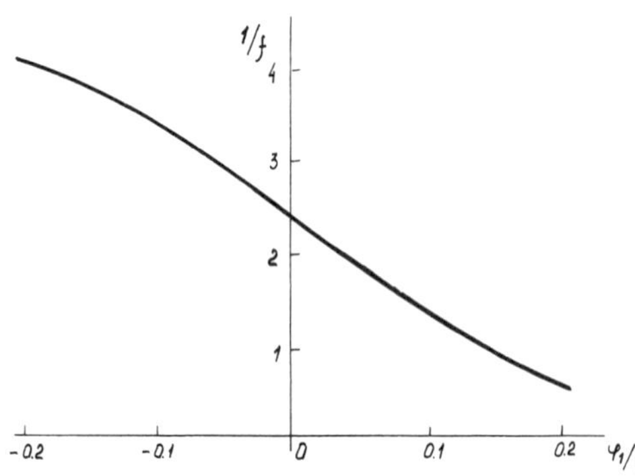

FIGURE 1-11. The lens power of an electron gun as a function of the ratio of potentials at the section (Wehnelt) φ_1 and anode φ_2. The mode in which $\varphi_1 < 0$ is used.

and the potentials must be chosen with due regard for this requirement: a low degree of magnification is achieved when the ratio of potentials is reduced, the diameter of the point in the grid is increased (in so doing, the focal length increases), and the distance between cathode and anode increases.

A special quantity called electron brightness and defined as the current density per unit solid angle is introduced as a characteristic of electron-beam systems. According to Langmuir,[34] the solid angle $\Delta\Omega$ in which electrons emitted by the cathode and accelerated by the potential φ move is determined by the ratio of thermal scatter of the electron's energy to the energy obtained due to acceleration in the field φ:

$$\Delta\Omega = \pi\Delta\varphi/\varphi = \pi kT/e\varphi \tag{1-86}$$

therefore, the electron brightness will be[34]

$$\beta = j_c e\varphi/\pi kT \tag{1-87}$$

where j_c is the thermionic current density at the cathode. The brightness calculated by Langmuir's formula, Eq. (1-87), is usually taken as the upper limit on this quantity; it is lower in actual guns. Brightness decreases because of the fact that the actual scatter of the electrons energy is greater than thermal scattering due to the influence of the space charge at the cathode. The space charge at the cathode can be reduced by raising the field at the cathode, which will be governed by the geometry of the gun electrodes (the cathode Wehnelt, and anode) as well as by the bias voltage on the Wehnelt (a negative potential) and the accelerating potential at the anode of the gun. By lowering the negative potential at the Wehnelt and raising the accelerating potential, we can increase the field at the cathode and lower the scattering of electron energy caused by the space charge. However, the surface area of the cathode having the accelerating potential increases at the same time; the effective area of the cathode increases and the amount of crossover increases. Experimental dependences of the brightness on the indicated factors are given in reference [3]. At a temperature of 2800 K the thermionic current density from a tungsten cathode is 3.5 A/cm^2, the scattering in electrons energy is 4–5 eV, and the brightness of a vee-shaped cathode is 1.5×10^5 A/cm^2 · steradian at an accelerating potential of 20 keV.

The experimental values of brightness can be described by[3]

$$\beta = j_{cr}/\pi\theta_s^2 \tag{1-88}$$

where j_{cr} is the current density at the center of the crossover, and θ_s is the spread of the divergent beam (see Fig. 1-10).

1-10. A GENERAL DESCRIPTION OF AN AXIALLY SYMMETRIC ELECTRIC FIELD[1]

In our previous discussion we used an approximate description of the electric field in an electrostatic lens via the field potential $\varphi(z)$ (and its derivatives) given in the system's axis of symmetry. In doing this the question of how to determine $\varphi(z)$ remained open. We will now return to the task of giving a complete description of the electric field in an axially symmetric system.

The potential of an electric field generated by a system of charges is determined from Poisson's equation:

$$\Delta\varphi = -4\pi\rho(x, y, z) \tag{1-89}$$

We have the following charges: the electrodes that produce a lens carry charges $q_1, \ldots,$ $q_i, \ldots$ on their surfaces and the electron current emitted from the cathode and focused by the lens creates a space charge $\rho(x, y, z)$. The charge q_i raises the potential of the corresponding electrode to a value φ_i determined by the equation $\varphi_i = q_i/c_i$, where

c_i is the electrode's capacitance. The electrode surface is equipotential. As a result, we need not consider the charges on the electrodes specifying instead their potentials $\varphi_1, \ldots, \varphi_i, \ldots$, especially since it is much easier in practice to set (and measure) the potentials by connecting the electrodes to their respective source of dc voltage. Then, instead of solving Poisson's equation, Eq. (1-89), with charges on the electrodes, we can solve the Laplace equation

$$\Delta\varphi = 0 \tag{1-90}$$

with boundary conditions whose meaning will become clear later.

When defining the optical properties of the lenses, we do not consider the field produced by the space charge of the current itself. However, when the current density in an electron beam is high, the field produced by this charge ejects charges from the beam, which increases the cross section of the beam. Clearly, this is extremely undesirable in electron lithography equipment in which, on the one hand, a fairly high current density is required in the electron beam (to reduce the exposure time and increase the equipment's throughput) and, on the other hand, the probe diameter must be less than a micron. The question of beam spreading due to the influence of the intrinsic field was examined in references [17,36].

Laplace's equation in cylindrical coordinates z, r, χ is

$$\frac{\partial^2\varphi}{\partial z^2} + \frac{1}{r}\frac{\partial\varphi}{\partial r} + \frac{\partial^2\varphi}{\partial r^2} + \frac{1}{r^2}\frac{\partial^2\varphi}{\partial\chi} = 0 \tag{1-91}$$

It is worthwhile to examine a few special cases of this equation. If the system electrodes extend a considerable distance along the optical axis in part of the space that is removed from the ends of the electrodes, the field can be described by a two-dimensional Laplace equation in rectangular coordinates:

$$\partial^2\varphi/\partial x^2 + \partial^2\varphi/\partial y^2 = 0 \tag{1-92}$$

The two-dimensional Eq. (1-92) is solved in an extremely elegant fashion using the theory of functions of a complex variable. A solution of Eq. (1-92) is any functions u and v such that a function $\varphi(\xi) = u + iv$ of the complex variable $\xi = x + iy$ is an analytic function, i.e., it satisfies the Cauchy–Riemann conditions:

$$\partial u/\partial x = \partial v/\partial y, \qquad \partial u/\partial y = -\partial v/\partial x \tag{1-93}$$

That the Cauchy–Riemann conditions are a simple alternate form of writing two-dimensional Laplace equations

$$\Delta_{x,y}u = 0, \qquad \Delta_{x,y}v = 0 \tag{1-94}$$

can easily be verified.

For example, the function

$$\varphi(\xi) = c\xi^2 = c(x^2 - y^2) + ic2xy \tag{1-95}$$

is analytic. Then the functions $u = c(x^2 - y^2)$ and $v = 2cxy$ describe families of hyperbolas that are cross sections of equipotential surfaces and the hyperbolas of the lines of force from the electric fields that are normal to them. We can fabricate electrodes in the form of hyperbolas (in cross section), combine them with four equipotentials and charge them to the potentials of these equipotentials. In doing this the field in the space between the electrodes does not change. In this matter we obtain a so-called quadrupole electrostatic lens used for the strong focusing of high-energy beams of charged particles in accelerators. The high focusing ability of quadrupole lenses is due to the fact that an intense focusing field E_r, normal to the z axis along which a beam of particles moves, can be produced in these lenses by applying a large difference of potentials between the lens electrodes (the maximal difference of potentials is governed by the breakdown effect). (In the axially symmetric lenses examined above the focusing component of the field E_r is a small portion of the $\bar{E}$ field directed almost along the lens axis.) We will not examine the optical properties of quadrupole lenses in detail; instead, we refer the interested reader to the monograph.[(2)]

The solution of a two-dimensional axially symmetric Laplace equation describes the field of a charged filament or the field in a tubular capacitor. This solution can be found by directly integrating the equation

$$\frac{\partial^2\varphi}{\partial r^2} + \frac{1}{r}\frac{\partial\varphi}{\partial r} = 0, \qquad \varphi(r) = c\ln r \tag{1-96}$$

which corresponds to the harmonic function $c\ln(x+iy) = c\ln r + i(\Phi + 2k\pi)$. We will return to Eq. (1-91) for an axially symmetric field. This case is the most interesting for us and must be examined in more detail. We shall seek a solution of Eq. (1-91) as a series:

$$\varphi(z,r) = \sum_{n=0}^{\infty} A_n(z) r^n \tag{1-97}$$

$$A_0(z) = \varphi(z,0) \tag{1-98}$$

The first term in the series ($n = 0$) is the function $\varphi(z,0)$ that we introduced earlier and describes the values of the potential on the optical axis. By substituting the series of Eq. (1-97) into Eq. (1-91) we obtain

$$\sum_{n=0}^{\infty} A^{\mathrm{II}}_{nzz} r^n + n^2 A_n(z) r^{n-2} = 0 \tag{1-99}$$

Because z and r are independent variables, Eq. (1-99) will be satisfied only when the coefficients are identically equal to 0 for the r^n in the equation. This gives a recursive relation for A_n:

$$n^2 A_n(z) + A^{\mathrm{II}}_{n-2}(z) = 0$$

$$A_0 \equiv \varphi(z,0), \qquad A_2 = -\frac{1}{2^2}A_0^{\mathrm{II}}(z), \qquad A_4(z) = \frac{1}{64}A_0^{\mathrm{IV}}(z), \ldots \tag{1-100}$$

Thus, the series for the potential and voltage of the focusing and lengthwise fields are

$$\varphi(z,r) = \varphi(z,0) - \frac{r^2}{4}\varphi^{\mathrm{II}}(z,0) + \frac{r^4}{64}\varphi^{\mathrm{IV}}(z,0) + \dots \tag{1-101}$$

$$E_r = -\frac{\partial\varphi}{\partial r} = \frac{r}{2}\varphi^{\mathrm{II}}(z,0) - \frac{r^3}{16}\varphi^{\mathrm{IV}}(z,0) + \dots \tag{1-102}$$

$$E_z = -\frac{\partial\varphi}{\partial z} = -\varphi^{\mathrm{I}}(z,0) + \frac{r^2}{4}\varphi^{\mathrm{III}}(z,0) - \frac{r^4}{64}\varphi^{\mathrm{V}}(z,0) + \dots \tag{1-103}$$

Needless to say, the function $\varphi(z,0)$ has remained undefined; it must be found by way of the boundary conditions given as constant potentials at the electrodes.

Because r and z are points on the electron trajectories, the maximal value of $r = r_m$ in an electron optic system is governed by a restricting diaphragm. In axial beams theory only the first terms of the series E_r and E_z are used to calculate the optical properties of the lenses. The smallness of subsequent terms, e.g., the E_r series, is found from a ratio of the type

$$\frac{E_r^{(3)}}{E_r^{(1)}} = \frac{r^2}{8}\frac{\varphi^{(\mathrm{IV})}(z,0)}{\varphi^{\mathrm{II}}(z,0)} \tag{1-104}$$

The changes in $\varphi(z,0)$ at the extremes of the lens are described by a characteristic length z_0 that has, obviously, the sense of a lens thickness. If we model, e.g., the field of a symmetric lens by the potential

$$\varphi(z,0) = A\exp(-z^2/z_0^2) \tag{1-105}$$

we can easily find

$$\frac{\varphi^{(\mathrm{IV})}}{\varphi^{\mathrm{II}}} = \frac{8z^4/z_0^8 - 56z^2/z_0^6 + 16/z_0^4}{4z^2/z_0^4 - 2/z_0^2} \tag{1-106}$$

or, when $z \ll z_0$, $\varphi^{(\mathrm{IV})}/\varphi^{\mathrm{II}} \simeq -8/z_0^2$ and

$$E_r^{(3)}/E_r^{(1)} \approx r_m^2/z_0^2 \tag{1-107}$$

Thus, the lens is nearly ideal if $r_m^2/z_0^2 \ll 1$. If the object image given by an ideal lens is taken to be an ideal image, then the terms for the fields E_r and E_z that are proportional to higher orders of r will produce distortions in the image, called aberrations.

Different methods have been developed for experimentally modelling the electric field in lenses.[1–3] The most popular of these is the electrolytic bath method. These methods provide a graphic representation and a qualitative analysis of the fields in the lenses. However, quantitative results and optimal lens design performed by computer methods are based, naturally, on solving the Laplace equation for specific electrode potentials.

We will now turn to investigating axially symmetric magnetic electron optical lenses.

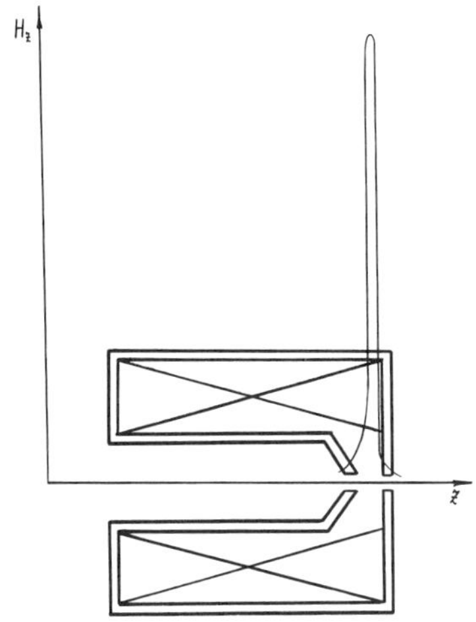

FIGURE 1-12. The lengthwise magnetic field H_z of the solenoid is concentrated in a narrow gap Δz by pole pieces.

1-11. A GENERAL DESCRIPTION OF AN AXIALLY SYMMETRIC MAGNETIC FIELD IN A LENS

Axially symmetric magnetic lenses are solenoids connected to a source of direct current. Magnetic line terminals made of soft iron (Fig. 1-12) are used to concentrate the field in a small interval Δz along the axis of symmetry. We can find an expression for this field as a series in r which is analogous to the series, Eq. (1-97), for an electrostatic field.

The initial equations for finding the magnetic field are the electrodynamics equations:[(1)]

$$\Delta \mathbf{A} = 1/c^2 \partial^2 \mathbf{A}/\partial t^2 - (4\pi/c)\bar{\jmath} \tag{1-108}$$

$$\mathbf{H} = \operatorname{curl} \mathbf{A} \tag{1-109}$$

Here the $\bar{\jmath}$ are the bulk current densities and $\mathbf{A}$ is the vector potential. We are interested in the magnetic field between the pole terminals, where $\bar{\jmath} = 0$. In addition, for contours that do not enclose a current, $\operatorname{curl} \mathbf{H} = 0$. Finally, the field is constant and $\partial^2 \mathbf{A}/\partial t^2 = 0$. Therefore, the system of equations that we must solve is

$$\Delta A = 0, \qquad \mathbf{H} = \operatorname{curl} \mathbf{A}, \qquad \operatorname{curl} \mathbf{H} = 0 \tag{1-110}$$

The boundary conditions are given by the relationship between the magnetic field and the number and distribution of the winding's ampere turns. If the winding is thin, the surface current density is the same as the number of ampere turns per unit length of the winding: $\bar{I} = n\bar{J}$; in turn, $\bar{I}$ is related to the discontinuity of the field in the current surface by Eq. (1-37):

$$\frac{4\pi}{c}\bar{I} = \bar{N} \cdot (\bar{H}_i - \bar{H}_e) \tag{1-111}$$

where $\bar{H}_i$, $\bar{H}_e$ is the field from different sides of the current surface, and $\bar{N}$ is a unit vector normal to the surface.* Assuming that the field outside the winding is $\bar{H}_e = 0$, we obtain the boundary condition for the field inside the winding immediately near the currents surface:

$$\bar{N} \cdot \bar{H}_i = (4\pi/c)n\bar{J} \tag{1-112}$$

It is difficult to give an analytic expression for the relationship between the ampere-turns distribution and the field at the surface when a winding with terminals has a complex geometry. We can assume, as we did for an electrostatic field, that by experience we know the values of the magnetic field on the axis of symmetry of a magnetic lens $H_z(z,0)$. In those cases where it is not possible to reduce the currents in the windings to surface currents we must solve the equation with bulk currents distribution: $\Delta\bar{A} = -(4\pi/c)\bar{\jmath}$.

We will introduce at the point z, r, χ a system of unit vectors for cylindrical system of coordinates $\bar{z}_0 \parallel \bar{z}$, $\bar{r}_0 \parallel \bar{r}$, $\bar{\chi} \perp \bar{r}_0$, and $\bar{z}_0$ and write the vector projections $\bar{H} = \text{curl } \bar{A}$ and curl $\bar{H} = 0$ in these directions:[1]

$$H_z = \frac{1}{r}\left[\frac{\partial(rA_\chi)}{\partial r} - \frac{\partial A_r}{\partial \chi}\right], \qquad H_r = \frac{1}{r}\frac{\partial A_z}{\partial \chi} - \frac{\partial A_\chi}{\partial z} \tag{1-113a}$$

$$H_\chi = \frac{\partial A_r}{\partial z} - \frac{\partial A_z}{\partial r} \tag{1-113b}$$

$$(\text{curl } \bar{H})_z = \frac{1}{r}\left[\frac{\partial(rH_\chi)}{\partial r} - \frac{\partial H_r}{\partial \chi}\right] = 0 \tag{1-114a}$$

$$(\text{curl } \bar{H})_r = \frac{1}{r}\frac{\partial H_z}{\partial \chi} - \frac{\partial H_\chi}{\partial z} = 0 \tag{1-114b}$$

$$(\text{curl } \bar{H})_\chi = \frac{\partial H_r}{\partial z} - \frac{\partial H_z}{\partial r} = 0 \tag{1-114c}$$

Because of the axial symmetry of the field, the derivatives of **A** and **H** with respect to χ are 0; in addition, $H_\chi = 0$. Then, by substituting

$$H_z = \frac{1}{r}\frac{\partial(rA_\chi)}{\partial r}, \qquad H_r = -\frac{\partial A_\chi}{\partial z} \tag{1-115}$$

in Eq. (1-114a) we obtain an equation for finding A_χ:

$$\frac{\partial^2 A_\chi}{\partial r^2} + \frac{1}{r}\frac{\partial A_\chi}{\partial r} - \frac{A_\chi}{r^2} + \frac{\partial^2 A_\chi}{\partial z^2} = 0 \tag{1-116}$$

With A_χ we can find the components H_z and H_r of the magnetic field according to Eq. (1-113).

We will seek a solution of Eq. (1-116) as a series in r:

$$A_\chi = \sum_{n=0}^{\infty} \alpha_n r^{n+1} H_z^{(n)}(z) \tag{1-117}$$

---

*For an infinitely thin and long solenoid having a surface current density $\bar{I} = n\bar{J}$, where n is the number of turns per unit length of winding and J is the current in the winding, the field inside the winding is uniform and equal to $H_i = (4\pi/c)nJ$, which follows directly from Eq. (1-111).

Substituting Eq. (1-117) into Eq. (1-116) and setting the coefficient to 0 when multiplying the independently varying quantities $r^{n-1}H_z^{(n)}(z)$, we obtain a recursive relation for the coefficients α:

$$\alpha_n = \alpha_{n-2}/n(n+2) \tag{1-118}$$

The relation

$$A_\chi^{(0)}(r,z) = 0.5rH_z(z,0) \tag{1-119}$$

determines the first two coefficients:

$$\alpha_0 = 0.5, \qquad \alpha_1 = 0 \tag{1-120}$$

Thus we obtain the series

$$A_\chi(z,r) = \frac{r}{2}H_z(z,0) - \frac{r^3}{16}H_z^{\mathrm{II}}(z,0) + \frac{r^5}{6\cdot 64}H_z^{(\mathrm{IV})}(z,0) + \ldots \tag{1-121}$$

$$H_z(z,r) = H_z(z,0) - \frac{r^2}{4}H_z^{\mathrm{II}}(z,0) + \frac{r^4}{64}H_z^{(\mathrm{IV})}(z,0) + \ldots \tag{1-122}$$

$$H_r(z,r) = -\frac{r}{2}H_z^{\mathrm{I}}(z,0) + \frac{r^3}{16}H_z^{\mathrm{III}}(z,0) - \frac{r^5}{6\cdot 64}H_z^{\mathrm{IV}}(z,0) + \ldots \tag{1-123}$$

$$H(r,z) = (H_r^2 + H_z^2)^{1/2} \tag{1-124}$$

for the A_χ and the components of the magnetic field. In a first approximation the axial component of the field does not depend on r and in the next approximation decays as $(-r^2)$. The radial component of the field increases in proportion to r, however, in a magnetic field the focusing action of a lens is governed not by the radial, but by the axial component of the field; the radial component rotates (turns) the electron (and the image) around the axial axis.

A solution was found in reference [38] for the magnetic field of a toroidal conductor carrying a current j. Let the conductor cross section be rectangular and limited by the cylindrical coordinates $\rho_a \le \rho \le \rho_b$ and $z_a \le z \le z_b$. This geometry models a solenoid that is $z_b - z_a$ units long and $\rho_b - \rho_a$ units thick with winding having a uniform cross section. The equation for the azimuthal component of the vector potential A_χ is[37]

$$\left(\frac{\partial^2}{\partial\rho^2} + \frac{1}{\rho}\frac{\partial}{\partial\rho} - \frac{1}{\rho^2} + \frac{\partial^2}{\partial z^2}\right)A_\chi = \begin{cases} 0, & b \in j = 0 \\ -\dfrac{4\pi}{c}j, & b \in j \neq 0 \end{cases} \tag{1-125}$$

under the boundary conditions $A_\chi \to 0$ as $\rho \to \infty$, $A_\chi \to 0$ as $\rho \to 0$. The integral representation of the solution, Eq. (1-125), is

$$A_\chi(\rho,z) = -\frac{j}{c}\int_{\rho_a}^{\rho_b}\int_{z_a}^{z_b}\int_0^{2\pi}\frac{\cos\chi\, d\chi\, dz_1\, \rho_1\, d\rho_1}{r_1} \tag{1-126}$$

$$r_1^2 = \rho^2 + \rho_1^2 - 2\rho\rho_1\cos\chi + (z_1 - z_2)^2 \tag{1-127}$$

Expanding the subintegral of Eq. (1-126) in a series yields

$$A_\chi(\rho, z) = 4\pi/c \sum_{n=0}^{\infty} \rho^{2n+1} I_n(z) \tag{1-128}$$

The null function $I_0(z)$ is found by direct integration of Eq. (1-126) when $\rho = 0$:

$$I_0(z) = j/4 \left[\left[w \ln |x + r| \right]_{\rho_a}^{\rho_b} \right]_{z_a}^{z_b} = j/4 \ln \frac{\left| \rho_b + \sqrt{\rho_b^2 + (z_b - z)^2} \right|^{z_b - z}}{\left| \rho_a + \sqrt{\rho_a^2 + (z_a - z)^2} \right|^{z_a - z}} \tag{1-129}$$

Substituting the series, Eq. (1-128) into Eq. (1-125) defines a recursive relation between the functions $I_n(z)$:

$$I_n(z) = -I_{n-1}^{\mathrm{II}}(z)/4n(n+1), \qquad n = 1, 2, \ldots \tag{1-130}$$

From here it follows that all functions I_n are given in terms of I_0:[38]

$$I_n(z) = \frac{(-1)^n I_0^{(2n)}(z)}{2^{2n} n!\,(n+1)!} \tag{1-131}$$

If we introduce other functions $A_{n+1}(z)$ via the relation $A_{n+1}(z) = (2/n!) I_0^{(n)}(z)$ and calculate the magnetic field components according to Eq. (1-113), it is easy to write the first three terms of the respective series:

$$A_\chi = \frac{4\pi}{c} \left(\frac{\rho}{2} A_1 - \frac{\rho^3}{8} A_3 + \frac{\rho^5}{16} A_5 + \ldots \right) \tag{1-132}$$

$$H_z = \frac{4\pi}{c} \left(A_1 - \frac{\rho^2}{2} A_3 + \frac{3\rho^4}{8} A_5 + \ldots \right) \tag{1-133}$$

$$H_\rho = \frac{4\pi}{c} \left(\frac{\rho}{2} A_2 - \frac{3\rho^3}{8} A_4 + \frac{5\rho^5}{16} A_6 + \ldots \right) \tag{1-134}$$

The functions $A_2, A_3, \ldots$ can easily be found by differentiating the expression $A(z) = 2I_0(z)$ in Eq. (1-129). We will not write the expressions obtained when we do this because of their awkwardness.

The example presented shows that solving Poisson's equation, Eq. (1-125), having boundary conditions that are set by the distribution of the currents in space makes it possible in principle to find the field for any optical system including, naturally, the deflection systems. When the geometry of the currents is complex it becomes necessary to resort to the computer to calculate the fields.

Numerical values obtained for ρ that are small in comparison with ρ_a and ρ_b ($\rho \leq 0.4$, $\rho_a = 90, \rho_b = 110$) considering only the three terms from the expansion [Eqs. (1-132)–(1-134)] gave results that almost matched the exact solution.[38]

1-12. THE EQUATION FOR THE TRAJECTORY OF ELECTRONS IN A MAGNETIC FIELD

A magnetic field acts on a moving electron with the Lorentz force

$$\bar{F} = -\frac{e}{c}\bar{v} \times \bar{H} \tag{1-135}$$

By writing $\bar{v}$ and $\bar{H}$ via projections in cylindrical coordinates, we find

$$\bar{v} \times \bar{H} = \bar{r}_0(v_\chi H_z) + \bar{\chi}_0(v_z H_r - v_r H_z) + \bar{z}_0(-v_\chi H_r) \tag{1-136}$$

for the projection of the vector product. An electron in an axially symmetric magnetic field moves in a complex manner: the $F_\chi = -(e/c)(v_z H_r - v_r H_z)$ component of the Lorentz force rotates (curls) the electron around the axis of symmetry; the radial component $F_r = -(e/c)v_\chi H_z$ alters the distance of the electron from the axis, i.e., governs the focusing action; the axial component $F_z = (e/c)v_\chi H_r$ of the force changes the lengthwise velocity component of the electron. (Although the magnetic field does no work and does not change the total energy of the electron, because the electron changes direction the energy of lengthwise motion may change after passing through a magnetic lens).

The equations for electron motion can be written in cylindrical coordinates as

$$\frac{d}{dt}(mr^2\dot{\chi}) = rF_\chi = -\frac{er}{c}(v_z H_r - v_r H_z) \tag{1-137a}$$

$$m\ddot{r} = -\frac{e}{c}(r\dot{\chi})H_z + mr\dot{\chi}^2 \tag{1-137b}$$

$$m\ddot{z} = \frac{e}{c}v_\chi H_r = \frac{er\dot{\chi}}{c}H_r \tag{1-137c}$$

The first equation was written as a moments equation. By substituting the first term of the series, Eq. (1-123), into it instead of H_r, we convert it to

$$\frac{d}{dt}(mr^2\dot{\chi}) = \frac{e}{c}\left(\frac{r^2}{2}\frac{dH_z}{dz}\frac{dz}{dt} + r\frac{dr}{dt}H_z\right) \equiv \frac{e}{c}\frac{d}{dt}\left(\frac{r^2 H_z}{2}\right) \tag{1-138}$$

Integrating Eq. (1-138) gives

$$mr^2\dot{\chi} = \frac{e}{c}\frac{r^2 H_z}{2} + \text{const} \tag{1-139}$$

The constant of integration is found from the initial conditions: at the input to the lens (when $H_z = 0$) there is no rotation, $\dot{\chi} = 0$, and const = 0. Thus, for the angular velocity of an electron we obtained the Larmor frequency:

$$\dot{\chi} = \frac{eH_z}{2mc} \tag{1-140}$$

Further integration of Eq. (1-139) proceeds easily, but holds little interest:

$$\dot{\chi} = \frac{d\chi}{dt} = \frac{d\chi}{dz}\frac{dz}{dt} = v_\chi \frac{\partial \chi}{\partial z} = \frac{eH_z}{2mc} \tag{1-141}$$

$$\Delta\chi_{AB} = \int_A^B d\chi = \frac{e}{2mc}\int_A^B \frac{H_z}{v_z}\,dz \tag{1-142}$$

The result, Eq. (1-142), determines the total rotation of an electron as it passes through a magnetic lens.

Substituting the value of $\dot{\chi}$ found from Eq. (1-140) into the equation for radial motion, we obtain

$$m\ddot{r} = -\frac{1}{4}\frac{e^2H_z^2r^2}{mc^2} \tag{1-143}$$

Using the identity $\ddot{r} = v_z^2 r''$ (the dots denote differentiation with respect to time, the primes denote differentiation with respect to z), we derive from Eq. (1-143) an equation for the trajectory of an electron in a magnetic lens:

$$\frac{d^2r}{dz^2} + \frac{r}{\lambda^2} = 0, \qquad \frac{1}{\lambda^2} = \frac{e^2H_z^2}{4m^2c^2v_z^2} \tag{1-144}$$

If at some distance along the z axis H_z and v_z can be considered constants, then the solution of Eq. (1-144) will be periodic (sinusoidal):

$$r(z) = c_1 \sin(z/\lambda) + c_2 \cos(z/\lambda) \tag{1-145}$$

the electron repeatedly crosses the system's axis of symmetry. However, in practice we deal with thin lenses in which these conditions are not valid; the field H_z is nonzero and changes rapidly over a narrow range of z values. It is important to note that the trajectory of a charged particle in a magnetic field depends on the ratio e/m; the scaling theorem is not valid for magnetic lenses.

1-13. THE FOCAL LENGTH OF A THIN, WEAK MAGNETIC LENS

We will present Eq. (1-144) as $d(dr/dz) = -(e^2H_z^2r/4m^2c^2v_z^2)dz$ and integrate from point A in the object plane to point B in the image plane (see Fig. 1-13):

$$\left.\frac{dr}{dz}\right|_B - \left.\frac{dr}{dz}\right|_A = -\frac{e^2}{4m^2c^2}\int_A^B \frac{H_z^2 r}{v_z^2}\,dz \tag{1-146}$$

We will make some simplifying assumptions when we integrate Eq. (1-146): we will say that within the lens the change in radial distance r and in the lengthwise component of velocity v_z can be neglected and take these quantities outside the integral on the right; we find the slopes of the trajectories on the left by the ratio of the distances:

$$\left.\frac{dr}{dz}\right|_B = -\frac{r_0}{b}, \qquad \left.\frac{dr}{dz}\right|_A = \frac{r_0}{a} \tag{1-147}$$

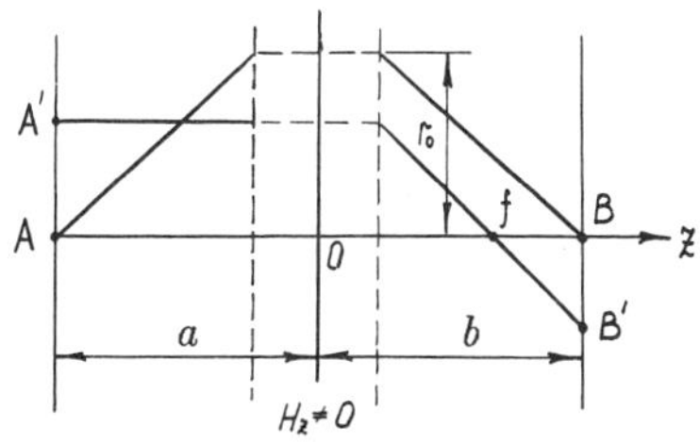

FIGURE 1-13. A thin, weak lens. The focal length of the lens is governed by the longitudinal component $H_z(z)$ of the magnetic field.

from which we obtain

$$\frac{r_0}{b} + \frac{r_0}{a} = \frac{e^2 r_0}{4m^2c^2v_z^2} \int_A^B H_z^2(z)\, dr = \frac{r_0}{f} \tag{1-148}$$

As we can see, we obtain the formula

$$\frac{1}{f} = \frac{e^2}{4m^2c^2v_z^2} \int_A^B H_z^2\, dz \tag{1-149}$$

for the focal length of a thin magnetic lens. Notice first of all that the focal length of a magnetic lens as an integral of H_z^2 is always positive; i.e., a magnetic lens is always a converging lens. Focal length is inversely proportional to the integral of the square of the lengthwise component of the magnetic field, which means that the relation

$$\frac{1}{f} = kJ^2 \tag{1-150}$$

between the focal length and current J in the lens winding is valid. On the other hand, f is proportional to the electron energy $mv_z^2/2 = e\varphi$. The established electron-beam focus will not be disrupted if the accelerating potential φ and current in the lens winding is varied in such a way that

$$J^2/\varphi = \text{const} \tag{1-151}$$

Obviously, concentrating the field in a narrow region Δz increases the integral of H_z^2 and decreases the focal length of the lens.

For a given lens winding geometry we can find an empirical formula for $H_z(z)$, e.g., of the type

$$H_z(z) = H_0 \exp(-z^2/a^2) \tag{1-152}$$

and calculate an explicit focal length. When we do this, we find that $f \sim a$, i.e., a less concentrated field provides less focusing.

More precise calculations of electron trajectories in a magnetic lens give us the ability to determine its aberrations. When performing these calculations, we must take how H_z depends on v_z and r into account.

We will point out how the focal length of a magnetic lens depends on a particle's charge-to-mass ratio e/m; at a fixed level of ion energy $mv_z^2/2 = e\varphi$, f is proportional to m. This leads to an excessive increase in the focal lengths of the lenses and the size of the optical column of a system when working with ion beams. For this reason electrostatic lenses are used primarily in ion-beam systems and the focal lengths of these lenses are independent of e/m.

1-14. THE DEFLECTION SYSTEM FOR ELECTRON BEAMS

In electron-beam lithography the resist is exposed according to a well-defined procedure. Two types of exposure systems have been distinguished: raster and vector. Raster systems operate on the principle of producing a television image: the deflection systems then move the beam over every point on the surface being scanned; when the beam passes over those areas on the surface that are not meant to be exposed, it is deflected by special "blanking" electrodes. Obviously, the major drawback to a raster system is the loss of operating time and reduced equipment throughput due to the unnecessary passing through unexposed areas by a switched "beam." This drawback is eliminated in vector systems: the electron beam touches only those surface areas of the resist that are to be exposed. The difficulties in producing such a system stem from the need for a precise beam extraction at the origin of the point where the next area on the resist surface is to be exposed. The software for a vector system is more complicated than that for a raster system.

A description of electron lithography systems as a whole will be given later; for now we will return to the relationship between beam deflection and the fields caused by this deflection.

A magnetic field deflects the electron beam in a direction perpendicular to the axis along which the electrons move in the system. Just as we did earlier, we will say that the electrons move along the z axis; the magnetic field H_x is directed along the x axis, and the electron beam is deflected along the y axis. The equation for electron motion in a deflecting field H_x is

$$m\ddot{y} = -\frac{e}{c} v_z H_x \tag{1-153}$$

Using the identity $\ddot{y} = v_z^2 y''_{zz}$ and integrating with respect to z (we say that the field H_z is uniform within the limits of integration from $z = 0$ to z and v_z = const), we obtain

$$y'_z = \frac{dy}{dz} = -\frac{eH_x}{mcv_z} z \tag{1-154}$$

If the deflecting field H_x is generated in the interval Δz, the deviation of the electron trajectory from the z axis at the point where it leaves the deflecting field H_x will be

$$\left.\frac{dy}{dz}\right|_{\Delta z} = -\frac{eH_x}{mcv_z} \Delta z \tag{1-155}$$

After leaving the deflection system the beam moves a distance L before it impinges upon the surface of a wafer with a resist and then the deflection Y of the beam will be

$$Y = L \left.\frac{dy}{dz}\right|_{\Delta z} = -\sqrt{\frac{e}{2mc}} \frac{H_x \, \Delta z \, L}{\varphi^{1/2}} \tag{1-156}$$

The sensitivity to deflection by a magnetic field is

$$S_H = \frac{Y}{H_x} = -\sqrt{\frac{e}{2mc}} \frac{\Delta z\, L}{\varphi^{1/2}} \tag{1-157}$$

As can be seen, sensitivity decreases when the energy of the electrons $e\varphi$ increases as $\varphi^{-1/2}$, i.e., rather slowly.

Performing a similar calculation for the sensitivity to deflection by an electric field, which we will not show here because of its simplicity, we obtain

$$S_E = Y/E_y = \Delta z\, L/2\varphi \tag{1-158}$$

A deflecting electric field is applied, naturally, along the deflection of y. In this case sensitivity decreases in inverse proportion to the energy of the electrons. The higher sensitivity to deflection by a magnetic field when the electron energy is high makes magnetic deflection the preferred method.

The sensitivity S_H to deflection in a magnetic field depends on the e/m ratio: $S_H \sim (e/m)^{1/2}$ because S_E is independent of e/m. In working with ion beams the smaller the S_H, the greater the ion mass m. For this reason electrostatic systems are, as a rule, used to deflect ion beams.

Nonuniformities in the deflecting magnetic field H_x are a source of additional aberrations. The field H_x is nonuniform in the O_{xy} plane as well as in the z direction (due to edge effects). Let the beam entering a deflecting field have a circular cross section. At those points on the beam cross section where H_x is larger, the electrons, in keeping with Eq. (1-155), undergo greater deflection. Thus the cross section of the beam ceases to be circular. Further, in producing a pattern of limited size we must force a ray to pass through various areas of the deflecting field; nonuniformities in the field $\Delta H_x(x, y, z)$ in these sections will be caused by nonuniformity in the sensitivity ΔS_H and a distortion-like alteration of the image. Aberrations introduced by the deflecting fields limit the extent to which a ray can be deflected. For example, if the beam diameter is 0.2 μm and we wish the resolution to be no worse than 0.2 μm over the entire surface of exposure then the maximal deflection of a ray must be such that the aberrations caused by deflection must remain much less than 0.2 μm.

1-15. ABERRATIONS IN ELECTRON-BEAM SYSTEMS[1,3]

The DeBroglie wavelength for electrons accelerated by a potential 10^4 eV is negligibly small in comparison with the wavelength of visible light 0.1220×10^{-8} cm. This wavelength determines the diffraction limit of the resolution obtained by electron beams. The sizes of the electron-beam cross sections that are actually attainable are substantially larger: they are governed by the effects of aberrations in electron lenses. Correcting aberrations in electron optical systems is much more complicated than in ordinary optical systems with glass lenses. (However, in the final analysis electron lenses give better resolution because the wavelength of electrons is 10^4 less than the wavelength of visible light.)

In electron lithography systems we are interested in the diameter of the electron beam that exposes the resist. Electron guns, described above, produce an electron beam whose minimal cross section, called crossover, has a diameter on the order of 100 μm and the divergence angle θ_s in this cross section is roughly 10^{-2} radian. Subsequent electron lenses form a reduced crossover image in the plane of the surface of the resist film being exposed. For this reason aberrations in lenses that create a crossover image will, in the final analysis, govern the beams cross section and the resolution of the device pattern being exposed.

The theory of aberrations in electron lenses is quite simple in principle and amounts to solving the problem of the electron's trajectory in lenses: in doing this the exact equations for the electrostatic field components, Eqs. (1-102) and (1-103), and for the magnetic field, Eqs. (1-122) and (1-123), must be used.

We will present the mathematical problem of aberrations in electron optics systems in a form that lends itself to machine computations.[39] We will introduce a Cartesian coordinate system in the form $w = x + iy,\ z$. The unknown equation of a ray is a function $w(z) = x(z) + iy(z)$. If the rays are not paraxial, i.e., they make a trace some distance from the optical axis of the system and form large angles with it, the object image created by these rays will be distorted because of geometric aberrations. Of course, we can introduce diaphragms into the system that will eliminate those rays that are some distance from the optical axis (and have large deflection angles). However, this leads to a reduction in beam intensity and, in addition, small diaphragms amplify the so-called diffraction blurring. In particle optics the latter effect is small and ordinarily not taken into account, although it does play a major role in visible light optics. Besides geometric aberrations, there are chromatic aberrations that are governed by the fact that a longer focal length in the lens corresponds to particles having greater energy. If the particles have some energy distribution, then the image of a point on the object will be a spot of limited size. Mechanical inaccuracies in fabricating and adjusting the lens electrodes relative to the optical system axis also distort the image of an object, and can be thought of as aberrations.

In a quantitative sense the extent of geometric aberrations is determined by solving the ray equation in a paraxial approximation: $u_g = w(z) - w_p(z)$, u_g is geometric aberration and w_p is the solution of the ray equation in a paraxial approximation. We will define chromatic aberration. Let $w(z)$ be a solution of the ray equation for some nominal value of the particles accelerating potential φ. We increment the accelerating field by an amount $\Delta\varphi$ and again find a solution $w_c(z)$ that corresponds to the field $\varphi + \Delta\varphi$. The difference $w_c - w$ then defines the chromatic aberration $u_c = w_c - w$. The total aberration is determined by the difference $u_{gc} = w_c - w_p = u_c + u_g$.

Perturbation methods are frequently used in aberrations theory, keeping terms of a specific order of smallness in the ray equations. In doing this the transverse coordinate w and its derivatives $w' = dw/dz$, i.e., the deflection angle of a ray to the z axis of the system, emerge as the small parameters. Because the initial coordinates of a particle when it leaves an object w_0, w_0' define its coordinates w, w' for any z along the system axis as it moves along a trajectory (ray), aberrations are frequently written as functions of the initial coordinates w_0, w_0' of the particle. Geometric aberrations that depend only on the deflection angle of the trajectory w_0, w_0' are called spherical aberrations. The same geometric aberrations that depend on the point at which the particle emerges are called field aberrations.

In machine computations it turns out that looking for aberrations as the difference in the solutions of the equations for nonparaxial and paraxial rays is inconvenient. The fact of the matter is that in this type of computation the values of the aberration may contain large errors because their small values will be close to the errors in computing the "large" values of w and w_p. Therefore, more accurate values of the aberrations can be obtained by solving special aberrations equations. We will write out the corresponding equation.[39]

A general equation for the rays $w(z) = x(z) + iy(z)$ is

$$w'' + \frac{1+2\varepsilon\varphi}{2\varphi(1+\varepsilon\varphi)}(1+w'\bar{w}')(E_w - w'E_z) + i\eta\left(\frac{1+w'\bar{w}'}{\varphi(1+\varepsilon\varphi)}\right)^{1/2} \times \left(H_w - \frac{w'^2\bar{H}'w}{2} + \frac{w'\bar{w}'H_w}{2} - w'H_z\right) = 0 \qquad (1\text{-}159)$$

Here the dependence of electron mass on its energy (in the nonrelativistic limit $\varepsilon\varphi = e\varphi/2mc^2 \to 0$); $\eta = (e/2m)^{1/2}$; E_w, E_z, H_w, H_z are, respectively, the transverse and lengthwise components of the electric and magnetic fields in an electron optic system; φ is the electrostatic potential; the prime denotes differentiation with respect to z, and the bar above a symbol denotes the complex conjugate of a number.

A ray equation in a paraxial approximation is obtained from Eq. (1-159) by replacing the field voltages with their values on the axis of the optical system and discarding the small squared w_p terms of the equation:

$$w_p'' + \frac{1+2\varepsilon\varphi_0}{2\varphi_0(1+\varepsilon\varphi_0)}(E_{w_1} - w_p'E_{z_0}) + i\eta(\varphi_0(1+\varepsilon\varphi_0))^{1/2}(B_{w_1} - w_p'B_{z_0}) = 0 \quad (1\text{-}160)$$

Here $\varphi_0 = \varphi(z;0)$ and H_{z_0} are the electrostatic potential and magnetic field on the axis, $E_{z_0} = -\varphi_0'$, and the transverse components of the fields are

$$\begin{aligned} E_{w_1} &= \frac{1}{2}w_p\varphi_0'' + E_{wd} \\ H_{w_1} &= -\frac{1}{2}w_pH_{z_0}' + H_{wd} \end{aligned} \qquad (1\text{-}161)$$

Here the components of the deflecting electric and magnetic fields E_{wd} and H_{wd} have been isolated and are respectively proportional to the potential V on the deflecting plates and the current I in the deflection coils:

$$E_{wd} = -\Phi_1 V(t), \qquad H_{wd} = \Psi_1 I(t)$$

The equation for a geometric aberration u_g can be obtained by substituting the expansion $w = w_g + u_g$ into Eq. (1-159) and subtracting Eq. (1-160) from it in a paraxial approximation. The result is[39]

$$u_g'' - (A_g E_z + i\eta C_g H_z)u_g' + A_g w'\bar{w}'(E_w - w'E_z) + A_g(E_w - E_{w_1})$$
$$+ B_g E_{w_1} - w_p'\{A_g(E_z - E_{z_0}) + B_g E_{z_0}\} + \frac{1}{2} i\eta C_g (w'\bar{w}' H_w - w'^2 \bar{H}_w$$
$$+ i\eta\{C_g(H_w - H_{w_1}) + D_g B_{w_1}\} - i\eta w_p'\{C_g(H_z - H_{z_0}) + D_g B_{z_0}\} = 0$$
$$A_g = (1 + 2\varepsilon\varphi)/2(1 + \varepsilon\varphi)\varphi$$
$$B_g = -(\varphi - \varphi_0)\{1 + \varepsilon(\varphi + \varphi_0) + 2\varepsilon^2\varphi\varphi_0\}[2\varphi\varphi_0(1 + \varepsilon\varphi_0)(1 + \varepsilon\varphi)]^{-1}$$
$$C_g = [(1 + w'\bar{w}')/(1 + \varepsilon\varphi)\varphi]^{1/2}$$
$$D_g = -\frac{(\varphi - \varphi_0)\{1 + E(\varphi + \varphi_0) + w'\bar{w}'\varphi_0(1 + \varepsilon\varphi_0)}{\varphi\varphi_0(1 + \varepsilon\varphi_0)(1 + \varepsilon\varphi)}$$
$$\times \left[\left(\frac{1 + w'\bar{w}'}{\varphi(1 + \varepsilon\varphi)}\right)^{1/2} + \left(\frac{1}{\varphi_0(1 + \varepsilon\varphi_0)}\right)^{1/2}\right]^{-1} \tag{1-162}$$

The total (geometric+chromatic) aberration is $u_{gc} = w_c - w_p$, and w_c is a solution of Eq. (1-159) that corresponds to the accelerating potential $\varphi = \Delta\varphi$ and satisfies Eq. (1-162). We find the chromatic aberration by subtracting the two solutions of Eq. (1-162): $w_c = u_{gc} - u_g$.

Computer programs that allow us to study the relationships between the various aberrations for specific optical systems have been written[39] to solve aberration problems according to the equations described.

We will present formulas that determine the magnitude of the most important types of aberrations.

Spherical aberration describes the image blurring caused by the dependence of the location of the image point on the angle that the rays emerge from the object (e.g., the object may be a crossover and the angles θ_c lie within the electron's cone of divergence from the crossover). This dependence is rewritten by the equations[3]

$$x_i - Mx_0 = M[c_s\alpha_0(\alpha_0^2 + \gamma_0^2)] \tag{1-163}$$
$$y_i - My_0 = M[c_s\gamma_0(\alpha_0^2 + \gamma_0^2)] \tag{1-164}$$

where $\alpha_0 = dx/dz|_{z_0}$ and $\gamma_0 = dy/dz|_{z_0}$ are the projected deflection angles of the trajectory of an electron in the plane of the object (crossover), c_s is the spherical aberration coefficient, and M is magnification due to the optical system. Because of spherical aberration, a point on the object will be imaged in the image plane by a circle of radius $r_s = Mc_s\theta_s^3$, where θ_s^3 is equal to the maximal values of α_0 and γ_0. Spherical aberration is strongly manifested in objective lenses in which the rays are moved at large angles α and γ to the optical axis.

The aberrations known as distortion are defined by a nonlinear dependence of the image point location on the location of a point in the object plane (D is the distortion coefficient):[3]

$$x_i - Mx_0 = M[Dx_0(x_0^2 + y_0^2)] \tag{1-165}$$
$$y_i - My_0 = M[Dy_0(x_0^2 + y_0^2)] \tag{1-166}$$

The initial coordinates, x_0 and y_0, of an object point uniquely define the position x_i and y_i of the image point. This means that distortion does not deteriorate the sharpness of an image, but causes a displacement that leads to image deformation (in our case to a change in the shape of the beams cross section). Distortion may lead to cushion distortion and barrel distortion of the square or coma distortion (the anisotropic distortion inherent to magnetic lenses). Distortion-like aberrations manifest themselves in projection lenses in which the electron's trajectory is almost parallel to the optical axis, but at substantial distances x_0 and y_0 from it.

Chromatic aberration in lenses is a result of the focal length being dependent on the electron energy: electron energy scattering $\Delta\varphi/\varphi$ leads to the following relationships between the x_0 and y_0 coordinates of an object point and the x_i and y_i coordinates of an image point:[3]

$$x_i - Mx_0 = M(c_c\alpha_0 + c_M x_0 - c_\theta y_0)\,\Delta\varphi/\varphi \tag{1-167}$$

$$y_i - My_0 = M(c_c\gamma_0 + c_M y_0 + c_\theta x_0)\,\Delta\varphi/\varphi \tag{1-168}$$

where c_c is the axial chromatic coefficient; c_M is the coefficient of chromatic aberration in magnification; and c_θ is the coefficient of anisotropic chromatic aberration, inherent to magnetic lenses. The aberration described by the terms $c_c\alpha_0$ and $c_c\gamma_0$ plays a major role in objective lenses and aberrations having the c_M and c_θ coefficients play a major role in projection lenses.

Aberration coefficients are the squares of complex functions of the fields at the system axis and of the transverse coordinates. By way of example we will present the formulas for the coefficients of spherical and chromatic aberrations in a magnetic lens, obtained from the theory of trajectories:[1]

$$c_s = \frac{\eta^2}{48\varphi}\int_{z_0}^{z_i}\left(\frac{4\eta^2}{\varphi}H_z^4 + 5H_z'^2 - H_zH_z''\right)r^4(z)\,dz \tag{1-169}$$

$$c_c = -\frac{\eta^2}{4\varphi}\int_{z_0}^{z_i} H_z^2 r^2(z)\,dz, \qquad \eta^2 = \frac{e}{2m} \tag{1-170}$$

The integration is performed from the object plane to the image plane; the energy $e\varphi$ of an electron near the lens is constant, and $r(z)$ is the deflection of an electron from the system's optical axis. The coefficients c_s and c_c are, respectively, functions of the fourth and second degrees of the distance r_z. They have the dimensions of length; the extent of their dependence on the fields H_z can vary from 1 mm to several cm.[1,3]

The lens system in electron-beam lithography equipment forms a reduced image of the electron gun crossover on the surface of the wafer being exposed. The dimensions of the image will be changed because of spherical, chromatic, and other types of aberration and diffraction. For any given probe current we can find the minimal probe diameter achievable at any level of magnification in the optical system.

By definition, the brightness of a virtual source of electrons at a crossover is

$$\beta = j_c/\pi\alpha_c^2 \tag{1-171}$$

where j_c is the current density at the crossover. We find the probe current by multiplying the current density by the cross-sectional area of the beam at the crossover:

$$j_s = j_c \cdot \pi r_c^2 = \beta\pi^2\alpha_c^2 r_c^2 \tag{1-172}$$

From the Helmholtz equation we have $\alpha_c r_c = \alpha r_i$, where r_i is the radius of the image and α is the beam convergence angle at the image plane. We can now easily write the formula for the probe diameter at the image plane:

$$d_i^2 = 4j_s/\beta\pi^2\alpha^2 \tag{1-173}$$

This relation is for a system without aberrations and it establishes the probe diameter as a function of probe current, gun brightness, and angle α.

We will now look at spherical aberration in the objective lens of a system. This type of aberration is predominant in systems having a large degree of linear reduction. As a result of spherical aberration every point on the object is imaged by a spot of diameter $d_s = 0.5c_s\alpha^3$, where c_s is the spherical aberration coefficient. We can assume that the probe diameter, with due regard for aberration, can be found from

$$d^2 = d_i^2 + d_s^2 \tag{1-174}$$

The squares of the diameters are additive quantities, i.e., they can be thought of as the variances in Gaussian distributions. It is easy to show that a minimum for d^2 occurs when $d_i = d_s$; in this case

$$\begin{gathered} d^2 = 2d_s^2 = 0.5c_s^2\alpha_{\text{opt}}^6, \qquad \alpha_{\text{opt}} = (2d^2/c_s^2)^{1/6} \\ j_s = k\beta d^{8/3}c_s^{-2/3}, \qquad k \equiv \pi^2/2^{5/3} \\ d = j_s^{3/8}c_s^{1/4}/(k\beta)^{3/8} \end{gathered} \tag{1-175}$$

As can be seen from Eq. (1-175), for any probe current j_s the probe diameter increases with the spherical aberration coefficient, but decreases as the brightness β of the virtual source increases.

When the contributions from chromatic aberration and diffraction are considered, we have[3]

$$\begin{gathered} d^2 = d_i^2 + d_s^2 + d_c^2 + d_d^2 \\ d_c = c_c\alpha\,\Delta\varphi/\varphi, \qquad d_d = 1.22\lambda/\alpha \end{gathered} \tag{1-176}$$

In keeping with Eq. (1-176) the beam diameter is given as a function of the convergence angle α by a function of the type $d^2 = a\alpha^{-2} + b\alpha^2 + c\alpha^6$. A minimum for $d(\alpha)$ occurs when α is[3]

$$\alpha = [-b/6c + (b^2 + 12ac)/6c]^{1/4} \tag{1-177}$$

The relationships obtained allow us to evaluate the different factors in producing electron beams having a given cross section.

What has been said relates to beams whose current in the cross section is Gaussian distributed. A method of producing a beam whose size and shape are variable is frequently employed in lithographic equipment. In this case it is desirable to have a constant probe current density on the target in the form of a figure spot and wherever possible a sharp decrease in current at the edge of the spot. In this case the spot edge will be determined by aberrations that cause a Gaussian decrease in the current density at the beam edge with parameters of the type given by Eq. (1-176). In conclusion we will point

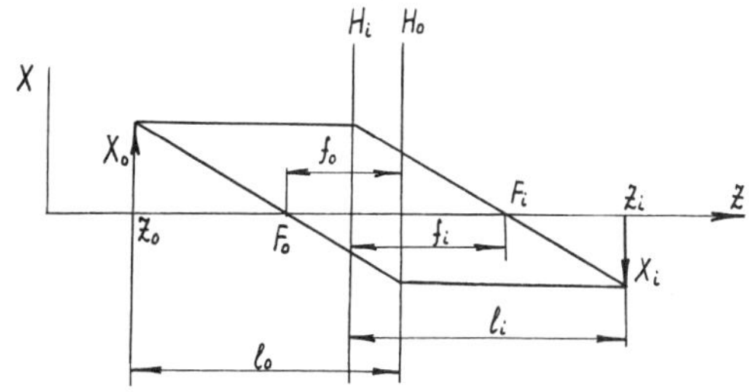

FIGURE 1-14. Cardinal elements of a lens: H_0, H_i are the principal planes, F_0, F_i are the focal planes in the object and its image spaces, and f_0, f_i are the respective focal lengths.

out some works that have been devoted to studying aberrations in electrostatic[40–42] and magnetic lenses.[43,44] The works[43,44] devoted to theoretical and experimental research into correcting solenoidal magnetic lenses by means of the space charge in electron beams are of particular interest.

1-16. THE CARDINAL ELEMENTS OF A LENS

Every lens is characterized by cardinal elements that completely describe its properties in an optical system. We will take two rays parallel to the optical axis of the system in the object and image spaces, respectively (Fig. 1-14). The rays in the object space and parallel to the axis are described by an equation $x_{\|}(z) = x_0 x_2(z)$ and another set of rays that are parallel but deflected from the axis (slope x_0') are described by an equation $x_1(z) = x_0' x_1(z) + x_0 x_2(z)$ (x_1, x_2—x-projections of the rays $r_1(z)$ and $r_2(z)$) (see Section 1-8, Fig. 1-5). For all rays $x_{\|}(z)$ when $x_2(z) = 0$ we have $x_{\|}(z) = 0$; i.e., they intersect the z axis at a single point z_{F_i}, called the focus in the images space. Further, for all deflected rays when $z = z_{F_i}$ we have $x_1(z_{F_i}) = x_0' x_1(z_{F_i})$; i.e., they all intersect at a single point $z = z_{F_i}$ in a plane perpendicular to the z axis. Consequently, parallel sets of rays are focused at points in this plane. For this reason, we call it a focal plane (in the images space).

The location of an image plane z_i is governed by the condition that all rays leaving a point $x_0 = \text{const}$, on the object intersect at its points, and x_0' is variable. This condition gives an equation $x_i(z) = x_0 x_2(z) + x_0' x_1(z) = \text{const}$ that defines the location of the image plane z.

By constructing asymptotes for a ray passing through F_i and x_i (in the image space) and parallel to the z axis (in the object space) until they intersect in the lens space, we find the location of the principal plane H_i in the image space. The point at which the principal plane intersects the z axis defines a principal point $z(H_i)$ (in the image space). In a similar manner the focal point z_{F_0} and the focal plane in the object space, as well as the principal plane H_0 and the principal point $z(H_0)$ in the object space (Fig. 1-14) are found. The focal lengths of a lens are determined by the equation

$$f_0 = z(H_0) - z(F_0), \qquad f_i = z(F_i) - z(H_i) \tag{1-178}$$

The distances from the object and image to the corresponding principal planes are

$$l_0 = z(H_0) - z_0, \qquad l_i = z(F_i) - z(H_i) \tag{1-179}$$

The quantities f_0, f_i, l_0, and l_i are associated with the lens equation $f_0/l_0 + f_i/l_i = 1$ and determine the lens magnification M:

$$M \equiv \frac{x_i}{x_0} = \frac{l_i}{l_0}\frac{f_0}{f_i} = \frac{f_0}{l_0 - f_0} = \frac{l_i - f_i}{f_i} \tag{1-180}$$

(see the similarity of the triangles in Fig. 1-14). If an object lies in a principal plane H_0 ($l_0 = 0$), its image lies in a principal plane H_i ($l_i = 0$) and, in this case, $M = 1$.

Lenses having axial symmetry have identical properties in any plane containing an axis of symmetry; therefore, all of the relationships for the Oxz plane discussed previously are also valid for the Oyz plane.

Recall also (see Section 1-9) that $f_i/f_0 = \sqrt{\varphi_i/\varphi_0}$ for immersion lenses, and $f_i = f_0 \equiv f$ for symmetric (singular) lenses.

The parameters of the cardinal elements of lenses can be used to compute the elements of matrices that describe the conversions made by a lens on a ray passing through it.

1-17. A MATRIX DESCRIPTION OF THE CONVERSION MADE BY A LENS

Using matrices to describe the conversions made by refracting surfaces is well known in geometric optics[45] and can also be successfully applied to electron optics.[46] We will give the trajectories $x_a(z)$ and $x_b(z)$ of two rays that enter (in the z_1 plane) the lens parallel to the axis and at an angle $\pi/4$ to it:

$$\begin{cases} x_a(z_1) = 1, \\ x'_a(z_1) = 0, \end{cases} \qquad \begin{cases} x_b(z_1) = 0 \\ x'_b(z_1) = 1 \end{cases} \tag{1-181}$$

The conversion made by a lens on the transverse bias $x(z_1) \equiv x_1$ and deflection $x'(z_1) \equiv x'_1$ of an arbitrary ray can (in a paraxial approximation), to obtain the displacement $x(z_2) \equiv x_2$ and deflection $x'(z_2) \equiv x'_2b$ in the emergence plane of the lens, be given as

$$\begin{aligned} x_2(z_2) &= x_1(z_1)x_a(z_2) + x'_1(z_1)x_b(z_2) \\ x'_2(z_2) &= x_1(z_1)x'_a(z_2) + x'_1(z_1)x'_b(z_2) \end{aligned} \tag{1-182}$$

Introducing the vector $\bar{x} = |x/x_1|$ and the matrix $T_x = \begin{vmatrix} x_a & x_b \\ x'_a & x'_b \end{vmatrix}$ allows us to write the conversion described by Eq. (1-182) as

$$\bar{x}(z_2) = T_x(z_2)x(z_1) \tag{1-183}$$

We can show that if the lens does not change the particle's energy the determinant of the matrix T_x is 1.

If the space between two lenses free of the field is S, the parameters of a ray in this space are converted as

$$x_2(z_2 + s) = x_1(z_2) + x_1'(z_2)s$$
$$x_2'(z_2 + s) = x_1'(z_2) \tag{1-184}$$

or, in matrix form,

$$\bar{x}(z_2 + s) = T(s)\bar{x}(z_2), \qquad T(s) \equiv \begin{pmatrix} 1 & s \\ 0 & 1 \end{pmatrix} \tag{1-185}$$

The total conversion of a ray's parameters by a system of several lenses and intervals will be described by the product of the respective matrices:

$$\bar{x}_f = T_{x_f} T(s_f) \cdots T(s_1) T_{x_1} \bar{x}_{in} \tag{1-186}$$

Here $\bar{x}_{in}$ and $\bar{x}_f$ are the ray's parameters at the input to the first lens and at the output of the last. We can add conversions at the intervals S_0 and S_i between the first and last lenses and the object and image planes and write

$$\bar{x}_i = T(s_i) T_{x_f} T(s_f) \cdots T(s_1) T_{x_1} T(s_0) \bar{x}_0 \tag{1-187}$$

The elements of a compound matrix that converts the parameters of a trajectory from the object plane to the image plane are given in terms of the magnification M and the focal length f of the system:

$$T = \begin{pmatrix} M & 0 \\ -1/f & 1/M \end{pmatrix} \tag{1-188}$$

The matrix elements for the conversion made by a one-lens system are given in terms of the cardinal elements of a lens:[46]

$$T = \begin{Bmatrix} -\frac{1}{f}[z(H_i) - z(F_i)] & f\left\{\frac{1}{f^2}[z(H_i) - z(F_i)][z(H_0) - z(F_0)] + 1\right\} \\ -\frac{1}{f} & \frac{1}{f}[z(H_0) - z(F_0)] \end{Bmatrix} \tag{1-189}$$

For a thin lens,

$$T = \begin{pmatrix} 1 & 0 \\ -1/f & 1 \end{pmatrix} \tag{1-190}$$

The matrix method gives the researcher a simple, but elegant formalism for analyzing and synthesizing optical systems of any given degree of complexity. It must be stated that for nonparaxial systems with aberrations the elements of the conversion matrix T_x will depend on the parameters of the ray being converted and the conversion will be nonlinear. Nevertheless, the ability to determine the properties of a complex optical system by multiplying the matrices will be preserved.

We can show that the conversion matrix determinants are unity. This property of conversion matrices is preserved even in complex systems that are the products of the matrices; it can be used to verify the correctness of the computations of the elements for complex systems matrices.

1-18. THE SPECTRAL PROPERTIES OF ELECTRON OPTICAL SYSTEMS

The ultimate purpose of electron-beam lithography systems is to create a specific distribution $I(x, y)$ of the electron current on the surface of the wafer being exposed. In probe diagnostic systems and lithographic equipment having a Gaussian beam the function $I(x, y)$ must be a Gaussian function for the given diameter. In projection (with reduction $M\times$) and shadow ($1\times$) systems as well as in image generators with variable beam cross sections the function $I(x, y)$ must be an aggregate of rectangles (or other figures) of specific size within which $I(x, y)$ = const and the transition to a value of $I(x, y) = 0$ in the neighborhood of the rectangle being exposed is made as sharply as possible (a step). When this happens, the function $I(\bar{r}_i)$ is the image (virtual or real) of an electron source (object) described by a brightness $B(\bar{r}_0, \bar{v}_0)$, $(r_0(x_0, y_0))$ is a point on the object, and $\bar{v}_0(x_0' y_0')$ is a unit vector oriented in the direction in which electrons leave the object plane. We need to find the relationship between the object $B(\bar{r}_0, \bar{v}_0)$ and its image $I(\bar{r}_i)$. A virtual source in the crossover of the electron gun may be the object in Gaussian beam systems, whereas in different situations the object is the electron current that passes through the points on diaphragms or masks.

Obviously, the relationship between an object $B(\bar{r}_0, \bar{v}_0)$ and image $I(\bar{r}_i)$ is established by the trajectory equations $r_i = r_i(\bar{r}_0, \bar{v}_0)$. Particles emitted from a point r_0 on the object into a solid angle dO are incident on an area $dx\, dy$ in the image plane:

$$i(r_i, r_0, v_0)\, dO = B(r_0, v_0)\, dx\, dy \tag{1-191}$$

Here $i(r_i, r_0, v_0)$ is the contribution to current density at the area $dx\, dy$ made by electrons emerging from the point r_0 on the object and moving in the v_0 direction. An element of the solid angle is $dO = dx_0'\, dy_0'/\cos\theta$, where θ is the angle between v_0 and the normal to the object plane. The relationship between $dx_0'\, dy_0'$ and $dx\, dy$ can be described by the Jacobian J of the conversions contained in the trajectory equations:

$$dx_0'\, dy_0' = J(r_i, r_0(v_0))\, dx\, dy \tag{1-192}$$

$$J = \det \begin{vmatrix} \partial x_0'/\partial x & \partial x_0'/\partial y \\ \partial y_0'/\partial x & \partial y_0'/\partial y \end{vmatrix} \tag{1-193}$$

In order to find J in analytic form we must solve the trajectory equation in terms of x_0', y_0': $v_0 = v_0(r_i, r_0)$. This procedure is elementary for trajectory equations in a paraxial approximation when the equations $r_i = r_i(\bar{r}_0, \bar{v}_0)$ are linear. When aberrations are considered, the equations $r_i = r_i(\bar{r}_0, \bar{v}_0)$ become nonlinear; in this case the Jacobian J can be found by numerical methods.

Using Eq. (1-192) instead of Eq. (1-191), we have

$$i(r_i, r_0, v_0)\, dx\, dy = B(r_0, v_0) J(r_i, r_0, v_0)\, dx\, dy/\cos\theta \tag{1-194}$$

Substituting the solution $v_0 = v_0(r_i; r_0)$ into Eq. (1-194), we reduce the number of parameters:

$$i(r_i, r_0)\,dx\,dy = B(r_0, r_i)J(r_0, r_i)/\cos\theta\,dx\,dy \tag{1-195}$$

We find the total current density at a point r_i by integrating Eq. (1-195) over the area of the object Σ_0:

$$I(r_i) = \left[\int_{\Sigma_0} B(r_0; r_i)J(r_0; r_i)/\cos\theta\,dx_0\,dy_0\right] \tag{1-196}$$

With Eq. (1-196) our problem has been resolved: we have found the image $I(r_i)$ of the object $B(r_0, v_0)$; the Jacobian $J(r_0; r_i)$ of the conversions contains in itself a description of the properties of an optical system that contains the object. We can, however, continue our study of the problem by striving for a clearer explanation of the properties of an optical system.

We find the total particle current from a point M_0 on an object by integrating the brightness with respect to the solid angle

$$B(r_0) = \int_0 B(r_0, v_0)\,dx_0'\,dy_0' \tag{1-197}$$

and find the fraction $h(r_0; r_i)$ of this current that impinges on a point r_i in the image plane:

$$h(r_0; r_i) = \frac{i(r_0; r_i)}{B(r_0)} = \frac{B(r_0; r_i)J(r_0; r_i)}{B(r_0)\cos\theta} \tag{1-198}$$

Equation (1-198) has the sense of a normalized distribution of the particles emitted from a point source at a point on the object over the image surface and bears the (very accurate) name of scattering function. As can be seen, the scattering function takes into account both the properties of the optical system (by way of the Jacobian $J(r_0; r_i)$, being a differential form of the trajectory equations) and the source (through the angular distribution of the emission current stated as $B(r_0; r_i)/B(r_0)\cos\theta$ and which determines the fraction of particles sent from a point r_0 on the object to a point r_i on the image plane).

If the scattering function is identical at every point r_0 on the object, then it depends only on the differences $\bar{r}_0 - M\bar{r}_i$:

$$h(\bar{r}_0; \bar{r}_i) = h(\bar{r}_0 - M\bar{r}_i) \tag{1-199}$$

and the conversion $h(\bar{r}_0 - M\bar{r}_i)$ is said to be uniform. Field aberrations that depend on the point r_0 violate the uniformity of a conversion; aberrations that depend only on the deflection of the rays (spherical and point chromatic aberrations and axial astigmatism) create a uniform conversion.

With the help of the scattering function, Eqs. (1-198) and (1-199), we can express the image current, Eq. (1-196):

$$I(\bar{r}_i) = \int B(\bar{r}_0)h(\bar{r}_0 - M\bar{r}_i)\,d\bar{r}_0 \tag{1-200}$$

Finding the function $B^*(\bar{r}_0)$ through

$$B^*(\bar{r}_0) = \begin{cases} B(\bar{r}_0) & \text{in } \Sigma_0 \\ 0 & \text{outside } \Sigma_0 \end{cases} \tag{1-201}$$

the image, Eq. (1-200), can be written as an indefinite integral:

$$I(\bar{r}_i) = \iint_{-\infty}^{+\infty} B^*(\bar{r}_0)h(\bar{r}_0 - M\bar{r}_i)\,d\bar{r}_0 \tag{1-202}$$

The integral in the right-hand side of Eq. (1-202) is the convolution of the functions $B(\bar{r}_0)$ and $h(\bar{r}_0 - M\bar{r}_i)$. Using the theorem for the Fourier transform of a convolution, we find

$$F_i(\nu_x, \nu_y) = F_0(\nu_x, \nu_y)H(\nu_x, \nu_y) \tag{1-203}$$

where

$$F_i(\nu_x, \nu_y) = \iint_{-\infty}^{+\infty} B(\bar{r}_i)e^{-2\pi i\bar{r}_i\bar{\nu}}\,d\bar{r}_i \tag{1-204}$$

$$F_0(\nu_x, \nu_y) = \iint_{-\infty}^{+\infty} B(\bar{r}_0)e^{-2\pi i\bar{r}_0\bar{\nu}}\,d\bar{r}_0 \tag{1-205}$$

$$H(\nu_x, \nu_y) = \iint_{-\infty}^{+\infty} h(\bar{r}_i - M\bar{r}_i)e^{-2\pi i(\bar{r}_0 - M\bar{r}_i)\nu}\,d(\bar{r}_0 - M\bar{r}_i) \tag{1-206}$$

$\bar{\nu}(\nu_x, \nu_y)$ is a two-dimensional vector of the spatial frequency, and $F_i(\bar{\nu})$, $F_0(\nu)$, and $H(\bar{\nu})$ are the two-dimensional spectra of the respective functions. The function $H(\nu)$ is called the transfer function of the optical system and it may be complex. The absolute value $|H(\nu)|$ is the modulation transfer function of the system.

Equation (1-203) establishes the fact that the optical system acts as a filter for the spatial frequencies; the spectral component $F_0(\nu)$ of an object is propagated with a coefficient of $|H(\nu)| \leq 1$, whereby distortion is created in the image of an object. The phases of the function $H(\bar{\nu}) = |H(\nu)|e^{i\varphi(\nu)}$ can play a major role in these distortions if the radiation has wave-like properties (diffraction of the electrons, electromagnetic radiation).

It will not be difficult to establish the physical sense of the $H(\nu)$ function. If the object is one emitting point of single cardinality, then $B(\bar{r}_0) = \delta(\bar{r}_0)$ and $F_0(\bar{\nu}) = 1$ at all frequencies. In this case $F_i(\nu) = H(\nu)$, i.e., $H(\nu)$ is the Fourier spectrum of the image of a δ-source. If the distribution of the intensities in the image of a δ-source can be described by a Gaussian curve of diameter d, the Fourier transform of this distribution will be a Gaussian function of the spatial frequency $\nu_x\nu_y$ having the frequency "diameter" $(2/\pi d)$ cm^{-1}. Thus, the larger the diameter d of the spot that images the δ-source, the narrower the band $\Delta\nu = (1/\pi d)$ of the spatial frequencies filter that is the optical system being investigated. In improving an optical system by reducing aberrations, we widen the band of spatial frequencies transmitted by the optical system. In an ideal aberration-free system the diameter of the spot that images a δ-source is governed by the diffraction in a circular diaphragm that limits the beam; these systems are called diffraction-limited because the bandwidth of the frequencies being transmitted is limited by diffraction effects.

The modulation transfer function $|H(\nu)|$ of a system can be determined experimentally by measuring the reduction in contrast of the periodic lattices. Let the emission from an object be described by a one-dimensional sinusoidal function:

$$B(r_0) = A_0 + A_\nu \sin(2\pi\nu x_0) \tag{1-207}$$

$A_0 \geq A_\nu$. The contrast of the object will be

$$k_0(\nu) = \frac{B(r_0)_{\max} - B(r_0)_{\min}}{B(r_0)_{\max} + B(r_0)_{\min}} = \frac{A_\nu}{A_0} \tag{1-208}$$

A spatial frequency filter that is equivalent to an optical system transmits the constant component A_0 in Eq. (1-207) without attenuation and the component at frequency ν with an attenuation coefficient of $|H(\nu)|$; therefore, the lattice image contrast will be

$$k_i(\nu) = \frac{|H(\nu)|A_\nu}{A_0} = k_0(\nu)|H(\nu)| \tag{1-209}$$

It is obvious that measuring $k_0(\nu)$ and $k_i(\nu)$ is simpler than measuring the distribution of intensities in a spot $h(r_i)$ of microscopic size.

The frequency methods of describing and characterizing optical systems that we have presented here are widely used in ordinary (electromagnetic) optics.

1-19. ELECTRON-BEAM SYSTEMS HAVING GAUSSIAN BEAMS

Electron-beam systems having a Gaussian current distribution over the cross section are currently thought of as machines for so-called nanometer lithography; i.e., they are designed for doing experimental work in creating devices having minimal sizes of from a few to 100 nm. Table 1-3 shows the parameters of several of the machines described in the literature. Figure 1-15 shows a simplified diagram of an optical system for a machine having a Gaussian beam. As can be seen from Table 1-3, thermionic cathodes made from LaB_6 and tungsten continue to be used with good success in electron guns. The requirement that nanometer dimensions be maintained by the beam over the diameter fairly strongly limits the field that can be scanned by the beam. For example, in the JBX-5DII equipment the scanning field is 80×80 μm^2 when using the machine in the nanometer mode for an electron energy of 50 keV and 160×160 μm^2 for an energy of 25 keV. In the submicron mode the scanning field increases to 750×750 μm^2 and 1500×1500 μm^2 for electron energies of 50 and 25 keV, respectively.[49]

The preferred mode for exposing specific portions of a resist layer is vector scanning, where the Gaussian ray scans only the exposed areas. The centers of the bands being scanned are a beam diameter apart; the total exposure is determined by the location of the respective Gaussian distributions. In order to correct the proximity effect (see Chapter 2) we must have the ability to expose each pixel with its own dose of radiation. The easiest way to provide this ability is to change the total exposure time of the pixel. At small beam diameters the current densities reach large values; e.g., when $d = 2.5$ nm and the current is 0.1 nA the current density is $\sim 10^3$ A/cm^2. At these levels of current density the necessary dosage must be applied by exposing the pixel several times in order to avoid intolerable heating in the resist layer and destroying the layer. The heating effect appears

TABLE 1-3. Parameters of Electron Beam Lithography Machines Having a Gaussian Beam

		Beam diameter, nm, minimum	Beam current, nA	Emitting electrode	Beam energy, keV	Electron gun brightnesss	Ref.
NSF-1	Univ. of Osaka	2.5	0.1	Zr/W	50	2×10^8	[47]
EBS-5	Hughes Res. Lab.						[48]
JBX-5DII	JEOL	$8–10^3$	10^{-2}	LaB_6 (100)	25–50		[49]
	IBM	5		LaB_6	10–100		[50]
VL-F1	VLSI Technol. Res. Association, Tokyo	100	40	Field emission	20		[51]
EB-F	NTT Electr. Comm. Lab., Japan	30	3	Ti/W	20, 50	10^8	[52]
EBES4	Bell Lab.	125	250	W/Zr	20	–	[52]

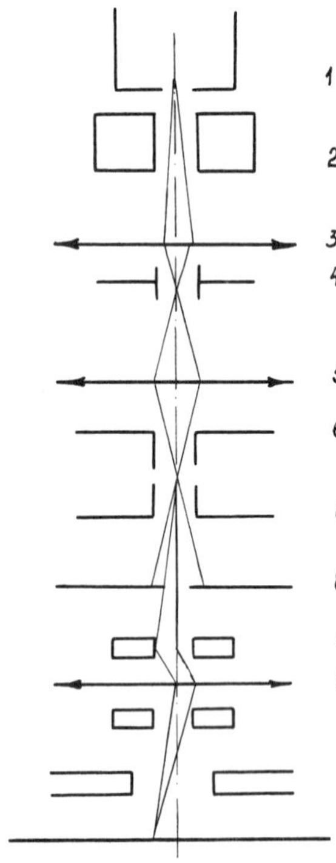

FIGURE 1-15. Diagram of the optical system for an electron-beam lithography machine having a Gaussian beam: 1) electron gun; 2) alignment system windings; 3) first condenser lens; 4) blanking plates; 5) second condensing plates; 6, 7) intermediate restricting apertures and blanking electrodes; 8) restricting aperture; 9, 11) deflection coils; 10) objective lens; 12) alignment system secondary electrons detectors; 13) wafer being exposed.

even at current densities of only 2 A/cm^2.[(53)] Exposing PMMA at this level of current density for a period of 20 μsec leads to melting in the resist layer.[(53)] Residual effects in the form of increased sensitivity in the resist were observed when it was exposed five times for a period of 4 μsec each time.

The throughput of an electron lithography system that uses a Gaussian beam in the nanometer mode is very low. We will estimate the time needed to generate a pattern in an electron resist that has a sensitivity of S = 10 μC/cm^2 and by a beam of diameter

TABLE 1-4. Optical Column Parameters for an FEPG Machine

Accelerating voltage, keV	20
Source (thermocathode)	W
Distance between source and wafer being exposed, mm	490
Size of the rectangular exposing dies, μm	0–10
Illumination scheme	Critical
Reduction of the rectangular aperture	1:50
Number of reduction stages	2
Beam deflection for shaping	Magnetic single
Focal length of the final lens, mm	15
Scanning deflection scheme	Dual magnetic
Scanning field, μm^2	162×162
Distance between the rectangular aperture and the specimen, mm	175

$d = 0.1$ μm. Let the data flow rate from the beam control system be $R = 20$ Mbit/sec, and the current density be $J = 10$ A/cm^2. The time needed for the subsequent exposure of a pattern of area L^2 is

$$T = \kappa((L/d)^2 \max(S/J, 1/R) + T_{\text{add}} \tag{1-210}$$

κ is the pattern filling coefficient, T_{add} is the additional time needed to move and position the table, to deflect the beam, etc. The pixel exposure time at given levels of resist sensitivity and beam current density is $\tau = S/J \simeq 10^{-6}$ sec, which is longer than the control signal arrival time. We will also take $\kappa = 0.1$, and $L = 1$ cm. Then the pure exposure time of a pattern is $T_{\min} = 10^3$ sec.

1-20. ELECTRON-BEAM LITHOGRAPHY SYSTEMS WITH VARIABLE BEAM SHAPES

To date, a large number of electron-beam lithography systems having variable beam shapes have been developed, produced, and put into use. These machines are the most universal in application, have high resolution, and have acceptable throughput when working on semiconductor wafers as well as when fabricating different types of masks. Figure 1-16 and Table 1-4 show a diagram and the parameters for one modern machine of the type being discussed.[54] Changing the size $a \times b$ of a rectangular beam is done by partially superimposing a rectangular beam formed by the S_1 aperture onto a second aperture S_2. The location relative to the S_2 aperture of the image from aperture S_1 produced by lenses L_4 and L_5 is controlled by the deflecting winding of deflector 5. A comparatively large number of lenses makes it possible to improve the system parameters: magnifying the image of the electron gun crossover at the S_1 aperture raises the uniformity of the current density in a rectangular beam (after the S_1 aperture). A large degree of reduction of the S_1 aperture image on the specimen makes it possible to reduce the effect of aperture edge defects on the shape of the edge of rectangular dies on the surface of the sample.

A wafer is exposed by 162 μm wide bands via the vector scanning method; scanning is done while the wafer (table) moves at a constant velocity (up to 20 cm/sec) along the

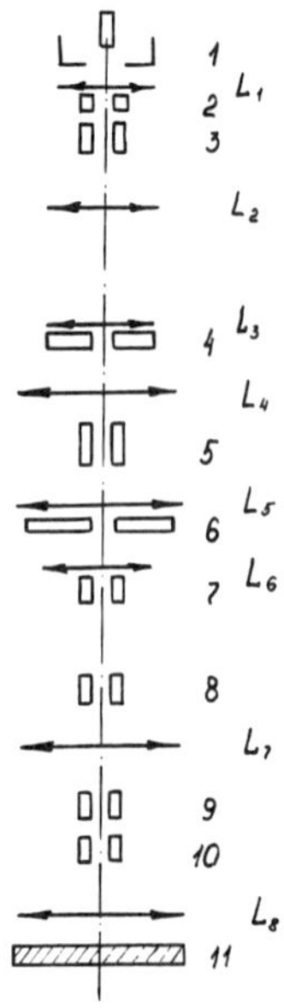

FIGURE 1-16. Diagram of a multibeam lithography system: 1) cathode; 2, 3) electron gun electrodes; 4) lens L_1; 5) blanking electrodes; 6) quadrature aperture; 7, 8) deflecting plates for the scanning system; 9) lens grid; 10) wafer being exposed.

y axis (write-on-the-fly mode). A dynamic deflection system takes the movement of the table during scanning and finding the coordinates of the next die into account. Between two dies (exposures) is a blanking ray; the blanking time can be varied from 0.4 to 2 μsec.

With the ability to operate at beam current densities of 1–10 A/cm^2 on a target and to vary the die sizes over wide ranges from a few tenths of a micron to 10 μm, machines with variable beam cross sections demonstrate substantial levels of throughput tens of square centimeters per minute or tens of wafers per hour. Gratings with line widths of 0.3 μm can be fabricated with these machines.[(53)]

The possibility of further improving machines whose beam cross section is variable by changing the exposure scheme was indicated in reference [55]. Critical exposure or Kohler exposure is used in the familiar machines. In both schemes a Gaussian beam is emitted by the virtual electron source in the electron gun crossover. In the critical exposure scheme the condensing lens projects a magnified image of the crossover onto the plane of the rectangular beam-shaping aperture. The case of a magnified Gaussian beam isolated by a diaphragm is an object the reduced image of which is produced on the target. The greater the diameter of the magnified Gaussian beam in comparison with the size of the diaphragm, the more uniform is the current distribution in the rectangular beam; however, this is attained at significant losses in the total current. In a Kohler exposure scheme the magnified image of a Gaussian beam in the crossover is produced at the input aperture of the final (objective) lens. The current distribution at the output of the electron (not in the crossover!) is projected in magnified form on the shaping rectangular aperture and the final lens projects the portion isolated by the diaphragm onto the target. Thus, a Kohler exposure scheme is preferred when the electron gun has a high degree of uniformity in the (angular) current distribution.

In the new exposure scheme suggested in reference [55] called a critical Kohler scheme, the gun's immersion lens projects the cathode surface rather than the current distribution at the output onto the shaping aperture. The portion of the magnified cathode surface image isolated by the diaphragm is projected onto the target. If the cathode's emission current distribution is fairly uniform, the current on the target (within the dies) will also have a high degree of uniformity. Uniform emission from a cathode will occur if the cathode has a smooth, plane surface, uniform heating, a uniform work function, and is also perpendicular and concentric to the optical axis. These requirements are not hard to meet when cathodes of monocrystalline LaB_6 are used. These cathodes have high emissivity and the size, orientation, and shape of the emitting area are easy to control. A hundredfold increase in the beam current stability, a reduction in edge blurring from rectangular electron current dies (owing to the reduction in the variance of the energy of the electrons in the beam), and a tenfold increase in exposure efficiency can be achieved by using a critical Kohler exposure scheme.

1-21. MULTIBEAM SCANNING SYSTEMS IN ELECTRON LITHOGRAPHY

Attempts to produce multibeam systems have been made in order to improve the throughput of electron beam lithography systems.[56–58] In an early approach to creating these systems[56] the patterns were produced on different areas of a wafer by different beams. All beams are scanned and blanked in parallel (simultaneously) so that identical patterns can be produced on each area of the wafer. The second approach[57,58] scans all beams in parallel (the raster method) but each beam is blanked individually, which allows the corresponding chip pattern to be produced by each beam. Obviously, these two approaches differ significantly in their demands on the beams grid: in the first case there may be only a few beams (equal to the number of areas into which the wafer surface is divided) and the distances between the beams are equal to, or less than, the chip dimension. In the second situation there may be a large number of beams but the distance between them must be small because each beam produces only a small portion of the chip pattern. We hasten to point out that neither system has been put into practical use. However, because of the importance attached to improving the throughput of electron-beam lithography systems, multibeam systems are worthy of examination.

Figure 1-17 shows a diagram of a parallel scanning system.[56] The electron gun accelerates the electrons to an energy of 1 keV and the beam current is approximately 2 mA. The rectangular aperture is uniformly exposed by the electron beam shaped by lens L_1. The half angle of beam divergence is 3°. Conducting plates having a grid of circular apertures of radius r_a from the side of the object and R from the side of the image are placed approximately 1 m from the aperture. These apertures form a grid of lenses, each of which shapes a separate beam in the space between the plane of the lenses and the wafer. Electrons passing through the apertures are focused by the field from the accelerating voltage V = 9 keV between the lens plates and the wafer. The cross section of the exposing beam on the surface of the wafer is a reduced image of the objective aperture. If the side of the aperture square is 50 μm, the cross section of each beam on the wafer may, e.g., be 0.5 μm. A large distance between the objective aperture and the plane of the lenses ensures that the input apertures of every lens in a circle 100 mm in diameter will be exposed. Two pairs of deflecting plates that ensure scanning by the ray are located in the space unaffected by the field between the objective aperture and the lenses. The control signals on these plates displace the location of the virtual objective aperture and consequently displace their images on the wafer's surface. Deflection aberrations limit the scanning area to 0.2 mm. For this reason the complete

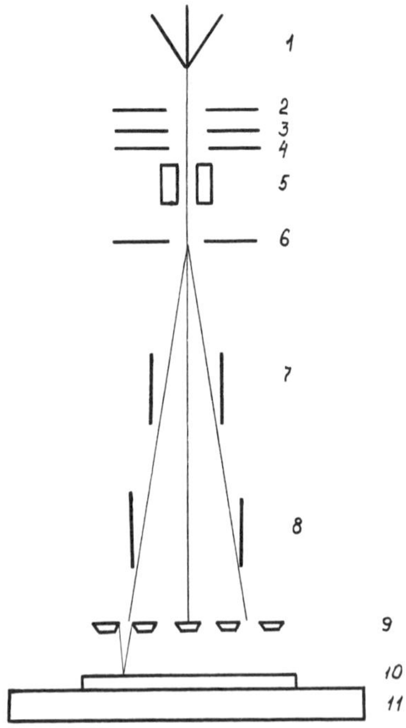

FIGURE 1-17. Diagram of a multibeam lithography system: 1) cathode; 2, 3) electron gun electrodes; 4) lens L_1; 5) blanking electrodes; 6) quadrature aperture; 7, 8) deflecting plates for the scanning system; 9) lens grid; 10) wafer being exposed; 11) controlled table.

pattern is produced on a section of the wafer by using the controllable table with a laser interferometer to move the wafer. When the ray is scanning in the x direction, the table moves at constant velocity in the y direction. The beams are simultaneously turned on and off, respectively, to the pattern on a portion of the wafer. An analysis of aberrations in the lens system and the experiments described in reference [56] make it possible to obtain a beam edge that is less than 0.2 μm wide and a current of more than 20 nA in the beam 0.5 μm. Estimates put the number of wafers produced by the machine at 60/h.

Multibeam systems whose beams are close to one another can be built on optical elements (lenses) into which are built gauze electrodes for splitting a wide beam into individual beams, focusing, and blanking them.(57) The diagram of a three-electrode single lens with a gauze built into the center electrode(57) is shown in Fig. 1-18. An ordinary three-electrode symmetric electrostatic lens has focusing properties. If a conductive foil that covers the entire aperture is introduced into the center electrode, the optical action of this electrode disappears because the tangential component of the field voltage on the surface of the foil is everywhere zero. These components will appear if an opening is made on the foil: every opening will act like a focusing lens. At the same time, the edge electrodes will act like independent defocusing lenses. If the focal plane of the individual beams can be arranged, a system of blanking electrodes controlled by commands that are individualized for each beam can be accommodated.(57,58)

A design for a system with 1024 = 32 $\times$ 32 rays, each of which is focused into a beam 0.1 μm in diameter, was built using the optical elements described. Each ray simultaneously exposes an area 3.2 $\times$ 3.2 μm^2 on the wafer. If the total current from every ray is 30 μA and the sensitivity of the resist is 10 $\mu C/cm^2$, the system has an exposure rate of 10^{10} pixels/h. A prototype of this system has probably not yet been

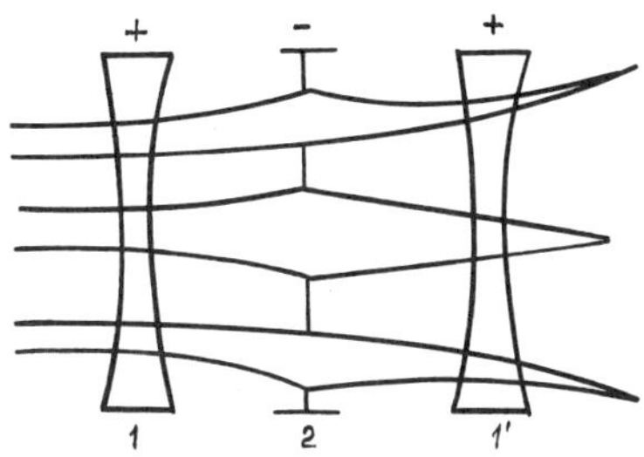

FIGURE 1-18. Diagram of a single lens converted by insertion into the central electrode of a grid with openings into a lens system: 1, 1') defocusing lenses; 2) focusing lenses grid.

built, although the optical elements have been modelled.[58] We can assume that the greatest difficulty in producing such a system is the individual blanking system for each ray, including the hardware and software.

Multibeam electron systems of the mirror type can also be constructed. In this case the fundamental system element must be a large focusing mirror divided into a grid of mirrors that are individually controlled.[57,58]

1-22. PROJECTION ELECTRON-BEAM LITHOGRAPHY SYSTEMS

The capabilities of an electron-beam projection multilens system in which a reduced image of a shadow mask (a metal foil with openings) is transferred to the wafer being exposed have been tested (Fig. 1-19). The main advantage of these systems is high throughput, because they make a parallel transfer of the image for an individual chip (in a version of stepwise align and expose) onto the entire wafer. However, the difficulties in fabricating large-area shadow masks, the difficulties in aligning to submicron accuracy, and the lack of application flexibility (in comparison with scanning machines) have hindered development of this type of machine. There have been no recent reports of further development of these systems.

Research into producing a projection electron-beam lithography system with a photocathode, a schematic diagram of which is shown in Fig. 1-20, is being much more actively pursued.[59] An additional coating of CsI photoemitting material is deposited onto a mask having a pattern on a chromium film. Photoemission of electrons occurs during exposure by ultraviolet radiation from the back side of a mask in the areas of the mask lacking the chromium coating. A uniform electric field applied between the mask (cathode) and plate (anode) accelerates the electrons to an energy of 10–20 keV. The solenoid creates a lengthwise magnetic field whose intensity focuses the electrons onto the wafer surface. The alignment marks on the wafer are made irom heavy metals which emit x-rays when irradiated by an electron beam. The respective detectors provide the necessary alignment by measuring the x-rays.

A mask containing the pattern for a layer of the entire silicon wafer is made in 1:1 scale. A prototype of the equipment prints 25 wafers/h with line widths of 0.5 μm.[60] However, printing the entire wafer limits the resolving power of transferring a mask image (because of surface roughness, irregularities in thickness, etc.). Recent work directed toward producing projection equipment with a photocathode pursued the goal of producing equipment having stepwise align and expose on each chip (steppers).[61,62] Figure 1-20 shows the diagram of a stepper with a photocathode. There are two gaps

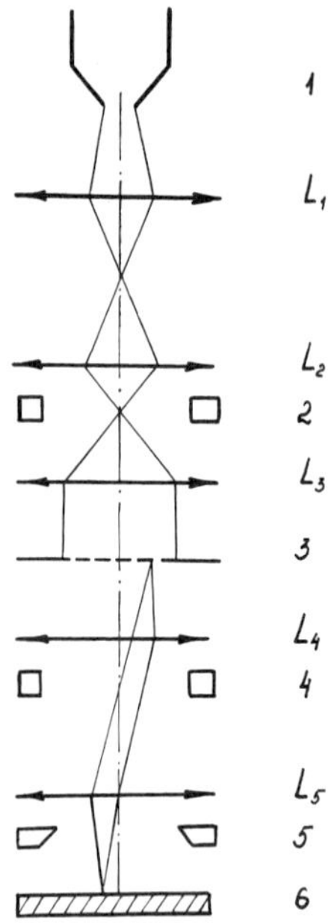

FIGURE 1-19. Diagram of an electron-beam projection lithography machine: 1) electron gun; L_1, L_2, and L_3 are condenser lens; 2) deflection windings; 3) mask with openings; L_4 and L_5 are projection lenses; 4) alignment system windings; 5) detectors for the electrons back-scattered from the alignment system; 6) wafer being exposed.

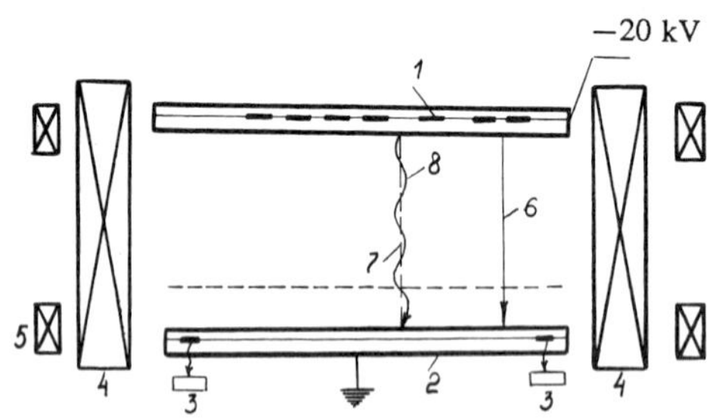

FIGURE 1-20. Diagram of an electron-beam projection machine with a photocathode: 1) photocathode–mask having a pattern on a chromium layer and a layer of photoelectric CsI; 2) wafer with a resist layer; 3) detectors for x-rays emitted from the alignment marks, 9; 4) solenoid for producing a uniform magnetic field, B; 5) alignment system windings; 6, 7) uniform electric and magnetic field; 8) electron trajectory.

between the electrodes. An accelerating voltage of 20 keV is applied in the first gap between the mask–photocathode and the gauze. The secondary-electron detectors for the alignment system are placed in the second gap, which is unaffected by the electric field. The lengthwise magnetic focusing field fills both gaps and both the planes of the gauze and the wafer are focal planes. In order that the gauze image not appear on the wafer the gauze moves at constant velocity in its own plane during exposure. The gauze comprises 100 μm wires and has a transmission of 50%. The allowable deviations in the placement of the gauze from the focal planes are 20 μm. Resolution of 0.3 μm was achieved in an experimental device. According to estimates, the throughput of the system can reach 60 wafers/h.

Another alignment system scheme was described in reference [62] and uses the signal from secondary (back-scattered) electrons. A thin film with a rectangular opening for exposing the wafer is used to collect the back-scattered electrons. The foil is placed parallel to the cathode (mask) and anode (wafer) in the interelectrode accelerating gap; it is placed in one of the equipotential planes, has the same potential as this plane, and, therefore, does not distort the accelerating field. The alignment marks are placed in the gaps between the chips and are etched-out grooves. The wafer and mask remain motionless throughout the entire exposure process; different areas of the wafer are successively exposed by relocating the foil with the opening. The signals at the back-scattered electrons detectors make it possible to achieve an alignment system sensitivity of 0.05 μm. An additional advantage of using a foil detector is that unexposed areas of the wafer are shielded from back-scattered electrons that appear on a wafer due to the action of the accelerating field.

The density of the photoemission current is not large—5–50 μA/cm. If the alignment time in every exposure field is no greater than 0.2 sec the process throughput may reach 50 wafers/h. In projection systems with a photocathode the possibility of using a photocathode mask for extended periods under production conditions remains unclear.

1-23. ELECTRON-BEAM SYSTEMS FOR SHADOW PRINTING WITH A GAP

The first works devoted to investigating methods of using broad electron beams to shadow print an image from stencil masks appeared in 1982.[63–65] The authors[63,64] used both a large-diameter Gaussian scanning beam (a defocused beam in a ZRM-12) as well as an immobile broad beam from an ÉM-14 transmission microscope for shadow printing an image from stencil masks. Nuclear filter films with circular openings, obtained by etching the film along the tracks made by heavy, high-energy ions, were used for the test stencil masks.[66] When there was no gap between the shadow mask–stencil and the resist layer on the silicon wafer, the resolution from the process was better than 0.1 μm. Silicon masks were used by the authors;[65] exposure was done by the beam scanning method, the beam having a hexagonal cross section (roughly 1 mm in diameter). A controllable (up to 500 μm) gap was established between the mask and the wafer and again a resolution of about 0.1 μm was achieved. Further work by the authors of reference [67,68] on developing shadow method in electron-beam lithography was directed toward explaining the possibilities of using high-energy (up to 100 keV) electron beams because the authors[69–72] had written works on producing the equipment and technology for using beams of nearly 10 keV energy.

Figure 1-21 shows the diagram of a machine for electron-beam printing with a gap in the scanning mode.[70] The mask contains an image of one circuit; therefore, stepwise align and expose methods are used. The electron gun accelerates the electrons to an energy of 10 keV. The collimating lens L_1 with a built-in aperture forms a parallel, hexagonal-shaped electron beam that moves along the system axis. The deflection windings move the beam over the screen within the confines of the chip by biasing the beam relative to the beam axis. Thin deflection windings can bend the beam relative to the axis, making it possible to compensate distortion in the masks. The amount of distortion in the masks, $D_x(x, y)$, $D_y(x, y)$, can be measured beforehand[71] and corrections made to the respective control program. The amount of distortion can be described by

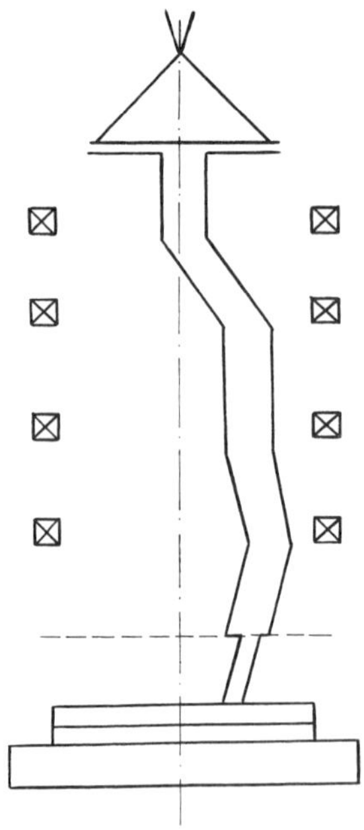

FIGURE 1-21. Diagram of an electron-beam machine for shadow printing with a gap: 1) electron gun; L_1 is the collimating lens; 2) hexagonal aperture; 3, 4) deflection windings for the scanning system; 5, 6) deflection windings for the beam deflecting system; 7) shadow stencil mask; 8) wafer with a resist layer; 9) controlled table.

nonlinear analytic equations in the form of polynomials:

$$D_x(x,y) = c_1 + c_2x + c_3y + c_4xy + c_5x^2y + c_6xy^2$$
$$D_y(x,y) = d_1 + d_2x + d_3y + d_4xy + d_5x^2y + d_6xy^2 \qquad (1\text{-}211)$$

Because of the smallness of the gap between the mask and the wafer (less than a millimeter), alignment is achieved through small changes in the directed current that flows through the silicon wafer: these changes are caused by variations in the reflection of primary electrons in the area of the alignment marks.(69) Marks are made on the wafer for global aligning the wafer and each chip individually. The errors in alignment are less than 0.1 μm.

The optical system of an electron beam machine for printing a stencil pattern on a wafer must generate a wide electron beam that has the most uniform possible distribution of electrons over the cross section of the beam and small angular divergence. The current density must be such that in a short time span (on the order of 100 msec) a resist having a sensitivity $d_0 = 5\ \mu\text{C/cm}^2$, common for electron resists, is exposed. From this emerges the current density requirement: $J \approx 50\ \mu\text{A/cm}^2$. On the other hand, the current density is limited from above by the allowable level of stencil heating due to absorption of the electrons: energy. Experiments show that stencils warp when heated beyond 120°.(65) On the other hand, at an exposure time of 350 msec stencil heating is no more than 20°.(65) Beam scanning ensures uniform exposure over a large area.

Designing an electron optical system that has the indicated requirements is not difficult at the current level of electron-beam systems design technology. For example, electron transmission microscopes can be used in experiments on transferring a stencil's image in a broad electron beam.[63,64]

It is obvious that the stencils used in printing via an electron beam must be quite different from those used in optical printing. A relatively thick substrate that is transparent to radiation and gives the stencil mechanical strength is used in optical stencils. Because no material is transparent to an electron beam, the stencil areas that pass the electron beam must simply be physical openings. In other words, stencils for electron-beam printing are stencils (stencil masks). A special technology has been developed for fabricating them which includes fabricating 3–20 μm thick silicon membranes, producing a resist mask having the respective pattern and etching through the membranes to obtain physical openings in the stencil. Obviously, the etching process must ensure that the dimensions of a resist mask be precisely reproducible and that the walls of the openings in the membrane are vertical. Ion-beam etching is the most appropriate (and universal) method for etching these openings. Choosing the ion energy and the composition of the plasma makes it possible to obtain good results when etching to a depth of several microns. It is to be hoped that this method can produce submicron stencils on membranes 5–10 μm thick. Etching the silicon with an anisotropic liquid is another suitable means of fabricating stencil masks on silicon membranes. Some etchants exhibit etching discrimination (selectivity) on the order of 1000 when etching silicon (the difference in etching rates for different crystallographic planes can be as much). Our investigations have shown the possibility of obtaining submicron (less than 0.2 μm) through cuts that have absolutely vertical (atomically smooth) channels in membranes up to 20 μm thick. Unfortunately, this etching method has limitations that lower the universality of the openings pattern obtained: 1) the channel walls are vertical in only one dimension and are sloped in the other, and 2) the direction of the sides of the openings is closely tied to the crystallographic directions in the crystal.[67,68] However, stencils obtained by this method have superior test qualities.

The main disadvantage of stencil masks is that each layer of an IC must be divided into two stencils when the topology is such that falling sections are produced when the stencil openings are formed. Fortunately, modern alignment technology provides totally satisfactory interlayer alignment between the two sections of a single pattern.[69] Attempts are also being made to suspend the falling sections by extremely thin ties so that their image is not formed on the surface of the resist because of their small size, electron scattering in the resist, or swinging the beam angle relative to the stencil. This method, however, does bring with itself substantially greater complications than does the method of dividing a pattern into two stencils. These two stencils are conveniently places on a single membrane.

The method of electron-beam shadow printing with a gap offers high resolution, reasonable throughput (50 wafers/h), and applications flexibility. The latter quality is reflected in the ability to make accurate alignments and to dynamically correct mask distortion. However, the main difficulty with the method currently lies in the technology for fabricating stencil masks. Additional complications arise as a result of dividing each topological layer into two masks. This factor not only doubles the amount of work that goes into fabricating masks and exposing the wafers, but introduces additional errors in interlayer alignment. In addition, the probability of defects in each layer is doubled. The number of defects can be reduced by performing special repair operations on the masks.[72] Despite the indicated shortcomings in the method, the results obtained to date allow us to optimistically evaluate the possibilities of using shadow printing to reproduce patterns whose minimal dimensions are at least 0.5 μm.[59]

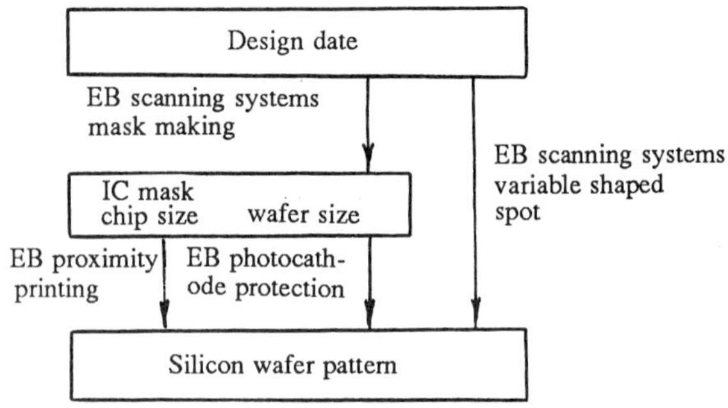

FIGURE 1-22. Diagram of "pure" electron lithography.

Scanning, projection, and shadow electron lithography methods make it possible in principle to perform every operation in producing masks and lithography on a wafer without resorting to other lithography methods (photo, x-ray, or ion lithography) which is shown in Fig. 1-22.[(59)] The relatively low throughput scanning systems give precise execution, starting from the designs for the layer topology pattern, the masks pattern in the area of the whole mask, a chip, or a group of chips. Systems with variable beam cross section have intermediate (between scanning and projection systems) throughput capabilities. They are designed for making LSI prototypes on a wafer by lithography, small batches of custom LSI, LSI on the entire silicon wafer, as well as fabricating all types of masks. The high throughput projection and shadow printing systems reproduce mask images (in the stepwise align and expose mode) on a chip or a group of chips. Projection systems with a photocathode can also print from masks containing patterns for the entire wafer.

1-24. A SCANNING TUNNEL MICROSCOPE AS AN ELECTRON LITHOGRAPH

The discussions in Section 1-5 concerning electrons tunneling through an energy barrier in connection with the subject of electrons leaving a cathode due to field emission can easily be extended to the case in which the distance d between cathode and anode becomes less than the height of the energy barrier. In this case electron tunneling from the cathode (point) will proceed not in a vacuum as it does during field emission, but directly from cathode to anode. The metallic cathode and anode can in this case be described as two potential wells separated by the tunneling energy barrier in the evacuated gap. Tunneling is possible because the wave functions of the conduction electrons from each electrode, decaying exponentially, reach the surface of the other electrode. At a distance of roughly 4 Å the attenuation of the electron wavefunctions is very strong and perturbations theory yields good accuracy for computing the tunneling current in a first approximation. For the current at low temperatures we can write[(73)]

$$I = \frac{2\pi}{h} e \sum_{\mu,\nu} f(E_\mu)[1 - f(E_\nu + eU)](M_{\mu\nu})^2 \delta(E_\mu - E_\nu) \tag{1-212}$$

Here the subscripts μ and ν are associated with quantities relating to the cathode (emitter) and anode, respectively; $f(E)$ is the Fermi distribution function; and $M_{\mu\nu}$ is a matrix

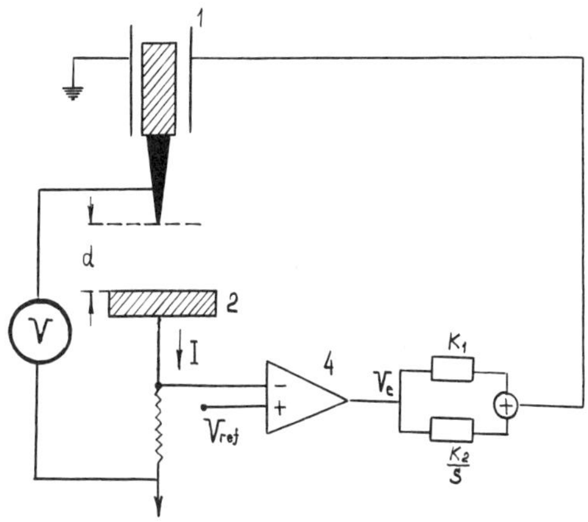

FIGURE 1-23. Fundamental diagram of a scanning tunneling microscope: 1) piezoelectric emitter vertical motion converter; 2) anode (sample being studied); 3) interelectrode source of potential difference; 4) electronic circuits for maintaining a constant current I by changing the interelectrode distance d.

element for the electron transition from state E_μ in the cathode to state E_ν in the anode. In keeping with[74] the matrix transition element can be represented as an integral over a surface S in the barrier (interelectrode) space of the electron current operator:

$$M_{\mu\nu} = \frac{\hbar^2}{2m} \int_S (\psi_\mu^* \nabla \psi_\nu - \psi_\nu \nabla \psi_\mu^*)\, d\bar{S} \tag{1-213}$$

Because of the exponential attenuation of the wavefunctions $\psi_\mu(r)$ and $\psi_\nu(r)$ at distances of atomic size, the matrix element and, correspondingly, the interelectrode tunneling current are exponential functions of the interelectrode distance d:

$$I \sim U \exp(-2\kappa d) \tag{1-214}$$

Here U is the difference of potentials applied between the electrodes; κ is the attenuation coefficient, dependent on the work function φ of the cathode and anode; and d is the interelectrode distance. Equation (1-214) was obtained for low temperatures and voltages: $eU \ll \varphi$. In tunneling through a vacuum $\kappa = \hbar^{-1}\sqrt{2m\varphi}$, where φ is the effective local work function. If $\varphi = 4$ eV, then $\kappa = 1$ Å^{-1}. This means that the current will change by an order of magnitude when the distance d is increased by only 1 Å. If the magnitude of the current I is held to within 2%, the distance d will be held constant to within 0.01 Å.[73]

The concepts presented above regarding interelectrode electron tunneling current is the basis for the so-called tunneling microscope.[73,75–77] Figure 1-23 shows a diagram of a tunneling microscope.[73] A difference of potentials U is applied between the electron's emitting point and the plane surface of the anode. The point scans the anode surface via two piezoelectric transducers that move the point in the x and y directions; the third transducer continually varies the distance $d(x, y)$ between the point and the anode surface, establishing a constant tunneling current I. Movements of the point $\Delta d(x, y)$ can easily be established according to the magnitudes of the control signals at the transducer that specifies vertical motion of the opening.

Because, by assumption, emission from the opening remains constant during scanning, the changes $\Delta d(x, y)$ are governed by the changes in electron density in the interelectrode space. Briefly, the apex of the opening will write out a surface during scanning that corresponds to a constant electron density $|\psi_\nu(r)|^2$. Because the electron density changes radically at distances on the order of an atomic radius, we can expect that the corresponding profiles will look like contour maps with atomic resolution, which has been shown in a large number of examples beginning with the pioneering works of Binnig *et al.*(75–77)

The achievement of atomic resolution in a tunneling microscope has given hope that a high (atomic) resolution process will be produced. The scanning field in a tunneling microscope is on the order of $10 \times 10\ \mu m^2$; scanning takes place along the conducting surfaces (of metals or semiconductors).(73) These conditions are far from those for electron-beam lithography in resist (dielectric) layers where electron beams having energies of 10–50 keV are used. Nevertheless, if the current flow is varied somewhat, lithographic processes that maintain high (possibly atomic) resolution become possible.(78,79) First, the accelerating difference of potentials U can be increased to levels of 10–1000 V: at these energies the electrons are capable of stimulating various physical and chemical reactions, and process on the target (anode). As U increases so does the distance d between the electrodes. When this happens, it is mainly the current flow mode that changes: instead of electrons tunneling between the electrodes, field tunneling emission of the electrons takes place in the vacuum; after they have passed through the tunneling region of the barrier further movement of the electrons takes place under the influence of the accelerating voltage U. Current limiting caused by the space charge does not occur, because of the smallness of the interelectrode distance. Thus, a tunneling microscope system differs from an electron lithography system with field emission at the cathode by the fact that optical elements (lenses, diaphragms, deflection windings) are completely eliminated. The entire optical system of the device comprises a cathode in the shape of a point and a planar anode and the distance between electrodes has been reduced to a few nanometers. The current density distribution at the anode when field emission takes place from the spherical surface of the point was modelled in [79]. It has been established that the radius of the current spot at the anode is nearly equal to the distance d between the electrodes. Thus, the spot current magnitude is the parameter that determines the resolution of the lithographic process.

The physical and chemical processes that take place on the surface of an anode being affected by electrons having an energy of $10–10^3$ eV can be extremely diverse: precipitation of material from the surrounding atmosphere, and reactions in layers whose thickness is comparable to the free path of the electrons. The processes by which masking films of carbon were precipitated from the surrounding atmosphere were demonstrated in [78] along with exposing ultrathin (three monomolecular layers) films of the Langmuir–Blodgett docosenoic acid, $CH_3(CH_2)_g{:}(CH_2)_nCOOH$. Chalcogenide and metal halide compounds can be used as resists. As a result of being activated by the electron collision of the molecules in metalloorganic compounds, precipitation of metallic films takes place. Finally, deformation (removal) of material from the anode surface is possible by simple scratching. Precipitation of material from the point onto the anode surface can also be observed. This process can be thought of as a process of recording information.(73)

In closing, we will note the main features of an electron lithography process that uses tunneling microscope type systems. Exposing the surface of a target is done by a low-energy, 10–1000 eV, electron beam. The magnitude of the probe current is not limited by the space charge, only by the processes taking place at the emitter (heating) and the target. For this reason the current can reach large values on the order of several mA. The beam diameter at the target is on the order of the interelectrode distance. Beam scanning proceeds with a high degree of precision, but the scanning frequency is low. The scanning area (~100 μm^2) is also small. The possibilities for improving these characteristics of a system must be studied. Nevertheless, a system of the tunneling microscope type can, probably, be thought of today as the prototype of the technological systems for atomic and molecular engineering. Revealing the possibilities of these systems is only just beginning.[73]

REFERENCES

1. V. E. Cosslett, *Introduction to Electron Optics* [Russian translation], Inostrannaya Literatura, Moscow (1950).
2. L. A. Artsimovich and S. Yu. Luk'yanov, *Charged Particle Motion in Electric and Magnetic Fields* [in Russian], Nauka, Moscow (1978).
3. P. W. Hawkes, *Electron Optics and Electron Microscopy* [Russian translation], Mir, Moscow (1974).
4. F. R. Brewer (editor), *Electron Beam Technology in Microelectronic Fabrication*, Academic Press, New York (1980).
5. S. Reed, *Electron Probe Microanalysis* [Russian translation], Mir, Moscow (1979).
6. L. A. Baranova and S. Ya. Yavor, *Electrostatic Electron Lenses* [in Russian], Nauka, Moscow (1986).
7. V. K. Popov and S. N. Yachmenev, *The Design and Analysis of Electron and Ion Lithography Equipment* [in Russian], Radio i Svyaz', Moscow (1985).
8. I. Brodie and J. J. Muray, *The Physics of Microfabrication*, Plenum Press (1982).
9. C. Kittel, *Introduction to Solid State Physics* [Russian translation], Nauka, Moscow (1978).
10. D. R. Herriot and G. R. Brewer, *Electron Beam Technology in Microelectronic Fabrication*, Academic Press, New York (1980), pp. 141–216.
11. D. M. Goebel, Y. Hirooka, and T. A. Sketchley, *Rev. Sci. Instrum.*, **56**, No. 9, 1717–1722 (1985); D. M. Goebel, Y. Hirooka, and G. A. Campbell, *Rev. Sci. Instrum.*, **56**, No. 10, 1888–1893 (1985).
12. R. L. Gordon, *J. Appl. Phys.*, **56**, No. 1, 810–813 (1984).
13. R. Chow and W. A. Tiller, *J. Appl. Phys.*, **55**, No. 5, 1339–1352 (1984).
14. K. W. Hadley and P. J. Donders, *J. Appl. Phys.*, **57**, No. 7, 2617–2625 (1985).
15. R. S. Khairnar, D. S. Joag, S. K. Kulkarni, A. S. Nigavecar, and P. L. Kanitkar, *Rev. Sci. Instrum.*, **55**, No. 9, 1505–1507 (1984).
16. Y. Furukawa, M. Yamabe, T. Ishizuka, and T. Inagaki, *J. Appl. Phys.*, **52**, No. 1, 533–534 (1981).
17. A. R. Shul'man and S. A. Fridrikhov, *Secondary Emission Methods in the Study of Solid State* [in Russian], Nauka, Moscow (1977).
18. L. N. Dobretsov, *Electron and Ion Emission* [in Russian], Gostekhizdat, Moscow–Leningrad (1952).
19. H. Boersch, *Z. Phys., B* **139**, 115–146 (1954).
20. W. Leger, *Z. Phys., B* **171**, 424–435 (1963).
21. V. A. Zhukov and K. A. Kalabushev, *Mikroelektronika*, **11**, No. 4, 316–325 (1982).
22. M. I. Elinson and G. F. Vasil'ev, *Field Emission of Electrons* [in Russian], Fizmatgiz, Moscow (1958).
23. A. A. Sokolov, Yu. M. Loskutov, and I. M. Ternov, *Quantum Mechanics* [in Russian], Uchpedgiz, Moscow (1962).
24. L. W. Nordheim, *Phyz. Z., B* **30**, No. 4, 177–196 (1929).
25. K. Fujii, S. Zaima, Y. Shibata, H. Adachi, and S. Otani, *J. Appl. Phys.*, **57**, No. 5, 1723–1728 (1985); H. Adachi, K. Fujii, S. Zaima, Y. Shibata, S. Oshima, S. Otani, and Y. Ischizawa, *Appl. Phys. Lett.*, **43**, 540–543 (1980).
26. J. C. Wolfe, *IEEE J.*, **SC-15**, No. 4, 540–543 (1980).
27. H. Pinna, K. Liang, M. Denizart, and B. Jouffrey, *Rev. Phys. Appl.*, **18**, 659–665 (1983).
28. B. V. Bondarenko, Yu. L. Rybakov, E. P. Sheshin, and A. A. Shchuka, *Electronic Technology, Ser. 4: High Vacuum Electronic and Gas Discharge Devices*, No. 4, 1–56 (1981).
29. B. V. Bondarenko, E. P. Sheshin, and A. A. Shchuka, *Foreign Electronic Technology*, No. 2, 3–43 (1979).
30. B. V. Bondarenko, *Electronic Technology, Ser. 4: High Vacuum Electronic and Gas Discharge Devices*, No. 6, 3–9 (1982).
31. B. V. Bondarenko, *Electronic Technology, Ser. 1: Very High Frequency Electronics*, No. 9, 3–9 (1980).
32. D. A. Kirkpatrick, R. E. Shefer, and G. Bekefi, *J. Appl. Phys.*, **57**, No. 11, 5011–5016 (1985).
33. K. H. Bayliss and R. V. Latham, *Vacuum*, **35**, No. 6, 211–217 (1985).
34. D. M. Goebel, Y. Hirooka, and G. A. Campbell, *Rev. Sci. Instrum.*, **56**, No. 10, 1888–1893 (1985).

35. D. B. Langmuir, *Proc. IRE*, No. 25, 971–991 (1937).
36. V. S. Lukoshkov, *Zh. Tekh. Fiz.*, **6**, 26–30 (1936).
37. I. E. Tamm, *Fundamentals of Electricity* [in Russian], Gostekhizdat, Moscow (1954).
38. J. Caldwell, *J. Appl. Phys.*, **57**, No. 1, 6–9 (1985).
39. K. Saito, T. Okubo, T. Takamoto, Y. Uno, and M. Kondo, *J. Vac. Sci. Technol.*, **A4**, No. 4, 1913–1917 (1986).
40. G. F. Rempfer, *J. Appl. Phys.*, **57**, No. 7, 2385–2401 (1985).
41. M. Szilagyi, S. J. Yakowitz, and M. O. Duff, *Appl. Phys. Lett.*, **44**, No. 1, 7–9 (1984).
42. M. Szilagyi, *Appl. Phys. Lett.*, **45**, No. 5, 499–501 (1985).
43. G. Hess and S. Humphries, Jr., *J. Appl. Phys.*, **60**, No. 5, 1569–1576 (1986).
44. P. Loschialpo, W. Namkung, M. Reiser, and J. D. Lawson, *J. Appl. Phys.*, **57**, No. 1, 10–17 (1985).
45. E. L. O'Neill, *Introduction to Statistical Optics*, Addison–Wesley, Palo Alto (1963).
46. V. I. Kotov and V. V. Miller, *Focusing and Isolation Relative to the Mass of High Energy Particles* [in Russian], Atomizdat, Moscow, (1969).
47. F. Emoto, K. Gamo, S. Namba, N. Samoto, and R. Shimizu, *Jap. J. Appl. Phys.*, **24**, No. 10, 809–811 (1985).
48. P. Riemenschneider, Q. Bui, A. Grohs, C. Jenzen, P. Landmeier, D. Morgan, D. Cumming, and M. Purvis, *J. Vac. Sci. Technol.*, **4**, No. 1, 68–72 (1986).
49. M. Hassel Shearer, H. Takemura, M. Isobe, N. Goto, K. Tanaka, and S. Miyauchi, *J. Vac. Sci. Technol.*, **B4**, No. 1, 64–67 (1986).
50. D. P. Kern, P. J. Coane. P. J. Houzego, and T. H. P. Chang, *Solid State Technol.*, 127–131, Feb. (1984).
51. S. Hosaka, M. Ichihashi, H. Hayakawa, S. Nishi, and M. Migitaka, *Jap. J. Appl. Phys.*, **21**, No. 3, 543–549 (1982).
52. K. Iwadate, R. Yamaguchi, K. Hirata, and K. Harada, 30th Intern. Symp. Electron, Ion, and Photon Beams, May 27–30, USA (1986).
53. E. de Chambost, B. Allanos, A. Frichet, J. Perrocheau, D. S. Alles, C. J. Biddick, J. H. Bruning, J. T. Clemens, *et al.*, *J. Vac. Sci. Technol.*, **B4**, No. 1, 73–77 (1986).
54. E. de Chambost, A. Frichet, M. Chartier, H. Ja The, and J. Trotel, *J. Vac. Sci. Technol.*, **B4**, No. 1, 78–82 (1986).
55. M. Essig and H. C. Pfeiffer, *J. Vac. Sci. Technol.*, **B4**, No. 1, 83–85 (1986).
56. I. Brodie, E. R. Westerberg, D. R. Cone, J. J. Muray, N. Williams, and L. Gasiorek, *IEEE Trans. Elect. Dev.*, **28**, No. 11, 422–428 (1981).
57. B. J. G. M. Roelofs, J. B. Le Poole, J. E. Barth, and C. B. de Gruyter, Proc. Microcircuit Eng. Cong., Cambridge, UK, Academic Press, New York (1983), pp. 91–98.
58. B. J. G. M. Roelofs, and J. E. Barth, *Microelectronic Engineering*, **2**, No. 4, 259–279 (1984).
59. W. Kulke, *Methods and Materials in Microelectronic Technology*, J. Barton (editor), Plenun Press, New York–London (1984), pp. 127–144.
60. R. Ward, A. R. Franklin, P. A. Gould, and M. J. Plummer, *J. Vac. Sci. Technol.*, **19**, No. 4, 966 (1982).
61. R. Ward, A. R. Franklin, I. H. Lewin, and P. A. Gould, and M. J. Plummer, *J. Vac. Sci. Technol.*, **B4**, No. 1, 89–93 (1986).
62. W. R. Livesay and L. B. Anderson, *J. Vac. Sci. Technol.*, **B4**, No. 1, 100–104 (1986).
63. K. A. Valiev, L. V. Velikov, R. Kh. Makhmutov, and A. V. Rakhov, *Dokl. Akad. Nauk SSSR*, **262**, No. 6, 1377–1380 (1982).
64. K. A. Valiev, L. V. Velikov, S. D. Dushenkov, R. Kh. Makhmutov, A. V. Rakhov, and N. Yu. Ustinov, *Mikroelektronika*, **11**, No. 5, 447–450 (1982).
65. H. Bohlen, J. Greschner, J. Keyser, W. Kulke, and P. Nehmiz, *IBM J. Res. Develop.*, **26**, No. 5, 568–579 (1982).
66. G. N. Flerov and V. S. Barashenkov, *Usp. Fiz. Nauk*, **114**, 351 (1974).
67. K. A. Valiev, L. V. Velikov, R. Kh. Makhmutov, and A. M. Prokhorov, *Pis'ma Zh. Tekh. Fiz.*, **11**, No. 17, 1048–1053 (1985).
68. K. A. Valiev, L. V. Velikov, R. Kh. Makhmutov, and A. M. Prokhorov, *Dokl. Akad. Nauk SSSR*, **284**, No. 3, 595–598 (1985).
69. P. Nehmiz, U. Behringer, H. Bohlen, and M. Kallmeyer, *J. Vac. Sci. Technol.*, **B1**, 1023 (1983).
70. P. Nehmiz, W. Zapka, U. Behringer, M. Kallmeyer, and H. Bohlen, *J. Vac. Sci. Technol.*, **B3**, No. 1, 136–139 (1985).
71. W. Zapka, P. Nehmiz, and H. Bohlen, *J. Vac. Sci. Technol.*, **B3**, No. 1, 140–143 (1985).
72. U. Behringer and P. Wettiger, *J. Vac. Sci. Technol.*, **B4**, No. 1, 94–99 (1986).
73. P. K. Hansma and J. Tersoff, *J. Appl. Phys.*, **61**, No. 2, R1–R23 (1987).

74. J. Bardeen, *Phys. Rev. Lett.*, **6**, 57 (1961).
75. G. Binnig, H. Rohrer, C. Gerber, and E. Weibel, *Phys. Rev. Lett.*, **49**, 57 (1982).
76. G. Binnig and H. Rohrer, *Surf. Sci.*, **126**, 236 (1983).
77. G. Binnig, H. Rohrer, C. Gerber, and E. Weibel, *Phys. Rev. Lett.*, **50**, 120 (1983).
78. M. A. McCord and R. F. W. Pease, *J. Vac. Sci. Technol.*, **B4**, No. 1, 86–88 (1986).
79. M. A. McCord and R. F. W. Pease, *J. Vac. Sci. Technol.*, **B3**, No. 1, 198–201 (1985).

2

The Physics of the Interactions between Fast Electrons and Matter

2-1. THE TASKS OF THE MICROSCOPIC THEORY OF ELECTRON LITHOGRAPHY

Just as in optical lithography, the task of electron lithography is to expose a pattern and remove (by etching) the exposed portion of the polymer resist film (the unexposed portion of the resist is removed when negative resists are processed). The result from exposing a positive resist is that the molecular linear chain is broken and the average molecular weight of the polymer in the exposed area is reduced. Reducing the molecular weight of the polymer is accompanied by an increase in the solubility (etching rate). Thus, the principal result from exposure is a pattern drawn in terms of the etching rate. It is desired that the contrast in the pattern, i.e., the ratio of the etching rate in the exposed region to the etching rate in the unexposed region be as large as possible. The contrast is largely dependent upon the etching method used; high when liquid (solvent) etching is used and lower when the etching is done in more active media in a plasma or beam of atoms and ions.[1-6]

It is clear from what has been said that the physical theory of electron lithography must have as the goal of computing the three-dimensional molecular weight distribution of the polymer molecules (or the equivalent distribution of the density of the number of disruptions in the polymer chains). With this kind of problem statement of all the events that take place when a fast electron interacts with a resist molecule, it is mainly the acts that disrupt the molecule that must be singled out. Because this procedure is so complicated we will replace it with something simpler; the problem of the three-dimensional distribution of the energy emitted by fast electrons (transmitted to the resist molecules) by assuming that there is a constant amount of energy absorbed when one act of breaking a molecule chain occurs, found experimentally to be 80–100 eV.

The sensitivity of electron resists is on the order of 10^5 C/cm^2 or 10^{14} e/cm^2. The energy of each electron in a monochromatic electron beam is 10–30 keV. The free path of electrons at this energy level in hydrocarbon polymers is 10 μm. The resist layer is 0.5–1 μm thick. Thus, it can be seen that electrons in a beam can easily reach the substrate material and that they are subsequently scattered in the substrate. After being

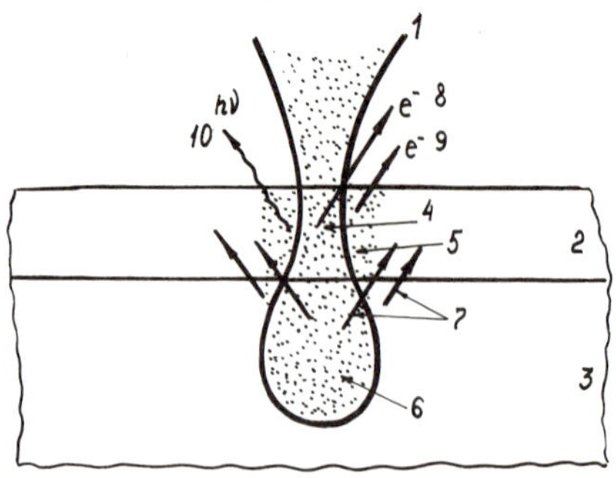

FIGURE 2-1. The physical processes in a resist–substrate system during exposure to an electron beam: 1) a focused electron beam; 2) resist layer; 3) substrate; 4, 5) resist layer regions exposed by electrons when the beam passes straight through and by electrons reflected from the substrate; 6) electron scattering area in the substrate; 7) electrons reflected from the substrate; 8) reflected and secondary electrons (including Auger electrons); $h\nu$ are x-ray and optical quanta.

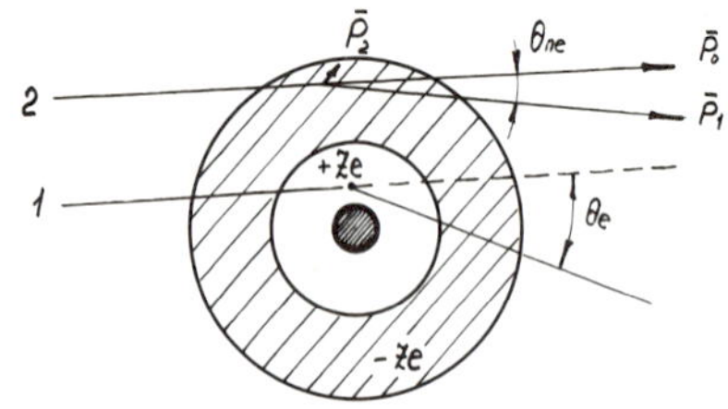

FIGURE 2-2. Elastic and inelastic scattering (collision) of a fast electron with an atom: 1) with elastic scattering the electron collides with an atom's nucleus and the scattering angle θ_e may be large; 2) with inelastic scattering a fast electron collides with an atom's electron and a portion of its energy $\Delta E = (p_0^2/2m - p_1^2/2m)$ is transmitted to the atom's electron. The inelastic scattering angle θ_{ne} is, as a rule, small; also, $\Delta E \ll p_0^2/2m$.

scattered in the substrate a fraction of the electrons return to the resist layer and again expose the layer. A few electrons, being scattered outward, i.e., reflected, leave the resist. Secondary electrons, optical, and x-ray radiation also occur. These processes are shown in Fig. 2-1. All these effects quite naturally influence the three-dimensional distribution of energy emitted by electrons in the resist that we are interested in. How do we go about finding this distribution?

To answer this question, we will investigate a fast electron in the act of penetrating a material (resist). There is a chance that this electron will be scattered when it encounters the first atom that lies in its path. This scattering may be elastic or inelastic. When elastic scattering occurs, the electron collides with the nucleus of the atom and retains its energy, but changes direction. When inelastic scattering occurs, the electron collides with the target atom's electrons and not only changes direction, but changes its energy. Because of the energy received from the fast electron the atom becomes excited (one of its electrons goes to an excited level) or is ionized (one of its electrons leaves the atom, creating a secondary electron that travels through the material). Because the target atom is part of a molecule in the resist, its excitation and ionization can disrupt a molecular chain in the resist. A diagram of elastic and inelastic electron collisions with an atom is shown in Fig. 2-2.

From what we have said it is apparent that we need to know what the scattering probabilities (the elastic and inelastic scattering cross sections of an electron in the atoms) are, which is the problem we will primarily be looking at in this chapter. The theory of electron scattering in atoms can be thought of as the theory of elementary actions that provide a foundation for constructing the trajectory of a fast electron.

If we study the motion of an electron as a series of elastic and inelastic collisions with a material's atoms we can derive an integrodifferential equation for calculating its trajectory. However, a more convenient approach is based on the concept of the mean free path length which is equal to $\lambda = 1/\sum_j N_j(\sigma_j^e + \sigma_j^{ne})$, where σ_j^e, σ_j^{ne} are the total elastic and inelastic scattering cross sections and N_j is the number of j atoms per cubic centimeter of material. If the total and differential scattering cross sections are known, an electron's trajectory can be modelled rather easily by the Monte Carlo method of statistical trials.[8–17] The trajectory of a single electron is not representative, i.e., it is impossible to evaluate the extent of the chemical conversions, the density of the emitted energy, etc. But, if we model 10^3–10^4 trajectories, the statistical processing for a group this large yields values for the density of the dissipated energy whose accuracy are quite adequate.

Thus, using the Monte Carlo method we obtain fairly accurate computer results for the density of the dissipated energy. It is also desirable to have an analytical solution for the problem, even if only approximate so that we can visualize the fundamental dependencies of the exposure parameters on the beam parameters (the energy of the electrons) and the parameters of the different layers of the material. Because most of the electrons in a material's atoms do not take part in the chemical bond, the interaction of an electron with the material is equivalent in a first approximation to the interaction with the respective group of free atoms. In an interactions model of this type the parameters of the material are simply the atomic numbers Z_j and the concentrations of the respective atoms N_j.

The smallness of the electron's mass and other properties of electron scattering require the use of a quantum mechanical scattering theory. In subsequent paragraphs we will present the respective theory.

2-2. THE QUANTUM MECHANICAL STATEMENT OF THE PROBLEM OF ELECTRON SCATTERING IN BORN'S METHOD (1926)

A quantum mechanical statement of the scattering problem means that the stationary Schrödinger equation is the basis for solving it.[18-20]

A fast particle (electron) far from a target's atom can move freely. This motion is described by the Schrödinger equation

$$-\frac{\hbar^2}{2m}\Delta\Psi = T\Psi \tag{2-1}$$

the solution of which, as we can easily satisfy ourselves by substitution, is the eigenfunction for a type of plane wave, $\Psi = e^{ikr}$, and k is the electron's wave number that gives the eigenvalues of the electron energy and momentum:

$$T = \frac{\hbar^2 k^2}{2m}, \qquad \bar{p} = \hbar\bar{k} \tag{2-2}$$

We will consider the energy levels W_α and the wave functions $\varphi_\alpha(\bar{r}_a)$ of the target atom's electrons that are a solution of the corresponding Schrödinger equation to be known:

$$\hat{H}\varphi_\alpha(\bar{r}_a) = W_\alpha \varphi_\alpha(\bar{r}_a) \tag{2-3}$$

Prior to a collision with a fast electron the atom is in the ground state (state having the least energy) W_0, $\varphi_0(\bar{r}_a)$. The system of fast electron + atom has total energy $E = T + W_0$ and is described by the wave function $\Psi_0 = e^{i\bar{k}\bar{r}}\varphi_0(\bar{r}_a)$ (the wave function of the noninteracting particles is equal to the product of the wave functions of the respective particles).

Now we include the interaction $V(\bar{r}, \bar{r}_a)$ of a fast electron at point $\bar{r}$ with the atom's electrons at points $\bar{r}_a$. We will say that $V(\bar{r}, \bar{r}_a)$ also includes the energy of the interaction between a fast electron and an atom's nucleus. The unknown wave functions can now be found from Schrödinger's equation:

$$\left(-\frac{\hbar^2}{2m}\Delta_r + \hat{H}_{r_a} + V(\bar{r}, \bar{r}_a)\right)\Psi(\bar{r}, \bar{r}_a) = E\Psi(\bar{r}, \bar{r}_a) \tag{2-4}$$

We will use the method of successive approximations to solve Eq. (2-4), writing the unknown solution as $\Psi = \Psi_0 + \Psi_1 + \ldots + \Psi_n + \ldots$, where Ψ_n has an order of smallness V^n. The first term $\Psi_0 = e^{i\bar{k}\bar{r}}\varphi_0(\bar{r}_a)$ describes the motion of the electron and the atom when there are no interactions between them and contains no indications of scattering. On the contrary, the term $\Psi_1(\bar{r}, \bar{r}_a)$ is given in terms of the interaction energy $V(\bar{r}, \bar{r}_a)$ and describes the change of state in the fast electron and the atom resulting from the collision, i.e., scattering. If we confine ourselves to calculating only Ψ_1 we will find the result of scattering in a first Born approximation.

Substituting $\Psi = \Psi_0 + \Psi_1 + \ldots$ into Eq. (2-4),

$$\left(-\frac{\hbar^2}{2m}\Delta_r + \hat{H}_{r_a} + V(\bar{r}, \bar{r}_a)\right)(\Psi_0 + \Psi_1 + \ldots) = E(\Psi_0 + \Psi_1 + \ldots) \tag{2-5}$$

and considering that the solution for Φ_0 satisfies the Schrödinger equation when there are no interactions,

$$\left(-\frac{\hbar^2}{2m}\Delta_r + \hat{H}_{r_a}\right)\Psi_0 = E\Psi_0 \tag{2-6}$$

we find an equation for correcting the Ψ_1 of the first approximation:

$$\left(-\frac{\hbar^2}{2m}\Delta_r + \hat{H}_{r_a} - E\right)\Psi_1 = -V(\bar{r}, \bar{r}_a)\Psi_0 \tag{2-7}$$

Subsequent corrections can be found by the method of successive approximations from equations of the type

$$\left(-\frac{\hbar^2}{2m}\Delta_r + \hat{H}_{r_a} - E\right)\Psi_n = -V(\bar{r}, \bar{r}_a)\Psi_{n-1} \tag{2-8}$$

that for our purposes are a sufficient first approximation.

We will seek the function $\Psi_1(\bar{r}, \bar{r}_a)$ in Eq. (2-7) as a linear combination of the eigenfunctions $\varphi(\bar{r}_a)$ of the target atom's electrons:

$$\Psi_1 = \sum_\alpha f_{\alpha,1}(\bar{r})\varphi_\alpha(\bar{r}_a) \tag{2-9}$$

the coefficients in the series are unknown functions of the fast electron's coordinates. Substituting Eq. (2-9) into Eq. (2-7), we obtain

$$\sum_\alpha \left(-\frac{\hbar^2}{2m}\Delta_r + W_\alpha - E\right) f_{\alpha,1}(\bar{r})\varphi_\alpha(\bar{r}_a) = -V(\bar{r}, \bar{r}_a)\Psi_0(\bar{r}, \bar{r}_a) \tag{2-10}$$

We multiply both sides of Eq. (2-10) by $\varphi^*_\beta(\bar{r}_a)$ and integrate with respect to the coordinates of the atom's electrons $\bar{r}_a$. Taking the orthogonality of the atomic wave functions $\int \varphi^*_\beta(\bar{r}_a)\varphi_\alpha(\bar{r}_a)\, d^3\bar{r}_a = \delta_{\alpha\beta}$ into account and introducing the notation $W_\alpha - E = -(\hbar^2/2m)k^2_\alpha$ we obtain an equation for the functions $f_{\alpha,1}(\bar{r})$:

$$(\Delta_r + k^2_\beta)f_{\beta,1}(\bar{r}) = \frac{2m}{\hbar^2}\int V(\bar{r}, \bar{r}_a)\Psi_0(\bar{r}, \bar{r}_a)\varphi^*_\beta(\bar{r}_a)\, d^3\bar{r}_a \tag{2-11}$$

The shortest route is to write a solution of Eq. (2-11) based on Green's functions.[21] Designating the right-hand side of Eq. (2-11) as $\Phi(\bar{r})$ and omitting the subscripts β and 1 on the left-hand side we find an equivalent expression for this equation:

$$(\Delta_r + k^2)f(\bar{r}) = \Phi(\bar{r}) \tag{2-12}$$

As we can see, the unknown functions $f(\bar{r})$ are the solution of the nonhomogeneous equation. Green's method introduces a nonhomogeneous equation having a δ right-hand side:

$$(\Delta_r + k^2)f(\bar{r}) = \delta(\bar{r} - \bar{r}') \tag{2-13}$$

whose solution is Green's function of Eq. (2-12):

$$G(\bar{r}, \bar{r}') = \frac{1}{4\pi}\frac{\exp(ik|\bar{r} - \bar{r}'|)}{|\bar{r} - \bar{r}'|} \tag{2-14}$$

The solution of a nonhomogeneous equation having an arbitrary right-hand side $\Phi(\bar{r})$ is written as Green's function and $\Phi(\bar{r})$:

$$f(\bar{r}) = -\int G(\bar{r}, \bar{r}')\Phi(\bar{r}')\, d\bar{r}' \tag{2-15}$$

Generally speaking, we are interested in the result of electron scattering in an atom at large distances from the atom, i.e., an asymptotic solution of $f(\bar{r} \to \infty)$. Because we can replace $|\bar{r} - \bar{r}'|$ with r in the denominator of Green's function, Eq. (2-14), and assume that $|\bar{r} - \bar{r}'| \simeq r|1 - \bar{r}\cdot\bar{r}'/r^2|$ in the exponent,

$$f(\bar{r} \to \infty) = \frac{e^{ikr}}{4\pi r}\int \Phi(\bar{r}')\, e^{-i\bar{k}\bar{r}'}\, d\bar{r}' \tag{2-16}$$

here we introduce the vector $\bar{k} = k\bar{r}/\bar{r}$. Replacing the subscripts and substituting the explicit form of $\Phi(\bar{r})$ into Eq. (2-14) we find

$$f_{\alpha,1}(\bar{r}) = \frac{e^{ik_\alpha r}}{r} \frac{m}{2\pi\hbar^2} \int d^3\bar{r}'\, e^{-i\bar{k}_\alpha \bar{r}'} \int d^3\bar{r}_a\, e^{i\bar{k}\bar{r}'} V(\bar{r}', \bar{r}_a)\varphi_0(\bar{r}_a)\varphi^*_\alpha(\bar{r}_a) \quad (2\text{-}17)$$

Finally we obtain

$$\Psi_1 = \sum_\alpha C_{0\alpha} \frac{e^{ik_\alpha r}}{r} \varphi_0(\bar{r}_a) \quad (2\text{-}18)$$

as a correction for the first approximation, where

$$C_{0\alpha} = \frac{m}{2\pi\hbar^2} \int d^3\bar{r}'\, e^{i(\bar{k}-\bar{k}_\alpha)\bar{r}'} \int \varphi^*_\alpha(\bar{r}_a) V(\bar{r}, \bar{r}_a)\varphi_0(\bar{r}_a)\, d^3\bar{r}_a \quad (2\text{-}19)$$

As we can see the correction for the first approximation is the superposition of the states in which after the collision the atom was in a state $\varphi_\alpha(\bar{r}_a)$ and the fast electron $\Psi_{\bar{k}} = e^{i\bar{k}\bar{r}}$ was a dissipating spherical wave $e^{ik_\alpha r}/r$ of amplitude $C_{0\alpha}(\bar{k} - \bar{k}_\alpha)$, depending on the difference in the electron's momenta before and after the collision $\bar{k} - \bar{k}_\alpha$.

Born's approximation is valid when $\Psi_1 \ll \Psi_0$.

2-3. THE TYPES OF COLLISIONS BETWEEN ELECTRONS AND ATOMS

According to their final states (after collision), different types of collisions between atoms and electrons can be distinguished.

1. *Elastic Scattering.* After colliding with an electron the atom remains in its ground state $\varphi_0(\bar{r}_a)$; therefore, $\alpha = 0$ and the electron conserves its energy but changes direction: $k^2 = k_0^2$, $\bar{k} \cdot \bar{k}_0 = k^2 \cos\theta$. The magnitude of elastic scattering of the electron at the atom is

$$C_{00} = \frac{m}{2\pi\hbar^2} \int d^3\bar{r}'\, e^{i(\bar{k}-\bar{k}_0)\bar{r}'} \int \varphi^*_0(\bar{r}_a) V(\bar{r}', \bar{r}_a)\varphi_0(\bar{r}_a)\, d^3\bar{r}_a \quad (2\text{-}20)$$

2. *Inelastic Scattering of an Electron That Excites the Atom.* In this case the energy level W_α of the atom is part of the atom's discrete spectrum.

3. *Inelastic Scattering of an Electron That Ionizes the Atom.* In this case the energy level W_α and the wave function $\varphi_\alpha(\bar{r}_a)$ of the atomic electron are part of the continuous spectrum.

4. If the atom was initially in an excited state $\varphi_\alpha(\bar{r}_a)$, W_α it can possibly go to a lower energy state $\varphi_{\alpha'}$, $W_{\alpha'}$ (e.g., its ground state) and transmit its excitation energy to the incident electron such that $k^2_{\alpha'} = (2m/\hbar^2)[T + (W_\alpha - W_{\alpha'})] > k^2$. This type of collision is called a collision of the second kind. Because in electron lithography a material's atoms are in their ground state collisions of the second kind do not occur.

5. The incident electron may be captured by the atom (recombination) and transmit all its energy to the atom's electrons (multielectron excitation). In this case $k^2_\alpha < 0$, $k_\alpha = i|k_\alpha|$, and the spherical dissipating wave $\exp(ik_\alpha r)/r$ becomes an exponentially decaying wave $\exp(-|k_\alpha|r)/r$ which, properly speaking, describes the electron capture

process. (Multielectron excitation of an atom leads, as a rule, ionizes the atom the Auger process (see [22]).

Processes 1–3 play a part in electron lithography. For this reason we will hereafter examine only these processes.

2-4. THE SCATTERING CROSS SECTION IN BORN'S THEORY

In keeping with the geometry used in experiments on particle scattering in targets a stream of scattered particles dj is set in the direction of the solid angle dO. The differential scattering cross section into the solid angle dO is governed by the ratio of the scattered and incident particle streams:

$$d\sigma = dj/j_f \tag{2-21}$$

In quantum mechanics a stream of particles described by a wave function $\Psi(\bar{r})$ is calculated from the formula[18]

$$\bar{j} = \frac{\hbar}{2mi}(\Psi^*\nabla\Psi - \Psi\nabla\Psi^*) \tag{2-22}$$

We have adopted $\Psi_{\bar{k}} = e^{i\bar{k}\bar{r}}$ as the wave function of the incident particles (electrons). Substituting the obvious equations $\nabla\Psi_{\bar{k}} = i\bar{k}\Psi_{\bar{k}}$ and $\nabla\Psi^*_{\bar{k}} = -i\bar{k}\Psi^*_{\bar{k}}$ into Eq. (2-22) yields $j_f = v$, where v is the electron's velocity. After scattering, when the atom has gone into the state W_α, the wave function of the electron is $f_{1\alpha}(\bar{r}) = C_{0\alpha}\, e^{i\bar{k}_\alpha r}/r$, where $C_{0\alpha}$ is independent of r. Then the stream of scattered particles through an area $ds = r^2\, dO$ is

$$dj = \frac{\hbar}{2mi}(f^*_{1\alpha}\nabla f_{1\alpha} - f_{1\alpha}\nabla f^*_{1\alpha})r^2\, dO = \frac{\hbar k_\alpha}{m}|C_{0\alpha}|^2\, dO \tag{2-23}$$

(here we say that $\nabla r^{-1} = -\bar{r}/r^3$ and $\nabla e^{ik_\alpha r} = (\bar{r}/r)\, e^{i\bar{k}_\alpha r}$).

Thus, the differential scattering cross section of an electron into a solid angle dO in the course of which the atom goes into the state W_α, φ_α is

$$d\sigma_\alpha = (k_\alpha/k)|C_{0\alpha}|^2\, dO \tag{2-24}$$

Calculating the differential cross section for elastic and inelastic processes reduces to calculating the amplitude $C_{0\alpha}$ of the scattered wave according to Eq. (2-19).

2-5. THE CROSS SECTION OF ELASTIC RUTHERFORD SCATTERING

As our first example we will use Eqs. (2-20) and (2-24) to calculate the electron scattering cross section in an infinitely heavy nucleus of an atom having charge Ze. We will consider the nucleus to be a point charge such that

$$\varphi^*_0(\bar{r}_a)\varphi_0(\bar{r}_a) = \delta(\bar{r}_a) \tag{2-25}$$

(the nucleus is made the origin of the coordinates system). Then the integral over the coordinates of the target atom's charges, Eq. (2-20), reduces to the Coulomb energy of interaction between the electron and the nuclear charge Ze:

$$\int V(\bar{r}, \bar{r}_a)\varphi_0^*(\bar{r}_a)\varphi_0(\bar{r}_a)\, d^3\bar{r}_a) = \int V\, \delta(\bar{r}_a)\, d^3 r_a = V(\bar{r}) = e^2 Z/r \tag{2-26}$$

The amplitude of the scattered electron's wave function is now

$$C_{00} = \frac{m}{2\pi\hbar^2}\int d^3\bar{r}'\, e^{i(\bar{k}-\bar{k}_0)\bar{r}'}\, V(\bar{r}') \tag{2-27}$$

We will introduce a spherical coordinates system whose z axis is parallel to the vector $\bar{k} - \bar{k}_0$; integrating over the angles we obtain

$$\begin{aligned} C_{00} &= \frac{m}{2\pi\hbar^2}\int_0^\infty r^2\, dr\, V(r)\int_0^\pi e^{i/\bar{k}-\bar{k}_0/r\cos\theta}\sin\theta\, d\theta\int_0^{2\pi} d\varphi \\ &= \frac{2m}{\hbar^2|\bar{k}-\bar{k}_0|}\int_0^\infty V(r) r \sin(|\bar{k}-\bar{k}_0|\bar{r})\, dr \end{aligned} \tag{2-28}$$

We make the substitution $V(r) = Ze^2/r$ and integrate over r by means of going to the limit

$$\begin{aligned} \int_0^\infty \sin x\, dx &= \lim_{\rho\to 0}\int_0^\infty e^{-\rho x}\sin x\, dx = \lim\frac{1}{1+\rho^2} = 1, \\ C_{00} &= \frac{2mZe^2}{\hbar^2|\bar{k}-\bar{k}_0|^2} = \frac{mZe^2}{2\hbar^2 k^2\sin^2\theta/2} \end{aligned} \tag{2-29}$$

Substituting the amplitude that we found for the wave function amplitude into Eq. (2-24) we obtain the elastic Rutherford scattering cross section of a fast electron in bare atomic nuclei:

$$d\sigma_{\text{el}} = |C_{00}|^2\, dO = \frac{Z^2 e^4}{4m^2 v^4}\frac{dO}{\sin^4\theta/2} \tag{2-30}$$

The basic physical meaning of this equation is this: the cross section of scattering into a given solid angle dO increases very quickly as the scattering angle θ decreases; as the particles' energy $E = mv^2/2$ increases the cross section decreases as $1/E^2$. It is easy to see by integrating Eq. (2-30) over a sphere that the total Rutherford scattering cross section goes to infinity.

2-6. SHIELDING EFFECTS DURING ELASTIC SCATTERING OF FAST ELECTRONS IN ATOMS

During elastic scattering of electrons in atoms the scattering will proceed differently than on bare nuclei—shielding of the nuclear charge from the fast electron by the charge of the atom's electron shells will manifest itself. We can expect the following as it relates to the role of electron shell charge shielding. If the scattering angle is large, it must mean that the electron has come close to the atomic nucleus during the scattering process so that the electrons charge remained outside the fast electron; in this case shielding must disappear or become very weak. Conversely, at small scattering angles the electron passes far from the atomic nucleus, which means that the atom's electron shell (distribution) is almost entirely between the electron being scattered and the nucleus; in this case shielding is strong. Thus, the shielding factor is a function of the scattering angle. We will formulate a theory of elastic scattering that takes shielding into account.

According to Eq. (2-20) the amplitude of the scattered electrons' wave function is, to within a factor, a Fourier component of the potential of the entire charge in the atom where the scattered electron is found:

$$C_{00} = \frac{m}{2\pi\hbar^2}\int d^3\bar{r}\, e^{i(\mathbf{k}-\mathbf{k}_0)\mathbf{r}}[-|e|\varphi(r)] = -\frac{me}{2\pi\hbar^2}\varphi_{\mathbf{q}}, \qquad \mathbf{q} = \mathbf{k} - \mathbf{k}_0 \tag{2-31}$$

Conversely, the field potential of the atom's charges is related to the charge distribution $\rho(r)$ by Poisson's equation:

$$\Delta\varphi(r) = -4\pi\rho(r) \tag{2-32}$$

Using Poisson's equation we can find the necessary Fourier component $\varphi_{\mathbf{q}}$ of the atom's potential. Instead of $\varphi(r)$ and $\rho(r)$ we substitute their Fourier expansions

$$\varphi(r) = \int \varphi_q\, e^{-i\mathbf{q}\mathbf{r}}\, d^3\mathbf{q}, \qquad \rho(r) = \int \rho_q\, e^{-i\mathbf{q}\mathbf{r}}\, d^3\mathbf{q} \tag{2-33}$$

into Eq. (2-31), which yields an algebraic relationship between $\varphi_{\mathbf{q}}$ and $\rho_{\mathbf{q}}$:

$$-q^2\varphi_{\mathbf{q}} = -4\pi\rho_{\mathbf{q}}, \qquad \varphi_{\mathbf{q}} = \frac{4\pi}{q^2}\rho_{\mathbf{q}} \tag{2-34}$$

From the definition

$$\rho_{\mathbf{q}} = \int \rho(r)\, e^{i\mathbf{q}\mathbf{r}}\, d^3\mathbf{r} \tag{2-35}$$

we can calculate $\rho_{\mathbf{q}}$ for any charge distribution.

We write the charge distribution in an atom as

$$\rho(r) = Ze\delta(r) - en(r) \tag{2-36}$$

where $n(r)$ is a spherically symmetric distribution of the number of electrons in the atom. Integrating Eq. (2-35) with Eq. (2-36) we find

$$\rho_{\mathbf{q}} = Ze - eF(\mathbf{q}), \qquad F(q) = \int n(r)\, e^{i\mathbf{q}\mathbf{r}}\, d^3\mathbf{r} \tag{2-37}$$

The function $F(q)$ is a Fourier component of the electron density in the atom and is called the atomic scattering form factor. The elastic scattering cross section in an atom is, with due regard for shielding,

$$d\sigma_{\text{el}} = \frac{e^4|Z - F(q)|^2}{4m^2v^4\sin^4\theta/2}\,dO \tag{2-38}$$

To find $F(q)$ we must first calculate $n(r)$.The most common tables of the $F(q)$ function are those obtained with a distribution $n(r)$ found from the statistical Thomas–Fermi model of the atom.[18–21] However, the behavior of $F(q)$ at large and small scattering angles can be analyzed directly using the definition of the form factor, Eq. (2-37). At large scattering angles $q = 2k\sin(\theta/2)$ is also large and $\mathbf{q}\cdot\mathbf{r} = qr_q \gg 1$. Then, within the volume of an atom (where $n(r)$ is nonzero) the function $\exp(i\mathbf{qr})$ changes sign several times and $F(q)$ will be small. According to the Thomas–Fermi model, for an atom of number Z most of the electrons have an orbit $a_0/Z^{1/3}$ and the condition for a large scattering angle (no shielding) will be

$$qa_0/Z^{1/3} \gg 1 \rightarrow 2\sin\theta/2 \gg Z^{1/3}/ka_0 \qquad (2\text{-}39)$$

The inverse inequality defines, obviously, small scattering angles when shielding is strong. To find the scattering cross section in this situation we expand $F(q)$ as a series in $\mathbf{q}\cdot\mathbf{r} = qr_q \ll 1$:

$$F(q) = \int n(r)\,d^3\mathbf{r} + iq\int r_q n(r)\,d^3\mathbf{r} - 0.5q^2\int r_q^2 n(r)\,d^3\mathbf{r} + \dots \qquad (2\text{-}40)$$

The first term in the expansion is the total number of electrons in the atom Z; the second term is zero because the $n(r)$ distribution is spherically symmetric. The third term is the most meaningful to us, because

$$Z - F(q) = 0.5q^2\int r_q^2 n(r)\,d^3\mathbf{r} \qquad (2\text{-}41)$$

The integral in Eq. (2-41) can be evaluated from the Thomas–Fermi model

$$\int r_q^2 n(r)\,d^3\mathbf{r} = \frac{1}{3}\int r^2 n(r)\,d^3\mathbf{r} \approx \frac{1}{3}\left(\frac{a_0}{Z^{1/3}}\right)^2 Z \qquad (2\text{-}42)$$

Then

$$d\sigma_{\text{el}} = \frac{e^4m^2}{9\hbar^4}\frac{a_0^4}{Z^{4/3}}Z^2\,dO = \frac{1}{9}a_0^2 Z^{2/3}\,dO \qquad (2\text{-}43)$$

As we can see, because of shielding the scattering cross section is independent of the scattered particles's energy and the scattering angle when the angles are small, and the cross section is governed by the geometric size of the orbits and the number of electrons in the atom.

Because the Thomas–Fermi model is so important to our purposes, we will present the most important results from this theory.[18,21,23] The potential of an atom at a point r is given by

$$\varphi(r) = \frac{Ze}{r}\chi\left(\frac{r}{r_s}\right), \qquad r_s = 0.885\frac{a_0}{Z^{1/3}} \qquad (2\text{-}44)$$

where $\chi(x)$ is a universal shielding function that is a solution of

$$\frac{d^2\chi}{dx^2} = x^{-1/2}\chi^{3/2} \qquad (2\text{-}45)$$

with boundary conditions $\chi = 1$ when $x = 0$ and $\chi = 0$ when $x = \infty$. The radial distribution of the electrons density $n(r)$ is given by

$$n(r) = Z^2 f\left(\frac{r}{r_s}\right), \qquad f(x) = \frac{32}{9}\pi^2\left(\frac{\chi(x)}{x}\right)^{3/2} \tag{2-46}$$

Integrating with Eq. (2-46), we obtain a new universal function $\varphi(x)$ for the scattering form factor:

$$F(q) = \int n(r)\, e^{i\bar{q}\bar{r}}\, d^3\mathbf{r} = Z \cdot \varphi(q \cdot r_s) \tag{2-47}$$

which has also been tabulated.[18] Finally, if we substitute Eq. (2-47) into the scattering cross section we obtain still another universal function $\Phi(x)$:

$$d\sigma_{\text{el}} = \frac{4e^4 m^2 Z^2}{\hbar^4 q^4}(1 - \varphi(qr_s))^2\, dO = Z^{2/3} a_0^2 \Phi(qr_s)\, dO \tag{2-48}$$

$$\Phi(qr_s) = 4 \cdot 0.0885^4 (1 - \varphi(qr_s))^2/(qr_s)^4 \tag{2-49}$$

The universal nature of these functions lies in the fact that they can be applied to any atom because they are functions of dimensionless arguments. The unit of length r_s in the dimensionless arguments is the radius of the electron orbit because it describes a given atom by means of Z.

The total elastic scattering cross section for electrons in an atom is finite if we take shielding of the nuclear charge by the atom's electrons into consideration. We will show this by integrating Eq. (2-48) over all angles. Considering the fact that small scattering angles make the major contribution to total scattering we can write $q = k\theta$, $dO = 2\pi\theta\, d\theta$ for small angles and use this expression over the entire range of angles:

$$\sigma = Z^{2/3} a_0^2 \int_0^{\pi} \Phi(kr_s\theta) 2\pi\theta\, d\theta = \frac{Z^{2/3} a_0^2}{k^2 r_s^2} \int_0^{\pi k r_s} \Phi(x) x\, dx \tag{2-50}$$

The upper limit of integration in Eq. (2-50) can be, without a great deal of error, assumed to be infinity. Using numerical methods to find a numerical value for $\int_0^{\infty} \Phi(x) x\, dx$,[18,20] we obtain

$$\sigma = 7.14\pi Z^{4/3}/k^2 \tag{2-51}$$

Obviously, the total elastic scattering cross section when shielding is accounted for is finite and inversely proportional to the electron's energy.

2-7. THE CROSS SECTION OF ELASTIC ELECTRON SCATTERING AT A CENTER HAVING AN EXPONENTIALLY SHIELDED COULOMB FIELD[20]

The equation found in Section 2-6 for the differential scattering cross section using the Thomas–Fermi model is given as a universal function that has been tabulated (it is a table for the generating function $\varphi(qr_s)$), but is unknown in an analytical form. For this reason it is still worth our while to find an analytical function for $d\sigma_{\text{el}}$. An extremely elegant formula is obtained if we assume that the field of the atomic nucleus is shielded by the atom's electrons as it is in a system of mobile charges, i.e., by the exponential Debye–Hückel law

$$V(r) = \frac{Ze^2}{r} \exp\left(-\frac{r}{r_D}\right), \qquad r_D = sr_s \tag{2-52}$$

Here the shielding radius r_D is given in terms of r_s in the Thomas–Fermi theory and the correction S must be found. The most appropriate way to find S is to require that the results for the total cross section σ_{el} obtained from a shielded Coulomb field, Eq. (2-52), on one hand and from the Thomas–Fermi model on the other hand, agree.

An elementary integration with the potential energy, Eq. (2-52), yields

$$C_{00} = \frac{2m}{\hbar^2}\frac{1}{q}\int_0^\infty \left(\frac{Ze}{r}\right)^2 e^{-r/r_D} \sin(qr)\, dr = \frac{2Zme^2}{\hbar^2}\frac{1}{q^2 + r_D^{-2}} \tag{2-53}$$

$$d\sigma_{\text{el}} = \frac{4m^2(Ze^2)^2}{\hbar^2}\frac{dO}{(q^2 + r_D^{-2})} = \left(\frac{Ze^2}{2mv^2}\right)^2 \left(\sin^2\frac{\theta}{2} + \alpha^2\right)^{-2} dO \tag{2-54}$$

$$\alpha = 0.56Z^{1/3}(e^2/\hbar v)S$$

for the amplitude of the scattered electron's wave function and the differential scattering cross section.

The formula for the total elastic scattering cross section at a center having a shield potential is obtained via an elementary integration over the angles:

$$\sigma_{\text{el}} = \int d\sigma_{\text{el}} = \left(\frac{Ze^2}{mv^2}\right)^2 \frac{\pi}{\alpha^2(1+\alpha^2)} \tag{2-55}$$

For fast electrons we can assume that $1 + \alpha^2 \approx 1$ in the denominator; then,

$$\sigma_{\text{el}} = 3.13\pi Z^{4/3}/k^2 S^2 \tag{2-56}$$

Equating the total cross sections of Eqs. (2-56) and (2-51) gives us $S = 0.66$.[20]

2-8. INELASTIC COLLISIONS BETWEEN FAST ELECTRONS AND ATOMS

In our previous discussions of the general theory of fast electrons scattering in Born's approximation we have written the inelastic electron scattering cross section as Eqs. (2-19) and (2-24):

$$d\sigma_\alpha = (k_\alpha/k)|C_{0\alpha}|^2\, dO \tag{2-57}$$

$$C_{0\alpha} = \left(\frac{m}{2\pi\hbar^2}\right) \int d^3\bar{r}_a\, \varphi^*(\bar{r}_a)\varphi_0(\bar{r}_a) \int e^{i\bar{q}\bar{r}}\, V(\bar{r}, \bar{r}_a)\, d^3\bar{r} \tag{2-58}$$

$$V(\bar{r}, \bar{r}_a) = -\frac{Ze^2}{r} + \sum_a \frac{e^2}{|\bar{r} - \bar{r}_a|} \tag{2-59}$$

These equations describe the intensity (probability) of the process in which an atom will remove energy $E_\alpha - E_0$ and momentum $\hbar\bar{q}$ from a fast electron, and go from state $\varphi_0(\bar{r}_a)$ to state $\varphi_\alpha(\bar{r}_a)$; the electron energy is reduced from $\hbar^2k^2/2m$ to $\hbar^2k_\alpha^2/2m$ and its momentum from $\hbar\bar{k}$ to $\hbar\bar{k}_\alpha$, so that the laws of conservation of energy and momentum for inelastic scattering are

$$(\hbar^2/2m)(k^2 - k_\alpha^2) = E_\alpha - E_0 \tag{2-60}$$

$$\hbar\bar{q} = \hbar(\bar{k} - \bar{k}_\alpha), \qquad q^2 = k^2 + k_\alpha^2 - 2kk_\alpha \cos\theta \tag{2-61}$$

An analysis of the laws of conservation, as we will see, is extremely useful when doing the calculations given by Eq. (2-58). We shall now begin these calculations. We will isolate the integral

$$e \int e^{i\bar{q}\bar{r}}/|\bar{r} - \bar{r}_a|\, d^3\bar{r} = \varphi_{\bar{q}}(\bar{r}_a) \tag{2-62}$$

which, obviously, has the sense of a Fourier component for the potential of an atom's electron at point $\bar{r}_a$. On the other hand, it can be thought of as the potential at point $\bar{r}_a$ produced by the charge distributed in space according to $\rho_q = e\exp(i\bar{q}\bar{r})$. This potential must then satisfy Poisson's equation that relates the potential and the charge density at the point

$$\Delta\varphi_q(\bar{r}_a) = -4\pi\rho_q(\bar{r}_a) = -4\pi e \exp(i\bar{q}\bar{r}_a) \tag{2-63}$$

We can easily satisfy ourselves (e.g., by substitution) that a solution of (2-63) is

$$\varphi_{\bar{q}}(\bar{r}_a) = \frac{4\pi e}{q^2} e^{i\bar{q}\bar{r}_a} \tag{2-64}$$

Equation (2-64) for the Fourier component of coulomb potential is widely used. Substituting Eq. (2-64) into Eq. (2-57) gives us

$$d\sigma_\alpha = 4\left(\frac{me^2}{\hbar^2}\right)^2 \frac{k_\alpha}{k}\frac{1}{q^4}\left|\int \sum_a e^{i\bar{q}\bar{r}_a}\varphi_\alpha^*(\bar{r}_a)\varphi_0(\bar{r}_a)\, d^3\bar{r}_a\right|^2 dO \tag{2-65}$$

Here the fact that the integrals of the energy of interaction between the fast electron and the nuclear charge fall out due to the orthogonality of the wave functions φ_α and φ_0 is taken into consideration.

To do further calculations, we must analyze the laws of conservation. First, we will use the fact that the energy lost by a fast electron is small so that

$$E_\alpha - E_0 = \frac{\hbar^2}{2m}(k + k_\alpha)(k - k_\alpha) \approx \hbar v(k - k_\alpha) \tag{2-66}$$

We can say the following about $E_\alpha - E_0$. If an atom is excited by colliding with an electron we can estimate $E_\alpha - E_0$ by the quantity $mv_a^2/2$, where v_a is the velocity of the atom's electrons. If ionization occurs, $E_\alpha - E_0 \approx mv_e^2/2$, where v_e is the velocity of the electron that was liberated due to ionization. The infinitesimality of the energy lost by a fast electron is given by

$$(k - k_\alpha) = \frac{E_\alpha - E}{\hbar v} \ll k, \qquad \frac{mv_a^2}{2\hbar v} \ll \frac{mv}{\hbar}, \qquad \frac{v_a^2}{2v^2} \ll 1 \tag{2-67}$$

The size of an electron orbit a and the velocity of an electron in that orbit v_a are tied together by an uncertainty relation $a \cdot mv_a \approx \hbar$. An electron that has been liberated by ionization is not localized in space and its velocity v_e may be large in comparison with v_a, but still less than that of a fast electron. On the other hand,

$$q^2 = k^2 + k_\alpha^2 - 2kk_\alpha\left(1 - \frac{\theta^2}{2}\right) = (k - k_\alpha)^2 + k^2\theta^2 \tag{2-68}$$

follows from the law of conservation of momentum when the scattering angles are small, $\theta \ll 1$. Substituting $k - k_\alpha = mv_a^2/2\hbar v$ and $k = mv/\hbar$ into Eq. (2-68) we obtain

$$q^2 = \left(\frac{mv_a^2}{2\hbar v}\right)^2 + \left(\frac{mv}{\hbar}\right)^2\theta^2 \tag{2-69}$$

In the range of the parameters for which Eq. (2-69) is valid the amount of momentum transmitted to the atom and the scattering angle of the fast electron increase simultaneously and are related by the hyperbolic Eq. (2-69). We rewrite Eq. (2-69) as

$$\frac{q^2}{k^2} = \theta^2 + (v_a^2/2v^2)^2 \tag{2-70}$$

According to Eq. (2-70) we can distinguish three ranges of values θ and q. The first range of very small angles is given by $\theta \ll v_a^2/v^2$. In this range the amount of transmitted momentum is independent of the scattering angle and is equal to

$$q_{\min} = k\left(\frac{v_a^2}{2v^2}\right) = \frac{(E_\alpha - E_0)}{\hbar v} \tag{2-71}$$

The second range of angle values is given by $(v_a^2/2v^2) < \theta < v_a/v$. In this range the deflection angle and the amount of momentum being transmitted are proportional to one another: $q = k\theta$. Considering the relationship between the velocity and the orbit size of the atom's electrons $a \cdot mv_a = \hbar$, we also have $qa \ll 1$ for the second range.

We define the third range of values for the quantities θ and q by $v_a/v \ll \theta \ll 1$. In this interval, $q = k\theta$ as usual, although $qa \gg 1$. At very large scattering angles a large amount of energy in comparison with the magnitude of the binding energy is transmitted to the atom's electron and the binding energy E_0 in the law of conservation of energy, Eq. (2-60), can be neglected, assuming that $E_\alpha = mv_e^2/2$. In this case the scattering is inelastic in the sense that the kinetic energy of the colliding particles is conserved.

The fourth range (large scattering angles) is given by $\theta \geq 1$. When $\theta \geq 1$ the energies of the scattered and scattering electrons are comparable after the collision and they cannot be distinguished by energy, which means that exchange effects must be considered in collisions having a large scattering angle.

Figure 2-3 shows θ as a function of q. The characteristic inequalities are given in Table 2-1.

An analysis of how much energy is transmitted to the atomic electrons in the scattering angle ranges is of interest. The transmitted energy is

$$\Delta E = E_\alpha - E_0 = \frac{\hbar^2}{2m}(k^2 - k_\alpha^2) \tag{2-72}$$

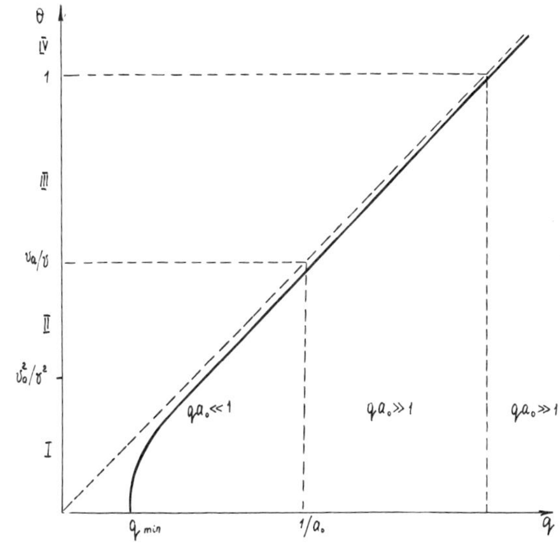

FIGURE 2-3. The scattering angle of a fast electron as a function of the momentum transmitted to the atom by the electron.

The momenta $\bar{k}$ and $\bar{k}_\alpha$ are related by the law of conservation $\bar{q} = \bar{k} - \bar{k}_\alpha$, which, for small scattering angles $\theta \ll 1$ (ranges I–III), is written

$$q^2 = (k - k_\alpha)^2 - k(k - k_\alpha)^2\theta^2 + k^2 \tag{2-73}$$

In the first range $k - k_\alpha = q_{\min}$ and

$$\Delta E = (E_\alpha - E_0) = \frac{\hbar^2}{m} k q_{\min} \tag{2-74}$$

The angular deflection of the scattered electron in the first range is extremely small and every change in its momentum is made in the lengthwise component of the momentum: $k - k_\alpha = q_{\min}$.

In ranges II and III we can assume that $q = k\theta$ and from Eq. (2-73) we obtain an equation for finding $(k - k_\alpha)$:

$$(k - k_\alpha)^2 - k(k - k_\alpha)\theta^2 = 0 \tag{2-75}$$

$$(k - k_\alpha) \approx k\theta^2 \tag{2-76}$$

Then, according to Eq. (2-72),

$$\Delta E_{\text{II}} = 2E\theta^2 \leq 2E\left(\frac{v_a}{v}\right)^2 \tag{2-77}$$

for range II and $\Delta E \gg |E_a|$ for range III. Even more so is $\Delta E \gg |E_a|$ for range IV. The continuous change in transmitted energy $\Delta E(\theta)$ describes the transition of an electron into states having a continuous spectrum, i.e., ionization. We will now give the scattering cross section into a solid angle dO in a specific interval of the transmitted momentum dq. The law of conservation of energy, Eq. (2-60), shows that if the energies

TABLE 2-1. Ranges of Angles for Inelastic Scattering and Characteristic Relationships between the Physical Quantities That Describe a Collision

	I	II	III	IV
Range of scattering angles	$\theta < v_a^2/v^2$	$v_a^2/v^2 \ll \theta < v_a/v$	$v_a/v < \theta < 1$	$\theta \geq 1$
Values of qa_0	$\ll 1$	$\ll 1$	$\gg 1$	$\gg 1$
Dependence of $q(\theta)$	$q = q_{\min} = \frac{E_\alpha - E_0}{\hbar v} = \text{const}$	$q = k\theta$	$q = k\theta$	$q^2 = k^2 + k_\alpha^2 - 2kk_\alpha \cos\theta$

of the atom's initial E_0 and final E_α states are defined and the initial energy of a fast electron is $E = \hbar^2 k_\alpha^2/2m$, then the final energy $\hbar^2 k_\alpha^2/2m$ of the fast electron will also be uniquely defined. Then the k and k_α of Eq. (2-61) can be considered constants and we take the derivative

$$q\,dq = kk_\alpha \sin\theta\, d\theta = \frac{kk_\alpha}{2\pi} dO \tag{2-78}$$

Substituting Eq. (2-78) into Eq. (2-65) yields

$$d\sigma_\alpha(q, q+dq) = 8\pi \left(\frac{e^2}{\hbar v}\right)^2 \frac{dq}{q^3} \left| \int \sum_\alpha e^{i\mathbf{q}\mathbf{r}_a} \varphi_\alpha^* \varphi_0\, d^3\mathbf{r}_a \right|^2 \tag{2-79}$$

We are now in a position to write the final equations for the differential scattering cross section of a fast electron, with due regard for the relationships between the quantities in the different ranges of scattering angles described above.

The amount of momentum transferred in range I is so very small that $\mathbf{q}\cdot\mathbf{r}_a \ll 1$. We make use of the expansion $\exp(i\mathbf{q}\mathbf{r}_a) = 1 + i\mathbf{q}\mathbf{r}_a$ and write $\mathbf{q}\mathbf{r}_a = qx_a$, taking the x axis as the direction in which the momentum is transmitted. In range I we will understand that $q = q_{\min} = (E_\alpha - E_0)/\hbar v = \text{const}$. Then, according to Eq. (2-65) we will have

$$\begin{aligned} d\sigma_\alpha^{(\mathrm{I})} &= 4\left(\frac{me^2}{\hbar^2}\right)^2 \frac{k_\alpha}{k} \frac{dO}{q_{\min}^4} \left| \int \sum (1 + iq_{\min} x_a) \varphi_\alpha^* \varphi_0\, d^3\mathbf{r}_a \right|^2 \\ &= 4\left(\frac{me}{\hbar^2}\right)^2 \frac{k_\alpha}{k} \frac{dO}{q_{\min}^2} |(d_x)_{0\alpha}|^2 \end{aligned} \tag{2-80}$$

where the atom's electric dipole moment operator $\mathbf{d} = \sum_a e\mathbf{r}_a$ and its projection $d_x = e\sum_a x_a$ in the direction $\mathbf{q}$ are introduced. We will now consider $\mathbf{q}\mathbf{r}_a \ll 1$ and $q = k\theta$ in range II. Expanding $\exp(i\mathbf{q}\mathbf{r}_a)$ and integrating as in Eq. (2-79), we find

$$d\sigma_\alpha^{(\mathrm{II})} = 8\pi \left(\frac{e}{\hbar v}\right)^2 \frac{dq}{q} |(d_x)_{0\alpha}|^2 \tag{2-81}$$

Making the substitution $q = k\theta$, we can rewrite Eq. (2-81):

$$d\sigma_\alpha^{(\mathrm{II})} = \left(\frac{2e}{\hbar v}\right)^2 |(d_x)_{0\alpha}|^2 \frac{dO}{\theta^2} \tag{2-82}$$

As we can see, for small scattering angles the differential scattering cross section $d\sigma/dO$ increases with decreasing θ as θ^{-2} and reaches a maximum value defined by Eq. (2-80). Because the cross sections in ranges I and II are given in terms of the matrix elements for the dipole moment of the atom's electron shell, the formulas obtained are called dipole approximation formulas.

We will do the calculations for the third range of angles $v_a/v \ll \theta \ll 1$. Now $\bar{q}\bar{r}_a \gg 1$ and we cannot use the expansion of $\exp(i\bar{q}\bar{r}_a)$ as we did in the dipole approximation. We must deal directly with the quantity $\sum_a \exp(i\bar{q}\bar{r}_a)$, which has the following physical meaning. We will introduce an operator for the density of the electrons in an atom:

$$\rho(\mathbf{r}) = \sum_\alpha \delta(\mathbf{r} - \mathbf{r}_a) \tag{2-83}$$

In fact, the sum of the δ-functions of Eq. (2-83) gives the number of electrons at a specific point r. Using

$$\rho_{\bar{q}} = \int \rho(\mathbf{r}) \exp(i\mathbf{qr})\, d^3\mathbf{r} \tag{2-84}$$

to define the Fourier component $\rho_{\bar{q}}$ of the electrons density, we find

$$\rho_{\bar{q}} = \sum_a \int \delta(\bar{r} - \bar{r}_a) \exp(i\bar{q}\bar{r})\, d^3\bar{r} = \sum_a e^{i\bar{q}\bar{r}_a} \tag{2-85}$$

Because, just as before, when $\theta \ll 1$ we can take $q = k\theta$, we find from Eq. (2-65) that

$$d\sigma_\alpha^{(\mathrm{III})} = \left(\frac{2me^2}{\hbar^2}\right)^2 \frac{k_\alpha}{k} \frac{1}{k^4\theta^4} |(\rho_{\bar{q}})_{0\alpha}|^2\, dO \tag{2-86}$$

This dependence corresponds to Rutherford's equation in which $2\sin\theta/2 \approx \theta$.

We will now investigate the quantity $|(\rho_{\bar{q}})_{0\alpha}|^2$. It follows from the quantum mechanical identity[18]

$$\sum_\alpha |f_{0\alpha}|^2 = \sum_\alpha f_{0\alpha} f_{0\alpha}^* = \sum_\alpha f_{0\alpha} (f^*)_{0\alpha} = (f f^*)_{00} \tag{2-87}$$

that

$$\sum_{\alpha \neq 0} |(\rho_{\bar{q}})_{0\alpha}|^2 = \left(|\rho_{\bar{q}}|^2\right)_{00} - |(\rho_{\bar{q}})_{00}|^2 \tag{2-88}$$

By the definition of the scattering form factor we have

$$(\rho_{\bar{q}})_{00} = \int \sum_a e^{i\bar{q}\bar{r}_a} |\varphi_0(\bar{r}_a)|^2\, d^3\bar{r}_a = \int n(r)\, e^{i\bar{q}\bar{r}}\, d^3\bar{r} = F(q) \tag{2-89}$$

Then, we have the identity

$$|\rho_{\bar{q}}|^2 = \left|\sum_{a=1}^{Z} e^{i\bar{q}\bar{r}_a}\right|^2 = Z + \sum_{a \neq b} e^{i\bar{q}(\bar{r}_a - \bar{r}_b)} \tag{2-90}$$

Substituting Eqs. (2-87) and (2-88) into (2-86) we find

$$d\sigma_\alpha^{(\mathrm{III})} = \left(\frac{2e}{mv^2}\right)^2 \frac{dO}{\theta^4} \left\{ Z - F^2(q) + \sum_{a \neq b} e^{i\bar{q}(\bar{r}_a - \bar{r}_b)} \right\} \tag{2-91}$$

At angles $v_0/v \ll \theta$ the scattering form factor $F(q)$ is small (see the analysis of $F(q)$ in Section 2-6) and the third term inside the brackets in Eq. (2-91) is also small. Discarding small terms we find

$$d\sigma_\alpha^{(\mathrm{III})} = \left(\frac{2e^2}{mv^2}\right)^2 Z \frac{dO}{\theta^4} \tag{2-92}$$

The inelastic scattering cross section comes about because of the contributions from the Z independently (incoherently) scattering electrons of the atom. The elastic scattering cross section is, as we have seen, proportional to Z^2.

We will now find the formula for the scattering cross section when $\theta \geq 1$ for interval IV. In fact, we have already found this formula (see Eq. (2-38) or (2-92)). If we use Eq. (2-38) its use can be justified by the fact that when large amounts of energy are transmitted we can neglect the electrons' binding energy and examine the scattering of a fast electron on free (quiescent) electrons. This scattering will be elastic in the sense that all of the kinetic energy of the colliding electrons is conserved. In the sense of the event we must consider scattering at each of an atom's electrons to be additive, i.e., assume that in Eq. (2-38) $Z = 1$ and $F = 0$. Then, in the calculation for a single electron

$$d\sigma_\alpha^{(\mathrm{IV})} = e^4/4m^2v^4\, dO/\sin^4\theta/2 \tag{2-93}$$

We obtain the same result if we rewrite Eq. (2-92) with the substitution $\theta \to 2\sin\theta/2$.

Equation (2-93) must, however, be corrected for exchange effects when collisions occur. The fact is that when large amounts of energy are transmitted from a fast electron we cannot say which of the two electrons produced because of the collision is the primary "fast" electron. Accounting for exchange effects reduces to adding the probabilities that after colliding, two electrons will be found in a given solid angle.

If we return to the amplitude $C(\theta)$ of a scattered wave $C(\theta)\, e^{ikr}/r$ we find by means of $d\sigma = |C(\theta)|^2\, dO$ that

$$C(\theta) = (e^2/2mv^2)(\sin^2\theta/2)^{-1} \tag{2-94}$$

According to the laws of conservation, when two particles of the same mass collide the scattering angles are θ and $\pi - \theta$ (in the center of mass system; see [24]). We denote the respective scattering amplitudes as $C(\theta)$ and $C(\pi - \theta)$. When the total spin of the colliding particles is even, $\bar{S} = \bar{S}_1 + \bar{S}_2 = 0$ the total amplitude, being the wave function of the colliding particles (to within $\exp(ikr)/r$), must be symmetric, i.e., it cannot change sign when the particles are transposed. This condition is satisfied if we take

$$C_s(\theta) = C(\theta) + C(\pi - \theta) \tag{2-95}$$

Similarly, when the total spin is odd, $|\bar{S}| = |\bar{S}_1 + \bar{S}_2| = 1$, we take the asymmetric wave amplitude

$$C_a(\theta) = C(\theta) - C(\pi - \theta) \tag{2-96}$$

The scattering cross section for these two cases can be written

$$d\sigma_s = |C_s(\theta)|^2\, dO \tag{2-97}$$

$$d\sigma_a = |C_a(\theta)|^2\, dO \tag{2-98}$$

When an unpolarized electron beam is scattered on an unpolarized target the total cross section is comes about because of the symmetric and asymmetric cross sections with their statistical weights: when $|\bar{S}| = 1$ there are three spin states $S_Z = -1,\ 0,\ +1$;

when $|\bar{S}| = 0$ there is one spin state, $S_Z = 0$. Accordingly, the statistical weights $d\sigma_a$ and $d\sigma_s$ of the cross sections are 3/4 and 1/4:

$$d\sigma = \frac{3}{4} d\sigma_a + \frac{1}{4} d\sigma_s \tag{2-99}$$

Substituting $C_s(\theta)$ and $C_a(\theta)$ into Eq. (2-99) we obtain

$$d\sigma^{(\mathrm{IV})} = \left[|C(\theta)|^2 + |C(\pi - \theta)|^2 - \frac{1}{2} C^*(\theta) C(\pi - \theta) - \frac{1}{2} C(\theta) C^*(\pi - \theta) \right] d\theta$$

$$= Z \left(\frac{e^2}{2mv^2} \right)^2 [\sin^{-4} \theta/2 + \cos^{-4} \theta/2 - \sin^{-2} \theta/2 \cos^{-2} \theta/2] \, d\theta \tag{2-100}$$

In a laboratory coordinates system, where one particle is at rest, the deflection angle is $\theta_1 = \theta/2$. Then,

$$d\theta = 2\pi \sin\theta \, d\theta = 4\pi \sin 2\theta_1 \, d\theta_1 = 4\cos\theta_1 2\pi \sin\theta_1 \, dO_1 = 4\cos\theta_1 \, dO_1$$

$$d\sigma^{\mathrm{IV}} = \left(\frac{e^2}{mv^2} \right)^2 [\sin^{-4} \theta_1 + \cos^{-4} \theta_1 - \sin^{-2} \theta_1 \cos^{-2} \theta_1] \cos\theta_1 \, dO_1 \tag{2-101}$$

2-9. THE DIFFERENTIAL IONIZING CROSS SECTION OF ATOMS

The theory of atomic ionization holds a great deal of interest for us because it is the most typical process that takes place when fast electrons act upon a target. Three particles are formed during ionization: a scattered electron, an ionized atom, and a departing (secondary) electron. Therefore, the initial state of the atom+electron system has an initial energy $E = \hbar^2 k^2/2m$, an electron momentum of $\hbar\bar{k}$, and a set of quantum numbers γ for the atom; the state of the system after a collision is determined by the energy $E' = \hbar^2 k'^2/2m$ and momentum $\hbar\bar{k}'$ of the scattered electron, the energy $E_f = \hbar^2 k_f^2/2m$ and momentum $\hbar\bar{k}_f$ of the secondary electron, and a set of quantum numbers γ' for the ion that was formed. The law of conservation of energy for this process is written

$$\frac{\hbar^2(k^2 - k'^2)}{2m} = \frac{\hbar^2 k_f^2}{2m} + E_{\gamma'} - E_\gamma \tag{2-102}$$

The quantity $\hbar\bar{q} = \hbar(\bar{k} - \bar{k}')$ is equal to the amount of momentum lost by the incoming electron.

Using perturbation theory we can do calculations for the ionization cross section that are similar to those in Section 2-2. In doing this, different quantities that denote the final state of the system may be chosen. For example, if a primary electron is scattered toward a solid angle dO and its final momentum is $\hbar\bar{k}'$, and the ionization electron moves in a direction $\bar{k}_f$ and its momentum is somewhere between $\bar{k}_f$ and $\bar{k}_f + d\bar{k}_f$, the differential cross section for this process is[24]

$$d\sigma = \frac{m^2}{(2\pi)^5 \hbar^4)} \frac{k'}{k} \left| \int e^{i\bar{q}\bar{r}} \, V(\bar{r}, \bar{r}_a)_{\gamma,\gamma',\bar{k}_f} \, d^3\bar{r} \right|^2 d\bar{k}_f \, dO \tag{2-103}$$

We can take both the change in the primary electron's energy W and the energy of momentum transfer $Q = \hbar^2 q^2/2m$ as the variables that describe this process. Using these variables gives the cross section a particularly simple form[25]:

$$d\sigma = \frac{\pi e^4}{E} \frac{1}{WQ} \frac{df(W,Q)}{dW} dW\, dQ \tag{2-104}$$

Here $df(W,Q)/dW$ is the density of the extended oscillator strength for an atom to make the transition from its initial to its final state. This quantity is extremely important in spectroscopy and in the theory of fast particle scattering; therefore, the question of defining them deserves a separate examination.

2-10. THE OSCILLATOR STRENGTHS OF TRANSITIONS IN ATOMS

The oscillator strength of a transition $0 \longrightarrow \alpha$ in an atom in the discrete spectrum is defined by the dimensionless quantity[26]

$$S_{0\alpha} = \frac{2m}{\hbar^2}(E_\alpha - E_0) \left| \left(\sum_a x^{(a)} \right)_{0\alpha} \right|^2 \tag{2-105}$$

For transitions into continuous spectrum states the density of the oscillator strengths per unit energy is introduced, as we have just seen.

Recall that the probabilities of spontaneous and forced transitions are stated in terms of oscillator strengths[26]:

$$A_{ik} = \frac{2e^2\omega_{ik}^2}{mc^2} S_{ik}, \qquad B_{ki} = \frac{\pi e^2}{m\hbar\omega_{ki}} S_{ki} \tag{2-106}$$

Many computations for hydrogen-like, one-electron ions can be done exactly rather than approximately. The oscillator strengths are also calculated exactly. For example, the oscillator strengths of transition that correspond to the Lyman series of spectral lines are[26]

$$S_{1s \to np} = \frac{2^8 n^5 (n-1)^{2n-4}}{3(n+1)^{2n+4}} \tag{2-107}$$

As can easily be seen these strengths are independent of the nuclear charge Z. This is easy to understand when we recall how the oscillator strengths of a transition, Eq. (2-105), are defined. For hydrogen-like ions the transition energy is $E_{np} - E_{1s} = Z^2 e^2/a_0(1 - 1/n^2)$, i.e., $E_\alpha - E_0 \sim Z^2$, because the radius of the electron orbit is inversely proportional to Z, and $x_{0\alpha}^2 \sim 1/Z^2$. As the principal quantum number n increases, the oscillator strengths decrease in proportion to $1/n^3$ for large n. The reason that the oscillator strengths decrease with increasing $n - n'$ is that the overlap in the electron wave functions in the n and n' shells is reduced. Tables of the oscillator strengths for a hydrogen atom transitions can be found in [26].

When calculating the eigenfunctions and eigenvalues of the electron energies in atoms having several electrons, the interactions between electrons must be considered. In this situation the problem becomes somewhat more complicated and an exact solution becomes impossible. To date, a number of approximation methods have been developed to solve the problems associated with multielectron atoms. We have already mentioned the Thomas–Fermi statistical method in which the atom's electrons are treated as a quantum gas. In the Thomas–Fermi approximation the atom's electrons lose some of their individuality. An atom is characterized by an electron radial distribution density $\rho(r)$ and a total energy that can be calculated via $\rho(r)$. The Thomas–Fermi approximation is inadequate for calculating the energies E_n of the individual electron shells and the oscillator strengths of transition $S_{nn'}$.

Several approximation methods for calculating the energy and the wave functions of an atom have been developed, in which the individual characteristics of the different electrons are preserved. As a rule, the individual electrons in an atom with several electrons are described by the same set of quantum numbers n, l, m, S_Z as in a one-electron atom. Slater's method[23] is a convenient approximation. A multielectron atom problem is solved by Slater's method as a one-electron atom problem and the potential energy of the atom's electron is described by

$$U(r) = -\frac{(Z-\gamma)e^2}{r} + \frac{e^2 a_0}{2r^2}(n^* - 1)n^* \tag{2-108}$$

The second term is the energy of the scattered electron interacting with all of the atom's other electrons. The energy of interaction between an electron and the nucleus is changed by the introduction of an effective nuclear charge $(Z-\gamma)e$, where γ is a coefficient that describes the shielding of a given electron from the nucleus and is generated by all of the other electrons. The numbers n^* and γ are in fact corrections; they are chosen so that theoretical values of energy match those found by x-ray spectroscopy, ionization energy, etc. as closely as possible.

Solving Schrödinger's equation for a one-electron atom having the potential energy given by Eq. (2-108) gives the following eigenvalues for energy and eigenfunctions for the electron:

$$E_{n^*\gamma} = -\frac{1}{2}\frac{(Z-\gamma)^2}{n^{*2}} Ry$$

$$\Psi_{n^*lm}(r) = A_{n^*} r^{n^*-1} \exp\left[-\frac{(Z-\gamma)r}{a_{n^*,\gamma}}\right] Y_{lm}(\theta,\varphi) \tag{2-109}$$

$$a_{n^*,\gamma} = a_0 n^*/(Z-\gamma) \tag{2-110}$$

The closest match between energy values calculated from Eq. (2-109) and those found from experiment is obtained when the following rules for choosing the numbers n^* and γ are used.

The true principal quantum numbers n = 1, 2, 3, 4, 5, 6 correspond to the effective quantum numbers n^* = 1, 2, 3, 3.7, 4.0, 4.2. To describe how the nucleus is shielded from one electron shell by another we combine the electrons into shells, starting with the innermost shell in the order in which they are filled:

$$(1s),\ (2s,2p),\ (3s,3p),\ (3d),\ (4s,4p),\ (4d,4f),\ (5s,5p)$$

1. Electron shells outside the one being examined do not contribute to shielding. 2. Other electrons in the shell to which the electron being examined belongs increase γ by 0.35.

TABLE 2-2. Calculation Done by Slater's Method for the Shielding Coefficients and Electron Energy Levels in a Silicon Atom

Shell	Shielding coefficient γ for shell electrons	$E_{n,\gamma}$, Ry = 13.6 eV
$1s$	$0.30 \cdot 1 = 0.30$	−187.7
$2s$	$0.85 \cdot 2 + 6 \cdot 0.35 + 0.3 = 4.10$	−24.5
$2p$	$0.85 \cdot 2 + 5 \cdot 0.35 + 2 \cdot 0.3 = 4.15$	−24.25
$3s$	$1 \cdot 2 + 0.85 \cdot 8 + 0.35 \cdot 2 + 0.3 \cdot 1 = 9.8$	−1.96
$3p$	$1 \cdot 2 + 0.85 \cdot 8 + 0.35 \cdot 1 + 0.3 \cdot 2 = 9.75$	−2.02

In this case the contribution from a $1s$ electron is 0.30. 3. If an electron belongs to a shell that is closer to the nucleus than the electron being examined and it is a (d, f) electron, its contribution to γ is 1.0. If it is an (s, p) electron and belongs to the shell that is closest (from within) to the shell being examined its contribution to γ is 0.85. If an electron belongs to a shell that is closer to the nucleus than is the nearest shell, its contribution to γ is 1.0.

As an example, the calculated values of γ and $E^*_{n\gamma}$ for the electrons of a silicon atom are shown in Table 2-2 (a silicon atom has 14 electrons in shells $(1s^2)$, $(2s^2 2p^6)$, $(3s^2 3p^2)$).

Using radial functions and Slater's energy eigenvalues, Eqs. (2-109) and (2-110), we can calculate the oscillator strengths of transitions. The rules for choosing according to quantum numbers l and m are given by integrating over the angular variables; they are identical to the selection rules in a hydrogen atom. The normalized factor for the radial component of Slater's wave function, as can easily be verified, is

$$A^*_n = \frac{(2/a_{n^*})^{(2n^*+3)/2}}{[(2n^*+2)!]^{1/2}} \tag{2-111}$$

when n^* is an integer and

$$A^*_n = \left(\frac{2}{a_{n^*}}\right)^{(2n^*+3)/2} \Gamma^{-1/2}(2n^*+3)$$

when it is not. The matrix element for the point r between states n and n' is

$$r_{n,n'} = \int_0^\infty R_n(r) R_{n'}(r) r^3\, dr = \frac{(2/a_n)^{(2n+3)/2}(2/a_n)^{(2n'+3)/2}}{(1/a_n + 1/a_{n'})^{n+n'+4}} \times \frac{\Gamma(n+n'+4)}{[\Gamma(2n+3)\Gamma(2n'+3)]^{1/2}} \tag{2-112}$$

(for simplicity we will omit the "asterisk" on the effective numbers). Substituting this result into the equation for the oscillator strengths yields

$$S_{ns\to n'p} = \frac{1}{3}[r_n, r_{n'}]^2 (E_n - E_{n'}) \frac{2m}{\hbar^2} = \frac{1}{3}\frac{1}{a_0^2}\left[\frac{(Z-\gamma_n)^2}{n^2} - \frac{(Z-\gamma_{n'})^2}{n'^2}\right] |r_{n,n'}|^2 \tag{2-113}$$

After substituting Eq. (2-112) for $r_{n,n'}$ into (2-113) and cancelling the a_0, the oscillator strength of a transition remains only a function of the numbers Z, n, n', γ_n, $\gamma_{n'}$:

$$S_{ns\to n'p} = \frac{1}{3}\left[\frac{(Z-\gamma_n)^2}{n^2} - \frac{(Z-\gamma_{n'})^2}{n'^2}\right]\frac{(2/\alpha_n)^{(2n+3)/2}(2/\alpha_{n'})^{(2n'+3)/2}}{(1/\alpha_n + 1/\alpha_{n'})^{n+n'+4}}$$
$$\times \frac{\Gamma(n+n'+4)}{\Gamma^{1/2}(2n+3)\Gamma^{1/2}(2n'+3)}; \qquad \alpha_n = \frac{a_n}{a_0} = \frac{n}{Z-\gamma_n} \tag{2-114}$$

Equation (2-114) serves as an example of a semiempirical formula for the oscillator strengths of transitions.

We will now take up the question of finding the oscillator strengths in a form that is useful for describing transitions into states with continuous energy (ionization). In keeping with Eq. (2-104) the ionization cross section is given in terms of the density of the generalized oscillator strengths $df(W,Q)/dW$. This function can be generalized so that Eq. (2-104) is also useful for describing the transition between states in the discrete spectrum (excitation of the atoms).[25] A function df/dW in a plane Q, W is called a Bethe surface. We will assume that a target is described by a finite number M of resonance transitions with energy W_i in the discrete spectrum state. The energy W_i contains, e.g., transitions in optically allowed states and the stimulation of collective (plasma) oscillations. During these transitions little momentum is transmitted. We can assume that the range of these transitions in the plane Q, W is limited by the conditions $W = W_i$ and $Q < W_i$. If in making a transition the amount of transmitted momentum is large so that $Q > W_i$ we can say that $W = Q$. These will be transitions in which resonant energies have not been considered; they correspond to large amounts of energy and momentum being transmitted to atomic electrons and the latter can be thought of as free particles. We will now relate the oscillator strength f_i to the transition energy W_i and write

$$\frac{\partial f(Q,W)}{dW} = \sum_{i=1}^{M} f_i \frac{dg_i(Q,W)}{dW} \tag{2-115}$$

$$\frac{dg_i(Q,W)}{dW} = \begin{cases} \delta(W-W_i), & Q < W_i \\ \delta(W-Q), & Q > W_i \end{cases} \tag{2-116}$$

In keeping Bethe's rule of sums, we can write

$$\sum_{i=1}^{M} f_i = Z \tag{2-117}$$

because all transitions are considered when Eq. (2-115) is integrated (both in discrete and continuous spectrum states). The quantities f_i and W_i are used to find the average ionization potential I of a target:

$$\sum_{i=1}^{M} f_i \ln W_i = Z \ln I \tag{2-118}$$

Equations (2-117) and (2-118) are inadequate to determine the $2M$ parameters of a target: W_i, f_i, $i = 1, \dots, M$; and, therefore, we must invoke additional physical considerations. We will assume that a single oscillator comprises all the electrons in an atom's shell (subshell) or energy band in the target. This condition defines f_i: $f_i = Z_i$, where Z_i is the number of electrons (in the atom) in the respective group.

Equation (2-118) gives the resonance energies the sense of the average ionization potential of the respective group Z_i of electrons. The energy W_i for an atomic shell must be related to the binding energy E_i in this shell. In the limit as $Q \to 0$ we move into the range of optical transitions: $df(W,0)/dW = df/dW$, in which the lower bound on the transitions begins at $W_i = E_i$. We can then write

$$\int_{E_i}^{\infty} \frac{df_{i\,\text{opt}}}{dW}\, dW = Z_i \tag{2-119}$$

$$\int_{E_i}^{\infty} \frac{df_{i\,\text{opt}}}{dW} \ln dW\, dW = Z_i \ln W_i \tag{2-120}$$

for the ith shell of the atom. Conversely, atomic photoeffect theory establishes an explicit (exponential) dependence for the density of the optical strengths of transition on the transmitted energy W:[27]

$$\frac{df_{i\,\text{opt}}(W)}{dW} \sim W^{-\alpha}; \qquad 2.5 \le \alpha \le 3.52 \tag{2-121}$$

By substituting Eq. (2-121) into Eqs. (2-119) and (2-120) we find the relationship between W_i and E_i:

$$W_i = E_i \exp[1/(\alpha - 1)] \tag{2-122}$$

whereby, when $2.5 \le \alpha \le 3.5$ then $W_i/E_i = 1.5$–2.0

Further activity in finding W_i is predicated upon the following assumptions.[25] 1. The ratio W_i/E_i is the same for all inner shells: $W_i/E_i = C$. 2. The W_i for valence electrons is known. 3. The E_i for the inner shells is known. We will introduce the number Z_{core} for the electrons in the inner shells of the atoms and calculate the quantity I' introduced by

$$Z \ln I' = \sum_{i,\text{val}} Z_i W_i + \sum_{i,\text{core}} Z_i \ln E_i \tag{2-123}$$

Using $W_i = cE_i$, we also write Eq. (2-118) as

$$Z \ln I = \sum_{i,\text{val}} Z_i W_i + \sum_{i,\text{core}} Z_i \ln(cE_i) \tag{2-124}$$

From Eqs. (2-123) and (2-124) we find C:

$$C = (I/I')^{Z/Z_{\text{core}}} \tag{2-125}$$

For silicon the energy W_i of all the valence electrons is made identical to the energy of the plasma oscillations: $W_{i\,\text{val}} = 16.6$ eV and for C and I we find: $C = 2.07$ and

TABLE 2-3. Binding Energy E_i and Partial Average Ionization Potential W_i for Electron Subshells in Silicon[25]; f_i Is the Number of Electrons in the Subshell

Shell	E_i, eV	W_i, eV	f_i
$1s$	1560	3099	2
$2s$	118	234	2
$2p$	74	146	6
$3s\,3p$	–	18	3

I = 161 eV. Values of W_i are shown in Table 2-3, where it is apparent that they are fairly close to the shell energy values calculated via Slater's method (Table 2-2). Similar calculations of W_i were done in [25] for a number of different metals (Be, Ae, Ni, Cu, Ag, Au). The values found for W_i were used to calculate the length of the free path of electrons and the deceleration S of fast electrons in a target.

The description of a target as a function of the density of the generalized oscillator strength, Eq. (2-116), given here is distinguished by its high degree of clarity and is based on knowing the energy spectrum of the target atoms E_i and the energy of the valence electrons.

2-11. THE TOTAL EXCITATION CROSS SECTION IN ATOMS

The length of a fast particle's free path in relation to some scattering process is $\lambda_j = 1/n\sigma_j$, where n is the number of scattering particles per unit volume and σ_j is the total scattering cross section. For this reason we must calculate the total inelastic scattering cross sections of electrons in an atom and free electrons in metals and semiconductors. We will begin with the total scattering cross section in atoms.

Our task is made more difficult by the fact that we have obtained different differential scattering cross section formulas for the different ranges of scattering angles (momenta q being transmitted). Four of these ranges must be introduced for scattering by an atom's electrons. In principle, we must calculate the contribution from each of these ranges and add the results. However, we come to the conclusion that one of these ranges makes a greater contribution and we consider the integral over this range equal to the total inelastic scattering cross section. This turns out to be the second range, where $v_0^2/v^2 < \theta < v_0/v$ or $q_{\min} < q < 1/a_0$. The first range, $\theta < v_0^2/v^2$, is very narrow and its contribution is small, even though the cross section in this range is large. The third and fourth ranges are large, but their differential cross section is small and, consequently, their contribution to the total cross section is also small. Straightforward calculations using Eqs. (2-80), (2-81), and (2-92) show that the contribution from range II is roughly $\ln(\beta_n \hbar v/e^2) \sim 3$ times that from ranges I and III.

We will first find the total excitation cross section for an atom's state. The differential cross section of the process in which an atom makes the transition from a state E_0 to

E_α and an electron goes from a state $\bar{k}$ to $\bar{k}'$ is, Eq. (2-81):

$$d\sigma_\alpha = 8\pi \left(\frac{e}{\hbar v}\right)^2 \left| \left(\sum_a x^{(a)} \right)_{0\alpha} \right|^2 \frac{dq}{q} \tag{2-126}$$

Equation (2-126) is valid over the range $q_{min} \leq q \leq 1/a$, $q_{min} = (E_\alpha - E_0)/hv$. We will say that the atom's final state is part of the discrete spectrum. Then an atom in its final state turns out to be excited into a state E_α. We will integrate Eq. (2-126) over all the possible values of momentum q that can be transmitted by a fast electron:

$$\sigma_\alpha = \int_{q_{min}}^{q_{max}} d\sigma_\alpha = 8\pi \left(\frac{e^2}{\hbar v}\right)^2 \left| \left(\sum_a x^{(a)} \right)_{0\alpha} \right|^2 \ln \left(\frac{q_{max}}{q_{min}} \right) \tag{2-127}$$

The logarithmic term is a function of the atom's excitation energy and can be given in different forms that are convenient for the evaluations. Assuming that $E_\alpha - E_0 = \beta_\alpha e^2/a$, where β_α is a coefficient on the order of unity, we obtain

$$\sigma_\alpha = 8\pi \left(\frac{e^2}{\hbar v}\right)^2 \left| \left(\sum_a x^{(a)} \right)_{0\alpha} \right|^2 \ln \left(\beta_\alpha \frac{\hbar v}{e^2} \right) \tag{2-128}$$

In another version we can assume that $E_\alpha - E_0 = \beta_\alpha m v_0^2$ and $m v_0 a \approx \hbar$, then

$$\sigma_\alpha = 8\pi \left(\frac{e^2}{\hbar v}\right)^2 \left| \left(\sum_a x^{(a)} \right)_{0\alpha} \right|^2 \ln \left(\beta_\alpha \frac{v}{v_0} \right) \tag{2-129}$$

Two comments must be made regarding Eqs. (2-128) and (2-129).

First, Eqs. (2-128) and (2-129) are, in the main, untrue in the extreme case of threshold energies of a fast particle. Indeed, if the energy of an incident electron is less than the excitation energy, a transition is impossible according to the law of conservation of energy and when this happens the cross section must go to zero. Consequently, the equations properly describe the dependence of the cross section on the energy of a fast particle when this energy is high and must be corrected when the energies are close to threshold energies.

Equations (2-128) and (2-129) can be rewritten, giving $(\sum_a x^{(a)})_{0\alpha}$ in terms of the oscillator strength of transition $S_{0\alpha}$ (see Eq. (2-105)). Also, we make the transition to using dimensionless values, having assumed the Rydberg energy $Ry = e^2/2a_0$ to be unity; then, instead of Eq. (2-128), we obtain

$$\sigma_\alpha = \frac{4\pi a_0^2}{\varepsilon} \frac{S_{0\alpha}}{\varepsilon_{0\alpha}} \ln(\beta_\alpha^2 \varepsilon), \quad \varepsilon = \frac{E}{Ry}, \quad \varepsilon_{0\alpha} = \frac{E_\alpha - E_0}{Ry}, \quad a_0 = \frac{\hbar^2}{me^2} \tag{2-130}$$

The oscillator strength of transition $S_{0\alpha}$ is the most important quantity in atomic spectroscopy and is calculated theoretically or found experimentally for many atomic transitions.[26,28]

In order to properly account for the cross section going to zero at the threshold energy $E = E_\alpha - E_0$, we introduce an empirical factor $\Phi(x)$ into Eq. (2-130) and write it[28]

$$\sigma_\alpha = 8\pi a_0^2 \frac{S_{0\alpha}}{\varepsilon_{0\alpha}^2} \Phi(x) c_1 \ln(c_2 x) \tag{2-131}$$

$$\Phi(x) = \frac{x^2}{(1+x^2)^2}, \qquad x^2 = \frac{E - E_{0\alpha}}{E_{0\alpha}} \tag{2-132}$$

The $\Phi(x)$ factor when $x \gg 1$ goes to $\varepsilon_{0\alpha}/\varepsilon$, and we obtain the previous formula, Eq. (2-130). In the range near the threshold $x \ll 1$, $\Phi(x) = x^2$ the cross section tends to zero according to $(\varepsilon - \varepsilon_{0\alpha})/\varepsilon_{0\alpha}$. Finally, a maximum is reached for σ_α when $x = 1$, i.e., when $E = 2E_{0\alpha}$. The constants c_1 and c_2 are on the order of unity and must be thought of as corrections when comparing theoretical and experimental results.

2-12. THE TOTAL IONIZATION CROSS SECTION IN AN ATOM

The total ionization cross section of an atom by electron collision is extremely interesting for us. As we have said, three particles are formed during ionization: the scattered electron, the ionized atom, and the departing electron. Therefore, the final state attained during an ionizing collision is described by the scattered electron's wave number $\bar{k}'$, the departing electron's wave number $\bar{k}_f$, and the atomic residue's (ion's) quantum numbers γ_0. The differential scattering cross section with ionization for the departing electron in the range of wave numbers $d\bar{k}_f$ was given in Section 2-9. We will not go through the calculation of Eq. (2-103), but merely present the result written in terms of the oscillator strength of transition $\partial S/\partial E_f$ that relates to a unit interval of the ionizing electron's energy[28] $E_f = \hbar^2 k_f^2/2m$:

$$d\sigma(\gamma, \gamma_0, E_f) = 8\pi a_0^2 \left(\frac{Ry}{E}\right) \left(\frac{Ry}{|E_\gamma| + E_f}\right) \sum_{\gamma'/\gamma_0} \frac{\partial S(\gamma, \gamma')}{\partial E_f} \ln(\beta^2 k a_0)\, dE_f \tag{2-133}$$

Here E_γ is the energy of the state from which the ionization occurs, and the summation over γ'/γ means summation over all quantum numbers which do not appear in γ_0 for the system's final state:

$$\beta^2 = \frac{e^2/a_0}{|E_\gamma| + E_f} \tag{2-134}$$

The total ionization cross section is obtained by integrating Eq. (2-133) over the energy E_f:

$$\sigma(\gamma, \gamma_0) = 8\pi a_0^2 \left(\frac{Ry}{E}\right) \sum_{\gamma'/\gamma_0} \langle f(\gamma, \gamma')\rangle \tag{2-135}$$

$$\langle f(\gamma, \gamma')\rangle = \int \frac{Ry}{|E_\gamma| + E_f} \frac{\partial S(\gamma, \gamma')}{\partial E_f} \ln(\beta^2 k a_0)\, dE_f \tag{2-136}$$

In order for us to do further calculations, we need an explicit form of the $\partial S/\partial E_f$ function and the dependence of the $\langle S(\gamma,\gamma')\rangle$ on the quantum numbers γ'.

Semiempirical formulas are often used for the total ionization cross section. These formulas are derived using the same considerations as in the case of excited atoms[28]:

$$\sigma_\gamma = 8\pi a_0^2 \frac{S_\gamma}{\varepsilon_\gamma^2} \frac{x^2}{(1+x^2)^2} c_1 \ln(c_2 x), \qquad S_\gamma = \sum_{\gamma'} \int \frac{\partial S(\gamma,\gamma')}{\partial E_f}\, dE_f \quad (2\text{-}137)$$

$$x^2 = \frac{E - |E_\gamma|}{|E_\gamma|} = \frac{\varepsilon - \varepsilon_\gamma}{\varepsilon_\gamma}, \qquad \varepsilon_\gamma = |E_\gamma|/Ry, \qquad \varepsilon = E/Ry \quad (2\text{-}138)$$

Here S_γ is equal to the sum of the oscillator strengths of transition from the E_γ level into all states in the continuous spectrum. If we add to the integral of S_γ in Eq. (2-137) the sum of the oscillator strengths of transition from level E_γ into all states in the atoms' discrete spectrum, then, by the rule of sums (see below) the total sum of the oscillator strengths will be equal to the number of electrons ξ_γ that occupy a given energy level (shell) in the atom:

$$\sum_\alpha S_{0\alpha} + S_\gamma = \xi_\gamma \quad (2\text{-}139)$$

Thompson's empirical formula for the total ionization cross section is

$$\sigma_\gamma = 4\pi a_0^2 \frac{\xi_\gamma}{\varepsilon_\gamma^2} \frac{x^2}{(1+x^2)^2} \quad (2\text{-}140)$$

and corresponds to the assumption that the sum of the oscillator strengths of transition into discrete levels is much less than the sum of the oscillator strengths of transition for the discrete spectrum (also, the logarithmic term is replaced by unity).

We pose the question: how does the total ionization cross section depend on the ionization energy of the respective level? The quantity $\Phi(x) = x^2 \ln(c_2 x)/\varepsilon_\gamma^2(1+x^2)^2$ gives the explicit dependence of σ_γ on ε_γ and ε:

$$\Phi(x) = \frac{1}{2} \frac{\varepsilon - \varepsilon_\gamma}{\varepsilon_\gamma \varepsilon^2} \ln\left[c_2^2 \frac{(\varepsilon - \varepsilon_\gamma)}{\varepsilon_\gamma}\right] \quad (2\text{-}141)$$

In addition, there is an implicit dependence by way of S_γ. If we say that $S_\gamma = \xi_\gamma$ according to Eq. (2-140), then only the explicit dependence indicated in Eq. (2-141) remains. In every case we can say that S_γ is weakly dependent on ε_γ and is independent of ε. We can assume that an electron at the beginning of its range has much more energy than it needs to ionize an atom's subshells: $\varepsilon \gg \varepsilon_\gamma$ (at least, for light atoms). Then the cross section σ_γ is described in the usual Born approximation:

$$\sigma_\gamma = 8\pi a_0^2 \frac{S_\gamma}{\varepsilon \varepsilon_\gamma} \frac{1}{2} c_1 \ln(c_2^2 \varepsilon/\varepsilon_\gamma) \quad (2\text{-}142)$$

As can be seen, the dependence of the cross section on ε_γ in this region becomes $\sigma_\gamma \sim (S_\gamma/\varepsilon_\gamma)\ln(c_2^2(\varepsilon/\varepsilon_\gamma))$; i.e., the cross section decreases fairly rapidly as the ionization energy, ϵ_γ, increases. This means that an electron collision ionizes mainly the outer subshells of the atoms. A still sharper dependence on ε_γ is seen in the threshold region when $\varepsilon - \varepsilon_\gamma \ll \varepsilon_\gamma$:

$$\sigma_\gamma = 4\pi a_0^2 \frac{S_\gamma}{\varepsilon_\gamma^2} \frac{\varepsilon - \varepsilon_\gamma}{\varepsilon_\gamma} c_1 \ln\left(c_2^2 \frac{\varepsilon - \varepsilon_\gamma}{\varepsilon_\gamma} \right) \tag{2-143}$$

A maximum in the ionization cross section occurs when the beam energy is $\varepsilon = 2\varepsilon_\gamma$. The total ionization cross section over several atomic shells is the superposition of several curves having maxima and thresholds that are characteristic for each shell.

We must point out that this kind of ionization cross section dependence on beam energy is similar to the trace of the spectral absorption curves for x-ray radiation and is caused by the photoeffect in the respective shells (see Chapter 4). There is, however, a significant difference between these effects. As we have already seen, the ionization cross section due to electron collision is maximal in the atom's outer shell, whereas for the photoeffect (ionization) by x-ray radiation the cross section is maximal for the atom's innermost shell (the k-shell).

The different formulas for the total ionization cross section of an atom's inner shells by electron collision are given by Powell.(29) Bethe's formula

$$\sigma_{nl} = \frac{\pi e^4 Z_{nl}}{E \cdot E_{nl}} b_{nl} \ln\left(c_{nl} \frac{E}{E_{nl}} \right) \tag{2-144}$$

is fundamental. Here E_{nl} is the binding energy of the electrons and Z_{nl} is the number of electrons in a shell having quantum numbers n and l. The b_{nl} and c_{nl} parameters are different for different shells and can detect the dependence on the atomic number Z. If we introduce the normalized energy of the incident electrons $U_{nl} = E/E_{nl}$, then $\sigma_{nl}E_{nl}^2$ (in $\mathrm{cm}^2 \cdot \mathrm{eV}^2$) is the same function U_{nl} for all elements:

$$\sigma_{nl}E_{nl}^2 = 6.51 \times 10^{-14}\, Z_{nl} b_{nl} \ln(c_{nl}U_{nl}) U_{nl}^{-1} \tag{2-145}$$

Mott and Massey(19) suggested the following constants for the K and L shells: $b_K = 0.35$, $b_L = 0.25$, $c_K = c_L = 2.42$.

We will also present some other modifications of Bethe's formula, as suggested by different authors (summarized by Powell(29) and Lotz(30)):

$$\sigma_{nl}E_{nl}^2 = a_{nl} Z_{nl} U_{nl}^{-1} \ln U_{nl} \{1 - b_{nl} \exp[-c_{nl}(U_{nl} - 1)]\} \tag{2-146}$$

and by Grisinsky(31):

$$\sigma_{nl}E_{nl}^2 = 6.51 \times 10^{-14}\, Z_{nl} g(U_{nl})$$
$$g(U_{nl}) = \frac{1}{U} \left(\frac{U-1}{U+1} \right)^{3/2} \left[1 + \frac{2}{3}\left(1 - \frac{1}{2U} \right) \ln[2.7 + (U-1)^{1/2}] \right] \tag{2-147}$$

The formulas above were compared with experimental results obtained by ionizing the K shell of C, Ne, N, and O atoms in Auger electron spectroscopy experiments on gaseous targets ionized by electron collision in the range of energies $U = 1$–25. The basic conclusions to be drawn from comparing theory with experiment are: Bethe's formula, Eq. (2-144), describes the experimental curves when $U_K \leq 4$ well, if $b_K = 0.9$ and $c_K = 0.65$. These values are more precise than those used earlier (see above). Grisinsky's formula, Eq. (2-147), is in good agreement with experiment near the threshold energies in the range $U_K = 1$–4. This good agreement is also exhibited by Lutz's formula if $a_K = 5 \times 10^{-14}$, $b_K = 0.75$, and $c_k = 0.5$.

The formulas presented here are used to compute electron trajectories and intensity characteristic of x-ray radiation and Auger electrons via the Monte Carlo method in problems of analyzing the composition of the targets.[8]

2-13. TOTAL EFFECTIVE DECELERATION OF FAST ELECTRONS

Total effective deceleration is the amount of energy lost by a fast particle per unit length of travel: dE/dx. The quantity dE/dx is obtained by summing the energy lost, $E_\alpha - E_0$, by an electron in exciting and ionizing the atoms of a material in per unit volume:

$$\frac{dE}{dx} = \sum_j N_j \sum_{q_{\min}}^{q_{\max}} (E_{\alpha j} - E_{0j}) \frac{d\sigma_{\alpha j}}{dq}\, dq \tag{2-148}$$

We divide the range of transmitted momenta $q_{\min} \leq q \leq q_{\max}$ into two ranges $q_{\min} \leq q \leq q_0$ and $q_0 \leq q \leq q_{\max}$ such that when $q = q_0$ the amount of transmitted momentum q ceases to be a function of the amount of energy transmitted $E_\alpha - E_0$.[18] Then the lower limit of integration in the first range $q_{\min}$–q_0 is a function of $E_\alpha - E_0$ and summing over α must be done after integrating over q:

$$\left(\frac{dE}{dx}\right)_1 = \sum_j N_j \sum_\alpha (E_{\alpha j} - E_{0j}) \int_{q_{\min}}^{q_0} \frac{d\sigma_{\alpha j}}{dq}\, dq \tag{2-149}$$

In the second range, $q_0 \cdots q_{\max}$, the q are independent of $E_\alpha - E_0$ and summing over α can be done before integrating over q:

$$\left(\frac{dE}{dx}\right)_2 = \sum_j N_j \int_{q_0}^{q_{\max}} dq \sum_\alpha (E_{\alpha j} - E_{0j}) \frac{d\sigma_{\alpha j}}{dq} \tag{2-150}$$

Choosing q_0 so that $q_0 a_0 \ll 1$ we can present the differential scattering cross section in Eq. (2-149) in a dipole approximation:

$$\left(\frac{dE}{dx}\right)_1 = 8\pi \left(\frac{e^2}{\hbar v}\right)^2 \sum_j N_j \sum_\alpha (E_{\alpha j} - E_{0j}) |x_{0\alpha}|^2 \int_{q_{\min}}^{q_0} \frac{dq}{q} \tag{2-151}$$

For the range q_0–$q_{\max}$, which includes larger values of transmitted momentum, we use a more general expression for the scattering cross section:

$$\left(\frac{dE}{dx}\right)_2 = 8\pi \left(\frac{e^2}{\hbar v}\right)^2 \sum_j N_j \int_{q_0}^{q_{\max}} dq/q^3 \sum_\alpha (E_{\alpha j} - E_{0j}) \left| \left(\sum_\alpha e^{iqr_a} \right) \right|^2 \tag{2-152}$$

Using the so-called rule of sums we have[18]

$$\sum_{\alpha}(E_\alpha - E_0)\left|\left(\sum_{\alpha} e^{iqr_a}\right)_{0\alpha}\right|^2 = \frac{\hbar^2 q^2}{2m} Z \tag{2-153}$$

A direct consequence of Eq. (2-153) in a dipole approximation $qr_a \ll 1$ is

$$\sum_{\alpha}(E_\alpha - E_0)\left|\left(\sum_{a} x_a\right)_{0\alpha}\right|^2 = \frac{\hbar^2}{2m} Z \tag{2-154}$$

Using Eq. (2-154) we can convert the sum in Eq. (2-151):

$$\sum_{\alpha}(E_\alpha - E_0)\left|\left(\sum_{a} x_a\right)_{0\alpha}\right|^2 \ln\left(\frac{q_0 \hbar v}{E_\alpha - E_0}\right) \equiv \frac{\hbar^2 Z}{2m} \ln\left(\frac{q_0 \hbar v}{I}\right) \tag{2-155}$$

The so-called average energy I that describes a specific type of atom is introduced here. From Eq. (2-155) and the expression for the oscillator strength of transition, Eq. (2-105), we have

$$I = \prod_{\alpha}(E_\alpha - E_0)^{S_{0\alpha}/Z} \tag{2-156}$$

Consequently, the energy loss (dE/dx) is given by

$$\left(\frac{dE}{dx}\right)_1 = \frac{4\pi e^4}{mv^2} \sum_j N_j Z_j \ln\left(\frac{q_0 \hbar v}{Ij}\right) \tag{2-157}$$

$$\left(\frac{dE}{dx}\right)_2 = \frac{4\pi e^4}{mv^2} \sum_j N_j Z_j \ln\left(\frac{q_{\max}^{(j)}}{q_0}\right) \tag{2-158}$$

$$\frac{dE}{dx} = \left(\frac{dE}{dx}\right)_1 + \left(\frac{dE}{dx}\right)_2 = \frac{4\pi e^4}{mv^2} \sum Z_j N_j \ln\left(\frac{q_{\max}^{(j)} hv}{Ij}\right) \tag{2-159}$$

As can be seen, the deceleration caused by j-type atoms is proportional to the concentration $N_j Z_j$ of electrons belonging to this atom. The logarithmic term weakens this dependence somewhat, because it is smaller for the heavier atoms. Indeed, we assume that $q_{\max} = 1/a$, where a is the average radius of an electron orbit in the atom. We can compare the energy I to the average frequency of motion for the electrons in the atom:

$$I = \hbar\omega = \hbar v_a / a$$

then, $q_{\max}\hbar v/I \approx v/v_0$. According to the Thomas–Fermi atom model $v_a \sim Z^{2/3}$, and the quantity in the logarithm decreases with increasing Z as $Z^{-2/3}$. Further, because $a \sim Z^{-1/3}$, $I = AZ$, $A = 13$ eV.

The deceleration equation can be written so that the logarithmic term contains a ratio of energies rather than a ratio of velocities. Because the momentum transmitted to an atom's electron is $q = k\theta$, we can assume for the range $q_{max}\theta \sim 1$ that $q_{max} = \beta k$, where β is a term on the order of unity. Then $q_{max}\hbar v/I = \beta m v^2/I$ and

$$\frac{dE}{dx} = \frac{4\pi e^4}{mv^2}\sum_j N_j Z_j \ln\left(2\beta\frac{E}{I_j}\right) \tag{2-160}$$

Bethe and others[32] have shown that for electrons, $2\beta = \sqrt{e/2} = 1.166$ (e is the base of natural logarithms), whereas we must assume for the energy losses of fast ions in matter that $\beta = 1$.

2-14. FAST ELECTRON SCATTERING IN METALS AND SEMICONDUCTORS

New scattering mechanisms present themselves when fast electrons move through metals and semiconductors. These mechanisms are connected with the existence of collective motions in the form of plasma oscillations in the electron system of a target. To examine scattering in targets such as these we must divide an atom's electrons into inner shell and outer (valence) shell electrons. Inner shell electrons that belong to different atoms are weakly bound to one another and the scattering that takes place in them is described by the equations given in previous paragraphs. In the last, however, we must understand that the number of electrons means the number of electrons in the atom's inner shells.

The collective properties of valence electrons are due to remote acting Coulomb interactions between the electrons and the fixed (positive) charges of the ionic skeletons that form a crystal lattice. The Coulomb repulsion of electrons by one another makes the electron system an elastic liquid. Conversely, if the equilibrium in a system of mobile electrons is disturbed (e.g., by the action of an external fast electron), so that their concentration in a small area of the solid is reduced, but increased in an adjacent area, the effect this has on the charges in the ionic skeletons of the lattice is to produce retentional forces that act on the electrons that have disturbed the equilibrium. The retentional forces and the electrons' inertias cause oscillations in the electron system, which will be plasma oscillations. Lacking an external stimulus, plasma oscillations are caused by thermal motion and have the form of thermal fluctuations in the system of mobile electrons.

However, plasma oscillations do not tell the whole story of thermal motion in a system of mobile electrons. If we conduct our analysis from the viewpoint of the energy losses (scattering) of a fast electron it turns out that one group of electrons—the distance to which is less than a distance called the shielding radius—will behave like a group of individual particles, whereas the other group of electrons, which are more than the shielding radius away, behave collectively—energy can be transmitted to the remote electrons only by stimulated plasma oscillations. Electrons near the exciting particle can acquire energy individually in the form of one-particle excitations. If we imagine a tube whose radius is equal to the shielding radius lying along the trajectory of a fast particle, then electrons inside the tube are excited in one-particle form, and outside the tube are excited as collective motions—plasma oscillations. This picture can be augmented by still one more statement. One-electron stimulations of an electron gas must, because they occur with close electron–electron collisions, be accompanied by the transmission of a large amount of momentum $\bar{q}$. In contrast, the stimulation of plasma waves occurs for remote collisions and the momentum that is transmitted in this case is relatively small. This means that we can incorporate some momentum $\bar{q}_c$ such that when the amount of momentum transmitted is $\bar{q} < \bar{q}_c$ plasma waves are stimulated and when the transfer of momentum is $\bar{q} > q_c$ one-electron excitations occur. We need to calculate the scattering cross section in one-electron excitations and in plasma oscillations; the corresponding effects will be additive.

2-15. COULOMB INTERACTION IN A SYSTEM OF FREE CRYSTAL ELECTRONS

The potential energy of interaction for the l electron with all other j electrons and charges of the e_I ionic skeletons is

$$U(r_l) = \sum_{j \neq l} \frac{e^2}{r} - \sum \frac{ee_I}{re_I} \tag{2-161}$$

Using Eq. (2-64) we can write

$$\sum_{j \neq l} \frac{e^2}{r_{l_j}} = \sum_q \frac{4\pi e^2}{q^2} \exp(i\bar{q}(\bar{r}_l - \bar{r}_j)) \tag{2-162}$$

The uniform electron charge distribution that corresponds to the term with $q = 0$ in the Fourier expansion of the potential energy must be compensated by positive charges in the ionic skeletons such that

$$C_0 = -\sum_I \frac{ee_I}{re_I} \tag{2-163}$$

as a result of which we obtain

$$U(r_l) = {\sum_q}' \frac{4\pi e^2}{q^2} e^{i\bar{q}\bar{r}_l} \sum_j e^{-i\bar{q}\bar{r}_j} \tag{2-164}$$

the lines in the summation mean that the term with $q = 0$ has been ignored.

In keeping with what was said earlier, plasma oscillations are caused by that portion of Coulomb interaction between particles when $q < q_c$:

$$U_p(r_{l_j}) = {\sum_{q<q_c}}' \frac{4\pi e^2}{q^2} \exp(i\bar{q}\bar{r}_{l_j}) \tag{2-165}$$

By making the transition from summation over q to integration in the q-space we can write

$$U_p(r) = \frac{1}{(2\pi)^3} \int \frac{4\pi e^2}{q^2} e^{i\bar{q}\bar{r}}\, d^3 q = \frac{2e^2}{\pi r} \int_0^{q_c r} \frac{\sin(qr)}{qr}\, d(qr) \tag{2-166}$$

For distances $r \gg 1/q_c$ the upper limit of integration can be replaced by ∞, in which case the integral is equal to $\pi/2$, and

$$U_p(r) = \frac{e^2}{r} \tag{2-167}$$

In fact, plasma waves are stimulated via an unshielded Coulomb potential.

We can now look at the interaction between charges at small distances which correspond to large q, such that $q > q_c$:

$$U_l(r_{l_j}) = \sum_{q>q_c} \frac{4\pi e^2}{q^2} e^{i\bar{q}\bar{r}_{l_j}} = \frac{2e^2}{\pi r} \int_{q_c r}^{\infty} \frac{\sin(qr)}{qr}\, d(qr) \tag{2-168}$$

At distances $r \ll 1/q_c$ we can write the approximation ($q_c r \ll 1$)

$$\begin{aligned} U_l(r) &= \frac{2e^2}{\pi r}\left[\int_0^{\infty} \frac{\sin x}{x}\, dx - \int_0^{q_c r} \frac{\sin x}{x}\, dx\right] \\ &\approx \frac{2e^2}{\pi r}\left[\frac{\pi}{2} - q_c r\right] \approx \frac{e^2}{r} \exp\left[-\frac{\pi}{2} q_c r\right] \end{aligned} \tag{2-169}$$

The result from a more precise calculation is

$$U_l = \frac{e^2}{r} \exp(-q_c r) \tag{2-170}$$

In fact, the Coulomb interaction at small distances turns out to be shielded and have a shielding radius $r_{\text{sh}} = 1/q_c$.

2-16. PLASMA OSCILLATIONS IN AN ELECTRON GAS

The maximum value of a wave number q_c is given naturally if we obtain equations that describe the behavior of the electronic charge density and define the conditions for which an oscillatory solution, i.e., plasma waves, occurs.

We introduce the density and the stream of a number of electrons by the equations

$$\rho(r) = \sum_l \delta(\bar{r} - \bar{r}_l), \qquad \bar{\jmath}(r) = \sum_l \mathbf{v}_l\, \delta(\bar{r} - \bar{r}_l) \tag{2-171}$$

The Fourier transform of these quantities is

$$\rho(r) = \sum_q \rho_q\, e^{i\bar{q}\bar{r}}, \qquad \bar{\jmath}(r) = \sum_q j_q\, e^{i\bar{q}\bar{r}} \tag{2-172}$$

$$\rho_q = \sum_l e^{-i\bar{q}\bar{r}_l}, \qquad \bar{\jmath}_q = \sum_l \mathbf{v}_l\, e^{-i\bar{q}\bar{r}_l} \tag{2-173}$$

The continuity equation for the electron charge density $\dot{\rho} + \operatorname{div} \bar{j} = 0$ for the respective Fourier component is

$$\dot{\rho}_q + \sum_l i\bar{q}\bar{v}_l \, e^{-i\bar{q}\bar{r}_l} = 0 \tag{2-174}$$

Taking the derivative of Eq. (2-174) with respect to time, again we find

$$\ddot{\rho}_q + \sum_l (\bar{q}\bar{v}_l)^2 \, e^{-i\bar{q}\bar{r}_l} + \sum_l i\bar{q}\dot{\bar{v}}_l \, e^{-i\bar{q}\bar{r}_l} = 0 \tag{2-175}$$

We find the electron acceleration term, $\dot{v}_l$, that appears in Eq. (2-175) from Newton's equation, $m\dot{\bar{v}}_l = \bar{F}_l$. We find the force acting on an electron by all the other particles by taking the derivative of the interactions energy, Eq. (2-164), with respect to $\bar{r}_l$:

$$m\ddot{\bar{r}}_l = -4\pi e^2 i \sum_{\bar{k},j}{}' \frac{\bar{k}}{k^2} \, e^{i\bar{k}\bar{r}_{lj}} = -4\pi e^2 ik \sum_{\bar{k}}{}' \frac{\bar{k}}{k^2} \, e^{i\bar{k}\bar{r}_{lk}} \, \rho_k \tag{2-176}$$

Substituting the $\ddot{\bar{r}}_l$ from Eq. (2-176) into Eq. (2-175) and making some conversions to form the quantity $\rho_{\bar{k}-\bar{q}}$, we obtain

$$\ddot{\rho}_q + \sum_l (\bar{q}\dot{\bar{r}}_l) e^{i\bar{q}\bar{r}_l} + \frac{4\pi e^2}{m} \sum_k \frac{\bar{q}\bar{k}}{k^2} \, \rho_{\bar{k}} \rho_{\bar{k}-\bar{q}} = 0 \tag{2-177}$$

We isolate the term with $k = q$ in the last sum and take the remaining terms to the right:

$$\ddot{\rho}_{\bar{q}} + \left(\frac{4\pi e^2 \rho_0}{m} \right) \rho_q = - \sum_l (\bar{q}\bar{v}_l)^2 \, e^{i\bar{q}\bar{r}_l} - \frac{4\pi e^2}{m} \sum_k \frac{\bar{q}\bar{k}}{k^2} \, \rho_{\bar{k}} \rho_{\bar{k}-\bar{q}} \tag{2-178}$$

By studying the equation obtained, Bom and Paines showed that the last sum is negligible due to the relationships between the phase factors $\rho_{\bar{k}}$ and $\rho_{\bar{k}-\bar{q}}$ and discarded it. This approximation is widely used in the theory of plasma oscillations and is called an approximation of random phases.

The remaining sum in the right-hand side of Eq. (2-178) attenuates the plasma oscillations. In order that the oscillatory process take place, the oscillatory force needed must be greater than the attenuating force:

$$\omega_p^2 \rho_q > \left| \sum (\bar{q}\bar{v}_l)^2 \, e^{-i\bar{q}\bar{r}_l} \right|, \qquad \omega_p^2 = \frac{4\pi e^2 \rho_0}{m} \tag{2-179}$$

An estimate from above for the sum that produces attenuation is obtained by replacing the electron velocity v_l by their maximal possible velocity v_F:

$$\omega_p^2 \rho_{\bar{q}} > \left| q^2 v_F^2 \sum e^{-i\bar{q}\bar{r}_l} \right| = q^2 v_F^2 \rho_{\bar{q}} \tag{2-180}$$

From this we find that plasma waves occur when $q < \omega_p / v_F \equiv q_c$. Because the plasma oscillation frequency and the Fermi velocity of free electrons are stated only in terms of the electron concentration in a material (along with the electron charge and mass) the

TABLE 2-4. Energy of Plasma Oscillations

Material	Number of valence electrons	$\hbar\omega_p$, eV experimental	$\hbar\omega_p$, eV calculated
Na	1	5.7	5.9
Li	1	7.12	8.02
K	1	3.72	4.29
Mg	2	10.5	10.9
Al	3	15.0	15.7
Be	2	18.9	18.8
Si	4	16.9	16.6
Ge	4	16.0	15.6
Sb	5	15.3	15.1
InSb	4	12.0–13.0	12.0

boundary wave number q_c and shielding radius $r_{\text{sh}} = q_c^{-1}$ acquire the natural definition $q_c = \omega_p/v_F$.

We will not study the attenuation mechanism described by the respective sum in Eq. (2-178) for plasma waves, called Landau damping, in detail. This means that electrons moving in the same direction as the propagating wave, but at a velocity slightly less than the phase velocity of the wave $v_{\text{ph}} = \omega_p/q$ acquire energy (see [35–38] for a more detailed discussion).

If we ignore the attenuation, Eq. (2-178) becomes an equation for the oscillations of a collective point $\rho_{\bar{q}}$:

$$\ddot{\rho}_{\bar{q}} + \omega_p^2 \rho_{\bar{q}} = 0 \tag{2-181}$$

having, if we approach it in classical fashion, a solution $\rho_q = a\cos(\omega_p t + \varphi)$. However, the classical solution is meaningful only for a charged gaseous plasma. In an electron plasma in a solid we must begin by writing the solution of the respective Schrödinger equation as eigenvalues of the energy $E_\alpha = \hbar\omega_p(\alpha+1/2)$ and the corresponding oscillator wave functions $\Psi_\alpha(\rho_{\bar{q}})$. In fact the particles density in a plasma in metals and the valence band of semiconductors is on the order of 10^{23} cm^{-3}, the plasma oscillations frequency is $\omega_p \approx 10^{16}$ Hz, and the quantum of energy is $\hbar\omega_p \approx 20$ eV. At normal temperatures $kT \approx 0.023$ eV and plasma wave oscillators are found in the ground state with an energy of $E_0 = 0.5\hbar\omega_p$. Under these conditions, the oscillators must be examined in only quantum mechanical terms.

Plasma oscillation frequencies for selected metals and semiconductors are shown in Table 2-4.(39,40)

We can now easily estimate the wave number to be $q_c = \omega_p/v_F \approx 10^8$ cm^{-1} and the shielding radius to be $r_{\text{sh}} = q_c^{-1} = 10^{-8}$ cm.

If we consider the elasticity of an electron gas to be caused by the motion of its particles (as in a standard gas), the frequency of the plasma oscillations displays a dependence on q , i.e., the dispersion(30,40)

$$\omega_q^2 = \omega_p^2 + \frac{3}{5} v_F^2 q^2 \tag{2-182}$$

A maximum value is $\omega_{\max} = \omega(q_c) = 1.2\text{–}1.3\omega_p$.

2-17. THE CROSS SECTION FOR STIMULATION OF PLASMA OSCILLATIONS BY A FAST ELECTRON

We can use the general theory of inelastic scattering developed from examining inelastic scattering in atoms to calculate the scattering cross section of fast electrons in plasma oscillations. We have established that if the electron system in a target described by a system of energy levels E_α and wave functions φ_α, the differential scattering cross section for a fast electron from state $\bar{k}$ to $\bar{k}'$ (in the solid angle dO) and the electron system from E_0, φ_0 to E_α, φ_α is (see [18])

$$d\sigma_\alpha = \left(\frac{2me^2}{\hbar^2}\right)\frac{k'}{k}\frac{dO}{q^4}\left|\int \sum_a e^{i\bar{q}\bar{r}_a}\varphi_0\varphi_\alpha^* \, d\tau\right|^2 \tag{2-183}$$

Here $\bar{q} = \bar{k} - \bar{k}'$ is the momentum transmitted to the target's electron system. Replacing the sum $\sum_a e^{i\bar{q}\bar{r}_a}$ in the integral with $\rho_{\bar{q}}$, we obtain (2-184). The integral Eq. (2-183) was written in the so-called configuration representation, in which the coordinates $\bar{r}_a$ of the electrons and $d\tau = \prod_a d^3 r_a$ are the arguments of the wave functions. Thus, we have calculated the integral when studying electron scattering in the atoms. By introducing the variable ρ_q into Eq. (2-183), we have converted to a representation in which the electron system coordinates are given in terms of the Fourier components of the electrons density.

Edwards showed(41,42) that the wave function for a system of valence electrons can be thought of as the state of free electrons that has been altered by the appearance of plasma oscillations:

$$\varphi = \varphi_{\text{col}}\varphi_f = \exp\left(-\sum_q \alpha_q |\rho_q|^2\right)\varphi_f$$

$$\alpha_q = \frac{2\pi e^2}{\hbar\omega_p q^2}, \qquad q < q_c; \qquad \alpha_q = 0 \quad \text{when } q > q_c \tag{2-185}$$

If we consider the wave function for free movement by the electrons φ_f to have been independently normalized, then the oscillatory part must also be normalized. The normalized oscillator wave function ρ_q in the ground state will be

$$\varphi_0(\rho_q) = \left(\frac{2\alpha_q}{\pi}\right)^{1/4} e^{-\alpha_q|\rho_q|^2} \tag{2-186}$$

Similarly, the wave function for the first excited state is written

$$\varphi_1(\rho_q) = 2\left(\frac{2\alpha_q^3}{\pi}\right)^{1/4} \rho_q e^{-\alpha_q|\rho_q|^2} \tag{2-187}$$

In calculating the matrix element ρ_q of Eq. (2-183) the wave functions for free movement and all other oscillations, except for the isolated ρ_q yield a factor of one, thanks to normalizing; the nonzero matrix element ρ_q is

$$\int \rho_q\varphi_0(\rho_q)\varphi_\alpha(\rho_q)d\rho_q = \frac{1}{2\sqrt{a_q}} = \left(\frac{\hbar\omega_p q^2}{8\pi e^2}\right)^{1/2} = \left(\frac{\hbar n_0 q^2}{2m\omega_p}\right)^{1/2} \tag{2-188}$$

where the inequality

$$\omega_p^2 = 4\pi e^2 n_0/m; \qquad |n_0 \equiv \rho_0|$$

is used. The scattering cross section is written

$$d\sigma_{0\to 1} = \left(\frac{2me^2}{\hbar^2}\right)^2 \frac{k'}{k} \frac{dO}{q^4} \frac{\hbar n_0 q^2}{2m\omega_p} \tag{2-189}$$

Further conversions of Eq. (2-189) are based on the conservation of energy and momentum in the process being examined:

$$q^2 = k^2 + k'^2 - 2kk' \cos\theta \tag{2-190}$$

$$\hbar\omega_p = (\hbar^2/2m)(k^2 - k'^2) \tag{2-191}$$

For fast electrons, $k \approx k'$ and $\theta \ll 1$. The angle between $\bar{k}$ and $\bar{q}$ is nearly $\pi/2$, which means that plasma waves stimulated by fast electrons will propagate in a direction that is roughly perpendicular to the direction of the electron beam.

For small scattering angles $\theta \ll 1$, which occur only when plasmons are generated, the conservation laws are written

$$q^2 = k^2 + k'^2 - 2kk'\left(1 - \frac{\theta^2}{2}\right) \approx (k - k')^2 + k^2\theta^2 \tag{2-192}$$

$$k - k' = \frac{m}{\hbar k}\left(\omega_p^2 + \frac{3}{5}q^2 v_F^2\right)^{1/2} \tag{2-193}$$

Substituting Eq. (2-193) into (2-192), we find the relationship between the momentum being transmitted and the scattering angle:

$$q^2 = \left(1 + \frac{3}{5}\frac{v_F^2}{v^2}\right) k^2(\theta^2 + \theta_E^2), \qquad \theta_E = \frac{\hbar\omega_p}{2E}, \qquad E = \frac{mv^2}{2} \tag{2-194}$$

Substituting Eq. (2-194) into the dispersion equation, Eq. (2-182), gives us the frequency of a plasmon being generated during electron scattering at an angle θ:

$$\omega^2(q) = \omega_p^2 + \frac{3}{5}q^2 v_F^2 = \omega_p^2 + \frac{3}{5}v_F^2 k^2(\theta^2 + \theta_F^2) \tag{2-195}$$

Differentiating Eq. (2-190) at fixed k and k', we find $dO = 2\pi q\, dq/kk'$ and

$$d\sigma_1 = 2\pi \frac{e^4 n_0}{(mv^2/2)\hbar\omega_p} \frac{q\, dq}{q^2} \tag{2-196}$$

Then, as in Eqs. (2-194),

$$q\, dq/q^2 = \theta\, d\theta/(\theta_E^2 + \theta^2) \tag{2-197}$$

and

$$d\sigma_1 = \frac{n_0 e^4}{E\hbar\omega_p} \frac{dO}{\theta_E^2 + \theta^2} \tag{2-198}$$

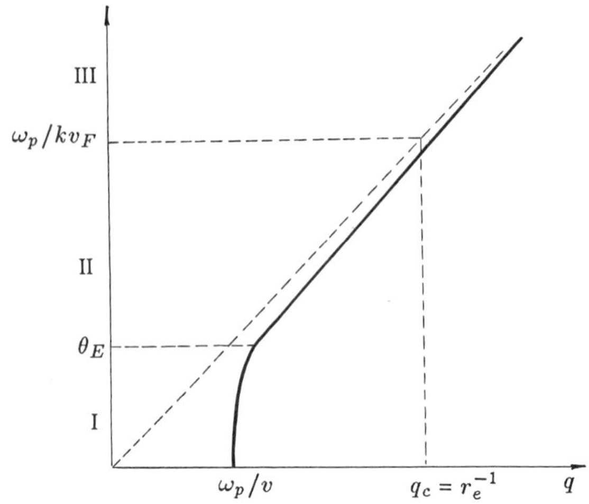

FIGURE 2-4. The scattering angle of a fast electron as a function of the momentum transmitted by plasma oscillations. In range I of the angles $\theta < \theta_E = (\hbar\omega_p/2E)q \simeq q_{\min} = \omega_p/v$; when $\theta_E < \theta \leq \omega_p/kv_F$ (range II), $q = k\theta$; in range III, $\theta > \omega_p/kv_F$ and electron scattering is governed by a single frequency perturbation of the metal's electrons.

Equation (2-194) shows that the amount of momentum transmitted q and the scattering angle θ of a fast electron are related by the equation of a hyperbola (Fig. 2.4). In the range $\theta \ll \theta_E$ the amount of momentum transmitted is independent of the scattering angle θ, and $q_{\min} = \omega_p/v$; when $\theta \gg \theta_E$, $q = k\theta$. This means that stimulation of high frequency plasmons requires a larger scattering angle. According to Eq. (2-195), as q and θ increase, so also does the quantum of energy $\hbar\omega_p$ emitted by a fast electron and the corresponding line in the electrons energy distribution is shifted. These results have been confirmed in experiments on fast electron scattering in thin metal films.[40,43]

We must now turn our attention to the scattering angle as a function of the energy of fast particles: $\theta = q/k$. As the energy E in an electron beam decreases, the angular scattering of the beam increases. We frequently deal with a beam of monochromatic electrons; as the electrons penetrate into a target their energy decreases and angular scattering increases. The maximal deflection angle for an electron when a plasmon having $q = q_c$ is generated is approximately 20 mrads when $E = 10^3$ eV. These deflection angles will be seen when fast electrons pass through films so very thin that the probability of several plasmons being generated by a single electron is negligibly small. In passing through films whose thickness is much greater than the free path length relative to generating plasmons, the observed angular deflection is the sum of many deflections and must be calculated using methods from multiple scattering theory.

2-18. THE ONE-PARTICLE CROSS SECTION OF FREE ELECTRON EXCITATION

As we have seen, one-particle excitation of free electrons occurs when fast electrons are scattered at angles $\theta > \theta_c$. This type of scattering is analogous to inelastic scattering in an atom's electrons on the assumptions that the binding energy of the atom's electrons can be neglected. Coulomb potential shielding when $r < r_c$ can be thought of as a new moment in comparison with inelastic scattering in an atom's electrons. However, when calculating the cross section of one-particle excitations Coulomb potential shielding can be ignored because these excitations correspond to scattering into finite angles $\theta > \theta_c$ and in this range of angles there are no peculiarities in the dependence of $d\sigma/dO$ on θ. Taking shielding into account may produce terms such as α^2 in the denominator of Eq. (2-54). These terms can be ignored for angles $\theta > \theta_c$.

Thus, if we ignore Coulomb shielding between fast electrons and the target material's electrons, we obtain (see Eqs. (2-92) and (2-103))

$$d\sigma = \left(\frac{2e^2}{mv^2}\right)^2 n_0 \frac{dO}{\theta^4} \qquad \text{when } \theta_c \leq \theta \leq 1 \tag{2-199}$$

$$d\sigma = n_0 \left(\frac{e^2}{2mv^2}\right)^2 (\sin^{-4}\theta/2 + \cos^{-4}\theta/2 - \sin^{-2}\theta/2\cos^{-2}\theta/2)\, dO \quad \text{when } \theta > 1 \tag{2-200}$$

Equations (2-199) and (2-200) are written for the number of electrons n_0 per unit volume of target material.

2-19. THE ENERGY LOST BY A FAST ELECTRON IN EXCITING PLASMA ELECTRONS

It is not hard to find the total cross section of plasma oscillation stimulation by a fast electron. Integrating Eq. (2-198) over the interval from $\theta_{\min} = \theta_E$ to $\theta_{\max} = \theta_c$ we find[(40)]

$$\begin{aligned}\sigma_p &= \frac{n_0 e^4}{E\hbar\omega_p}\int_{\theta_E}^{\theta_c}\frac{dO}{\theta_E^2 + \theta^2} = \frac{\pi n_0 e^4}{E\hbar\omega_p}\ln\frac{\theta_c^2 + \theta_E^2}{\theta_E^2}\\ &= \frac{2\pi n_0 e^4}{E\hbar\omega_p}\ln(\theta_c/\theta_E), \qquad \frac{\theta_c}{\theta_E} = \frac{v}{v_F}\end{aligned} \tag{2-201}$$

Because we have calculated the cross section for n_0 electrons per unit volume, the free path length for an electron relative to that for plasmon stimulation is $\lambda_p = 1/\sigma_p$. The probability that n plasmon stimulations will occur on a path l is given by the Poisson distribution

$$I_n = \left(\frac{l}{\lambda_p}\right)^n \frac{\exp(-l/\lambda_p)}{n!} \tag{2-202}$$

where l/λ_p is the average number of plasmons stimulated by a fast electron in a layer l. The number of plasmon stimulations per unit path length will be $1/\lambda_p$, because the energy lost (deceleration) per unit path length in stimulating plasmons is

$$\left|\frac{dE}{dS}\right|_{\text{pl}} = \hbar\omega_p\frac{1}{\lambda_p} = \frac{2\pi n_0 e^4}{E}\ln(\theta_c/\theta_E) \tag{2-203}$$

It is not hard to find how much deceleration will occur due to one-particle excitations of the plasma electrons in close collisions. To do this we use the inelastic scattering cross section, Eq. (2-199), and the amount of energy transmitted to a target material electron as a function of the scattering angle[24]:

$$\Delta E(\theta) = \frac{mv^2}{2}\sin^2\theta \approx \frac{mv^2}{2}\theta^2 \tag{2-204}$$

at angles $\theta_c \leq \theta \leq 1$. Integrating the total losses per unit path length we find

$$\left|\frac{dE}{dS}\right|_e = \int \Delta E(\theta)\frac{d\sigma}{dO}\,dO = \frac{n_0 e^4}{(mv^2/2)}\int_{\theta_c}^{1}\frac{dO}{\theta^2} = \frac{2\pi n_0 e^4}{E}\ln(1/\theta_c) \tag{2-205}$$

The total losses are the sum of the losses due to stimulating plasmons, one-particle excitation of the plasma electrons, and exciting and ionizing the inner shell electrons:

$$\left|\frac{dE}{dS}\right| = \left|\frac{dE}{dS}\right|_{\text{pl}} + \left|\frac{dE}{dS}\right|_{\text{single}} + \left|\frac{dE}{dS}\right|_{\text{core}} \tag{2-206}$$

$$\left|\frac{dE}{dS}\right|_{\text{pl}} + \left|\frac{dE}{dS}\right|_{\text{single}} = \frac{2\pi n_0 e^4}{E}\ln\left(\frac{1}{\theta_E}\right) = \frac{2\pi n_0 e^4}{E}\ln\left(\frac{2E}{\hbar\omega_p}\right) \tag{2-207}$$

The energy losses in an atom's inner electron shells are given by a formula similar to Eq. (2-160):

$$\left|\frac{dE}{dS}\right|_{\text{core}} = \frac{2\pi e^4 N(Z - Z_B)}{E}\ln(1.166E/I) \tag{2-208}$$

Here the average ionization energy I must be calculated without regard for the valence electrons that will form the plasma and the losses that have already been considered. The inner shell losses will be small in comparison with the losses in the valence electrons, because $I \gg \hbar\omega_p$. As we have already seen, the ionization energy increases rapidly when transitions in a multielectron atom are made to more interior shells (see Slater's calculation of I in Section 2-11). The path length of an electron with energy of E = 30 keV relative to that for plasmon stimulation in aluminum is λ_p = 600 Å.

2-20. THE ENERGY SCATTERING MECHANISMS FOR LOW-ENERGY ELECTRONS IN SOLIDS[44]

Our examination of electron scattering processes will be incomplete if we ignore the scattering processes in low-energy electrons. Born's theory of inelastic atom–electron collisions is based on the assumption that the energy of a fast electron is much greater than the binding energy of the atom's electrons. This energy is roughly 10 eV for the valence electrons and 10^2–10^4 eV for the inner shell electrons. How do the electron scattering processes proceed when the electron's energy falls below the ionization energy of a material's valence electrons?

The subject of low-energy electron scattering also arises when the scattering of the secondary electrons liberated due to the atoms (molecules) of a material being ionized is examined. It is well known that the majority of secondary electrons have low energy, in the range $E_s \geq 30$ eV. This means that a fast electron with energy 10^4 eV produces somewhere on the order of 10^3 secondary electrons in ionizing a material's atoms. It is mainly these secondary electrons that, by dissipating their energy, convert the energy of the primary fast electrons into heat.

Finally, low-energy electrons arise when atoms (molecules) are photoionized during photolithography. Due to the internal photoeffect, ultraviolet or soft x-ray radiation produces relatively low-energy photoelectrons and Auger electrons whose energy is related to that of a primary photon by: $E_s \leq h\nu - I$, where I is the energy of the photoelectron. In most cases the energy of these electrons lies in the $\leq 10^2$–10^3 eV range.

At room temperature the thermal energy of electrons is E_T, which is much less than the energy of those electrons that we have placed in the low-energy category. In semiconductor physics, electrons with energy $E \gg E_T$ are called hot electrons. They are obtained by injection through heterojunctions and their energy rarely exceeds 1 eV. Thus, the electrons we are examining and which have energy $\sim$ 10 ev are, from the standpoint of hot electron physics, "superheated." Nevertheless, the processes that occur when they are decelerated will be similar to what happens when hot electrons are decelerated in semiconductors.

The differential scattering cross section for a primary electron with energy E in an atom that emits an ionization electron having energy $\Delta E = \varepsilon E$, $\varepsilon < 1$ can easily be obtained from Eq. (2-100) if we use the relation $\Delta E = E\sin^2(\theta/2)$ between the scattering angle and transmitted energy:

$$\frac{d\sigma}{d\varepsilon} = \frac{B}{E}\left(\frac{1}{\varepsilon^2} + \frac{1}{(1-\varepsilon)^2} - \frac{1}{\varepsilon(1-\varepsilon)}\right), \qquad B = \frac{2\pi e^4}{mv^2} \tag{2-209}$$

For small values of transmitted energy $\varepsilon \ll 1$ the cross section increases as ε^{-2}, i.e., low-energy secondary electrons are generated with greater probability. Eggarter(45) and Grisinsky(33) suggested semiempirical formulas that contain in explicit form the binding energy I_i of the electron being emitted (in units of E):

$$\left(\frac{d\sigma}{d\varepsilon}\right)_E = \frac{B}{E}\left(\frac{1}{\varepsilon^2} + \frac{1}{(1-\varepsilon+I_i)^2} - \frac{1}{\varepsilon(1-\varepsilon+I_i)}\right) \tag{2-210}$$

$$\left(\frac{d\sigma}{d\varepsilon}\right)_G = \frac{\pi e^4}{E^2}(1+\varepsilon_{\rm nl})^{-3/2}\,\frac{(1-\varepsilon)^{\varepsilon_{\rm nl}/(\varepsilon+\varepsilon_{\rm nl})}}{\varepsilon^2} \times \left\{(1-\varepsilon_{\rm nl}) + \frac{4}{3}\frac{\varepsilon_{\rm nl}}{3}\ln\left[2.7 + \left(\frac{1-\varepsilon}{\varepsilon_{\rm nl}}\right)^{1/2}\right]\right\} \tag{2-211}$$

In these formulas the minimal transmitted energy is $\varepsilon \gg I_i$, $\varepsilon_{\rm nl}$.

If we assume that the formulas from the inelastic scattering theory for fast electrons can be used for electron energies of as much as $E \approx I_v$, then electrons having energy $E \geq I_v$ must belong to low-energy electrons.

Figure 2-5 shows an energy diagram that can be used to model inelastic scattering for low-energy electrons.(44) A polymer resist layer has been patterned as a dielectric having a forbidden band E_g. In this model ionization of a resist molecule is shown as filling the conduction band with an electron from the valence (ionization of the valence electrons) or deeper bands that correspond to the energy levels of the atoms' inner shells. Below the bottom of the conduction band lie the molecule exciton energy levels that

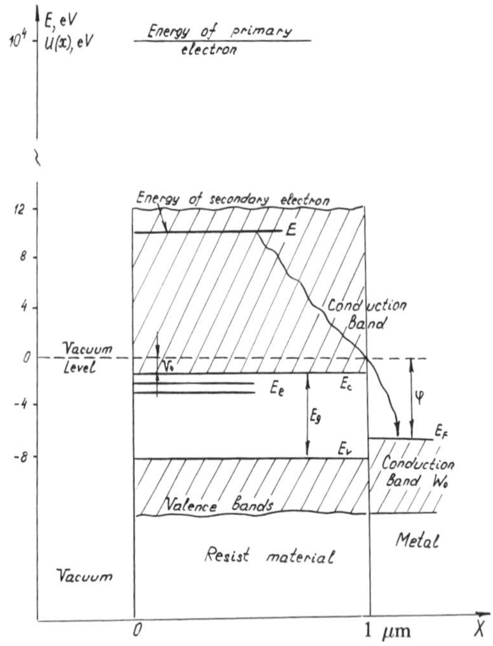

FIGURE 2-5. Energy diagram for constructing a model of inelastic scattering of low-energy electrons.

correspond to excited states of the resist molecules (without ionization). The resist layer in a metal film is described by a work function φ.

The location of the bottom of the conduction band V_0 in a dielectric relative to the zero-energy level in a vacuum is an important parameter. Two factors contribute to the energy of an electron at V_0 in a dielectric: 1) the polarization energy of the dielectric when acted upon by an electron:

$$U_p = \frac{3\alpha e^2}{2r_s^2}\left[\frac{8}{7} + \left(1 + \frac{8\pi\alpha N}{3}\right)^{-1}\right] \tag{2-212}$$

α is the molecule's susceptibility to polarization, cm^3; N is the number of molecules per cubic centimeter, and; $r_s = ((3/4)\pi N)^{-1/3}$ is the radius of a Wigner–Seitz sphere, calculated for one molecule of the medium; 2) the electron's kinetic energy $T_0 = \hbar^2 k_0^2/2m^*$ that corresponds to the motion of an electron in a potential well that is a spherical layer of thickness $\Delta r = r_s - a$, where a is the radius of action for short-range Hartree–Fock electron exchange repulsion forces from the atom's electron shells.(46) According to the uncertainty principle, $\Delta r \cdot k_0 \sim 1$; if $\Delta r \sim 1$ Å then $k_0 \approx 10^8$ cm^{-1} and $T_0 \approx 3$ eV. Conversely, the polarization binding energy is of roughly the same magnitude: $|U_p| \approx 3$–4 eV. However, because most dielectrics have finite work functions we must assume that $|U_p| > T_0$, $V_0 = T_0 - |U_p| < 0$. Numerical computations in [44] for benzene yielded values of $U_p = -3.40$ eV, $T_0 = 3.1$ eV, and $V_0 = -0.3$ eV.

Now, it is not hard to describe the energy relaxation process in low-energy electrons. These electrons have roughly 10 eV above the ground energy level in the conduction band of a polymer dielectric. This is enough energy for electrons to make transitions into molecule ionizing states, which normally lies in the range 4–8 eV (from the top of the valence band). A second path of electron energy relaxation is the transmission of energy via intramolecular oscillations; the energies of these transitions lie in the range 0.1–0.5 eV; for a specific material they can be found via IR and combination spectroscopy.

Finally, a third path is energy transmission via low-frequency "lattice" oscillations in the medium. The energy of the phonons in these oscillations is less than 0.1 eV.

Experiments in the transmission spectra of low-energy electrons through thin dielectric layers (4–10 monolayers or 1.8–4.5 nm for benzene) show that while the electron has enough energy for electrons to make transitions into conduction band states, it is mainly this path that is dominant in the relaxation of a hot electron's energy. Then, at energies $E < E_g$, energy relaxation proceeds by stimulating intramolecular oscillations and then, phonons.[44] At the end of the energy relaxation process the electrons are at the bottom of the conduction band and provide the polymer's radiation conductance. A fraction of the electrons pass through to the metal, whose Fermi energy is much less than the energy at the bottom of the polymer's conduction band. The passage of electrons into the metal prevents a space charge being formed in the polymer during exposure to an electron beam. If the polymer film has been deposited onto a dielectric (e.g., SiO_2) space charges can form in the polymer film and in the dielectric, which can create significant disturbances when the polymer resists are exposed to electron beams and distort the lithographic pattern. Obviously, the best way to prevent this type of distortion is to remove the charge from the polymer film (put the dielectric into contact with a grounded metal).

2-21. THE ENERGY SPECTRUM OF ELECTRONS PENETRATING A TARGET

The beam energy $E(S)$ in a target after travelling a path S can be calculated from Bethe's formula for electron deceleration:

$$-\frac{dE(S)}{dS} = \frac{k}{E(S)} \ln\left[\frac{aE(S)}{I}\right] \tag{2-213}$$

$$k = 2\pi N_A e^4 \rho Z/A \tag{2-214}$$

$a = 0.1658$, ρ is the target density, and the other quantities have the usual designations. For a small path S we can assume that

$$E(S) = E_0 - kS, \qquad S_0 \equiv \frac{E_0}{R_B} \tag{2-215}$$

Substituting Eq. (2-215) into (2-213) and integrating, we find[47]

$$E(S) = E_0\left\{1 - \frac{kS_0}{2E_0}([\ln \xi_0]^2 - [\ln \xi_0(1 - S/R_B)]^2\right\}, \qquad \xi_0 = \frac{aE_0}{I} \tag{2-216}$$

Equation (2-216) does not satisfy the boundary condition $E = 0$ when $S = R_B$, and gives a result that diverges to ln(0). To eliminate this divergence, we make the substitution $S \to \gamma_S$ and impose the condition that $E \to 0$ when $S \to R_B$. This condition defines the quantity γ:

$$\gamma = 1 - \exp\left[\sqrt{(\ln \xi_0)^2 - \frac{2E_0}{S_0 K}} - \ln \xi_0\right] \tag{2-217}$$

For the energy of a beam travelling a path S we will have[47]

$$E(S) = E_0 \left[1 + \frac{\Lambda}{2\xi_0^2} \ln(1 - \gamma_S/R_B)\right] [2 \ln \xi_0 + \ln(1 - \gamma_S/R_B)] \tag{2-218}$$

$$\Lambda = (a/I)^2 k R_B \tag{2-219}$$

Although Eq. (2-218) yields the value of $E(S)$, it contains no information about the statistical distribution of the electron energy in the beam. The problem of the electron energy distribution in a beam entering a target was first examined by Landau in 1944.[48] We will present results from Landau's theory.

The energy distribution of a beam entering a target is

$$f(E, 0) = \delta(E_0 - E) \tag{2-220}$$

We will introduce the probability density $f(\Delta, S)$ that an electron on a path S loses an amount Δ of energy. Electron energy losses occur because of inelastic collisions with the target atoms' electron. The inelastic scattering cross section for an electron by an electron when large amounts of energy are transmitted is given by Rutherford's formula, Eq. (2-93). By multiplying Eq. (2-93) by the number of electrons per unit volume $n = \rho(N_A/A)Z$, we obtain the scattering cross section per unit length of the electron's path:

$$d\sigma' = \rho \frac{N_A}{A} Z \left(\frac{e^2}{2mv^2}\right)^2 \frac{dO}{\sin^4 \theta/2} \tag{2-221}$$

The velocity that an electron acquires due to a collision in the target is[24]

$$v_2' = \frac{2m_1 v}{m_1 + m_2} \sin \theta/2, \qquad m_1 = m_2 = m \tag{2-222}$$

Consequently, the energy lost by the fast electron (and acquired by the target's electron) is

$$\varepsilon = \frac{m_2 v_2'^2}{2} = \frac{mv^2}{2} \sin^2 \theta/2 \tag{2-223}$$

Making the conversion to the energy ε in Eq. (2-221) we find the differential cross section for transmitting energy in the interval from ε to $\varepsilon + d\varepsilon$ per unit length of the electron's path:

$$\frac{d\sigma'}{d\varepsilon} \equiv \omega(\varepsilon) = \pi\rho \frac{N_A Z}{A} \frac{e^4}{E} \frac{1}{\varepsilon^2} \tag{2-224}$$

We use the result obtained to compose a kinetic equation for $f(S, \Delta)$. The change df in this function over the path from S to $S + dS$ is governed by the difference between the number of electrons at S with an energy $E_0 - \Delta + \varepsilon$ and going into a state with an energy $E_0 - \Delta$ due to losing energy ε over the path dS:

$$\delta f_+ = \int_0^\infty \omega(\varepsilon) f(\Delta - \varepsilon, S)\, d\varepsilon\, dS \tag{2-225}$$

and by the number of electrons at the point S in a state with energy $E_0 - \Delta$ and leaving this state due to losing energy ε over the path dS:

$$\delta f_- = \int_0^\infty \omega(\varepsilon) f(\Delta, S)\, d\varepsilon\, dS \tag{2-226}$$

Thus, we find an integral equation for the distribution function $f(\Delta, S)$:

$$\frac{df}{dS} = \int_0^\infty \omega(\varepsilon)[f(\Delta - \varepsilon, S) - f(\Delta, S)]\, d\varepsilon \tag{2-227}$$

In order to solve Eq. (2-227) we use the Laplace transformation

$$\Phi(S, p) = \int_0^\infty e^{-p\Delta} f(S, \Delta)\, d\Delta; \qquad f(S, \Delta) = \frac{1}{2\pi i} \int_{-i\omega+\sigma}^{i\omega+\sigma} e^{p\Delta} \Phi(S, p)\, dp \tag{2-228}$$

Calculations yield

$$\Phi(S, p) = \exp\left[-S \int_0^\infty \omega(\varepsilon)(1 - e^{-p\varepsilon})\, d\varepsilon\right] \tag{2-229}$$

for the Laplacian of $\Phi(S, p)$. With due regard for the properties of the $\omega(\varepsilon)$ function (I is the average ionization energy for the target atoms)

$$\omega(\varepsilon) = 0 \qquad \text{when} \begin{cases} \varepsilon < I \\ \varepsilon > \varepsilon_{\max} = mv^2/2 \end{cases}$$

we can find[47]

$$\int_0^\infty \omega(S)(1 - e^{-p\varepsilon})\, d\varepsilon = \frac{\pi N_A e^4 Z}{E_0 A} \rho \left(1 - C - \ln\left(\frac{I^2 p}{4E_0}\right)\right) p \tag{2-230}$$

Returning to the distribution function $f(S, \Delta)$ we find, according to Eq. (2-228),

$$f(S, \Delta) = \frac{p(\lambda)}{\xi}, \qquad p(\lambda) = \frac{1}{2\pi i} \int_{-i\infty+\sigma}^{+i\infty+\sigma} e^{u(\ln u + \lambda)}\, du \tag{2-231}$$

where

$$\xi = \frac{\pi N_A e^4}{E_0} \frac{Z}{A} ps, \qquad \lambda = \frac{\Delta - \xi[1 - C + \ln(4E_0\xi/I^2)]}{\xi} \tag{2-232}$$

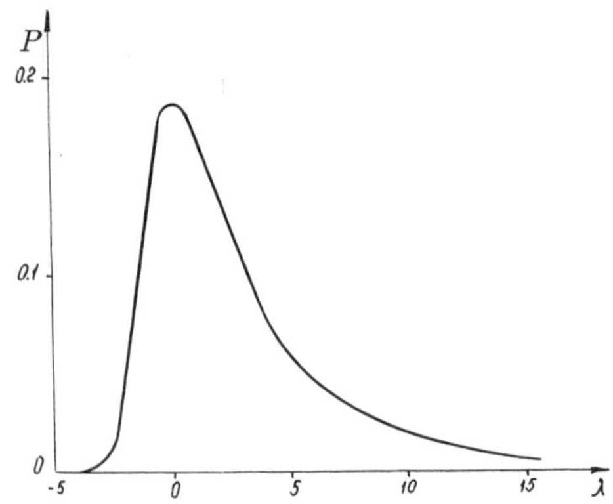

FIGURE 2-6. The amount of energy lost by an electron as a function of beam penetration depth into the target; λ is the reduced value of energy loss.

We can show[47] that the average energy Δ lost on a path S is

$$\Delta = 2\xi \ln\left(1.166\frac{E_0}{I}\right) = \left[2\pi \frac{N_A e^4}{E_0}\frac{Z}{A}\rho \ln\left(1.166\frac{E_0}{I}\right)\right] \cdot S \qquad (2\text{-}233)$$

The distribution $P(\lambda)$ has been tabulated and a graph of this distribution is shown in Fig. 2-6. As we can see, the distribution is markedly asymmetric. The probability of losses greater than Δ falls off slowly, whereas the probability of losses less than Δ falls off rapidly.

2-22. ELECTRON-BEAM SCATTERING WHEN EXPOSING A RESIST

In previous sections we have discussed the physics of the elementary electron scattering processes. How can we now use this knowledge in our study of electron lithography processes?

First of all, we will define some functions that are characteristic of the lithographic process. A thin (0.1–1 μm) polymer layer (electron resist) is exposed by an electron beam whose charge (current) distribution is considered to be Gaussian,

$$N(r) = N_0 \frac{a}{\pi} e^{-ar^2}, \qquad \int_0^\infty N(r) 2\pi r \, dr = N_0 \qquad (2\text{-}234)$$

The beam diameter is $d = 2/\sqrt{a}$; N_0 is the total number of electrons entering the film per second and is related to the beam current: $N_0 = qit$, i is the beam current in amperes, and q is the number of electrons per Coulomb of charge, $q = 10^{19}/1.6$.

Figure 2-7 shows the cross section of isoenergy surfaces in the exposed region when a single line passes through it.[49] As can easily be seen, a fairly wide area of the resist is exposed. The maximal width of this area is equal to the beam width plus the distance from the beam to both sides, roughly equal to the mean free path of electrons in the beam. At a beam energy of 20 keV the mean free path in a PMMA resist is 7.5 μm. At

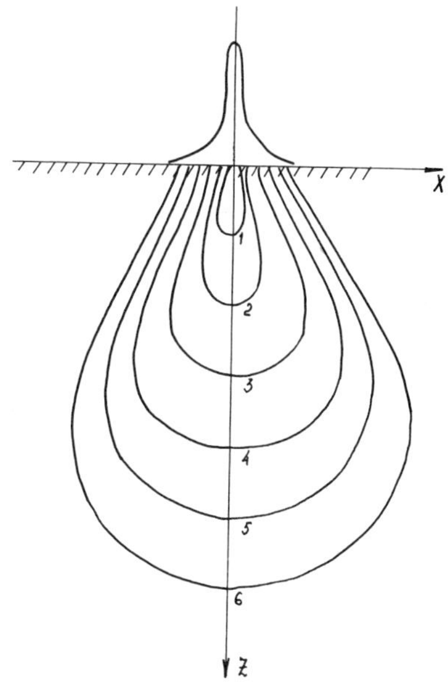

FIGURE 2-7. The resist exposed by an electron beam is tear-shaped. 1, 2, 3, 4, 5, 6 are the cross sections of the isoenergy surfaces for a given energy density.

first glance, it appears that we cannot obtain a high degree of resolution even when the beam itself is less than a micron wide.

To obtain a more realistic evaluation of the resolving power of electron-beam exposure we must know more precisely what the energy density of the beam being absorbed by the resist is. We will designate the coordinates of points in the area being exposed as $r_T(x_T, y_T, z_T)$, and the density of the energy being absorbed at a point r_T as $D(r_T)$. For a given etchant and etching procedure there is a threshold of absorbed energy density D_0 such that only that quantity of resist which has absorbed an amount of energy $D \geq D_0$ is etched. Therefore, the quantity of polymer bounded by the isoenergy surface $D(r_T) = D_0$ is said to be etched.[50] The width of this band is approximately that of the beam and may be less than a micron for beams having submicron cross section.

The second question deals with the substrate's role. How much of the beam is reflected from the substrate? What is its energy spectrum, and what is the density of the energy dissipated in a resist layer by the reflected beam? Qualitative answers to these questions are the following. The free path of electrons in a resist is much greater than the layer thickness so that the layer is "thin." Almost all the electrons pass through the resist and enter the substrate. At energies of about 5 keV the number of electrons that penetrate the substrate is reduced by as much as 50% for layers 0.4 μm thick. The majority of beam electrons passing through the resist layer are scattered at only small angles; the so-called large-angle scattering cross section is relatively small. Therefore, we can distinguish two components of the electron beam in the resist layer: electrons undergoing small-angle scattering (or no scattering at all) and electrons undergoing large-angle scattering.

A large number of electrons are reflected from the substrate. The higher the substrate material's atomic number and the lower the beam energy, the larger is the reflection coefficient. For a beam energy 20 keV it is roughly 0.2 for a silicon substrate and 0.5 for a gold substrate. Because electrons are reflected from the substrate by scattering on the substrate atoms, the reflected beam has broad three-dimensional and energy distributions.

Therefore, the energy dissipated by the reflected beam in a resist layer covers a large area in all directions from the initial Gaussian beam that entered the resist layer. We will examine the questions that have been raised in more detail. It is obvious that our main task now is to study the three-dimensional distribution of the electron beam and the energy that it dissipates into the target.

2-23. THE ANGULAR DISTRIBUTION OF THE SCATTERED PARTICLES

We will introduce a function $f(\bar{r}, \vartheta, \varphi, t)$ for the probability density of finding a particle at point $\bar{r}$ in the target at time t and moving in the ϑ, φ direction. The function's rate of change is given by the Boltzmann kinetic equation[20]

$$\delta f = -\bar{v}\,\mathrm{grad}\, f + \delta f^{+} - \delta f^{-} \tag{2-235}$$

where δf^{+} and δf^{-} are increments governed by the scattering processes

$$\begin{aligned} \delta f^{+} &= N \iint v\sigma(\zeta, v) f(\theta, \Phi, t) \sin\zeta\, d\zeta\, d\chi \\ \delta f^{-} &= N \iint v\sigma(\zeta, v) f(\vartheta, \varphi, t) \sin\zeta\, d\zeta\, d\chi \end{aligned} \tag{2-236}$$

Here $\sigma(\zeta, v)$ is the differential scattering cross section into a solid angle $\sin\zeta\, d\zeta\, d\chi$, N is the number of target atoms being scattered per unit volume, and

$$\cos\zeta = \cos\theta\cos\vartheta + \sin\theta\sin\vartheta\cos(\Phi - \varphi) \tag{2-237}$$

Thus, we find ($\bar{v}_0 = \bar{v}/|\bar{v}|$)

$$\frac{\partial f}{\partial s} = -\bar{v}_0\,\mathrm{grad}\, f + N \iint \sigma(\zeta, v)[f(\bar{r}, \theta, \Phi, t) - f(r, \vartheta, \varphi, t)] \sin\zeta\, d\zeta\, d\chi \tag{2-238}$$

The initial condition for an electron impinging along the normal to the target's surface at the coordinates system origin is

$$f(\bar{r}, \bar{\vartheta}, \varphi, 0) = \delta(\bar{r})\delta(1 - \cos\vartheta)/2\pi \tag{2-239}$$

In order to simplify our problem of finding the three-dimensional electrons distribution, we will only consider acts of elastic scattering, so that the particle's velocity will remain constant.

We now introduce a function $F(\vartheta, \varphi, s) = \int f\, d^3\bar{r}$ for the angular distribution of the particles. This function defines the probability that an electron will move in the ϑ, φ direction after it has traversed a path s in the target. We obtain the function F for the angular scattering of particles by integrating Eq. (2-238) over all points in the space $\bar{r}$. Averaging Eq. (2-238) over $\bar{r}$ yields

$$\frac{\partial F}{\partial s} = N \iint \sigma(\zeta, v)[F(\theta, \Phi, s) - F(\vartheta, \varphi, s)] \sin\zeta\, d\zeta\, d\chi \tag{2-240}$$

because

$$\int d^3\bar{r}\bar{v}_0 \,\mathrm{grad}\, f = 0$$

Equation (2-240) is solved for the initial condition

$$F(\vartheta, \varphi, 0) = \delta(1 - \cos\vartheta)/2\pi \tag{2-241}$$

The symmetry of our problem emphasizes the fact that the angular distribution F is independent of the angle φ. We will seek $F(\vartheta)$ as a series of Legendre polynomials $P_l(\cos\vartheta)$:

$$F(\vartheta, s) = \sum_l F_l(s, v) P_l(\cos\vartheta); \qquad F(\theta, s) = \sum_l F_l(s, v) P_l(\cos\theta) \tag{2-242}$$

Substituting Eq. (2-242) into Eq. (2-240) yields

$$\sum_l \frac{\partial F_l}{\partial s} P_l(\cos\vartheta) = \sum_l N \iint \sigma(\zeta, v) F_l(s, v)[P_l(\cos\theta) - P_l(\cos\vartheta)] \sin\zeta \, d\zeta \, d\chi \tag{2-243}$$

We determine the function $P_l(\cos\theta)$ according to the rule for multiplying Legendre polynomials:

$$P_l(\cos\theta) = \left[P_l(\cos\zeta) - 2 \sum_{m=l}^{l} \frac{(l-m)!}{(l+m)!} P_l^m(\cos\theta) P_l^m(\cos\vartheta) \right] / P_l(\cos\vartheta) \tag{2-244}$$

We multiply Eq. (2-243) by $P_l(\cos\vartheta)$ and integrate with respect to $\sin\vartheta\, d\vartheta\, d\varphi$. Keeping the orthogonality of the polynomials in mind, we obtain

$$\frac{\partial F_l}{\partial s} = -k_l F_l, \qquad k_l = N \iint \sigma(\zeta, v)(1 - P_l(\cos\zeta)) \sin\zeta \, d\zeta \, d\chi \tag{2-245}$$

Taking the initial conditions, written

$$\delta(1 - \cos\vartheta) = \sum_{l=0}^{\infty} \frac{2l+1}{2} P_l(\cos\vartheta) \tag{2-246}$$

into account, we find

$$F(\vartheta, s, v) = \frac{1}{4\pi} \sum_{l=0}^{\infty} (2l+1) P_l(\cos\vartheta) e^{-k_l(v)s} \tag{2-247}$$

Before we look at the properties of the solution of Eq. (2-247) we must calculate the "attenuation coefficients" $k_l(v)$. Because $P_0 = 1$, $k_0 = 0$; which means that the isotropic term in the solution does not attenuate as the depth of penetration into the target increases, whereas all nonisotropic terms $P_l(\cos\vartheta), l \neq 0$ attenuate at distances $1/k_l$. At a depth $s \gg 1/k_l$ the motion of the beam electrons becomes completely isotropic and diffusive and the electrons completely forget what their original direction of motion was.

If elastic scattering is described by a Rutherford differential cross section with shielding,

$$\sigma(\vartheta, v) = \frac{Z^2 e^4}{4m^2 v^4} \frac{1}{(\sin^2 \vartheta/2 + \alpha)^2}, \qquad \alpha^{1/2} = 0.56 Z^{1/3} \kappa (e^2/\hbar)/v, \qquad \kappa = 0.66$$

the total cross section is

$$\sigma_0 = 2\pi \int_0^{\pi} \sigma(\vartheta, v) \sin\vartheta \, d\vartheta = \frac{\pi Z^2 e^4}{m^2 v^4} \frac{1}{\alpha(1+\alpha)} \tag{2-248}$$

The integral

$$\begin{aligned} \sigma_1 &= 2\pi \int_0^{\infty} \sigma(\vartheta, v)(1 - \cos\vartheta) \sin\vartheta \, d\vartheta \\ &= 2\pi \frac{Z^2 e^4}{m^2 v^4} \left[\ln\left(1 + \frac{1}{\alpha}\right) - \frac{1}{1+\alpha} \right] \sim 4\pi \frac{Z^2 e^4}{m^2 v^4} \ln(367 Z^{-1/3} v/c) \end{aligned} \tag{2-249}$$

describes the electron's loss of velocity in its initial direction of motion and can therefore be called the cross section of momentum losses.

The attenuation constants k_l can be rewritten as

$$\begin{aligned} k_l &= N\sigma_0 \iint f(\vartheta)(1 - P_l(\cos\vartheta)) \sin\vartheta \, d\vartheta \, d\varphi = N\sigma_0(a_0 - a_l) \\ a_l &= \iint f(\vartheta) P_l(\cos\vartheta) \sin\vartheta \, d\vartheta \, d\varphi, \qquad f(\vartheta) = \sigma(\vartheta)/\sigma_0 \end{aligned} \tag{2-250}$$

When scattering is described by the Rutherford formula with shielding, Eq. (2-248), we can find recurrence formulas for calculating a_{n+1} in terms of a_n and a_{n-1}.[51] The form factor for the differential scattering cross section is

$$f(\vartheta) = \sigma(\vartheta)/\sigma_0 = \frac{\alpha(1+\alpha)}{\pi} \frac{1}{(1 - \cos\vartheta + 2\alpha)^2} \tag{2-251}$$

The quantity a_n can be written as an integral:

$$\begin{aligned} a_n &= -2\alpha(1+\alpha) \int_0^{\pi} \frac{P_n(\cos\vartheta) \, d(\cos\vartheta)}{(1 - \cos\vartheta + 2\alpha)^2} \\ &= 2\alpha(1+\alpha) \int_{-1}^{+1} \frac{P_n(u) \, du}{(1 + 2x - u)^2} \end{aligned} \tag{2-252}$$

Setting $x = 1 + 2\alpha$ and using the identity $(x - u)^{-2} = -d(x - u)^{-1}/dx$ we rewrite Eq. (2-252):

$$\frac{a_n}{2\alpha(1+\alpha)} = -2\frac{d}{dx} \int_{-1}^{+1} \frac{P_n(u)}{2(x - u)} du = -2\frac{dQ_n(x)}{dx} \tag{2-253}$$

where

$$Q_n(x) = \int_{-1}^{+1} \frac{P_n(u)\, du}{2(x - u)} \tag{2-254}$$

are second-order Legendre functions. Recurrent relations occur between them (the prime denotes differentiation)[50]:

$$nQ_n = xQ'_n - Q'_{n-1}; \qquad (n+1)Q_n = Q'_{n+1} - xQ'_n \tag{2-255}$$

Eliminating Q_n from Eq. (2-255), we find

$$nQ'_{n+1} = (2n+1)xQ'_n - (n+1)Q'_{n-1} \tag{2-256}$$

or

$$na_{n+1} = (2n+1)(1+2\alpha)a_n - (n-1)a_{n-1} \tag{2-257}$$

Values of a_0 and a_1 can now easily be calculated:

$$\begin{gathered} Q_0(x) = \frac{1}{2} \ln \frac{x+1}{x-1}, \qquad Q'_0(x) = -\frac{1}{(x+1)(x-1)} = -\frac{1}{4\alpha(1+\alpha)} \\ Q_1(x) = \frac{x}{2} \ln \frac{x+1}{x-1}, \qquad Q'_1 = -\frac{1 + 2\alpha + 2\alpha(1+\alpha)\ln[\alpha/(1+\alpha)]}{4\alpha(1+\alpha)} \\ a_0 = 1, \qquad a_1 = 1 + 2\alpha + 2\alpha(1+\alpha)\ln[\alpha/(1+\alpha)] \end{gathered} \tag{2-258}$$

Noting that $N\sigma_0 = 1/\lambda$, where λ is the free path length, and measuring the path travelled by an electron in units of the free path length we obtain

$$sk_l = N\sigma_0(a_0 - a_l)s = p(1 - a_l), \qquad p = \frac{s}{\lambda} \tag{2-259}$$

As the final result, we write the angular distribution function for electrons travelling a path $s = p\lambda$ in the target, having isolated the unscattered portion of the beam $\delta(\vartheta)\exp(-p)$ in explicit form as

$$\left.\begin{aligned} F(\vartheta, p) &= e^{-p} \left[\delta(\vartheta) + \frac{1}{4\pi} \sum_{n=0}^{\infty} (2n+1)(e^{pa_n} - 1) P_n(\cos\vartheta) \right] \\ \delta(\vartheta) &= \frac{1}{4\pi} \sum_{n=0}^{\infty} (2n+1) P_n(\cos\vartheta) \end{aligned}\right\} \tag{2-260}$$

We can also introduce the integral function of the angular distribution as the probability that a scattered electron will, after travelling a path $s = p\lambda$ in the target, be found in a cone bounded by the angles $0 \le \varphi \le 2\pi$, $\vartheta \le \theta$:

$$G(\vartheta \le \theta, p) = \int_0^{2\pi} d\varphi \int_0^{\theta} F(p, \vartheta) \sin\vartheta \, d\vartheta \tag{2-261}$$

Starting with Eq. (2-260), the probability G can be written as a series which is convenient for calculating, e.g., by the Monte Carlo method. We write Eq. (2-260) as

$$F(\vartheta, p) = \sum_{n=0}^{\infty} (2n+1) b_n P_n(\cos\vartheta), \qquad b_n = e^{-p(1-a_n)} \tag{2-262}$$

Direct integration of the Legendre polynomials and using recurrence formulas for them yields[51]

$$G(\theta, p) = \frac{1}{2}\left[b_0 + b_1 P_0(u) - \sum_{n=1}^{\infty}(b_{n-1} - b_{n+1}) P_n(u)\right], \qquad u = \cos\theta$$

Substituting $b_n = e^{-p(1-a_n)}$, we obtain[51]

$$G(\theta, p) = \frac{1}{2}(1 + e^{-p(1-a_1)}) - \sum_{n=1}^{\infty} \frac{1}{2}[e^{-p(1-a_{n-1})} - e^{-p(1-a_{n+1})}] P_n(\cos\theta) \tag{2-263}$$

2-24. AN ASYMPTOTIC REPRESENTATION OF THE ANGULAR DISTRIBUTION OF ELECTRONS IN A TARGET

It is clear from general considerations that the probability of an angular distribution, Eq. (2-260), written as a series in Legendre polynomials must converge to a normal (Gaussian) distribution when $p = s/\lambda$ is large. This was shown in Moliere's work.[52]

For a given target thickness s electrons will experience a larger number of scattering actions when the scattering angle is smaller. We can expect that the series given by Eq. (2-126) will be Gaussian in nature for some range of angles and a given target thickness and will be distinctly non-Gaussian for large angles, becoming the function for the angular dependence of the probability of an individual collision (Rutherford cross section).

Summation of the series proceeds on the assumption that the scattering angles are small and the number of collisions p is large.

We will use the approximation $P_l(\cos\theta) \simeq J_0(l\theta)$, valid for small angles θ, and calculate $k_l s (a \equiv \sqrt{\alpha})$:

$$k_l s = N\sigma_0 s[1 - a_l] = N\sigma_0 s\left[1 - a^2(1+a^2)\int_0^{\pi} \frac{J_0(l\theta)\theta\, d\theta}{(\theta^2 + 4a^2)^2}\right] \tag{2-264}$$

Using the integral form of the Bessel function $J_0(l\theta)$,

$$J_0(\lambda, y) = \frac{2}{\pi}\int_0^{\infty} \sin(\lambda y \cos ht)\, dt \tag{2-265}$$

Moliere calculated the integral that appears in Eq. (2-264), extending the limit to ∞:

$$\frac{2}{\pi}\int_0^{\infty}\int_0^{\infty} \frac{y \sin(\lambda y \cos ht)\, dt}{(1+y^2)^2} = \frac{\lambda}{2}\int_0^{\infty} e^{-\lambda \cos ht} \cos ht\, dt = -\frac{\lambda}{2} K_1(\lambda) \tag{2-266}$$

so that

$$k_l s = p[1 + 2laK_1(2la)] \tag{2-267}$$

Replacing summation over l in Eq. (2-260) with integration, we write the unknown distribution as an integral:

$$F(\theta, p) = \frac{1}{2\pi} \int_0^\infty x J_0(x\theta) \exp[-p\{1 + 2axK_1(2ax)\}]\, dx \tag{2-268}$$

The presence of an exponential term having large p means that the main contribution to the integral comes from the region in which the values of the $J_0(x\theta)$ and $K_1(2ax)$ function arguments are small. Using the expansions of these functions for small values of the arguments and integrating we can write the unknown distribution as[52]

$$F(\theta, p) = \frac{1}{2\pi\theta_2^2}\left\{2e^{-\Psi^2} + \frac{F_1(\Psi)}{B} + \frac{F_2(\Psi)}{B^2} + \dots\right\}, \qquad \Psi = \theta/\theta_2$$

$$F_n(\Psi) = \frac{1}{n!}\int_0^\infty x J_0(\Psi x)\left(\frac{x^2}{4}\log\frac{x^2}{4}\right)^n e^{-x^2/4}\, dx \tag{2-269}$$

$$B - \log B = \log p + 1 - 2C, \qquad C = 0.5772$$

$$\theta_2^2 = 4a^2 Bp \simeq 4a^2 p \log p = 8\pi N \frac{Z^2 e^4}{m^2 v^4} s \ln\{603 Z^{2/3} e^2 (Ns)^{1/2}/mv_c\}$$

A remarkable feature of this result is that for small angles $\theta \leq \theta_2$ we have a Gaussian distribution with variance θ_2^2, whereas for large angles $\theta \gg \theta_2$, $F(\theta, p)$ becomes the cross section for one-time Rutherford scattering. In fact, the Gaussian term in the distribution can be neglected for angles $\theta \gg \theta_2$. For the expansions of F_1 and F_2 in powers of $1/\Psi$ we obtain by integrating over the expansion $\exp(-x^2/4)$:

$$F_1(\Psi) \sim 2/\Psi^4 + 8/\Psi^6 + \dots; \qquad F_2(\Psi) \sim (16/\Psi^6)\log(\Gamma\Psi - 3/2) + \dots$$

In this manner, the principal term in the expansion of $F(\theta, p)$ when $\theta \gg \theta_2$ is

$$F(\theta, p) = \frac{1}{2\pi\theta_2^2 B}\frac{2\theta_2^4}{\theta^4} \simeq \frac{4Z^2e^4N}{m^2v^4}\frac{1}{\theta^4}s \tag{2-270}$$

i.e., the Rutherford scattering cross section.

The angular distribution function given by Eq. (2-260) describes how the shape of the beam evolves as it penetrates into the target. The quantities a_l, as can be seen from the values of a_0 and a_1 and the recurrence formula given by Eq. (2-257) for the a_l is a decreasing series and the attenuation constants k_l are an increasing series. When $s = 0$ the angular distribution is a δ-function; when $s = \infty$, $f = 1/4\pi$; i.e., it becomes a uniform distribution over angles. It is well known that a uniform (spherically symmetric) distribution corresponds, e.g., to a solution of the equation for particles diffusing from a point source. In fact, for large values a Boltzmann type equation for the three-dimensional distribution function can be converted to a differential equation for the diffusion particles.

2-25. THE EQUATIONS FOR THE THREE-DIMENSIONAL DISTRIBUTION FUNCTION FOR BEAM PARTICLES IN A TARGET

Our main concern in electron lithography is the three-dimensional distribution of the energy absorbed by the target. This quantity is closely related to the probability density $f(\bar{r}, \bar{v}_0, t)$ of finding a particle at point $\bar{r}$ in the target at time t and moving in a direction $\bar{v}_0(\vartheta, \varphi)$. The function $f(\bar{r}, \bar{v}_0, t)$ is, as we have seen, subordinate to the equation

$$\frac{\partial f}{\partial s} = -\bar{v}_0 \nabla f + N \int \sigma(|\bar{v}_0 - \bar{v}_0'|)[f(\bar{r}, \bar{v}_0', s) - f(\bar{r}, \bar{v}_0, s)]\, d\bar{v}_0 \tag{2-271}$$

The boundary conditions correspond, e.g., to the perpendicular entry of the beam into the target: $f(\bar{r}, \bar{v}_0, 0) = \delta(\bar{r})\delta(\bar{v} - \bar{v}_0)$. As Lewis did,[53] we will seek a general solution of Eq. (2-271) as a series of spherical functions of the angles ϑ and φ that define the electron's direction of motion:

$$f(\bar{r}, \bar{v}_0, s) = \sum_{l,m} f_{lm}(\bar{r} s) Y_{lm}(\bar{v}_0) \tag{2-272}$$

If we integrate Eq. (2-272) over the angles $\bar{v}_0(\vartheta, \varphi)$ we obtain the three-dimensional distribution for the electrons:

$$\int f(\bar{r}, \bar{v}, s)\, d\bar{v}_0 = f_{00}(\bar{r}, s) \tag{2-273}$$

Thus, in order to describe the three-dimensional distribution of electrons in a target we must know a unique coefficient in the expansion of Eq. (2-272). Unfortunately, the $f_{lm}(\bar{r}, s)$ coefficients are interconnected among themselves by a chain of linking equations, which makes it impossible to define $f_{00}(\bar{r}, s)$ explicitly. We substitute the expansion given by Eq. (2-272) into (2-271), multiply by $Y_{lm}^*(\bar{v}_0)$ and integrate over the sphere. The result is

$$\begin{aligned} \frac{\partial f_{lm}}{\partial s} = &-\sum_{\lambda,\mu} \nabla f_{\lambda\mu} Q_{lm}^{\lambda\mu} + N \sum_{\lambda,\mu} f_{\lambda\mu} \iint Y_{lm}^*(\bar{v}_0)\sigma \\ &\times (|\bar{v}_0 - \bar{v}_0'|)[Y_{\lambda\mu}(\bar{v}_0') - Y_{\lambda\mu}(\bar{v}_0)]\, d^3\bar{v}_0\, d^3 v_0' \end{aligned} \tag{2-274}$$

$$Q_{lm}^{\lambda\mu} = \int Y_{lm}^*(\bar{v}_0)\bar{v}_0 Y_{\lambda\mu}(\bar{v}_0)\, d^3\bar{v}_0$$

By converting the angular portion of the collisions integral in the right-hand side of Eq. (2-274) more accurately, as we did in Section 2-23 for the angular distribution, we obtain

$$\frac{\partial f_{lm}}{\partial s} = -k_l f_{lm} - \sum_{\lambda,\mu} \nabla f_{\lambda\mu} Q_{lm}^{\lambda\mu} \tag{2-275}$$

$$k_l = 2\pi N \int_0^{\pi} \sigma(\vartheta)[1 - P_l(\cos\vartheta)] \sin\vartheta\, d\vartheta \tag{2-276}$$

The boundary conditions for f_{lm} are

$$f_{lm}(r,0) = \delta(r)\delta_{m0}Y_{l0}(0) = \left(\frac{2l+1}{4\pi}\right)^{1/2} \delta_{m0}\delta(r) \tag{2-277}$$

Thus, the f_{lm} coefficients prove to be interconnected among themselves through the matrix elements $Q_{lm}^{\lambda\mu}$ of a unit velocity vector $\bar{v}_0$. This relationship states the fact that the change in an electron's direction of motion due to collisions with a target's atoms is a direct cause of beam spreading. The matrix elements $Q_{lm}^{\lambda\mu}$ will be zero if $|\lambda - l| > 1$, $|\mu - m| > 1$.

In general, the system of Eqs. (2-275) cannot be solved. We will discuss an approximate solution that corresponds to the asymptote for small angle scattering. Lewis[(53)] showed the possibility of computing the so-called moments of the particles' coordinates using Eq. (2-275) as a basis. We will introduce a special function for the electron distribution about the z coordinate:

$$g_{lm}(z,s) = \int f_{lm}(r,s)\,dx\,dy \equiv \delta_{m0}g_l(z,s) \tag{2-278}$$

and moments of the z coordinate:

$$H_{ln} = \int_{-\infty}^{+\infty} z^n g_l(z,s)\,dz \tag{2-279}$$

According to [53], the function g_l and moments H_{ln} are subordinate to

$$\begin{gathered}\frac{\partial g_l}{\partial s} + k_1 g_l + \frac{\partial}{\partial z}[\alpha_l g_{l-1} + \alpha_{l+1}g_{l+1}] = 0\\ \frac{\partial H_{ln}}{\partial s} + k_l H_{ln} - n[\alpha_l H_{l-1,n-1} + \alpha_{l+1}H_{l+1,n-1}] = 0\\ \alpha_l = l/(4l^2 - 1)^{1/2}\end{gathered} \tag{2-280}$$

In principle, finding the moments for all degrees is equivalent to finding a solution for the system of Eqs. (2-275). However, the method of moments has not been widely adopted because of the difficulty encountered putting the information obtained to practical use.

2-26. A SMALL-ANGLE APPROXIMATION FOR THE THREE-DIMENSIONAL DISTRIBUTION FUNCTION OF A BEAM

The scattering cross section for fast particles becomes large only for scattering at small angles. We can therefore assume that the scattering angles ζ that appear in the differential scattering cross section and the angles ϑ between the electron's velocity vector **v** and the Oz axis are also small relative to unity, although the angles ϑ are the sum of a large number of angles ζ that have come about because of multiple scattering. With these assumptions we can convert Eq. (2-271).

It is worth our while to introduce a representation of the so-called particle deflection angles ϑ_x and ϑ_y. Let a single vector $\bar{v}_0$ be defined by the angles ϑ and φ: $\bar{v}_0(\sin\vartheta\cos\varphi, \sin\vartheta, \sin\varphi, \cos\vartheta)$. When $\vartheta \ll 1$ we can write

$$\bar{v}_0\left(\vartheta\cos\varphi, \vartheta\sin\varphi, 1-\frac{\vartheta^2}{2}\right) \equiv \bar{v}_0\left(\vartheta_x, \vartheta_y, 1-\frac{\vartheta_x^2+\vartheta_y^2}{2}\right) \tag{2-281}$$

ϑ_x is the angle that the projection of $\bar{v}_0$ onto the plane Oxz makes with the z axis; ϑ_y is defined in the same manner. An element of the solid angle $\sin\vartheta\, d\vartheta\, d\varphi$ can be replaced through $d\vartheta_x\, d\vartheta_y$, because $d\vartheta_x\, d\vartheta_y$ coincides with an element of area $dx\, dy$ on a plane perpendicular to the z axis and tangent to a single sphere.

We now return to the Boltzmann equation, Eq. (2-271), to consider the angular arguments of the projected angles ϑ_x and ϑ_y. We will expand the difference $f(r, \theta_x, \theta_y, s) - f(r, \vartheta_x, \vartheta_y, s)$ in the collisions integral as a series in small scattering angles ζ_x and ζ_y:

$$\begin{aligned} f(r, \theta_x, \theta_y, s) - f(r, \vartheta_x, \vartheta_y, s) &= f(r, \vartheta_x+\zeta_x, \vartheta_y+\zeta_y, s) - f(r, \vartheta_x, \vartheta_y, s) \\ &= \frac{\partial f}{\partial\vartheta_x}\zeta_x + \frac{\partial f}{\partial\vartheta_y}\zeta_y + \frac{1}{2}\left(\frac{\partial^2 f}{\partial\vartheta_x^2}\zeta_x^2 + 2\frac{\partial^2 f}{\partial\vartheta_x\,\partial\vartheta_y}\zeta_x\zeta_y + \frac{\partial^2 f}{\partial\vartheta_y^2}\zeta_y^2\right) + \dots \end{aligned} \tag{2-282}$$

Because of the evenness of the scattering cross section $\sigma(\zeta_x, \zeta_y)$ relative to ζ_x, ζ_y, the linear terms of the expansion become zero during integration and we obtain

$$\frac{\partial f}{\partial s} = -\bar{v}_0 \operatorname{grad} f + \left(\frac{\partial^2 f}{\partial\vartheta_x^2} + \frac{\partial^2 f}{\partial\vartheta_y^2}\right) N \int \frac{1}{2}\zeta_x^2 \sigma(\zeta_x, \zeta_y)\, d\zeta_x\, d\zeta_y \tag{2-283}$$

Here the angles ζ_x and ζ_y are said to be equal. If we recall the definition of the momentum losses cross section, Eq. (2-249), we see that for small scattering angles the coefficient of the last term in the equation obtained can be written as the reciprocal of the length of the momentum losses:

$$N\int \sigma(\zeta_x, \zeta_y)\frac{1}{2}\zeta_x^2\, d\zeta_x\, d\zeta_y = N\int \sigma(\zeta)\frac{1}{4}\zeta^2\, dO = N\int \sigma(\zeta)\frac{1}{2}(1-\cos\zeta)\, dO = 1/\Lambda \tag{2-284}$$

Thus, we come to the differential equation

$$\frac{\partial f}{\partial s} = -\bar{v}_0 \operatorname{grad} f + \frac{1}{\Lambda}\left(\frac{\partial^2 f}{\partial\vartheta_x^2} + \frac{\partial^2 f}{\partial\vartheta^2 y}\right) \tag{2-285}$$

It must be noted that the transition from the integro-differential Boltzmann-type equation to the Fokker–Planck differential equation, Eq. (2- 285), is a widely accepted means for finding an approximate solution for the problem.[36]

The ϑ_x and ϑ_y, and x and y variables appear quite regularly in Eq. (2-285). We will also assume that $s = z$, and therefore we can say that

$$f(r, \vartheta_x, \vartheta_y, s) = f_1(x, \vartheta_x; z) f_2(y, \vartheta_y; z)$$

and in the stationary case we find by eliminating variables that

$$\frac{\partial f_1}{\partial z} + \vartheta_x \frac{\partial f_1}{\partial x} = \frac{1}{\Lambda} \frac{\partial^2 f_1}{\partial \vartheta_x^2} \tag{2-286}$$

and a similar equation for f_2. The initial conditions for Eq. (2-286) will have the form $f_1(x, \vartheta_x, 0, 0) = \delta(x)\, \delta(\vartheta_x)$.

The simple conclusion drawn from Eq. (2-285) is contained in the work of Rossi and Greizen.[(54)] A solution of Eq. (2-286) is, as we can easily see by substitution,

$$f_1(x, \vartheta_x, z) = \frac{3^{1/2}\Lambda}{2\pi z^2} \exp\left[-\frac{\Lambda}{z}\left(\vartheta_x^2 - \frac{3\vartheta_x x}{z} + \frac{3x^2}{z^2}\right)\right] \tag{2-287}$$

Integrating Eq. (2-287) over all x we obtain a function for the angular distribution of electrons at a depth z in the target, which, as we would expect, is a Gaussian function:

$$F(\vartheta_x; z) = \frac{1}{2\sqrt{\pi}} \sqrt{\frac{\Lambda}{2}} \exp\left[-\frac{\Lambda}{4z}\vartheta_x^2\right], \qquad \bar{\vartheta}_x^2 = \frac{2z}{\Lambda} \tag{2-288}$$

Similarly, integration over all angles gives an electron distribution function relative to the three-dimensional deflection to the side from the point at which the particle enters the target:

$$H(x; z) = \frac{\sqrt{3}}{2\sqrt{\pi}} \sqrt{\frac{\Lambda}{z^3}} \exp\left[-\frac{3\Lambda}{4z^3} x^2\right], \qquad \overline{x^2(z)} = \frac{2z^3}{3\Lambda} \tag{2-289}$$

It remains to show that according to Eq. (2-287) electrons having a deflection x are not Gaussian distributed, but inversely and that electrons moving in the ϑ_x direction are not Gaussian distributed relative to the displacements x.

The question arises as to what is the range of applicability for the distribution given by Eq. (2-287) and its derivative distributions given by Eq. (2-288) and Eq. (2-289). First, integrating the distribution $f_1 f_2$ over the plane Oxy yields a value of unity; i.e., there is no reflection. Thus, in principle, it cannot describe the reflection of beam electrons from a target because it is based on the assumption that every electron is scattered at only a small angle. Nevertheless, scattering does occur at large angles $\vartheta \leq \pi/2$ and at angles $\vartheta > \pi/2$, which results in electrons being reflected from the target; these cross sections are much smaller than is the small-angle cross section. If we examine the beam at some point in the target then a Gaussian distribution over the angles remains valid for angles $\theta \leq \theta_2$, as was shown in our analysis of Moliere's distribution, Eq. (2-269). At large angles $\theta \gg \theta_2$ the distribution becomes the Rutherford cross section for an elastic collision. In a similar fashion we can assume that the Gaussian distribution, Eq. (2-289), of electrons relative to the linear deflection r from the beam axis will be fairly accurate for $r^2 \leq 4z^3/3\Lambda$. For large x and y the probability of finding beam electrons will be defined by large-angle scattering.

Conversely, the question arises as to what is the range of applicability for Eqs. (2-288) and (2-289) when the beam penetration depth z increases. Until now we have not considered inelastic scattering and particle deceleration in our examination. Clearly, deceleration (energy loss) determines the maximal and average depths to which particles can penetrate into a target and makes them finite. Because the mean free path is finite the three-dimensional distribution functions will be nonzero in a finite space in the target. The linear dimensions of this space are the same as the magnitudes of the mean free paths.

2-27. THE DIFFUSIVE STAGE OF BEAM PROPAGATION THROUGH A TARGET

Using the qualitative analysis discussed above we will assume that starting from a depth s to which a beam has penetrated into the target the unknown three-dimensional particles distribution function can be written as[20]

$$f(r, \theta; s) = \rho(r, s) + J(r, s) \cos\theta \tag{2-290}$$

where the ρ and J are given by

$$\rho(r, s) = \frac{1}{2} \int_0^{\pi} f(r, \theta, s) \sin\theta \, d\theta \tag{2-291}$$

$$\frac{4\pi}{3} J(r, s) = 2\pi \int_0^{\pi} \cos\theta \, f(r, \theta, s) \sin\theta \, d\theta \tag{2-292}$$

which respectively have the meaning of the particle density and current in the original direction of the propagating beam.

Substituting Eq. (2-290) and (2-285) and integrating over all angles we find that

$$\frac{\partial \rho}{\partial s} = -\frac{1}{3} \frac{\partial J}{\partial z} \tag{2-293}$$

If we multiply Eq. (2-285) by $\cos\theta$, substitute Eq. (2-290), and integrate over the angles, we find

$$\frac{\partial J}{\partial s} = -\frac{\partial \rho}{\partial z} - \frac{2}{\Lambda} J \tag{2-294}$$

We can say that $2J/\Lambda \gg \partial J/\partial s$.[20] Combining Eqs. (2-293) and (2-294), we obtain an equation for one-dimensional diffusion into the target:

$$\frac{\partial \rho}{\partial s} = \frac{\partial \rho}{v \, dt} = \frac{\Lambda}{6} \frac{\partial^2 \rho}{\partial z^2} \tag{2-295}$$

The quantity $\Lambda v/6$ emerges as the diffusion constant. The length Λ of the momentum free path is a function of the particle's energy which changes because the particle decelerates as it penetrates more deeply into the target. Therefore, the quantity Λv is an implicit function of t or s.

Unfortunately, Eq. (2-295) gives us very little useful information about how the particles are distributed in space. It does tell us that, starting from some depth to which the particles have penetrated into the target, the translational motion of the particles according to the law of uniformly decelerated motion $s = \int_0^t v(t') \, dt'$ ceases and the motion becomes diffusive. A group of particles injected into the target at time $t = 0$

and having δ-like initial conditions and initially moving according to $s = \int_0^t v\,dt'$ at the end of the motion moves ahead at only the diffusion rate. The amount of displacement Δz changes with time as $\Delta z = \sqrt{D\Delta t}$, if we introduce the instantaneous value of the diffusion constant $D = \Lambda v/6$. The transition from uniformly decelerated motion to diffusive motion takes place gradually and we do not have an intermediate solution. Bethe *et al.*[(55)] assumed that for values $\overline{\cos}\,\theta > 0.5$ the motion can be considered translational, and diffusive when $\overline{\cos}\,\theta < 0.5$. Bethe's condition allows us to introduce the so-called diffusion depth at which the diffusive stage of beam propagation begins. Using the angular distribution function, Eq. (2-247), we find

$$\overline{\cos}\,\theta = 2\pi \int_0^\pi F(\theta; s)\cos\theta \sin\theta\, d\theta = e^{-K_1 s} \tag{2-296}$$

Considering that $K_1 = \Lambda^{-1}$ is a function of the path traversed, s, we can write

$$\overline{\cos}\,\theta = \exp\left(-\int_0^s ds'/\Lambda(s')\right) \tag{2-297}$$

The transition to diffusive motion takes place, according to Bethe,[(55)] when $s = s_1$ is such that

$$\overline{\cos}\,\theta = \exp\left(-\int_0^{s_1} ds'/\Lambda(s')\right) = \frac{1}{2}, \qquad \int_0^{s_1} ds'/\Lambda(s') = \ln 2 \tag{2-298}$$

Stating these quantities in terms of particle energy, we find, with the help of the continuous losses function, that

$$\int_{E_1}^{E_0} \frac{dE}{\Lambda(E)|dE/ds|} = \ln 2 \tag{2-299}$$

In travelling the path s_1 or when the energy falls to E_1 the particle moves ahead diffusively. We define the diffusion depth as

$$R_D = \int_0^{s_1} \cos\theta(s)\, ds \tag{2-300}$$

and it will be less than s_1. If we introduce the variables $\tau(s)$ and $\tau(E)$ via

$$\tau(s) = 1/6 \int_0^s \Lambda(s)\, ds \tag{2-301a}$$

$$\tau(E) = \int_E^{E_0} \frac{\Lambda(E)\, dE}{|dE/ds|} \tag{2-301b}$$

then the diffusion equation, Eq. (2-295), can be converted to

$$\partial\rho/\partial\tau = \partial^2\rho/\partial z^2 \tag{2-302}$$

At this point we will conclude our discussion of the diffusion stage of beam motion in a target.

2-28. THE RANGE OF AN ELECTRON IN A TARGET

The range of an electron in a given material is determined by the electron's initial energy E_0 and the amount of energy it loses per unit length of the path $s = |dE/ds|$ (deceleration):

$$R = \int ds = \int_{E_0}^{0} \frac{dE}{|dE/ds|} = \frac{1}{2\pi n_e e^4} \int_0^{E_0} \frac{E\,dE}{\ln[1.166E/I]} \tag{2-303}$$

The logarithmic term inside the integral is a slowly varying function of energy. We take this term outside the integral, replacing E with some value $\tilde{E}$ for the interval 0–E_0:

$$R = (2\pi n_e e^4 \ln(1.166\tilde{E}/I))^{-1} \int_0^{E_0} E\,dE \tag{2-304}$$

In this approximation we obtain the dependence of the range as the square of the initial energy E_0

$$R = AE_0^2, \qquad A = 1/4\pi Z N_a e^4 \ln[1.166\tilde{E}/I], \qquad R \sim E_0^2/Z \tag{2-305}$$

If we integrate Eq. (2-305) up to the energy E at a point on the path s and rewrite the result as a function of the electron's velocity we obtain the so-called Thomson–Whiddington law for the velocity of an electron as a function of the path travelled in a target[(55)]:

$$s = A[E_0^2 - E^2(s)] \tag{2-306}$$

$$v^4 = v_0^4 \left(1 - \frac{s}{AE_0^2}\right) = v^4 \left(1 - \frac{s}{R}\right) \tag{2-307}$$

If we use

$$n_e = ZN_a = Z\rho\frac{N_A}{A}$$

$$n_e = \sum_j Z_j N_{aj} = N_A \sum \frac{Z_j}{A_j}\rho_j \equiv N_A\rho\left(\frac{\tilde{Z}}{A}\right) \tag{2-308}$$

$$\left(\frac{\tilde{Z}}{A}\right) = \sum \frac{Z_j}{A_j}\frac{\rho_j}{\rho}$$

for one-component and multicomponent targets and say that the ratio Z_j/A_j remains approximately constant, varying from 0.5–0.4 for all the elements in the periodic table from helium (Z = 2) to gold Z = 79, the range R can be written as a function of the electron's initial energy E_0 and the density of the target, ρ:

$$R = [4\pi e^4 \ln(1.166E_0/I)]^{-1}\frac{E_0^2 A}{\rho Z N_A} = \frac{v_0^4}{c_T\rho} \tag{2-309}$$

An empirical value of the constant c_T is, according to [56], $c_T = 5.05\times10^{42}$ cm^6/g·sec^4.

Range can be calculated more precisely on the basis of the logarithmic term in the integral of Eq. (2-302). Because the deceleration, Eq. (2-160), is only valid for electron energies $E \gg I$, we can replace the lower limit of integration with $E_1 \approx I$; introducing the variable $x = 2\ln(1.166E/I)$, we obtain[57]

$$\begin{aligned} R &= \frac{1}{2\pi n_e e^4} \int_{E_1}^{E_0} \frac{E\,dE}{\ln(1.166E/I)} \\ &= \frac{I^2}{1.166^2 \cdot 2\pi n_e e^4} \int_{x_1}^{x_2} \frac{e^x}{\ln x}\,dx \\ &= \frac{I^2}{1.166^2 \cdot 2\pi n_e e^4} [Ei(x_2) - Ei(x_1)] \end{aligned} \qquad (2\text{-}310)$$

We can choose E_1 so that $Ei(x_1) = 0$; then,

$$R = \frac{1}{1.166^2 \rho} \frac{(A/Z)I^2}{2\pi e^4 N_A} Ei(2\ln(1.166E_0/I)) \qquad (2\text{-}311)$$

because the integral exponential function $Ei(x)$ has been tabulated,[58] we can use Eq. (2-311) to calculate range R for any material described by the parameters ρ, I, A, and Z. We can assume that $I = 11.5Z$ eV.[57]

2-29. THE TRANSMISSION COEFFICIENT FOR A BEAM OF ELECTRONS THROUGH A THIN FILM

The ranges of the individual electrons in a beam are determined by the sequence of electron collisions with a target's atoms and are therefore random (statistical) quantities. The values that we have calculated for R can be thought of as the result of averaging. A rigid theory of scattering effects must, as, we have seen, be built upon a kinetic equation. We must solve this equation for the probability distribution of the coordinates of the electron being scattered and use this distribution and averaging to find the value of any quantity we are interested in that is also a function of the coordinates. However, the mathematical difficulties encountered in this method are significant, and simpler models are often used. One such model is the model for one-time scattering. We first calculate the transmission coefficient for electrons through thin layers of material and within the confines of one-time scattering theory. Beam attenuation as the electrons penetrate into the target is caused by an electron being scattered at a large angle θ from its original direction of motion, its reflection, or its cessation of motion. We will adopt the convention that when calculating attenuation, we can confine ourselves to considering only one-time scattering at large angles θ. Large-angle deflections due to multiple scattering are not considered because our computations of transmission will be increased and the reflection coefficient decreased. As we will see, the transmission and reflection coefficients in the model for one-time scattering will be functions of only one parameter and we can incorporate a correction into this value, choosing it so that the equations obtained will properly describe the experimental results quantitatively.

We will calculate the transmission coefficient for a beam through a layer of material as in [56,59]. The number of electrons scattered into the solid angle $2\pi \sin\theta \, d\theta$ in a layer of the material

$$dn = n(z)\frac{Z^2 e^4}{4m^2 v^4}\frac{2\pi \sin\theta \, d\theta}{\sin^4 \theta/2} N_a \, dz \tag{2-312}$$

where $n(z)$ is the number of electrons that reach the layer z, $z+dz$, and N_a is the number of atoms per unit volume of material. Notice that in this model small-angle electron scattering is not considered: the electrons scattered at small angles are considered, as usual, to be moving in the original direction of motion along the normal to the target's surface. Therefore, the path s in the Thomson–Whiddington equation can be set equal to the depth z:

$$v^4 = v_0^4(1 - z/R) \tag{2-313}$$

If we then write $N_a = N_A\rho/A$ and introduce the dimensionless coordinate $\xi = z/R$, instead of Eq. (2-312) we obtain

$$dn = -\frac{1}{2}\beta\frac{n(\xi)\, d\xi}{(1-\xi)}\frac{\sin\theta \, d\theta}{\sin^4 \theta/2}, \qquad \beta = \frac{\pi Z^2 e^4 N_A}{m^2 c_T A} \tag{2-314}$$

The minus sign is introduced here to denote the reduction in the electron beam's intensity. Now we must find the range of scattering angles for which the scattered electrons cannot reach a given depth ξ (Fig. 2-8). Let scattering have occurred to a depth of $\xi < \xi_0$. If $\theta > \pi/2$, the electron is reflected backward and does not reach the depth ξ_0. However, when the scattering angle is $\theta > \theta_m$, $\cos\theta_m = (\xi_0 - \xi)/(1 - \xi)$, the electron will also not reach the layer ξ_0, because of the limited range in the target. Integrating Eq. (2-314) over the angle θ from $\theta_m = \arccos[(\xi_0 - \xi)/(1 - \xi)]$ to π and then over ξ from 0 to ξ_0 gives us the transmission through a target of thickness ξ_0:

$$\tau(\xi_0, \beta) = \frac{n(\xi_0)}{n_0} = (1 - \xi_0)^{-\beta} \exp\left(-\frac{2\beta\xi_0}{1 - \xi_0}\right) \tag{2-315}$$

The transmission coefficient for a thin target can also be written in terms of the electron's initial energy E_0, ξ_0, and β:

$$\tau = [1 - (E_{\xi_0}/E_0)^2]^{-\beta} \exp\left[-\frac{2\beta(E_{\xi_0}/E_0)^2}{1 - (E_{\xi_0}/E_0)^2}\right] \tag{2-316}$$

where E_{ξ_0} is the energy of an electron whose range is equal to the thickness of the film, ξ_0. It is interesting to compare the result obtained when beam attenuation is due only to reflection; i.e., scattering at angles $\theta \geq \pi/2$. Now by integrating Eq. (2-314) over θ from $\pi/2$ to π yields

$$\int_{\pi/2}^{\pi} \frac{\sin\theta \, d\theta}{\sin^4 \theta/2} = 2, \qquad dn = -\beta\frac{n(\xi)}{(1-\xi)} d\xi \tag{2-317}$$

Separating variables and integrating Eq. (2-317) over the layer thickness ξ_0, we find the beam attenuation caused by scattering at angles $\theta \geq \pi/2$:

$$\tau^*(\xi_0, \beta) = (1 - \xi_0)^{\beta} \tag{2-318}$$

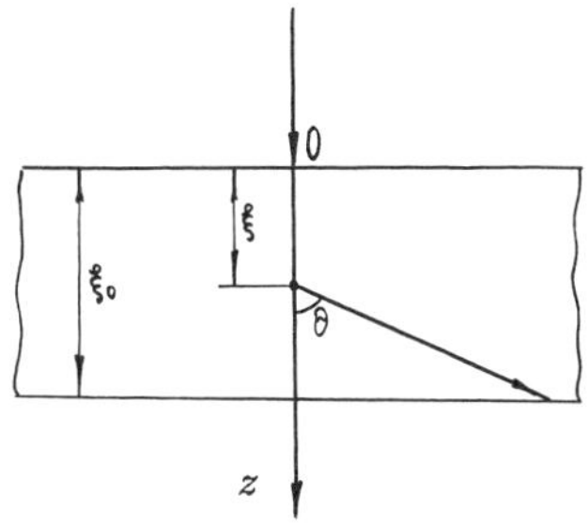

FIGURE 2-8. For calculating the beam transmission coefficient through a thin layer. If the dimensionless path is $\xi + (\xi_0 - \xi)/\cos\theta \geq 1$, an electron will stop in the layer.

The different exponent on the $(1 - \xi_0)$ term in Eqs. (2-315) and (2-318) is confusing. However, we can easily satisfy ourselves of the closeness of these equations by taking their asymptotic form for a thin target, $\xi_0 \ll 1$:

$$\begin{aligned} \tau(\xi_0 \ll 1) &\approx (1 + \beta\xi_0)\exp(-2\beta\xi_0) \simeq \exp(-\beta\xi_0) \\ \tau^*(\xi_0 \ll 1) &\approx (1 - \beta\xi_0) \simeq \exp(-\beta\xi_0) \end{aligned} \tag{2-319}$$

In this form the result obtained is close to Leonard's empirical formula[57] for the coefficient of beam transmission:

$$\tau(z_0) = \exp(-\sigma\rho z_0) \tag{2-320}$$

where ρ is the target material density, $\sigma = 1.4 \times 10^{10}\, E_0^{-1.5}$, E_0 is in eV, and z_0 is the target thickness.

The experimental dependencies of the electron-beam transmission coefficient through thin films of metallic aluminum are shown in Fig. 2-9.[60] The values of this coefficient, obtained through Monte Carlo calculations, describe the experimental curve very well. Values of the transmission coefficient through a 0.4-μm-thick layer of PMMA resist were obtained by Hawryuluk *et al.* by the Monte Carlo method are shown in Table 2-5. Leonard's formula gives much lower values of the transmission coefficient than those obtained by the Monte Carlo method. Dispersion is particularly high when the beam energies are low. Apparently, Leonard's empirical formula can be used to estimate beam transmission through thin layers when beam energies are high.

2-30. THE COEFFICIENT OF ELECTRON REFLECTION FROM A TARGET

The reflection of electron beams, which plays a major role in lithography, has been extensively studied, theoretically and experimentally. Electron beams used in lithography have an energy of 10–30 keV and will pass almost completely unattenuated through a resist layer 0.3–1 μm thick and into the substrate. Electrons reflected in the substrate return to the resist layer, reexposing the resist at rather large distances from the beam axis. Obviously, this process plays a key role in reducing the resolving power of the electron lithography process and it is for just this reason that electron reflection deserves a thorough investigation.

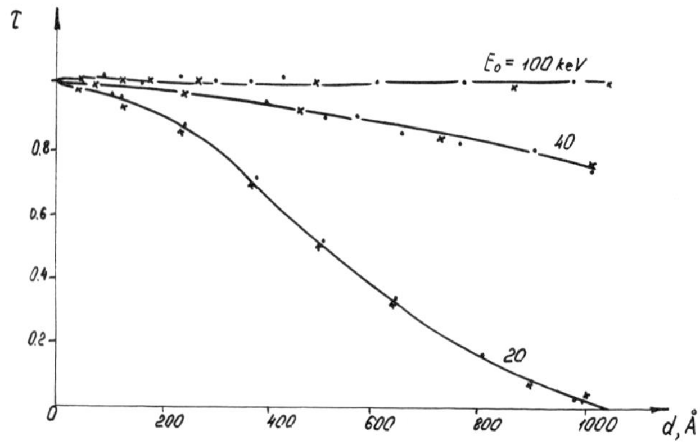

FIGURE 2-9. Coefficient of electron transmission through an Al target as a function of target thickness d. + are experimental values, • are calculated values.

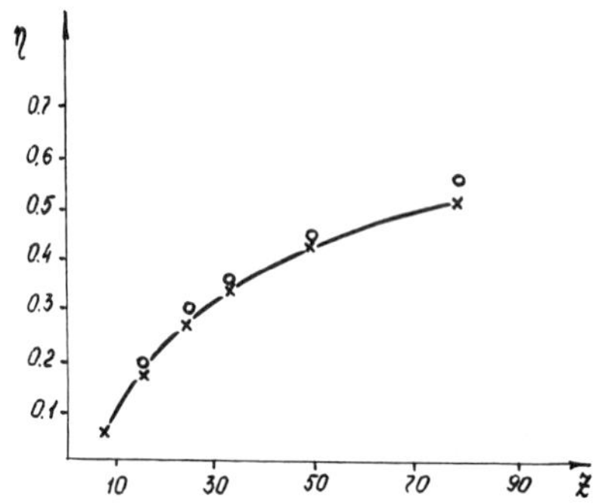

FIGURE 2-10. The electron reflection coefficient as a function of the target material's atomic number.

TABLE 2-5. Values of the Transmission Coefficient Calculated by the Monte Carlo Method for a Beam Passing through a 0.4-μm-Thick Layer of PMMA Resist

Electron energy, E_0, keV	5	10	20
Transmission coefficient, % τ	43.9	95.9	99.5

The fundamental dependencies for the electron reflection coefficient are discussed in [60]. The electron reflection coefficient increases as the atomic number Z of the target material increases (Fig. 2-10). This dependence is not hard to understand when we consider the fact that electrons are reflected because of elastic back scattering and the elastic scattering cross section is proportional to Z^2. At the same time inelastic scattering, whose cross section is proportional to Z, affects the reflection coefficient, because the total path for beam electron motion in a target is limited by its range $R \sim 1/Z$. Consequently, those electrons that have been reflected at too great a depth in the target cannot return to the surface and exit. The probabilities of electrons having 30 keV energy escaping an aluminum target as a function of the depth of electron penetration were found by the Monte Carlo method for a computer experiment and are presented in [60]. The electrons are reflected at a depth that is roughly one-third their maximal range.

A weak dependence of the reflection coefficient on the energy of the primary beam

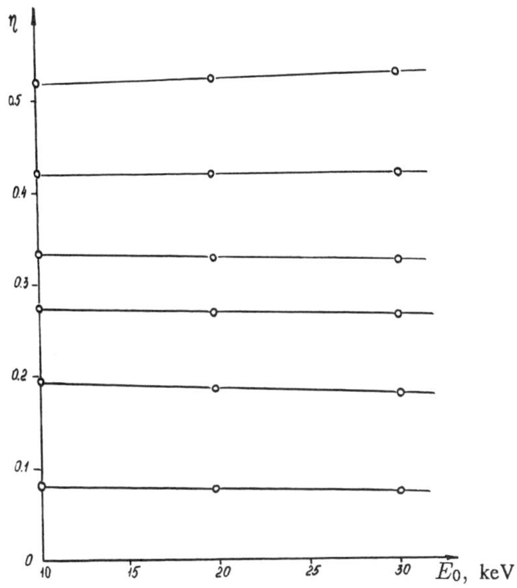

FIGURE 2-11. The electron reflection coefficient (backscattering) as a function of energy.

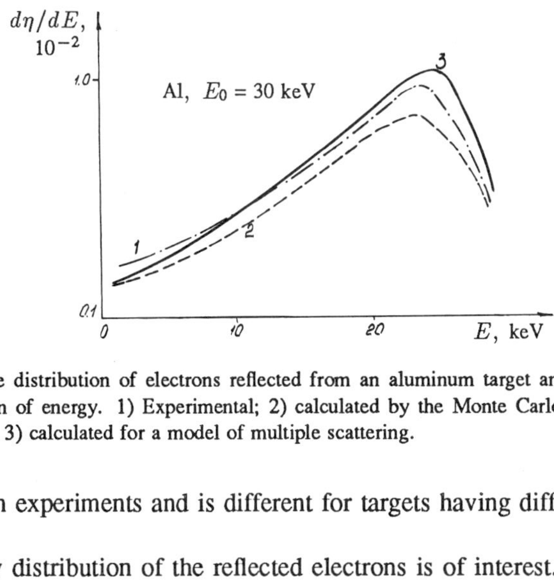

FIGURE 2-12. The distribution of electrons reflected from an aluminum target and having E_0 = 30 keV energy as a function of energy. 1) Experimental; 2) calculated by the Monte Carlo method for a model of one-time scattering; 3) calculated for a model of multiple scattering.

has been seen in experiments and is different for targets having different Z (see Fig. 2-11[57]).

The energy distribution of the reflected electrons is of interest. Figure 2-12 shows the experimental and calculated dependencies $d\eta/dE$ for an aluminum target at a beam energy of 30 keV. The calculated values of $d\eta/dE$ are found by the Monte Carlo method in models for one-time and multiple scattering. As can be seen, the energy spectrum of the reflected electrons is fairly wide. This correlates with the fact that electrons are reflected at various depths in the target, which extend over a wide range.

Along with the primary beam electrons, secondary electrons also escape from the target. In those cases where the recording systems add the current due to reflected primary and secondary electrons, difficulties arise in comparing the experimental and calculated values of the reflection coefficient. However, the energy spectrum of the secondary electrons lies within a rather narrow range, from zero to a few tens of eV. In

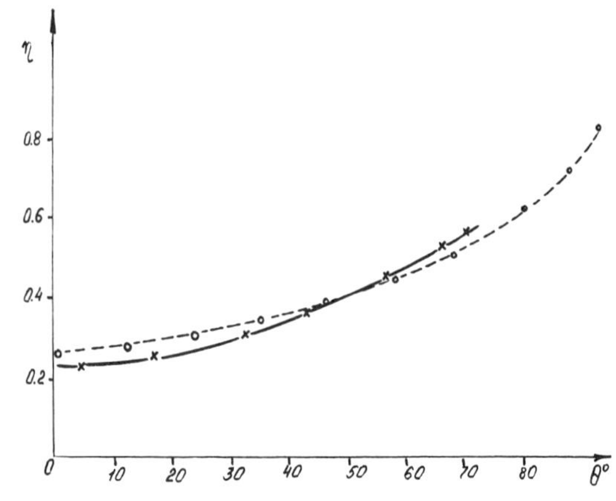

FIGURE 2-13. The reflection coefficient of an electron beam as a function of the angle of incidence θ ($\theta = 0$ corresponds to incidence normal to the surface; $E_0 = 30$ keV; the target is iron containing 3.22% Si). + are experimental values for Fe–3.22Si; o are values calculated by the Monte Carlo method.

this range the intensity of the reflected electrons from the primary beam is negligible. Therefore, in an experiment we can distinguish the secondary electrons by applying a decelerating voltage of about 50 eV to the devices that record reflected electrons. The three-dimensional distribution of reflected electrons can be studied by observing the extent to which the resist is exposed by reflected electrons, as well as by the Monte Carlo method.[(60)] These studies show that reflected electrons escape the target at distances ~ 1 μm from the point of entry.

In electron lithography the electron reflection coefficient as a function of the beam's angle of incidence on the target surface is also significant. Until now we have only examined beams that are incident from the normal and in electron-lithography machines the electron beams are incident from the normal to the surface of the wafer being exposed. However, these wafers do have relief caused by previous processing (e.g., etching through the resist mask) so that during exposure in a raster scan the beam's angle of incidence on the film's surface changes with the position of the beam. It has been suggested in a number of works[(13,16,61)] that special notches be created on a wafer's surface to serve as alignment marks so that, by detecting the change in reflected electrons current caused by a change in the beam's angle of incidence in the area of the alignment marks, these marks will be detected and the pattern being exposed will be aligned with them.

When the beam is incident from the normal, the angular distribution of reflected electrons is nearly conical[(60)]: $f(\theta) \sim \cos\theta$, where θ of the reflected electron's motion measured from the normal. Monte Carlo calculations lead to the same result.[(60)] When the beam is incident from an angle, the reflection coefficient increases (see Fig. 2-13[(60)]).

These are the fundamental experimental dependencies for fast electrons reflecting from a substrate. In principle, they can be explained on the basis of the kinetic equation, Eq. (2-271), that takes elastic and inelastic scattering in target atoms into account, and by finding values of the distribution function $f(r, t)$ when $z < 0$, i.e., the region outside the target. Dashen has discussed the problems of electron reflection using a kinetic equation similar to Eq. (2-271).[(7)] The various model theories of reflection are distinguished by the multiplicity of the elastic scattering processes that they consider. For simplicity and clarity, one-time scattering theory emerges as a first approach. Here the reflection coefficient is thought of as the probability of an electron being backscattered, i.e., at

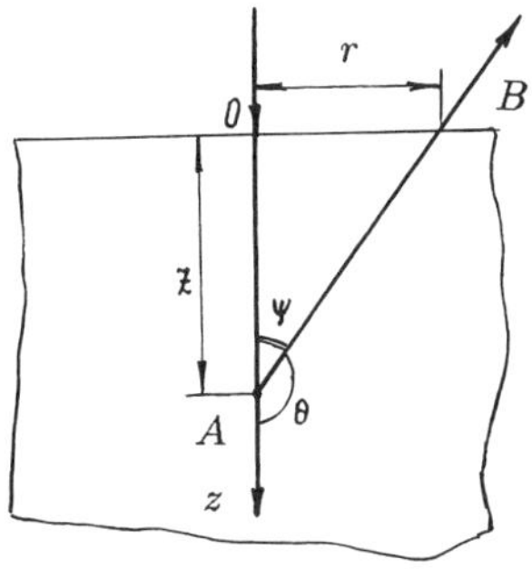

FIGURE 2-14. Trajectory of an electron reflected due to scattering. Elastic scattering occurs at point A at an angle θ; the electron leaves the target at point B.

an angle $\theta \geq \pi/2$ on the condition that the electron can escape the target. It is clear that one-time scattering theory yields low values for the reflection coefficient because the possibility of electrons being reflected at angles less than $\pi/2$ is not considered. For this reason it is conventional to make a correction to the formulas used in one-time scattering theory.(56) By studying the solution of the kinetic equation Dashen came to the conclusion(56) that the contributions from small-angle scattering and one-time scattering $\theta \geq \pi/2$ are roughly equal. Thus, electron reflection can be adequately described only within the framework of multiple scattering theory. Nevertheless, we will confine ourselves to discussing the theory of reflection for one-time scattering because of its simplicity, assuming that this theory correctly describes the basic features of the event.

We will assume that an electron has entered the target from the normal to its surface at the coordinates origin, penetrated a distance z into the target, undergone elastic scattering at an angle $\theta > \pi/2$, and been redirected toward the target's surface (see Fig. 2-14). If the path $s = z + \sqrt{z^2 + r^2}$ for an electron to the surface is less than or equal to the range R in the given material, the electron can escape the target (be reflected). We will calculate the total probability of these events, which will be the reflection coefficient.(56)

From Eqs. (2-314)–(2-318) we can write the number of electrons reflected into a solid angle $2\pi \sin\psi\, d\psi$ by a target layer $d\xi = dz/R$ at a depth $\xi = z/R$:

$$dn = \frac{1}{2}\beta n_0(1-\xi)^{\beta-1}\frac{\sin\psi\, d\psi}{\cos^4\psi/2}\, d\xi \tag{2-321}$$

(all distances are measured in units of the range R). The condition for an electron to escape the target is

$$\xi + \sqrt{\xi^2 + k^2} \leq 1, \qquad \xi = z/R, \qquad k = r/R \tag{2-322}$$

From here it follows that at a specific elastic scattering depth ξ, the electron will escape the target if k lies in the interval 0–$\sqrt{1-2\xi_0}$. Returning to the variables k and ξ in Eq. (2-321), we integrate over k from zero to $\sqrt{1-2\xi}$ and then over ξ from zero to $\xi_{max} = 1/2$. Because

$$\begin{gathered} \tan\psi = k/\xi, \qquad d\psi = \cos^2\psi\, dk/\xi \\ \cos\psi = (1+\tan^2\psi)^{-1/2}, \qquad \sin\psi = \tan\psi(1+\tan^2\psi)^{-1/2} \\ \cos^2\psi/2 = 0.5[1+(1+\tan^2\psi)^{1/2}](1+\tan^2\psi)^{-1/2} \end{gathered} \tag{2-323}$$

then

$$dn(\xi, k) = 2\beta n_0 (1-\xi)^{\beta-1} \left(1+\frac{k^2}{\xi^2}\right)^{-1/2} \left(1+\left(1+\frac{k^2}{\xi^2}\right)^{1/2}\right)^{-2} \frac{k\,dk}{\xi^2} d\xi \quad \text{(2-324)}$$

We calculate the integral over k by making the substitution $1 + k^2/\xi^2 = t^2$.

$$\int_0^{\sqrt{1-2\xi}} (1+k^2/\xi^2)^{-1/2}(1+(1+k^2/\xi^2)^{1/2})^{-2}\frac{k\,dk}{\xi^2} = \int_1^{(1-\xi)/\xi} \frac{dt}{(1+t)^2} = \left(\frac{1}{2}-\xi\right) \quad \text{(2-325)}$$

It is now easy to integrate over the thickness of the specimen from 0 to $\xi = 1/2$:

$$\eta = \int \frac{dn}{n_0} = 2\beta \int_0^{1/2} (1-\xi)^{\beta-1}\left(\frac{1}{2}-\xi\right) d\xi = \frac{\beta - 1 + (0.5)^\beta}{1+\beta} \quad \text{(2-326)}$$

A natural parameter that defines the reflection coefficient η turns out to be the β parameter that describes the scattering processes:

$$\beta = \frac{\pi e^4 N_A}{m^2 c_T}\left(\frac{Z}{A}\right) Z = \alpha Z \quad \text{(2-327)}$$

Ordinarily, we assume that $\alpha = 0.045$.[56]

N. G. Nakhodkin *et al.*[59] obtained fairly general results from electron scattering theory within the framework of the model for one-time scattering. We will present some of these results, which can then be compared with experimental data.

When the beam is incident at an angle θ, measured from the normal to the target's surface, the reflection coefficient is[59]

$$\eta(\beta, p = \cos\theta) = (\beta + p^2)^{-1}(\beta + 2p^2)^{-1} \times \left[\beta^2 + \beta p^2 - 2p^6 + 2p^3(\beta + p^2 + p^3)\left(\frac{p}{1+p}\right)^{\beta/p^2+1}\right] \quad \text{(2-328)}$$

When $\beta > 1$ (a material having a large Z), we obtain the approximate formula, Eq. (2-158)

$$\eta(\beta, p) = \left(1 + \frac{2}{\beta}\cos^2\theta\right)^{-1} \quad \text{(2-329)}$$

When the beam is incident from the normal, $p = 1$ and Eq. (2-328) becomes Eq. (2-326).

In experiments on free films (not attached to a substrate) of thickness ξ we must use

$$\eta(\xi, \beta, p = \cos\theta) = \frac{(\beta + p^2)\beta - 2p^6}{(\beta + 2p^2)(\beta + p^2)} - \beta(1 - \xi/p)^{\beta/p^2} \times \left[\frac{1-p^2}{\beta + 2p^2}(1-\xi/p)^2 + \frac{2p^2}{\beta + p^2}(1 - \xi/p) - p^2/\beta\right] \quad \text{(2-330)}$$

to find the reflection coefficient,[59] where $\xi = z/R$ is the reduced thickness of the film and θ is the angle of beam incidence measured from the normal.

A reflection coefficient was also derived in [59] for a two-layer target:

$$\eta_{1+2} = \eta(\xi_1 \beta_1, p) + (1 - \xi_1/p)^{(\beta_1 - \beta_2)/p^2} \times \left[\eta\left(\frac{p}{1+p}, \beta_2, p\right) - \eta(\xi_1, \beta_2, p)\right] \tag{2-331}$$

where β_1 and β_2 describe the thin upper layer and the substrate, $\xi_1 = d/R_1$ is the normalized thickness of the thin layer, the functions $\eta(\xi, \beta, p)$ are found from Eq. (2-330), and $\eta(p/(1+p), \beta_2, p)$ is the substrate's reflection coefficient. When $\xi_1 \ll 1$, then[59]

$$\eta_{1+2} = \eta\left(\frac{p}{1+p}, \beta_2, p\right) - (\beta_1 - \beta_2)(\eta_2 - 1)\xi_1/p^3 \tag{2-332}$$

We can also find the energy distribution of the reflected electrons in the model for one-time scattering.[59] If an electron (when the beam is incident from the normal) escapes the target because of large angle scattering at a depth z, the path s that it travels determines the energy with which it escapes the target:

$$E^2(s) = E_0^2(1 - s/R) = E_0^2\left(1 - \frac{z + (z^2 + r^2)^{1/2}}{R}\right)$$
$$\varepsilon = \frac{E^2(s)}{E_0^2} = 1 - \xi - \frac{\xi}{\cos\psi} \tag{2-333}$$

Converting to the variables ε and ξ, with Eq. (2-321),

$$\cos\psi = \xi/(1 - \xi - \varepsilon), \qquad \sin\psi\, d\psi = -\xi\, d\varepsilon/(1 - \xi - \varepsilon)^2$$

we find

$$dn = -2\beta n_0 \frac{d\varepsilon}{(1-\varepsilon)^2}(1 - \xi)^{\beta - 1}\xi\, d\xi \tag{2-334}$$

We establish the energy $\varepsilon = E^2/E_0^2$ of the reflected electrons. Electrons reflected maximally from a depth $\xi = 1/2(1 - \varepsilon)$ when the beam is incident from the normal and reflected along the normal will have this much energy. This sets the limit of integration over ξ needed to find the energy distribution of the reflected electrons for a specific energy:

$$\frac{d\eta}{d\varepsilon} = \frac{2}{(1+\beta)(1-\varepsilon)^2}\left\{1 - \left[1 + \frac{\beta}{2}(1 - \varepsilon)\right]\left(\frac{1+\varepsilon}{2}\right)^{\beta}\right\} \tag{2-335}$$

Converting to the energy of the reflected electrons, we obtain

$$\frac{d\eta}{d(E/E_0)} = \frac{4}{1+\beta}\frac{E/E_0}{1 - E^2/E_0^2}\left\{1 - \left[1 + \frac{\beta}{2}\left(1 - \frac{E^2}{E_0^2}\right)\right]\left(\frac{1 + E^2/E_0^2}{2}\right)^{\beta}\right\} \tag{2-336}$$

The formula obtained fairly accurately describes the experimental curves for the energy spectrum of reflected electrons, excluding the unlimited increase near $E = E_0$.[59]

The average energy of reflected electrons,

$$E = \int_0^{E_0} E\frac{d\eta}{dE}\,dE \Big/ \int_0^{E_0} \frac{d\eta}{dE}\,dE = (0.632 + 1.26 \times 10^{-3}\ Z)E_0 \qquad (2\text{-}337)$$

was also calculated in [59]. Experiments yielded the result[59]

$$E_{\text{exp}} = E_0(0.45 + 2 \times 10^{-3}\ Z) \qquad (2\text{-}338)$$

We will now use the model for one-time scattering at large angles to find the angular dependence of the distribution of reflected electrons. The number of electrons reflected from a layer $\xi, \xi + d\xi$ into a solid angle $dO = 2\pi \sin\psi\, d\psi$ is

$$dn = \frac{1}{2}\beta n_0(1 - \xi)^{\beta - 1}\frac{\sin\psi\, d\psi}{\cos^4\psi/2}\,d\xi \qquad (2\text{-}339)$$

ψ is the angle between the normal to the surface and the direction of the reflected electron's motion. The beam is incident on the target from the normal. We must now specify the solid angle dO and integrate with respect to the depth of the target, allowing for the fact that the total path travelled by the electron cannot exceed the range R:

$$s = z + \frac{z}{\cos\psi} \leq R, \qquad \xi\left(1 + \frac{1}{\cos\psi}\right) \leq 1 \qquad (2\text{-}340)$$

From here it follows that the maximal depth $\varepsilon_{\max}$ for a reflected electron moving in the direction ψ is $\xi_{\max} = \cos\psi/(1 + \cos\psi)$. Integrating over ξ between the limits given, we find the function for the angular distribution of reflected electrons:

$$f(\psi) = \frac{dn}{n_0 2\pi \sin\psi\, d\psi} = \frac{1}{4\pi}\frac{[1 - (1/2\cos^2\psi/2)^{\beta}]}{\cos^4\psi/2} \qquad (2\text{-}341)$$

The maximum for the distribution corresponds to reflection along the normal: $f(0) = (4\pi)^{-1}(1 - 0.5^{\beta})$; when $\psi = \pi/2$, $f(\psi)$ is zero.

Antolak and Williamson performed generalized Monte Carlo calculations for electrons reflected from a homogeneous target[62] and for a substrate–film target.[63] A SANDYL program was used on a CRAY 1 computer. The homogeneous targets used were Al, Cu, Ag, and Au; the energy of the incident electrons was varied from 10 keV to 1 MeV; and the angles of incidence were 0°, 30°, 45°, and 60° from the normal. The values calculated for the reflection coefficient η as a function of the target's atomic number Z were slightly less than the experimental values and those found from Arnal's *et al.* empirical formula[64]: $\eta = 2^{-9/\sqrt{Z}}$. A great deal of attention has been devoted to the energy parameters of reflection: in every case the magnitude of the reflected energy, as well as the energy spectrum of the reflected electrons (for some parameter values) have been reported. The amount of energy reflected increases as the target's atomic number Z and the beam's angle of incidence increases. From the data obtained, we can analyze the coefficient of beam energy reflection as a function of atomic number and

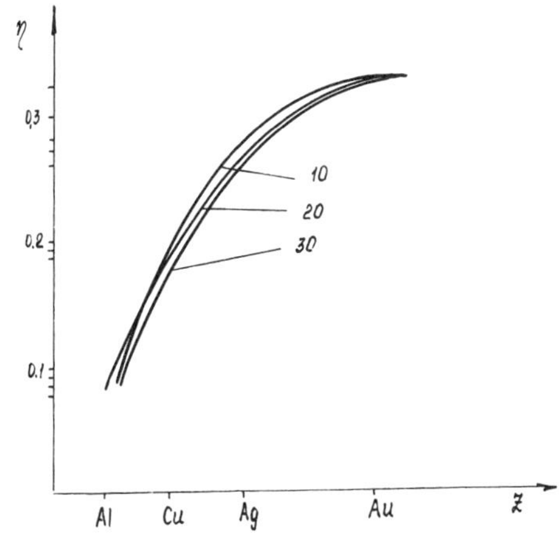

FIGURE 2-15. Beam energy reflection coefficient as a function of the target's atomic number for different values of beam energy.

beam energy, as well as the angle of beam incidence. The dependencies $\eta_E \sim Z$ are shown in Fig. 2-15 for several values of beam energy (which are close to those used in lithography). As can be seen, η_E and η increase as the target's atomic number increases. This $\eta_E \sim Z$ dependence is significant for targets with small Z (Al), and η_E is almost constant for heavy targets (Au). The energy spectra of reflected electrons are narrower for heavy targets; they can be described by the beta distribution

$$\frac{d\eta}{dx} = \eta \frac{\Gamma(\alpha_1 + \alpha_2 + 2)}{\Gamma(\alpha_1 + 1)\Gamma(\alpha_2 + 1)} x^{\alpha_2}(1 - x)^{\alpha_1}, \qquad x = \frac{\Delta E}{E_0} \tag{2-342}$$

Values of the α_1 and α_2 parameters are found by adjusting (by the method of least squares fit) for every target and for every value of energy.[62]

Calculations were done in [63] for η and for the reflected energy from a film–substrate system for Al/Pt, Ag/Pt, Ag/Al, and Au/Al at energies of 20, 40, 60, and 100 keV and the angles of beam incidence mentioned above. The results obtained are not unexpected: the η and η_E have intermediate values somewhere between the values of these quantities for the film and the substrate material; as film thickness increases η and η_E tend toward the η_f and η_{Ef} for the film material, and as film thickness decreases they tend toward the values η_s and η_{Es} for the substrate material.

2-31. COMPUTER EXPERIMENTS IN ELECTRON LITHOGRAPHY BY THE MONTE CARLO METHOD

The Monte Carlo method is a computational procedure for solving a given problem by statistical methods for a large number of machine (numerical) experiments. In the last decade the Monte Carlo method has been put to unusually widespread use, mainly to solve problems associated with actual random processes. One example of these processes is charge transfer (transport) in a scattering medium. As we have seen, finding an analytical solution for problems of this type is mathematically very difficult and can only be done after a great deal of simplification. Because of this an analytical solution has only a qualitative nature, and, what is important, we do not know how much error has been introduced through the simplifications. In every case the Monte Carlo method proves to be an essential investigative tool.

The main shortcoming of the Monte Carlo method is its nonanalytical statistical nature. Statistical reliability in the results obtained can only be achieved by conducting tens of thousands of random tests. The quantities we are interested in are obtained in the Monte Carlo method by averaging the results from random tests. When the initial conditions vary (e.g., electron-beam scattering in a target, the electrons energy and angle of incidence, the composition of the target, etc.), the computations must be iterated, which takes a great deal of machine time. Nevertheless, the Monte Carlo method is widely used because in many cases an analytical solution of the problem is unknown and Monte Carlo is the only method available to the researcher. When an approximate analytical solution is known, the Monte Carlo method makes it possible to verify the accuracy of the analytical method, which is also quite valuable. In the final analysis, being essentially a method for computer experimentation the Monte Carlo method allows us to achieve a major economy by replacing actual experiments with those performed on a computer.

We will describe briefly the essence of the Monte Carlo method.

In the Monte Carlo method we model a real random process using random numbers. Modern computers have built-in random number generators from which random numbers γ uniformly distributed over the interval $(0, 1)$ and having a mean value and variance of $E_\gamma = 1/2$ and $D_\gamma = 1/12$ are taken. Let a physical random variable ξ be defined in the interval (a, b) and the probability density of finding a specific value x of this variable in this interval be $P(x)$. Because the probability density $p(x)$ has been normed to unity, $\int_a^b p(x)\,dx = 1$, the values of the integral are

$$I(\xi) = \int_a^\xi p(x)\,dx \tag{2-343}$$

where ξ lies within the interval (a, b), and coincide with one of the random numbers. Thus, the actual random physical quantity ξ that we are interested in can be associated with the random numbers γ by the relation

$$\int_a^\xi p(x)\,dx = \gamma \tag{2-344}$$

If the integral in Eq. (2-344) is calculated and the equation obtained can be solved in terms of $\xi = g(\gamma)$, a value of the random number ξ that corresponds to the "experiment" is found when the number γ is obtained from the random number generator: $\xi = g(\gamma)$. If the integral in Eq. (2-344) cannot be calculated in a straightforward manner, we must turn to transform methods.

If the random number ξ assumes discrete values $x_1, \ldots, x_n$ with probabilities $p_1, \ldots, p_n$,

$$\xi = \begin{pmatrix} x_1, x_2, \ldots, x_n \\ p_1, p_2, \ldots, p_n \end{pmatrix} \tag{2-345}$$

a vector having components $p_1, p_1 + p_2, \ldots, p_1 + p_2 + \cdots + p_n$ is introduced. We will assume that the random number γ falls within the interval $p_1 + \cdots + p_{i+1}, p_1 + \cdots + p_i$. Then the value of the random number that corresponds to γ is x_i.

Thus, in order to do the computations by means of the Monte Carlo method we must first find, from theory or experiment, a possibly more accurate distribution $p(x)$ and its corresponding function $\xi = g(\gamma)$. Using these in the Monte Carlo method we can mathematically create a specific particle trajectory if we are talking about transport processes. Iterating the procedure gives a large number (about 10^4) of trajectories. Then the quantities we are interested in (the probability density $f(r)$ of finding a particle at a specific point in the target, the density of the energy dissipated, $D(r)$) will be found by methods of mathematical averaging.

2-32. USING THE MONTE CARLO METHOD TO MODEL ELECTRON TRAJECTORIES

We will show how to perform a procedure for Monte Carlo calculations when examining the motion of an electron in a target. We must first find the random variables x that define the process and their distribution $p(x)$, as well as their random occurrences $\xi = g(\gamma)$.

An electron moving through a target changes direction because of elastic collisions with the target's atoms. The differential cross section of elastic collision is found from Rutherford's equation with shielding:

$$\sigma(\theta) = \frac{Z(Z+1)e^4}{4E^2}(1 - \cos\theta + 2\beta)^{-2} \tag{2-346}$$

The total cross section for elastic collision is[20]

$$\sigma_t = \frac{\pi e^4 Z(Z+1)}{4E^2\beta(1+\beta)} \tag{2-347}$$

The total cross section σ_t defines the average length of the free path λ relative to elastic scattering:

$$\frac{1}{\lambda} = N\sigma_t = (N_A/A)\rho\sigma_t \tag{2-348}$$

In a compound material we have

$$\frac{1}{\lambda} = N_A\rho\sum_i C_i\sigma_{ti}/A_i \tag{2-349}$$

where the C_i are the weighted fractions of the respective elements in the target.

An electron also experiences inelastic collisions and loses energy in the course of exciting or ionizing the atoms in the target and changes its direction of motion. In the model for continuous losses the role of inelastic collisions comes down to electron deceleration, as described by Bethe's formula, Eq. (2-160), which can be written

$$\frac{dE}{ds} = -\frac{2\pi e^4\rho N_A}{E}\sum C_i\frac{Z_i}{A_i}\ \mathrm{In}(1.166E/I_i) \tag{2-350}$$

for a compound material. Changes in the direction due to inelastic collisions are taken care of by replacing the Z^2 in Eq. (2-346) with $Z(Z+1)$.

These formulas define the probability density functions $p(x)$ and the occurrences $\xi = g(\gamma)$ that we need for the respective random processes.

The probability density that an elastic collision will not occur on a path ds if the average length of the mean free path is λ can be found from

$$\frac{dp}{p} = -\frac{ds}{\lambda(r)} \tag{2-351}$$

and is equal to

$$p(s) = A \exp\left[-\int_0^s ds/\lambda(r)\right] \tag{2-352}$$

The dependence $\lambda(r)$ suggests the possibility, e.g., of a layered target in which each layer is described by its own λ. If $\lambda(r)$ = const, the normalized probability density that the length of the mean free path will be s is

$$p(s) = \frac{1}{\lambda} e^{-s/\lambda}, \qquad s = 0 \ldots \infty \tag{2-353}$$

Using Eq. (2-344), we can write

$$\int_\xi^\infty p(s)\, ds = \int_\xi^\infty \frac{1}{\lambda} e^{-s/\lambda}\, ds = \gamma \tag{2-354}$$

Solving Eq. (2-354) for ξ, we obtain

$$\xi = -\lambda \ln \gamma \tag{2-355}$$

Thus, using the first random number γ_1 generated, we find the coordinates of the point at which the first elastic collision occurred. If the initial conditions were $r(x, y, 0)$, $\theta = 0$, i.e., the electron has entered the target at the point x, y and had a velocity normal to the target surface, the first collision occurred at the point $x, y, -\lambda \ln \gamma_1$.

If the target comprises different atoms the question arises as to which atom was collided with. The probability that a randomly chosen atom will be a i-type atom is, obviously,

$$p_i = (C_i\sigma_i/A_i)/\sum_j (C_j\sigma_j/A_j) \tag{2-356}$$

For a $C_5H_8O_2$ PMMA resist, we can find the corresponding table of probabilities of discrete events.

Having chosen the second random number γ_2, we compare it with the components of the p_H, $p_H + p_C$, $p_H + p_C + p_O$ vector. If the random number γ_2 falls, e.g., within the interval between p_H and $p_H + p_C$, we conclude that in this case the electron has collided with a carbon atom.

We must now find the result of the elastic collision with the carbon atom, i.e., find the scattering angle θ. The probability density $p(\theta)$ can be found by dividing the differential cross section by the total cross section:

$$p(\theta) = \sigma(\theta)/\sigma_t = \frac{\beta(1+\beta)}{\pi}(1 - \cos\theta + 2\beta)^{-2} \tag{2-357}$$

In keeping with Eq. (2-344) we compare the third random number γ_3 to the scattering angle θ that occurred in a given collision:

$$\int_\theta^\pi p(\theta)\sin\theta\,d\theta\int_0^{2\pi} d\varphi = \gamma_3 \tag{2-358}$$

Calculating the integral in Eq. (2-358), we obtain

$$\frac{\beta}{1+\beta}\,\frac{1-\sin^2\theta/2}{\beta+\sin^2\theta/2} = \gamma_3 \tag{2-359}$$

Solving Eq. (2-359) for $\cos\theta$, we find

$$\cos\theta = \frac{(1+2\beta)\gamma_3 - \beta}{\beta+\gamma_3} \tag{2-360}$$

We can find the azimuthal angle φ from the equation $\varphi = 2\pi\gamma_4$ because of scattering symmetry, because

$$p(\varphi) = 1/2\pi \tag{2-361}$$

Finally, we must find the point x, y at which the electron impinges on the target. If the electron beam has a Gaussian cross section, we have

$$p(x) = (a/\pi)^{1/2}\exp(-ax^2) \tag{2-362}$$

for the probability density $p(x)$ (and similarly for $p(y)$). Using Eq. (2-344), we obtain

$$\int_x^\infty p(x)\,dx = \pi^{-1/2}\mathrm{Erfc}(\sqrt{a}\,x) = \gamma_5 \tag{2-363}$$

$$\int_y^\infty p(y)\,dy = \pi^{-1/2}\mathrm{Erfc}(\sqrt{a}\,y) = \gamma_6 \tag{2-364}$$

$$\mathrm{Erfc}(z) = \int_z^\infty e^{-t^2}\,dt \tag{2-365}$$

Obviously, we have encountered a situation in which we need a special transform in order to find the random variables $x = g(\gamma_5)$ and $y = g(\gamma_6)$. A simpler procedure was used in [56]. If the point of incidence is given in polar coordinates ρ, χ, the Gaussian function for the beam current will be

$$p(\rho) = (a/\pi)\exp(-a\rho^2) \tag{2-366}$$

Because of the problem's symmetry the azimuthal angle is found from the uniform distribution $p = (\chi) = 1/2\pi$ in the interval $(0, 2\pi)$:

$$\chi = 2\pi\gamma_5 \tag{2-367}$$

then we use

$$\exp(-a\rho^2) = \gamma_6 \rightarrow \rho = [-(1/a)\ln\gamma_6]^{1/2} \tag{2-368}$$

to find the radial distance p.

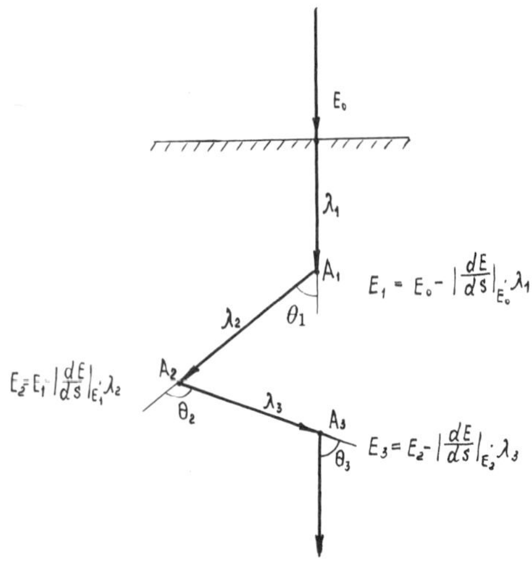

FIGURE 2-16. The trajectory constructed according to the model for elastic collisions and continuous deceleration for an electron moving through a target. The electron is decelerated in the λ_1, λ_2 region of the trajectory and elastic collisions that change the electron's direction of motion occur at points A_1, A_2.

Following this procedure the random functions x and y can be modelled in terms of the random numbers γ_5 and γ_6 through the relations

$$
\begin{aligned}
x = \rho \cos \chi &= \left[-\left(\frac{1}{a}\right) \ln \gamma_6\right]^{1/2} \cos(2\pi\gamma_5) \\
y = \rho \sin \chi &= \left[-\left(\frac{1}{a}\right) \ln \gamma_6\right]^{1/2} \sin(2\pi\gamma_5)
\end{aligned}
\tag{2-369}
$$

The energy of the electron falls off gradually as it travels between elastic collision points. At the point of the $n + 1$ collision the electron energy is

$$E_{n+1} = E_n - |dE/ds|_{E_n} \lambda_n \tag{2-370}$$

Figure 2-16 shows a trajectory calculated by the Monte Carlo method.

The calculation procedure described above allows us to find not only the trajectory of electrons, but even more complex quantities such as, e.g., the density of the electron beam energy absorbed by the polymer. In fact, by knowing an electron's trajectory and its energy $E(r)$ at any point on the trajectory we can easily find the amount of energy dissipated $\Delta E = |dE/ds|\Delta s$ in any interval of the trajectory by using, e.g., Bethe's formula, Eq. (2-160) to calculate the deceleration $|dE/ds|$. Calculating the density of the dissipated energy then reduces to the following procedure. Near the point r at which we wish to find the density of the dissipated energy, we isolate a volume ΔV. The quantity ΔV is small, but it is macroscopic (e.g., $\Delta V = \Delta x \, \Delta y \, \Delta z = 40 \times 40 \times 40$ nm). The choice of dimensions for the volume element is dictated by the requirements on the accuracy of the calculations. We will compose a program that fixes the trajectory for the i electron through the volume ΔV and calculates the amount of energy ΔV dissipated in the volume $\Delta E_i = |dE/ds|_i \, \Delta s_i$. Then we let the program add all contributions

ΔE_i from the N_0 electrons that impinge on the target during the numerical experiment. Dividing $\sum_i \Delta E_i$ by the volume ΔV and the number of incident electrons N_0, we find the universal density function for the energy dissipated by a single electron, $I(r)$, that was introduced above:

$$I(r) = \sum_{i=1}^{N_0} |dE/ds|_i \, \Delta s_i / N_0 \, \Delta V \tag{2-371}$$

We find the dosage (the density of the dissipated energy) as $D(R) = N_0 I(r)$.

The shape of the volume ΔV is chosen according to the symmetry of the problem. If, e.g., a point on the surface of a resist is exposed by a Gaussian beam the problem has cylindrical symmetry and ΔV is conveniently chosen as a ring with rectangular cross section $\Delta r \, \Delta z$ and radius r located a depth z in the resist layer. Then[56]

$$D(r) = N_0 I(r, z) = \left(\sum_{i=1}^{N_0} \Delta E_i \right) / 2\pi r \, \Delta r \, \Delta z \tag{2-372}$$

When scanning is linear, we use a linear electron density Q [electrons/cm] = $i(A) q / v$ [(electrons/C)/(cm/sec)$^{-1}$]. We then choose as the volume ΔV a parallelepiped $\Delta x \, \Delta y \, \Delta z$ centered at the x, y, z. (The coordinates system is fixed and the beam moves along the x axis.) Over the interval $x_b, x_b + dx_b$ the beam conveys $dN = Q \, dx_b$ electrons to the target. These electrons dissipate an amount of energy:

$$dE = I(x - x_b, y, z) \Delta V \, Q \, dx_b = \sum_{i=1}^{dN} \Delta E_i \tag{2-373}$$

Here $I(x - x_b, y, z)$ is the contribution made by dN electrons to the density of the energy dissipated in the volume ΔV. Adding these contributions from each interval dx_b on the path, we find

$$E(y, z) = \int_{-\infty}^{+\infty} I(x - x_b, y, z) \, dx_b \qquad \frac{\text{eV/cm}^3}{(\text{electrons/cm})^{-1}} \tag{2-374}$$

Multiplying Eq. (2-374) by the number q of electrons per coulomb of charge, we find the density of the energy dissipated by a line charge of 1 C/cm:

$$E_c(y, z) = q E(y, z) \qquad \frac{\text{eV/cm}^3}{(\text{C/cm})^{-1}} \tag{2-375}$$

All of the quantities used here can be found in the literature.

This brief discussion of the Monte Carlo method and its application to electron lithography adequately describes the method and the value of the results it yields. The attractiveness of the method is its uncommon universality. Its main shortcoming is that the results obtained are nonanalytical. We can show some ways to improve Monte Carlo calculations of electron beam interactions with matter. Our discussion will center around the use of more accurate equations for the scattering cross section in these calculations. In the calculations discussed above, inelastic scattering processes were described by Bethe's formula in the model for continuous losses. More accurate calculations may include the inelastic scattering cross section as a function of the energy losses ΔE.

It was pointed out in Section 2-12 that the inelastic scattering cross section is adequately described by theory in an approximation and several semiempirical formulas for the inelastic scattering cross section have been suggested. Using one of these formulas, we can organize the process of calculating the energy lost by electrons in a resist. In doing this the opportunity to study both the trajectory and the scattering of the secondary electrons energy presents itself, but the number of calculations will increase significantly. We can point to the work done by Murata *et al.*(9) in which the affect on electron lithography processes from secondary electrons created by fast electrons ionizing the atoms via inelastic scattering was studied. Using the Monte Carlo method to calculate the electrons energy spectrum also assumes that analytical formulas are used for the inelastic scattering cross sections.(8)

We will make numerical estimates for a number of physical quantities that describe electron beam actions on a resist. Deceleration (the average energy losses per unit path length) can easily be calculated from Bethe's formula:

$$\left|\frac{dE}{ds}\right| = \frac{2\pi e^4 n_e}{E} \ln(1.166E/I) \tag{2-376}$$

Polymers used as electron resists are characterized by a relatively small effective atomic number $Z_{\text{eff}} \approx 4$ and the amount of energy lost by an electron is on the order of 10^8 eV/cm (see Table 2-5 in [65]). In a 1-μm-thick resist film electrons having about 10^4 eV energy lose about 3 keV of energy (when passing directly through):

$$\Delta E = \left|\frac{dE}{ds}\right| \Delta s = 2.7 \times 10^7 \,\frac{\text{eV}}{\text{V}} \times 10^{-4}\,\text{cm} = 2.7\ \text{keV} \tag{2-377}$$

For 1-μm-thick substrate from materials having large Z these losses are on the order of 10 keV.

A fundamental experimental quantity in electron lithography is the so-called sensitivity of a resist, measured in C/cm^2. This sensitivity in C/cm^2 is a function of the energy of the incident electrons and the etchant used to develop the resist (etching the exposed portion of the resist). A typical value of sensitivity for a PMMA-type electron resist is $Q = 1.5 \times 10^{-5}$ C/cm^2. When the sensitivity is known, a relationship between the parameters in the point-by-point mode of exposure can easily be found:

$$Q = i\tau/s \tag{2-378}$$

where i (C/sec) is the beam current, s (cm^2) is the beam cross section, and τ (sec) is the exposure time of the point. For a cross section of 10^{-10} cm^2 (beam diameter of 0.1 μm and beam current of 1.5 nA), the exposure time for the point is 10 μsec.

A line (linear) charge density

$$Q_l = \frac{i}{v}\,\frac{\text{(C/sec)}}{\text{cm/sec}} = \frac{i}{v}\ \text{C/cm} \tag{2-379}$$

is used in a number of works for the linear beam scanning mode. In a nominal (sensitivity of 10^{-5} C/cm^2) exposure mode, v = 1 cm/sec, the line charge density is $Q_l = 10^{-9}$ C/cm.

We can easily convert exposure given in C/cm^2 to the number of electrons per cm^2: 1 C/cm^2 = 0.625 $\times$ 10^{19} electrons/cm^2.

For us the most interesting quantities are the energies. Let a resist be exposed by a charge density $Q = 1.5 \times 10^{-5}$ C/cm^2 $\simeq 10^{14}$ electrons/cm^2 at an electron energy of 10^4 eV. An electron beam of this type will take into a target a stream of energy whose density is

$$S_0 = E_0 Q \simeq 10^{18}\ \text{eV/cm}^2 = 0.16\ \text{J/cm}^2 \tag{2-380}$$

As we saw earlier, a substantial amount of energy is reflected (taken out of the target by reflected electrons); considering the reflection coefficient for the energy being absorbed by the target the flow of energy is

$$S = (1 - \eta_E) S_0 = 0.1\ \text{J/cm}^2 \tag{2-381}$$

If we assume for estimating purposes that the electron beam energy remaining in the target is absorbed uniformly in a layer whose thickness is equal to the electron range in the resist, the density of the absorbed energy can be estimated from

$$D = (1 - \eta_E) S_0 / R_B \tag{2-382}$$

in a PMMA-type resist the range of electrons having 10 keV of energy is 2.18 μm, then

$$D = \frac{0.1\ \text{J/cm}^2}{2.18 \times 10^{-4}\ \text{cm}} = 500\ \text{J/cm}^3 \tag{2-383}$$

The density of the energy absorbed in the resist can be estimated in another way. In a 1-μm-thick resist an electron having 10^4 eV energy loses about 2.7 keV of energy in passing straight through. Multiplying this quantity by the electron density in the beam, we find

$$D_f = (2.7\ \text{keV/electron} \times 10^{14}\ \text{electrons/cm}^2) : 10^{-4}\ \text{cm} = 432\ \text{J/cm}^3 \tag{2-384}$$

which agrees with our previous estimate.

It is interesting to estimate the density of the energy dissipated by a single electron. The physical process through which a fast electron loses energy includes, as we have examined in detail, the energy lost in exciting and ionizing the target's atoms. Ionization creates secondary electrons whose range is R_{Bs}. We can say that the volume of target material into which an electron dissipates its energy is equal to the volume of a pipe whose length is equal to the range of the primary (fast) electron R_B and whose radius is equal to the range of the secondary electron R_{Bs}:

$$I_{\max} = E_0 / \pi R_{Bs}^2 R_B \tag{2-385}$$

When E_0 = 10 keV, R_B = 2.18 μm in PMMA. We assume for estimating purposes that $R_{Bs} = 10^{-6}$ cm; then, $I_{\max} = 1.4 \times 10^{19}$ eV/cm^3 $\simeq 1$ J/cm^3.

It is interesting to estimate the amount of heat produced in a resist film by electron exposure. Only that amount of energy absorbed from the beam is converted to heat; the remainder is stored by the resist as the potential energy of the ruptured molecular bonds.

We will assume that all of the energy absorbed from the beam goes into heating the film. Then we can estimate the amount of heat expected to go into the film by

$$\Delta T = D/C \cdot \rho \tag{2-386}$$

where C is the specific heat of the resist and ρ is its density. If the exposure time for a point is measured in microseconds we can ignore the effect of thermal conductivity on heating. For PMMA $\rho = 1.2$ g/cm^3 and the specific heat of PMMA-type polymers is $C = 1.73$ J/g $\cdot$ deg.[66] Earlier, we estimated the amount of energy absorbed to be $D = 500$ J/cm^3; then, the amount of heat expected is $\Delta T \simeq 250°$. Heating resist films to 200–300° produces irreversible changes in their properties (e.g., vitrification). The exposure modes must ensure that heat is removed via thermal conductivity by increasing the exposure time or multiple exposure.

According to data from radiation chemistry, the average amount of energy that will be absorbed by a polymer and break one chemical bond is 50–80 eV.[67] From here we can easily estimate the concentration (density) of broken bonds:

$$N = \frac{D}{\Delta E} = 10^{20}\ \text{cm}^{-3} \tag{2-387}$$

The number of resist molecules per cubic centimeter is

$$N_R = N_A \rho_R / M_R = 6 \times 10^{23} \times 1.2\ \text{g/cm}^3/9 \times 10^5\ \text{AEM} = 10^{18}\ \text{cm}^{-3} \tag{2-388}$$

so that the calculated number of breaks in a single molecule is roughly 100. This means that exposure reduces the average molecular weight by a factor of 100. However, a large number of the bond breaks take place in the side radicals and terminate by forming volatile fragments of low molecular weight. The number of times the average molecular weight of the skeletal polymer chains is reduced is, apparently, much less than 100. Nevertheless, it does produce a substantial change in the solubility of the exposed polymer.

Polymer solubility as a function of the dosage (density) of the energy absorbed is given by different empirical exponential formulas, e.g.,[49]

$$R(D) = R_1(C_m + D/D_0)^{\alpha} \tag{2-389}$$

where R_1, C_m, D_0, and α are empirical parameters. Formulas of the type given by Eq. (2-389) are used to model the polymer etching process in the respective etchant on a computer.

2-33. USING THE MONTE CARLO METHOD TO MODEL THE ELECTRON ENERGY SPECTRUM

Several authors have used the Monte Carlo method to calculate the electron energy spectrum and the reflection and transmission coefficients in thin films.[8,10–12,55] The total and differential scattering cross sections for both elastic and inelastic collisions of a fast electron with an atom are needed in these computations. Grisinsky's formula[31] has been used in recent works for calculating the differential inelastic scattering cross section:

$$\frac{d\sigma}{d\Delta E} = \frac{\pi e^4}{\Delta E^3}\frac{E_B}{E_0}\left(\frac{E_0}{E_0+E_B}\right)^{3/2}\left(1-\frac{\Delta E}{E_0}\right)^{E_B/(E_B+\Delta E)}$$
$$\times\left\{\frac{\Delta E}{E_B}\left(1-\frac{E_B}{E_0}\right)+\frac{4}{3}\ln\left[2.7+\left(\frac{E_0-\Delta E}{E_B}\right)^{1/2}\right]\right\} \qquad (2\text{-}390)$$

where E_0 is the electron's energy before the collision, ΔE is the energy lost in the collision, and E_B is the binding energy of the atom's electron that acquires the energy ΔE. The scattering angle of the electron during the collision is determined by the amount of energy lost, ΔE:

$$\sin\theta/2 = (\Delta E/E_0)^{1/2} \qquad (2\text{-}391)$$

The total inelastic scattering cross section obtained from Eq. (2-388) is[10]

$$\sigma_{\text{inel}} = 6.51\times10^{-14}\frac{N_S}{E_B^2 u}\left(\frac{u-1}{u+1}\right)^{3/2}\left[1+\frac{2}{3}\left(1-\frac{1}{2u}\right)\ln(2.7+(u-1)^{1/2})\right] \qquad (2\text{-}392)$$

where $u = E_0/E_B$, and N_s is the number of electrons in the respective shell. The binding energy of the electrons in the atom's inner shells can be found, e.g., by Slater's method.[23] Slater's procedure cannot be used for the valence electrons. For these electrons the binding energy E_l is, according to Shimizu and Everhart,

$$\int_{E_B}^{E_0}\Delta E\frac{d\sigma}{d\Delta E}\,d\Delta E = \left(\frac{dE}{ds}\right)_{\text{Bethe}} - \left(\frac{dE}{ds}\right)_{\text{core}} \qquad (2\text{-}393)$$

This equation means that the deceleration $(dE/ds)_{\text{core}}$ due to scattering in the inner electrons and the total deceleration $(dE/ds)_{\text{Bethe}}$, calculated from Grisinsky's and Bethe's formulas, respectively, have the required degree of accuracy and allow us to find the binding energy E_B of the valence electrons as a correction if Grisinsky's formula is extended to the valence electrons (in the integral in the left-hand side of Eq. (2-393) the cross section $d\sigma/d\Delta E$ is given by Grisinsky's formula, Eq. (2-390)). The valence electron binding energies found were 4.0, 4.0, and 10.0 eV, respectively, for Al, Si, and PMMA.

As for the cross section for elastic collisions, the Rutherford formula with shielding,

$$\frac{d\sigma}{dO} = \frac{Z^2e^4}{4E^2}\frac{1}{(1-\cos\theta+2\beta_N)^2}, \qquad \int d\sigma = \sigma_{\text{tot}}^{\text{el}} = \frac{\pi e^4 Z^2}{4\beta_N(1+\beta_N)E^2} \qquad (2\text{-}394)$$

is used to calculate the electron spectra. However, the shielding parameter $\beta_N = 5.43Z^{2/3}/E$ (E is the electron's energy in eV before the collision) can also be thought of as a correction if precise experimental or calculated (computer) data regarding the scattering cross section are available. Specifically, the authors[10] showed that the Rutherford cross section with shielding $\beta^* = 0.5\beta_N$ is a good representation of the results obtained

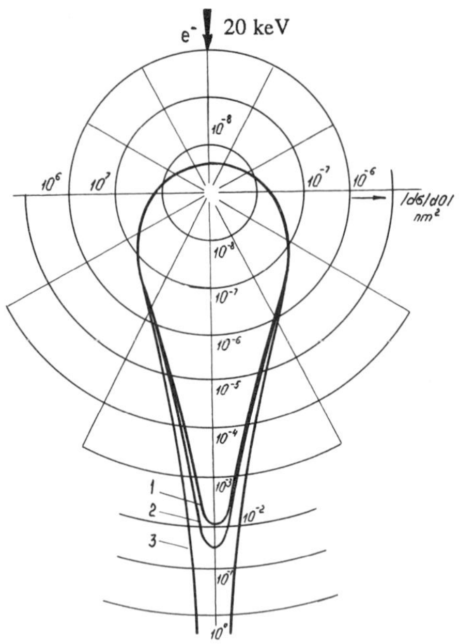

FIGURE 2-17. Polar diagrams of the angular dependence of the differential scattering cross section for electrons having an energy $E_0 = 20$ keV in PMMA. 1) Calculated with a shielding coefficient of $\beta_N = 5.43 Z^{2/3}/E$, E is in eV; 2) with a shielding coefficient of $0.5\beta_N$; 3) without shielding.

from a numerical computation of the cross section by the method of partial waves expansion. Figure 2-17 shows the angular dependence of the differential elastic scattering cross section.[10]

Knowing the total cross section of a process allows us to give the length of a fast electron's mean free path in relation to that process:

$$\frac{1}{\lambda_{\text{el}}} = N\sigma_{\text{tot}}^{\text{el}}, \qquad \frac{1}{\lambda_{\text{inel}}} = N\sigma_{\text{tot}}^{\text{inel}} \tag{2-395}$$

For compound materials we can give the average values as the atomic percentages of the material's composition:

$$\frac{1}{\lambda_{\text{el}}} = \sum_i N_i(\sigma_{\text{tot}}^{\text{el}})_i, \qquad \frac{1}{\lambda_{\text{inel}}} = \sum_i N_i(\sigma_{\text{tot}}^{\text{inel}})_i \tag{2-396}$$

The average free path length λ_{tot} in relation to any process is found from

$$\frac{1}{\lambda_{\text{tot}}} = \frac{1}{\lambda_{\text{el}}} + \frac{1}{\lambda_{\text{inel}}} \tag{2-397}$$

We will now outline a Monte Carlo procedure for calculating the energy spectrum. We choose a random number γ_1 and find the path the particle takes from the time it crosses the target's boundary until the first collision:

$$s = -\lambda_{\text{tot}} \ln \gamma_1 \tag{2-398}$$

We now choose a second random number γ_2 and compare it with the ratio $\lambda_{\text{el}}^{-1}/\lambda_{\text{tot}}^{-1}$; if $\gamma_2 \leq \lambda_{\text{el}}^{-1}/\lambda_{\text{tot}}^{-1}$ an elastic collision occurs and the result (scattering direction) is found, as was described earlier. The type of atom in which scattering occurs (when the target is multiatomic) and then the scattering angle are subsequently revealed.

If $\gamma_2 > \lambda_{\text{el}}^{-1}/\lambda_{\text{tot}}^{-1}$, inelastic scattering takes place. Depending upon which inequality

$$\frac{\lambda_{\text{el}}^{-1}}{\lambda_{\text{tot}}^{-1}} < \gamma_2 \leq \frac{\lambda_{\text{el}}^{-1} + \lambda_L^{-1}}{\lambda_{\text{tot}}^{-1}} \quad \text{or} \quad \frac{\lambda_{\text{el}}^{-1} + \lambda_L^{-1}}{\lambda_{\text{tot}}^{-1}} < \gamma_2 \leq 1 \tag{2-399}$$

is satisfied, we decide that inelastic scattering occurs from the atom's L-shell electrons or valence electrons.

To find how much energy is lost by a fast electron through inelastic scattering, ΔE, we choose the next random number γ_3 and find ΔE from

$$\gamma_3 = \int_{E_B}^{\Delta E} d\sigma(\Delta E') \Bigg/ \int_{E_B}^{E_0} d\sigma(\Delta E') \tag{2-400}$$

where E_B, $d\sigma(\Delta E')$ pertain to the respective group of electrons (L-shell or valence electrons; we consider the scattering cross section in K-shell electrons to be negligible). The angular deflection when inelastic scattering occurs is calculated from Eq. (2-391).

In Shimizu *et al.*,[12] the electron energy spectrum in metals was calculated with regard for a special inelastic scattering cross section in the conduction electrons, Grisinsky's formula for L-shell electrons, and the formulas in Section 2-17 for the stimulation cross section for plasma oscillations. Values calculated for the free path lengths Al of electrons having λ of energy are respectively 15 keV, and λ_L = 1426 Å, λ_c = 1260 Å, λ_p = 297 Å, λ_{el} = 162 Å for inelastic scattering from the L-shell electrons, the conduction electrons (one-particle stimulations), in plasma oscillations, and elastic scattering. The rates of energy loss were 0.189 eV/Å, 0.063 eV/Å, and 0.004 eV/Å, respectively, for losses in L-shell electrons, conduction electrons, and plasmons. The sum of all these losses was 0.306 eV/Å in comparison with the 0.314 eV/Å obtained by Bethe's formula, Eq. (2-160), for electrons having an energy of E_0 = 15 keV.

Figures 2-18 and 2-19 show the data obtained in [10] for the reflection and transmission coefficients in thin PMMA resist films and the energy spectrum calculated by the Monte Carlo method for the electrons that pass through a film of the respective thickness. The points in Fig. 2-18 show the experimental values of the transmission coefficient for a beam having an energy of E = 20 keV.

Reflection coefficients obtained by the Monte Carlo method for electrons having 10 and 20 keV of energy from a thick PMMA target were 0.09 and 0.098, respectively. The experimental value of η_B for Be was 0.05. The Monte Carlo electron energy distribution after passing through a PMMA layer was wider than in Bethe's model for continuous losses. The calculated results are in good agreement with experimental results obtained for aluminum (Fig. 8 in [10]).

2-34. THE DISTRIBUTION DENSITY OF THE ENERGY DISSIPATED BY ELECTRONS IN A RESIST LAYER. THE PROXIMITY EFFECT

We can apparently divide, with sufficient accuracy, the contributions made to energy dissipation at a given point in a resist layer into two parts: the contribution from electrons

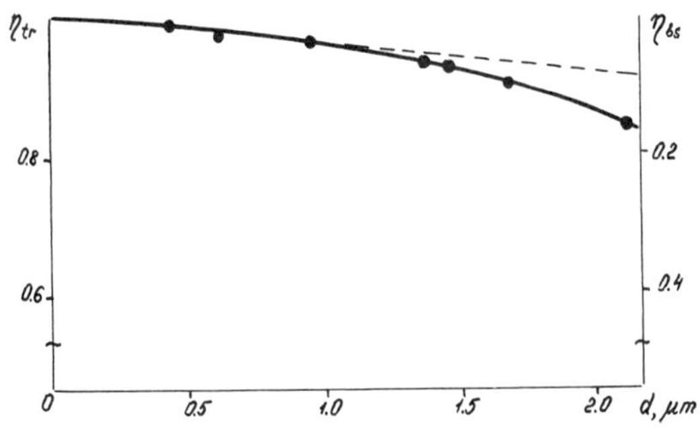

FIGURE 2-18. The reflection η_{Bs} (dotted line) and transmission η_{tr} (solid line) coefficients for a beam of electrons having an energy of E_0 = 20 keV as functions of PMMA layer thickness.

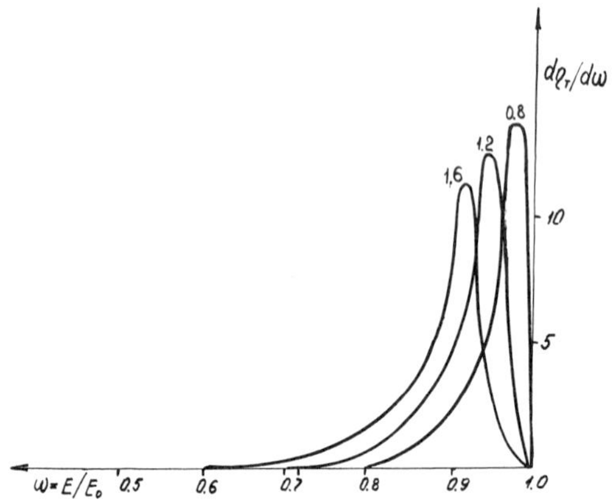

FIGURE 2-19. The energy distribution of electrons passing through PMMA layers of different thickness.

that intersect the resist when passing straight through, and the contribution from electrons that are reflected from the substrate and the deeper resist layers. When this happens, the basic part of the reflected contribution are electrons reflected from the substrate, because the resist layer is relatively thin and, as a rule, consists of light atoms. Therefore, we usually speak of reflection from the substrate. Our task is to find the density of the energy dissipated by these electron streams in the resist later.

We will use Bethe's formula,

$$\left|\frac{dE}{ds}\right| = 2\pi e^4 n_e E^{-1} \ln(1.116E/I) \tag{2-401}$$

to calculate the energy dE dissipated by an electron as it travels a path ds.

The resist layer is thin in comparison with the range of electrons having an energy of 10–20 keV. Therefore, in order to calculate beam energy losses in a resist layer during direct penetration we can make two simplifying assumptions: we consider the energy E to be constant and equal to E_0, and we say that $s = z$.

We will examine a point in the resist being exposed by a Gaussian beam for a period of time τ:

$$N(x, y) = \frac{\tau q i a}{\pi} \exp[-a(x^2 + y^2)] \tag{2-402}$$

The probability that an electron entering a target at a point with coordinates x, y on the surface will pass through a point with coordinates x_T, y_T, z_T in the target is given, according to Eq. (2-139), by the distribution

$$P(x_T, y_T, z_T; x, y) = \frac{b}{\pi} \exp[-b((x - x_T)^2 + (y - y_T)^2)], \qquad \frac{1}{b} = \overline{r^2} = 4z_T^3/3\Lambda \tag{2-403}$$

The density of the energy dissipated at point x_T, y_T, z_T during the exposure time τ is found by integrating the product of the functions in Eqs. (2-401), (2-402), and (2-403):

$$D(x_T, y_T, z_T; \tau) = 2\pi e^4 n_e E_0^{-1} \ln(1.166 E_0/I) \times \iint_{-\infty}^{+\infty} N(x, y) P(x_T, Y_T, z_T; x, y)\, dx\, dy \tag{2-404}$$

Performing this elementary integration, we find

$$D(r_T) = \left|\frac{dE}{ds}\right|_{E_0} \times \frac{q_i \tau c}{\pi} \exp[-c(x_T^2 + y_T^2)] \tag{2-405}$$

where $1/c = 1/a + 1/b$; and D is measured in electrons $\cdot$ J/cm^3. Thus, the density of the energy dissipated when a beam passes straight through is Gaussian distributed in every plane perpendicular to the beam axis. As the beam penetrates more deeply into the target the width of the exposed region increases.

Calculating the contributions made to absorbed energy by a stream of reflected electrons is much more complicated. Reflected electrons traverse complex paths and finding the energy of the reflected electrons at a specific point r_T requires that the entire path the electron travels prior to arriving at the point r_T be considered. The necessary calculations were done by Hawryluk[56] in approximating one-time scattering at large angles. However, the formulas he derived are cumbersome and the results they yield are inaccurate. In practice, simple functions based on an analytical approximation of the results from statistical modelling of electron trajectories by the Monte Carlo method are used.[68–77]

First of all, we introduce a proximity function $f(|\bar{r}|)$ as the energy density distribution produced by a δ-shaped electron beam. From Eq. (2-405) we can easily find the density distribution when the electrons pass straight through. For the δ-shaped source $a = \infty$ and the width of the Gaussian distribution is determined by small-angle scattering when the beam passes straight through: $\bar{r}^2 = 1/b = 4z_T^3/\Lambda$. Graphs of the Monte Carlo generated distribution of the absorbed beam energy density can be approximated by the sum of two Gaussian functions:

$$f(|\bar{r}|) = \varkappa \left[\exp\left(-\frac{r^2}{\beta_f^2}\right) + \eta_E^* \left(\frac{\beta_f}{\beta_b}\right) \exp\left(-\frac{r^2}{\beta_b^2}\right)\right] \tag{2-406}$$

The widths of the distribution for the Gaussian functions β_f and β_b are quite different: $\beta_b \gg \beta_f$. Therefore, the width of the first Gaussian function is assumed to be associated with beam spreading as it passes straight through, whereas the width of the second Gaussian function is equal to the width of the beam reflected from the deeper resist layers and the substrate. The parameter η_E that describes the ration of the total energy dissipated by reflected electrons in the resist to that dissipated by electrons when passing through a resist layer can be physically measured, and $\varkappa$ is a normalizing factor.

The proximity function for a real (not δ-shaped) beam is found by taking the convolution of the proximity function given by Eq. (2-406) with the distribution function for electrons over a transverse section of the beam. Convolution with a Gaussian beam of radius r_b again produces a function of the type given by Eq. (2-406) and having parameters

$$\beta_f^* = (\beta_f^2 + r_b^2)^{1/2}, \qquad \beta_b^* = (\beta_b^2 + r_b^2)^{1/2} \tag{2-407}$$

For simple shapes such as stripes and rectangles, the resulting absorbed energy density distribution is given as an error function, which is calculated on a computer for more complex shapes.

We will obtain a latent image produced in a resist layer by a rectangular die. Let an electron beam produce uniform exposure d_0 in the rectangle $P(x_1 < x < x_2, y_1 < y < y_2$, i.e.,

$$d(x,y) = \begin{cases} d_0, & (x,y) \in p \\ 0, & (x,y) \notin p \end{cases} \tag{2-408}$$

Taking the convolution of Eq. (2-408) with the proximity function, Eq. (2-406), we easily find[78]

$$\begin{aligned} D(x,y) = \frac{\varkappa d_0}{e} \Bigg\{ &\frac{\pi\beta_f^2}{4}\left[\operatorname{erf}\left(\frac{x_2 - x}{\beta_f}\right) - \operatorname{erf}\left(\frac{x_1 - x}{\beta_f}\right)\right] \\ &\times \left[\operatorname{erf}\left(\frac{y_2 - y}{\beta_f}\right) - \operatorname{erf}\left(\frac{y_1 - y}{\beta_f}\right)\right] + \eta\frac{\beta_b^2\pi}{4}\left[\operatorname{erf}\left(\frac{x_2 - x}{\beta_b}\right)\right. \\ &\left. - \operatorname{erf}\left(\frac{x_1 - x}{\beta_b}\right)\right]\left[\operatorname{erf}\left(\frac{y_2 - y}{\beta_b}\right) - \operatorname{erf}\left(\frac{y_1 - y}{\beta_b}\right)\right]\Bigg\} \end{aligned} \tag{2-409}$$

Here, $\eta = \eta_E(\beta_f/\beta_b)$.

The actual electron current density distribution over the beam cross section in lithography machines with a rectangular die displays some edge fuzziness. Attempts have been made[76] to approximate the decay in current at the edge of the die by a linear dependence. However, this approximation is rather coarse for small dies. An approximation of the electron current distribution in a rectangular die was used in [77]. This approximation is equivalent to exposing an area P on a rectangular die to uniform scanning by a Gaussian beam having a variance σ:

$$d(x,y) = \frac{d_0}{\pi\sigma^2}\int_P \exp\left[-\frac{(x - x')^2 + (y - y')^2}{\sigma^2}\right] dx'\, dy' \tag{2-410}$$

The width W of the decay region for the function given by Eq. (2-410) is defined by the σ parameter. If σ is small in comparison with the size of the die, the W found for the level 0.9–0.1 is 1.85σ. The integral in Eq. (2-410) over a rectangle is written as an errors integral:

$$d(x,y) = \frac{d_0}{4}\left[\operatorname{erf}\left(\frac{x_2 - x}{\sigma}\right) - \operatorname{erf}\left(\frac{x_1 - x}{\sigma}\right)\right] \times \left[\operatorname{erf}\left(\frac{y_2 - y}{\sigma}\right) - \operatorname{erf}\left(\frac{y_1 - y}{\sigma}\right)\right] \tag{2-411}$$

Obviously, Eqs. (2-410) and (2-411) do not consider effects described by the proximity function, Eq. (2-406). The latent image of a rectangular die, with regard for the proximity effect, can also be obtained in analytical form.[78] We choose a proximity function in the form $I(r) = \varkappa(I_f + \eta I_b)$ and take the convolution of the I and d functions:

$$D(x,y) = \frac{\varkappa}{e}[I_f * d + \eta I_b * d] \tag{2-412}$$

We introduce a unit rectangle function

$$g(x) = \begin{cases} 1, & x_1 \le x \le x_2 \\ 0, & x < x_1, x > x_2 \end{cases} \tag{2-413}$$

For brevity in our equations we will use the notation $\exp(-x^2/\sigma^2) \equiv \tilde{e}(x/\sigma)$. With the unit rectangle function the beam current density, Eq. (2-410), can be written as the product of the convolutions of the variables with respect to x and y:

$$d(x,y) = \frac{d_0}{\pi\sigma^2}\left[g(x) * \tilde{e}\left(\frac{x}{\sigma}\right)\right]\left[g(y) * \tilde{e}\left(\frac{y}{\sigma}\right)\right] \tag{2-414}$$

Taking the convolution of Eq. (2-414) with the proximity function $I_f(r)$, we find

$$\begin{aligned} I_f * d(x,y) &= \frac{\varkappa d_0}{\pi\sigma^2}\left[\tilde{e}\left(\frac{x}{\beta_f}\right) * \left(g(x) * \tilde{e}\left(\frac{x}{\sigma}\right)\right)\right]\left[\tilde{e}\left(\frac{y}{\beta_f}\right) * \left(g(y) * \tilde{e}\left(\frac{y}{\sigma}\right)\right)\right] \\ &= \frac{\varkappa d_0}{\pi\sigma^2}\left[\left(e\left(\frac{x}{\beta_f}\right) * \tilde{e}\left(\frac{x}{\sigma}\right)\right) * g(x)\right]\left[\left(\tilde{e}\left(\frac{y}{\beta_f}\right) * \tilde{e}\left(\frac{y}{\sigma}\right)\right) * g(y)\right] \\ &\times \frac{\beta_f^2}{\beta_f^2 + \sigma^2} d_0 \int_{x_1}^{x_2}\int_{y_1}^{y_2} \exp\left[-\frac{(x - x')^2 + (y - y')^2}{\sigma^2 + \beta_f^2}\right] dx'\,dy' \end{aligned} \tag{2-415}$$

We find a similar result for the convolution $I_b(r) * d$. Writing Eq. (2-415) in terms of Eq. (2-411), we obtain a formula for the latent image of a rectangular die that accounts for proximity effects:

$$D(x,y) = \frac{\varkappa d_0}{e}\left\{\frac{\pi\beta_f^2}{4}\left[\operatorname{erf}\left(\frac{x-x_1}{\sigma_f^*}\right) - \operatorname{erf}\left(\frac{x-x_2}{\sigma_f^*}\right)\right]\right.$$
$$\times\left[\operatorname{erf}\left(\frac{y-y_1}{\sigma_f^*}\right) - \operatorname{erf}\left(\frac{y-y_2}{\sigma_f^*}\right)\right]$$
$$+\,\eta\frac{\pi\beta_b^2}{4}\left[\operatorname{erf}\left(\frac{x-x_1}{\sigma_b^*}\right) - \operatorname{erf}\left(\frac{x-x_2}{\sigma_b^*}\right)\right]$$
$$\left.\times\left[\operatorname{erf}\left(\frac{y-y_1}{\sigma_b^*}\right) - \operatorname{erf}\left(\frac{y-y_2}{\sigma_b^*}\right)\right]\right\} \qquad (2\text{-}416)$$

$$\sigma_f^* = (\sigma^2 + \beta_f^2)^{1/2}; \qquad \sigma_b^* = (\sigma^2 + \beta_b^2)^{1/2} \qquad (2\text{-}417)$$

Our result can be stated as follows. Considering proximity effects, a rectangular die with Gaussian edges is converted to a composition of dies with Gaussian edges whose widths are increased according to Eq. (2-417).

A characteristic of the proximity function given by Eq. (2-406) is that the absorbed energy distribution for reflected electrons is fairly wide and reaches 2–3 μm (for the beam energies typically used in electron lithography). Therefore, the small gaps between exposed figures also acquire some of the energy absorbed from reflected electrons. Reciprocal exposure also occurs within the exposed figures themselves. Reciprocal exposure effects appear at distances β_b and have been given the name proximity effects. The proximity effect ultimately defines the resolving power of the electron lithography process and has been painstakingly studied by several authors. The main purpose in these investigations is to develop methods for correcting the proximity effect.

2-35. METHODS FOR CORRECTING THE PROXIMITY EFFECT

The ultimate goal of electron lithography is to obtain a resist mask of specific size on the surface of a workable wafer (a blank or a silicon wafer). The lithography process comprises two steps: exposing the resist layer (creating a latent image) and removing the exposed regions from the resist layer (developing the latent image). (We will assume that a positive resist is being processed.) The theory of electron lithography must examine both steps. To date, latent image production processes that are completely described by the density $D(r)$ of the absorbed beam energy (or by the density of the radiation-chemical conversions that are proportional to $D(r)$) at a point r on the resist layer have been the most thoroughly developed from a theoretical standpoint. The transition is made to the latent image developing step by using empirical dependencies of the developing (etching) rate in the exposed resist layer on the absorbed energy: $v_{\text{sol}} = v(D(r))$.

The main limitation to producing high quality resist masks having submicron dimensions is the proximity effect, caused by the beam electrons being scattered in the resist and reflected from the substrate.[69,74,79] We will now concentrate on methods for correcting this proximity effect.

The purpose of correction is to obtain a roughly uniform absorbed energy density level in an exposed area, rapid decay of the energy absorbed at the edges of the areas being exposed, and minimal quantities of absorbed energy in the unexposed areas of the resist layer. These conditions (goals) of correction must provide a high quality mask: a high decay gradient $\nabla D(r)$ at the edges of the shapes being exposed produces a resist mask having minimal deviation from the specified size and almost vertical walls, whereas minimal values of $D(r)$ in the unexposed areas of the resist layer ensure more complete conservation (over thickness) of the resist mask. Thus, from the very beginning the stated goals of proximity effect correction consider the properties of the latent image developing processes (etching the exposed layer).

Methods studied to date for reducing the proximity effect's influence on mask pattern quality have included using multilayer resists,[68,79] choosing the resist or substrate thickness, increasing the energy of the electrons in the beam,[68,86] and step-by-step exposure of the resist.[81,82] Under given conditions (beam energy, resist layer thickness, substrate material, etc.) the means for reducing the impact of the proximity effect consists of changing the dimensions[83] and radiation dosages in the different areas being exposed[76,84] (or combining these two possibilities[76,85]).

The approaches to solving the problem of proximity effect correction in electron lithography can conditionally be divided into two groups. The first group of methods is a rigid mathematical statement of the problem which is then solved on powerful computers with effective numerical methods: matrix algebra,[84] iterative methods,[85,86] Fourier transform methods,[87,88] and linear programming.[70,89] Approaches in which simple computing algorithms based on empirical reference data tables,[90] hierarchical rules,[91] and dosage compensation curves[92,93] are used belong to the second group. In this case the correction algorithms can be executed in real time. Solutions for the correction problem based on rigid methods for each topological layer of an integrated circuit must compose an exposure program for an electron-lithography machine and at the same time make it possible to verify the accuracy of, and augment the reference data used in the second group of methods.

We will not discuss the second group of methods in any detail, but instead concentrate on the rigid methods of formulating and solving the problem of correcting the proximity effect. Notice first of all that an absorbed energy density averaged over the thickness of the resist layer $D(x, y) = \langle D(x, y, z) \rangle$ is used in each case. This approximation is sufficiently rigid when the beam energy is such that the electron range is many times greater than the thickness of the resist layer. Very often, methods for correcting the proximity effect are demonstrated for one-dimensional lattices. Although these specific results can be used in fabricating lattices, the utility of formulating the problem as it applies from the very beginning to two-dimensional patterns is obvious. We will now discuss the rigid statements that have been developed for the problem of correcting the proximity effect.

1. The Matrix Method (Parikh[84])

We divide a region of a surface G_1 into n fragments of area s_i, $i = 1, 2, \ldots, n$, and assign an exposure dosage d_i to each area. The dosages d_i are unknown and must be determined.

The density of the energy dissipated at the points having coordinates $\bar{r}_i(x_i; y_i)$ on the surface of the resist because of uniform (within each fragment) exposure of the n

fragments is

$$D(\bar{r}_i) = \frac{1}{e}\sum_{j=1}^{n} d_j \int_{s_j} f(|\bar{r}_j - \bar{r}_i|\, d\bar{s}_j, \qquad d\bar{s}_j \equiv dx_j\, dy_j \tag{2-419}$$

where $f(|\bar{r}|)$ is a proximity function that describes the three-dimensional energy density distribution during point exposure by a single incident electron, obtained by averaging over the resist layer thickness.

The average energy dissipated per unit volume in the ith fragment of area s_i is found by integrating over this area:

$$\bar{D}_i = \frac{1}{e} s_i^{-1} \sum_{j=1}^{n} d_j \int_{s_i}\int_{s_j} f(|\bar{r}_i - \bar{r}_j|)\, ds_i\, ds_j \tag{2-420}$$

We require that the resulting average dosage in every exposed fragment be the same: $\bar{D}_i = D_0$, $i = 1, \ldots, n$. This gives us a system of linear algebraic equations with constant coefficients for the n unknown d_j:

$$\sum_{j=1}^{n} \bar{a}_{ij} d_j = D_0, \qquad i = 1, 2, \ldots, n \tag{2-421}$$

$$\bar{a}_{ij} = \frac{1}{e} s_i^{-1} \int_{s_i}\int_{s_j} f(|\bar{r}_i - \bar{r}_j|)\, ds_i\, ds_j \tag{2-422}$$

The d_j are calculated in two steps: first, the n^2 coefficients $\bar{a}_{ij}$ are calculated, and then the unknown d_j are calculated by the methods of matrix algebra. We use the results from Section 2-33 to find the a_{ij}.

The contribution to the resulting dosage at point r_k from the ith area exposed by a uniform dosage d_i is

$$D_i(x_k, y_k) = a_{ik} d_i \tag{2-423}$$

Making the $D(x, y)$ in Eq. (2-409) identical to $D_i(x_k, y_k)$, d_0 to d_i, and x, y to x_k, y_k, we find the values for the a_{ik} coefficients:

$$\begin{aligned} a_{ik} = \frac{\pi}{4}\frac{\varkappa}{e}\Bigg\{ &\beta_f^2 \left[\operatorname{erf}\left(\frac{x_k - x_1^i}{\beta_f^*}\right) - \operatorname{erf}\left(\frac{x_k - x_2^i}{\beta_f^*}\right)\right] \\ &\times \left[\operatorname{erf}\left(\frac{y_k - y_1^i}{\beta_f^*}\right) - \operatorname{erf}\left(\frac{y_k - y_2^i}{\beta_f^*}\right)\right] \\ &+ \eta\beta_b^2 \left[\operatorname{erf}\left(\frac{x_k - x_1^i}{\beta_b^*}\right) - \operatorname{erf}\left(\frac{x_k - x_2^i}{\beta_b^*}\right)\right] \\ &\times \left[\operatorname{erf}\left(\frac{y_k - y_1^i}{\beta_b^*}\right) - \operatorname{erf}\left(\frac{y_k - y_2^i}{\beta_b^*}\right)\right]\Bigg\} \end{aligned} \tag{2-424}$$

in doing this the Gaussian current distribution at the edge of the die is accounted for by using the parameters given by Eq. (2-407).

Note the deficiencies in the method described, from which more highly developed methods must not suffer.

1. The division into fragments, introduced into the problem from the outside, is arbitrary, is unrelated to the nature of the respective areas of the topology, and is unformalized in any respect.

2. For the correction to be accurate the region must be divided into very small fragments, which correspondingly increases the number of computations needed to solve the system given by Eq. (2-421).

3. When the system of equations is solved, negative dosages may be obtained, which is physically impossible.

4. The radiation dosage and its rate of decay at the edges of the figure are uncontrolled, i.e., the deviation in the size of the figures from nominal and the quality of the mask profile after developing are uncontrolled. When $\bar{D}_i = D_0$, the dosage at the figure edges D_r is less than D_0, and the smaller the D_r, the larger the figure.[68] To ensure that the condition $D_r = D_0$ is satisfied, which corresponds to complete development of the exposed figures, the dosages found by solving Eqs. (2-421) are usually multiplied by a coefficient $k > 1$. A consequence of this procedure is that small figures are overexposed and the size of the developed figure increases. We can differentiate the values of k as a function of the figure size, which increases still further the number of parameters arbitrarily introduced into the problem.

5. The dosage in the unexposed regions is uncontrolled. If the dosage D is controlled at auxiliary working points chosen, e.g., at the edges of a figure or the areas between them, we come to a mathematically incorrect problem of solving a redefined system of equations:

$$a \cdot d = D \tag{2-425}$$

where d and D are matrices of size $n \times 1$ and $N \times 1$, $N > n$, and $a - N \times n$ is the matrix of a_{ij} elements. Generally speaking, the redefined system of equations has no solution.

2. *The Iterative Method*[76,84]

The redefined system of equations, Eq. (2-425), can be solved by iterative methods starting from the requirement that the norm be minimized:

$$\|ad - D\| \to \min$$

A solution of the system, Eq. (2-421), can be used as a zero-order approximation under the auxiliary condition that the dosages be positive, $d_i > 0$. A solution found in the iterative method depends on both the choice of the zero-order approximation and on the method used to divide the figures being exposed into fragments. The iterative method will not always yield a solution that satisfies the condition that $D_r = D_0$ at the edges of adjacent figures being exposed and that $D_{G_2} < D_{\min}$ at points chosen in the unexposed area G_2 between them. When we do this, we must simultaneously change the size of the figures being exposed,[71,82] keeping in mind those cases in which this method will not produce the desired result.

3. *Fourier Analysis*[87,88]

Fourier analysis reveals a high degree of similarity between the problem of creating a latent image in electron beam exposure and the problem of creating an image $i(\bar{r})$ of an object $o(\bar{\rho})$ in an optical system (see Chapter 4). In the general case, not dividing the surface to be exposed into figures and fragments, the dosage of absorbed energy at a point $\bar{r}(x, y)$ on the plane can be represented by the convolution of the functions $d(\bar{r})$ and $f(\bar{r})$:

$$D(\bar{r}) = \frac{1}{e} \int d(\bar{r}') f(|\bar{r} - \bar{r}'|)\, d\bar{r}' \tag{2-426}$$

The illumination of an optical image of an object having an intensity distribution $o(\bar{\rho})$ can also be given by a convolution. When this is done, the proximity function $f(|\bar{r} - \bar{r}'|)$ is replaced by a "scattering function" $s(|\bar{r} - \bar{\rho}|)$ that describes how light from a point source in the object plane is distributed in the image plane. In optics a scattering function describes the properties of the optical system and not the medium in which the image is recorded, whereas the proximity function describes the interactions (scattering of a point electron beam in a resist layer). The scattering process emerges as an "optical system," converting the exposure dosage distribution $d(\bar{r})$ into a lower quality absorbed energy density distribution $D(\bar{r})$. In striving to obtain the desired shape of an "image" $D(\bar{r})$, we change the "object" $d(\bar{r}')$ because the "optical system" that converts an object into an image is described by a proximity function $f(|\bar{r} - \bar{r}'|)$ that has been specified for the given electron beam exposure conditions. In optics, efforts are constantly being made to improve optical systems in order to make the "scattering function" an almost δ-shaped function (diffraction limited systems). In order to obtain a given image $i(\bar{r})$ we can also distort an object $o(\bar{\rho})$. The electron-optic analog mentioned here will not be examined further.

From a mathematical viewpoint Eq. (2-426) reduces the problem of correcting the proximity effect to solving an integral equation for finding the unknown function $d(\bar{r}')$ for a given proximity function $f(|\bar{r} - \bar{r}'|)$ and absorbed beam energy density, $D(\bar{r})$. When the limits of integration can be extended over the entire x, y plane, the integral equation will have an exact solution that can be found via Fourier transforms.[87] Indeed, taking the Fourier transform of Eq. (2-426), we obtain

$$D_\varkappa = d_\varkappa f_\varkappa \tag{2-427}$$

where the $D_\varkappa$, $d_\varkappa$, and $f_\varkappa$ are the Fourier transforms of the functions $D(\bar{r})$, $d(\bar{r})$, and $f(\bar{r})$, respectively. After taking the inverse Fourier transform of $d_\varkappa = D_\varkappa / f_\varkappa$ we can find the unknown function for the exposure dosage $d(\bar{r}')$ that corresponds to the desired function $D(\bar{r})$ for a latent image.

Fourier analysis makes it possible, in principle, to find an exact solution for the proximity effect correction problem. However, when we examine this beam more closely, the difficulties that arise become obvious. Assume that the desired function $D(\bar{r})$ has the form

$$D_c(\bar{r}) = \begin{cases} D_0, & \bar{r} \in G_1, \Gamma; \\ 0, & \bar{r} \in G_2 \end{cases} \tag{2-428}$$

Discontinuities in the function $D_c(\bar{r})$ at the edges of the figures being exposed correspond to infinitely large spatial frequencies in the Fourier spectrum of $D_\varkappa$, i.e., the left-hand side of Eq. (2-427) is nonzero for any frequency $\varkappa$, no matter how high. Conversely, the proximity function is a bell-shaped curve having a finite width (e.g., a Gaussian function with limited variance); therefore, its spectrum is also limited: $f_{\varkappa} \to 0$, $\varkappa \to \infty$. This means that at high frequencies $\varkappa$ it is impossible to find physically realizable values of $d_{\varkappa}$ that would satisfy Eq. (2-427). Thus, for the latent image given by Eq. (2-428) and finite width of the proximity function $f(\bar{r})$ the problem of correcting the proximity effect has no solution. Spectral analysis of the problem reveals why there is no solution in this case. Equation (2-427) also reveals maximal capabilities in terms of the sharpness of a latent image produced by electron lithography. Assume that a δ-shaped beam is used to make the exposure: $d(\bar{r}) = d_0\,\delta(\bar{r})$. Then, $d_{\varkappa} = \text{const}$, $D_{\varkappa} = \text{const} \cdot f_{\varkappa}$. As we can see, the sharpness of the edge of the latent image is equal to the sharpness of the edge of the proximity function in this case.

In solving the problem of correcting the proximity effect through Fourier analysis special processors that can rapidly take Fourier transforms are used.[88] As can be seen from the examples given here, a solution found for $d(\bar{r})$ might not be physically realizable. In this case we must smooth the $D(\bar{r})$ function somehow and iterate the procedure until we can find a solution that is physically realizable. The $D(r)$ function can be smoothed by using sloped (rather than vertical) boundaries, rounding sharp angles,[87,88] etc.

4. *The Optimization Method*[70,73,78,89,94,95]

An optimization approach to solving the proximity effect correction problem was suggested in [70, 73]. Later, the optimization method was formulated in [78, 89, 94, 95] in the rigid form that we will use as the basis for our discussion.

In the general statement of the problem we will consider the exposure dosages d_i, $i = 1, 2, \ldots, N$, which are constant in the rectangular areas and the α_i, $i = 1, 2, \ldots, 4N$ parameters that define the sizes and locations of these areas to be unknown. The number of rectangular exposure areas N is limited by the size of the database for controlling the exposure process (e.g., N may be an order of magnitude larger than the number of figures in the topology). We will say that the coordinates of the rectangular areas being exposed (which means their dimensions) can be varied independently of each other and of the coordinates of the figures in the topology. When the problem is stated this way, some areas of the figures may remain unexposed (they are exposed indirectly due to the proximity effect) and other areas may be exposed several times due to the location of the rectangular areas being exposed. Because of this possibility we will say "covering the topological figures" instead of the familiar "separating the topological figures" into the rectangular exposure areas.

With optimization methods the unknown dosages d_i and coordinates α_i of the areas being exposed must be found from a system of inequalities that impose requirements on the exposure process that will produce a latent image having "optimal" parameters, or, if we consider the developing processes, requirements on an optimal resist mask (see below). We will formulate these requirements for the two-dimensional case.

1) The exposure dosages in the rectangular regions must be positive:

$$d_i \geq 0, \qquad i = 1, \ldots, N \tag{2-429}$$

2) The resulting dosage $D(\bar{r})$ within the pattern elements must be greater than (equal to) the critical value D_0 which ensures that the resist will be completely developed in

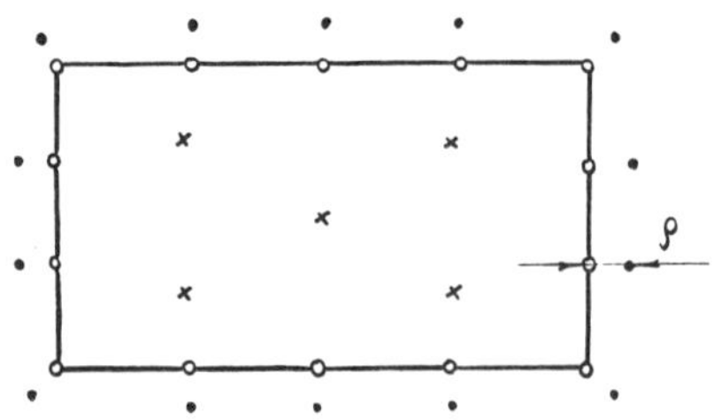

FIGURE 2-20. Arrangement of the points in the sets $\{\varkappa_{in}\}$, $\{\varkappa_\Gamma\}$, and $\{\Omega\}$ for an isolated rectangular element. The arrangements are shown by ×, o, and •, respectively.

any developing mode (liquid or plasma etching):

$$D(\bar{r}_k) \geq D_0, \qquad \bar{r}_k \in \{\varkappa_{\text{in}}\} \tag{2-430}$$

where $\{\varkappa_{\text{in}}\}$ is a finite set of interior points in the figure being exposed. The functions $D(\bar{r}_k)$ are linearly dependent on the unknown exposure dosages $D(\bar{r}_k) = \sum_i a_{ik} d_i$, which makes it possible for us to reduce the optimization problem to a linear programming problem for which effective standard methods, e.g., the simplex method, exist for solving the problem.

3a) Of particular significance are the conditions on the resulting dosage $D(r)$ at the edge of the figures being etched and in the adjacent outside area in which $D(r)$ decays rapidly. These conditions define the accuracy with which the dimensions of the figures can be reproduced in a resist mask and the profile of the mask's "edge." In the two-dimensional case, points on the edge of the figure are not equivalent and setting the condition that $D(r_\Gamma) = D_0$ for all edge points would lead to an infinite system of equations. In order to avoid this we will extend the inequalities of Eq. (2-430) to a finite set $\{\varkappa_\Gamma\}$ of edge points (see Fig. 2-20):

$$D(\bar{r}_\Gamma) \geq D_0, \qquad \bar{r}_\Gamma \in \{\varkappa_\Gamma\} \tag{2-431}$$

and impose the condition

$$P_1 = \sum_j D(\bar{r}_j) \to \min, \qquad \bar{r}_j \in \{\varkappa_\Gamma\} \tag{2-432}$$

on the $D(\bar{r}_\Gamma)$ at the edge of the figure. Because we do not require that $D(\bar{r}_\Gamma) = D_0$, the number of points in the set $\{\varkappa_\Gamma\}$ can be greater than the number of unknown α_i, d_i, which, on the one hand, ensures that a given figure contour can be reproduced fairly accurately and on the other hand, does not increase the number of equations in comparison with the number of unknowns (does not redefine the system of equations).

The values of $D(r)$ at points r_Γ on the edge are limited from below by the relationship, Eq. (2-431), and therefore the condition, Eq. (2-432), deprives us of a solution for which the average deviation $D - D_0$ at the edge points is minimal. The method described can therefore be called a method of minimal deviation from the critical value of dosage on the edges of the figures.

3b) We can define an objective function in such a way as to guarantee the fastest possible decay of the $D(r)$ function at the edges of the figures. This minimizes the average drift in the sizes (contours) of the figures during subsequent etching and makes the edge of the resist mask nearly vertical. We select a set $\{\Omega\}$ of points M, each of which lies a small distance $\bar{\rho}$ outside the figures being exposed (along a perpendicular to the edge) from the respective point in the set of edge points $\{\varkappa_\Gamma\}$. We require the function which is linear with respect to the unknown dosages d_i have minima,

$$P_2 = \sum_{i=1}^{M} D(\bar{r}_\Gamma + \bar{\rho}) \rightarrow \min, \qquad (\bar{r}_\Gamma + \bar{\rho}) \in \{\Omega\} \tag{2-433}$$

It is intuitively clear that a minimum P_2 will occur at the maximal rate of decay (gradient) of the function $D(\bar{r})$. In fact, we can write

$$\frac{P_2}{M} = \overline{D(r_\Gamma)} - \overline{|\nabla_\rho D|} \cdot \rho \tag{2-434}$$

where $\overline{D(r_\Gamma)}$ and $\overline{|\nabla_\rho D|}$ are the average values over the M edge points of the dosages D and the gradient of the dosage $|\nabla_\rho D|$. Considering the condition imposed on an edge, Eq. (2-431), we can also write

$$\overline{D(\bar{r}_\Gamma)} = D_0 + \overline{\Delta D} \tag{2-435}$$

where $\overline{\Delta D}$ is the average value of the deviation $D(r_\Gamma) - D_0$ over these points. Thus, a minimum in the functional P_2 of Eq. (2-433) ensures both a maximum in the gradient of the $D(r)$ function and minimal deviations of the dosage from D_0 at the figures' edge. The method described is called the method of maximal gradient.

3c) We choose a set of points $\{\Omega'\}$ that occupy an intermediate position between the figures being exposed and require a minimum in the functional

$$P_2 = \sum_{\Omega} D(\bar{r}_{\text{out}}) \rightarrow \min, \qquad \bar{r}_{\text{out}} \in G_2 \tag{2-436}$$

We call the solution found under this condition (and the conditions given by Eqs. (2-429)–(2-432) a solution by the method of minimal background exposure.

Note that all three methods can be performed by a single computer program, because the only difference between the methods is the coordinates of the points in sets $\{\varkappa_{\text{in}}\}$, $\{\varkappa_\Gamma\}$, $\{\Omega_\rho\}$, and $\{\Omega'\}$. Choosing the coordinates of the points in these sets is still an open question. We need to specify a procedure for finding both how many of these points there are and where they are located. We will give the results from some computations in which the number and the location of these points have been chosen by the operator. This choice can be automated through the use of a heuristic procedure. However, the number of points may remain an operator-specified function when this is done, beginning from some heuristic criterion that is based upon the acceptable level of accuracy in the solution.

We must also consider the question of how many figures (described by a system of equations) are being examined simultaneously. It is clear from physical considerations that two figures separated by a distance that is greater than the range of electrons in the material cannot affect one another and therefore the respective equations can be examined independently. However, there is an effect from this moderated "interaction": both figures, thanks to the proximity effect, will affect a figure located between them a distance that is less than the radius of the proximity effect and the respective equations are chained together. This kind of chaining in a topology that comprises hundreds of thousands of figures produces an immense system of equations. However, it is easy to decouple these equations because the actual influence of the proximity effect manifests itself only at distances equal to the range of the electrons. Taking a group of figures that lie within these distances, we set up and solve the system of equations. Solutions for the edge figures will be corrected by allowing for the fact that this figure will appear in the adjacent system of equations. Even simpler is the procedure in which each figure is divided into fragments, whereas the adjacent figures are considered to be uniformly exposed over their entire area. A detailed procedure for decoupling the systems of equations and subsequent joining of the solutions for the "neighboring" systems of equations has not yet been described in the literature.

We now return to the procedure for solving the entire system numerically

$$d_i \geq 0, \qquad i = 1, \dots, N \tag{2-429a}$$

$$\sum a_{ik} d_i \equiv D(\bar{r}_k) \geq D_0, \qquad \bar{r}_k \in \{\varkappa_{\text{in}}\} \tag{2-430a}$$

$$D(\bar{r}_\Gamma) \geq D_0, \qquad \bar{r}_\Gamma \in \{\varkappa_\Gamma\} \tag{2-431a}$$

$$P_2 = \sum_{i=1}^{M} D(\bar{r}_\Gamma + \bar{\rho}) \to \min, \qquad (\bar{r}_\Gamma + \bar{\rho}) \in \{\Omega\} \tag{2-433a}$$

We denote a minimum in the function P_2 for fixed $\alpha_1, \dots, \alpha_{4N}$ as P_{2m}. The magnitude of P_{2m} (as well as the corresponding dosages d_i) is uniquely determined by the method of maximal gradient discussed earlier and is a function of the α_i parameters:

$$P_{2m} = P_{2m}(\alpha_1, \alpha_2, \dots, \alpha_{4N}) \tag{2-437}$$

It remains for us to describe a method for varying the α_i parameters in order to find the d_i that correspond to the "global" minimum in the P_2 function that occurs when the α_i parameters are varied. The function $P_2(\alpha_1, \alpha_2, \dots, \alpha_{4N})$ is continuous, but its partial derivatives may contain discontinuities and therefore gradient methods cannot be used to find a minimum. A cyclic method for changing the arguments was suggested in [78]. In modern lithography machines the α_i are varied discretely in specific minimal increments. The size of this increment is taken as the minimal increment in the cyclic procedure for varying the α_i parameters and is considered in the calculation algorithm. Into this algorithm are also included restrictions that consider the minimal dimensions of the constant exposure areas.

Figure 2-21 shows the results from calculating the $D(r)$ and the arrangement of the uniformly exposed rectangular dies for a topology that comprises two rectangles with a 0.3-μm gap.(78) The number of dies N was taken as 20. The decay distance in the electron current density at the edges of a rectangular die was characterized by a parameter $\sigma_2 = 0.02\ \mu\text{m}^2$. The isolines $D(r) = \text{const}$ are shown for values of D_0 (solid line), $0.8D_0$, $0.5D_0$, and $0.25D_0$ (dashed lines). The isolines for $D(r) = D_0$ contain closed loops within the exposed figures.

First, we notice that the isoline for $D(r) = D_0$ practically merges with the edge of the figures and the isolines $0.8D_0$ and $0.5D_0$ are very close to it. This is the result from the method of maximal gradient given by Eq. (2-433). Because the individual exposure areas (dies) can be placed up against one another the number of resist areas that receive

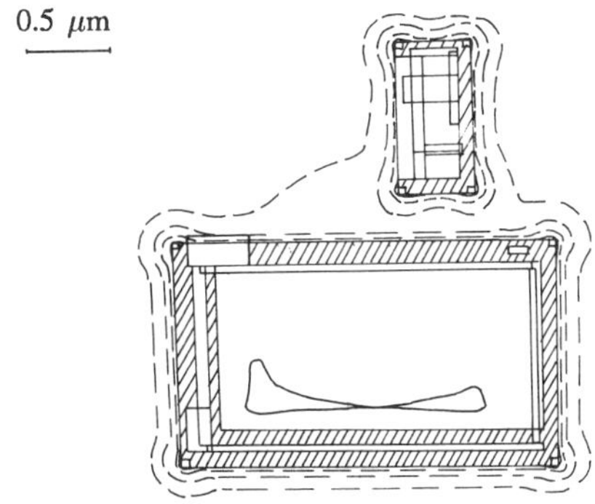

FIGURE 2-21. Isolines $D(x, y)$ = const and an optimal arrangement of dies for two rectangular elements with a 0.3-μm gap. The isoline $D(x, y) = D_0$ coincides almost completely with the element edges. Those portions of the elements that are not covered by one die are cross-hatched.

different dosages is much greater than N (for the structure shown in Fig. 2-21 there are 39 areas when $N = 20$). When the figures are optimally covered, the dies might not cover the entire area of the figure (those areas that are not covered by a single die are shown by dashed lines). The dosage parameters and coordinates of the dies are shown in Table 2-6. Along with the d_i parameters that correspond to Eq. (2-416), which are proportional to the exposure time and equal to the dosages at the centers of dies whose dimensions are much larger than σ (see Eq. (2-410)), the values $\tilde{d}_i$ of actual exposure dosages for the center of dies are also shown in Table 2-6. These values were calculated with regard for the reduction in current density in small dies (see Eq. (2-410)). Naturally, $\tilde{d}_i \simeq d_i$ for large dies (normed values of d_i/d_0 and $\tilde{d}_i/d_0$ are shown in Table 2-6, where d_0 is the exposure that yields a dosage D_0 at the center of an infinite square). The coordinates are measured from the lower left-hand corner of a large rectangle.

A comparison of the results obtained for the test structures shows that the best results are obtained via the methods of maximal gradient and minimal background exposure.[78,94,95] The method of maximal gradient automatically ensures minimal exposure in the small gaps between the figures. In light of this we may assume that the method of maximal gradient has additional advantages over the method of minimal background exposure because it optimizes the resulting dosage $D(r)$ in the edge area of each figure, thereby ensuring the least drift in this figure's size. It must be noted that by varying the (ρ, D_Γ) parameters that appear in the method of maximal gradient we can find solutions that guarantee optimal quality in a mask obtained under given conditions of developing.

2-36. THE TRANSFER FUNCTION FOR SHADOW MASKS

We can describe the ability of a system to transfer a mask pattern (object) onto a resist layer (image) by means of a system transfer function, similar to what is done in optics. We call a function that describes the profile of the electron current density distribution on the surface of a resist layer when the mask surface is uniformly irradiated a stencil transfer function. The profile of the latent image produced in a resist layer can be obtained in terms of the proximity effect by taking the convolution of the mask transfer function with the proximity function.

TABLE 2-6

N	d_i/d_0*)	$\tilde{d}_i/d_0$	x_1**)	x_2	y_1	y_2
1	0.360	0.311	0.10	0.40	0.05	0.20
2	17.751	9.701	0.15	1.35	1.13	0.20
3	0.076	0.072	1.35	1.55	0.10	0.50
4	16.000	6.124	0.10	0.15	0.18	2.40
5	1.431	1.431	0.25	1.35	0.28	2.40
6	29.694	11.365	1.30	1.35	0.18	2.35
7	19.079	7.303	0.15	1.30	2.33	2.38
8	1.121	0.338	1.40	1.45	2.20	2.33
9	9.839	4.583	1.80	1.90	1.50	1.60
10	0.867	0.798	1.90	2.70	1.50	1.60
11	0.504	0.343	1.80	1.90	1.60	1.90
12	6.733	6.713	1.90	2.70	1.60	1.90
13	6.628	2.537	2.25	2.68	1.85	1.90
14	37.095	5.437	0.00	0.05	0.00	0.05
15	36.347	5.327	1.45	1.50	0.00	0.05
16	36.267	5.315	1.45	1.50	2.45	2.50
17	32.884	4.819	0.00	0.05	2.45	2.50
18	35.835	5.252	2.75	2.80	1.50	1.55
19	34.259	5.021	2.75	2.80	1.95	2.00
20	33.703	4.939	1.80	1.85	1.95	2.00

*The quantity d_0 is equal to the exposure dosage for which the density of the energy dissipated at the center of an infinite square is equal to D_0.
**The origin of the coordinate system is the lower left-hand corner of a large rectangle.

We will calculate the transfer function for a shadow mask containing a slit a wide and $b \gg a$ long. Let the mask thickness be h and the gap between the mask and the surface of the resist layer be H (Fig. 2-22). We choose the coordinates system as in Fig. 2-22b: the illumination system of the machine forms a Gaussian electron beam of radius R that impinges on the mask. We will consider the variance in the lengthwise component of the electrons velocity to be zero and the transverse component to be Maxwellian distributed with a variance $v_{\perp 0}$. The distribution function of the electron beam on the surface of the mask is given by[97]

$$f_0(x, y, v_\perp, v_z) = \varkappa \exp\left(-\frac{x^2+y^2}{R^2}\right)\exp\left(-\frac{v_\perp^2}{v_{\perp 0}^2}\right)\delta\left(\frac{v_z}{v_0}-1\right) \qquad (2\text{-}438)$$

$$\varkappa = I/(q v_0^3 \pi^2 R^2 \theta_0^2), \qquad \theta_0 \approx \tan\theta_0 = v_{\perp 0}/v_0 \qquad (2\text{-}439)$$

I is the total beam current and q is the electronic charge. When $R \gg a, b$ the electron distribution in the area of the slit can be considered uniform.

It is well known that the transfer function for optical masks in a printing scheme with a gap is determined exclusively by diffraction effects, because the geometric dimensions of the transparent slits in the mask are on the order of the wavelength λ of the illumination that transfers the image: $\lambda \leq a, b$. Conversely, when transferring an image by an electron beam having an energy of E = 10–100 keV the de Broglie wavelength of the electrons is $\lambda_e \ll a, b$ and diffraction does not play a defining role in forming the mask transfer function.

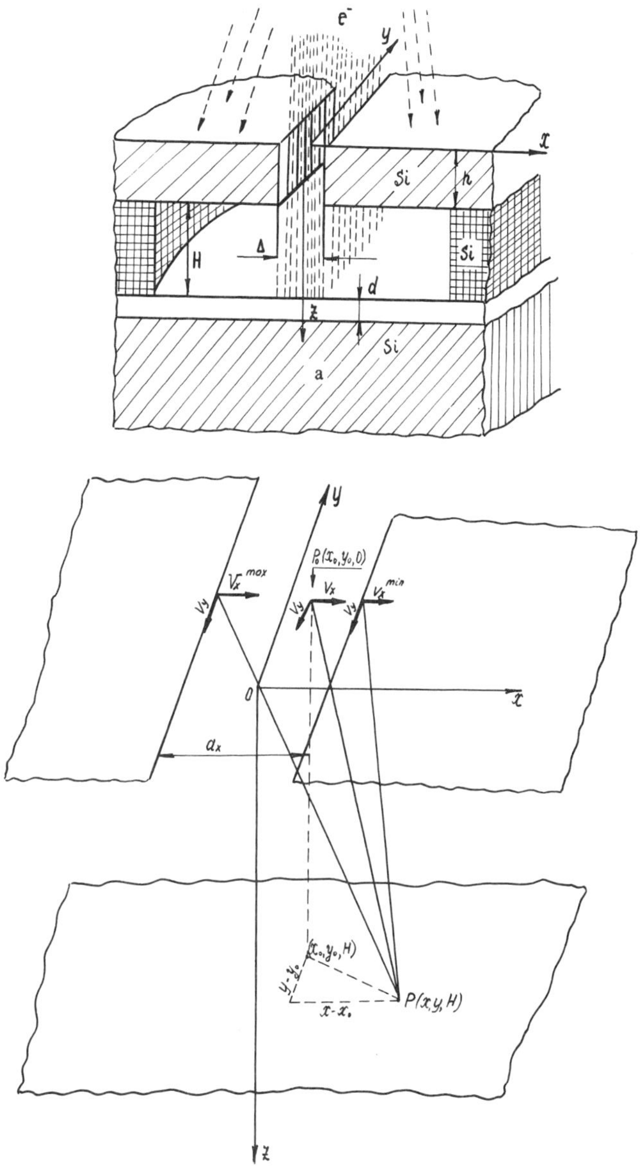

b

FIGURE 2-22. a) Scheme for electron lithography using shadow masks; b) possible electron trajectories; scheme for analytically calculating the transfer function of a mask having an isolated slit.

An electron current passing through a mask to a point $\bar{r}$ on the resist layer surface can be divided into three components: the current $I_0(r)$ that does not interact with the mask as it passes through a window, the current $I_s(r)$ from electrons scattered by the walls of the slit, and the current I_f that initially impinges on the shaded part (background) of the mask.

The intensity of the background I_f is, obviously, governed by the transmittance of the h-thick film (see Section 2-29). The number of electrons that pass through a film of given thickness decreases fairly monotonically (see Fig. 2-9). When the mask thickness h is much less than the magnitude of the projected electron range, the mask cannot provide a high degree of transmission contrast in the image of the mask slits. However, we must consider the three-dimensional (Eq. (2-291)) and the angular (Eq. (2-290)) spread of a beam propagating through the material of the mask. The dispersion in the scattering angle is 1 radian for a thickness of $h = \Lambda/2$, where Λ is the electron momentum loss length (roughly equal to the length of the projected electron range). This means that when $h = \Lambda/2$ the background current electrons leave the mask with a large angular distribution and, when the windows in the mask surface are arranged more or less uniformly and the gap H is finite, produce uniform illumination on the mask surface. Homogenization of the background electron distribution over the surface of the resist layer lowers the reduction in contrast electrons that initially impinge on the dark area of the mask surface. The average intensity of the background is

$$I_f = \tau(h) \cdot S \tag{2-440}$$

where $\tau(h)$ is the beam transmission coefficient under specific conditions (defined by beam energy, thickness h, and substrate material), and S is the dark area of the mask. Hereafter, we will consider the intensity of the background current to be small.

The number of electrons impinging on the side walls of a slit in the mask will be smaller as the angular divergence of the beam θ_0 and the mask thickness h: $I_s \sim \theta_0 \cdot h$ are smaller. Under the condition that I_f and I_s be small the mask can be thought of as an infinitesimally thin diaphragm that is impenetrable by particles in the dark area.

We will find the distribution of the electrons intensity $I(x, y, H)$ on the surface of a resist $z = H$, considering the electron distribution on the surface of the mask $z = 0$ to be described by Eq. (2-438).

We can easily see that an electron starting from a point $P(x_0, y_0, 0)$ in the plane of the mask surface impinges on a point $P(x, y, H)$ in the plane of the resist's surface if the v_x, v_y components of its transverse velocity satisfy the conditions

$$(x - x_0) = \left(\frac{v_0}{H}\right)^{-1} v_x, \qquad (y - y_0) = \left(\frac{v_0}{H}\right)^{-1} v_y \tag{2-441}$$

The probability density of this event is

$$f_{P_0 \to P}(x, v_x; y, v_y) = \varkappa \exp\left(-\frac{x_0^2 + y_0^2}{R^2}\right) \exp\left(\frac{v_x^2 + v_y^2}{v_0^2 \theta_0^2}\right) \delta\left(\frac{v_z}{v_0} - 1\right) \tag{2-442}$$

where the dependencies $x_0(x, v_x)$, $y_0(y, v_y)$ are given by Eq. (2-441).

The total electron current at the point $P(x, y, H)$ will be

$$I(x, y, H) = \int_{-\infty}^{+\infty} dv_y \int_{-\infty}^{+\infty} dv_z \int_{v_{x\min}}^{v_{x\max}} dv_x\, f_1(x, y, v_x, v_y) \tag{2-443}$$

$$v_{x\min} = \left(x - \frac{a}{2}\right)\frac{v_0}{H}, \qquad v_{x\max} = \left(x + \frac{a}{2}\right)\frac{v_0}{H} \tag{2-444}$$

The limits of integration over v_y account for the condition that $b \gg a$. Performing the integration, we obtain[97]

$$I(x, y, H) = j_0 \exp\left(-\frac{x^2 + y^2}{R_1^2}\right) F(x, H) \tag{2-445}$$

$$F(x, H) = \frac{1}{2}[\operatorname{erf}(\xi_+) - \operatorname{erf}(\xi_-)] \tag{2-446}$$

$$\operatorname{erf}(\xi) = \frac{2}{\sqrt{\pi}} \int_0^{\xi} \exp(-\xi'^2)\, d\xi' \tag{2-447}$$

$$\xi_\pm = \frac{x \pm (a/2)(R_1/R)^2}{H\theta_0/(R_1/R)}, \qquad R_1 = (R^2 - H^2\theta_0^2)^{1/2} \tag{2-448}$$

$$j_0 = I_0/(\pi q R_1^2) \tag{2-449}$$

We assume that the electron current impinging on the mask is uniform: $R_1 \to \infty$. Then, according to Eq. (2-445), $I(x, y, H) = j_0 F(x, H)$; i.e., the function $F(x, H)$ is the mask's transfer function. In the extreme case of no gap

$$\lim_{H \to 0} F(x, H) = \begin{cases} 1, & \text{when } |x| \leq a/2 \\ 0, & \text{when } |x| > a/2 \end{cases} \tag{2-450}$$

which corresponds to our assumption that the dark area of the mask is impenetrable.

The width of the slit can be conveniently measured in units of $H \cdot \theta_0$: $\xi = a/H\theta_0$. At a fixed H the function $F(x, H)$ has a maximum at $x = 0$ and points of inflection at $x = \pm x_n$, which are given by

$$\frac{a}{2x_n} \ln\left(\frac{1 + a/2x_n}{1 - a/2x_n}\right) = \left(\frac{a}{H\theta_0}\right)^2 \tag{2-451}$$

As $H \to 0$, $x_n \to a/2$, and when $\xi > 1$, $x_n > a/2$. Figure 2-23 shows the transfer function profiles for a mask with a slit for different values of ξ obtained by numerical modelling and the analytical expression, Eq. (2-446). The high degree of steepness in decay at the edge of the slit is preserved when $\xi \geq 10$ or $H \leq 0.1a/\theta_0$. We can describe the steepness of the transfer function decay by β, the angle of deflection to the ordinate axis at the transfer function's point of inflection on the transfer function's graph:

$$\beta = \operatorname{arccot}(|dF/d(x/a)|x_n) = \xi(1 - \exp(-\xi^2)) \tag{2-452}$$

Values of $\beta \leq 5°$ occur when $\xi \geq \xi_0 = 20$. Assuming that $\xi > \xi_0$ and specifying for H and θ_0 the minimal practically attainable values $H = 5\ \mu$m and $\theta_0 = 1$, we can estimate the maximal resolving power of a mask: $a_{\min} = 0.07\ \mu$m.

The mask transfer function in the two-dimensional case is obtained by taking the product of the one-dimensional transfer functions:

$$F(x, y, H) = F(x, H) F(y, H) \tag{2-453}$$

where $F(y, H)$ is given by Eq. (2-446) with the substitutions $x \to y$ and $a \to b$. For a mask having a large number of windows (arranged in regular or irregular fashion)

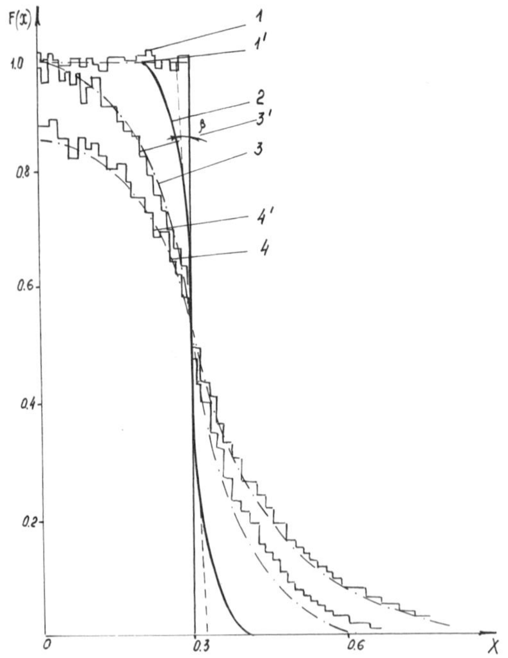

FIGURE 2-23. Profiles of the transfer function for a mask having an isolated slit $a = 0.6\ \mu$m wide when $\xi = \infty$ (1, 1′), 10 (2), 4 (3, 3′), 2 (4, 4′), obtained by numerical modelling, (1′–3′), and from the analytical expression, Eq. (2-446), (1–4).

the mask transfer function is obtained by simply adding the transfer functions of the individual windows, considering the location of each window relative to the axis of the beam, which is measured from the origin of the coordinate system:

$$F_M(x, y, H) = \sum_{i=1}^{M} F(x - x_{i0}, H)F(y - y_{i0}, H) \tag{2-454}$$

$$I_M(x, y, H) = j_0\, e^{-(x^2+y^2)/R_1^2} F_M(x, y, H) \tag{2-455}$$

x_{i0}, y_{i0} are the coordinates of the center of the ith rectangle, and M is the number of windows in the mask.

The contribution from any window to the transparency of the mask, owing to the absence of diffraction and interference effects (they are small), in exposing an arbitrary point is calculated in a similar manner to calculating the contribution from an isolated window, as in Eq. (2-448). In doing this only the contribution from windows in close proximity to the chosen point is significant. Calculations for a regular structure with windows of submicron size ($a = b = 0.6\ \mu$m) show that when $\xi > 10$, the overlap from the transfer functions of adjacent windows can be neglected. When $\xi = 4$, weak overlapping from the transfer functions of adjacent windows occurs. As ξ is reduced further, strong overlapping from the transfer functions of adjacent windows occurs, and when $\xi \leq 1$ an almost uniform level of illumination is achieved.[97] When windows of different size are present on the mask, the value of the $H \cdot \theta_0$ parameter must be chosen for the condition that the image from the smallest window be transferred with high quality.

It is also interesting to examine the case in which the beam impinges on the mask at an oblique angle. Let α be the angle between the beam axis and the normal to the mask. Calculations similar to those for the case of normal incidence show that the transfer function during oblique incidence coincides with the transfer function during normal incidence if we make the substitutions

$$x \to x_\alpha = (x - H \tan \alpha)\left(1 - B\frac{\theta_0^2 H^2}{R^2}\right) \qquad \text{(2-454a)}$$

$$H \to H_\alpha = H(1 + B), \qquad B \equiv \left(x + \frac{a}{2}\right)\frac{\tan \alpha}{H} \qquad \text{(2-456)}$$

It follows from Eq. (2-454a) that when the beam impinges at an oblique angle the image from the slit is displaced a distance $\Delta x = H \tan \alpha$ and a change in the image scale of the slit occurs; the scale factor is $(1 - B(x)(\theta_0^2 H^2/R^2)$. In transferring an image in the two-dimensional case (windows instead of a slit) we can introduce the projection of the unit vector **e** along a ray on the coordinate axis: $\mathbf{e}(\sin \alpha \cos \varphi, \sin \alpha \sin \varphi, \cos \alpha)$. Taking the smallness of the angle α into account, we can make the substitutions $\sin \alpha \cos \varphi = \alpha \cos \varphi \equiv \alpha_x$, $\sin \alpha \sin \varphi = \alpha \sin \varphi \equiv \alpha_y$. The transfer function for a beam impinging a rectangular window at an oblique angle will coincide with Eq. (2-446) when the substitutions $x \to x_{\alpha_x}$, $H \to H_{\alpha_x}$ and $y \to y_{\alpha_y}$, $H \to H_{\alpha_y}$ are made.

The angular divergence of the beam leads to electrons impinging on the lateral sides of the windows in the mask and reduced beam intensity. It is not hard to estimate these losses. Starting from Eq. (2-445) we calculate the number of particles that intersect a slit of width a when $z = h$. To do this, we must integrate Eq. (2-445) over the width of the slit. Then we find the relative number of electrons lost through impinging on the (h high) walls of the slit:

$$\frac{\Delta N(h)}{N(0)} = \frac{N(0) - N(h)}{N(0)} = 1 - \frac{a \cdot \operatorname{erf}(\xi_1) - \dfrac{2}{\sqrt{\pi}}\dfrac{a}{\xi_1}(1 - \exp(-\xi_1^2)}{\sqrt{\pi}\, R \operatorname{erf}(a/2R)} \qquad \text{(2-457)}$$

Considering the conditions $\xi_1 = (a/h\theta_0) \gg 1$ and $(a/2R) \ll 1)$ to be satisfied, we can use the asymptotic representations $\operatorname{erf}(\xi)$ when $\xi \gg 1$ and find

$$\frac{\Delta N(h)}{N(0)} = \frac{2h\theta_0}{\sqrt{\pi}\, a} + O(\xi_1^{-2}) \qquad \text{(2-458)}$$

In the situation that we are interested in, $h \geq a$, the beam electrons lost is defined by the angular divergence; when $h = a$, $\theta_0 = 5'(\Delta N/N) < 0.2\%$.

REFERENCES

1. G. R. Brewer (editor), *Electron Beam Technology in Microelectronic Fabrication*, Academic Press, New York (1980), pp. 1–59.
2. J. N. Helbert, C. F. Cook, and E. H. Poindexter, *Pol. Eng. and Sci.*, **17**, No. 6, 414–419 (1977).
3. E. Gipstein, A. C. Ouano, D. E. Johnson, and O. U. Need, *J. Res. Develop.*, **21**, No. 2, 143–153 (1977).
4. M. G. Rosenfield and A. R. Neureuther, *IEEE Trans.*, **ED-28**, No. 11, 1289–1294 (1981).
5. K. Harada, O. Kogure, and K. Murase, *IEEE J.*, **SC-17**, No. 2, 148–154 (1982).

6. S. Hattori, J. Tamano, and M. Yamada, *Thin Solid Films*, **83**, 189–194 (1981).
7. R. F. Dashen. "Theory of electron backscattering," *Phys. Rev.*, **134**, No. 4A, A1025–1032 (1964).
8. "The use of Monte Carlo calculations in electron probe microanalysis and scanning microscopy," Proc. of a Workshop held at the NBS, Gaithersburg, MD, 1975, NBS-81 (1976).
9. K. Murata D. F. Kyser, and C. H. Ting, *J. Appl. Phys.*, **52**, No. 7, 4396–4405 (1981).
10. I. Adesida, R. Shimizu, and T. E. Everhart, *J. Appl. Phys.*, **51**, No. 11, 5692–5969 (1980).
11. I. Adesida, R. Shimizu, and T. E. Everhart, *Appl. Phys. Lett.*, **33**, No. 10, 849–850 (1978).
12. R. Shimizu, Y. Kataoka, and T. Ikuta, *J. Appl. Phys.*, **9**, 101–114 (1976).
13. D. Stephani, *J. Vac. Sci. Technol.*, **16**, No. 6, 1739–1742 (1979).
14. D. F. Kyser and R. Pyle, *IBM J. Res. Develop.*, **24**, No. 4, 426–437 (1980).
15. I. Adesida and T. E. Everhart, *J. Appl. Phys.*, **51**, No. 11, 5994–6005 (1980).
16. N. Aizaki, *Jap. J. Appl. Phys.*, **18**, 319–325 (1979).
17. A. F. Akkerman, Yu. M. Nikitushev, and V. A. Botvin, *Solving Problems in the Transmission of Fast Electrons in Matter by the Monte Carlo Method* [in Russian], Nauka, Alma Ata (1972).
18. L. D. Landau and E. M. Lifshits, *Quantum Mechanics* [in Russian], Fizmatgiz, Moscow (1963).
19. N. Mott and H. Massey, *Theory of Atomic Collisions* [Russian translation], Inostrannaya Literatura, Moscow (1951).
20. H. S. W. Massey, *Adv. in Electronics*, **4**, No. 4, 2–68 (1952).
21. A. S. Davydov, *Quantum Mechanics* [in Russian], Fizmatgiz, Moscow (1963).
22. E. S. Parilis, *The Auger Effect* [in Russian], Izd-vo. AN UzbSSR, Tashkent (1969).
23. P. Gombash, *The Problem of Multiple Particles in Quantum Mechanics* [Russian translation], Inostrannaya Literatura, Moscow (1953).
24. L. D. Landau and E. M. Lifshits, *Mechanics* [in Russian], Fizmatgiz, Moscow (1958).
25. D. Liljequist, *J. Appl. Phys.*, **57**(3), 657 (1985).
26. M. A. El'yashevich, *Atomic and Molecular Spectroscopy* [in Russian], Fizmatgiz, Moscow (1962).
27. H. Bethe and E. Salpeter, *Quantum Mechanics of One- and Two-Electron Atoms* [Russian translation], Fizmatgiz, Moscow (1960).
28. I. I. Sobel'man, *Introduction to the Theory of Atomic Spectra* [in Russian], Fizmatgiz, Moscow (1963).
29. C. J. Powell, Proc. of a Workshop held at the NBS, Gaithersburg, MD, 1975. NBS (1976), pp. 97–104.
30. W. Lotz, *Z. Phys.*, **232**, No. 2, 101–107 (1970).
31. M. Grisinsky, *Phys. Rev.*, **138**, A336–A358 (1965).
32. H. A. Bethe and J. Ashkin, *Experimental Nuclear Physics*, J. Wiley, New York (1953).
33. A. S. Davydov, *Theory of Solids* [in Russian], Nauka, Moscow (1976).
34. D. Paines, *Elementary Excitations in Solids* [Russian translation], Mir, Moscow (1965).
35. L. D. Landau, *Zh. Eksp. Teor. Fiz.*, **16**, 574 (1946).
36. E. M. Lifshits and L. P. Pitaevskii, *Physical Kinetics* [in Russian], Nauka, Moscow (1979).
37. L. A. Artsimovich and R. Z. Sagdeev, *Plasma Physics for Physicists* [in Russian], Atomizdat, Moscow (1978).
38. V. L. Ginsburg, *Theoretical Physics and Astrophysics* [in Russian], Nauka, Moscow (1975).
39. C. Kittel, *The Quantum Theory of Solids* [Russian translation], Nauka, Moscow (1967).
40. P. M. Platzman and P. A. Wolff, *Waves and Interactions in a Solid Plasma* [Russian translation], Mir, Moscow (1975).
41. S. F. Edwards, *J. Proc. Sci.*, **72**, Pt. 5, No. 467, 685–693 (1958).
42. R. Feynman, *Statistical Mechanics* [Russian translation], Mir, Moscow (1978).
43. A. R. Shulman (editor), *Characteristic Electron Energy Losses in Solids. Selected Articles* [Russian translation], Inostrannaya Literatura, Moscow (1959).
44. T. Goulet and J. Jay-Gerin, *P. Radiat. Phys. Chem.*, **27**, No. 3, 220–239 (1986).
45. E. Eggarter, *J. Chem. Phys.*, **62**, 833 (1975).
46. B. E. Springet, J. Jorther, and M. H. Kohen, *J. Chem. Phys.*, **48**, 2720 (1968).
47. J. S. Greeneich and T. Van Duzer, *IEEE Trans. Electron Devices*, 598–600 (1973).
48. L. D. Landau, *The Energy Lost by Fast Particles in Ionization. Collected Works, Vol. 1* [in Russian], Nauka, Moscow (1969), pp. 482–490.
49. G. Shinoda, T. E. Everhart, R. F. Herzog, M. S. Chung, and W. J. Devore, Proc. of the Sixth Intern. Conf. on X-ray Optics and Microanalysis. University of Tokyo Press (1972), pp. 81–86.
50. A. R. Neureuther, D. F. Kyser, and C. H. Ting, *IEEE Trans.*, **ED-26**, No. 4. 686–692.
51. J. Henos and F. Maurice, Proc. of a Workshop held at the NBS, Gaithersburg, MD, 1975, NBS (1976), pp. 61–69.
52. G. Moliere. *Z. Naturforsch.*, **2a**, No. 3, 133–145 (1947); **3a**, No. 2, 78–97 (1948).

53. H. W. Lewis, *Phys. Rev.*, **78**, 526–529 (1950).
54. B. Rossi and K. Greisen, *Rev. Mod. Phys.*, **13**, No. 4, 262–268 (1941).
55. H. A. Bethe, M. E. Rose, and L. P. Smith, *Proc. Amer. Phil. Soc.*, **78**, 573–585 (1938).
56. R. J. Hawryluk, A. M. Hawryluk, and H. I. Smith, *J. Appl. Phys.*, **45**, No. 6, 2551–2566 (1974).
57. S. Reid, *Electron Probe Microanalysis* [Russian translation], Mir, Moscow (1979).
58. E. Jahnke and F. Emde, *Tables of Functions with Formulae and Curves* [Russian translation], Mir, Moscow (1949).
59. N. G. Nakhodkin, A. A. Ostroukhov, and V. A. Romanovskii, *Fiz. Tverd. Tela*, **4**, 1514–1523 (1962); **7**, 210–216 (1965).
60. D. E. Newbury and H. Yakowitz, "The use of Monte Carlo calculations in electron probe microanalysis and scanning electron microscopy," Proc. of a Workshop held at the NBS, Gaithersburg, MD, 1975, NBS (1976), pp. 15–44.
61. Lin Yi-Ching and A. R. Neureuther, *IEEE Trans.*, **ED-28**, No. 11, 1397–1404 (1981).
62. A. J. Antolak and W. Williamson, *J. Appl. Phys.*, **58**(1), 526–534 (1985).
63. A. J. Antolak and W. Williamson, *J. Appl. Phys.*, **58**, No. 10, 3687–3691 (1985).
64. F. Arnal, P. Verdier, and P. D. Vinsensini, *C. R. Acad. Sci.*, **268**, 1526 (1969).
65. A. F. Akkerman and A. L. Gibrekhterman, *Mikroelektronika*, **12**, No. 1, 78–80 (1983).
66. G. W. C. Kaye and T. H. Laby, *Tables of Physical and Chemical Constants* [Russian translation], Fizmatgiz, Moscow (1962).
67. A. Charlesby, *Nuclear Radiation and Polymers* [Russian translation], Inostrannaya Literatura, Moscow (1962).
68. M. Parikh, *IBM J. Res. Develop.*, **24**, No. 4, 438–451 (1980).
69. M. Nakase and M. Yoshimi, *IEEE J.*, **SC-15**, No. 4, 525–530 (1980).
70. A. M. Carrol, *J. Appl. Phys.*, **52**, No. 1, 434–437 (1981).
71. N. Aizaki, *J. Vac. Sci. Technol.*, **16**, No. 6, 1726–1733 (1979).
72. T. H. P. Chang, *J. Vac. Sci. Technol.*, **12**, 1271 (1975).
73. M. J. Parikh, *J. Vac. Sci. Technol.*, **15**, 931 (1978).
74. R. J. Hawryluk, *J. Vac. Sci. Technol.*, **19**, No. 1, 1–17 (1981).
75. M. Parikh and D. F. Kyser, *Appl. Phys.*, **50**, No. 2, 1104–1111 (1979).
76. N. Sigiyama, K. Saitoh, K. Shimizu, and Y. Tarui, *IEEE Trans., Electron. Dev.*, **ED-27**, No. 8, 1466–1474 (1980).
77. T. Okubo and K. Takemoto, *Jap. J. Appl. Phys.*, **22**, No. 8, 1335–1341 (1983).
78. K. A. Valiev, B. N. Kovtun, and T. M. Makhviladze, Preprint, IOFAN, No. 194, 60; No. 209, 26 (1987).
79. J. S. Greeneich, 9th Intern. Conf. on Electron and Ion Beam Science and Technology, R. Bakish (editor), The Electrochemical Society, Princeton (1980), pp. 282–303.
80. A. R. Neureuther, *TIIER*, **71**, No. 1, 149–158 (1983).
81. M. G. Rosenfield and A. R. Neureuther, *IEEE Trans. Electr. Dev.*, **B3**, No. 1, 153–164 (1985).
82. G. Owen, P. Rissman, and M. F. Long, *J. Vac. Sci. Technol.*, **B3**, No. 1, 153–158 (1985).
83. J. Swell, *J. Vac. Sci. Technol.*, **15**, 927 (1978).
84. M. Parikh, *J. Appl. Phys.*, **50**, No. 6, 4371–4387 (1979).
85. N. D. Wittels and Youngman, 8th Intern. Conf. on Electron and Ion Beam Science and Technology, R. Bakish (editor), The Electrochemical Society, Princeton (1978), p. 361.
86. V. P. Derkach, A. A. Mervinskii, and L. V. Starikova, *Mikroelektronika*, **14**, No. 6, 467–477 (1985).
87. D. Kern, 9th Intern. Conf. on Electron and Ion Beam Science and Technology, R. Bakish (editor), The Electrochemical Society, Princeton (1980), pp. 326–339.
88. D. G. L. Chow, J. F. McDonald, D. C. King, W. Smith, K. Molnar, and A. J. Steckl, *J. Vac. Sci. Technol.*, **B1**, No. 4, 1383–1390 (1983).
89. A. M. Carrol, *J. Vac. Sci. Technol.*, **19**, 1296–1299 (1981).
90. P. Hendy and M. E. Jones, Microcircuit Engineering '82 International Conf. on Microlithography, Grenoble. France, 5–8 October 1982, pp. 22–27.
91. W. T. Lynch, T. E. Smith, and W. Fichtner, *ibid.*, pp. 309–314.
92. J. S. Greeneich, *J. Vac. Sci. Technol.*, **19**, No. 4, 1269–1274 (1981).
93. V. D. Verner, Yu. N. D'yakov, A. N. Kononov. D. P. Polikarpov, I. Ya. Smolyanitskii, A. F. Urakhchin, and M. Yu. Shcherbakova, Works from the international Conference on Electron Beam Technologies (Varna, May 26–June 2, 1985), Izd-vo IE BAN, Sofia (1985), pp. 510–515.
94. K. A. Valiev, A. N. Kirillov, B. N. Kovtun, T. M. Makhviladze, and M. M. Mkrtchyan, *Problems in Microlithography (Works of the IOFAN, Vol. 8)* [in Russian], Nauka, Moscow (1987).

95. K. A. Valiev, A. N. Kirillov, B. N. Kovtun, T. M. Makhviladze, and M. M. Mkrtchyan, *Mikroelektronika*, No. 6, 122–130 (1987).
96. D. B. Yudin and E. G. Gol'shtein, *Linear Programming, Theory, Methods, and Applications* [in Russian], Nauka, Moscow (1969).
97. K. A. Valiev, A. N. Kirillov, T. M. Makhviladze, and M. M. Mkrtchyan, Preprint, IOFAN, No. 215 (1987).

3

The Physics of Ion-Beam Lithography

3-1. THE FUNDAMENTAL PROCESSES IN ION LITHOGRAPHY

Ion beams are used for very diverse purposes in electronics and can be divided into three categories: 1) micromilling by removing material, 2) controlling changes in the properties of a material by doping, and 3) analyzing the structure and composition of materials. Figure 3-1 shows the physical processes that occur when an ion beam $E <$ 1 MeV interacts with a solid target and how they are used in electronic technology and diagnostics. The parameters of the beams used in technology and diagnostics work are shown in Table 3-1, and the energy and dose ranges of the ion beams are shown pictorially in Fig. 3-2.[(2)] Thus, the physical theory presented in this chapter may be thought of as the theoretical foundation for not only ion-beam lithography, but for ion-beam technology and diagnostics as well. As can be seen, the energies of the beams used in electronics range from a few keV to 100 MeV and the doses are 10^9–10^{19} ions/cm^2. A surface is machined by sputtering the material with low-energy ions. A controlled change in a material's properties is accomplished by implanting fixed amounts of impurities in the material. Familiar examples of changing the properties of a semiconductor material are the creation of "buried layers," isolating transistor regions in an integrated circuit, and altering the charge in the SiO_2 oxide of a SiO_2–Si structure in order to change the threshold voltage of a field-effect transistor, etc.

Interest in ion beams as tools for microlithography is comparatively recent, and the number of experiments and amount of work done in ion-beam lithography is still quite small.[(3,4)] The physical processes that are the basis of ion-beam lithography, however, have been developed to a significant level both theoretically and experimentally. This is because wide ion beams have been used for a long time in electronics to implant impurities. A great deal of research in ion implantation has been done and the results obtained make it possible for us to evaluate the capabilities of ion-beam lithography and understand the physical processes that are the basis of this method. In turn, ion implantation has come about because of research in nuclear physics that advances our understanding of how high-energy atomic and ionic beams interact with matter. Therefore, ion-beam lithography is based on widespread physical experimentation and extensively developed

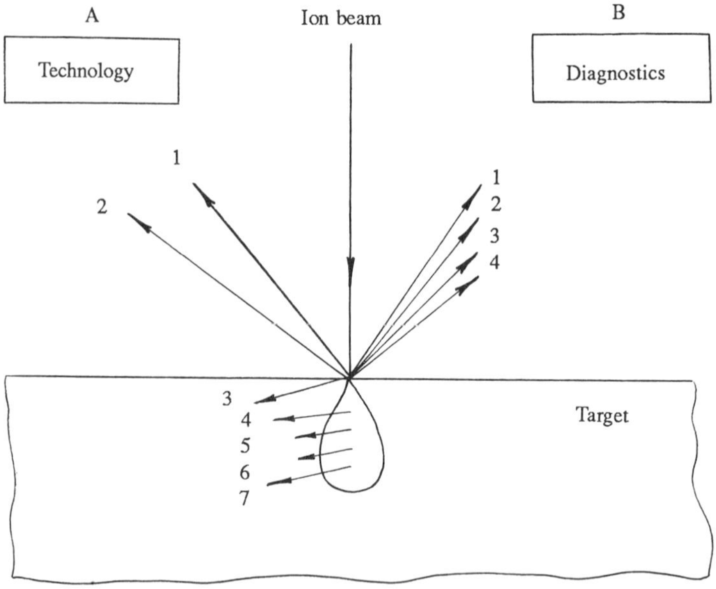

FIGURE 3-1. Ion beams in technology and diagnostics. A) Technology: 1) sputtering a target for micromilling or for depositing sputtered atoms onto substrates; 2) aligning a pattern on a target and mask by secondary electrons; 3) depositing beam ions on targets; 4) implanting active dopants into solid targets; 5) lithography on inorganic films; 6) lithography on organic films; 7) mixing the materials in multilayer devices. B) Diagnostics: 1) using Rutherford backscattering to analyze the structure and chemical composition of a target; 2) secondary ion mass spectrometry; 3) chemical analysis by electron spectrometry (ESCA) and by Auger electron spectrometry (AES); 4) using x-rays to analyze the chemical composition.

theory. It is to be hoped that all of this will inspire the rapid development and widespread use of ion-beam microlithography in the realm of microelectronics.

As we did in discussing the physical problems encountered in electron-beam lithography, we can relate the topics of ion sources that form ion beams of submicron cross section and the physics of ion-beam interactions with matter to the physics of ion-beam lithography. Our knowledge of electron beams greatly simplifies our discussion of these topics. For example, because the theory of electron optics is totally applicable to ion beams the need to discuss ion optics is obviated. However, there are areas in which ion beams are quite different from electron beams and a detailed discussion of ion-beam physics is essential. This has mainly to do with the physics of ion-beam interactions with matter. Two factors distinguish the physics of ion beam interactions with matter from those of electron beams: the much greater mass of an ion relative to that of an electron, and the fact that an ion is essentially a multielectron system, which leads us to the fact that the energy of interaction between an ion and the atoms of a target is a more complex function of the distance between the ion and the target's atoms than can be described by Coulomb's law. The large mass of an ion produces a number of significant results, chief among which is, perhaps, the fact that a heavy ion has a great deal of momentum $m_1 v_1$. Transmitting this momentum to a target's atoms during a collision may dislodge the latter from the lattice sites and produce a vacancy–interstitial atom type of defect (Frenkel defects). The dislodged atom may acquire enough energy from the ion that it too may dislodge other atoms during a collision, producing a series of collisions that create a large number of Frenkel defects. At the same time, a fast electron having $E \leq 10^5$ eV of energy has so little momentum that it is incapable of dislodging atoms

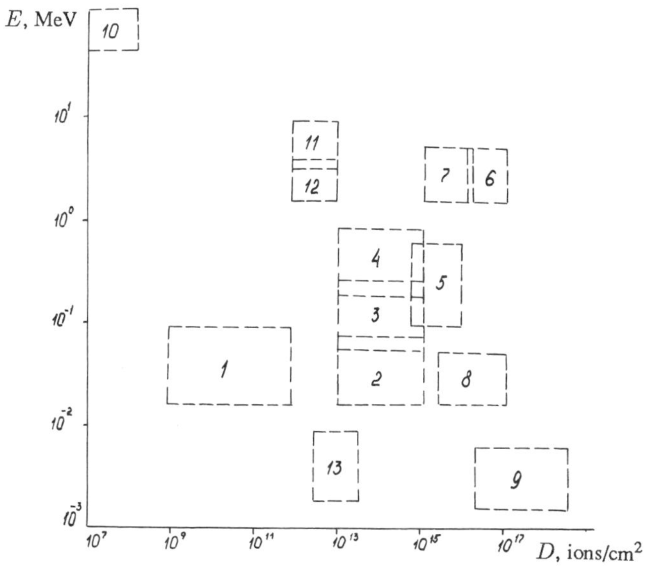

FIGURE 3-2. Ranges of energies and doses of ion radiation in electronics: 1) varying the charge in an oxide to adjust the threshold voltage in MOS transistors; 2) corralling a dopant prior to diffusion; 3) ion lithography; 4) doping the collector of an IIL bipolar transistor; 5) implanting dopants into A_3B_5 group semiconductors; 6) implanting buried layers in a bipolar transistor; 7) implanting protons in GaAs to produce isolation; 8) fabricating planar photoconductors; 9) ion etching; 10) fabricating nuclear filters in polymer films; 11) using ion-induced nuclear reactions to analyze chemical composition; 12) using ion backscattering and ion-induced x-ray radiation to analyze the structure and chemical composition of a material; 13) using beams of low-energy ions to analyze the structure of a surface.

from the lattice sites. Further, for a given energy level, the velocity of the particles in a beam is inversely proportional to the square root of the mass: $v = \sqrt{2E/m}$. Thus, at $E = 10^4$ eV of energy the velocity of an electron is $v_e = 0.58 \times 10^9$ cm/sec, whereas the velocity of argon ions is only 2.15×10^6 cm/sec. This difference between the velocities of ions and electrons creates a substantial difference in the ionizing ability of ions and electrons. The fact is that the probability that an impinging particle will ionize an atom is high only when the velocity of the impinging particle is comparable to (or greater than) the velocity of the atom's electron, roughly $v_a = 10^8$ cm/sec. If the velocity of the impinging particle is $v_i \ll v_a$, the probability that the atom will be ionized is sharply reduced and the collision will become, as they say, adiabatic. When the energies of the ion beam are relatively low the basic losses in the energy of fast particles will come from dislodging atoms from the lattice sites, not by ionizing and exciting a target's atoms as with an electron beam. For intermediate beam energies ($v_i \approx v_a$) the ionizing losses must be considered along with the losses caused by dislodging atoms.

The ion–atom collisions that result from dislodging target atoms are called nuclear collisions and the energy losses incurred are called nuclear losses. These collisions are also called elastic because in (almost) all of them the kinetic energy of the particles is maintained up to and after the collision. The energy that fast ions lose in ionizing and exciting a target's atoms is called electron loss.

From the physical point of view, an atom–atom or an ion–atom collision can be described in semiclassical terms, describing the atoms and ions by their trajectories and the collision by the impact parameter, which makes the theory more comprehensible and, sometimes, simpler.

TABLE 3-1. Ion Beam Process Characteristics[1]

Beam parameters	Lithography	Ion implantation	Etching	Deposition	Surface modification	Diagnostics
Size	0.1–0.3 μm (focused beams), 1–2 cm (wide beams)	1–1.5 cm	1–10 cm	1–10 cm	1–2 cm	0.1 μm
Ion type	H^+, He^+, O^+, Ga^+	Doping materials	F, Cl, O	Metals, dielectrics, semiconductors	All ions	All ions
Energy (keV)	1–50	1–1000	10^{-3}–10	10^{-3}–10	1–1000	1–30
Current (A)	10^{-9}–10^{-6} (focused beams), 10^{-3}–10^{-2} (wide beams)	1–10^{-3}	Large	Large	1–10^{-3}	10^{-11}–10^{-6}
Purity	Low	High	Medium	High	Medium	High
Angle of incidence, deg	90	Variable	Variable	Variable	90	Variable
Dose, ions/cm^2	10^9–10^{16}	10^{12}–10^{19}	10^{14}–10^{18}	Maximum possible	10^{13}–10^{19}	Minimal

We will now return to the subject of ion-beam lithography. A thin ion beam has much weaker angular scattering in a target than does an electron beam and produces much higher resolution in the lithography process. Reflection of an ion beam is also much less pronounced and the effects due to the reflected ions exposing a polymer layer are insignificant. In a polymer layer the total nuclear and electron losses from an ion beam are much (roughly 100 times) greater than the energy lost from an electron beam; therefore, polymer resists are 10–100 times more sensitive to an ion beam than to an electron beam. This gives us hope that a thin ion beam will yield a faster process of exposure. However, the exposure rate will be limited by the beam control process. Visualize a beam being deflected onto a given point by an electric field produced by parallel, plane plates Δz long. The ions travel a distance Δz in a period of time $t_i = \Delta z / v_i$, where v_i is the velocity of the ions. The time required to move the beam from one exposure point to the next is $t > t_i$. If the exposure time for the point is $\tau < t_i$, then it is the time t_i that will determine the overall exposure time for a wafer.

The creation of Frenkel (vacancy–interstitial atom) defects changes the solubility (for given solvents) of the inorganic materials that are typically used in microelectronics the metals and dielectrics.[4–6] Therefore, the possibility of a lithographic process without a polymer resist presents itself, since in ion-beam lithography the layers of the materials (that make up the structure of a device) manifest themselves as inorganic resists. The ratio of the etching rate v_T in a material exposed by ions to the etching rate v_G of the unexposed material can be as much as five, which is completely acceptable.

In terms of productivity, we can assume that the use of a thin (submicron) beam of ions to create a pattern permits only the production of masks (stencils). Stencils having submicron dimensions may serve as the medium for transferring a pattern onto a wafer by means of a wide beam of x-ray or ultraviolet photons,[5–7] electrons,[8,9] or ions.[10,11] Photon lithography uses transparent substrates with a pattern made from an opaque material. Masks with slots are suggested for projection and contact lithography with electron and ion beams.[8,9,11] A continuous substrate (without slots) made from a monocrystalline material can be used with ion beams. The "transparency" of the monocrystalline substrate to the ion beam is due to channeling,[10] which consists in reducing scattering and increasing the range of the ions that move strictly along the crystallographic axes or the planes in the crystal.[12]

Ion current sources are an important topic in ion lithography. In our discussion of electron lithography we dwelled at length on the thermionic and field emission of electrons that define the parameters of electron guns, which can have a brightness of as much as 10^6 A/cm^2 · steradian in practical applications. Naturally, sources having the same level of brightness are desirable for ion lithography. The brightness of sources that extract ions from a gaseous plasma is many orders of magnitude less than the aforementioned and for this reason plasma sources are not used in thin ion-beam machines. (Although they can be used in systems for ion lithography through a mask). A fundamental type of source for ion lithography are the so-called liquid-metal sources. The physical processes that take place in these sources have been studied in detail in connection with field-ion microscopy.[13,14] The surface of the thin tip used in liquid-metal sources is made from a refractory material (tungsten) wetted with a liquid metal (gallium, gold, alloys). A difference of potentials that will produce a voltage of about 10^8 V/cm over the very small radius of curvature of the surface of the liquid film on the tip is applied to the tip. Evaporation and subsequent ionization of the neutral atoms near the tip by the strong electric field is the most likely reason for an ion current to appear in this type of source. In every case, neutral atoms are a part of the evaporating particles and their presence can be detected by the optical radiation emitted by the excited neutral atoms in the small region near the tip.[15,16] It is likely that field-induced evaporation and ionization play the same role in liquid-metal ion sources.[15,16] Because the area from which the ion current is emitted is so small, the density of the ion current is rather large and the space charge due to the current affects the volt–ampere characteristics of the ion sources.[17] The brightness of liquid-metal ion sources is on the order of 10^6 A/cm^2 · steradian, i.e., comparable to the brightness of field emission electron sources.

We will now discuss the physical problems that form the basis of ion-beam lithography.

3-2. THE ELECTRIC FIELD OF A CHARGED TIP AND THE SHAPE OF THE LIQUID-METAL ELECTRODE

It was pointed out in Chapter 1 that a voltage $E = V_0/r_s$ will appear near the surface of a metal sphere of radius r_s and charged to a potential V_0. The sources actually used in ion lithography have, by assumption, a spherical surface of radius r_s at the apex of the tip; this spherical surface is then connected with the surface of a cone whose opening angle is θ.[18] The voltage of the field E near a tip with a radius of curvature r_s for a spherical apex is said to be described by the one-parameter equation[14]

$$E = V_0/(kr_s) \tag{3-1}$$

The geometric factor k is assigned numerical values ranging from 4 to 5. Obviously, with such a rough description of the field the actual values of field voltage E may be quite different from those calculated in Eq. (3-1). Generally speaking, the coefficient k will be a specific, unknown value for each sample, since it depends on the geometry of the particular ion source being used.

An exact value of the electric field $E(r) = -\text{grad}\,\varphi(r)$ can be found from Poisson's equation, $\Delta\varphi = 0$. A solution of Poisson's equation for various tip geometries and electrodes was found[18] by numerical methods, where it was shown that the results obtained for a field E at a point r_s on the apex of the tip can be described by the two-parameter empirical equation

$$E = (V_0/(k_1 r_s))\exp(-k_2\theta) \tag{3-2}$$

Here, r_s is the radius of curvature of the spherical segment at the end of the tip and θ is the opening angle associated with this segment on the cone of rotation (Fig. 3-3). The rationale behind isolating mainly the cone opening angle θ from all of the geometric "variables" is that finding a particular value of θ is difficult when the tip is being produced, but the randomly obtained θ can be measured.

The constant k_2, which describes how the opening angle of the conical portion of the tip surface affects the field E, remains constant when θ is varied within 0–9° and changes only slightly when the polar angle is varied by as much as 60°.[18] At the same time, the constant k_1 is strongly dependent upon the tip's radius of curvature r_s and the angles θ and φ, varying between 6.8 to 14.2. Values of k_1 have been tabulated.[18]

Analytical expressions are used in a number of works for the potential and the field voltage on the surface of and near the solid tip. Thus, for the model of a diode in which

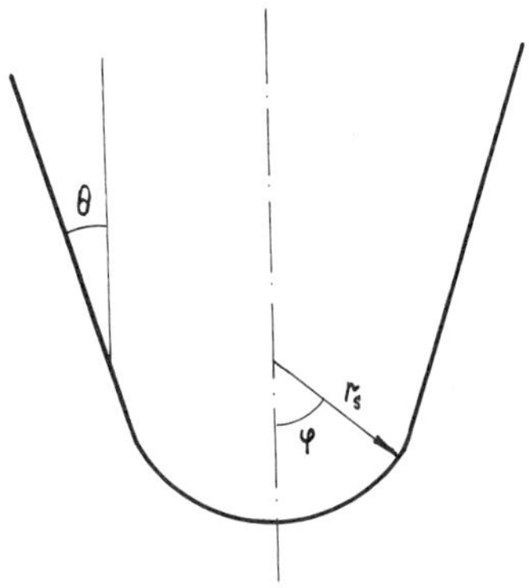

FIGURE 3-3. The shape of the metallic tip used in ion microscopy. The spherical surface of radius r_s at the apex of the tip is connected with the surface of a cone having an opening angle θ. The polar angle φ defines the size of the spherical segment.

one electrode is a sphere on an orthogonal cone and the other is a plane located a distance R from the center of the sphere, the field potential at an arbitrary point r, θ is[19]

$$V(r,\theta) = V_R \left(\frac{a}{R}\right)^n \left[\left(\frac{r}{a}\right)^n - \left(\frac{a}{r}\right)^{n+1}\right] P_n(\cos\theta) - V_0 \qquad (3\text{-}3)$$

$$a = 0.5r_0; \qquad n = 0, 15$$

The center of the sphere was chosen as the origin for the spherical coordinate system and the cone's axis to the outside was chosen as the polar axis; r_0 is the radius of the spherical tip. When $r = r_0$ (at the surface of the tip) the field is given by

$$V(r_0,\theta) = V_R \left(\frac{a}{R}\right)^n \left[2^n - \frac{1}{2^{n+1}}\right] P_n(\cos\theta) - V_0 \qquad (3\text{-}4)$$

$$r = R : V(R,\theta) \simeq V_R P_n(\cos\theta) - V_0$$

Gomer[20] and Ishitani *et al.*[21] used this diode geometry and a somewhat different formula for the field potential, from which the field voltage at the axis of the cone follows[22]:

$$E(z) = \frac{E_a}{(1 + 3a_0^2/r_0^2)} \left(\frac{r_0}{z + r_0}\right)^{1/2} \left(1 + \frac{3a_0^2}{(z + r_0)^2}\right) \qquad (3\text{-}5)$$

The maximum value of $E(z)$ occurs at the apex of the tip: $E(0) = E_a$.

The calculated distributions of potential $V(r,\theta)$ near a tip covered with a liquid metal require that two essential factors be considered: 1) the change in the shape of the surface of the tip covered with a liquid metal when acted upon by an electric field, and 2) the effect due to the space charge field created by the ion current, i.e., shielding effects. A complete system of equations for a current-carrying diode includes electron hydrodynamic equations for defining the shape of the liquid metal surface, Poisson's equation for finding the field potential when the space charge is taken into account, and an equation for the ion current. To date, the complete system of equations has not been solved. We will present some of the results that have been obtained through efforts made in this direction.

The solution of the electrohydrodynamic equations in a linear approximation of a planar diode, in which one electrode is covered by a liquid metal a thick, produces a dispersion equation that relates the frequency ω and the wave number k of the surface waves[23]:

$$\omega^2(x) = \frac{T}{\rho b^3} x \tan(\alpha x) \left(x^2 + \frac{\rho g b^2}{T} - \frac{\varepsilon_0 E_0^2 b}{T} x \cot(x) \right) \tag{3-6}$$

Here T, ρ are the surface tension and density of the liquid metal; $E_0 = \varphi/b$ is the voltage of the electric field in the planar diode (φ is the difference of potentials in the diode and b is the distance between the electrodes); g is the acceleration due to gravity; ε is the permittivity in a vacuum; and $\alpha \equiv a/b$, $x \equiv kb$.

It follows from Eq. (3-6) that conditions for the liquid surface to be unstable arise in fields E_0 stronger than some critical value E_c. These conditions are revealed by the fact that the square of the frequency is a negative number over some range of wave numbers: $\omega^2(x) < 0$. Thus, for a field $E_0 = 2$ MV/cm these conditions arise in the range $x < 0.9\ \mu\text{m}^{-1}$.[24] When the field is $E_0 \geq 10$ MV/cm the instability wavelength is $\lambda \approx 1\ \mu$m. We may assume that in the two-dimensional case, a two-dimensional lattice from tips having periods λ_x, $\lambda_y < 1\ \mu$m will appear on the surface of the liquid metal. It is assumed that the emission of a 10 μA ion current from every tip will produce a current of density 5 kA/cm^2.[2] An examination of the electrohydrodynamic equations in a nonlinear approximation confirms the results from the linear theory. The time needed for instabilities to become established is estimated to be less than 1 nsec.

The notion of a teardrop-shaped liquid, the so-called Taylor cone, forming on the end of a metal tip wetted with a liquid metal, is prevalent in the literature on liquid-metal ion sources.[18] In this situation the electrohydrodynamic equations are given as the elementary condition for the liquid surface to be in static equilibrium when acted upon by the forces of surface tension and electrostatic forces acting on the charge at the surface of the liquid metal:

$$T\left(\frac{1}{r_1} + \frac{1}{r_2}\right) = \frac{1}{2}\varepsilon_0 E^2 \tag{3-7}$$

Here, r_1, r_2 are the principal radii of curvature of the surface and $E(r)$ is the field at the surface of the liquid metal.

The radii of curvature for a conical surface are $r_1 = R/\cot\theta_0$, $r_2 = 0$, where θ_0 is the half angle of the cone's opening and R is the radial coordinate on the surface of a cone whose origin is at the apex of the cone (Fig. 3-4). Since the surface of the liquid-metal cone must be an equipotential surface the equilibrium condition, Eq. (3-7), is due to the normal component of the electric field $E_N = dV/dn$; the tangential component is zero. In the coordinate system we have chosen, $dn = R\,d\theta$, and instead of Eq. (3-7) for the cone we will have[25]

$$T\cot\theta_0/R = 1/8\pi(dV/R\,d\theta)^2 \tag{3-8}$$

If we choose V the function

$$V(R,\theta) = V_0 + AR^{1/2}P_{1/2}(\cos\theta), \qquad \theta = \pi - \theta_0 \tag{3-9}$$

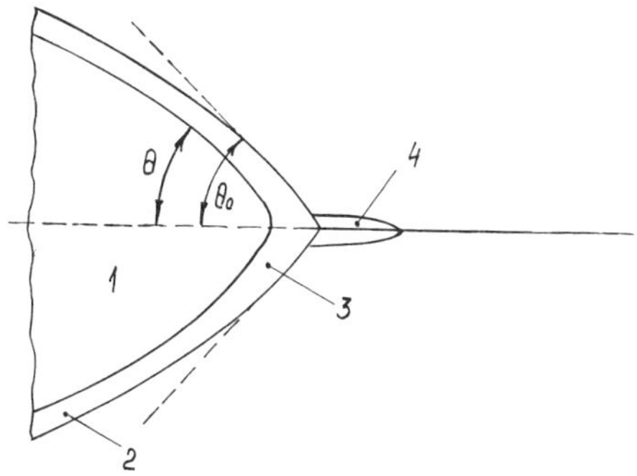

FIGURE 3-4. Shape of a liquid-metal tip in a strong electric field: 1) needle of solid material; 2) liquid-metal film; 3) node shaped like a Taylor cone and in static equilibrium in the electric field; 4) Kingham and Swanson's streamlike liquid-metal node.(22) $0z$ is the axis of the polar coordinate system used in calculating the shape of the Taylor cone.

for the potential we can satisfy both the equilibrium condition given by Eq. (3-8) and the condition for an equipotential surface on the cone. The latter condition requires that

$$P_{1/2}(\cos\theta) = 0 \tag{3-10}$$

which, according to tables of the $P_{1/2}(\cos\theta)$ function, will occur for only one value of the angle, $\theta = 130.7°$or $\theta_0 = 49.3°$. Thus, the condition of equipotentiality defines the angle at the apex of the liquid-metal cone. On the other hand, the choice $V(R) \sim R^{1/2}$ satisfies the equilibrium condition given by Eq. (3-8), from which we find the constant A:

$$A = [8\pi T \cot\theta_0]/0.974 = 1.432 \times 10^3\ T^{1/2}\ (\text{V}\cdot\text{cm}) \tag{3-11}$$

Here we use the equality $dP_{1/2}/d\theta|_{\theta=\theta_0} = -0.974$.(25)

We can use Eq. (3-9) to find the critical difference of potentials needed to form a Taylor cone. Assume that the potential is V_0 = at the surface of the cone. When $\theta = 0$, $R = R_c$, where R_c is the distance between the anode and cathode, we will have a cathode potential of $V_c = AR_c^{1/2}P_{1/2}$ (1). Substituting the A found from Eq. (3-11) gives us

$$V_c = 1.4 \times 10^3\ T^{1/2} R_c^{1/2}\ (\text{V}) \tag{3-12}$$

The curvature at the apex of the cone theoretically becomes infinite.

The concept of an unusually small Taylor cone apex does not give a coherent physical picture of a diode with an ion current. Specifically, when the diameter of the emitting surface is several nanometers and the current is 10 μA, the ion space charge density near the tip is so large that the anode field must be strongly shielded. Consequently, the field will become insufficient to produce field evaporation, which has a threshold nature and requires ultrahigh field voltages on the order of 100 MV/cm (see below).

Kingham and Swanson(22) suggested that additional factors in determining the shape of the liquid-metal teardrop at the end of the tip be considered, namely the streamlike flow of liquid to the apex of the liquid teardrop that compensates the loss of mass due to the outflow of the ion current. The streamlike movement of the liquid creates a negative Bernoulli pressure inside the liquid stream. The equilibrium condition on the liquid surface now has the form

$$T\left(\frac{1}{r_1}+\frac{1}{r_2}\right)+\frac{1}{2}\rho v^2=\frac{1}{2}\varepsilon_0 E^2 \tag{3-13}$$

Equation (3-13) leads to the formation of a protrusion that is thinner than the Taylor cone, and which is described by length, apex radius, and a shape factor S. The value $S = 1$ corresponds to the formation of a thick protrusion, and $S = 0$ to a thin protrusion. The authors(22) used an iterative procedure to solve the system of equations that includes Eq. (3-13) and Poisson's equation. A protrusion shape on the Taylor cone was used as a zero-order approximation and the field E_s needed to produce the given protrusion shape was found. Poisson's equation was then solved to find the field E_p on the surface of the cone. Consistency between the solutions of Eq. (3-13) and Poisson's equation was achieved by imposing the constraint that $E_S = E_p$.

The investigations showed that the radius of the apex of the stream protrusion increases in roughly direct proportion to the magnitude of the ion current, and that the length of the protrusion also increases with current (e.g., from 40 nm at 5 μA to 500 nm at 30 μA for a gallium source). Self-consistency between the shape of the tip and space charge effects leads us to conclude that the stream is longer for heavy elements whose ions are more strongly affected by the space charge (because of their lower velocity). A longer stream reduces the density of the space charge and compensates the space charge effect. The length of the protrusion was 40, 120, 230, 550 nm for liquid-metal sources from Al, Ga, In, Bi, respectively, at a current of 10 μA.

A review of the literature on liquid-metal sources of ion current can be found in [26].

3-3. THE EFFECT FROM FIELD IONIZING ATOMS NEAR THE SURFACE OF A METAL

The effect from ionizing atoms placed into an electric field has been studied theoretically with hydrogen atoms.(27) The potential energy of the hydrogen atom's electron in the field of the nucleus is described by a Coulomb potential $V(r) = -e^2/r$ that forms a symmetric potential well for the electron. The electron energy levels that correspond to the associated state form a discrete, negative-valued series (Fig. 3-5). If the atom is placed in a strong electric field E_x the potential energy of the atom in the field of the nucleus is augmented by the energy in the external electric field:

$$V(x) = -\frac{e^2}{|x - x_i|} + eE_x \cdot x \tag{3-14}$$

This energy is shown graphically in Fig. (3-5). As we can see, the potential well in which the electron moves has now become noticeably asymmetric: a barrier of finite width has arisen on the negative x side. The electron can penetrate the space beyond and escape the atom. This process has been called field ionization. In order for an external field E_x to produce a noticeable deformation of the electron's potential well, the external field must be comparable in magnitude to the interatomic electric field $E_a = e/a_{\text{B}}^2$, where a_{B} is the Bohr radius of the orbit. The intraatomic field is $E_a = 5.142 \times 10^9$ V/cm and therefore, electric fields on the order of 10^8 V/cm are needed for field ionization to occur. The electron energy levels for an atom placed into an electric field are not the same as those for an atom without a field; this shift in energy levels is called the Stark effect. One reason for this shift in energy levels is the polarization of the atom: when an atom is placed into an electric field the atom's electron shell is displaced relative to the nucleus in a direction opposite to that of the voltage vector of the electric field (the electron charge is negative). Consequently, the atom has an electric dipole moment $d_x = \alpha E_x$ that has been induced by the external field. The quantity α is called the polarizability of the atom. The dipole moment energy of an atom in a field E_x is

$$W(E_x) = \frac{1}{2}\alpha E_x^2$$

The polarizability of a hydrogen atom in a state having quantum numbers n, m is[27]

$$\alpha_{n,m} = a_{\text{B}}^3 \frac{n^4}{8}(17n^2 - 9m^2 + 19)$$

It is important to our purposes that the field near the tip be nonuniform. A nonuniform field produces a force that acts on a neutral atom via the atom's induced dipole moment

$$F_{x_i} = -\frac{\partial W}{\partial x_i} = -\alpha E_{x_i} \frac{\partial E_{x_i}}{\partial x_i}, \qquad x_i = x, y, z$$

which results in the directed motion (attraction) of the atoms to the tip, if the tip is located in a charged gaseous medium. Some of the atoms being attracted to the tip are ionized near its surface, forming ions that have a large force of repulsion from the tip (the tip potential is positive). If a fluorescent screen is placed in the path of the repelled ions we will see an image of the tip surface, the contrast of which is, as we shall see, due to the difference in the intensities of the field ionization processes at different points on the surface of the tip. Such a device is called a field-ion microscope, and many of the physical problems to be discussed here have been developed in the context of the field-ion microscope.[13,14]

The probability per unit time that a hydrogen atom will be ionized is, from perturbation theory,[27]

$$P(E_x) = \frac{32I^2}{\hbar(e^3 E_x)} \exp\left[-\frac{8}{3}\frac{I^2}{e^3 E_x}\right] \tag{3-15}$$

Here $I = me^4/2\hbar^2 = e^2/2a_{\text{B}} = 13.6$ eV is the ionization energy of a free hydrogen atom. As we will see, the probability that an atom will be field-ionized in an electric field is given by the ratio of the voltage of the external field to the voltage of the atom's electric field:

$$e^3 E_x/4I^2 = e^3 E_x/(e^4/a_{\text{B}}^2) = E_x/(e/a_{\text{B}}^2)$$

Equation (3-15) holds only auxiliary significance for us. In point of fact, the potential of an electron in an atom near the surface of a metal is given by a more complex function than Eq. (3-14):

$$V(x) = -e^2/|x - x_i| + eE_x x - e^2/4x + e^2/(x + x_i) \tag{3-16}$$

$V(x)$ includes the Coulomb interaction between an electron at a point x; an ion at x_i; the electron's image in the metal, $(-e^2/4x)$; the ion's image, $e^2/(x + x_i)$; and the energy in

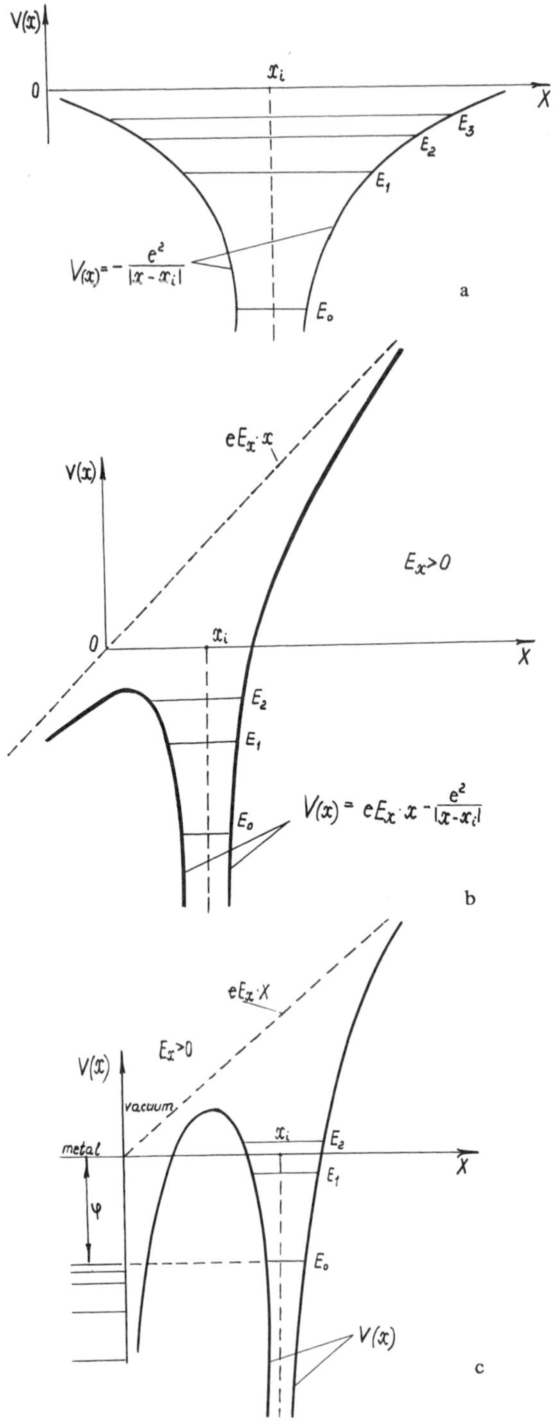

FIGURE 3-5. The potential energy of an electron and its energy levels when in the field produced by an ion.

the field, E_x. Equation (3-16) is shown graphically as Fig. (3-5c). Here φ is the work function of the metal, and E_0, I are the ground state energy and ionization energy of the atom. As a result of field ionization the electron enters the metal by tunneling through the energy barrier. Obviously, this process is possible if and only if $E_0 > E_F$, since in this case the corresponding energy levels of the electron in the metal are free. When $E_0 < E_F$ the electron cannot enter the metal because of the Pauli exclusion principle, since all energy levels below E_F in the metal are filled.

The probability that an electron will tunnel through a potential barrier is calculated from the quantum mechanical formula[27]

$$D(E, V(x)) = \exp\left\{-\left(\frac{8m}{\hbar}\right)^{1/2} \int_{x_1}^{x_2} [V(x) - E]^{1/2} dx\right\} \qquad (3\text{-}17)$$

where $V(x)$ is the electron's potential energy, E is the electron's energy level in the atom (total energy), and x_1 and x_2 are the points at which the function $V(x)$ intersects the level E on the graph.

Equation (3-17) with the function $V(x)$ can only be calculated by numerical methods. We can attempt to represent the function $V(x)$ by a model function. To do this, it is worthwhile to analyze the total energy in the atom + metal system, $V_{\text{a+Me}}$. It is not a simple task to define the function $V(x)_{\text{a+Me}}$. We will initially define it for the ion + metal system, A^+ + Me^-. We place a neutral atom A an infinite distance from the surface of the metal and ionize it, which requires that we do an amount of work equal to the atom's ionization energy I. We move the electron that was produced into the metal. In doing this the electron will occupy a free energy level on the Fermi surface, giving up an amount of energy equal to the metal's work function φ. Thus, the system A^+ + Me^- has $V = I - \varphi$ of energy when the distance between the ion A^+ and the metal Me^- is large. We bring the ion and metal to within a distance x and apply the electric field E_x, which reduces the ion's potential energy by

$$\Delta V = -e^2/4x - \frac{1}{2}\alpha_i E_x^2 - exE_x$$

if we consider the contributions made by the ion's image forces $(-e^2/4x)$, the polarization $(-1/2)\alpha_i E_x^2)$ of the ion, and the energy of the ion's charge in the electric field $(eE_x x)$. At small distances, exchange forces of repulsion arise which have a positive energy of $V_{\text{rep}}^{(i)}$. Thus, the function $V(x)$ for the Me^- + A^+ system is

$$V(x)_{i+\text{Me}} = I - \varphi - \frac{e^2}{4x} - \frac{1}{2}\alpha_i E_x^2 - exE_x + V_{\text{rep}}^{(i)} \qquad (3\text{-}18)$$

We will repeat our analysis for a neutral atom + uncharged metal. We will say that the energy of this system is zero when the atom and the surface of the metal are far from each other. We bring the metal and atom to within a distance x of each other and apply the electric field E_x. When this happens the energy of the neutral system Me + A is

$$V(x)_{a+\text{Me}} = -\frac{1}{2}\alpha_a E_x^2 + V_{\text{rep}}^{(a)}$$

Figure 3-6 shows diagrams of the potentials for the systems A + Me and A^+ + Me^- (the field has not been applied, Fig. 3-6a, and has been applied, Fig. 3-6b). As can be

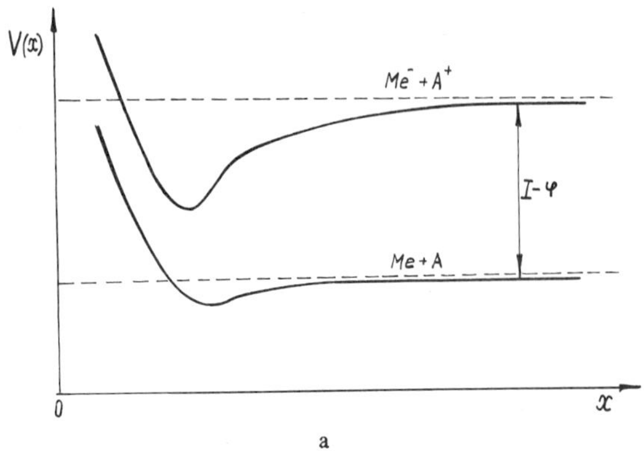

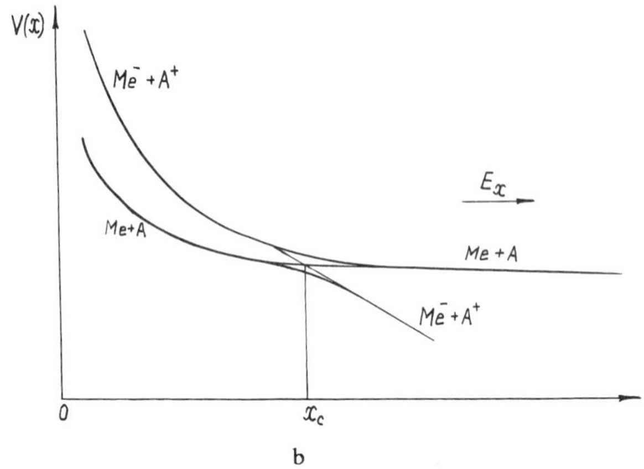

FIGURE 3-6. Energy of the atom + metal and ion + charged metal systems as functions of the distance between them.

seen, with no field applied the energy of the A + Me system is much less than that of the $A^+ + Me^-$ system at all distances x and ionization cannot occur. When the electric field E_x is applied the ionized state of the atom is energetically more useful at distances $x > x_c$: the energy of the $A^+ + Me^-$ system is less than that of the neutral A + Me system. At distances x that are much greater than x_c the energy barrier to an electron going into the metal is wide and the probability of tunneling is low and will become significant only in the narrow region $\Delta x = x - x_c$ in the immediate vicinity of x_c. It is in mainly this narrow region near the surface of the metal that field ionization plays a role. The x_c point is found from the equation

$$I - \varphi - \frac{e^2}{4x_c} - \frac{1}{2}\alpha_i E_x^2 - e x_c E_x + V_{\text{rep}}^{(i)} = -\frac{1}{2}\alpha_a E_x^2 + V_{\text{rep}}^{(a)}$$

If we can ignore the energies due to image and polarization forces, and repulsion forces (the latter appear only when the atom and metal are in direct contact), we obtain

$$x_c = (I - \varphi)/(e \cdot E_x) \tag{3-19}$$

Using Eq. (3-19) we find for a helium atom (I = 24.5 eV) on a tungsten (φ = 4.5 eV) surface that x_c = 0.4 nm, if we assume that $E_x = 5 \times 10^8$ V/cm.

Our analysis allows us to model the barrier as an isosceles triangle whose base is at x_c, apex is at x_c, and has a height of $I - 2(e^3 E_x)^{1/2}$ [$\Delta V = 2(e^3 E_x)^{1/2}$, the reduction in barrier height is due to image forces and the electric field, the Schottky effect]. For the transparency of a triangular barrier having the parameters indicated we find that[13]

$$D(x_c) = \exp\left[-4.55 \times 10^7 (I - 7.60 \times 10^{-4} E_x^{1/2})^{1/2} \frac{I - \varphi}{E_x}\right]$$

where I and φ are in eV and E_x is in V/cm.

The total probability that an atom will be ionized in time t can be written

$$P(t) = 1 - \exp(-t/\tau) \tag{3-20}$$

where τ is the atom's lifetime in the preionized state, equal to $\tau = (1/D)\nu$, and ν is the frequency of collisions between the atom's electron and the energy barrier (ν is roughly equal to an electron's orbital frequency in an atom, $\nu = 10^{16}$ sec^{-1}). Because an atom near the surface of the metal follows a random trajectory that is defined by its initial thermal velocity and by the electric forces that attract the atom to the tip, we can, instead of Eq. (3-20), calculate[13]

$$P(t) = 1 - \exp\left(-\nu \oint \frac{D(x)}{v(x)} dx\right) \tag{3-21}$$

where the integration is done along the atom's trajectory (a line integral). Using Eq. (3-21) we can calculate the ion current from any given region near the surface of the metal, the distribution of the ions with respect to their energy (the energy of an ion will depend on the point at which it was formed, since it will be accelerated from this point by the field from the ion gun), and other characteristics of the current. We will now discuss these topics.

3-4. AN ION CURRENT PRODUCED BY THE THERMAL EVAPORATION OF ATOMS AND THEIR SUBSEQUENT FIELD IONIZATION

We will, in the manner of Kingham and Swanson,[22] discuss the theory of an ion current obtained by evaporating atoms from a liquid metal and subsequently field-ionizing these atoms. The probability that during one second an atom will be thermally emitted from a surface is

$$P_t = \nu_t \exp(-\Lambda/kT) \tag{3-22}$$

Here Λ is the atom's binding energy and ν_t is its thermal vibrational frequency: $\nu_T \approx 10^{12}$ sec^{-1}. The number of atoms emitted from a surface area of 1 cm^2 during one second will be $N_t = N_0 \cdot P_t$, where N_0 is the number of atoms per unit area of the material's surface. When the solid and the saturated vapor are in equilibrium the number of ions being emitted is equal to the flux of particles deposited onto the surface from the vapor:

$$N_0 \cdot \nu_t \exp(-\Lambda/kT) = \gamma(2\pi mkT)^{-1/2} \cdot P \tag{3-23}$$

where γ is the accommodation (adhesion) coefficient, and P is the pressure of the saturated vapors. Equation (3-23) is well satisfied for Al, Ga, In, and Bi liquid-metal sources if the values chosen for λ are reduced by an amount equal to the melting heat.[22] In order to find the total number of atoms that have been emitted from the surface of a conical tip, we must first find the temperature profile on the surface of the tip. We will say that we can confine ourselves to calculating the evaporation from the sides of the cone, considering the area of the cone's apex to be small. The energy that goes into heating the tip reaches the apex of the tip as the energy from the electrons from the field-ionized atoms. Each of these electrons carries 10–100 eV of energy. We denote the incident energy as W. Then the temperature profile on the side of the cone is[28]

$$T(r) = T_0 + W/[2\pi k(1 - \cos\theta_0)r] \tag{3-24}$$

where k is the thermal conductivity of the liquid metal, r is the distance from the cone's apex to the point on its surface, and θ_0 is the half angle of the opening at the cone's apex. The maximum temperature at the apex of the tip is

$$T_a = T_0 + \frac{W\tan\theta_0}{2\pi k(1 - \cos\theta_0)r_a} \tag{3-25}$$

Calculations show that when $W = 10^{-4}$ W the temperature at the apex of the tip may reach 1000–4000°K, depending on the opening θ_0 and the radius of the sphere at the apex.

Substituting Eq. (3-24) into Eq. (3-22) and integrating over the lateral surface of the cone, we can write

$$N_t = \nu_t N_0 \int_{r'_a}^{\infty} 2\pi r \sin\theta \exp\left(-\frac{\Lambda}{kT(r)}\right) dr, \qquad r'_a = r_a/\tan\theta_0 \tag{3-26}$$

Integrating Eq. (3-26) with the condition that $T_a \gg T_0$ gives us

$$N_t = \frac{2\pi N_0 \nu_t r_a^2 \sin\theta_0}{\tan^2\theta_0}\left[\frac{\Lambda}{kT_a} - 1\right]^{-1} \exp(-\Lambda/kT_a) \tag{3-27}$$

The probability that a neutral atom will be captured in a region of the strong field near the tip is defined by the ratio of the polarization energy $W_p = (1/2)\alpha \cdot E_a^2$ and the thermal energy kT:

$$P_c(r) = 1 - \exp(-\alpha E_a^2/2kT(r)) \tag{3-28}$$

We will also introduce the probability P_i that an atom will be ionized in a region of the strong field as in Eq. (3-21); the ion current obtained by emitting atoms and then field ionizing them will be

$$I_{EI} = q\nu_t N_0 \int_{r'_a}^{\infty} 2\pi r \sin\theta \exp\left(-\frac{\Lambda}{kT(r)} - \frac{\alpha E_a^2}{2kT(r)}\right) P_i \, dr \tag{3-29}$$

where q is the charge on the ion. We can say that the probability P_i is unity when a critical value $E_a \geq E_0$ is reached. Then, in the same approximations as for Eq. (3-27) and assuming that $(1/2)\alpha E_a^2 \ll \Lambda$, we obtain

$$I_{EI} = qN_t(1 - \exp(-\alpha E_a^2/2kT_a)) \tag{3-30}$$

Numerical calculations show that the magnitude of an ion current obtained by the processes of thermal evaporation and subsequent ionization is, as a rule, a smaller part of the current produced by field evaporation. Indeed, the current is an exponential function of the heat of evaporation Λ, which is much higher than the field evaporation barrier $Q(E)$. At relatively high tip heating temperatures and small amounts of polarization energy for neutral atoms $E_p = (1/2)\alpha E_a^2$, the probability P_c that an atom will be captured in a region of the strong field may also become much less than unity. According to Eq. (3-29) this also serves to reduce the magnitude of the current I_{EI}.

3-5. THE EFFECT DUE TO FIELD EVAPORATION. AN ION CURRENT PRODUCED BY FIELD EVAPORATION

As was stated in the paragraph that introduced this chapter, the effect due to field evaporation plays an important role in liquid-metal ion sources. Physically, this effect is very similar to the field emission of electrons that was examined in Chapter 1. We will visualize a strong electric field of opposite polarity to that used for the field emission of electrons, applied to a metal. Now the force extracted from the metal as viewed from the field is applied not to electrons, but to positive ions that can tunnel through the potential barrier just as electrons do in field emission. However, since the transparency of the barrier, Eq. (3-17), is a function of the particle's mass m the probability that ions will tunnel is much lower than for electrons (under strict, equivalent conditions $D_{\text{ion}} = (D_{\text{el}})^{(M_i/m_e)^{1/2}}$). A noticeable probability of tunneling will occur only when thermal populating of those vibrational energy levels that are higher than the ground level are accounted for in an atom bound on the surface of the crystal. In any event however, tunneling plays a subordinate role. The basic ion flux is due to the thermal current of ions whose energy is greater than the barrier: the role of the field reduces to lowering the barrier height (the Schottky effect).

We will consider the field evaporation of an ion that has been adsorbed on the surface of a metal. Figure 3-7 shows the potential energy curves for this situation. The curve for the $A^+ + Me^-$ system is everywhere below the curve for the A + Me system, which means that the ion is evaporated without changing the electron state. The distance x_0 corresponds to the distance at equilibrium to which an ion is removed from a metal surface. The energy levels corresponding to the different vibrational states of the atom are shown in a potential well. The height of the energy barrier that keeps an ion on the metal surface is[13]

$$Q = \Lambda + I - \varphi - (e^3 E_x)^{1/2} + ex_0 E_x$$

where Λ is the ion's heat of sublimation.

The ion will be found in the ground vibrational state when the temperature is low. Thermal excitation plays a negligible role; field evaporation proceeds by way of an ion of mass M tunneling through the Schottky barrier. If we approximate this barrier with a triangle of height Q and having slopes of S and S', the integral in the exponential term

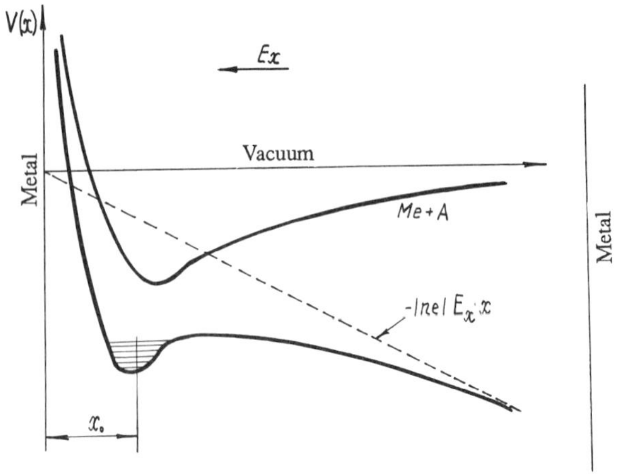

FIGURE 3-7. Energy diagram for an atom on the surface of a metal. A number of vibrational levels exist for an ionized atom in a potential well. A thermal ion current appears when the energy E of thermal motion is greater than the barrier height u_b. Tunneling from the vibrational levels through a relatively wide barrier is less likely than is thermal excitation to a level $E > u_b$.

for the transparency coefficient, Eq. (3-17), is easily evaluated and we find that

$$P_e = \nu_i \exp\left[-\frac{4}{3}\left(\frac{2M_i}{\hbar^2}\right)^{1/2}\left(\frac{1}{S}+\frac{1}{|S'|}\right)Q^{3/2}\right] \tag{3-31}$$

for the probability that an ion will be evaporated per unit time.[13] Here ν_i is the vibrational frequency of an ion bound on the metal surface and corresponds to the number of occasions per unit time the ion impinges on the barrier, and M_i is the ion's mass.

At high temperatures evaporation due to tunneling can be neglected; evaporation occurs only by way of thermal activation as in Arrhenius' equation:

$$P_e = \nu_i \exp(-Q/kT) \tag{3-32}$$

The temperature T_c, below which evaporation due to tunneling prevails, can be found by equating Eqs. (3-31) and (3-32):

$$T_c = \frac{3}{4}\frac{\hbar}{k}(2M_iQ)^{-1/2}(1/S + 1/|S'|)^{-1} \tag{3-33}$$

Using Eq. (3-33) to calculate T_c gives a value of T_c in the range 70–430° K for different metals.[13]

At intermediate temperatures the total probability of evaporation can be found by adding the probabilities of tunneling from all of the vibrational levels in the potential well and the probability of thermal excitation, as in Eq. (3-32) (see [13]).

Equations (3-31) and (3-32) are useful for examining evaporation of ions attached to a metal by a metallic bond, in doing which[13]

$$Q = \Lambda + I - \varphi - (e^3 E_x)^{1/2} + \frac{1}{2}(\alpha_a - \alpha_i)E_x^2 \tag{3-34}$$

If we allow for the possibility of n-tuple ionization of the ions, we must then make the replacements $V(x)$ in the formulas for the energy barriers and potentials $e \rightarrow ne$, $\varphi \rightarrow n\varphi$, $I \rightarrow I_n$ where I_n is the energy for an n-tuple ionization of an atom. Setting $Q = 0$ in Eq. (3-34) we can estimate the voltage of the evaporating electric field E_{ev}. If we neglect polarization effects in Eq. (3-34) we find that

$$E_{\text{ev}}^{(n)} = (ne)^{-3}(\Lambda + I_n - n\varphi)^2$$

for n-tuple ionized ions.

The magnitudes of the evaporating fields calculated for various metals are in the range of a few 10^8 V/cm,[13] which is in good agreement with the values E_{ev} found experimentally for several metals.[13] It remains, nevertheless, to point out that the potentials, Eq. (3-18), that form the basis of the calculations, are very approximate in nature. The fact of the matter is that we use a function, Eq. (3-18), for such small distances between an atom and the metal (right down to distances smaller than interatomic) so that representing the metal as a homogeneous medium in which an image charge appears is a major idealization. Therefore, a shape for a Schottky barrier obtained via Eq. (3-18) is hardly close to the actual shape of a barrier. Simplifying the shape of the barrier to a triangle does not, in light of what has been said, weaken the theory. That the calculated and measured values of E_{ev} are in agreement shows that, as a whole, the representations discussed here correctly describe the processes by which atoms are evaporated by a field and subsequently field-ionized.

According to [22], in fields that satisfy the condition $Q(E) > 0.5$ eV for the barrier height as a function of the field the nonlinear dependence given by Eq. (3-34) can be replaced by a linear dependence

$$Q_L(E) = A - BE \tag{3-35}$$

Empirical values for the A and B coefficients for liquid Al, Ga, In, Bi sources have been tabulated in [22]. The linear relation, Eq. (3-35), makes it easier (by integrating over the surface) to calculate the ion current produced by field evaporation from a liquid metal if the distribution of the field E over the surface is known. Unfortunately, analytical expressions for the field rarely take the shielding created by the space charge of the current into consideration. In our model of a right-angled cone we may assume that, e.g.,[21]

$$E = E_a(1 - C\theta^2), \qquad C = 0.353, \qquad \theta \leq \theta_c \tag{3-36}$$

where E_a is the maximum field voltage at the apex of the liquid-metal tip and θ is the angle between the axis of the cone and the point on the surface being examined. Using Eqs. (3-35) and (3-36) we can write the equation for an ion current obtained by field evaporating atoms as an integral over the surface of the tip's apex:

$$I = \int N_a \nu_t \exp\left[-\frac{A - BE_a(1 - C\theta^2)}{kT(\theta)}\right] 2\pi r_0^2 \sin\theta \, d\theta \tag{3-37}$$

Assuming that $T(\theta) \equiv T_a = \text{const}$ and $r_0 = (2/3)r_a$, we find from Eq. (3-37) that[22]

$$I = 10.3 N_a \nu_t r_a^2 \left(\frac{kT_a}{BE_a}\right) \exp(-Q_L(E_a)/kT_a) \tag{3-38}$$

(in SI units). Because the actual field $E(\theta)$ is unknown (it changes due to shielding by the space charge), we can approximate the voltage within the limits of the surface of the apex by a constant: $E(\theta) \equiv E_a = \text{const}$. We then find

$$I = 2\pi r_a^2 N_a \nu_t \exp[-Q(E_a)/kT_a] \tag{3-39}$$

for the ion current obtained by field evaporation.

It must be noted that the quantity $Q(E_a)/kT_a$ in the exponential term in Eq. (3-39) changes slightly when the difference of potentials φ_a in the diode is changed. Increasing φ_a increases the current, the space charge density in the interelectrode area, and shielding of the liquid-metal electrode increases, which causes E_a to increase much more slowly in comparison with φ_a. We can say that, approximately, $Q(E_a)kT_a \simeq$ const, and the ion current increases because of an increase in the area on the emitting surface of the stream protrusion.[22]

A comparison of the ion currents obtained by field evaporation, Eq. (3-39), and by thermal evaporation and subsequent ionization, Eq. (3-30), shows that in weak fields, $E < E_1$, the current produced by thermal evaporation may prevail, and in strong fields $E > E_0$ the current produced by field evaporation prevails. In a relatively narrow range $E_1 < E < E_0$ the currents produced by field and by thermal evaporation are comparable.[22] Magnitudes of the E_1 and E_2 fields are given in [22].

The energy spectrum of an ion current is an interesting subject. The method of decelerating potential[29] and magnetic or electrostatic spectrometers[30] can be used to study these spectra experimentally. Measurements show that the spread in the energy of $\Delta E_{1/2}$ ions is Ga^+ at low values of current 1 μA and increases rapidly as the current increases because of the Boersch effect.[22] Another feature typical of the spectrum is the so-called energy deficit 4.5 eV of ions in comparison with the energy they would have in passing through an accelerating field $E_{i\,\max} = eV_0$. This deficit is obvious in the case of field evaporation V_0: $E = eV_0 - \Delta E$. It is simply equal to the work done in forming an evaporating ion[22]:

$$\Delta E = \Lambda + I - \varphi - Q(E) \tag{3-40}$$

When $Q(E) \ll \Lambda + I - \varphi$ the values of $\Delta E \simeq \Lambda + I - \varphi$ calculated according to Eq. (3-40) are 3–5 eV for different ions. These values may be compared with the peak values in the energy distribution of the ions in the current (the shift in the peaks and the width of the distribution are comparable, which makes it difficult to locate the peaks). Within measurement accuracy, the locations of the peaks and the energy deficits calculated according to Eq. (3-40) are in good agreement.[22]

Ions produced by thermal evaporation and subsequent ionization at some distance from the tip must have a broader spectrum and greater energy deficit. For example, if a neutral atom is ionized 3 nm from the surface of the tip and the field voltage is 10 V/nm, the energy deficit is ΔE = 30 eV. With due regard for the distribution $E(z)$ along the z axis, we can write

$$\Delta E(z) = \int_0^z eE(z')\,dz' \tag{3-41}$$

for ΔE, where z is the distance from the apex of the tip at which ionization occurred. Numerical calculations show that ΔE and the widths of the peaks increase rapidly with E_a: when E_a is increased from 5 to 15 eV/nm, ΔE increases from 5 to 90 eV.[22] Since both components in the current, the field evaporation component and the thermal evaporation (with subsequent ionization) component, can be seen at the same time, two peaks are seen in the energy spectrum of the ion current. The large spread in ion energies causes large chromatic aberration in the optical components of ion-beam systems. For this reason it is desirable to eliminate from the ion current those components caused by thermal evaporation and subsequent ionization. Various measures can suppress this component: elements or alloys that have low vapor pressures at the melting temperature, and high thermal conductivity will enhance evaporation by efficiently removing the heat.

3-6. VOLT–AMPERE CHARACTERISTICS OF ION SOURCES

The volt–ampere characteristics of liquid-metal ion sources are calculated with regard for the effects that arise due to the current's space charge (see Section 1-3) and electrohydrodynamic effects. For the most part, the latter determine the critical (threshold) voltage V_{0c} at which the liquid deforms and creates an emitting protrusion and the ion current appears[31]

$$V_{0c} = (r_a \cdot T/\varepsilon_0)^{1/2} \ln(2h/r_a) \tag{3-42}$$

Here T is the surface tension of the liquid metal, r_a is the radius of curvature of the apex of the needle (tip), and h is the distance between the needle and the parabolic elongating electrode. As the interelectrode voltage V_0 is lowered the ion current disappears at a somewhat lower voltage $V_{0x} < V_{0c}$ (see below). For the situation in which the ion emitting protrusion of liquid metal is created in the opening of a capillary of radius R, Mair[32] found the $I - V$ characteristic for a liquid-metal ion source as

$$I = 3\pi(2e/M)^{1/2} T \cdot R \cos\varphi \left[\left(\frac{V_0}{V_{0x}} - 1\right) + \frac{1}{2}\left(\frac{V_0}{V_{0x}} - 1\right)^2\right] V_0^{-1/2} \tag{3-43}$$

Here e/M is the charge-to-mass ratio of the ions being emitted, and φ is the contact angle of the liquid metal to the capillary wall ($\varphi = 43.3°$ [31]). As we can see, when $V_0 \gg V_{0x}$ Langmuir's three-halves law for a space-charge-limited current is valid. However, a high threshold voltage V_{0x} allows us to make Eq. (3-43) linear. In observations of the ion current from liquid-metal sources the voltage changes satisfy the condition that $(V_0/V_{0x} - 1) \leq 0.1$. Discarding the squared small term in Eq. (3-43) and making the substitution $V_0^{1/2} \simeq V_{0x}$ we find that, to within 1%,

$$I = 3\pi(2e/M)^{1/2} T \cdot R \cos\varphi V_{0x}^{-1/2} \left(\frac{V}{V_x} - 1\right) \tag{3-44}$$

The linear (above threshold V_{0x}) increase in current is well satisfied in experiments.[31,32] The slope of the straight lines $I \sim V_0$ depends on the afflux impedance of the liquid metal to the tip: at low impedances (tips with grooves, large radius capillaries) the slope of the straight lines is much steeper.

The threshold at which an ion current in a capillary source disappears is defined[32] by

$$V_{0x} = (2kTR\cos\varphi/\varepsilon_0)^{1/2} \tag{3-45}$$

k is a logarithmic term that depends on the ratio h/R. Taking Eq. (3-45) into consideration, we can write Eq. (3-44) as

$$I \sim V_{0x}^{3/2}\left(\frac{V_0}{V_{0x}} - 1\right) \tag{3-46}$$

As can be seen, $dI/dV_0 \sim V_{0x}^{1/2}$, i.e., the higher the threshold voltage, the steeper are the slopes of the current curves. A faster increase in current with voltage will be seen for lighter ions ($I \sim M^{-1/2}$) and ions having greater multiplicity of charge ($I \sim e^{1/2}$).

Experimental data for ion sources are discussed in [33–35].

3-7. PLASMA ION SOURCES

We will now briefly discuss plasmas as sources of ions. A large variety of plasma ion sources have been developed for different applications. In plasma ion sources a discharge is ignited in a low-pressure gas. A high-temperature chamber is needed to vaporize materials whose normal state is solid and the vapors from the material flow into the discharge chamber. The ions are extracted through the hole (holes) in the electrode to which an extraction voltage V_g has been applied. A shell in the form of a dual layer is formed at the edge of the due to the escape of the more mobile electrons whose temperature is $T_e \gg T_i$, where T_i is the ion temperature. The difference of potentials in the dual layer is roughly $\Delta V = 5(kT_e/e)$ and therefore, the density of the ion current through the dual layer is proportional to the electron temperature T_e in the plasma[36]:

$$J_i = 1.45 \times 10^{-15} n_i \sqrt{T_e/M_i} \tag{3-47}$$

Here n_i is the density of the ions, cm^{-3}; and M_i is the mass number of the ions. The ion density in a low-pressure plasma is 10^{10}–10^{11} cm^{-3}. Using the ordinary cold plasma value of electron temperature $T_e = 10^5$ and $M = 100$, we can easily estimate the current density through the shell in the plasma region as: J_i = 1–5 mA/cm^{-2}. In this case, the space charge imposes no restrictions on the ion current.

The brightness of plasma sources can be estimated from the theoretical relation $\beta = J_i V_a/\pi V_i$, where $eV_i = (3/2)kT_i$ is the energy of the ions in the plasma, and V_a is the accelerating voltage. At $T_i = 1500°$ K, $V_a = 10^4$ V and $M = 40$, we find that $\beta = 10^3 \text{A} \cdot \text{cm}^2 \cdot \text{steradian}^{-1}$. In practice the brightness of plasma sources is never greater than 10–100 $\text{A} \cdot \text{cm}^{-2} \cdot \text{steradian}^{-1}$. The reasons for the reduction in brightness are diverse. If a plasma source has been designed for use in machines with fine focusing the diameter of the hole in the extracting electrode must be on the order of 0.1 mm, in which case ion losses on the walls of the hole can play a major role. The aperture in the accelerating electrode has the same properties as a defocusing lens, resulting in the virtual size of the hole being less than the actual size.

Another important characteristic of plasma sources is the spread ΔE in the energy of the ions. Values of ΔE seen in practice range from 10 eV (microwave ion sources) to 50–100 eV (radio frequency ion sources). Obviously, these values are much greater than the thermal spread $eV_i = kT_i \simeq 0.1$ eV. Also, obviously, the observed spread in the ion energies is connected with the spread in values of the field potential inside the plasma, which causes ions produced at different points in the plasma to have different

TABLE 3-2

Ion source	Brightness, $A \cdot cm^{-2} \cdot ster^{-1}$	Energy spread ΔE, eV	Floating potential in the plasma, V	Beam diameter, mm	Beam current, mA
Microwave	10–100	5	10–20	3–60	3–400
Electron cyclotron resonance in the plasma		10–20	10	3–40	100
Radiofrequency		50–100	100–300		1
Duoplasmatron	10^2	10–20		0.1	1–10
Hot cathode and electron bombardment		4–70	4–6	50	10–50

energies. Associated with this phenomenon is the so-called surplus of energy in ions extracted from a plasma: the energy of ions at the extracting electrode may exceed the energy of acceleration $E_i = eV_a + \Delta E_{ex}$. The surplus energy is associated with the so-called floating potential in the plasma: $e\Delta V = \Delta E_{ex}$. It follows from plasma theory that[37]

$$\Delta V = kT_e/e \log(M_i T_e/m_e T_i)^{1/2} \tag{3-48}$$

for the floating potential. As can be seen, ΔV is proportional to T_e (ΔV is logarithmically dependent on the remaining terms). When the discharge conditions in the plasma are changed ΔV, the changes in T_e are fixed. According to [38], in a microwave discharge in Ar the floating potential increases from 10 to 25 V as the gas pressure is lowered from 2 Pa to 0.04 Pa. The potential ΔV also increases as the input power W is increased.

Fluctuations in the floating potential due to the plasma being unstable in a magnetic field may explain the spread ΔE in the energy of the ions in the current from the plasma.[38] Measurements of ΔE in a microwave plasma of Ar show that it does not change when the accelerating potential V_a is changed. Very minor changes in ΔE are seen when the gas pressure in the discharge and the input microwave power are increased.[38]

The most important characteristics of ion beams are, as we have already mentioned, the spread in their energies, the brightness of the source (or ion current density), and the geometric dimensions. Large dimensions in a source are important in projection and shadow ion lithography. Plasma sources make it possible to obtain uniform beams of collimated ions up to 50 cm in diameter by extracting ions through a hole in a grid. Large current densities during stationary (over an extended period of time) exposure may produce a great deal of heat in the target, which limits the maximal current densities that can be allowed.

Plasma ion current sources are classified basically according to the methods used to create the plasma. Typical values of ion current for the various sources are shown in Table 3-2.[38]

In radio frequency ion sources, a plasma discharge is stimulated in a gas at a pressure of 10^{-3}–10^{-2} torr from an electrodeless, 10–30 MHz RF generator having an output power of a few hundred watts. The gas is held in a dielectric container (glass, quartz). Electrons from the background ionization of the gas are accelerated by the RF field to an energy above the gas atoms' ionization threshold, the gas is ionized, and electrons are multiplied. Eventually the density of the ions and electrons reaches a steady-state value that is determined by the level of RF energy being absorbed; this level becomes saturated because of field shielding.(39) The ions are extracted through a hole in the extracting electrode. The ion current is accompanied by a flux of free neutrals that leave the plasma. The use of RF sources is limited by the broad energy spectrum of the ions in the current (as much as 100 eV). The use of a mass separator makes it possible to isolate ions whose energies lie in a narrow band; however, source efficiency falls off sharply when this is done. Efficiency in the process of ionizing by emitting primary electrons from a cathode (in sources that use a cathode) increases sharply if the discharge medium is placed into a magnetic field. The magnetic field increases an electron's path in the gas (prior to impinging on the vessel walls or recombining in the plasma) due to the magnetic twisting of the trajectory. This is why a magnetic field is used in all high-efficiency ion current sources. The absorption of RF energy is particularly efficient during electron cyclotron resonance, which occurs only at low gas pressures (for resonant absorption to occur, the mean free path length of the electrons must be comparable with the radius of the cyclotron orbit of an electron in a magnetic field). Consequently, electron cyclotron resonance ion sources have a low level of brightness.

The gas ions are maintained at a pressure of $P < 10^{-2}$ torr in microwave sources. Microwave energy (e.g., at a frequency $\nu = 2.45$ MHz) is fed to the discharge chamber through a coaxial waveguide; a resonator may also be used.(40) The size of the extracting grid (diameter d of the beam) is governed by the size of the resonator, e.g., $d = 6$ cm for a TE_{211} resonator.(40) When lower frequencies are used ($\nu = 915$ MHz) the beam diameter can be increased up to 50 cm.(38) Some microwave source designs make it possible to obtain beams of chemically active ions: F^+, O^+, Cl^+, H^+. The relatively small spread in ion energies $\Delta E \leq 5$ eV with microwave sources makes it possible to use them with high resolution optics: ion projectors or shadow exposure machines.

In plasmatrons a DC voltage source stimulates a discharge in a chamber with a hot cathode.(39) The magnetic field concentrates the plasma flux that emerges from the discharge chamber into an intermediate chamber through a gap in the anode (a duoplasmatron). The ion current is extracted from the intermediate chamber by an electrode-extractor. A duoplasmatron produces a fairly high brightness ion current. The broad energy spectrum of the ions ($\Delta E \simeq 10$–20 eV) and the fact that it can only be used with gas-producing materials limits the use of a plasmatron. The attachment of a vaporizing chamber to a duoplasmatron makes its design overly complicated.

Ion sources in which the plasma discharge is ignited by a beam of fast electrons can be used in systems with focusing. These sources of metal ions were studied in [41]. Metal vapors are produced in a high-temperature chamber and flow from there into a storage chamber shaped like a long tube. At one end of the tube is a system of electrodes which simultaneously serves as a gun for producing electrons (having an energy of about 5 keV) along the entire tube and an ion gun that extracts ions from the discharge in the tube. The discharge is ignited in the tube at a vapor pressure of 10^{-4}–10^{-2} torr. The electrons in the beam are gathered by a collector located at the opposite end of the storage tube. The brightness of these sources is unusually high: $\beta \approx 10^7\ \mathrm{A \cdot cm^{-2} \cdot rad^{-2}}$ for a current of silver ions on the order of $I_i \approx 1$ mA.

3-8. INTERACTIONS BETWEEN ION BEAMS AND MATTER

The energy that fast atomic particles lose in matter is due to two distinct processes. Electron losses are due to exciting and ionizing a target's atoms without dislodging them from the crystal lattice. Nuclear losses are due to dislodging atoms in the target from the lattice sites and creating Frenkel defects (vacancy + interstitial atom). It is namely these processes that we must study in detail. The processes of decelerating a fast atomic particle may be complicated by a change in the state of the ion itself while it is decelerating. Because of the interaction between a fast particle and the target's atoms the charge on the atom may change, i.e., it may gain or lose charge. When a fast ion begins moving in a target, it will be stripped, losing electrons and increasing charge at the expense of its own kinetic energy. When it reaches the end of its path and slows down, the reverse process, i.e., electron capture and charge reduction, will take place. In this manner the number of electrons Z_1 will be a function of the depth of beam penetration into the target: $Z_1(x)$. Ordinarily, we consider the number of electrons on the ion to be constant, ignoring the change in the ion's charge. The basis for this assumption is the fact that the potentials of an ion–atom interaction are only weakly sensitive to the change in the fast ion's charge when the number of electrons Z_1 on the fast ion changes only slightly. In addition, ion stripping occurs at energies $E_i > 1$ MeV, whereas ion beams having energies of only 50–200 keV are used in ion-beam lithography.

The processes of changing an ion's charge can play a major role in the sputtering (ion etching) of materials by a beam of relatively slow ($E \leq 10^4$ eV), but multiply ionized, atoms. We will assume that the atoms have been ionized ten times. If the probability that these ions will capture electrons is high, then this capture will take place in the uppermost layers of the solid. As electrons are captured an amount of potential energy equal to the sum of the ionization energies of all the captured electrons will be released: $\Delta E = \sum_j I_j$. This energy is on the order of the ion's kinetic energy, but in contrast to it, is dissipated as heat in a small region of the target at the beginning of the ion's path. The target atoms that have acquired energy as a result of changing the amount of charge leave this area, producing a microexplosion that leads to the simultaneous separation and dispersion of entire groups of atoms.[42] Usually, in ion-beam lithography only singly charged ions having a small reserve of potential energy of ionization are used; therefore, this energy is usually not taken into account.

Finally, the channeling effect, which reduces scattering and increases the range of ions moving strictly parallel to the crystallographic axes of the planes of a monocrystal, is usually ignored. Clearly, these effects are absent if the targets are amorphous (e.g., polymeric) or polycrystalline materials and when collimated beams act on monocrystalline targets or when the axes and planes of the crystal are oriented at angles in excess of the channeling angle.[12] The channeling effect can be used in ion-beam lithography to ensure the "transparency" of contact masks during exposure in a wide, collimated beam of ions.[10] We will now turn to the subsequent examination of the problems of ion scattering in the targets.

3-9. THE CONSERVATION OF MOMENTUM AND ENERGY IN THE THEORY OF THE ELASTIC SCATTERING OF FAST IONS IN ATOMS

We will outline the collision process between a fast particle and a target's atoms. As a fast particle and target atom approach one another an interaction described by an energy $V(r)$ and force $\mathbf{F} = -\mathrm{grad}\, V$ arises. Because of the collision the momentum

$p_2 = m_2 v_2$ and kinetic energy $T = m_2 v_2^2/2$ will be transmitted to the target atom. If the energy T that the target atom acquires is below some threshold T_d the atom will not move from the lattice site it occupied prior to the collision and the energy it acquired T will be manifested as vibrations that are stronger than those of the surrounding atoms. As it vibrates it will transmit its excess energy to its neighboring atoms and the lattice will again enter an equilibrium state. If $T > T_d$, the target atom will leave the lattice site and occupy an interstitial location at some point in the target, leaving a vacancy at the lattice site. The velocity of an atom when T = 20 eV is roughly $v_2 \approx 10^6$ cm/sec. The time needed to move to an interstitial location is roughly $\tau = (l/v_i) = 10^{-14}$ sec, which is much less than the period of vibration $\tau_{\text{vib}} = 10^{-12}$ sec for the atoms in the lattice, so that while target atoms are being dislodged from their lattice sites the lattice appears to be immobile. The stresses that occur near the vacancy and interstitial atom relax by dissipating the vibratory energy within the vibratory (phonon) relaxation time $\tau_{\text{ph}} \gg \tau_{\text{vib}}$. Such are the time scales of the events that occur. As to the energy T, only a small amount, about 5 eV is conserved as the potential energy of the vacancy + interstitial atom defect; the remainder is dissipated through the thermal vibrations of the lattice. If the Frenkel pair recombines (the interstitial atom returns to the vacancy in the lattice site) the remaining energy is converted to heat.[43]

The scattering processes must be examined in both a center of mass coordinate system (CMS) and in a laboratory coordinate system (LCS). The theory of collisions is more easily constructed in CMS; however, experimental results have been related to a LCS. Therefore, we must be able to convert the results calculated in a CMS to a LCS.

The trajectories, scattering angles, momenta, and velocities of the impinging (1) and scattering (2) particles in LCS and CMS are shown in Fig. 3-8. We are interested in the velocities and momenta after a collision has occurred. The coordinates and velocities for these in a CMS are, by definition,

$$\mathbf{R} = (m_1\mathbf{r}_1 + m_2\mathbf{r}_2)/(m_1 + m_2); \qquad \dot{\mathbf{R}} = (m_1\mathbf{v}_1 + m_2\mathbf{v}_2)/(m_1 + m_2) \tag{3-49}$$

we can write

$$\mathbf{v}_1 = (m_1\mathbf{v}_1 + m_2\mathbf{v}_2)/(m_1 + m_2) + \mathbf{v}_1', \qquad \mathbf{v}_2 = (m_1\mathbf{v}_1 + m_2\mathbf{v}_2)/(m_1 + m_2) + \mathbf{v}_2' \tag{3-50}$$

Solving the system, Eq. (3-50), for $\mathbf{v}_1'$ and $\mathbf{v}_2'$, we obtain

$$\mathbf{v}_1' = (\mu/m_1)\mathbf{v}_r, \qquad \mathbf{v}_2' = (-\mu/m_2)\mathbf{v}_r \tag{3-51}$$

here $\mu = m_1 m_2/(m_1 + m_2)$ is the reduced mass and the relative velocity is $\mathbf{v}_r = \mathbf{v}_1 - \mathbf{v}_2$. From the velocities triangle shown in Fig. 3-8d we have

$$\frac{\mathbf{v}_1' \sin\theta}{v_1' \cos\theta + |\dot{\mathbf{R}}|} = \tan\theta_1 \tag{3-52}$$

and we find $|\dot{\mathbf{R}}_1|$ from the law of conservation of momentum:

$$m_1\mathbf{v}_0 = (m_1 + m_2)|\dot{\mathbf{R}}| \tag{3-53}$$

In addition, because there are no external forces acting on the system of two particles, the magnitude of the relative velocity is unchanged, $|\mathbf{v}_r| = |\mathbf{v}_0|$, and, according to Eq. (3-51),

$$\bar{v}_1' = (\mu/m_1)\mathbf{v}_0 \tag{3-54}$$

Substituting Eqs. (3-53) and (3-54) into Eq. (3-52) yields

$$\tan\theta_1 = \sin\theta/(\cos\theta + m_1/m_2) \tag{3-55}$$

Similarly, from the velocities triangle (Fig. 3-8d) we find

$$\tan\theta_2 = v_2'\sin\theta/(v_2'\cos\theta + |\dot{\mathbf{R}}|) \tag{3-56}$$

Substituting the v_2' and $|\dot{\mathbf{R}}|$ from Eqs. (3-51) and (3-53) readily yields $\tan\theta_2 = \cot\theta/2$, i.e.,

$$\theta_2 = \pi/2 - \theta/2 \tag{3-57}$$

Equations (3-55) and (3-57) define the scattering angles θ_1 and θ_2 in a LCS if the scattering angle has been specified in a CMS. To derive these relations only the law of conservation of momentum is needed. The sum of angles $\theta_1 + \theta_2$ is the particles' scattering angle after collision; it follows from the equations we have obtained that when $m_1 > m_2$, $\theta_1 + \theta_2 < \pi/2$, and $\theta_1 + \theta_2 > \pi/2$ when $m_1 < m_2$. The special cases $m_1 = m_2$ and $m_1 \ll m_2$ are of interest. When $m_1 = m_2$ we easily find from Eqs. (3-55) and (3-57) that

$$\theta_1 = \theta/2, \qquad \theta_1 + \theta_2 = \pi/2$$

From here it follows that particles of equal mass will scatter at right angles when they collide with one another elastically. This corresponds, e.g., to the scattering of fast electrons by electrons: secondary electrons depart in a direction that is roughly perpendicular to the beam of primary electrons. When $m_1 \ll m_2$ (scattering of an electron by an atom) $\theta_1 \approx \theta$ (the scattering situations for a primary particle in a CMS and a LCS are roughly the same) and the scattering angle is $\theta_1 + \theta_2 = (\pi + \theta)/2$. When $\theta = \pi$ we have $\theta_1 + \theta_2 = \pi$; i.e., the lighter particle is reflected backwards.

Using similar calculations it is not hard to find the velocity of the particles after collision has occurred, v_1 and v_2, and the energy transmitted to the target atom, $T = m_2 v_2^2/2$:

$$\begin{aligned} v_1(\theta) &= v_0(m_1^2 + m_2^2 + 2m_1 m_2\cos\theta)^{1/2}(m_1+m_2)^{-1} \\ v_2(\theta) &= v_0 2m_1(m_1+m_2)^{-1}\sin\theta/2 \\ T(\theta) &= \frac{4m_1 m_2}{(m_1+m_2)^2}E_1\sin^2\frac{\theta}{2}, \qquad E_1 = \frac{m_1 v_0^2}{2} \end{aligned} \tag{3-58}$$

The formulas above are based on only the laws of conservation and no assumptions regarding the potential of interaction were used in deriving these formulas. The probability (cross section) of scattering at a specific angle can be found only if the $V(r)$ law of interaction between an ion and a target atom is known.

3-10. THE SCATTERING CROSS SECTION FOR ATOMIC PARTICLES

We can use the concept of a trajectory and an impact parameter b to introduce the concept of a scattering cross section for atomic particles (atoms, ions). Let the potential of interaction $V(r)$ for the scattered (1) and the scattering (2) particles be known. Then, by solving the trajectory problem, we can establish a one-to-one correspondence between

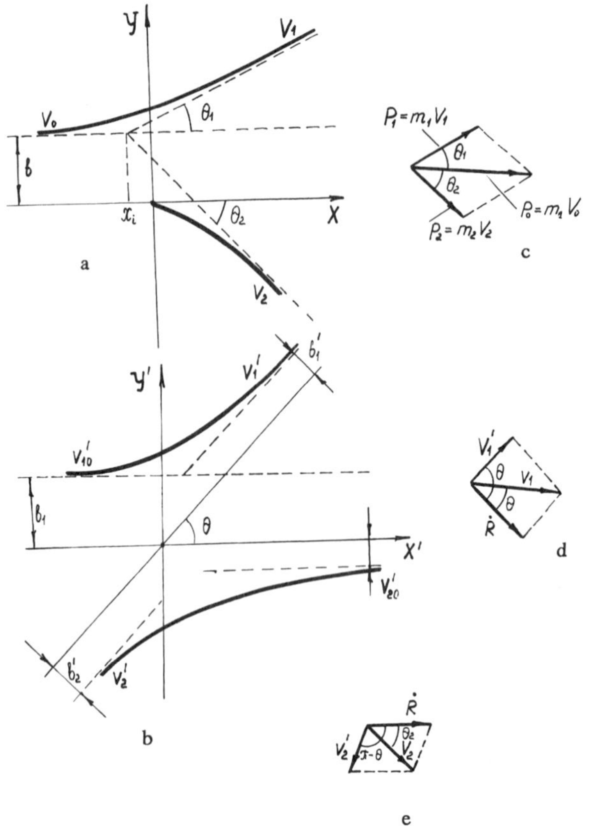

FIGURE 3-8. Trajectories of scattered and scattering particles in a LCS (a) and a CMS (b); parallelogram of momenta (c) and velocity triangles $\mathbf{v}_1 = \dot{\mathbf{R}} + \mathbf{v}_1'$ (d) and $\mathbf{v}_2 = \dot{\mathbf{R}} + \mathbf{v}_2'$ (e).

the scattering angle θ and the impact parameter b in a given collision: $\theta = \theta(b)$ and $b = \theta^{-1}(\theta)$. We designate the density of the particle flux as n (cm^{-2}). Into a ring of width b to $b + db$ near the scattering particle enter $dN = (2\pi b\, db)n$ particles which are scattered into the solid angle defined by the function $\theta = \theta(b)$.

We define the effective scattering cross section as the ratio

$$d\sigma = dN(\theta, \theta + d\theta)/n$$

of the number of particles scattered into the solid angle θ, $\theta + d\theta$ to the density of the particle flux. Using the identity $db = |db/d\theta|\, d\theta$, we easily find that

$$d\sigma = 2\pi b \left|\frac{db}{d\theta}\right| d\theta = \frac{b(\theta)}{\sin\theta} \left|\frac{db}{d\theta}\right| dO = \sigma(\theta)\, dO$$

We can find formulas for making the conversion to a LCS. If the given solid angle in the CMS and the LCS is, respectively, written $dO = 2\pi \sin\theta\, d\theta$ and $dO_1 = 2\pi \sin\theta_1\, d\theta_1$, the equation for the particle flux then has the form

$$\sigma(\theta)\, dO = \sigma(\theta_1)\, dO_1$$

or

$$\sigma(\theta_1) = \sigma(\theta)\frac{\sin\theta\, d\theta}{\sin\theta_1\, d\theta_1} = \sigma(\theta)\left|\frac{d\cos\theta}{d\cos\theta_1}\right|$$

Similarly, for $\sigma(\theta_2)$

$$\sigma(\theta_2) = \sigma(\theta)\left|\frac{d\cos\theta}{d\cos\theta_2}\right| \tag{3-59}$$

Using the conversion formula $\theta_2 = \pi/2 - \theta/2$ instead of Eq. (3-59) we easily obtain

$$\sigma(\theta_2) = \sigma(\theta(\theta_2))4\cos\theta_2$$

Using the relationship between θ_1 and θ, Eq. (3-55), we find[43] that when $m_1 \leq m_2$ and $0 \leq \theta_1 \leq \pi$,

$$\sigma(\theta_1) = \sigma(\theta_+)\left\{2\frac{m_1}{m_2}\cos\theta_1 + \frac{1 + (m_1/m_2)^2\cos 2\theta_1}{(1 - (m_1/m_2)^2\sin^2\theta_1)^{1/2}}\right\}$$

when $m_1 > m_2$, $\theta \leq \theta_1 \leq \arcsin(m_2/m_1)$,

$$\sigma(\theta_1) = \sigma(\theta_+)\left\{2\frac{m_1}{m_2}\cos\theta_1 + \frac{1 + (m_1/m_2)^2\cos 2\theta_1}{[1 - (m_1/m_2)^2\sin^2\theta_1]^{1/2}}\right\}$$
$$- \sigma(\theta_-)\left\{2\frac{m_1}{m_2}\cos\theta_1 - \frac{1 + (m_1/m_2)^2\cos 2\theta_1}{[1 - (m_1/m_2)^2\sin^2\theta_1]^{1/2}}\right\}$$

The angles θ_+ and θ_- are found[43] from

$$\cos\theta_\pm = -(m_1/m_2)\sin^2\theta_1 \pm \cos\theta_1\{1 - (m_1/m_2)^2\sin^2\theta_1\}^{1/2}$$

Finally, the cross sections $\sigma(\theta_1)$ and $\sigma(\theta_2)$ can be written as functions of the angle[43] θ:

$$\sigma(\theta_1(\theta)) = \sigma(\theta)\frac{[1 + (m_1/m_2)^2 + 2(m_1/m_2)\cos\theta]^{3/2}}{|1 + (m_1/m_2)\cos\theta|}$$
$$\sigma(\theta_2(\theta)) = \sigma(\theta)4\sin(\theta/2)$$

3-11. A GENERAL STATEMENT OF THE PROBLEM OF THE SCATTERING CROSS SECTION FOR ATOMIC PARTICLES. THE POTENTIALS OF AN ATOM–ATOM INTERACTION

We will say that the final solution for the trajectory problem in scattering an atomic particle in an atom is known[44]:

$$\theta = \pi - 2\int_{r_m}^{\infty}\frac{b}{r^2}\left[1 - \frac{V(r)}{E_r} - \frac{b^2}{r^2}\right]^{-1/2} dr \tag{3-60}$$

This result is not hard to derive (see, e.g., [44]). Here θ is the scattering angle in a CMS, b is the impact parameter for the collision, r_m is the closest approach distance of the particles during the collision, found from the condition that $\dot{r} = 0$:

$$\frac{\mu \dot{r}^2}{2E_r} = 1 - \frac{V(r_m)}{E_r} - \frac{b^2}{r_m^2} = 0 \tag{3-61}$$

E_r is the kinetic energy of motion in the CMS:

$$E_r = \frac{\mu}{2} v_\infty^2 = \frac{m_2}{m_1 + m_2} E_1, \qquad E_1 = \frac{m_1 v_\infty^2}{2}$$

and $V(r)$ is the potential of the central forces of the interaction between atoms.

The interactions between a multielectron ion and an atom can be thought of as the Coulomb interaction between their nuclei and the charges $Z_1 e$ and $Z_2 e$. This interaction is shielded by the electron shells of the atoms. We can therefore seek $V(r)$ in the form

$$V(r) = \frac{Z_1 Z_2 e^2}{r} \chi(r/a) \tag{3-62}$$

where $\chi(r/a)$ is the shielding function and a is the characteristic shielding length. We have already encountered the shielding function in the potential found by using the statistical Thomas–Fermi atom model. Firsov showed[(45)] that the shielding function in the Thomas–Fermi theory may be used for the interaction between two multielectron atoms (or ions) if we choose the quantity

$$a_{\mathrm{F}} = \left(\frac{9\pi}{128}\right)^{1/3} a_{\mathrm{B}}(Z_1^{1/2} + Z_2^{1/2})^{-2/3} = 0.47(Z_1^{1/2} + Z_2^{1/2})^{-2/3} \times 10^{-8}\ \mathrm{cm} \tag{3-63}$$

as the shielding length. A simpler version of the shielding function is the exponential term in Bohr's potential

$$V(r) = \frac{Z_1 Z_2 e^2}{r} \exp(-r/a), \qquad a = a_B/(Z_1^{2/3} + Z_2^{2/3})^{1/2} \tag{3-64}$$

Simple forms of the interatom potential are the exponential potentials of Born and Meier

$$V(r) = A \exp(-r/a_{\mathrm{BM}}) \tag{3-65}$$

or the exponentiated potentials

$$V(r) = C_n/r^n, \qquad C_n > 0 \tag{3-66}$$

The parameters of the Thomas–Fermi and Bohr potentials are calculated in terms of the Z_1 and Z_2 parameters of the atoms (from first principles); the parameters A and a_{BM} in the Born–Meier potential, and C_n and n in the exponentiated potential must be chosen from a comparison of theory and experiment; i.e., these potentials are empirical. Attempts at choosing, e.g., C_n and n in the exponentiated potential show that the values of these parameters must be changed, depending on the approach distance r of the atoms.

A measure that describes the degree of approach of the atoms in a given collision is the amount of energy transmitted, Eq. (3-58), which is determined by the impact parameter of the collision and by the energy of the relative motion of the colliding particles.

We will denote the reduced value of the energy of relative motion as ε:

$$\varepsilon = \frac{E_r}{(Z_1 Z_2 e^2/a_F)} = \frac{m_2}{m_1 + m_2}\frac{E_1}{(Z_1 Z_2 e^2/a_F)} \tag{3-67}$$

taking the Coulomb (not the shielded) energy of interaction between the nuclei of the colliding particles at a distance equal to the shielding distance a_F as unity and introduce the approach parameter t for atoms in a given collision by means of

$$t = \varepsilon^2 \sin^2 \theta/2 = (m_2/m_1)E_1 T/(Z_1 Z_2 e^2/a_F)^2 \tag{3-68}$$

As can be seen, t is equal to the product of the reduced values of the initial energy of the impinging particle and the transmitted energy (to within the factor m_2/m_1).

A comparison of the calculated and the measured scattering cross sections shows that the Coulomb potential $V(r) = Z_1 Z_2 e^2/r$ can be used when $t \gg 1$, which physically corresponds to a high-energy collision and a small impact parameter b. When this happens the nuclei of the colliding particles come so close to one another that the electron shells are actually outside both nuclei and their shielding action disappears. As t decreases, the n parameter in the exponentiated potential increases[(12)]:

$$V(r) \sim \begin{cases} r^{-2} & \text{when } 10^{-2} \le t \le 1 \\ r^{-3} & \text{when } 10^{-3} \le t \le 10^{-2} \\ r^{-4} & \text{when } t < 10^{-3} \end{cases}$$

The so-called approximation by the potential of solid spheres

$$V(r) = \begin{cases} \infty & \text{when } r \le R = r_1 + r_2 \\ 0 & \text{when } r > R \end{cases}$$

is used to describe collisions at low energies E_r and small impact parameter when scattering leads to the impinging particle being reflected backwards ($\theta \approx \pi$).

The potential given by Eq. (3-62) with the Thomas–Fermi shielding function can be used for all t, i.e., it is the most universal. This allows us to introduce a universal function for the scattering cross section of an atom by an atom. (We saw in Chapter 2 that new universal functions appear every time in computations using the Thomas–Fermi potential.)

We will assume that we have calculated the scattering angle θ as a function of E_r and b as in Eq. (3-60) and have written the result as a function of the reduced energy and impact parameter:

$$\theta = \theta(\varepsilon, b/a_F) \tag{3-69}$$

We introduce the inverse function of Eq. (3-69):

$$b/a_F = \varphi(\theta, \varepsilon) \tag{3-70}$$

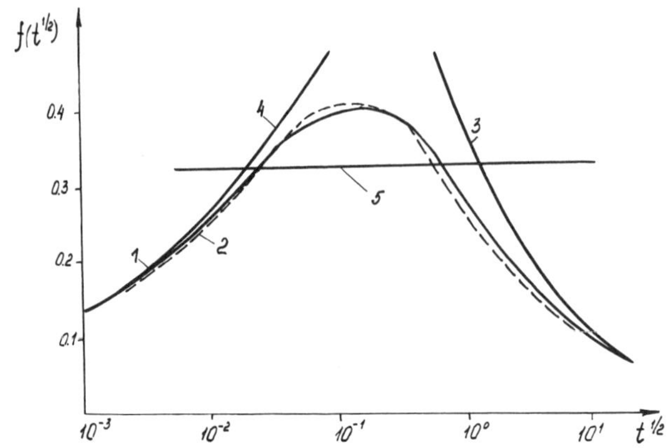

FIGURE 3-9. The function $f(t^{1/2})$ in the scattering cross section given by Eq. (3-71): 1) in the Thomas–Fermi atom model; 2) as in Eq. (3-71a); 3) as in Eq. (3-72) when $n = 3$; 4) as in Eq. (3-72) when $n = 3$; 5) as in Eq. (3-72) when $n = 2$.

Using Eq. (3-70) the differential scattering cross section can be written as

$$d\sigma = d(\pi b^2) = 2\pi a_F^2 \cdot \varphi \frac{\partial \varphi}{\partial \theta}\, d\theta \equiv \pi a_F^2 F(\varepsilon, \theta)\, d\theta$$

It can be shown that the function $F(\varepsilon, \theta)$ does in fact depend on the single complex argument $t^{1/2} = \varepsilon \sin \theta/2$; therefore, we can introduce a new function $f(t)$ via the equality

$$F(\varepsilon, \theta)\, d\theta = f(t^{1/2})\, dt/2t^{3/2}$$

and rewrite the expression for the scattering cross section:

$$d\sigma = \pi a_F^2 f(t^{1/2})\, dt/2t^{3/2} \tag{3-71a}$$

The function $f(t^{1/2})$ has been tabulated for collisions between neutral atoms according to the Thomas–Fermi theory. The tabulated values of $f(t^{1/2})$ are in good agreement with those obtained via the empirical formula

$$f(t^{1/2}) = 1.309t^{1/6}[1 + (2.618t^{2/3})^{2/3}]^{-3/2} \tag{3-71b}$$

For exponentiated potentials with $n = 1, 2, 3$, the function $f(t^{1/2})$ has the form[12]

$$f(t^{1/2}) = \lambda_n t^{(n-2)/n} \tag{3-72}$$

for values of $\lambda_1 = 0.5$; $\lambda_2 = 0.327$; $\lambda_3 = 1.309$. Values of $f(t^{1/2})$ calculated according to Eq. (3-72) coincide with those calculated according to the Thomas–Fermi model in a limited range of values for t: for $n = 1$ in the range $t \gg 1$; for $n = 3$ in the range $t \ll 10^{-2}$; for $n = 2$, $f(t^{1/2})$ remains constant and can, in a first approximation, be used to describe scattering in the range $10^{-2} \leq t \leq 1$ (see Fig. 3-9).

3-12. THE SCATTERING CROSS SECTION FOR ATOMIC PARTICLES FOR SPECIFIC KINDS OF POTENTIALS

A scattered particle's angle of deflection θ as a function of the impact parameter b can be calculated exactly according to Eq. (3-60) only for certain types of potentials, e.g., for exponentiated potentials when $n = 1$ (Coulomb interaction), $n = 2$, and the potential of solid spheres. We will begin by discussing results from calculating the function $\theta(b)$ according to Eq. (3-60), the differential scattering cross section, and the differential energy transmission cross section. We will then examine an extremely important method for calculating approximate values of these quantities that allows us to obtain analytical formulas for significantly more complex potentials, including potentials exponentiated to an arbitrary power n. This method is called the method of momentum approximation.

1. Coulomb interaction: $V(r) = C_1/r$, $C_1 = Z_1 Z_2 e^2$.

For this case we can easily find the closest approach distance according to Eq. (3-61):

$$r_m = b[C_1/(2bE_r) + (1 + C_1/(2bE_r))^{1/2}] \tag{3-73}$$

Evaluating the integral, Eq. (3-60), with the lower limit given by Eq. (3-73) we find

$$\theta = 2 \arcsin \left[\frac{C_1/(2bE_r)}{1 + C_1/(2bE_r)} \right] \tag{3-74}$$

Equation (3-74) can easily be converted to

$$\tan \frac{\theta}{2} = C_1/(2bE_r) = \frac{Z_1 Z_2 e^2}{E_1} \frac{m_1 + m_2}{2m_2} \frac{1}{b} \tag{3-75}$$

because $E_r = E_1 m_2/(m_1 + m_2)$. We find the differential scattering cross section $\sigma(\theta)$ by means of Eq. (3-75):

$$d\sigma = \sigma(\theta)\, dO = (C_1/4E_r)^2 \sin^{-4} \theta/2\, dO, \qquad 0 \leq \theta \leq |\pm \pi| \tag{3-76}$$

Using Eq. (3-58) we can convert the cross section given by Eq. (3-76) to the cross section for the transmitted energy:

$$d\sigma = \sigma(T; E_1)\, dT = \frac{2\pi C_1^2}{m_2 v_1^2} \frac{dT}{T^2}, \qquad 0 \leq T \leq T_m \tag{3-77}$$

The cross sections given by Eqs. (3-76) and (3-77) are the well-known Rutherford cross sections for elastic scattering in a Coulomb potential. Earlier, we used a quantum mechanical approach to derive this formula for electron scattering (Section 2-5, Eq. (2-30)) based on, just as it is now, the laws of classical mechanics that govern the motion of heavy ions. It is instructive to examine how the closest approach distance r_m and the scattering angle θ depend on the impact parameter b and the energy of relative motion E_r. These relationships are purely classical in nature and have no place in the theory of quantum mechanics (Chapter 2).

The quantity r_m when $b = 0$ (head-on, central collision) has an absolute minimum $r_m = C_1/2E_r$ (for a given E_r): as b increases r_m increases, and $r_m \to b$ when $b \gg C_1/2E_r$. Equation (3-75) shows that the scattering angle θ increases rapidly as the impact parameter b decreases. When $b \gg C_1/2E_r$, $\theta \sim 1/b$; when $b = C_1/2E_r$, $\theta = \pi/2$, and $\theta \to \pi$ as $b = 0$.

2. The potential of solid spheres $V(r) = \begin{cases} \infty & \text{when } r \leq R \\ 0 & \text{when } r > R \end{cases}$

This potential is close to the actual when a collision takes place with low relative energy E_r and is often used in computer experiments using the Monte Carlo method in problems in the theory of gases. Indeed, the condition that E_r be small is well satisfied when the molecules move at thermal velocities.

For solid spheres the closest approach distance is $r_m = R$. Integrating Eq. (3-60) with the lower limit R gives us

$$\theta = \pi - 2\int_R^\infty \frac{b}{r^2}\left(1 - \frac{b^2}{r^2}\right)^{-1/2} dr = \pi - \arcsin(b/R)$$

or

$$\cos\theta/2 = b/R$$

Using the formulas mentioned we can find the differential scattering cross section via elementary calculations:

$$\sigma(\theta) = \frac{b}{\sin\theta}\left|\frac{db}{d\theta}\right| = \frac{R^2}{4}, \qquad 0 \leq \theta \leq \pi$$

which is independent of the scattering angle and the energy of the collision E_r. The differential energy transmission cross section is

$$d\sigma = \pi R^2\, dT/T_m, \qquad 0 \leq T \leq T_m$$

i.e., it is independent of the amount of energy transmitted. The total scattering cross section coincides with the geometric cross section of a sphere whose radius is R:

$$\sigma_{\text{tot}} = \int d\sigma = \pi R^2$$

We will also introduce an expression for finding the escape angle of the recoil atom in a LCS:

$$\sin\theta_2 = b/R$$

and the scattering cross section for the recoil atom at the angle θ_2: $\sigma(\theta_2) = R^2\cos\theta_2$.

3. The potential $V(r) = C_2/r^2$, $C_2 > 0$.

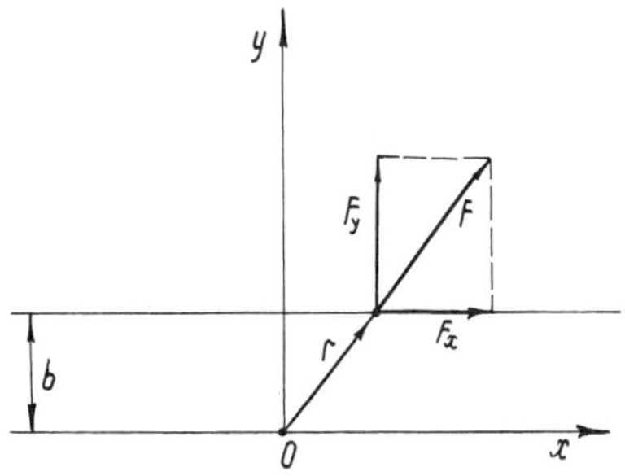

FIGURE 3-10. Schematic of collisions in a momentum approximation.

In this case $r_m = (b^2 + C_2/E_r)^{1/2}$, and

$$\theta(b) = \pi - 2\int_{r_m}^{\infty} (b/r^2)\left[1 - \frac{C_2}{E_r r^2} - \frac{b^2}{r^2}\right]^{-1/2} dr = \pi\left(1 - \frac{b}{r_m}\right)$$

$$\sigma(\theta) = \frac{C_2}{E_r}\frac{\pi^2(\pi - \theta)}{\theta^2(2\pi - \theta)^2}\frac{1}{\sin\theta}$$

$$d\sigma = \frac{2C_2}{E_r}\frac{[\tau(1-\tau)]^{-1/2}\left(1 - \frac{2}{\pi}\arcsin\sqrt{\tau}\right)d\tau}{\left[1 - \left(1 - \frac{2}{\pi}\arcsin\sqrt{\tau}\right)^2\right]^2}$$

$$\tau = T/T_m$$

3-13. MOMENTUM APPROXIMATION

The name of this method comes from the fact that the change in a particle's momentum during a collision is used to calculate its scattering angle. The approximation works very well for small scattering angles (forward scattering). When this is done we consider the motion of the scattered particle in the relatively small area of strong interaction between the colliding particles to be rectilinear and its y coordinate to be constant and equal to the impact parameter b (Fig. 3-10). The scattering particle remains immobile. We will conduct our examination in a LCS. The initial momentum of the impinging particle is $p_x = m_1 v_1$, $p_y = 0$. If the impinging particle acquires momentum Δp_y during the collision its escape angle will be $\theta_1 = \Delta p_y / p_x$. Thus, we must calculate the change in the impinging particle's momentum during the collision.

By definition,

$$\Delta p_x = \int_{-\infty}^{+\infty} F_x(t)\,dt, \qquad \Delta p_y = \int_{-\infty}^{+\infty} F_y(t)\,dt \tag{3-78}$$

and we say that the potential of the interaction between the colliding particles $V(r)$ is known, so that we can calculate the forces acting on the impinging particle:

$$F_x = -\frac{\partial V}{\partial r}\frac{x}{r}, \qquad F_y = -\frac{\partial V}{\partial r}\frac{b}{r}$$

Because the F_x component of the force changes sign when the particle moves from the first half of its trajectory to the second half, $\Delta p_x = 0$. Because $x = v_1 t$, we can easily make the change in Eq. (3-78) to integrating over the coordinate:

$$\Delta p_y = 2\int_0^\infty F_y(t)\,dt = \frac{2b}{v_1}\int_b^\infty \left(-\frac{\partial V}{\partial r}\right)\frac{dr}{(r^2 - b^2)^{1/2}}$$

Now we find the impinging particle's angle of deflection:

$$\theta_1 = \frac{\Delta p_y}{p_x} = \frac{b}{E_1}\int_b^\infty \left(-\frac{dV}{dr}\right)\frac{dr}{(r^2 - b^2)^{1/2}} \tag{3-79}$$

We can easily write this result in a CMS, using Eq. (3-55) for small θ_1 and θ, as $\theta = \theta_1(m_1 + m_2)/m_2$:

$$\theta = \frac{b}{E_r}\int_b^\infty \left(-\frac{dV}{dr}\right)\frac{dr}{(r^2 - b^2)^{1/2}}, \qquad E_r = \frac{m_2}{m_1 + m_2}E_1 \tag{3-80}$$

3-14. THE SCATTERING CROSS SECTION FOR EXPONENTIATED POTENTIALS IN A MOMENTUM APPROXIMATION

The integral of Eq. (3-80) can easily be evaluated for the exponentiated potential $V(r) = C_n/r^n$. We make the substitution $dV/dr = -nC_n/r^{n+1}$ and convert the integral, introducing the new variable $x = r/b$:

$$\theta = \frac{nC_n}{b^n E_r}\int_1^\infty \frac{dx}{x^{n+1}(x^2 - 1)^{1/2}} \tag{3-81}$$

We convert the integral given by Eq. (3-81) again, introducing the variable $y = 1/x$:

$$I = \int_1^\infty \frac{dx}{x^{n+1}(x^2 - 1)^{1/2}} = \int_0^1 \frac{y^n\,dy}{(1 - y^2)^{1/2}}$$

The integral obtained corresponds to the so-called beta function, defined[47] by

$$B(x, y) = \frac{\Gamma(x)\Gamma(y)}{\Gamma(x + y)} = 2\int_0^1 t^{2x-1}(1 - t^2)^{y-1}\,dt$$

As can be seen, in this case we have $I = (1/2)B((n+1)/2, 1/2)$ and

$$\theta = \frac{n\lambda_n}{E_r b^n}, \qquad \lambda_n = \frac{C_n}{2} \frac{\Gamma\left(\frac{n+1}{2}\right)\Gamma\left(\frac{1}{2}\right)}{\Gamma\left(\frac{n}{2}+1\right)} \tag{3-82}$$

We calculate the scattering cross section, using $\sin\theta \approx \theta$ for small angles θ when the momentum approximation is valid:

$$\frac{d\sigma}{dO} = \frac{b}{\sin\theta}\left|\frac{db}{d\theta}\right| = \frac{b^2}{n\theta^2}$$

because

$$\left|\frac{d\theta}{db}\right| = n^2\lambda_n/(b^{n+1}E_r) = n\theta/b$$

Replacing b^2 as in Eq. (3-82),

$$b^2 = (n\lambda_n/(\theta E_r))^{2/n}$$

we finally come to

$$\frac{d\sigma}{dO} = \frac{1}{n\theta^2}\left(\frac{n\lambda_n}{\theta E_r}\right)^{2/n} \tag{3-83}$$

Using the relationship $T = T_m \sin^2\theta/2$ we can easily convert to the magnitude of the cross section of transmitted energy T:

$$d\sigma = \frac{\pi}{n}\frac{dT}{T}\left(\frac{n\lambda_n}{E_r(4T/T_m)^{1/2}}\right)^{2/n} \tag{3-84}$$

As the exponent n on the potential function increases, the atoms become more "rigid." It is interesting to see how the respective cross sections as functions of the angles and amounts of transmitted energy change when this happens:

n	1	2	3	4
$d\sigma/dO$	$1/\theta^4$	$1/\theta^3$	$1/\theta^{8/3}$	$1/\theta^{5/2}$
$d\sigma/dT$	$1/T^2$	$1/T^{3/2}$	$1/T^{4/3}$	$1/T^{5/4}$

As we can see, the more "rigid" the atom, the weaker the dependence.

The method of momentum approximation enables us to find solutions in analytical form for not only exponentiated, but even more complex forms of the interatom potential. For the potential given by Eq. (3-64) with an exponential shielding function,

$$\theta = \frac{b}{a}K_1\left(\frac{b}{a}\right)$$

was obtained in reference [12], where K_1 is a modified second-order Bessel function.

With momentum approximation more accurate results can be obtained for the function $\theta(b, E_r)$ in the form of a series.

3-15. INELASTIC SCATTERING OF ATOMIC PARTICLES IN SOLIDS

We examined elastic atom–atom scattering in preceding paragraphs, considering the total kinetic energy of the colliding particles to be the same before and after the collision occurred. However, electron excitation, ionization, an exchange or change in the amount of charge may occur during the collision and consume a portion of the colliding particles' kinetic energy. This kind of collision is called inelastic. If the scattering atom in the solid is immobile and the energy of the impinging atom (ion) is E_1, the law of conservation of energy can be written

$$E_1 = T_1 + T_2 + Q$$

where T_1 and T_2 are the kinetic energies of the particles after the collision and Q is the amount of energy lost during the electron processes that occurred during the collision. As we said, these losses cause electron deceleration.

When atoms collide they penetrate one another and this penetration is described by the closest approach distance r_m of their nuclei. The quantity r_m is found from Eq. (3-61) and is small when the collision impact parameter is small and the relative energy E_r is high. In head-on collisions having high energy E_r the nuclei of the colliding particles come to within 0.001 nm of one another, which is hundreds of times smaller than the radii of the atoms. At the moment of closest approach the atoms form a quasiatom that has a nuclear charge of $(Z_1 + Z_2)e$ and a total of $Z_1 + Z_2$ electrons. When the distance r_m is roughly equal to the radius of the atom the colliding atoms form a quasimolecule instead. Be that as it may, the fact that the atoms penetrate one another increases the likelihood that electron processes, including collective processes, will proceed. It then becomes necessary to solve the collision problem for two multielectron atoms, taking the interactions between their electrons into account.

Solving the atomic collisions problem stated in multielectron form runs into significant obstacles. We will show a way to solve the problem of direct excitation of an electron (transmitting energy to the electron) during a knock-on collision between an ion and an electron. Solving this limited problem is made easier by the fact that the motion of the nuclei of the colliding particles can be examined on the basis of classical mechanics, preserving the concept of a trajectory with a given impact parameter, just as before. Only when we examine the electron transitions do we use Schrodinger's quantum mechanical equation. The problem is made especially simple if we solve the trajectory problem in a momentum approximation (see Section 3-10).

Using the laws of conservation of energy and momentum we can solve the kinematic problem of scattering angles for a given inelastic loss of energy Q. We can neglect the momentum transmitted to the electrons, because it is so small. The solution of the kinematic problem in a laboratory coordinates system is[43]

$$\frac{Q}{E_1} = \frac{2}{\alpha}\left(\frac{T_1}{E_1}\right)^{1/2}\cos\theta_1 + \frac{\alpha - 1}{\alpha} - \frac{\alpha+1}{\alpha}\frac{T_1}{E_1}, \qquad \alpha = \frac{m_2}{m_1} \tag{3-85}$$

$$\frac{Q}{E_1} = 2\left(\alpha\frac{T_2}{E_1}\right)^{1/2}\cos\theta_2 - (1+\alpha)\left(\frac{T_2}{E_1}\right) \tag{3-86}$$

The formulas for converting from a LCS to a CMS are

$$\sin\theta = \left(\frac{\alpha+1}{\alpha}\frac{T_1}{E_1}\right)^{1/2}\left(\frac{\alpha}{1+\alpha} - \frac{Q}{E_1}\right)^{1/2}\sin\theta_1$$

$$\sin\theta = \left((1+\alpha)\frac{T_2}{E_1}\right)^{1/2}\left(\frac{\alpha}{1+\alpha} - \frac{Q}{E_1}\right)^{-1/2}\sin\theta_2$$

Equations (3-85) and (3-86) are important because they allow us to find the inelastic energy loss Q experimentally if we measure, e.g., the kinetic energy after the collision has occurred, T_1, and the escape angle θ_1 of the impinging particle (Eq. (3-85)). In this approach the energy E_1 is assumed to be known. These experiments showed that the amount of inelastic energy $Q(\theta_1(b))$ lost in a collision increases rapidly with decreasing closest approach distance r_m (or impact parameter b).

3-16. USING CLASSICAL CONCEPTS TO EVALUATE INELASTIC ENERGY LOSSES

We can use classical mechanics to evaluate the amount of inelastic energy loss Q as a result of being transmitted to a target atom's electron, calculating it as the work done by the ion on the electron during a collision having a specific impact parameter. The collision situation corresponds to the conditions for a momentum approximation (see Section 3-12, Fig. 3-10). The F_x component of force does an amount of work $A_x = \int F_x\,dx = 0$ on the electron because F_x changes sign, but not magnitude, at $x = 0$. The F_y component does not change sign and has its maximum value at $x = 0$. The amount of work $A_y = Q$ can be found as $Q = \Delta p_y^2/2m_e$, evaluating Δp_y as the product of the maximal force $F_y(b)$, which occurs at $x = 0$, and the duration of the collision, which can be equated to $\Delta t = 2b/v_1$. We say that the interaction between the ion and the target atom's electron is Coulomb and set the ion's charge equal to its nuclear charge, not the degree to which it has been ionized. This means that the nuclear charge is unshielded by the electron shells, which happens when the impact parameter and closest approach distance are small quantities. Under these conditions

$$\Delta p_y = F_y\Delta t = \frac{Z_1 e^2}{b^2}\frac{2b}{v_1}$$

$$Q = \frac{2Z_1^2 e^4}{mv_1^2}\frac{1}{b^2} \tag{3-87}$$

As can be seen, inelastic losses decrease with increasing impact parameter b as $1/b^2$.

We can construct a much more accurate classical theory of inelastic collision when we consider the electron's frequency motion ω_0 in the atom. This frequency accounts for the ionization energy $I = \hbar\omega_0$. The force that the ion exerts on the target atom's electron can be expanded in a spectrum $F_x(\omega)$, $F_y(\omega)$. The amount of work done on the electron is[12]

$$Q = |F(\omega_0)|^2/m = \frac{2Z_1^2e^4}{mv_1^2b^2}\left[K_1^2(z) + \frac{1}{\gamma^2}K_0^2(z)\right]z^2, \qquad z = \frac{b\omega_0}{\gamma v_1} \tag{3-88}$$

Here the K_n are modified Bessel functions, and $\gamma = (1 - v^2/c^2)^{-1/2} = 1$ for the nonrelativistic case we are examining.

Equation (3-88) contains what is essentially a new, in comparison with Eq. (3-87), result which we will present. Large values of the argument $z = \omega_0 b/v_1$ of the $K_n(z)$ functions correspond to the duration b/v_1 of a collision much longer than the period of the electron's motion $2\pi/\omega_0$. Using an asymptotic representation of the modified Bessel functions for large[12]

$$K_n(z) = (\pi/2z)^{1/2}e^{-z}, \qquad z \gg 1$$

we obtain

$$Q = \frac{2Z_1^2e^4}{mv_1^2b^2}\left(\frac{\pi b\omega_0}{v_1}\right)\exp\left(-\frac{b\omega_0}{v_1}\right) \tag{3-89}$$

According to Eq. (3-89), the amount of energy transmitted decreases exponentially and becomes negligible when the impact parameters are large. In a quantum mechanical formulation of the theory, this corresponds to an exponential decrease in the probability (cross section) of inelastic scattering. Collisions with exponentially small energy transmission (scattering cross section) are called adiabatic.

Using the expansion of the Bessel functions for small $z \ll 1$ (small impact parameters),[12]

$$K_0(z) = \ln\frac{2}{cz}, \qquad K_1(z) = -K_0'(z) = 1/z$$

we obtain the result given by Eq. (3-87).

3-17. THE QUANTUM THEORY OF INELASTIC COLLISIONS WITH A SPECIFIC IMPACT PARAMETER[48]

A formalism that corresponds to a Born first approximation (see Section 2-2), but in which the motion of the impinging ion is examined in classical terms, e.g., on the assumptions of a momentum approximation (see Section 3-22) is widely used in the quantum theory of atomic particle collisions. The motion of the electrons in the colliding particles is described in quantum mechanical terms.

We will say that a solution of Schrödinger's equation for the motion of the target atom's electrons has been found

$$\left\{\frac{\hbar^2}{2m}\sum_a \Delta_a + V_A(r_a)\right\}\varphi_\alpha(r_a) = E_\alpha\varphi_\alpha(r_a) \tag{3-90}$$

and that the energy levels E_α and the wave functions φ_α for the atom's electrons have been defined.

At time t the impinging particle is at a point $\mathbf{R}$ and we will let its energy of interaction with the atom's electron at point r_a be $V(\mathbf{R} - \bar{r}_a)$. To determine the consequences for the state of the target atom's electron due to a collision, we must solve Schrödinger's temporal equation:

$$\left\{ -\frac{\hbar^2}{2m} \sum_a \Delta_a + V_A(r_a) + V(\mathbf{R} - \bar{r}_a) \right\} \Phi = i\hbar \frac{\partial \Phi}{\partial t} \tag{3-91}$$

We will seek a solution of Eq. (3-91) as an expansion in the functions $\varphi_a(r_a)$:

$$\Phi = \sum_n a_{0n}(t) \varphi_n(r_a) \exp(-E_n t/\hbar) \tag{3-92}$$

In a momentum approximation the trajectory of the ion is a straight line, $y = b$. The energy transmitted from the ion to the electrons in one collision is small, so that the velocity $v_x = v_1$ is also constant. We choose the moment from which time is to be measured such that $x = v_1 t$, $t = x/v_1$. Substituting Eq. (3-92) into Eq. (3-91) and taking Eq. (3-90) into consideration yields

$$i\hbar \sum_n \frac{da_{0n}(x)}{dx} \varphi_n(r_a) \exp(-iE_n x/(v_1 \hbar)) = V(\mathbf{R} - \bar{r}_a) \sum_n a_{0n}(x) \varphi_n(r_a) \times \exp(-E_n x/(\hbar v_1)) \tag{3-93}$$

We multiply Eq. (3-93) by $\varphi_\beta^*(r_a) \exp(iE_\beta x/(\hbar v_1))$ and integrate with respect to $d^2 \bar{r}_a$, which yields

$$i\hbar \frac{da_{0\beta}}{dx} = \frac{1}{v_1} \sum_n a_{0n}(x) V_{\beta,n}(R) \exp\left(-\frac{ix}{\hbar v_1}(E_n - E_\beta) \right) \tag{3-94}$$

$$V_{\beta,n} = \int \varphi_\beta^*(r_a) V(\bar{R} - \bar{r}_a) \varphi_n(\bar{r}_a)\, d^3 \bar{r}_a \tag{3-95}$$

The initial conditions are $a(-\infty)_{0n} = \delta_{0n}$ because the atom is in its ground state prior to the collision. Substituting the zero-order approximation $a_{0n}(x) = \delta_{0n}$ into Eq. (3-94), we obtain

$$\begin{aligned} a_{00}(x) &= 1 - \frac{i}{\hbar v_1} \int_{-\infty}^{x} V_{00}(R)\, dx \\ a_{0\beta}(x) &= -\frac{i}{\hbar v_1} \int_{-\infty}^{x} V_{0,\beta}(R) \exp\left[-\frac{ix}{\hbar v_1}(E_0 - E_\beta) \right] dx \end{aligned} \tag{3-96}$$

The dimensionless probability that an atom's electron will make the transition from a state $\varphi_0(r_a)$ to a state $\varphi_\beta(r_a)$ due to colliding with a cast ion (atom) will be

$$P_{0\beta} = |a_{0\beta}(\infty)|^2 \tag{3-97}$$

This probability is a function of the impact parameter b for the collision. Adding the probabilities Eq. (3-97) for all impact parameter values gives us the cross section $\sigma_{0\beta}$ of the transition $0 \to \beta$

$$\sigma_{0\beta} = 2\pi \int_0^{\infty} P_{0\beta}(b) b \, db \tag{3-98}$$

In order to obtain specific physical results, we must find the energy of interaction between the impinging atom and the electron in the target and calculate the matrix elements of Eq. (3-95) in an explicit form. We will assign the nucleus of the impinging atom a charge $Z_1 e$. Then its energy of interaction with the target atom's electrons is

$$V(\bar{R} - \bar{r}_a) = \sum_a \frac{Z_1 e^2}{|\mathbf{R} - \mathbf{r}_a|} \tag{3-99}$$

We make the center of the target atom the origin of the coordinate system and expand Eq. (3-99) as a series in multipoles; the first two terms in the expansion are

$$V(\bar{R} - \bar{r}_a) = \frac{Z_1 e \sum_a e_a}{R} + \frac{Z_1 e \mathbf{dR}}{R^3} + \ldots$$

In a momentum approximation (Fig. 3-7) the dipole term is rewritten as

$$V_d(R) = \frac{Z_1 e}{R^3}(x d_x + b d_y)$$

$$d_x = e \sum_a x_a, \qquad d_y = e \sum_a y_a, \qquad R = (x^2 + b^2)^{1/2}$$

We can now easily write the matrix elements for the energy of interaction

$$V_{0\beta} = \frac{Z_1 e}{R^3}[x(d_x)_{0\beta} + b(d_y)_{0\beta}] \tag{3-100}$$

Substituting Eq. (3-100) into Eq. (3-96) gives us

$$a_{0\beta}(\infty) = -iZ_1 e \int_{-\infty}^{+\infty} \exp(-i\omega_{0\beta} x / v_1)[x(d_x)_{0\beta} + b(d_y)_{0\beta}] R^{-3} \, dx \tag{3-101}$$

for the amplitude of the $a_{0\beta}$ state. The subintegral function $R^{-3} = (x^2 + b^2)^{-3/2}$ is large only when $x \leq b$; therefore, the index for the exponential curve must be evaluated when $x = b$. If $\omega_{0\beta}/v_1 \ll 1$ the exponential curve can be replaced by unity. The function xR^{-3} in the remainder of the integral is odd and the integral between symmetric limits is zero. The remainder of the integral can easily be evaluated:

$$\int_{-\infty}^{+\infty} R^{-3} \, dx = 2/b^2$$

and

$$a_{0\beta}(b) = -\frac{2iZ_1 e(d_y)_{0\beta}}{\hbar v_1 b}$$
$$P_{0\beta}(b) = \frac{4Z_1^2 e^2 |(d_y)_{0\beta}|^2}{\hbar^2 v_1^2 b^2} \tag{3-102}$$

For values $\omega_{0\beta} b/v_1 \gg 1$ the same exponential decrease is seen in the probability of a transition and corresponds to the same exponential decrease that was found in Section 3-16 by the classical model for the amount of energy transmitted to an electron by an impinging atom in an adiabatic collision. Considering the evenness and oddness of the subintegral functions in Eq. (3-101), we can write

$$I_1 = \int_{-\infty}^{+\infty} \exp(-i\omega x/v_1) R^{-3}\, dx = 2\int_0^{\infty} R^{-3} \cos(\omega x/v_1)\, dx \tag{3-103}$$

$$I_2 = \int_{-\infty}^{+\infty} \exp(-i\omega x/v_1) x R^{-3}\, dx = 2i\int_0^{\infty} x R^{-3} \sin(\omega x/v_1)\, dx \tag{3-104}$$

The integrals, Eqs. (3-103) and (3-104), are given as modified Bessel functions[47] according to the relations

$$\int_0^{\infty} (t^2 + a^2)^{-\nu - 1/2} \cos(ut)\, dt = \frac{\pi^{1/2} u K_\nu(au)}{2a\Gamma(\nu + 1/2)}, \qquad Re\nu > -\frac{1}{2} \tag{3-105}$$

$$\int_0^{\infty} (t^2 + a^2)^{+\nu - 3/2} t \sin(ut)\, dt = \frac{\pi^{1/2}(2a)^{\nu} u^{1-\nu} K_\nu(au)}{\Gamma(3/2 - \nu)} \tag{3-106}$$

Using Eqs. (3-105) and (3-106) we find the integral of I_1 and I_2 and write them in asymptotic form for large $z = b\omega/v_1 \gg 1$ (see Section 3-16) as

$$I_1 = \frac{2\omega}{bv_1} K_1\left(\frac{b\omega}{v_1}\right) \approx \frac{2\omega}{v_1 b}\left(\frac{2b\omega}{\pi v_1}\right)^{-1/2} \exp\left(-\frac{b\omega}{v_1}\right) \tag{3-107}$$

$$I_2 = \frac{2i\omega}{v_1} K_0\left(\frac{b\omega}{v_1}\right) \approx \frac{2i\omega}{v_1}\left(\frac{2b\omega}{\pi v_1}\right)^{-1/2} \exp\left(-\frac{b\omega}{v_1}\right) \tag{3-108}$$

Substituting Eqs. (3-107) and (3-108) into Eqs. (3-101) and (3-97) we find

$$a_{0\beta}(\infty) = -\frac{iZ_1 e^2}{\hbar^2 v_1}\left[i(d_x)_{0\beta} K_0\left(\frac{b\omega}{v_1}\right) + (d_y)_{0\beta} K_1\left(\frac{\omega b}{v_1}\right)\right]\frac{2\omega}{v_1}$$
$$P_{0\beta} = \frac{4(Z_1 e\omega_{0\beta})^2}{\hbar^2 v_1^4}[(d_x)_{0\beta}^2 + (d_y)_{0\beta}^2]\left(\frac{2b\omega}{\pi v_1}\right)^{-1} e^{-2b\omega/v_1} \tag{3-109}$$

for the magnitude and the probability of an electron transition during a collision.

We found an exponential decrease in the probability of a transition when the duration of the collision was much longer than the inverse frequency of the transition. This result is in complete agreement with the result calculated from the classical representation (see Eq. (3-89)).

Using Eqs. (3-102) and (3-109) for the probabilities of a transition we can calculate the average amount of energy transmitted to an electron by an ion in one collision as a result of the $0 \to \beta$ transition. To do this, we need only multiply the probability of the transition by the energy of the transition:

$$Q_{0\beta} = \hbar\omega_{0\beta} P_{0\beta} = \frac{4Z_1^2 e(d_y)}{\hbar^2 v_1^2} \frac{|(d_y)_{0\beta}|^2}{b^2} \hbar\omega_{0\beta} \tag{3-110}$$

The absolute values of the matrix element $(d_y)_{0\beta}$ and the transition frequencies $\omega_{0\beta}$, which have a specific value for a specific transition, appear in Eq. (3-110). Their magnitudes can be estimated by assuming them to be equal to the atom length (Bohr radius) $ea_{\mathrm{B}} = e\hbar^2/mc^2$ and atomic frequency $\omega_a = v_a/a_{\mathrm{B}} = me^4/\hbar^3$, so that

$$|(d_y)_{0\beta}|^2 \hbar\omega_{0\beta} \approx e^2\hbar^2/m$$

Because the quantity $Q_{0\beta}$ in this estimate has no unique characteristics, the transition subscripts may be omitted:

$$Q = \frac{4Z_1^2 e^4}{mv_1^2} \frac{1}{b^2}$$

which agrees with the classical estimate of the amount of energy transmitted, Eq. (3-87).

3-18. ION DECELERATION IN A TARGET

The energy lost by a fast atom (ion) as it moves through a target can be described by the concept of decelerating ability (deceleration) if the average energy lost in one collision is small in comparison with the particle's energy. We have seen that this condition is well satisfied for losses incurred through exciting the electrons or ionization. It is usually assumed that this condition is also satisfied for the nuclear losses that describe the transmission of energy from the fast atom to the recoil atom in elastic collisions.

If there are N atoms per unit volume of the target material and the differential cross section $d\sigma$ for scattering a fast atom on them is known, then the probability that the indicated scattering will occur when the fast atom passes through a layer dx in the target is

$$dP = N\, dx\, d\sigma$$

The average amount of energy lost in a layer dx due to the indicated processes is

$$\overline{dE} = \int T\, dP = N\, dx \int_0^{T_m} T \frac{d\sigma}{dT}\, dT$$

Going to losses per unit path length, we can write

$$\frac{\overline{dE}}{dx} = -NS(E), \qquad S(E) = \int_0^{T_m} T \frac{d\sigma(E;T)}{dT}\, dT \tag{3-111}$$

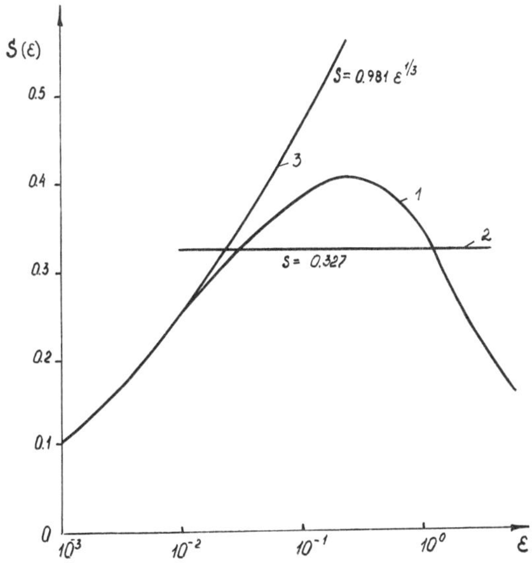

FIGURE 3-11. Nuclear deceleration as a function of energy: 1) for the Thomas–Fermi screened potential; 2, 3) for the exponentiated potentials with $n = 2$ and $n = 3$.

Using the cross section of the nuclear losses in an exponentiated potential, for the energy of interaction, Eq. (3-84), we find

$$S_n(E) = \frac{\pi}{n-1}\left(\frac{n\lambda_n}{2E_r}\right)^{2/n} T_m \tag{3-112}$$

If the energy of interaction is the Coulomb shielding potential, Eq. (3-71), the nuclear deceleration is

$$S_n(E) = \pi a_{\mathrm{F}}^2 T_m \varepsilon_1^{-1} \int_0^{\varepsilon} f(t)^{1/2}\, d(t^{1/2}) \tag{3-113}$$

If we normalize the path length to $l = 1/(N\pi a_{\mathrm{F}}^2 \gamma)$, $\gamma = 4m_1 m_2/(m_1 + m_2)^2$ and the energy is, as before, normalized to $Z_1 Z_2 e^2/a_{\mathrm{F}}$ we obtain a reduced value for the nuclear decelerating ability:

$$S_n(\varepsilon) = -\frac{d\varepsilon}{d\rho} = \int_0^{\varepsilon} f(t^{1/2})\, dt^{1/2}$$

The function for the nuclear decelerating ability increases at small $\varepsilon \ll 1$ as $\varepsilon^{1/3}$ and decreases for large $\varepsilon \gg 1$ as ε^{-1}. This is because the exponent on the potential of interaction changes from $n = 3$ for small relative energies of interaction to $n = 1$ for large (Rutherford scattering). Because $S_n \sim Z_1^2$, the magnitude of nuclear deceleration is a few keV per angstrom for heavy ions. Figure 3-11 is a graph of the function $S_n(\varepsilon)$.[48]

A particularly simple expression for the decelerating ability is obtained when $n = 2$ and is useful in estimating its order of magnitude. According to Eq. (3-112), when $n = 2$ we have

$$S_n(E) = \frac{\pi}{2}\lambda_2\gamma = -1.308\pi a_{\mathrm{F}}^2 Z_1 Z_2 e^2 m_1 (m_1 + m_2)^{-1} \tag{3-114}$$

Equation (3-114) enables us to analyze decelerating ability as a function of the relationship between the masses of the atoms in the beam and the atoms in the target. For light ions $m_1 \ll m_2$ and deceleration is only slightly dependent on the target material (since $Z_2 \sim m_2$, the dependence on the target material is maintained only through the a_F parameter). At the same time, the decelerating ability for light ions has a square law dependence on their mass: $S_n(E) \sim Z_1 m_1 \sim m_1^2$.

If, on the contrary, $m_1 \gg m_2$, the nuclear decelerating ability is proportional to the mass density of the target $N m_2$ and the atomic number of the atoms in the beam Z_1.

3-19. ION DECELERATION BY ELECTRONS IN THE TARGET

Ion deceleration caused by their energy being transmitted to the target atoms' electron can easily be calculated for fast ions whose velocity exceeds that of the atoms' electrons: $v_1 \gg v_a$, which occurs when the ions have about 1 MeV of energy per nucleon. Beams used in ion implantation and ion lithography have much less energy per nucleon, about 1-10 keV. The velocity of the ions in these beams is so low that $v_1 < v_a$. Calculating the amount of electron deceleration for beams of slow ions presents a much more difficult problem. Firsov(49) and Lindhard and Scharff(50) performed these calculations.

We will use Eq. (3-87) for the amount of energy transmitted from a fast ion to the electrons when $v_1 \gg v_a$ and write an integral that determines the extent to which fast ions are decelerated by electrons:

$$S_e(v_1) = N Z_2 \int_{b_{\min}}^{b_{\max}} T(b) 2\pi b \, db = \frac{4\pi Z_1^2 e^4 N Z_2}{m v_1^2} \ln\left(\frac{b_{\max}}{b_{\min}}\right) \tag{3-115}$$

Here N is the number of atoms per unit volume of target material, and Z_2 is the number of electrons per atom. The lower limit of integration $b_{\min}$ is determined from the condition that $T(b_{\min}) = T_{\max}$. According to Eq. (3-58), when $m_1 \gg m_2$ (m_1 is the mass of the fast ion and $m_2 = m$ is the electron mass), $T_{\max} = 2 m v_1^2$ and $b_{\min} = Z_1 e^2/(m v_1^2)$. Equation (3-87) for the amount of energy transmitted is valid up to those values for which a collision can be considered adiabatic, $b\omega_0/v_1 > 1$; therefore, we may use $b_{\max} = v_1/\omega_0$. Using this value we find that the extent to which ions are decelerated by electrons is

$$S_e(v_1) = \frac{4\pi Z_1^2 e^4 N Z_2}{m v_1^2} \ln\left(\frac{m v_1^2}{Z_1 e^2 \omega_0}\right) \tag{3-116}$$

A characteristic of electron deceleration of fast ions is the reduction in deceleration with increasing ion velocity as v_1^{-2} (if we neglect the weak dependence of the logarithmic term on the velocity v_1). As we will see, the electron deceleration of slow ions increases in proportion to the velocity of the ions as v_1.

Equation (3-116) has several variations, which are distinguished in the logarithmic term by their dependence on the properties of that group of target atoms to which it relates. Thus, the electron deceleration caused by the transmission of energy to the

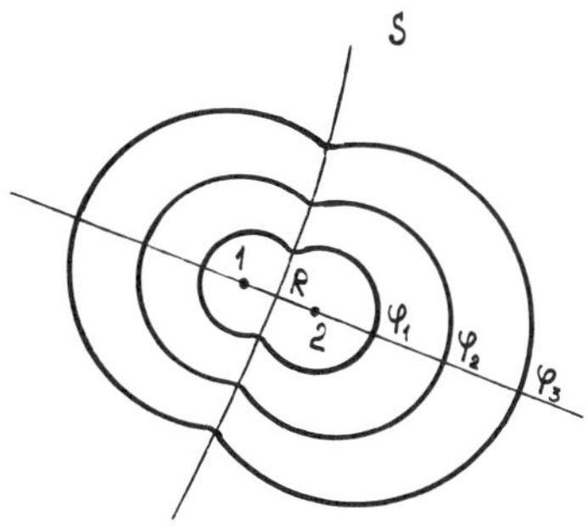

FIGURE 3-12. Equipotential surfaces and a surface S for separating the atoms in the quasimolecule created by colliding atoms.

conduction or valence electrons in a semiconductor is given[12] by

$$S_e(v_1) = \frac{4\pi Z_1^2 e^4 n_0}{m v_1^2} \ln\left(\frac{2mv_1^2}{\hbar\omega_p}\right)$$

The formula embraces the characteristics of the electron plasma: the number of electrons per unit volume n_0 and the plasma frequency ω_p. The derivation of this formula is similar to that for Eq. (2-205) for the energy lost by fast electrons in the electrons of a solid plasma.

It would seem that we could calculate the electron deceleration of slow ions from Eq. (3-115) using the decreasing function given by Eq. (3-89) for adiabatic collisions as $T(b)$. However, the predictions arrived at through this calculation have not been confirmed through experiment. It is evident that the premises upon which Eq. (3-89) is based are violated when collisions having large impact parameters and low velocities are involved. Indeed, in deriving Eq. (3-89) shielding was not considered and it was assumed that the energy levels in a target atom (the transition frequency ω_0) did not change.

Firsov,[49] using his own theory of inelastic collisions between slow, heavy ions and the atoms of a target, suggested a physical situation that is essentially different from that found in the theory of fast ion collisions. Imagine the atoms approaching one another during collision and their electron shells penetrating each other. We will introduce a procedure for establishing an electron's affiliation to one atom or the other when that electron is in the region in which the shells overlap. We calculate the potential of the quasimolecule created by the collision of these atoms. This potential will be a minimum at some point on the line connecting the atoms' nuclei. Through this point we draw a surface S to which the equipotential lines are everywhere perpendicular (Fig. 3-12). This surface will be a plane when the atoms are identical. The surface S separates the potential regions of the atoms, and therefore also the affiliation regions of the electrons: the electrons on one side of the surface belong to the atom on that side.

During a collision the distance R between the atoms changes, causing the electrons from atom 1 to move through the surface S to atom 2 and back again. This process of electron exchange causes momentum and energy to be transmitted from the moving ion to the target atom. This model is very reminiscent of the calculations done for frictional forces (viscosity) in the laminar flow of a gas or liquid: the exchange of atoms and molecules in the flow transfers the momentum.

Let an impinging atom transfer dn/dt electrons to the target atom per unit time; these electrons carry away an amount of momentum

$$m\mathbf{u}\frac{dn}{dt} = \frac{d\mathbf{P}}{dt} = \mathbf{F} \tag{3-117}$$

The target atom's electrons replace the departing electrons, but, because they have no directed motion, they transmit no momentum to the impinging atom. Thus, Eq. (3-117) describes the frictional forces between the impinging atom and the target atom's electrons.

In a coordinate system that moves with the impinging atom, the velocities of the atom's electrons $\mathbf{v}$ can be considered isotropic. Then the flow of electrons through the surface s per unit time is

$$\frac{dn}{dt} = \int_s \frac{nv}{4}\,ds$$

The amount of work done by frictional forces during a collision will be equal to the amount of energy transmitted to the target atom's electrons during the collision:

$$T(b) = \int_l \mathbf{F}\,d\mathbf{l} = m\int_l \left(\int_s \frac{nv}{4}\,ds\right)\mathbf{u}\,d\mathbf{l} \tag{3-118}$$

The electron density in the atom, $n(r)$, and their velocity, $v(r)$, can be stated in terms of the potential $\varphi(r)$ at a given point. In the Thomas–Fermi atom model[51]

$$n = 2^{3/2}(me\varphi)^{3/2}/3\pi^2\hbar^3$$

$$v = \frac{3}{4}(3\pi^2)^{1/3}\hbar n^{1/3}/m$$

Then instead of Eq. (3-118), we can write

$$T(b) = \frac{m^2e^2u}{4\pi^2\hbar^3}\int\left(\int_s \varphi^2\,ds\right)dx, \qquad dx = \frac{\mathbf{u}\,d\mathbf{l}}{|\mathbf{u}|} \tag{3-119}$$

We use a formula from the Thomas–Fermi–Firsov theory[49] to find the potential φ at a given point within the quasimolecule:

$$\varphi(r) = \frac{Z_1 + Z_2}{r}\chi(1.13(Z_1 + Z_2)^{1/3}r) \tag{3-120}$$

Integrating Eq. (3-119) with Eq. (3-120) yields

$$T(b) = \frac{0.35(Z_1 + Z_2)^{5/3}\hbar u/a_0}{[1 + 0.16(Z_1 + Z_2)^{1/3}b/a_0]^5} = \frac{(Z_1 + Z_2)^{5/3}4.3\cdot 10^{-8}\,u}{[1 + 3.1(Z_1 + Z_2)^{1/3}10^7\,b]^5} \tag{3-121}$$

for the amount of energy transmitted during a collision having an impact parameter b,[51] where b is in cm, u is in cm/sec, and T is in eV. To find the decelerating ability we must average $T(b)$ over all possible values of the impact parameter from 0 to ∞ and multiply by the number of atoms per cm^3; the result[12] is

$$S_e(u) = 2.34 \cdot 10^{-23} N(Z_1 + Z_2)u, \qquad u \ll v_a \tag{3-122}$$

According to Eq. (3-122), deceleration increases in proportion to the velocity of the slow ion.

A similar result without the derivation was published by Lindhard and Scharff in 1961.[50] We can assume that every part of the electron shell of an impinging atom experiences a frictional force $F(r)$ so that the decelerating ability can be thought of as the sum of these forces:

$$\frac{dE}{dx} = N \int_{\Delta V} \mathbf{F}(r)\, d^3 r \tag{3-123}$$

(the argumentation is Kumakhov's[12]). In this case the integration must be done over the volume of the impinging atom, ΔV. We can put the function $\mathbf{F}(r)$ in concrete form as follows. Obviously, $\mathbf{F}(r)$ must first be proportional to the electron density of the impinging atom: $\mathbf{F}(r) \sim n(r)$. Next, the decelerating force is governed by the elastic scattering of the impinging atom's electrons by the target atom's electrons, namely the momentum lost by the impinging atom's electrons. The processes by which momentum is lost are described by a special momentum loss cross section (or transport cross section, see Section 2-23):

$$\sigma_{tr} = \int_0^{\pi} \frac{d\sigma}{dO}(1 - \cos\theta) 2\pi \sin\theta\, d\theta$$

Therefore, we may assume that

$$F(r) = mu^2 n(r) \sigma_{tr}$$

For slow ions the transport cross section is $\sigma_{tr} \sim u^{-1}$; therefore, $F \sim u$. Using the Thomas–Fermi electron density function and integrating Eq. (3-123) gives us, via numerical integration,

$$\left(\frac{dE}{dx}\right)_e = \xi \frac{8\pi e^2 a_0 Z_1 Z_2 N}{(Z_1^{2/3} + Z_2^{2/3})^{3/2}} \frac{u}{v_0} \tag{3-124}$$

where ξ increases as $Z_1^{1/6}$ and has a value that lies within $\xi = 1$–2. The electron losses as a function of the impinging ion's energy can be given as

$$\left(\frac{dE}{dx}\right)_e = aE_1^{1/2} \tag{3-125}$$

or, in reduced units,

$$\frac{d\varepsilon}{d\rho} = -S_e(\varepsilon) = -k\varepsilon^{1/2} \tag{3-126}$$

$$k = \xi \frac{0.0793 Z_1^{1/2} Z_2^{1/2} (A_1 + A_2)^{3/2}}{(Z_1^{2/3} + Z_2^{2/3})^{3/4} A_1^{3/2} A_2^{1/2}} \tag{3-127}$$

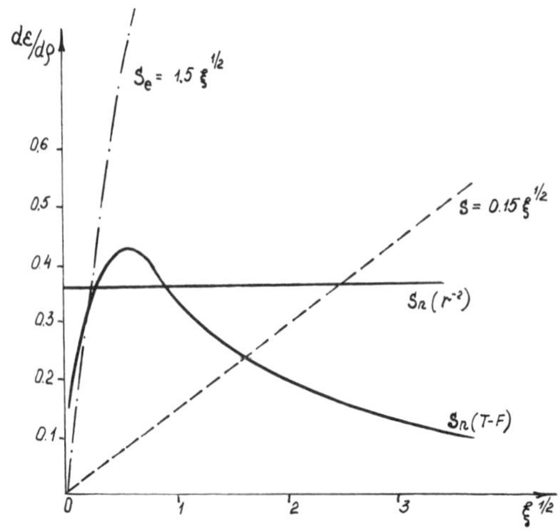

FIGURE 3-13. Electron decelerating ability (dotted lines) and nuclear decelerating ability (solid lines) as functions of beam energy.

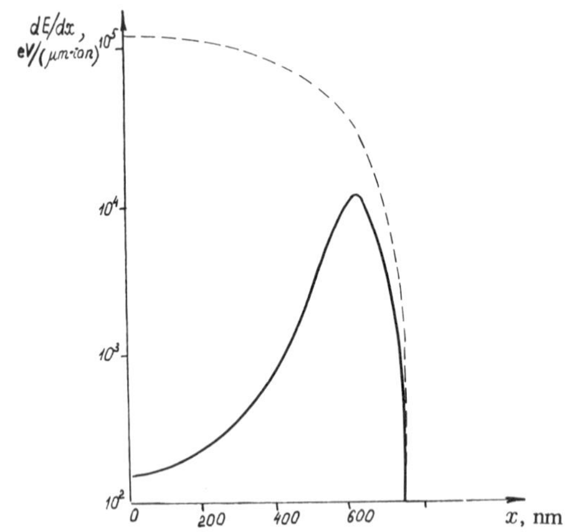

FIGURE 3-14. Electron (dotted line) and nuclear (solid line) proton energy losses.

Figure 3-13 shows the relationship between electron and nuclear energy losses as a function of the impinging atom's reduced energy. At very low ion energies $\varepsilon < 0.2$, we have $s_e \ll s_n$, and $s_n \sim \varepsilon^{1/3}$ (for an exponentiated potential $n = 3$). For light ions $Z_1 \ll Z_2$ the inequality $s_e < s_n$ is satisfied at even lower energies ε. Near the region $\varepsilon \approx 1$ we can expect that the total deceleration $s_e + s_n$ is approximately constant. At high energies $\varepsilon \gg 1$ electron losses prevail: $s_e \gg s_n$.

The relationships described have great significance for ion-lithography processes. Figure 3-14 shows the electron and nuclear losses for protons with initial energy E = 60 keV penetrating a layer of a PMMA-type polymer electron resist.[52] The path of the protons (depth of beam penetration into the target) is about 0.75 μm and at almost all depths electron losses prevail. As the protons decelerate, nuclear losses increase, although they are still less than electron losses.

3-20. A KINETIC EQUATION FOR A BEAM OF IONS IN MATTER

The probability that at time t an ion will be found within a differential volume $d^3\mathbf{r}$ and moving at $d^3\mathbf{v}_0$ in the velocities space if at time $t = 0$ it was at $r = 0$ moving at velocity $\mathbf{v}$ is[48]

$$-\frac{1}{v}\frac{\partial}{\partial t}F(\mathbf{v},\mathbf{v}_0;\mathbf{r},t) = \mathbf{e}\nabla F + N\int d\sigma[F(\mathbf{v},\mathbf{v}_0;\mathbf{r},t) - F(\mathbf{v}',\mathbf{v}_0;\mathbf{r},t)]$$
$$+ NS_e(E)\frac{\partial}{\partial E}F(\mathbf{v},\mathbf{v}_0;\mathbf{r},t) \qquad (3\text{-}128)$$

Equation (3-128) for ions is identical to Eq. (2-271) for electrons in a target. It has the most general form and is the starting point when analyzing problems in multiple scattering of an ion, ion energy losses, etc. It is however, impossible to find a general solution for Eq. (3-128) with realistic functions for the scattering cross section $d\sigma$ and electron deceleration $S_e(E)$. As in the theory of electron scattering (transfer), we can introduce special distribution functions for which we can find approximate solutions in analytical form. An example of these functions is the spatial distribution function

$$F_r(\mathbf{r},t) \equiv \int F(\mathbf{v},\mathbf{v}_0;\mathbf{r},t)\, d^3\mathbf{v}_0 \qquad (3\text{-}129)$$

and the angular distribution function

$$F_\Omega(\Omega,T) \equiv \int F(\mathbf{v},\mathbf{v}_0,\mathbf{r},t)\, d^3\mathbf{r} \qquad (3\text{-}130)$$

If in Eq. (3-129) we go to the limit as $t \to \infty$ and designate the distribution function obtained in going to this limit as

$$F(\mathbf{v},\mathbf{v}_0,\mathbf{r},\infty) \equiv F(E,\mathbf{e},\mathbf{r}) \qquad (3\text{-}131)$$

we obtain a distribution function for the directed ranges of the ions. Indeed, at $t = \infty$ the ions' velocity is zero: $v_0 = 0$ so that the tip of the vector $\mathbf{r}$ is at the ion's stopping point (with the initial condition that $\mathbf{r} = 0$ when $t = 0$). The quantities E and $\mathbf{e}$ in Eq. (3-131) denote the ion's initial energy and velocity. An equation for $F(E,\mathbf{e},\mathbf{r})$,

$$-\mathbf{e}\nabla F(E,\mathbf{e};\mathbf{r}) = N\int d\sigma[F(E,\mathbf{e};\mathbf{r}) - F(E-T,\mathbf{e};\mathbf{r})] + NS_e(E)\frac{\partial}{\partial E}F(E,\mathbf{e};\mathbf{r}) \qquad (3\text{-}132)$$

follows immediately from Eq. (3-128). A unique solution for Eq. (3-132) is obtained when the normalizing condition

$$\int F(E,\mathbf{e},\mathbf{r})\, d^3\mathbf{r} = 1$$

is taken into consideration. Let e' be the unit vector for the ion's velocity after a collision such that

$$\mathbf{e}\mathbf{e}' = \cos\theta_1$$

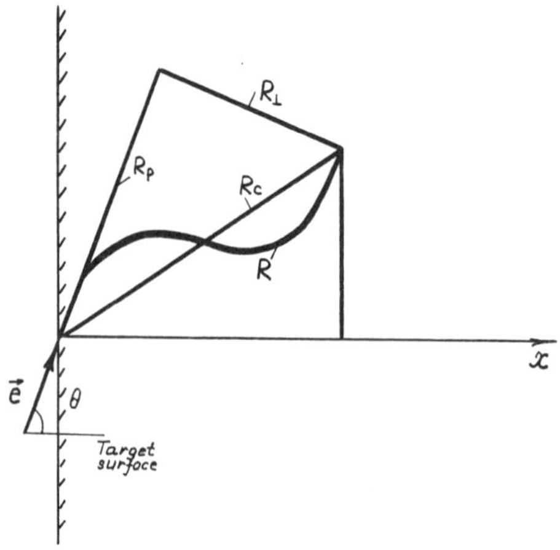

FIGURE 3-15. Characteristic values of ion range in a target.

the angle θ_1 is the ion's scattering angle in a laboratory coordinate system and is related to the energy T lost during the collision by

$$\cos\theta_1 = \left(1 - \frac{T}{E}\right)^{1/2} + \frac{1}{2}\left(1 - \frac{m_2}{m_1}\right)\left(\frac{T}{E}\right)\left(1 - \frac{T}{E}\right)^{-1/2} \tag{3-132a}$$

(see Eqs. (3-55) and (3-58)). Then, introducing integration with respect to $d\mathbf{e}'$ on the unit sphere, we can write

$$d\sigma = \frac{d\sigma}{dT}\, dT\, \frac{d\bar{e}'}{2\pi}\, \delta(\mathbf{e}\cdot\mathbf{e}' - \cos\theta_1(T))$$

Solving Eq. (3-132) we find the distribution of the stopped ions over depth, the projected range of the ions and the transverse range (see Fig. 3-15).

An interesting example of a simple distribution function is the distribution $P(E, R)$ of the ions over their total path length in a material R. We will use the deduction from Linhard's equation[48,50] for $P(E, R)$. Using similar arguments we can derive Eq. (3-128) and equations for other special distribution functions.

We will examine the trajectories of a beam of ions at the point where the beam enters the target. Let every ion traverse an interval Δx on its path. The probability that an ion will be elastically scattered on this path Δx is $dP = N\,\Delta x\, d\sigma$, and the number of ions scattered will be $\Delta N = N\,\Delta x \int d\sigma$. The distribution comprises, as is evident, the probability density that the ions scattered on the path Δx and the ions that were not scattered on this path interval will have a range R. The fraction of unscattered ions is $(1 - N\,\Delta x \int d\sigma)$. Her we must take the change in the ions' energy due to electron deceleration and the shift in range on Δx into consideration, which gives us

$$\begin{aligned} P(E, R) = N\,\Delta x \int d\sigma P(E - T - NS_e\,\Delta x, R - \Delta x) \\ + P(E - NS_e\,\Delta x, R - \Delta x)(1 - N\,\Delta x \int d\sigma) \end{aligned} \tag{3-133}$$

Equation (3-133) is based on the assumption that ion–electron collisions which can be described by the electron decelerating ability S_e are uncorrelated with nuclear (elastic) collisions and have not been subject to fluctuations. A correlation between nuclear and electron losses does in fact occur, particularly when the inner shells of the atoms are excited and ionized.

Expanding Eq. (3-133) in small increments and going to the limit as $\Delta x \to 0$ gives us

$$-\frac{\partial P}{\partial R} = N \int d\sigma[P(E,R) - P(E-T,R)] + NS_e \frac{\partial P(E,R)}{\partial E}$$

$$\int_0^{\infty} P(E,R)\, dR = 1 \tag{3-134}$$

The distribution functions associated with the recoil atoms are very significant. In every elastic collision a moving recoil atom arises. When the energy of a recoil atom is sufficient to dislodge atoms, its elastic collisions produce secondary recoil atoms, etc. Therefore, every primary ion that enters a target causes an avalanche of moving atoms. The motion of all the atoms involved in the avalanche can be called atom motion that occurs in contrast to the electron motion that occurs as a result of the atoms' electrons being excited and ionization. Let the average number of atoms set into motion due to an avalanche be $F(E, E_0)\, dE_0$. When these atoms begin moving, they have an amount of energy that lies in the range $E_0, E_0 + dE_0$. Every atom that goes into motion is reckoned only once. The mass of the ion is assumed to be the same as that of the target atom. With the same arguments that were used in deriving Eq. (3-133), we can write

$$F(E,E_0) = N\,\Delta x \frac{d\sigma(E,E_0)}{dE_0} + N\,\Delta x \int d\sigma[F(E-T,E_0) + F(T,E_0)]$$

$$+ F(E - NS_e\,\Delta x, E_0)(1 - N\,\Delta x \int d\sigma) \tag{3-135}$$

The first term on the right-hand side is the number of recoil atoms that have $E_0, E_0 + dE_0$ of energy and were produced in collisions at a depth Δx in the target. The second term is the number of atoms that have $E_0, E_0 + dE_0$ of energy and were produced by ions scattering on the layer at Δx losing energy T, and recoil atoms that have T ($T \neq E_0$) in their subsequent motion. Finally, the third term describes the contribution of the atoms that were not scattered on the layer at Δx. The impinging atom and the recoil atom are assumed to be independent immediately after the collision. The target is also assumed to contain a fairly large number of atoms that can go into motion. Physically, these assumptions are more justified at high particle energies. After expanding Eq. (3-135) in small increments and going to the limit as $\Delta x \to 0$ we obtain

$$\int d\sigma[F(E,E_0) - F(E-T,E_0) - F(T,E_0)] + S_e(E)\frac{dF(E,E_0)}{dE} = \frac{d\sigma(E,E_0)}{dE_0} \tag{3-136}$$

The boundary condition that defines a unique solution for Eq. (3-136) is

$$F(E,E_0) = 0 \qquad \text{when } E_0 > E$$

which follows from the law of conservation of energy in a collision. Equation (3-136) is useful in dealing with the energy distribution of the reflected ions and other problems.

We will now introduce a spatial distribution function for the recoil atoms. Let a primary ion having energy E and velocity direction $\mathbf{e}$ begin moving into a target at $\mathbf{r} = 0$. We use $F(E, \mathbf{e}; E_0, \mathbf{r})\, d^3\mathbf{r}\, dE$ to denote the average number of recoil atoms going into motion in a volume $(\mathbf{r}, d^3\mathbf{r})$ and having initial energies in the range $E_0, E_0 + dE_0$ in the avalanche that results from the motion of the primary ion. The equation for $F(E, \mathbf{e}; E_0, \mathbf{r})$ differs from Eq. (3-136) by a gradient term[48]:

$$\delta(\mathbf{r})\frac{d\sigma(E, E_0)}{dE_0} - \mathbf{e}\nabla F(E, \mathbf{e}; E_0, \mathbf{r}) = N \int d\sigma[F(E, \mathbf{e}; E_0, \mathbf{r}) - F(E - T, \mathbf{e}', E_0, \mathbf{r}) - F(T, \mathbf{e}'', E_0, \mathbf{r})] + N S_e \frac{\partial F(E, \mathbf{e}, E_0, \mathbf{r})}{\partial E} \tag{3-137}$$

Here $\mathbf{e}'$ and $\mathbf{e}''$ are unit vectors for the velocity of the scattered ion and the atom. In keeping with the law of conservation of energy,

$$F(E, \mathbf{e}, E_0, \mathbf{r}) \equiv 0 \qquad \text{when } E_0 > E$$

The function $F(E, \mathbf{e}, E_0, \mathbf{r})$ is a generalization of $F(E, E_0)$:

$$\int F(E, \mathbf{e}, E_0, \mathbf{r})\, d^3\mathbf{r} \equiv F(E, E_0)$$

The intergrodifferential equations presented above can be solved in rare, special cases. Useful information can be obtained from them in the form of the moments of the quantities on which the respective distribution functions depend. The moments of a distribution function can be measured experimentally. We will now turn to specific problems of the types described.

3-21. THE PATH LENGTH OF AN ION IN A TARGET

We will return to Eq. (3-134) for the distribution $P(E, R)$ for an ion's path length R in a target. Before we put the functions $d\sigma$ and S_e into a specific form in an attempt to find an analytical solution for the equation, we will make one more simplification in Eq. (3-134) by assuming that the energy T transmitted to the recoil during a collision is small in comparison with that of the scattered ion, $T \ll E$. We may then use the expansion

$$P(E - T, R) = P(E, R) - T\frac{\partial P(E, R)}{\partial E}$$

and, instead of Eq. (3-134), obtain

$$-\frac{\partial P(E, R)}{\partial R} = N \int d\sigma T \frac{\partial P(E, R)}{\partial E} + N S_e \frac{dP(E, R)}{dE} \tag{3-138}$$

Noting the fact that $\partial P(E, R)/\partial E$ is independent of the variable of integration over $(d\sigma/dT)\, dT$, we can take it outside the integral, and the integral $\int T\, d\sigma = S_n$ then has the sense of nuclear deceleration, Eq. (3-111). Eliminating the function $P(E, R)$ from the equation obtained, we find the relationship between the particle's initial energy E and its path length R in a target:

$$R = \int_0^R dR = -\int_E^0 \frac{dE}{N[S_e(E) + S_n(E)]} \tag{3-139}$$

Equation (3-139) is usually written directly from the definition of nuclear and electron deceleration. The derivation of Eq. (3-139) from the kinetic equation allows us to state more precisely the assumptions for which it is valid.

An ion's path length in a target as a function of its initial energy is determined by the relationship between electron and nuclear decelerations. If we use normalized values of the length and the energy, we can write

$$|\rho| = \int_0^\varepsilon \frac{d\varepsilon}{s_e(\varepsilon) + s_n(\varepsilon)}$$

An analysis of the dependencies of $s_e(\varepsilon)$ and $s_n(\varepsilon)$ at very low ion energies $\varepsilon \ll 1$ shows us that we can ignore electron losses $s_e \ll s_n$ and $s_n \sim \varepsilon^{1/3}$, $\rho \sim \varepsilon^{2/3}$. At energies $\varepsilon \sim 1$ the sum $s_n + s_e$ is roughly constant so that $\rho \sim \varepsilon$. At high energies $\varepsilon \gg 1$ electron losses prevail, $s_e \gg s_n$, $s_e = k\varepsilon^{1/2}$, and $\rho \sim \varepsilon^{1/2}$. When the ion's velocity is greater than that of the atom's electron, the electron losses begin to decrease with increasing velocity and the quantity R increases more rapidly than $E^{1/2}$.

If we ignore electron deceleration at low ion velocities and use Eq. (3-111) for s_n we obtain

$$R(E) = \frac{(m_1 + m_2)E}{1.308\pi N a_F Z_1 Z_2 m_1 e^2} \tag{3-140}$$

the path length $n = 2$ of an ion, which, when R, corresponds to an exponentiated interatom potential.

According to Eq. (3-139) the range of a beam of light ions is $R \sim E/Z_1^2$, whereas $R \sim E/(Z_1 Z_2)$ when $m_1 \gg m_2$.

We return to the distribution $P(E, R)$, Eq. (3-134), again. The easiest way to obtain useful information from an equation like Eq. (3-97) is to calculate the moments for the quantities that are arguments of the distribution function, which in our case is the ion's path length R in a target. We multiply Eq. (3-134) by R^q and integrate from 0 to ∞, which gives us

$$qR^{q-1}(E) = N \int d\sigma[R^q(E) - R^q(E - T)] + N S_e \frac{dR^q(E)}{dE} \tag{3-141}$$

where $R^q(E)$ is the R-moment of q:

$$R^q(E) = \int_0^\infty R^q P(E, R)\, dR$$

Equation (3-139) follows from Eq. (3-141) if Eq. (3-141) is written for $q = 1$ and the expansion

$$R^1(E - T) = R^1(E) - T \frac{dR^1(E)}{dE}$$

is used.

We will seek a solution for the moments, Eq. (3-141), in the form

$$R^q(E) = A^q \left(\frac{E^{2/n}}{CN}\right)^q, \qquad A^0 = 1$$
$$C = \frac{\pi}{2}\lambda_n a_F^2 \left(\frac{m_1}{m_2}\right)^{1/n} \left(\frac{2Z_1Z_2e^2}{a_F}\right)^{2/n} \tag{3-142}$$

Eq. (3-142) into Eq. (3-141) produces a recurrence relation between A^q and A^{q-1}:

$$qA^{q-1} = A^q[-n\gamma^{-1/n} - B_\gamma(-1/n, 1 + 2q/n)]$$
$$B_\gamma(x, y) = \int_0^\gamma d\tau\tau^{x-1}(1-\tau)^{y-1}$$
$$\tau = T/E, \qquad \gamma = T_m/E = 4m_1m_2/(m_1+m_2)^2$$

In the case of an exponentiated potential the quantity $L_n = E^{2/n}/(NC)$ is a measure of length. From here it follows that the distribution function can be written

$$P(E, R) = P\left(\frac{R}{L_n}\right)\frac{1}{L_n} = \frac{NC}{E^{2/n}}P\left(\frac{RCN}{E^{2/n}}\right)$$

Because the moments $R^q(E)$ with the distribution $P(R, E)$ are known, the problem of restoring a $P(R, E)$ function from its moments arises. This is, obviously, a purely mathematical problem. The most obvious way to solve it is as follows. We represent $P(R, E)$ as a test function having unknown parameters. Calculating the moments $R^q(E)$ via the test function, we can state these moments in terms of the test function parameters. Equating the moments obtained via the test function to the moments, Eq. (3-142), calculated via the integrodifferential equation for $P(R, E)$ gives us an equation we can use to define the test function parameters in terms of the moments of the distribution.

For example, if the graph $P(R, E)$ is a curve with a maximum, similar in form to a Gaussian function and a polynomial,[48]

$$P(E, R) = \exp\left(-\frac{(R - R^1(E))^2}{2\langle(R - R^1(E))^2\rangle}\right)\cdot[A + BR + CR^2 + \ldots]$$

The procedure described[48] can be used to find the length $R(E)$, the variance $\langle(R - R')\rangle$, and the coefficients $A, B, C\ldots$ of the polynomial in terms of the moments R^2.

As to an ion's path length R in a target we must make an essential stipulation. The fact is that R is not, in general, a measurable quantity because the ion's trajectory cannot be made visible (with the exception of high energy ions whose tracks may be observable). For this reason the theoretical study of R that we have presented has a rather methodical nature. The depth x to which an ion penetrates a target is a measurable quantity and it coincides with the projected range R_p when the ion beam impinges on the target from the normal. We will therefore turn to investigating the respective distribution for x (or R_p).

3-22. THE PROJECTED ION RANGE IN A TARGET

According to the definitions shown by Fig. 3-14 we can introduce and use an ion's range in its initial direction—the projected range R_p, along with its transverse range $R_\perp$ and depth of penetration x. An equation for the range along a chord r was derived earlier (see Eq. (3-132)). We define the distribution function with respect to depth of penetration into a target as

$$F(E, \mathbf{e}, x) \equiv \int F(E, \mathbf{e}, \mathbf{r})\, dy\, dz = F(E, \cos\theta, x)$$

Integrating Eq. (3-132) over y and z gives us an equation for $F(E, \cos\theta, x)$:

$$-\cos\theta \frac{\partial}{\partial x} F(E, \cos\theta, x) = N \int d\sigma [F(E, \cos\theta, x) - F(E - T, \cos\theta', x)] + N S_e(E) \frac{\partial}{\partial E} F(E, \cos\theta, x) \quad (3\text{-}143)$$

We must normalize $F(E, \cos\theta, x)$, taking into account the fact that the ion may be reflected as a result of being scattered in the target; i.e., the appearance of $x < 0$ in the left-hand (empty) half space:

$$\int_{-\infty}^{+\infty} F(E, \cos\theta, x)\, dx = 1$$

We define the moments $F^n(E, \cos\theta)$ of the distribution $F(E, \cos\theta, x)$ by

$$F^n(E, \cos\theta) = \int_{-\infty}^{+\infty} x^n F(E, \cos\theta, x)\, dx$$

Multiplying Eq. (3-143) by x^n and integrating over x from $-\infty$ to $+\infty$ gives us

$$\cos\theta\, n F^{n-1}(E, \cos\theta) = N \int d\sigma [F^n(E, \cos\theta) - F^n(E - T, \cos\theta)] + N S_e \frac{\partial}{\partial E} F^n(E, \cos\theta) \quad (3\text{-}144)$$

Obviously, $F^0 = 1$ and F^1 has the sense of an average depth of ion penetration x into a target. We will adopt the convention that a beam impinges on a target at the perpendicular to its surface so that $\cos\theta = 1$; $x = R_p$ and $F^1(E - T, \cos\theta) = \cos\Phi' F'(E - T)$, and the equation for the first moment becomes

$$1 = N \int d\sigma [R_p(E) - \cos\Phi' R_p(E - T)] + N S_e(E) \frac{dR_p(E)}{dE} \quad (3\text{-}145)$$

where $\cos\Phi'$ is a function of T as in Eq. (3-132). We can seek a solution of Eq. (3-145) when $S_e = 0$ as[48]

$$R_p(E) = A_p E^{2/n} / (NC)$$

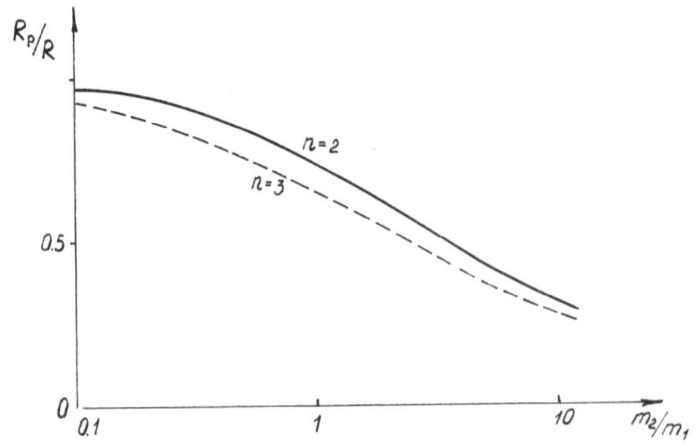

FIGURE 3-16. Ratio of projected range R_p to total range R of an ion in a target as a function of the ratio of masses m_2/m_1 for powers of n = 2 and n = 3 of the ion–atom potential of interaction.

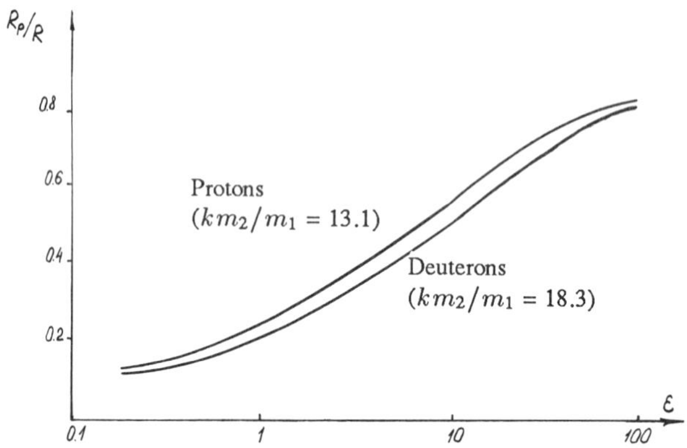

FIGURE 3-17. Ratio R_p/R for protons and deuterons in a target composed of heavy atoms as a function of the normalized energy ε (k is a constant of electron deceleration, $s_e = k\varepsilon^{1/2}$).

The latter equation means that an ion's path length R and projected range R_p are proportional to one another. The ratio R_p/R is a function of the ratio of the masses m_2/m_1 and the exponent on the ion–atom interaction potential.

Figure 3-16 shows the ratio R_p/R, obtained from Eq. (3-145) by numerical methods, as a function of the ratio of the masses m_2/m_1; Fig. 3-17 shows it as a function of the normalized energy ε for deuterons and protons.[48] The dependencies obtained have an extremely intuitive explanation. When heavy ions impinge on a target composed of light ions (the area $m_2/m_1 \ll 1$ in Fig. 3-16), the ions' trajectory is nearly rectilinear and $R_p/R \leq 1$.

Conversely, when $m_2/m_1 > 1$ (light ions impinge on a target composed of heavy atoms) the angular scattering of the ions is prominent and the projected range (or ion's depth of penetration into the target) is roughly one-tenth of the ion's path length in the target $R_p/R \sim 0.1$. Scattering effects grow weaker as the energy of the beam increases: for protons and deuterons $R_p/R \to 1$ as $\varepsilon \to \infty$.

All of these facts hold a great deal of meaning for ion-beam lithography. The resolution of ion-beam lithography and its throughput are governed by the beam diameter and current, by the deceleration $S(E) = S_e(E)+S_n(E)$ of the beam in the target, and by angular scattering. As we saw earlier, the Boersch effect, which defines the energy spread

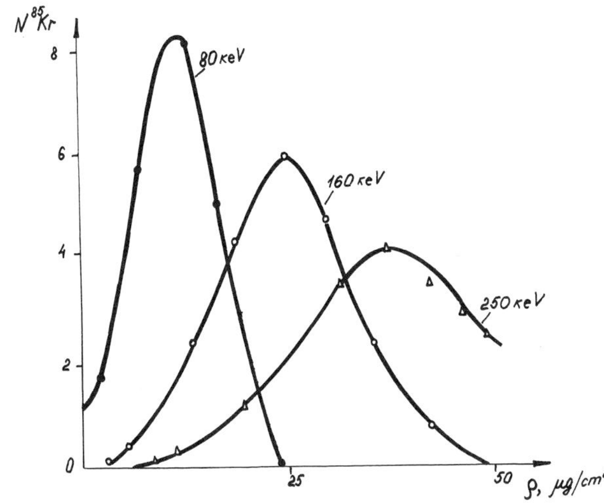

FIGURE 3-18. Distribution of ^{85}Kr from a beam in Al_2O_3 over the target thickness. This distribution is equivalent to the distribution for the projected range $R_p = x$.

in the beam and arises because of Coulomb interaction within the beam, is proportional to $m_1^{1/2}$. For this reason it is desirable from the viewpoint of obtaining ultrathin ion beams to work with light ions (e.g., protons). On the other hand, light ions undergo strong angular scattering in a target which spreads the beam and lowers the resolution of the process. Angular scattering can be compensated (reduced) if we use high energy beams; however, using a high energy beam may in turn create undesirable effects: a reduction in the decelerating ability $S_e(E) + S_n(E)$ of the resist and the beam penetrating into the structural layers of the device below the resist layer due to the overall increase in the projected range $R_p \sim E$. This creates defects in the structural layers which must be eliminated by special methods (e.g., annealing). All of these comments allow us to contemplate ways to optimize ion-beam lithography processes and to define the physical problems that must be studied to achieve this optimization.

By calculating the moments of the distribution according to Eq. (3-144) we can find the spread in the respective quantities, e.g., $\Delta R_p^2 = R_p^2 - (R_p)^2$. It turns out that the ratio $\Delta R_p^2 / R_p^2$ is small when $m_2/m_1 < 1$ and increases sharply to magnitudes greater than unity when $m_2/m_1 > 1$. The quantity $R_\perp^2 / R_p^2$ behaves in a similar manner.[48]

Figure 3-18 shows the measured distributions of the density of radioactive impurities removed from a beam at different beam energies E. The shape of these curves is almost Gaussian. The width of the distribution increases with beam energy (depth of penetration).

3-23. THE REFLECTION OF AN ION BEAM FROM A TARGET

We can use the same techniques to examine the reflection of an ion beam as we did in Chapter 2 when we studied electron beam reflection. Ion reflection is the exclusive result of angular deflection during elastic ion–atom scattering, because inelastic scattering of an ion by electrons contributes almost nothing to angular scattering, changing only the ion's energy (electron deceleration). The question then arises as to what contributions are

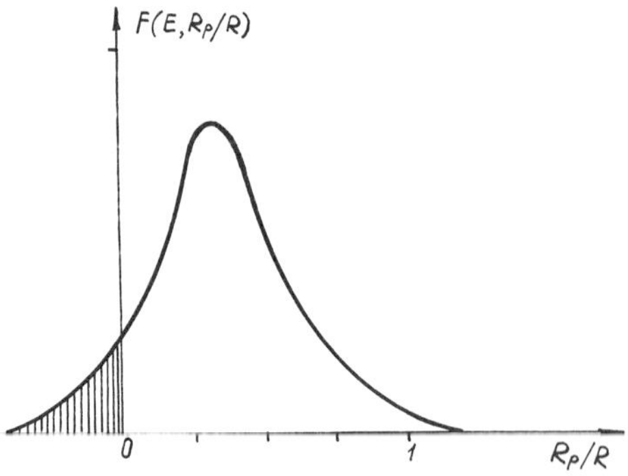

FIGURE 3-19. Distribution of ranges $F(E, R_p/R)$. (The area under the distribution curve in the region $R_p/R < 0$ shows the fraction of reflected particles.)

made to the reflections coefficient by one-time and multiple scattering at small angles. It is relatively easy to calculate one-time scattering at large angles if the cross section for elastic ion–atom scattering is known. It is however, impossible to derive a more or less general formula even for one-time scattering in the case of an ion beam, because the scattering cross sections are different for different energy ranges and mass ratios m_2/m_1. We can naturally expect that the reflection coefficient will increase as the ratio of masses m_2/m_1 increases. Obviously, the question of reflection coefficients of a beam in terms of the number of particles η and their energy η_E deserves as exhaustive an examination as for an electron beam. To date however, this has not been done as thoroughly for ion beams as it has for electron beams.

General relationships for reflection can be established via numerical methods using the kinetic equation for the distribution of the ranges, Eq. (3-143). The distribution function $F(E, x, \cos\theta)$ is defined over all space from $x = -\infty$ to $+\infty$. Impinging ions that have an initial energy of E intersect the surface of a target at the plane $x = 0$. Those ions that stop in the region $x < 0$ must be considered to have been reflected and the reflection coefficient is

$$\eta(E) = \int_{-\infty}^{0} F(E, \cos\theta, x)\, dx$$

(the function $F(E, x, \cos\theta)$ has been normalized to unity).

An example of the distribution $F(E, x)$ calculated for the case in which $m_2/m_1 = 4$, $n = 2$, and $S_e = 0$ is shown in Fig. 3-19.[49] The ranges are given in units of the maximal path length $R(E)$. There are three distinctive features of the curve: 1) the appreciable value of $F(E, x)$ when $x < 0$ defines a substantial reflection coefficient for light ions in a heavy target: $m_1/m_2 = 1/4$; 2) the average depth of ion penetration (nearly x_m when $F(E, x)$ has a maximum) is quite far from $R(E)$, i.e., $x_m/R(E) \leq 0.5$; 3) the shape of the distribution is asymmetric, having higher values at the greater penetration depths than a Gaussian curve has. This "tail" describes those ions that have a long range because they have not experienced elastic deflecting scattering over the entire path. At high ion energies the "tail" of the distribution is higher than Gaussian for values $x < x_m$. This corresponds to ions that have been reflected backward because of elastic backscattering.

Measurements of the refection coefficient η for a beam with relatively low energy $\varepsilon = 0.2$ show a strong dependence on the ratio of masses m_2/m_1. When $m_2/m_1 < 1$, η is less than 0.01, but increases rapidly to 0.2 and higher when $m_2/m_1 = 8$. Thus, reflection may be significant when a low energy beam of light ions is used for ion lithography. At beam energies of $\varepsilon \gg 1$, beam reflection falls to insignificant levels.

The study of the energy spectrum of reflected light ions when the initial energy is high, E = 1–3 MeV, has become quite important. At this level of energy, shielding of the nuclear charges can be neglected and the energy of an ion–atom interaction considered equal to that of the Coulomb interaction of the nuclei, $V(R) = Z_1 Z_2 e^2/R$. Therefore, the scattering cross section is given by the equation for the Rutherford scattering of an ion at a Coulomb center and the reflection of the ions under the indicated conditions is called Rutherford backscattering (RBS). RBS spectroscopy is extremely informative in studying the chemical composition and structure of the surface layers of materials. We will briefly discuss the theory behind this effect.

3-24. RUTHERFORD BACKSCATTERING SPECTROSCOPY

We will examine those ions that were reflected because of one-time Coulomb scattering at the nuclei of target atoms. Let such a collision have occurred for a given ion at a depth x. Entering the target with an initial energy of E_0, the ion loses energy due to electron deceleration so that by the time the reflecting collision occurs, it will have an energy of

$$E(x) = E_0 - \int_0^x \frac{dE}{dx}\, dx$$

If the depth x is small in comparison with the ion's path length R in the target, the deceleration dE/dx may be considered a constant that corresponds to the initial energy E_0 and write

$$E(x) = E_0 - \left.\frac{dE}{dx}\right|_{E_0} x$$

If scattering occurred at an angle $\theta_1 > \pi/2$ the path travelled by the ion from the scattering point to the surface is $l = x/\cos\theta_1$, and the amount of energy lost by the ion as it travels this path is

$$\Delta E = \left.\frac{dE}{dx}\right|_{E(x)-T} \frac{x}{\cos\theta_1}$$

Here we will also say that $l \ll R$, since we are considering a scattering angle of $\theta_1 \leq \pi$.

Using Eq. (3-55) and (3-58), it is not hard to find the energy of an ion scattered at an angle θ_1:

$$E(x) - T = E(x) \left[\frac{\left(1 - \frac{m_1}{m_2}\sin^2\theta_1\right)^{1/2} + \frac{m_1}{m_2}\cos\theta_1}{1 + m_1/m_2} \right]^2 \tag{3-146}$$

As a result, the ion leaves the target with an energy of

$$E_{m_2} = E_0 - \left.\frac{dE}{dx}\right|_{E_0} x - \left.\frac{dE}{dx}\right|_{E(x)-T} \frac{x}{\cos\theta_1} - T(\theta_1) \tag{3-147}$$

Solving this equation for x, we can, from the measured energy of the ion E_{m_2} and the angle θ_1, find the depth in the target at which a collision with an atom of mass m_2 occurred. We can imagine that at a depth x in the target there are atoms of different elements, which have masses $m_2^{(1)}$, $m_2^{(2)}$, and $m_2^{(3)}$. We also fix the angle θ_1 for observing reflected ions. Then ions reflected from the depth x and having energies of $E_{m_2^{(1)}}$, $E_{m_2^{(2)}}$, and $E_{m_2^{(3)}}$ will appear. The ratio of the number of ions having these energies and reflected from a thin layer x, $x + \Delta x$ is given by the product of the scattering cross section into a solid angle Ω (the angle of the reflected ions detector) and the number of atoms of the corresponding element in the layer:

$$I_{BS}^{m_2^{(i)}} = IN_{m_2^{(i)}}\Delta x \int_\Omega d\sigma$$

In a LCS the Rutherford scattering cross section at an angle θ_1 is[2]

$$\frac{d\sigma}{dO} = \left(\frac{Z_1 Z_2 e^2}{2E(x)\sin\theta_1}\right)^2 \frac{\left\{\cos\theta_1 + \left[1 - \left(\frac{m_1}{m_2}\right)^2 \sin^2\theta_1\right]^{1/2}\right\}^2}{\left[1 - \left(\frac{m_1}{m_2}\right)^2 \sin^2\theta_1\right]^{1/2}} \tag{3-148}$$

In observing backscattered ions the scattering angle is approximately 170°(the angle between the normal to the surface and the velocity of a backscattered ion is $\Psi = \pi - \theta_1 = 0.17$ radians). Assuming that $m_1 \leq m_2$ and $\Psi \ll 1$, the cross section given by Eq. (3-148) can be put into the simple form

$$\frac{d\sigma}{dO} = \left(\frac{Z_1 Z_2 e^2}{2E(x)}\right)^2 \frac{\Psi^2}{4}$$

As we can see, the cross section decreases in the region $\theta_1 \sim \pi$ as the square of the angle $\Psi = \pi - \theta_1$. In a similar manner we can use Eq. (3-146) to evaluate the fraction of energy that the ion will retain when reflected backwards from a collision with the nucleus of an atom of mass m_2,

$$k = \frac{E(x) - T}{E(x)} = \left(\frac{m_2 - m_1}{m_2 + m_1}\right)^2$$

When reflected from a heavy atom, an ion retains almost all of its energy and loses almost all of its energy when reflected from an atom having roughly the same mass. Table 3-3 shows values of k obtained by reflecting helium ions and protons off various target atoms.[2] The data from this table locate the "lines" of the individual elements in the energy spectrum of the reflected light ions. The width and shape of the lines is determined by how the impurities are distributed relative to depth. Mixtures of heavy atoms implanted at a depth $\sim$ 100 nm are preferred for analyzing the distribution relative

TABLE 3-3. Fraction of Energy That an Ion of Mass m_1 Retains When Reflected Backwards from an Atomic Nucleus of Mass m_2 ($m_1 = H^1,\ He^4$)

Atomic nucleus	B^{11}	C^{12}	N^{14}	O^{16}	Al^{27}	Si^{28}	P^{31}	Sb^{121}	Au^{197}
$K_H^{180°}$	0.6948	0.7161	0.7516	0.7789	0.8622	0.8670	0.8790	0.9674	0.9961
$K_{He}^{180°}$	0.2223	0.2500	0.3086	0.3600	0.5502	0.5625	0.5951	0.8762	0.9605

to depth in light materials. At greater depths the spectra of the reflected ions include the contributions from several scatterings and the accuracy of the determinations deteriorates.

Detecting crystal defects from RBS spectra is related to the channeling and shading effects that occur when ions are reflected from a crystal. If a highly collimated beam of ions impinges on a crystal from a direction that is strictly parallel to the crystallographic axis or plane, two phenomena will be observed: ions impinging in the gap between the atomic rows or planes will have less scattering and deceleration and greater range than in an amorphous substance and will be grouped closer to the middle of the gap, which corresponds to "channeling"; other ions impinging on an atomic row (plane) "head-on" will be reflected from the first surface atom in the row, which shades the atoms that are deeper in the row. It is intuitively clear that the shading effect in ion backscattering allows us to analyze the composition of the monatomic layer on the surface of a crystal, whereas the channeling effect makes it possible for us to establish that there are atoms in the gap between the rows of atoms or between the planes. Interstitial impurities are usually found in this space (substitutional impurities are found at the nodes of the crystal lattice).

Kumakhov and Shirmer(12) discussed the theory of the channeling effect. The methods of the kinetic equation and momentum approximation are used extensively to describe scattering in the theory of channeling.

3-25. THE PROCESSES BY WHICH ATOMS IN A SOLID ARE DISLODGED(43)

When a fast ion collides with an atom in a solid the atom receives an amount of kinetic energy T. What then happens with the atom that received this energy?

Two things can happen. 1) The energy T received can be manifested as intensified vibrational motion in the atom; the neighboring atoms have the vibrational energy corresponding to the temperature of the solid. Therefore, gradually, during the background relaxation period, the energy T is converted to heat and dissipated uniformly throughout the crystal. 2) The energy that the atom received is sufficient to dislodge the atom from the site it occupied in the solid and move it to a new location, leaving a vacancy behind. It turns out that the dislodging process has a threshold nature: an atom can be dislodged if, and only if, T is greater than the threshold dislodging energy T_d that is characteristic of the specific material. The energy T_d is several times higher than the binding energy and is on the order of 20 eV. It is natural to assume that the threshold energy T_d in monocrystals will be a function of the direction in which the atom was dislodged, relative to the crystallographic axes: $T_d(\theta, \varphi)$. The probability P that an atom will be dislodged in a given direction when energy is transmitted to it is

$$P(\theta, \varphi, T) = \begin{cases} 1 & \text{if } T > T_d(\theta, \varphi) \\ 0 & \text{if } T \leq T_d(\theta, \varphi) \end{cases} \tag{3-149}$$

Dislodging an atom from the site it occupies in a crystal takes very little time: $\tau \approx a/v_i = 10^{-14}$ sec, and the lattice can be thought of as immobile when an ion collides with its atom. The dislodged atom is deposited not far from the vacancy it left behind. During relaxation, which is the next immediate event following the collision with and dislodging of the atom, the dislodged atom may return to the vacancy and recombine with it. The geometric points around the vacancy from which the interstitial atom and the vacancy recombine is the recombination region. If recombination occurs while the crystal is relaxing from the ion–atom collision, the process must be type 1 and the threshold energy T_d must relate to dislodging the atom beyond the limits of the recombination region. The recombination region covers roughly a hundred lattice sites and increases with temperature.[43] The recombination region in monocrystals is anisotropic: a vacancy and an interstitial atom recombine when the distance between them is greater, if they are located in an atomic row that corresponds to a crystallographic axis having a low index.

We can introduce a cross section σ_d for dislodging an atom from a lattice site by an impinging particle having energy E:

$$\sigma_d(E, T_d) = \int_0^{T_{\max}} P_d(T) \frac{d\sigma(E,T)}{dT}\, dT \tag{3-150}$$

which is the sum of the products of the cross section of every T transmitted to the atom and the probability $P_d(T)$ of dislodging the atom when an amount of energy T has been received. If $P_d(T)$ is given by Eq. (3-149) for an isotropic energy T_d we have, instead of Eq. (3-150),

$$\sigma_d(E, T_d) = \int_{T_d}^{T_{\max}} \frac{d\sigma(E,T)}{dT}\, dT$$

The dislodging cross section σ_d depends on the threshold energy T_d as a parameter. For this reason T_d can be found experimentally by comparing the measured defect density in a crystal after irradiation with the calculated density according to

$$n_d = I \cdot N_a \cdot \sigma_d$$

Here I is the ion flux into the target and N_a is the number of target atoms per unit volume.

The physics of point defects in crystals relies on such geometrically obvious objects as a vacancy and an interstitial atom. The concept of these objects becomes less obvious when we look at amorphous solids and polymers. There is no long-range order in an amorphous solid, only short-range order; although the arrangement of the atoms even in short-range order is not rigidly defined and is described by a radial distribution function.[54] Under these conditions the concepts of vacancy and interstitial atom become vague. However, in amorphous solids with covalent bonds, in polymers, molecular crystals, etc. ruptured unsaturated chemical bonds are formed in both the molecule that lost the atom and in the atom dislodged by the ion–atom collision. In this case the concept of recombination between a vacancy and an interstitial atom takes on the meaning of a chemical recombination, the concept of a vacancy—that of a ruptured bond in the molecular structure of the solid, and the concept of an interstitial atom—that of an unsaturated bond in the atom. In describing the physics of interaction between a polymer and an ion beam we will adhere to the "chemical" definitions of the defects introduced here.

In ion- and electron-beam lithography the spatial distribution of Frenkel-like defects (for inorganic resists) and the density of the energy dissipated by the ions (for organic resists) are quite meaningful. Because an ion's energy in a target is converted into atom and electron motion, we must find and study each one of these distributions.

3-26. THE DISTRIBUTION OF THE DEFECTS CREATED BY COLLISION AVALANCHES

We introduce a function for the spatial distribution of the recoil atoms $F(E, \mathbf{e}, E_0, \mathbf{r})$ so that $F(E, \mathbf{e}, E_0, \mathbf{r})\, d^3\mathbf{r}\, dE_0$ is the average number of atoms set into motion in a region $(\mathbf{r}, d^3\mathbf{r})$ at an initial energy of E_0, $E_0 + dE_0$ by a collision avalanche caused by a primary ion having energy E and initial velocity $\mathbf{e}$ and at an initial location $\mathbf{r} = 0$.

Obviously, the function is the distribution of the recoil atoms relative to energy:

$$F(E, E_0) = \int d^3\mathbf{r} \cdot F(E, \mathbf{e}, E_0, \mathbf{r})$$

In a similar manner we can introduce a spatial distribution function for the density of the energy dissipated by an ion $F_\nu(E, \mathbf{e}, \mathbf{r})$ into atom motion so that $F_\nu(E, \mathbf{e}, \mathbf{r})\, d^3\mathbf{r}$ is the energy of atom motion in a volume $d^3\mathbf{r}$. This motion is the result of an avalanche of collisions caused by a primary ion having the indicated parameters. If we integrate $F_\nu(E, \mathbf{e}, \mathbf{r})$ over the volume we find the amount of the primary ion's energy that went into atom motion:

$$\nu(E) = \int F_\nu(E, \mathbf{e}, \mathbf{r})\, d^3\mathbf{r}$$

Similarly, we can introduce a spatial distribution for the density of the energy of electron motion $F_e(E, \mathbf{e}, \mathbf{r})$ and the portion $e(E)$ of primary ion energy that goes into electron motion:

$$e(E) = \int F_e(E, \mathbf{e}, \mathbf{r})\, d^3\mathbf{r}$$

It is obvious that

$$\nu(E) + e(E) = E$$

The distributions introduced are subject to the constraints that $F(E, \mathbf{e}, E_0, \mathbf{r}) = 0$ when $E_0 > E$, and $F(E, E_0) = 0$ when $E_0 > E$, which simply means that an ion cannot transmit more energy than it has.

The average number of atoms dislodged as the result of a collision avalanche will be

$$N(E) = \int_0^E F(E, E_0) P_d(E_0)\, dE_0$$

or when there is a distinct dislodging threshold T_d,

$$N(E) = \int_{T_d}^E F(E, E_0)\, dE_0$$

It is obvious that between the number of atoms set into motion $F(E, E_0)$ or the number of dislodged atoms $N(E)$ and the energy of the primary atom going into motion $\nu(E)$, there must be a proportionality relation, which is

$$F(E, E_0), \quad N(E) \sim \nu(E)$$

The explicit form of the expected relationships is found by solving the respective equations for the functions.

The equation for $F(E, E_0)$ was presented earlier (see Eqs. (3-135) and (3-136)). We will now derive an equation for $\nu(F)$. The energy $\nu(E)$ comprises the energy of the recoil atoms that appear in a layer Δx, the energy of going into atom motion, and the energy transmitted into atom motion beyond the layer Δx:

$$\nu(E) = N\,\Delta x \int d\sigma[\nu(E-T)+\nu(T)]+\nu(E-NS_e\,\Delta x)(1-\int N\,\Delta x\, d\sigma) \tag{3-151}$$

A simple conversion of Eq. (3-151) gives

$$\int_0^E d\sigma[\nu(E) - \nu(E-T) - \nu(T)] + S_e(E)\frac{d\nu(E)}{dE} = 0 \tag{3-152}$$

A solution for Eq. (3-136) can be found in analytical form if we assume that $S_e = 0$ and make use of the cross section for exponentiated potentials for the scattering cross section. Then, along with Eq. (3-152), we have

$$\int_0^E \frac{dT}{T^{1+1/n}}[F(E, E_0) - F(E-T, E_0) - F(T, E_0)] = \frac{1}{E_0^{1+1/n}} \tag{3-153}$$

Using the Laplace transform, we find an asymptotic relation for Eq. (3-153)[55]:

$$F(E, E_0) \sim \frac{1/n}{\Psi(1) - \Psi(1-1/n)}\frac{E}{E_0^2} \qquad \text{when } E_0 \ll E$$

$$F(E, E_0) \sim \frac{1}{\Gamma(1/n)\Gamma(1-1/n)}\frac{1}{E_0}\left(\frac{E}{E_0} - 1\right)^{1/n} \qquad \text{when } E_0 \leq E$$

$$\Psi(x) = \frac{d}{dx}\ln\Gamma(x)$$

Comparing this with the solution of Eq. (3-152) shows that

$$F(E, E_0) \sim \frac{1/n}{\Psi(1) - \Psi(1-1/n)}\frac{\nu(E)}{E_0^2} \qquad \text{when } E_0 \ll E$$

When small amounts of energy E_0 are transmitted the atoms may be modelled as solid spheres; then, $n = \infty$, and

$$F(E, E_0) \approx \frac{6}{\pi^2}\frac{\nu(E)}{E_0^2} \tag{3-154}$$

We find the average number of dislodged atoms as

$$N(E) \approx \frac{6}{\pi^2}\frac{\nu(E)}{E_d} \tag{3-155}$$

in a solid spheres approximation. Thus, the density of the recoil atoms is zero when $E_0 = E$ and varies in this range as $(E - E_0)^{1/n}/E_0^{1+1/n}$. At the other extreme, when $E_0 \ll E$ the density of the moving atoms increases to infinity as E_0^{-2}. However, the number of dislodged atoms is finite, owing to the threshold energy T_d. As we saw earlier, the amount of an ion's energy that goes into atom motion, $\nu(E)$, is, at low ion energies when the reduced energy of the ion is $\varepsilon \ll 1$, roughly equal to E. If $T_d = 20$ eV, then a primary ion having $E = 10^5$ eV of energy generates about 4000 dislodged atoms in a target. If we assume that each Frenkel defect has a potential energy of 5 eV, the amount of surplus energy retained in the crystal is about 20 keV, or 20% of the primary ion's energy. The other 80% of the primary ion's energy is dissipated as heat during vibrational relaxation. The surplus potential energy after a collision avalanche in amorphous solids and polymers is, for the most part, retained as the energy of ruptured chemical bonds. This potential energy makes the material more chemically active: irradiated crystals, amorphous inorganic materials, and polymers have a higher rate of dissolution.

We now turn to the spatial distribution of the number of recoil atoms $F(E, \mathbf{e}, E_0, \mathbf{r})$ and the density of the energy that goes into atom motion $F_\nu(E, \mathbf{e}, \mathbf{r})$. The integral equations for the $F(E, \mathbf{e}, E_0, \mathbf{r})$ functions were given earlier (see Eq. (3-137)); a similar equation for $F_\nu(E, \mathbf{e}, \mathbf{r})$ has the form[55]

$$-\mathbf{e}\nabla F_\nu(E,\mathbf{e},\mathbf{r}) = N\int d\sigma[F_\nu(E,\mathbf{e},\mathbf{r}) - F_\nu(E-T,\mathbf{e},\mathbf{r}) - F_\nu(T,\mathbf{e}'',\mathbf{r})] + NS_e(E)\frac{\partial F_\nu(E,\mathbf{e},\mathbf{r})}{\partial E} \tag{3-156}$$

Under certain assumptions ($E_0 \gg E$, etc.), there is a proportionality relation between Eqs. (3-137) and (3-156),

$$F(E,\mathbf{e},E_0,\mathbf{r}) = \frac{6}{\pi^2}\frac{F_\nu(E,\mathbf{e},\mathbf{r})}{E_0^2} \tag{3-157}$$

which is analogous to Eq. (3-117). Integrating Eq. (3-157) with respect to the energy of the recoil atom we obtain

$$\int_{T_d}^{E} F(E,\mathbf{e},E_0,\mathbf{r})\,dE_0 \equiv N(E,\mathbf{e},\mathbf{r}) = \frac{6}{\pi^2}\frac{F_\nu(E,\mathbf{e},\mathbf{r})}{T_d} \tag{3-158}$$

i.e., the density of the dislodged atoms and the density of the energy of atom motion have identical spatial distributions.

With the help of Eqs. (3-137) and (3-156), we can easily find equations for the moments $\langle x^n \rangle$ of the respective distributions. These distributions can be restored from known moments by the methods described in Section 3-21. A comparison of the moments and the distributions reveals a number of relationships. The ratio of the projected range R_p to the average depth of penetration R_D is shown in Fig. 3-20a (the ratio of the

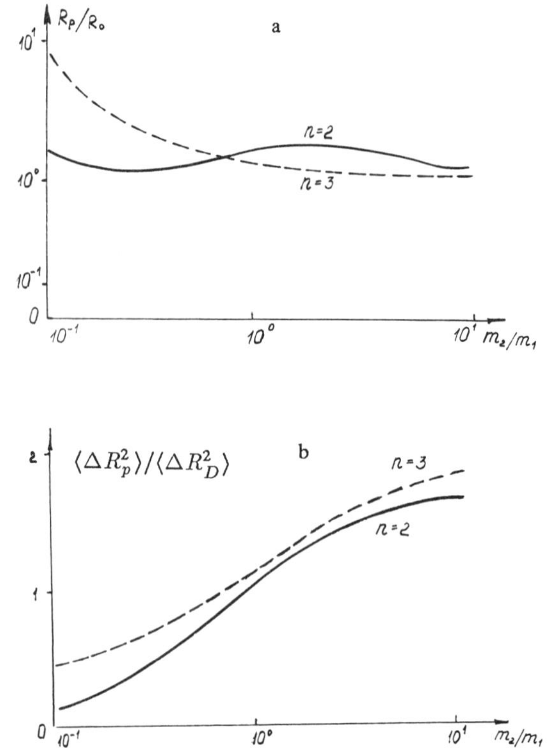

FIGURE 3-20. Ratio of projected range R_p to the average depth of defects R_D (a) and their variance as a function of the ratio of masses m_2/m_1 (b).

first moments of the respective spatial distributions); Fig. 3-20b shows the ratio of the variances of these distributions as functions of the ratio of masses. Electron losses have been ignored, which corresponds to low ion energies: $\varepsilon \ll 1$, which is typical when $m_2/m_1 \ll 1$, as well as when $m_2/m_1 < 1$, $R_p/R_D > 1$. This means that the peak of the density of implanted impurities is always deeper than the peak of the defects distribution (or the energy that goes into atom motion). The variance of the defects density distribution is much greater than the variance of impurity density distribution when $m_2/m_1 < 1$; the peak of the impurity density distribution is narrower than the peak of the defects distribution. This means that heavy ions entering a light target create defects along the entire path and that almost all of these ions stop at a depth roughly equal to the average range R_p. When $m_2/m_1 > 1$, i.e., when light ions enter a heavy target, the peak of the defects distribution is narrower than the peak of the distribution of the implanted impurities density. Trajectories of light ions in heavy metals fluctuate a great deal and defects are produced in the denser region that has been penetrated by the initial sections of the ion trajectories.

The distribution of defects relative to depth can be found by direct electron microscope observation of the defect clusters via, e.g., RBS. Figure 3-21 shows experimental and theoretical distributions of defects and impurity atoms relative to depth.[53] Prominent are the deeper location of the peak of the impurity density distribution in comparison with the peak of the defects distribution, as well as the difference in the width of the distributions.

In an exponentiated potential approximation of an ion–atom interaction and neglect-

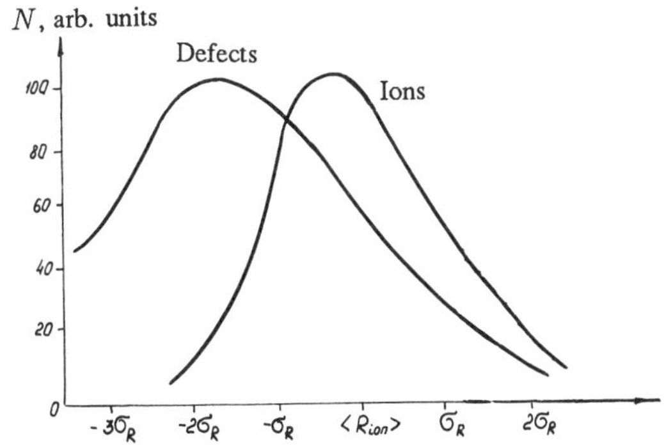

FIGURE 3-21. Distribution of a concentration N (in arbitrary units) of Bi ions having an initial energy of E = 50 keV in Si and the defects they produce relative to depth.

ing electron deceleration, we can find the energy dependencies of the avalanche area and the density of the energy dissipated in the avalanche.[55] Because the projected range in this case is $R_p \sim E^{2/n}/NC$, the avalanche area can be estimated as

$$\Omega \sim R_p^3 \sim E^{6/n}/(NC)^3$$

and we obtain

$$E/\Omega \sim E^{1-6/n}$$

for the amount of energy dissipated in the avalanche. When $n < 6$ the avalanche area increases with energy faster than the first degree of E and, correspondingly, the density of the energy dissipated decreases as the ion's energy increases. It is likely that at high ion energies an avalanche is broken up into subavalanches that have been produced by the primary recoil atoms that have acquired so much more energy that they can now generate subavalanches; when this happens fluctuations in the density of the defects within a cluster increase.

We can introduce a density similar to Eq. (3-158) for the atoms dislodged in an avalanche. Assuming that at low ion energies $\varepsilon \ll 1$ and $\nu(E) = E$, we find from Eqs. (3-155) and (3-159) that

$$C_D = \frac{N(E)}{\Omega} = \frac{6}{\pi^2}\frac{E}{T_d\Omega} \sim E^{1-6/n}$$

which coincides with the dependence of the density of the energy dissipated in the avalanche. The density of the dislodged atoms turns out to be quite high: on the order of 10^{18}–10^{19} cm^{-3} (10^{-2}–10^{-3} at.%).

The process of collision avalanches and defect creation has been studied several times by the Monte Carlo method.[56] Figure 3-22 shows the spatial distribution of vacancies (light circles) and interstitial atoms (shaded circles) generated in copper by an interaction with an ion having energy of 2 keV.[56] New relationships that were not seen in our previous examination are distinguished: 1) the distance between the vacancies and interstitial atoms, and 2) the formation of clusters (tight groups) of vacancies. The vacancies are highly concentrated at the core area of the avalanche, whereas the interstitial atoms form a cloud around this area (core). The calculations were done for two sets of parameters: Cu(I): T_d = 25 eV, and the recombination radius is r_ν = 1.5a, where

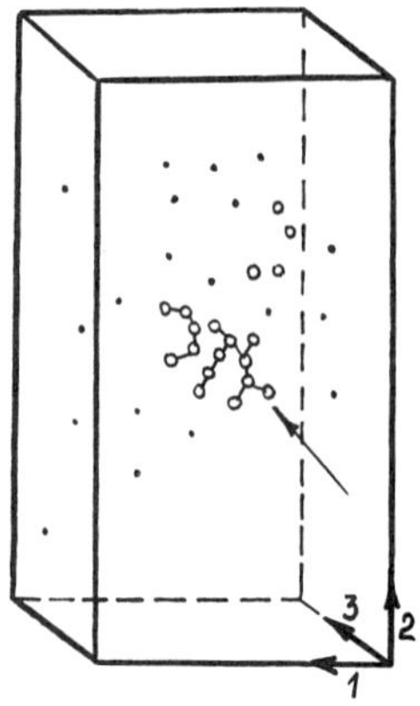

FIGURE 3-22. Vacancies (light circles) and interstitial atoms (shaded circles) in an avalanche created in a Cu(II) crystal by an ion having an initial energy of 2 keV.

TABLE 3-4. Average Number of Frenkel Defects $\langle N \rangle$ as a Function of Ion Energy E (calculated by the Monte Carlo method for the parameters shown in the last row of the table[56])

	$\langle N \rangle$			
E, keV	Cu(I)	Cu(II)	Al	Au
1	7	11	6	5
2	13	21	12	–
5	31	48	28	–
10	–	90	54	–
20	110	167	–	84
50	225	–	223	–
100	472	–	–	–
	$T_d = 25$ eV, $r_\nu = 1.5$	$T_d = 5$ eV, $r_\nu = 2.2$	$T_d = 25$ eV, $r_\nu = 1.5$	$T_d = 35$ eV, $r_\nu = 1.5$

a is the lattice constant; Cu(II): $T_d = 5$ eV, $r_\nu = 2.2a$. Moliere's potential was used for the potential of the ion–atom interaction. Table 3-4 shows the number of Frenkel defects obtained in [56] for copper. If we compare the values given in Table 3-4 with the number of dislodged atoms estimated from Eq. (3-155) we find that the difference is between these values roughly a factor of 10. One reason for this discrepancy is that theory does not consider the fact that Frenkel defects may recombine. Whereas in the Monte Carlo method it turned out that with the first set of parameters for Cu(I) 60% of the defects recombine and with the second Cu(II) 88% recombine.

The creation of defects when platinum was irradiated by Kr^+ ions having 20 keV of energy was modelled in [57]; these results were compared with results from experimental studies of the defects created by avalanches generated by a single ion by means of a field-ion microscope. A tip of radius 30 nm was fabricated from a Pt monocrystal which was then irradiated at 60°K. The density of vacancies in the avalanche was found by flash field evaporating the atomic layers of the tip, layer by layer; a photograph was taken after every flash of the evaporating field. On the average, one ion having 20 keV of energy created 184 vacancies, which is in good agreement with the theoretical estimate obtained from

$$\langle N_\nu \rangle = \eta \frac{\nu(E)}{2T_d} \tag{3-159}$$

where η is a coefficient that describes the efficiency of energy loss by the ion in the elastic collisions that created the defects (vacancies) $\eta = 0.84$ and the threshold energy T_d of defect creation is 36 eV for Pt. The average measured diameter of the avalanche (in a plane perpendicular to the direction of the ion's motion) was $\langle \lambda \rangle = 26, 36 \pm 12$ Å; the distribution of the vacancies extended to a depth 60 Å. The average density of vacancies in the avalanche was 6 at.%. The results obtained from modelling are very close to the measured avalanche parameters.

3-27. THE ANGULAR AND SPATIAL DISTRIBUTIONS OF AN ION BEAM IN A TARGET

From the viewpoint of the physics of ion-beam lithography, it is worth our while to examine the angular scattering functions for ions at a depth x in the target and the spatial distribution functions in a plane perpendicular to the beam. The thicknesses of the resists being exposed lie in the range of 0.5–1 μm. The beam energies used in lithography are chosen so that the projected range R_p is not much greater than the thickness of the layer being exposed, which means that the energy of the ions will change a great deal within the resist layer. Consequently, the theory of angular and spatial distributions of beam particles in a target must be constructed with due regard for the changes in the energy of the particles in the beam. The theory is much simpler if we assume that the beam energy remains constant, an approximation that is satisfied with reasonable accuracy if $R_p \gg d$, where d is the thickness of the layer.

Experiments were conducted for the purpose of observing angular scattering of ions by bombarding thin, suspended targets under the condition that the number of ion collisions over the thickness of the target be small. Accordingly, a great deal of attention has been given to one-time and repeated ($n \leq 20$) scattering. Multiple scattering corresponds to $n > 20$. In ion-beam lithography we are concerned with one-time and repeated scattering when the resist layers are roughly 1 μm thick and the beam energy is ~100 keV at shallow depths in the layer, and with multiple scattering in deeper areas of the resist.

The results discussed in Chapter 2 that were obtained from calculating the distribution of electrons in a beam can be applied directly to ion distributions only when their energy is on the order of 100 MeV; when we can neglect shielding of the nuclei by the electron shells and consider scattering to proceed at the Coulomb potential $V(R) = Z_1 Z_2 e^2/R$. For slower ions, having $E < 1$ MeV of energy, the potential of an ion–atom interaction is better approximated, as we have said, by the exponentiated functions, Eq. (3-29). The most general results are obtained when a Coulomb potential with the Thomas–Fermi, or the Thomas–Fermi–Firsov shielding function[(12)] is used. Meyer has developed a theory of angular scattering[(58)] that uses this potential.

The following assumptions are the foundation of Meyer's calculations: the scattering centers in a target (amorphous material) are randomly located; binary collisions at the independent atoms of the target, and small scattering angles, $\theta_1 \ll 1$; the energy of the ions is constant; the electrons of the medium are not scattered (electrons participate in the scattering process only to the extent that they shield the nuclei of the colliding atoms). The most restrictive application of the results obtained from the theory is the assumption that the energy of the ions is constant; the formulas obtained are valid only for shallow depths in a layer of the material.

We rewrite the ion scattering cross section at an atom for a Coulomb potential with Thomas–Fermi shielding, Eq. (3-71), introducing the variable $\eta = t^{1/2} = \varepsilon \sin\theta/2$:

$$d\sigma = \pi a^2 \frac{f(t^{1/2})}{2t^{3/2}}\, dt = \pi a^2 \frac{f(\eta)\, d\eta}{\eta^2} \tag{3-160}$$

We introduce the reduced quantities:

energy,

$$\varepsilon = E/\varepsilon_0, \qquad \varepsilon_0 = 2Z_1 Z_2 e^2/a$$

depth of beam penetration into the target,

$$\tau = x/x_0, \qquad x_0 = 1/(\pi a^2 N)$$

and deviation in the ion's direction of travel from its initial direction,

$$\tilde{\vartheta} = \theta/\theta_0, \qquad \theta_0 = 4Z_1 Z_2 e^2/aE = 2\varepsilon, \qquad \tilde{\vartheta} = \frac{1}{2}\varepsilon\theta$$

In these formulas the shielding length a is defined in the Thomas–Fermi–Firsov approximation as

$$a_{\mathrm{F}} = 0.885 a_0 (Z_1^{1/2} + Z_2^{1/2})^{-2/3}$$

or in the Thomas–Fermi–Lindhard approximation as

$$a_{\mathrm{L}} = 0.885 a_0 (Z_1^{2/3} + Z_2^{2/3})^{-1/2}$$

Meyer hypothesized[(58)] that each target atom scatters an ion according to Eq. (3-160) within a cross section πr_0^2 defined by the size of the atom in the target: $r_0 \approx 1/2N^{1/3}$, where N is the number of atoms per cubic centimeter. The number of collisions n that the ion will experience in a layer x of the target is

$$n = \pi r_0^2 N x = \frac{\pi}{8}\frac{x}{r_0}$$

The minimal "deviation" η_0 is found by means of Eq. (3-160) in terms of r_0:

$$\int_{\eta_0}^{\varepsilon} \frac{d\sigma}{d\eta}\, d\eta = \pi r_0^2$$

We rewrite Eq. (3-160), converting to the scattering cross section into a solid angle dO in a laboratory coordinate system:

$$\frac{d\sigma}{dO} = \frac{\varepsilon}{8} a^2 \left(\frac{m_1 + m_2}{m_2}\right)^2 \frac{f(\eta)}{\eta^3}$$

We now write the probability $F_1(\vartheta)\, dO$ that an ion will move within a solid angle dO as a result of one-time scattering:

$$F_1(\vartheta)\, dO = \begin{cases} \dfrac{1}{\pi r_0^2} \dfrac{d\sigma}{dO}\, dO & \text{when } \eta \geq \eta_0 \\ 0 & \text{when } \eta < \eta_0 \end{cases}$$

After the second collision we find, in keeping with the rule for multiplying probabilities, the distribution:

$$F_2(\vartheta)\, dO = dO \int_0^\infty \int_0^{2\pi} F_1(\vartheta - \chi) F_1(\chi)\, dO_\chi$$

Here the angles ϑ and χ are thought of as vectors in a plane located a unit distance away from the surface of the target and perpendicular to the initial direction of the beam.

We can write

$$F_n(\vartheta)\, dO = dO \int_0^\infty \int_0^{2\pi} F_{n-1}(\vartheta - \chi) F_1(\chi)\, dO_\chi \tag{3-161}$$

for the distribution that appears due to n collisions in a layer x. Using a Fourier-like transform with a zero-order Bessel function, instead of the $2(n-1)$-tuple integral of Eq. (3-161), we obtain[58]

$$F_n(\vartheta d)\, dO = \frac{dO}{2\pi} \int_0^\infty G_1^n(y) J_0(\vartheta y) y\, dy \tag{3-162}$$

$$G_1(y) = 2\pi \int_0^\infty F_1(x) J_0(yx)\, dx \tag{3-163}$$

By simplifying Eqs. (3-162) and (3-163) and discarding small terms, Meyer obtained a function for the scattering of ions at a reduced depth τ in a target as a universal (Thomas–Fermi) function $f_1(\tau, \tilde{\vartheta})$:

$$F_n(\vartheta)\, dO = \frac{dO}{2\pi} \frac{\varepsilon^2}{4} \left(\frac{m_1 + m_2}{m_2} \right)^2 f_1(\tau, \tilde{\vartheta}) \tag{3-164}$$

$$f_1(\tau, \tilde{\vartheta}) = \int_0^\infty \exp[-\tau \Delta(z)] J_0(\tilde{\vartheta} z) z\, dz \tag{3-165}$$

$$\Delta(x) = \int_0^\infty \frac{f(n)}{\eta^2} [1 - J_0(x\eta)]\, d\eta$$

$$\tilde{\vartheta} = \vartheta \frac{\varepsilon}{2} \frac{m_1 + m_2}{m_2} = \frac{\varepsilon \theta}{2}$$

The universal form-function $f_1(\tilde{\vartheta}, \tau)$ in Eq. (3-164) is mapped as a curve having a maximum at $f_1(0, \tau)$. The angular scattering half-width, $\tilde{\vartheta}_{1/2}$, is, evidently, another (Thomas–Fermi) universal function of the reduced depth of beam penetration into a target τ: $\tilde{\vartheta}_{1/2} = g_1(\tau)$. Meyer has tabulated the functions $f_1(\tilde{\vartheta}, \tau)$ and $g_1(\tau)$.[58]

TABLE 3-5. Parameters of the Function That Describes the Ability of Various Elements to Decelerate Protons: $dE/dx = \alpha y(1 - \beta y)$ eV/Å, $y = v_1/v_0$, $v_0 = 10^8$ cm/sec

Element	Z	α	β	Element	Z	α	β
Be	4	7.4	0.137	Ni	28	9.7	0.092
C	6	8.1	0.124	Cu	29	8.0	0.081
Al	13	7.5	0.134	Ge	32	8.5	0.129
Si	14	2.9	0.080	Ag	47	7.7	0.083
Ti	22	12.3	0.145	Sn	50	6.1	0.090
Cr	24	7.9	0.077	Sb	51	6.5	0.111
Mn	25	6.4	0.065	Au	79	7.3	0.065
Fe	26	7.3	0.076	Bi	83	7.2	0.106
Co	27	7.3	0.060				

When there is a large number of collisions ($n > 20$, multiple scattering) the numerical values for Meyer's angular scattering function agree with those obtained via the formulas used in Moliere's theory.[(59)] Notice once again that in both theories the energy of the particles is considered to be constant. In Gott's work[(60)] formulas were obtained for the angular distribution function for particles passing through a target of thickness d when scattering ions in an exponentiated potential having the exponent $n = 2$, and in a special case of ion energy loss established empirically for protons ($E < 100$ keV):

$$\frac{dE}{dx} = -\alpha v(1 - \beta v)$$

Here $\alpha(Z_2)$ and $\beta(Z_2)$ are constants that describe the deceleration of protons in a target of atomic number Z_2, and v is the proton velocity in 10^8 cm/sec.

Table 3-5 gives the values of α and β for various Z.

Beginning with the general formula for the angular distribution function[(61,62)],

$$f(\theta, d) = \frac{1}{2\pi} \sum_{l=0}^{\infty} \left(l + \frac{1}{2}\right) \exp\left[-N \int_0^d \sigma_l \, dx\right] P_l(\cos\theta)$$

$$\sigma_l = 2\pi \int_0^{\pi} [1 - P_l(\cos\alpha)] \frac{d\sigma}{dO} \sin\alpha \, d\alpha$$

and taking the differential scattering cross section to be

$$\frac{d\sigma}{dO} = \frac{4.1}{32\pi} \frac{Z_1 Z_2 e^2 a}{E \sin^2(\alpha/2)} \frac{m_1 + m_2}{m_2}$$

Gott[(60)] found

$$f(\theta, d) = \frac{1}{4\pi} \frac{1 - \exp(-2p)}{1 - 2\cos\theta \exp(-p)} \tag{3-166}$$

where

$$p = N \int_0^d \sigma_1 \, dx = \frac{2.56 \times 10^4 Z_1 Z_2 e^2 a N (m_1 + m_2)}{\alpha m_2} \times \left[\frac{1}{v_d} - \frac{1}{v_0} + \beta \ln \frac{v_0 (1 - \beta v_d)}{v_d (1 - \beta v_0)} \right]$$

$$v_d = \beta^{-1} [1 - (1 - \beta v_0) \exp(1.6 \times 10^{-20} \alpha \beta d / m_1)]$$

In a physical sense, p is the "attenuation" caused by the "transport" cross section σ_1 in a target of thickness d, and v_d is the velocity of a proton passing through this target d. Gott's results are of interest in the theory of ion lithography that uses proton beams.

We will indicate an asymptotic form of the angular distribution of an ion beam in a very thin target, $\tau \ll 1$:

$$F(\vartheta, x) \, dO = N x \frac{d\sigma}{dO} \, dO \tag{3-167}$$

which simply corresponds to the probability of one-time scattering into the respective solid angle in a layer; for a Coulomb potential we have, instead of Eq. (3-167),

$$F(\vartheta, x) \, dO = 4 N x \left(\frac{Z_1 Z_2 e^2}{pv} \right)^2 \frac{dO}{\vartheta^4} \tag{3-168}$$

For thoroughness' sake, we will also present a Gaussian angular distribution function in a Coulomb scattering potential, which is valid when $\tau \gg 1$:

$$F(\vartheta, x) \, d\vartheta = \frac{2\vartheta}{\langle \vartheta^2 \rangle} \exp \left[-\frac{\vartheta^2}{\langle \vartheta^2 \rangle} \right] d\vartheta \tag{3-169}$$

$$\langle \vartheta^2 \rangle = 4\pi \left(\frac{Z_1 Z_2 e^2}{pv} \right)^2 N x \ln[4\pi Z_1^2 Z_2^2 N x (\hbar / p)^2]$$

Marwick and Sigmund[63] studied the scaling law between the spatial and angular distribution functions for ion beams penetrating a target. If (in a LCS) an ion has been scattered at an angle ϑ on the surface of the target, its distance from the beam axis at a depth x will be

$$\rho = (y^2 + z^2)^{1/2} = x \tan \vartheta \approx x \cdot \vartheta \tag{3-170}$$

If the scattering at the angle ϑ occurred at a depth $x' < x$ or the deviation of ϑ was caused by many (several) acts of scattering, then it is intuitively obvious that in this case, the ion's deviation at the depth x will be less than the maximal deviation given by Eq. (3-170)

$$\rho(\vartheta, x) = \Gamma \cdot x \vartheta \tag{3-171}$$

Thus, a scale factor $\Gamma < 1$ must exist. Marwick and Sigmund[63] used a spatial distribution function

$$G(\rho, x)\, d^2\rho = \rho\, d\rho \int_0^\infty k\, dk\, J_0(k\rho) \exp\left[-N \int_0^x \sigma_0(kx')\, dx'\right]$$

$$2\pi \int_0^\infty G(\rho, x)\rho\, d\rho = 1$$

obtained as a solution of the kinetic equation when $\rho \ll x$, as their initial function for analyzing the scaling question. Here,

$$\sigma_0(k) = \int_0^\infty d\sigma(\alpha)[1 - J_0(k\alpha)]$$

The scaling law

$$\rho = \frac{x}{\Gamma_n}\vartheta, \qquad \Gamma_n = (1 + 2/n)^{n/2} \tag{3-172}$$

was established for exponentiated potentials of ion–atom interaction. A similar relationship exists between the half-widths of the distributions:

$$\rho_{1/2} = (x/\Gamma_n)\vartheta_{1/2}$$

As is evident from Eq. (3-172), $n = 1$ when $\Gamma_1 = \sqrt{3}$, and $\Gamma_\infty = l = 2.7183$ when $n = \infty$.

A scaling law was also established in [63] for scattering at the Thomas–Fermi potential. If we use the relation

$$G(\rho, x)\, d^2\rho = g_1(\tilde{\rho}, \tau)\tilde{\rho}\, d\tilde{\rho}, \qquad \tilde{\rho} = \frac{\rho\varepsilon}{2\tau_0}, \qquad \tau = \frac{x}{\tau_0} \tag{3-173}$$

$$\tilde{\vartheta} = \frac{\varepsilon\tilde{\vartheta}}{2}, \qquad \tau_0 = \frac{1}{\pi a^2 N}, \qquad \varepsilon = E/(Z_1 Z_2 e^2/a)$$

to introduce a spatial distribution function $g_1(\rho, \tau)$ for a beam of particles, we can then establish that

$$g_1(\tilde{\rho}, \tau) \approx \left[\frac{\Gamma(\tau)}{\tau}\right]^2 f_1\left(\tau, \Gamma\frac{\rho}{\tau}\right) \tag{3-174}$$

where $\Gamma(\tau)$ is weakly dependent on the reduced target thickness (it varies from 2.18 to 1.78 when τ varies from 10^{-3} to 10^3). The half-widths of the distributions are related as $\tilde{\rho}_{1/2}/\tilde{\vartheta}_{1/2} = \tau/\Gamma(\tau)$. This means that the dependence of $\tilde{\rho}_{1/2}$ on τ is one degree stronger than is the dependence of $\tilde{\vartheta}_{1/2}(\tau)$. When $n = 1$ (a Coulomb potential) $\tilde{\vartheta}_{1/2} \sim x^{1/2}$, $\tilde{\rho}_{1/2} \sim x^{3/2}$.

The half-width of angular scattering $\vartheta_{1/2}$ may, naturally, be associated with the value of nuclear deceleration $S_n(E)$.[48] In fact, for small scattering angles according to

Eq. (3-55), $\theta_1 = [m_2/(m_1 + m_2)]\theta$, and

$$T = \frac{4m_1 m_2}{(m_1 + m_2)^2} E \sin^2 \frac{\theta}{2} \approx \frac{m_1}{m_2} E\theta_1^2$$

$$\langle\theta_1^2\rangle = \frac{m_2}{m_1}\frac{\langle T\rangle}{E} = \frac{m_2}{m_1}\frac{NxS_n(E)}{E} \tag{3-175}$$

This means that we can find the magnitude of the nuclear deceleration S_n by measuring the variance of the angular distribution of ions passing through a target of given thickness.[48]

Good agreement is found between values obtained experimentally and those obtained from the theoretical formulas for angular and spatial distributions.

3-28. EXPOSING RESISTS WITH ION BEAMS

We can approach the theory of ion beam exposure in much the same way that we did in Chapter 2 when we constructed the theory for exposing resists with electron beams. Let a Gaussian beam

$$N(x, y) = \frac{qia}{\pi} \exp[-a(x^2 + y^2)] \qquad \frac{\text{ions}}{\text{cm}^2 \cdot \text{sec}} \tag{3-176}$$

be used to expose a resist.

We introduce the function $I(r_T; x, y)$ to describe the density of the energy dissipated by an ion penetrating the resist layer at a point x, y in a region of the resist layer located at $\mathbf{r}_T$. The density of the energy dissipated by the Gaussian beam at $\mathbf{r}_T$ during the exposure period t will then be

$$D(\mathbf{r}_T) = t\frac{qia}{\pi} \iint_{-\infty}^{+\infty} I(\mathbf{r}_T; x, y) e^{-a(x^2+y^2)} \, dx \, dy$$

When the pattern is linearly scanned by a Gaussian beam we must evaluate a more complex convolution of the functions, similar to Eq. (2-175):

$$D(\mathbf{r}_T) = \frac{1}{v_s} \sum_i \int_{x_{c_{1i}}}^{x_{c_{2i}}} dx_{ci} \iint_{-\infty}^{+\infty} N(x - x_{ci}, y - y_c) I(x_T - x, y_T - y, z_T) \, dx \, dy$$

From the results of the preceding, we can define a function $I(\mathbf{r}_T; x, y)$. We isolate a volume $dV_T = \rho \, d\rho \, d\varphi \, dz$ in a layer z_T, $z_T + dz_T$ of the resist. The probability that an ion penetrating the resist at the point $x, y, 0$ passes through this volume is, according to Eq. (3-173),

$$G(z_T; \rho_T) \, d^2\rho = g_1 \left(\tau_T; \frac{\sqrt{(x_T - x)^2 + (y_T - y)^2}}{\tau_0} \right) \frac{dx \, dy}{2\pi\tau_0^2}$$

We will neglect the difference between the path s that the ion travelled in the resist and the depth z_T because angular scattering is small. Then the energy dissipated by the ion into a volume dV_T is

$$dE = \left(\frac{dE}{dz}\right)_{z_T} dz_T \cdot g_1\left(\frac{z_T}{\tau_0}, \frac{\sqrt{(x_T - x)^2 + (y_T - y)^2}}{\tau_0}\right) \frac{dx\,dy}{2\pi\tau_0^2} \tag{3-177}$$

The density of the energy dissipated in the volume dV_T by a beam having a distribution of ions $N(x, y)$ is

$$D(\mathbf{r}_T) = \frac{t}{2\pi\tau_0^2}\iint_{-\infty}^{+\infty} N(x,y) g_1\left(\frac{z_T}{\tau_0}, \frac{\sqrt{(x_T - x)^2 + (y_T - y)^2}}{\tau_0}\right)\left(\frac{dE}{dz}\right)_{z_T} dx\,dy \tag{3-178a}$$

$$\left(\frac{dE}{dz}\right)_{z_T} = N[S_n(E(z_T)) + S_e(E(z_T))] \tag{3-178b}$$

In order to perform further calculations, we must have the functions $g_1(r_T; x, y)$ and dE/dz in explicit form. Assuming that the ion's path s to the point $\mathbf{r}_T$ can be equated to the depth z_T at which the point $\mathbf{r}_T, (dE/dz)_{z_T}$ is located, it turns out that $(dE/dz)_{z_T}$ is dependent only upon the coordinates of z_T and can be taken outside the integral in Eq. (3-178a). Solving

$$\int_{E(0)}^{E(z_T)} \frac{dE}{S_n(E) + S_e(E)} = N\int_0^{z_T} dz = N z_T \tag{3-179}$$

gives us the ion's energy $E(z_T)$, which we can use to find $(dE/dz)_{z_T}$ according to Eq. (3-178b). For large $\tau_T \gg 1$ the function g_1 can be approximated by the Gaussian function

$$\frac{1}{2\pi\tau_0^2} g_1(\tau, \tilde{\rho}) = \frac{1}{2\pi\sigma_1^2}\exp\left[-\frac{(x_T - x)^2 + (y_T - y)^2}{2\sigma_1^2}\right]$$

Integrating Eq. (3-178a) with Eq. (3-176) yields a Gaussian function whose variance is equal to the sum of the variances of the functions whose convolution is being evaluated

$$D(r_T) = \frac{qiat}{\pi}\left(\frac{dE}{dz}\right)_{z_T} \exp\left[-\frac{x_T^2 + y_T^2}{2\sigma^2}\right]$$
$$\sigma^2 = \sigma_0^2 + \sigma_1^2, \qquad \sigma_0^2 = 1/2a \tag{3-180}$$

If we define $\rho_{1/2}$ as the half-width at the half-height of a Gaussian curve, we find the following relationship between the variance σ of the curve and the half-width of $\rho_{1/2}$:

$$\rho_{1/2} = \sqrt{2\ln 2}\,\sigma = 1.177\sigma$$

We will now examine a specific case of exposing a PMMA-type resist with a beam of protons that have $E = 60$ keV of energy. We say that the resist layer is $d = 0.4\ \mu$m thick and make the convert to reduced quantities. The average atomic number $Z_{2\text{eff}}$ and atomic weight A_{eff} for a $C_5H_8O_2$ PMMA compound are 3.6 and 6.66, respectively. Hence,

$$a_F = \frac{0.885a_0}{(Z_1^{1/2} + Z_2^{1/2})^{2/3}} = 0.442a_0$$

We find the number of atoms per cubic centimeter from the density of PMMA ρ = 1.2 g/cm^3 and $A_{\text{eff}} = 6.6$: $N = 10^{23}$ cm^3. We now find the unit measure of length

$$\tau_0 = \frac{1}{\pi a_F^2 N} = 5.8 \text{ nm}$$

and the reduced thickness of the PMMA layer is $\tau = d/\tau_0 = 70$. If we define repeated scattering as $1 \leq n \leq 20$ and multiple scattering as $n > 20$, then ions having 60 keV of energy undergo one-time scattering in the upper layer of the resist, which is roughly 6 nm thick; undergo repeated scattering in the layer between 6 and 120 nm; and undergo multiple scattering in the layer from 120 to 400 nm. The distribution of the ions over the cross section will be nearly Gaussian in the multiple scattering region.

The unit of energy measure ε_0 when $a_F = 0.442a_0$ is

$$\varepsilon_0 = \frac{2Z_1 Z_2 e^2}{a_F} = 443 \text{ eV}$$

and the reduced energy of a beam of protons having $E = 60$ keV of energy is $\varepsilon = 135.4$. Consequently, the potential of the proton–atom interaction in collision may be thought of as Coulomb, right up to those depths at which the energy ε falls to values $\varepsilon \lesssim 10$, i.e., to the value of the order of 4 keV. Because protons having $\varepsilon > 10$ of energy pass through the entire resist layer, the potential of interaction may be considered to be Coulomb at every level throughout the thickness of the resist. As a result, the variance σ_1 of the Gaussian curve g_1 can be found using the Coulomb potential of interaction. In addition, when $\varepsilon > 10$, nuclear deceleration can be approximated by the function $S_n = kaE^{-1}$ and electron deceleration by $S_e = kE^{1/2}$; then the integration foreseen in Eq. (3-179) is evaluated to the end point:

$$\frac{1}{k}\int_{E(0)}^{E(z)} \frac{dE}{E^{1/2} + aE^{-1}} = \frac{2}{k}\left[x - \frac{a^{1/3}}{3}\left\{\frac{1}{2}\ln\frac{(x + a^{1/3})^2}{(x^2 + a^{2/3} - ax)} + \sqrt{3}\arctan\left(\frac{2x - a^{1/3}}{\sqrt{3}\,a^{1/3}}\right)\right\}\right|_{E^{1/2}(0)}^{E^{1/2}(z)} = Nz \qquad \text{(3-181)}$$

The energy $E(z)$ of an ion at a depth z can be found from Eq. (3-181) by numerical calculations.

The results from the analytical theory may be compared with the data obtained via a numerical experiment by the Monte Carlo method.[50] The lithographic process used to expose a PMMA resist with protons having 60 keV of energy was evaluated. The elastic scattering cross section for the protons at the atoms was calculated from Eq. (3-71), where $f(t^{1/2})$ was chosen as

$$f(t^{1/2}) = \lambda t^{1/2}[1 + (2\lambda t^{1-m})^2]^{-1/q}$$

with $\lambda = 2.54$, $m = 0.25$ and $q = 0.475$ (Winterbone's formula). The contributions to the scattering of the various atomic components of a compound material are considered on the basis of Bragg's rule of adaptivity.

Figure 3-23 shows the trajectories of fifty protons in a layer of PMMA attached to different substrates. Of these fifty protons, not one was reflected from the silicon substrate, whereas at least four of them were reflected from the gold substrate. On the whole, we can say that, under the conditions of a given experiment, reflection from the substrate plays no significant role. Figure 3-24 shows the distribution of the energy dissipated in a plane perpendicular to the beam at depths of 120, 200, and 400 nm. At 400 nm the width of the distribution curve is roughly 33 nm at the level where a tenfold reduction in the amount of energy absorbed occurs. Figure 3-25 shows the distribution of the energy dissipated by the ions when exposing a resist with five bands 0.1 μm wide and 0.2 μm between centers. The energy distribution is shown for up to 400 nm deep. As can be seen, ion lithography produces distinct lines 0.1 μm wide. This conclusion is supported when modelling the process of etching an exposed resist layer according to the equation for the etching rate $R(D)$:

$$R = R_1(C_m + D/D_0)^\alpha \tag{3-182}$$

where D is the density of the energy dissipated in J/cm^3, D_0 = 174 J/cm^3, R_1 = 0.1 nm/sec, $\alpha = 1.9$, and $C_m = 0$ are empirical parameters that describe the etching of an exposed layer in an enchant consisting of a mixture of methyl isobutyl ketone and isopropyl alcohol in 1:1 proportion. Evidently, nuclear and electron processes produce a different radiation yield per 100 eV of dissipated energy and a natural extension of Eq. (3-181) for the etching rate is

$$R(D) = R_1(C_m + D_n/D_{0n} + D_e/D_{0e})^\alpha \tag{3-183}$$

where the constants D_{0n} and D_{0e} have been introduced to describe the chemical efficiency of converting the ions' energy into nuclear and electron motion, respectively.

Figure 3-26 shows the developing curves for a resist exposed by five 0.1 μm wide bands at doses of 0.6 and 2.0 C/cm^2 after 15, 30, 45, 60, and 75 seconds of etching. It is apparent that the walls of the dissolved region are closer to the vertical when the exposure dose is higher. Obviously, increasing the beam energy as well as the exposure dose corresponds to "more vertical" walls.

Figure 3-27 shows the change in electron and nuclear energy losses with depth for a beam of protons having E = 60 keV of energy.[64] The situation is reversed for heavy ions; e.g., gallium ions having E = 250 keV of energy transmit almost 85% of their energy into nuclear motion.

3-29. ION BEAMS FOR SPUTTERING (ETCHING) SOLIDS

One practical application of ion beams in microelectronics is etching.[65] This process is widely used to analyze the composition of materials by secondary ion spectroscopy.[66] Ion etching offers hope for obtaining vertical walls in the etching profile.[65] A fundamental property of ion etching is the sputtering yield (coefficient), equal to the number of target atoms sputtered per implanted ion. We can assume that the sputtering yield will depend on the ratio of the ion mass to the mass of the target atoms, the binding energy of

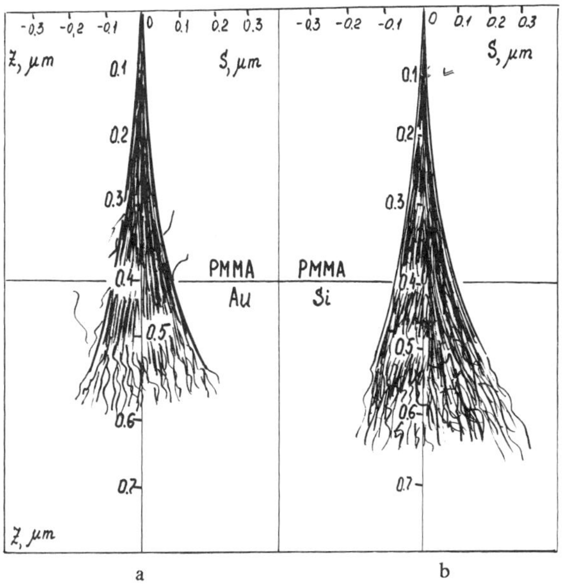

FIGURE 3-23. Trajectories calculated by the Monte Carlo method for protons in a PMMA layer on a gold (a) and a silicon (b) substrate (S is the proton scattering radius).

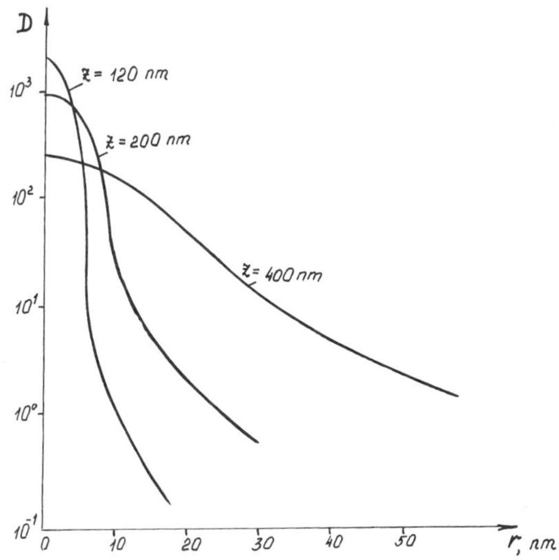

FIGURE 3-24. Radial distribution density of the energy dissipated by an electron beam in a resist. The infinitesimal beam enters at point $r = 0$ along the normal to the surface of the PMMA.

the target atoms, the angle of incidence, and the energy of the ion beam. Anisotropy in the atomic structure of a monocrystalline target may play a major role in the sputtering processes.

The flux of sputtered target atoms can be described by its own energy and angular distributions.

Experiments in ion sputtering in its purest form are conducted by making a monochromatic beam of inert gas ions act on a target in a vacuum. When a target is exposed to a beam containing ions of different elements, a chemical reaction between an ion

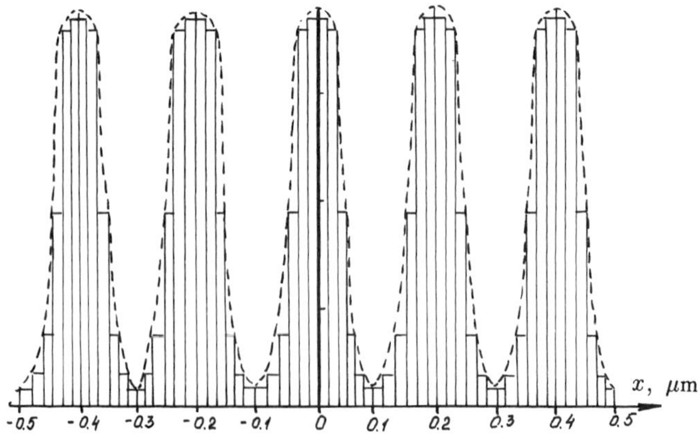

FIGURE 3-25. Distribution of the energy dissipated by protons at a depth of E_0 = 60 keV in a resist layer when exposing five Z = 400 nm wide bands through a 0.1-μm to 0.2-μm gap.

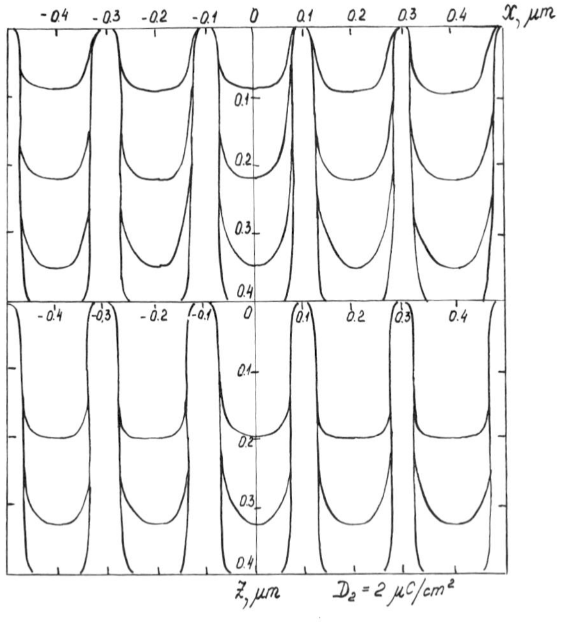

FIGURE 3-26. Developing profiles for a resist in a 1:1 methyl isobutyl ketone–isopropyl alcohol solvent. a) Etching times of 1, 3, 5, 7, 9 min; b) etching times of 15, 30, 45, 60, 75 sec.

and the target atoms which affects the sputtering process may be suspected. However, the sputtering of targets with ions of different elements having energies on the order of keV and higher can be confined to general relationships, which, apparently, attests to the physical nature of the effect under the conditions indicated. There is no question that heterogeneous chemical reactions on the surface of a target in a gas discharge plasma containing active ion-radicals play a role in etching.[65,67] In these situations we must speak of ion-chemical (or plasma chemical) etching when physical sputtering, caused by avalanches of atom–atom collisions due to an impinging ion, may play a subordinate role. Here we will examine only physical sputtering in the sense indicated.

In microlithography ion etching is accomplished by placing a target in a wide homogeneous beam of monochromatic ions. The beam impinges on the surface from the

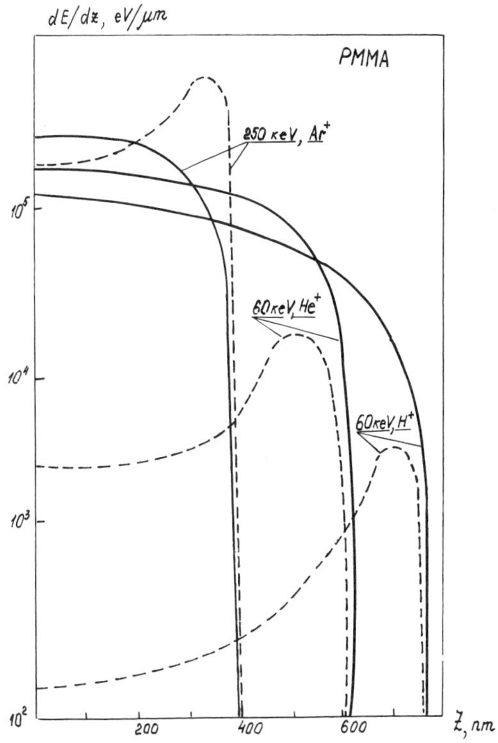

FIGURE 3-27. Change in electron (solid lines) and nuclear (dotted lines) energy losses for H^+, He^+, and Ar^+ ions in a PMMA resist as a function of penetration depth.

normal. The structural layers of the materials are etched through a mask containing windows. A polymer layer of resist material or a layer of inorganic, e.g., metallic, materials may serve as the mask. Ideally, the mask must not be subject to etching, whereas etching in a structural layer must proceed at a sufficient rate (on the order of 1 nm/sec). Unfortunately, for specific beam parameters the etching rates in different materials are not so different; these differences are often as much as a factor of five.[68]

Avalanche collisions theory makes it possible for us to explain the relationships that exist in ion etching.[55,68,69] A fast ion impinging on a target causes an avalanche of moving primary, secondary, etc. recoil atoms. Equation (3-157) describes the spatial density and energy spectrum of the moving atoms in the avalanche theory. We multiply Eq. (3-157) by the number of ions impinging a target, I_0 ions/sec, and integrate both sides over y and z:

$$I_0 \int F(E, \mathbf{e}, E_0, \mathbf{r})\, dy\, dz \equiv I_0 F(E, \mathbf{e}, E_0, x) = \frac{6}{\pi^2} \frac{F_\nu(E, \mathbf{e}, x)}{E_0^2} I_0$$

$$F_\nu(E, \mathbf{e}, x) = \int F_\nu(E, \mathbf{e}, \mathbf{r})\, dy\, dz$$

The quantity obtained is the spectral density of the moving atoms as a function of depth x in the target (the surface of the target is the plane $x = 0$). We will say that the velocities $\mathbf{v}_0$ of the recoil atoms are isotopically distributed, in which case the flux of atoms impinging on the surface of the target from within is

$$dI_a = I_0 F_\nu(E, \mathbf{e}, E_0, 0)\, dE_0\, v_{0x} \frac{dO}{4\pi} \tag{3-184}$$

However, this flux must be reduced by a factor of R_p/v_0 because the source of moving atoms in a target is limited by R_p. The need for this factor can be supported by the arguments that: atoms from a depth v_0 will pass through the plane of the surface during one second in an infinite source, whereas when the source is $R_p < v_0$ thick, only atoms from a depth R_p will pass through the surface and therefore, will be a factor of R_p/v_0 less than in an infinite source.

Introducing a factor into Eq. (3-184) that accounts for the finiteness of the source and, because $v_{0x} = v_0 \cos\theta$, gives us

$$dI_a = \frac{6F_\nu(E, \mathbf{e}, 0)}{E_0^2} \frac{R_p}{v_0} I_0 v_0 \cos\theta \sin\theta\, d\theta\, d\varphi\, dE_0 \tag{3-185}$$

Integration is done with respect to the escape angles of the atoms within the left hemisphere. The limits of integration with respect to the atoms' energy E_0 are defined by the fact that an atom must overcome an energy barrier U_0 in order to pass through the surface. To do this the component of the atom's energy that is associated with motion perpendicular to the surface must be greater than U_0: $m_2(v_{0x}^2/2) \geq U_0$. Therefore, the total ion flux out of the target is

$$I_a = \frac{6I_0}{4\pi^3} R_p F_\nu(E, \mathbf{e}, 0) \int_0^{2\pi} d\varphi \int_0^{\pi} \cos\theta \sin\theta\, d\theta \int_{U_0/\cos^2\theta}^{E} \frac{dE_0}{E_0^2}$$

Simple integrations yield

$$I_a = \frac{3I_0}{4\pi^2} \frac{F_\nu(E, \mathbf{e}, 0)}{U_0} R_p \tag{3-186}$$

Dividing Eq. (3-186) by the number of primary ions I_0 gives us the sputtering coefficient S_a:

$$S_a = \frac{3}{4\pi^2} \frac{F_\nu(E, \mathbf{e}, 0)}{U_0} R_p \tag{3-187}$$

Sigmund[(55,69)] obtained this same result by solving the kinetic equation for the atoms' energy distribution function E_0:

$$\frac{I_0}{v_0} \delta(E - E_0) = N \int d\sigma [G(E, E_0) - G(E - T, E_0) - G(T, E_0)] + N S_e(E)\, \partial G(E, E_0)/\partial E \tag{3-188}$$

The solution of Eq. (3-188) for exponentiated potentials and when $S_e = 0$ [55, 69] has the form

$$G(E, E_0) = \frac{n^{-1}(1 - n^{-1})}{\Psi(1) - \Psi(1 - n^{-1})} \frac{E}{E_0^2} \frac{I_0}{v_0} \left(\frac{E_0^{2/n}}{NC_n} \right) \tag{3-189}$$

If we write the last equation for the potential of solid spheres ($n = \infty$), replace the energy E by the fraction $\nu(E)$ that goes to the moving atoms, and extend the entire relation, Eq. (3-189), to the ions' spectral distribution in a space x we obtain[55]

$$dI_a = \frac{6}{\pi^2} \frac{F_\nu(E, \mathbf{e}, 0)}{E_0^2} \frac{I_0}{v_0} \frac{1}{NC_0} v_0 \cos\theta \, dE \, \frac{dO}{4\pi}$$

which is analogous to Eq. (3-185).

We can simplify the equation for the etching coefficient if we consider the fact that the quantity $F_\nu(E, \mathbf{e}, 0)$ has the dimensions of the derivative of the energy with respect to x, which means that $F_\nu(E, \mathbf{e}, 0)$ must be proportional to the nuclear decelerating ability $NS_n(E)$:

$$F_\nu(E, \mathbf{e}, 0) = \alpha(m_2/m_1) N S_n(E) \qquad (3\text{-}190)$$

The coefficient α is a function of the ratio of the masses m_2/m_1. Also, by defining the range as $R_p = 1/NC_0$, where C_0 is, according to [69],

$$C_0 = \frac{\pi}{2} \lambda_0 a^2, \qquad \lambda_0 = 24 \qquad a = 0.219 \text{ Å}$$

we obtain[69]

$$S_a(E) = 0.042 \frac{\alpha S_n(E)}{U_0} \qquad (3\text{-}191)$$

This relationship reflects the energy dependence of the sputtering coefficient very well. An expression that somewhat reveals the subtle dependence on the parameters of the ion and target atoms in Eq. (3-191) is[51]

$$S_a(E_0) = \frac{3.56}{U_0} \frac{Z_1 Z_2}{(Z_1^{2/3} + Z_2^{2/3})^{1/2}} \frac{M_1}{M_1 + M_2} \alpha \left(\frac{M_2}{M_1}\right) S_n(\varepsilon) \qquad (3\text{-}192)$$

Here $S_n(\varepsilon)$ is a reduced function of the nuclear decelerating ability, ε is the reduced energy of the ion, and $\alpha(M_2/M_1)$ is a weak function of M_2/M_1 which, according to [70], can be put in the form $\alpha = 0.15 + 0.13 M_2/M_1$. Considering this fact, Eq. (3-192) can be written as an approximation[70]

$$S_a(E_0) = \frac{(Z_1 Z_2)^{5/6}}{3U_0} S_n(E) \qquad (3\text{-}193)$$

The normalized yield of sputtered atoms $3S_a(E_0)U_0/(Z_1 Z_2)^{5/6}$ must agree with the reduced nuclear decelerating ability $S_n(\varepsilon)$, which is satisfied with sufficient accuracy in experiments in sputtering with inert gas ions (Figs. 5 and 6 in [70]).

A simple function $S_n(\varepsilon)$ was given in [71] and describes the experimental data for the nuclear decelerating ability very well:

$$S_n(\varepsilon) = \frac{1/2 \ln(1 + \varepsilon)}{\varepsilon + 0.14\varepsilon^{0.42}} \qquad (3\text{-}194)$$

It must be noted that $S_n \to \varepsilon^{0.58}$ as $\varepsilon \to 0$ and, accordingly, the yield of sputtered atoms $S_a(E_0)$ tends to zero only when the energy of the impinging ion is zero, which makes no physical sense since there is a barrier U_0 for the sputtering process and the formulas that can be used in the range of energies $E_0 \geq U_0$ must have factors that make the yield $S(E_0)$ go to zero as $E_0 \to U_0$. Such factors were introduced, e.g., in [72] as

$$S^*(E_0) = S(E_0)\left(1 - \sqrt{\frac{E_{\text{th}}}{E_0}}\right) \tag{3-195}$$

and as

$$S^*(E_0) = Q_{tp} \cdot 0.0085 \left(\frac{E_0}{E_{\text{th}}}\right)^{1/4} \left(1 - \frac{E_{\text{th}}}{E_0}\right)^{1/2} \tag{3-196}$$

in [73] for sputtering heavy targets with light ions. Here E_{th} is the threshold energy which must be determined, and Q_{tp} is a factor that depends on the M and Z parameters of the ions and the target. The threshold energy E_{th} has the sense of a statistical parameter for which the expression

$$E_{\text{th}} = \frac{U_0}{\gamma(1-\gamma)}, \qquad \gamma = \frac{4M_1M_2}{(M_1+M_2)^2} \tag{3-197}$$

was suggested in [73]. The expression

$$E_{\text{th}} = [0.214 + 4.77(M_1/M_2)^{0.567} + 0.256(M_2/M_1)]U_0 \tag{3-198}$$

was obtained in [74] from computer experiments.

Sputtering was examined in [74] on the basis of the mechanics of binary collisions, considering no more than one or two collisions. A surface atom was emitted when $E_\perp > U$, where $E_\perp = (mv_1^2/2)$ is the energy of the target atom's motion, acquired by colliding with a fashion, normal to the surface of the solid. The energy $E_\perp$ is a complex function of collision parameters such as the angle of incidence (angle from the normal) θ_0 and the scattering angle γ. For example, when a fast ion is reflected into the solid because of colliding with an atom in the first monolayer of the target and the target atom is reflected outward and only one such collision takes place, the threshold energy is

$$E_{\text{th}} = \frac{U_0}{\gamma}\sin^{-4}(\theta_0/2) \tag{3-199}$$

where θ_0 is the angle between the ion's motion v_0 and the normal to the surface. It was found in [74] for normal incidence of the ions that

$$\begin{aligned} E_{\text{th}} &= (4/3)^6 U_0/\gamma, \qquad M_1 \geq M_2 \\ E_{\text{th}} &= [(2M_1 + 2M_2)/(M_1 + 2M_2)]^6 U_0/\gamma, \qquad M_1 \leq M_2 \end{aligned} \tag{3-200}$$

The reduced threshold energy of sputtering E_{th}/U_0 changes rather markedly as the ratio M_1/M_2 changes, reaching values of about 50 when $M_1/M_2 = 5 \times 10^{-3}$ (Fig. 6 in [74]). When the target comprises several elements we can use the Bragg rule to find the binding energy U_0 of a surface atom(75):

$$U_0 = \sum_i C_{is} U_{0i} \tag{3-201}$$

where C_{is} are the atomic percentages of the elements in the surface monolayer and U_{0i} is the binding energy in the ith pure element.

By choosing one potential of atomic interaction or another, we can use the Monte Carlo method to model the processes of sputtering materials. The electron losses of the ions' energy are accounted for by means of the formulas for electron deceleration in an approximation of the continuous losses. The sputtering of uranium by uranium ions (self sputtering) was studied in [76] and the sputtering of copper, gold, copper–gold alloys by copper ions as well as TiC targets by Ar^+ ions having 3 keV of energy was studied in [75]. The calculations show that a large fraction of the sputtered atoms come from the first atomic layer; as the ion energy increases atoms are also emitted from the second and third monolayers. The average energy of the reflected primary U^+ ions is proportional to their initial energy: $\langle E_R \rangle = (R_E/R_N)E_0$, where R_N and R_E are the reflection coefficients for the ions relative to their number and their energy. When impingement is at the normal the function $R_N(E_0)$ has a maximum at $E = 300$ eV. Interestingly, the reflection coefficient for low-energy U^+ ions is negligible. The sputtering coefficient $S(E_0)$ for uranium is a weak function of the energy of the ions when $E_0 > 2$ keV; when $E_0 < 2$ keV, $S(E_0)$ falls off rapidly to small values. The average energy of the sputtered atoms depends on the ion energy E_0 roughly as $\langle E_S \rangle \sim E_0^{1/3}$. The sputtering coefficient S as a function of the angle of incidence can be satisfactorily mapped by a symmetric curve having a broad maximum near $\theta_0 = 60°$. The reflection coefficient for primary U^+ ions increases with the angle θ_0 and reaches a maximum at 80°.

We will now discuss the emission of electrons when solids are irradiated by ions. Secondary electrons are used in ion-beam systems to signal the alignment marks produced, e.g., as etched grooves. A review of the literature on electron emission during ion irradiation can be found in [77]. The fundamental law in this phenomenon is that the coefficient of electron emission γ (the number of electrons emitted per ion) is proportional to the amount of electron deceleration at the initial energy of the primary ion:

$$\gamma = \beta N S_e(E_0) \tag{3-202}$$

where N is the number of atoms per unit volume of the target. The ratio $\gamma/S_e(E_0)$ for metallic targets is constant over a very wide range of energies E_0 from 10^3 to 10^7 eV.[78] This relationship is easy to substantiate. Indeed, on a portion dx of its path an ion loses an amount of energy $dE = S_e(E_x, x)\,dx$ in exciting the electrons of the target atoms. The number of electrons that are produced on the path interval dx will be $dn = dE/J$, where J is the average energy expended in producing a single electron–hole pair. Because the probability that the electron flux will decrease along the path x to the target surface is $\exp(-x/L)$, where L is the attenuation distance (nearly equal to the mean free path of the electrons λ) we can write

$$\gamma = \frac{P}{2J}\int_0^\infty S_e(E_x, x)\exp(-x/L)\,dx \tag{3-203}$$

for the number of electrons emitted by the interaction with one ion.[79] The coefficient P statistically considers the electrons' distribution over energy and the direction of motion as well as the process by which they overcome the surface energy barrier. Since L is small we can say that $S_e(E_x, x) = S_e(E_0)$ and write

$$\gamma = \frac{P \cdot L}{2J} S_e(E_0) \equiv \beta N S_e(E_0) \tag{3-204}$$

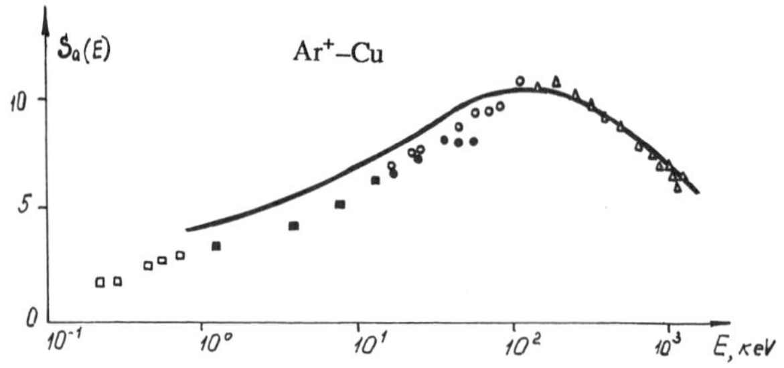

FIGURE 3-28. Energy dependence of the sputtering coefficient $S_a(E)$ for sputtering copper with argon ions.

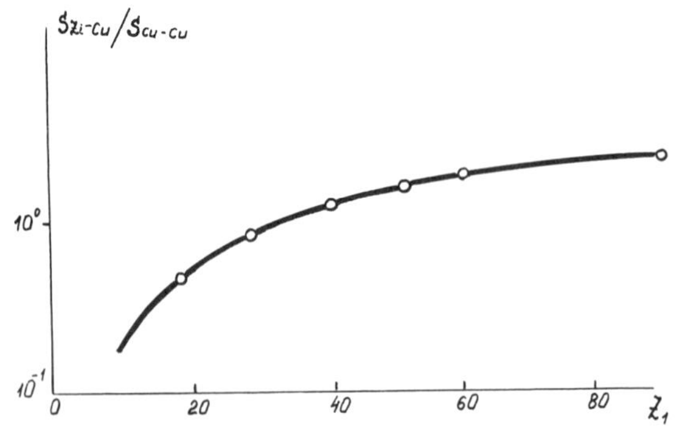

FIGURE 3-29. Ratio of the sputtering coefficient for polycrystalline copper for different elements Z_1 to the sputtering coefficient for copper ions.

Without going into further analysis of the formula, we will recall that the relationships for the change in $S_e(E_0)$ with energy and as function of the ion and target parameters were discussed in Sections 3-16. The γ may reach values of 1–5 at ion energies of 10^4 eV and depend rather strongly on the status of the target surface.[77]

Figure 3-28 shows measurements of the sputtering coefficient from sputtering polycrystalline copper with argon ions. As is evident, the function $S_a(E)$ passes through a maximum near 100 keV. At this level of energy the projected range of the ions is roughly 100 nm, i.e., the material being sputtered experiences a high degree of doping. We try to avoid doping the materials being etched by an ion beam with the sputtered material during microlithography, and for this reason etching is done with an ion beam having an energy of only a few hundred eV. Obviously, etching efficiency is insignificant when this is done. We can suggest a procedure for optimizing the ion etching process according to the materials used in the mask and the substrate: the energy of the etching ions must be chosen so that it is close to the energy at which a maximum is achieved in the sputtering coefficient, whereas the mask material must be chosen so that the energy of the etching beam is far from the maximum of the etching coefficient for the mask. We repeat that such optimization is possible if the number of etching ions implanted to a depth $R_p(E)$ in a layer of the structural material is insignificant.

Figure 3-29 shows the etching coefficient as a function of the atomic number of the etching ions. For a given beam energy the sputtering coefficient increases fairly rapidly in the range Z_1 = 10–20 and remains almost constant in the range Z_1 = 40–80. The experimentally observed dependence of $S_a(Z_1)$ can also be explained on the basis of Eq. (3-149).

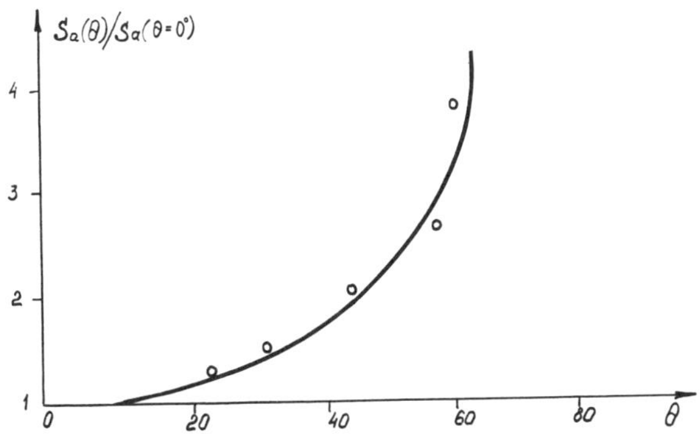

FIGURE 3-30. The sputtering coefficient $S_a(\theta)$ for polycrystalline copper for Ar^+ ions as a function of the angle θ between the direction of the beam and the normal to the target surface. (Solid line—theory; points—experiment).

The avalanche theory of sputtering also explains in what regard the sputtering coefficient depends on the beam's angle of deviation from the normal to the surface (see Fig. 3-30).

The angular dependence of the etching coefficient is, obviously, contained in the function for the density of the energy dissipated into atom motion $F_\nu(E, \mathbf{e}, 0)$ in Eq. (3-186). Sigmund's analysis of this dependence[69] reveals a close correspondence to the experimental dependence of $S_a(\theta)$, which is only slightly sensitive to the energy of the ions and the ratio of the masses and can be approximated by $(\cos\theta)^{-5/3}$. When $m_1 \ll m_2$ however, $S_a(\theta)$ is closer to a dependence of the $(\cos\theta)^{-1}$ type.

Finally, we can subject the predicted energy spectrum of the atoms being sputtered to experimental verification. According to Eq. (3-154), the dependence that we would expect from the avalanche theory for the energy spectrum of the atoms being sputtered is $dI_a/dE_0 \sim E_0^{-2}$ when $E_0 \ll E$. This result closely approximates the particles' energy spectrum seen in experiments (see Fig. 3-31).[55]

Everything we have said relates basically to amorphous or polycrystalline targets. Researchers have devoted a great deal of attention to studying how a crystalline target structure affects the process of ion sputtering a monocrystalline material. In this case the angular sputtering of the target atoms being sputtered is correlated with the direction of the crystallographic axes that intersect the plane of the target surface. These correlations are explained by (a) the channeling effect and by (b) the increase in the possibility of transmitting a pulse of energy over relatively long distances along chains of elastically coupled atoms (similar to the transmission of momentum along a chain of elastic spheres that abut one another). We refer the reader to Pleshivtsev,[68] Mashkova and Molchanov,[80] Veksler,[81] and Kumakhov and Shirmer[12] for a thorough discussion of the theory and experiments of these phenomena.

3-30. TOPICS IN THE CREATION AND CONTROL OF ION BEAMS IN LITHOGRAPHY MACHINES

The velocities of the particles in ion and electron beams having the same amount of energy are different: $v_{\text{ion}} = v_{\text{el}}(m/m_i)^{1/2}$. We will briefly examine a number of topics in ion optics that are related to this difference in velocities.

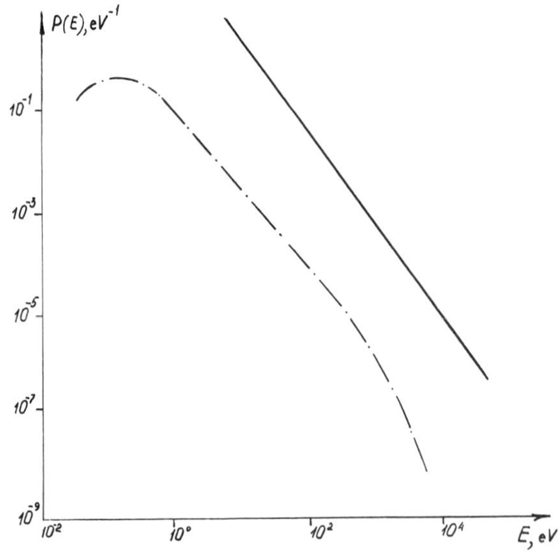

FIGURE 3-31. Energy spectrum for the atoms of polycrystalline gold sputtered with Xe^+ ions having E = 45 keV of energy at an angle of 45° from the normal to the surface.

We examined the principles of operation behind electrostatic and magnetic lenses in Chapter 1 and established that the focal length of an electrostatic lens is independent of the mass of the particles being focused, whereas the focal length of a magnetic lens is inversely proportional to the mass of the beam particles (see Chapter 1). For this reason the focal length of a magnetic lens when its geometry, the current in its windings, and the beam energy are given is m_i/m_e times greater for an ion beam than for an electron beam. From here it follows that the use of magnetic optics in ion-beam systems is impractical. The optics of electrostatic lenses, for which there is no relationship between focal length and particle mass, is the only possibility. The parameters of electrostatic lenses are defined by the geometry and the potentials of the electrodes that make up the lens. Obviously, the construction of complex, multielectrode electrostatic lenses is a more complex engineering problem than is constructing magnetic lenses from many (primary and correcting) windings. In electrostatic lenses the electrodes must be precisely fabricated and mounted and the geometry of the electrodes maintained throughout the operating period. Electric breakdown in both the atmosphere of residual gases as well as in the dielectric materials used in the structure of an electrostatic lens are serious problems. The appearance of breakdown effects may enhance the contamination of the electrodes and the surfaces of the dielectric details while the device is operating.

The use of a magnetic deflection system to control (scan) the ion beam is also a limitation. The sensitivity of magnetic and electric deflection systems is given by Eqs. (1-75) and (1-76), respectively. Under strict, equal conditions the deflection of an ion beam by a magnetic winding is a factor of (m_i/m_e) less than that for an electron beam, whereas in an electrostatic system the deflection is the same for both ion and electron beams. The advantages of an electric deflection system disappear as the energy of the ions increases; however, at beam energies of $E \leq 10^5$ eV electric method is, evidently, the method of choice.

The velocity of the ions in the control system limits the frequency at which the beam can be addressed to different points on the target. In fact, if Δz is the length of the region in which a deflection field has been created, the time it takes an ion to travel through the deflecting region is

$$t_i = \Delta z / v_i$$

We will say that $\Delta z = 3$ cm and $v_i = 3 \times 10^6$ cm/sec; then $t_i = 10^{-6}$ sec. If we need a fivefold reserve of time for readdressing relative to the travel time t_i, including the exposure time, then the readdressing frequency cannot be greater than 200 kHz. Obviously, for low-energy beams this is a serious limitation on the speed of an ion-beam system. The formula for the deflection of an ion in an electric system is

$$Y_E = \frac{E_y \, \Delta z \, L}{2\varphi}$$

which suggests ways of making an ion's travel time through the control field shorter; i.e., reduce Δz and at the same time increase the voltage of the control field E_y and the distance L between the surface being exposed and the deflection system.

The readdressing frequency can be made an insignificant limitation if we stop using a Gaussian beam of submicron cross section to expose a target, which was carried over into beam lithography from electron and ion scanning microscopy, and instead use shaped spots of particles.[83,84] In creating shaped spots the throughput of the system may increase by about twenty-five times, which corresponds to a 25-fold increase in the addressing frequency over a Gaussian beam.

The low velocity of ions in comparison with that of electrons manifests itself in yet another phenomenon—an increase in the energy spread of longitudinal ion motion in a beam (the Boersch effect). The method of the kinetic equation, which considers Coulomb collisions in a beam of particles (electrons, protons) was used in [84, 85] to find the mean-squared value of the spread in the energy of a beam after it had passed through an optical system:

$$\langle \Delta E^2 \rangle \sim r_0 \beta^{1/2} / (E/m)^{1/2} \qquad (3\text{-}205)$$

here r_0 is the beam radius at a crossover, β is the brightness of the source, and E and m are the energy and mass of the beam particles.

Because $v_z = (2E/m)^{1/2}$ we can write

$$\langle \Delta E^2 \rangle \sim r_0 \beta^{1/2} / v_z$$

instead of Eq. (3-205), i.e., the mean-squared increase in the energy spectra of the particles in a beam is inversely proportional to the velocity at which the particles pass through the optical system.

When the energy of an ion beam and an electron beam is the same, when the brightness of the sources and the radii of the crossovers is the same, the width of the energy spectrum for a beam of ions $\langle \Delta E^2 \rangle^{1/2}$ is $(m_i/m_e)^{1/4}$ times greater than that of an electron beam. The chromatic diameter of the spot, $d_c = C_{c\infty} \theta (\Delta E/E)$, also increases accordingly.

TABLE 3-6. Applications of Focused Ion Beams in Technology

Technology using a focused ion beam	Elements used	Ion energy, keV	Ion flux dose, cm^{-2}
Exposing polymer resists			
a) Wet developing	H, He, Be, B, Si	50–200	10^{-8}–10^{-5} C
b) Dry developing	Ga, Si		10^{-4}–10^{-2} C
Exposing inorganic materials	N, Si, Ga	50–200	10^{14} ions
Maskless etching			
a) Sputtering	Ar, Ga	1–10	10^{-3}–1 C
b) Ion–chemical etching	F, Cl, O	1–10	10^{-6}–10^{-3} C
Maskless ion implantation	B, Sb, As, P, Si, Be, Zn	30–200	10^{11}–10^{16} ions
Maskless deposition		1–5	10^{16} ions
Scanning ion microscopy	Ga, O, Al	10–50	10^{11}–10^{12} ions
Altering the chemistry of a surface layer	All ions		10^{18} ions

3-31. FOCUSED-BEAM ION SYSTEMS

Three types of ion-beam lithography machines have been constructed and are presently being used for research(1,86): 1) machines with a thin, focused ion probe (FIBL machines), 2) projection machines that produce an ion-beam image of a mask with slits (IPL machines), and 3) machines for shadow printing mask images in a scheme with a gap (MIBL machines).

Table 3-6 shows some applications of machines with a thin focused ion probe. The opportunities for widespread usage of the enumerated processes is governed by the dose of ion flux needed to obtain the desired result. The smallest doses are required for exposing polymer resists for wet developing (10^{-8}–10^{-5} C/cm^2). Interesting possibilities are being discovered when using dielectric inorganic films, e.g., SiO_2, as resists for ion-beam exposure.(5,87–92) The wet etching rate for films exposed to ion radiation doses of $\sim 2 \times 10^{14}$ ions/cm^2 increases by a factor of 3–5 in comparison with the etching rate in a film that has not been irradiated. As we can see, the sensitivity of dielectric inorganic films is roughly 1% that of polymer resists, which makes using them a problematic situation. In equal measure, the widespread use of sputter etching, ion implantation, film deposition, and altering the chemistry of the surface layer of a substrate by means of highly focused beams is limited because of the low throughput of these processes. Nevertheless, the possibility of using all of the processes shown in Table 3-6 to obtain devices having submicron dimensions (0.25–0.5 μm [1, 86, 92–104]) has been demonstrated under experimental conditions.

Figure 3-32 shows a diagram of a lithography machine that uses a highly focused ion probe.(86) The characteristics of a triode ion gun with a liquid metal ion source were thoroughly studied in [94]. The geometric parameters of the gun are its aperture diameter D of the grid electrode and the length H to which the emitter tip emerges from this aperture (Fig. 8 in [94]). Other parameters are the source temperature, the bias voltage V_g on the grid electrode (0–2 keV) and the accelerating voltage V_a (20–50 keV). At specific values of gun current i (= 5 μA) the voltages V_g and V_a can be varied between wide limits according to the relation $V_a = V_0 + kV_g$, $k > 0$, $V_g < 0$. The values of the constants V_0 and k decrease with increasing H. The mean-squared deviation in the ion energy $\Delta E_{1/2}$ increases (in a roughly linear fashion) from 5 eV to 30 eV when the ion current increases from 1 μA to 30 μA. The magnitude of $\Delta E_{1/2}$ is greater as the temperature of the liquid metal is higher. A nonlinear increase (with saturation) in the

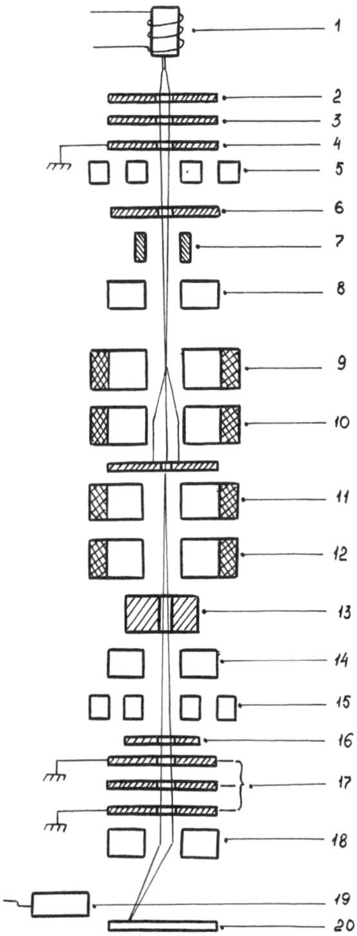

FIGURE 3-32. Schematic of a scanning ion lithography machine that uses a focused beam of ions: 1) liquid metal electrode with indirect heating; 2) extractor; 3, 4) first and second accelerating electrodes; 5) octupole stigmator; 6) gun aperture; 7) deflecting plates in the alignment system; 8) deflection blanking plates; 9, 10, 11, 12) magnets for the ion filter; 13) beam current sensor (set off the system axis); 14, 15) octupole stigmator; 16) lens aperture; 17) single lens for producing a focused ion probe; 18) deflection plates for the scanning system; 19) secondary electrons detector for the alignment system; 20) wafer being exposed.

mean-squared value of the ions' angular distribution is also seen: from 0.3 radian at small currents of about 1 μA up to 0.8 radian for currents of 50 μA. The diameter of the focused ion probe is defined by the input aperture, and the chromatic and spherical aberrations of the lens. The properties of the focusing optical systems in ion-beam machines can be studied via modelling with the Monte Carlo method.[105]

If the spread in ion energies is fairly large we may, in a first approximation, say that the beam diameter in systems with a focused beam is equal to the diameter of the chromatic aberration disc of a point image. We will examine a simplified focusing system that comprises a liquid metal source, an extractor electrode at a potential V_g, and an accelerating electrode at a potential V_a (Fig. 3-33).[31] From the Helmholtz–Lagrange theorem for the linear magnification in a system,

$$M = (\alpha_0/\alpha_i)(V_g/V_a)^{1/2} \tag{3-206}$$

V_g and V_a are potentials in the object and image spaces, respectively, and α_0 and α_i are the angular dimensions of a beam in these spaces. The ion probe current I is

$$I = \pi\alpha_0^2 J_\Omega^0 = \pi\alpha_i^2 J_\Omega^i \tag{3-207}$$

where J_Ω^0 and J_Ω^i are the angular intensities of the source and relate to the source and image planes. According to Eq. (3-206), J_Ω^0 and J_Ω^i are related by

$$J_\Omega^i = J_\Omega^0 M^2(V_a/V_g)$$

The diameter of the chromatic disc will be[31]

$$d_c = 2MC_c\alpha_0(\Delta V/V_g) = 2MC_c(V_a/V_g)\alpha_0(\Delta V/V_a) \tag{3-208}$$

As we can see from Eq. (3-208), the disk diameter can be given in terms of the spread in ion energies with respect to the beam energy in the source (object) or image plane. Combining the formulas presented we can write

$$I = \frac{\pi d_c^2 V_a^2}{4M^2C_c^2(V_a/V_g)^2} \frac{J_\Omega^0}{(\Delta V)^2} \tag{3-209}$$

for the ion probe current. Having set the beam diameter d_c (e.g., $d_c = 0.1\ \mu$m) we wish to obtain the largest possible probe current. To do this, the structure of the system must ensure that the quantities in the denominator of Eq. (3-209) are as small as possible. Typical values of the optical parameters are $C_c = 25$ mm, $(V_a/V_g) = 10$ and $M = 1$. It is interesting to examine the behavior of the so-called chromatic angular intensity $J_\Omega^0/(\Delta V)^2$. From measurements taken on a gallium liquid metal source, the function $J_e^0/\Delta V$ as a function of the source current i has a sharp maximum at a current of about 2 μA.[31] As the current increases beyond i the angular divergence of the beam at the source increases rapidly (J_Ω^0 falls off); at the same time the energy spectrum of the ions increases rapidly (due to the Boersch effect). This behavior is, evidently, typical of all liquid-metal sources. Thus, the operating mode of a source that corresponds to a maximum in the chromatic angular intensity will be optimal.

In general, we must examine the diameter of the probe in which the contributions from a Gaussian image of the source are considered (d_g is the geometric diameter of the image of the virtual source) and the diameter d_s of the disc caused by spherical aberration:

$$d = (d_g^2 + d_s^2 + d_c^2)^{1/2} \tag{3-210}$$

The contributions made by spherical and chromatic aberrations to the diameter of the scattering disk become comparable when ion sources having small relative energy spreads are used. Typical values of $\Delta V/V$ for liquid metal, duoplasmatron, and surface ionization sources are 10^{-3}, 5×10^{-4}, and 2×10^{-5}, respectively.[38]

The coefficients of spherical and chromatic aberrations, C_s and C_c, fall off as the focal length of the lens is decreased. However, in practical systems the focal length is limited from below by the structural requirements on arranging the deflection system, the secondary electrons detector, and the devices for observing and changing the wafers being exposed. Therefore, it becomes necessary to minimize the probe diameter (Eq. (3-210)) for a given value of the working gap d_w. The possibility of this kind of optimization

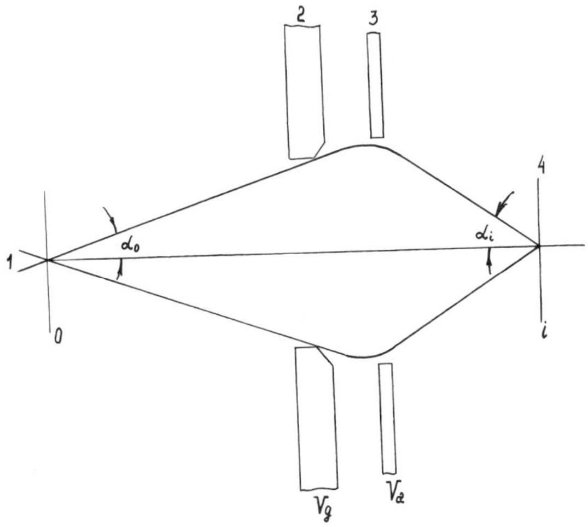

FIGURE 3-33. Simplified diagram of a focusing ion-beam machine: 1) liquid-metal ion source; 2) electrode-extractor at a potential V_g; 3) accelerating electrode at a potential V_0; 4) target (wafer being exposed).

has been shown in work on single electrostatic lenses. An optimum is achieved when, roughly, $d_s = d_c$, i.e., when

$$(C_s/C_c)_{\text{opt}} = 2(\Delta V/V)/\alpha^2 \tag{3-211}$$

In practice, single-lens optical systems are awkward. Their magnification $M = d_g/d_0$ (d_0 is the diameter of the virtual source and d_g is the geometric diameter of its image) is given in terms of the distances between the source and the lens a and between the lens and the image b (Fig. 3-34a)

$$M = \frac{b}{a}\sqrt{\frac{V_1}{V_2}} \tag{3-212}$$

where V_1 and V_2 are potentials in the space of the source and its image. For a single, symmetric lens, $V_1 = V_2$. Obviously, changing the magnification requires that the source or the subject being exposed be moved relative to the lens. When a large amount of reduction is needed (($M \ll 1$) and the working gap $d_w = b$ has been specified by the requirements on moving the auxiliary systems (see above) then the gap a must be increased drastically. And last: at a fixed beam aperture the probe current can only be changed by changing the operating mode of the gun.[36]

A two-lens system (Fig. 3-34b) is described by four gaps a_1, b_1, a_2, b_2; and $c = b_1 + a_2$ is the distance between lenses. In this case the magnification is

$$M = \frac{b_2/a_1}{c(b_1 - 1)} \tag{3-213}$$

As we can see, the magnification M in a two-lens system can be controlled by electrical means: namely by changing the distance b_1 between the condenser lens and the intermediate image d_1 (the distances a_1 and b_1 are fixed). The magnification M can be varied over a wide range of values, right up to strong reduction, for small overall system dimensions since c/b_1 can be made quite large.[36]

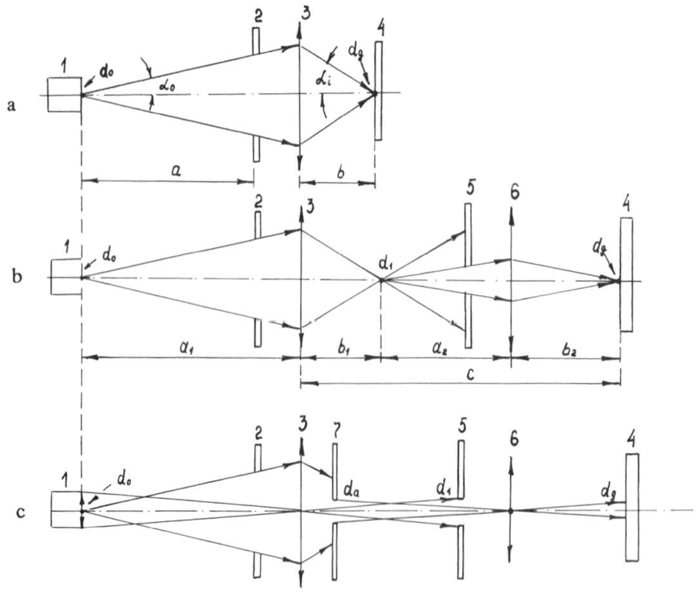

FIGURE 3-34. a) A single-lens system that creates an image of a virtual source of ions on a target (d_0, d_g are the diameters of the source and its image); b) a two-lens system (d_0, d_2, d_g are the diameters of the virtual source, the intermediate image, and the image on the target); c) a two-lens system that uses Kohler illumination: an aperture of diameter d_a is imaged on the target by a condenser lens. 1) Ion source; 2, 5, 7) aperture diaphragms for the lenses; 3, 6) lenses; 4) target.

Just as in electron-beam systems, it is convenient to use Kohler illumination in ion-beam systems.[36] It is well known that with this type of illumination an image of the source is produced by the condenser lens at the input aperture of the objective lens, and the objective lens projects the aperture of the condenser lens onto the surface of the target being exposed (Fig. 3-34c). The probe diameter at the target is defined by the diameter of the condenser lens aperture:

$$d_g = d_a b_2 / c \tag{3-214}$$

According to Eq. (3-214), the probe diameter is independent of the size and shape of the source and the parameters of the source can be optimized to maximize the probe current and its uniformity. If the source has a Gaussian current distribution cross section, the aperture of the objective lens in a Kohler scheme extracts the central portion from this distribution (on condition, of course, that the image of the source at the aperture is much larger than the size of the aperture).

Mass separators are built into ion-beam systems to extract ions of a particular type from the probe current. The optical properties of a system with a mass separator, deflection devices, and secondary radiation detectors are described in [36]. In these cases, systems that create an image of the source on the target being exposed and systems using Kohler illumination become multilens systems.

The intensity of secondary radiation (electrons, ions, photons) is proportional to the probe current, Eq. (3-209). Luminous radiation can be used for microscope observation of that portion of a target surface being exposed by ions in situ. To improve the efficiency of collecting secondary charged particles, they can be accelerated when they pass back through the objective lens. It is important to note that secondary electrons can be used to find alignment marks fabricated, e.g., as etching grooves. Modelling the emission of secondary electrons by the Monte Carlo method has shown that when irradiating GaAs with an ion probe of Si^+ having 160 keV of energy and a diameter of $d = 0.2\ \mu m$, the maximal contrast to secondary electrons occurs when the slope of a lateral wall is $\theta = 85°$ and the depth of the channel is $h = 2\ \mu m$.[107] The alignment accuracy is

$$\delta = w/(S/N) \tag{3-215}$$

where w is the half-width of the secondary electrons peak, and S/N is the signal-to-noise ratio.[107] Calculations according to the Monte Carlo method and experiments show that for a given wall slope θ in a groove the dependence of S/N on the depth of the groove is given by curves with a maximum: $(S/N)_{max} \simeq 10$ occurs at $\theta = 54.7°$ and $h = 0.2\ \mu m$, and when $\theta = 85°$, $h = 2\ \mu m$, $(S/N)_{max} \simeq 30$. The alignment accuracy obtained by secondary electrons in actual situations is $0.1\ \mu m$.[107]

3-32. PROJECTION ION-BEAM SYSTEMS

Ion projectors operate on the same principles as do optical projection steppers: a $10\times$ reduced image of a stencil is created on the surface of the wafer being exposed and the exposure on each circuit (chip) is done according to the principle of align and expose. There are reports[108–111] of several types of improved machines. A very thorough review of the questions involved in using projection methods of lithomicro machining was given in [1]. Our discussion will follow these sources.

The preferred ion source in the ion projectors to be discussed is the duoplasmatron. The diameter of the anode aperture in the source is 250 μm; considering angular divergence, the diameter of the virtual source is 50 μm. The source generates a flux of ions from H^+, He^+, N^+, Ne^+, Ar^+, and Xe^+ gases. The ion current reaches magnitudes of 200 μA and the beam divergence angle is 4°. The source brightness is as much as $10^4\ A \cdot cm^{-2} \cdot steradian^{-1}$. The energy of ions emitted from the source can be varied over a range of 5–10 keV.

Figure 3-35 shows the optical scheme of the projector. The mask is placed in the region outside the field and is a membrane with slits (a stencil mask). The technology of masks with slits is capable of producing masks with minimal dimensions of 2 μm, which produces a minimal pattern dimension of 0.2 μm (after $10\times$ reduction) on the wafer surface. A flux of ions having as much as 10 keV of energy irradiates a mask 30 mm in diameter with a current of up to 70 μA; the power density is 50 mW/cm^2 under these conditions. To avoid thermal deformation caused by mask heating, the mask is stretched on a frame. As the mask is heated tension is lost, but the mask retains its planar form.

The immersion two-electrode lens L_1 accelerates the ions to their final energy of 60–90 keV. The optical elements that ensure precise electronic alignment of the mask image with the pattern on the wafer are located outside the field, between the immersion accelerating lens L_1 and the objective lens L_2. The octupole electrostatic lens moves the mask image $\pm 1\ \mu m$ in the x and y directions; the axial magnetic field rotates the image. In addition, if nonalignment between the mask image and wafer pattern scales is detected the scale can be changed by varying the voltage on the projection lens.

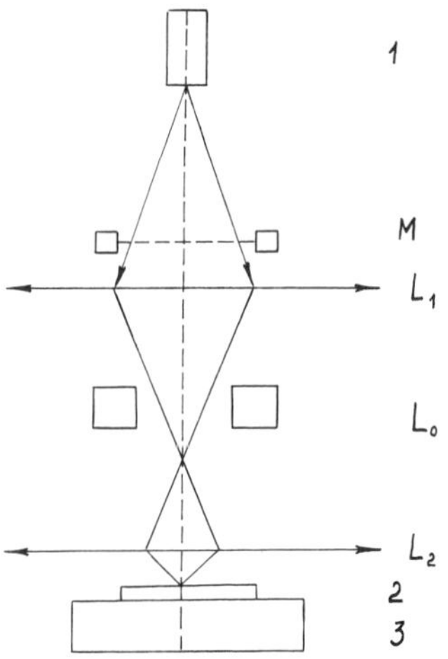

FIGURE 3-35. Simplified diagram of the optical system in an ion projection lithography machine: 1) Ion source; M is a mask with slits; L_1 is an immersion accelerating lens; L_0 is an octupole deflection lens for the alignment system; L_2 is the objective lens; 2) wafer being exposed; 3) controllable table.

Alignment detection is accomplished by special slits in the mask. Ion beams pass through these slits to the wafer surface and seek the alignment marks. The presence or absence of alignment is recorded by sensors that detect particles that have been backscattered by the alignment marks on the wafer (V-shaped grooves). Channeltrons serve as the particle detectors. Coarse alignment to within ± 1 μm is done by the system responsible for moving and positioning the table. The accuracy of the electronic fine alignment is better than 0.05 μm.

The projection electrostatic lens L_2 has a focal length of about 100 mm and creates a 10× reduced image of the mask on the wafer. The numerical aperture of the optical system is small and corresponds to a paraxial approximation, which reduces the influence of aberrations. The depth of focus is more than 100 μm for a projection lens. The optical system provides resolution of 0.1 μm in a field 3 mm in diameter.

If a reticle is used in the capacity of a mask, the number of pores will be equal to the number of beams by which the wafer will be scanned simultaneously when a deflecting voltage is applied to the octupole lens. The multibeam scanning mode can be used to fabricate submicron grids, devices on surface acoustic waves and quantum wells, integrated optical devices, and in graphoepitaxy.

Figure 3-36 shows graphs of wafer exposure times as functions of the required doses (in ions/cm^2 or C/cm^2) for three different ion current densities. The limits of the ion doses required for different technological processes are shown in Table 3-7. For a current density of 1 mA/cm^2, which is entirely within reach (thanks to the 10× reduction, the current on the wafer increases by a factor of 100 in comparison with the current density on the mask), the exposure time for polymer resists is a few milliseconds and is several tens of seconds for inorganic resists. At a current density of 1 mA/cm^2 and ion energy of 50 keV the power density is 50 W/cm^2. Prolonged exposure at this level of power will lead to rapid heating in the target. Heating can only be avoided by lowering the power density, i.e., the ion current density (for a given level of ion energy).

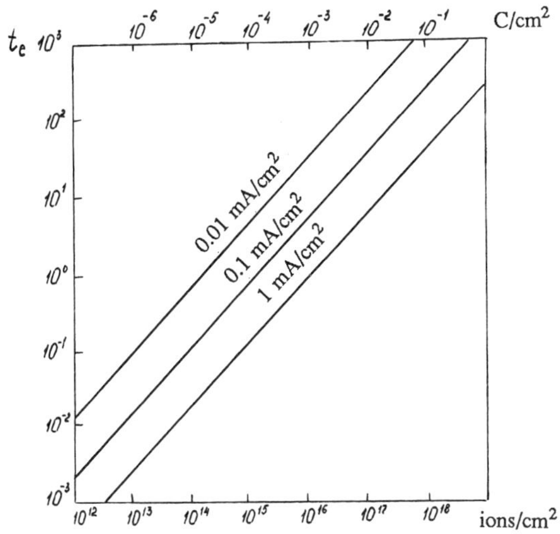

FIGURE 3-36. Exposure time as a function of exposure dose required on the wafer. The lower abscissa is the dose in photons/cm^2 and the upper abscissa is the dose in C/cm^2. The graphs show three different values of ion dose on the target.

TABLE 3-7. Dosages and Exposure Times Needed to Perform Different Tasks on an Ion Projection Machine (the current density in the target is measured in mA/cm^2)

Application for the ion projection machine	Dose, ions/cm^2	Exposure time, sec
Exposing organic resists	10^{12}–10^{14}	10^{-3}–10^{-2}
Exposing inorganic resists	10^{14}–10^{16}	10^{-2}–10^{0}
Implantation	10^{12}–10^{16}	10^{-3}–10^{0}
Ion beam mixing	10^{14}–10^{16}	10^{-2}–10^{0}
Ion etching	10^{15}–10^{18}	10^{-1}–10^{2}
Ion-actuated etching	10^{15}–10^{18}	10^{-1}–10^{2}
Altering the physical and chemical properties of the surface layer of materials	10^{12}–10^{18}	10^{3}–10^{2}

3-33. ION-BEAM SYSTEMS FOR PROXIMITY PRINTING

A prototype align-and-expose ion-beam system (Fig. 3-37) has been described in [112]. The machine is designed for the shadow transfer of a mask image onto a wafer in a 1:1 scale when the gap between mask and wafer is about 25 μm. The overall scheme of the machine is identical to that used in optical or x-ray machines for exposing through a mask and across a gap. The optical system of the machine has been designed to produce a highly collimated beam of protons having as much as 300 keV of energy. The beam divergence is $\leq 0.1°$, the size of the beam is ≥ 2 cm^2, and the uniformity of the ion flux in the beam is better than 95%. Since the mask is irradiated by a beam of high-energy ions, the requirement for a low degree of mask heating during the exposure period limits the current density to a magnitude of $i = 0.5$–1 μA/cm^2.

A mass separator that screens out ions other than protons is placed between the protons source and the collimating optics. This reduces heating and the generation of defects in the mask.

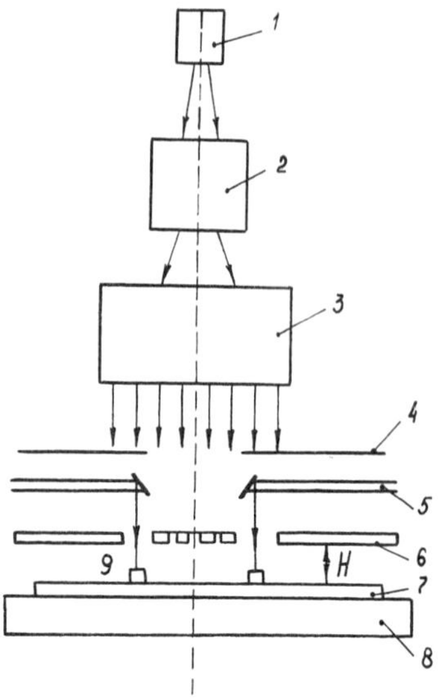

FIGURE 3-37. Simplified diagram of an ion-beam proximity shadow printing system: 1) ion source; 2) mass separator; 3) collimator; 4) diaphragm; 5) wafer with mirrors for the laser alignment system; 6) stencil mask with auxiliary slits for the alignment system; 7) wafer with a resist; 8) controllable table; 9) alignment marks. The gap between mask and wafer is $H = 25\ \mu m$.

The system elements have been designed to print minimal dimensions of 0.5 μm on the mask to within 0.1 μm. It follows from these conditions that the angular divergence of the beam after passing through the mask membrane is not greater than 0.3°. This requirement is met by channeling silicon masks having a 0.5 μm thick membrane at a proton energy of 200 keV. An angular beam divergence of $\leq 0.3°$ ensures that the width of an exposure line edge is 0.1 μm when the gap between the mask and the wafer is 25 μm. The conditions for channeling the collimated beam are attained by varying the inclination of the beam (fine tuning the optical system) without moving the mask. The "dark" areas of the mask are the thicker areas (all-silicon masks) and the contrast in the masks is 4:1. Thermal changes (distortion) in the mask will be less than 0.1 μm if the density of the power absorbed by the mask during the exposure period does not exceed 0.1 mW/cm^2 and the mask area is less than 1.6 cm^2. In this fashion, the chip area is limited to being no larger than 12.5×12.5 mm^2.

The laser alignment system (an interferometer on a grating) provides an alignment accuracy of 0.1 μm. Alignment is achieved by moving the table, including rotation, and is insensitive to changes in the distance between the mask and the wafer and to changes in the reflectance of the wafer. The system's detection sensitivity is ± 20 nm.

The machine throughput can reach 60 wafers per hour for a diameter of 100 mm and 40 exposure fields (chips). To accomplish this, the work pace must be: 0.5 sec to expose the chip, 0.5 sec to move the table, and 0.4 sec for alignment. The resists used must have a sensitivity of $\leq 2 \times 10^{12}$ protons/cm^2 in order to ensure an exposure time of 0.5 sec. A large number of positive and negative resists have this kind of sensitivity.[(112)]

When the protons have 200 keV of energy, the thickness of the resist layer can typically be about 1 μm. The ranges of the secondary electrons in the resist are not greater than 5 nm and the proximity effect is practically nonexistent. Periodic structures with 0.08 μm spacing obtained by exposure to a beam of protons through shadow masks have been demonstrated.[(113)]

Results obtained from experiments in off-contact shadow printing for stencil and grid masks placed in a proton flux are described in [114, 115]. For both mask types the slits in the membrane may be obtained by etching with reactive ions. The geometry of the pattern for stencil masks is limited by the requirement that there be no portions that fall away; this limitation does not exist for grid masks since the grids are created within those portions of the mask that correspond to the slits in a stencil mask. The mask must be rocked during the exposure period in order that the grid image not be transferred to the wafer. The presence of a grid in the slits reduces the ion flux and extends the exposure period; the noise due to scattered ions also increases. The contrast for stencil masks in ion lithography was modelled in [116] by the Monte Carlo method.

The changes in the width of a developed line in a resist as a function of the dose of protons having 50 keV of energy and exposing a PMMA resist was studied in [114]. For doses in the range of $D = 2 \times 10^{13}$–2×10^{14} ions/cm^2, the change in line width is

$$L = 0.8662 + 0.04063 \log_{10} D \tag{3-216}$$

for a 0.65 μm wide slit in the mask. According to Eq. (3-216), the width of the developed line varies slightly with dose: by 10% when the dose was changed by a factor of ten. This increase in line width may be caused by background irradiation from the ions scattered by the mask, the angular divergence of the beam, and irradiation by secondary electrons. A much stronger relationship between the width of the developed line and dose is seen in x-ray and optical lithography.

We must briefly discuss the question of mask degradation caused by prolonged usage. Degradation caused by ion sputtering and swelling when ions are absorbed by (implanted in) the mask are obvious. The coefficient of ion sputtering for silicon by protons having $\sim$1 keV of energy does not exceed 0.1. When the resist sensitivity is 2×10^{13} ions/cm^2 a single exposure removes 2×10^{12} Si atoms from 1 cm^2 and one atomic layer will be sputtered when 500 chips have been exposed. This sputtering rate can, evidently, be considered weak. Swelling in the material is seen when the density of implanted impurities is $\sim 10^{20}$; this means that changes in the dimensions of the slits in a mask due to swelling will be seen after 10^7 exposures.

3-34. THE SENSITIVITY AND RESOLUTION OF HIGH-MOLECULAR-WEIGHT ORGANIC RESISTS IN ION BEAM LITHOGRAPHY

Ions ranging in initial energy from a few tens to several hundred keV are used in practically every published work dealing with ion-beam lithography. These particles are decelerated in a material due to the energy lost in electron and nuclear collisions. Nuclear collisions prevail when the velocity of a particle is much less than the orbital velocity of a hydrogen atom's electron, ($v_H = e^2/\hbar = 2 \times 10^8$ cm $\cdot$ sec^{-1}). As the velocity of the impinging particle (v_1) increases, nuclear deceleration plays a decreasing role and the energy losses due to electron deceleration increase. Table 3-8 shows rounded values of the ratio v_1/v_H for a number of ions ranging in energy from 5 to 500 keV.

It is evident from this table that $v_1/v_H > 1$ occurs only for protons having $E_0 \geq$ 30 keV of energy, for a helium ion having $E_0 > 100$ keV of energy, and for a nitrogen ion having $E_0 \geq 500$ keV of energy. The range of energies 5–100 keV corresponds to the condition $v_1/v_H < 1$ for practically every ion except protons. Nuclear deceleration

TABLE 3-8. Ratio v_1/v_H for a Number of Ions Ranging in Energy E_0 = 5–500 keV

	v_1/v_H								
E_0, keV	^1H	^4He	^{11}B	^{14}N	^{20}Ne	^{31}P	^{40}Ar	^{70}Ga	^{84}Kr
5	0.44	0.22	0.14	0.12	0.10	0.08	0.07	0.05	0.05
10	0.63	0.32	0.19	0.16	0.14	0.11	0.10	0.075	0.07
50	1.40	0.70	0.42	0.37	0.32	0.25	0.22	0.16	0.15
100	1.99	0.98	0.60	0.53	0.44	0.36	0.32	0.23	0.22
500	4.38	2.19	1.44	1.19	0.99	0.79	0.70	0.53	0.48

prevails in a target bombarded by ions at this level of energy or when the amount of energy lost due to nuclear and electron deceleration is comparable.

All the modern high-molecular-weight resists used in electron-beam lithography can be used in ion-beam lithography.[64,117,118] As with an electron beam, positive resists have elevated dissolution rates during ion irradiation because of the disruption of polymer chains (reduction in the average molecular weight) and negative resists become insoluble due to the "knitting" of polymer chains. The processes of disrupting and "knitting" polymer chains when resists are ion irradiated is due to nuclear and electron collisions between the impinging particles and the polymer atoms. The more energy an impinging particle loses per unit path length in the resist during collision, the more numerous are the chemical acts of disrupting or knitting the polymer molecules in the irradiated region.

Figure 3-27 shows calculated curves of the energy lost by fast charged particles in a PMMA (polymethyl methacrylate) resist.[64] The solid lines correspond to the energy lost in inelastic (electron) collisions and the dashed lines to the energy lost in elastic (nuclear) collisions. We see from the figure that the amount of energy electrons having an initial energy of 20 keV lose is tens of times less than that lost by ions. In this context, it is natural to expect that the sensitivity of electron resists to ion irradiation will increase markedly. When this happens it is assumed that the radiation chemistry efficiency, i.e., the number of chemical acts of rupturing or knitting in the polymer chains per 100 eV of energy lost does not change significantly.

The fundamental characteristics of high-molecular-weight organic resists are sensitivity, contrast range, and resolution. These characteristics determine the throughput of a process and the limits on creating a microtopological pattern. In addition, those resist characteristics such as microdefects density, adhesion to other materials used in microelectronics and their chemical, plasmachemical, and thermal stability in the etching processes are important. The first three characteristics are highly dependent on how the radiation and the resist physically interact.

Just as we did for an electron resist, we will understand the sensitivity of an ion resist to be the minimal dose of radiation needed to produce a radiation chemistry conversion of a polymer throughout the thickness of the irradiated section as [119]. The minimal dose indicated is determined experimentally in the course of developing an irradiated resist according to the increase in the etching rate over the entire depth. However, different doses may produce a specific change in the dissolution rate, depending on the composition of the developer, the developing time, and the thickness of the resist. For this reason, comparing the sensitivity of a resist to the different types of radiation must be done for a prescribed thickness, exposure time, and developer composition.

Contrast is defined of an ion resist just as it was for an electron resist: the reciprocal

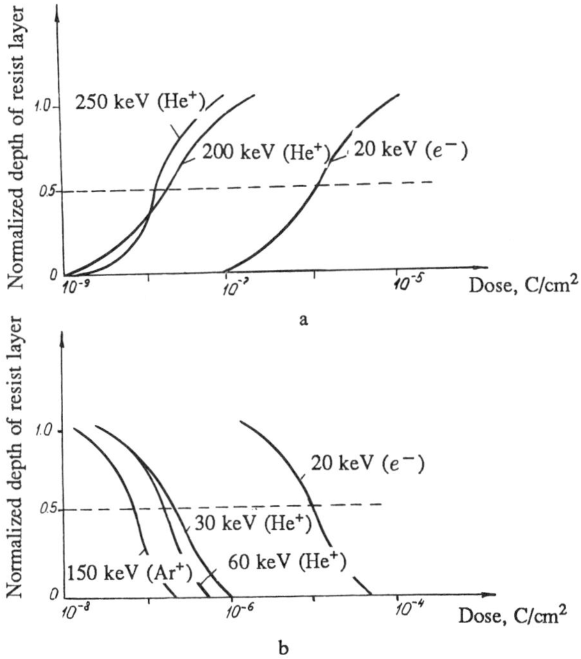

FIGURE 3-38. Characteristic curves for PDMS (a) and for PMMA (b) resists.

TABLE 3-9. Sensitivity of Some Positive Resists to Ion (120 keV) and (20 keV) Beams

Resist	Average molecular weight	Developer	Sensitivity, μC/cm² $^{1}H^+$	$^{40}Ar^+$	e^-
Polymethyl methacrylate	350,000	Methyl isobutyl ketone in isopropyl alcohol (1:1) 20°C	0.58	0.034	40–80
Polymethyl methacrylate + ethyl acetate	600,000	Isoamyl ethyl acetate 20°C	0.53	0.077	50
Polymethyl isobutyl ketone + anone	75,000	Butyl ethyl acetate 20°C	0.96	–	60

of the logarithm to base ten of the ratio of sensitivity to the threshold dose needed to initiate developing. We will examine the characteristics of resolution later, but for now we will turn our attention to analyzing the familiar experimental results.

Figure 3-38 shows the normalized residual thickness as a function of the radiation dose from electron and ion beams (a) is a negative PDMS (polydimethyl siloxane) resist and (b) is a positive PMMA resist.[64] It is apparent from the figure that these resists, e.g., at the point $D_{0.5}$, are tens of times more sensitive to ion irradiation than to electron irradiation, which is in good agreement with the change in the amount of energy lost for these types of radiation. Table 3-9 gives the sensitivity for a few positive resists when irradiated by 20 keV electrons, 120 keV protons, and argon ions.[117] The resist thickness when irradiated by the protons was 1 μm, and 0.2 μm when irradiated by Ar^+ ions. It is clear from the table that the resists may be hundreds of times more sensitive to ion irradiation than to electron irradiation.

Comparing the experimental and calculated relationships between the thickness of

TABLE 3-10. Radiation Chemistry Efficiency G (number of acts per 100 eV) When Irradiating PMMA and PDMS with Ions and Electrons

	Ions		
Resist	G_e	G_n	Electrons
PMMA	1.7	0.9	1.9
PDMS	2.4	0.4	2.2

the resist layer removed and its developing time for different radiation doses makes it possible to find the radiation chemistry efficiencies G_n and G_e that indicate the average number of chemical acts of disruption or knitting in polymer chains per 100 eV of energy lost during elastic (nuclear) and inelastic (electron) collisions, respectively. Thus, for example, the calculation was done for a positive resist according to formulas where the etching time and thickness of the layer removed are related to the change in the average molecular weight, which, in turn, depends on the quantities G_n and G_e, the energy losses in elastic and inelastic collisions, and the radiation dose. Three empirical parameters, found from the experimental relationships between the dissolution rate of the polymer and its average molecular weight, are present in these formulas.

The G_n and G_e calculated for PMMA and PDMS resists subjected to irradiation by $^4He^+$ and $^{40}He^+$ ions having different initial energies showed that the indicated radiation chemistry efficiencies are independent of the type of ions and its initial energy.[(64)] For comparison, Table 3-10 gives the results and the radiation chemistry efficiencies when the resists were irradiated by an electron beam.

We can see that G, G_n, and G_e are approximately equal. The closeness of these radiation chemistry efficiencies in ion and electron irradiated resists supports the conclusions drawn earlier that the increase in electron resist sensitivity is fundamentally related to the increase in energy losses. Radiation chemistry efficiency is higher with inelastic collisions than for elastic, however, the difference is small for a positive resist: the ratio G_e/G_n is roughly two, which allows us to speak of the essential contribution made by nuclear collisions to the chemical acts of disrupting polymer PMMA chains. Thus, it is apparent from Fig. 3-25 that about the same amount of energy is lost when PMMA films are irradiated by 60 keV He^+ ions and 150 keV Ar^+ ions due to inelastic collisions. At the same time the energy lost in nuclear collisions is a few tens of times higher for Ar^+ than for He^+. As follows from Fig. 3-38b, the sensitivity of PMMA to radiation by Ar^+ is roughly an order of magnitude higher than to irradiation by He^+. This experimental result supports the conclusion regarding the contribution made by nuclear collisions to resist sensitivity.

We will now pause to discuss resolution in resists, usually defined as the minimal line width that can be developed in a resist layer of specified thickness. When a resist is irradiated with ions having average energy, the range of the secondary electrons and ions backscattered from the substrate have almost no effect on resolution. The basic influence on resolution is the elastic collisions that occur between the atoms in the resist and the impinging particles and recoil nuclei, when their direction of motion is changed markedly. Also, the angular divergence of the ion beam impinging on the resist may reduce resolution.

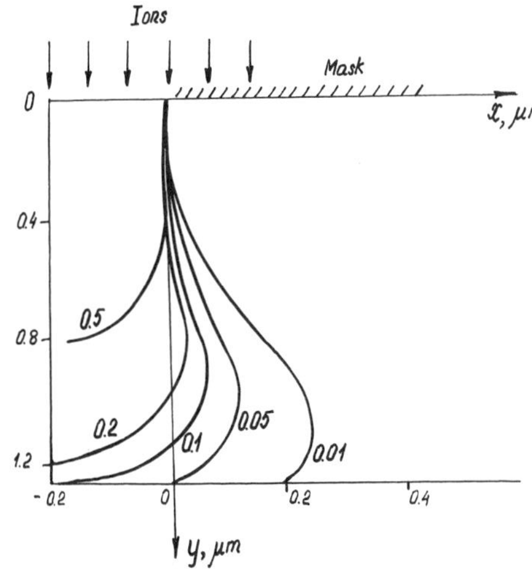

FIGURE 3-39. Cross section of isoenergy surfaces corresponding to the exposing PMMA by He^+ ions having 200 keV of energy.

In a practical sense, the problem of evaluating resolution in a resist reduces to calculating the isoenergy surfaces in the irradiated region.

Figure 3-39 shows calculated curves corresponding to the cross section of the isoenergy surfaces for the energy that a He^+ ion having an initial energy of 200 keV loses in a PMMA resist film screened by a mask that is impenetrable to ions.[64] The calculation was done within the framework of collisions theory and backscattering from the substrate was not considered. The initial direction of the ion beam was taken to be strictly perpendicular to the resist surface. The energy losses on the isoenergy lines were normalized relative to the losses at the resist surface which, for the given situation, were 25×10^8 eV/cm. For a specific process we can use a PMMA resist 0.8 μm thick. By developing the resist over a period of time corresponding to the dissolution of a 1.2 μm thick layer, we obtain a profile with a drift under the mask, which corresponds to an isoenergy line having normalized losses of 0.2. This drift from the base of the resist does not exceed 0.8 μm. This kind of resolution can be obtained in a 20 nm thick layer of PMMA. Similar values of resolution in PMMA were given in other works.[60,118] To ensure high resolution in resist layers 1.0 μm thick an ion beam impinging on the resist cannot diverge more than 1 μm (0.025 rad.).

An extremely graphic explanation of high resolution in a PMMA resist for ion lithography is given by calculating the spatial distribution of the energy the resist absorbs from impinging ions having an initial energy of 60 keV.[60] The calculation was done by the Monte Carlo method for three situations: the thickness of the resist layer is greater than the average projected range of the protons ($R_p \simeq 700$ nm); a resist less than R_p thick on a gold substrate, and a resist R_p thick on a silicon substrate. A distribution was obtained for the energy absorbed in a plane parallel to the surface and 400 nm deep from an infinitesimally narrow beam of ions impinging from the normal in all three cases. It turned out that most of the energy from the ions was absorbed within 25 nm of the point at which the beam impinged on the surface and was more than an order of magnitude higher than the losses occurred at distances farther than 25 nm from the point of impingement. This order-of-magnitude reduction in the amount of energy absorbed corresponds to a roughly 100-fold reduction in its developing rate, which means that the resolution in 400-nm-thick PMMA irradiated by protons is 20–25 nm.

Exposing a resist layer by an ion beam can be used not to accelerate the etching rate of the irradiated material but, conversely, to stop the etching. In this case the resist surface is irradiated by ions having an energy of from a few to tens of keV and the thin subsurface layer containing the implanted ions becomes an insoluble mask during the subsequent etching process. In this lithography method the pattern being developed after ion bombardment is similar to that in a negative resist irradiated by electrons. It has been shown experimentally, e.g., that a thin subsurface layer into which In^+ ions at a dose of 2×10^6 ions/cm^2 and initial energy of from 3 to 200 keV have been implanted can be used as the mask for reactive ion etching. The actual resolution achieved in this process is characterized by lines.[87]

3-35. CHARACTERISTICS OF EXPOSING ORGANIC RESISTS WITH ION BEAMS

We will note some additional distinctions between exposing resists with ion beams and electron beams. We have already discussed the higher sensitivity to ion beams (by 1 to 2 orders of magnitude) of the resists and their high resolution.

1. Several studies[120–122] have shown that the proximity effect is either absent or very weak in ion-beam lithography. Formulas from the analytical theory of how an ion beam spreads as it passes through the resist along with the reflection from the substrate due to elastic scattering at large angles allow us to estimate the magnitude of the intraproximity and interproximity effects in ion-beam lithography. Quantitative data relating to the ion beam proximity effect can be obtained by computer (Monte Carlo) and full scale experiments.

A computer (Monte Carlo) experiment was performed using a beam of H^+ protons on a PMMA layer in [120]. The modelling showed that there were virtually no ions reflected from the substrate and no reciprocal exposure in adjacent figures separated by more than 0.1 μm. Full scale experiments were performed to study the ion beam proximity effect in [121]. PMMA (positive resist) films 0.12 μm thick and polytrimethyl silistyrene with chloromethyl styrene P($SiSt_{90}$–CMS_{10}) (negative resist) 0.15 μm thick were exposed to a focused beam of Ga^+ ions having 100 keV of energy. The positive PMMA resist was exposed to doses ranging from 0.3 to 2 μC/cm^2, and the dosages used on the negative resist ranged from 0.03 to 0.3 μC/cm^2. As a function of dose the width of a developed line under given conditions increases almost linearly with the logarithm of the dose: $\Delta L \sim \ln D$. A nonlinear increase in the width of a developed line in PMMA exposed by Be^{++} ions having 100 keV of energy as a function of dose (C/cm) was also observed in [121]. (It is a simple matter to show that for a Gaussian beam $D(x) = D_0 \exp(-x^2/a^2)$ the width of a developed line with respect to dose D^* will be $x = a[\ln(D_0/D^*)]^{1/2}$. The experiment showed that line width (within the accuracy of measurement) did not depend on the type of substrate that the resist was placed on, polymer photoresist, silicon, or a layer of gold, and shows that there was no proximity effect due to ions backscattered from the substrate. It must be pointed out that the thickness of the resist layers (0.12 and 0.14 μm) was chosen with due regard for the range of the Ga^+ ions in the resists: 74 and 90 nm, respectively.

The absence of an interproximity effect was seen in experiments performed to measure the width of developed lines in a system of parallel, exposed bands. At a fixed dose the width of the developed lines is independent of the lattice period for values up to 0.3 μm in PMMA and 0.4 μm in the copolymer P($SiSt_{90}$– CMS_{10}). Notice that the widths of the developed lines lie in the 0.1–0.2 μm range for PMMA and 0.15–0.35 μm for P($SiSt_{90}$–CMS_{10}).

The continuity of lines 0.1 μm wide was destroyed at $(1.2–0.8) \times 10^{-8}$ C/cm^2 doses for P($SiSt_{90}$–CMS_{10}), and 0.7×10^{-7} C/cm^2 for PMMA. The lines retained continuity when the number of exposing ions per each pixel (0.1 μm)^2 in size was 44 for PMMA and 7.5 for P($SiSt_{90}$–CMS_{10}). Even when there are fewer ions per pixel, line continuity can be destroyed because of nonuniformities in the properties of the resists (spot noise), shot noise from the current, and, possibly, the additional noise from the ion current due to fluctuations in the emitting protrusion of a liquid-metal field source.[121] Similar noises may be revealed during etching and ion implantation by means of focused ion beams.[121]

2. In the 50–200 keV range of energies the projected range of an ion is only a few tenths of a micron. The distribution of the ions implanted into the resist layer is distinctly nonuniform and can be described by curves having a maximum. A summary of the data on the distributions of the ions, their ranges, and the variances in these ranges can be found in [123]. The distribution of the energy dissipated by the ions, which is the sum of the electron and nuclear (elastic) losses is also nonuniform with depth and is also seen to vary with ion energy and the ratio of the masses of the ions and target atoms. Accordingly, the density of radiation defects, which governs the rate at which the resists will dissolve during developing will also be a function of depth in the resist layer. Apparently, the nonuniformity of the distribution of the dissipated energy is small, since it is practically unnoticed during the developing process. We can therefore use a simplified (incremental) model for the exposure of a polymer resist by an ion beam, saying that the rate of dissolution is constant up to a depth equal to the thickness of the exposed layer, h. This depth can be associated with the depth at which the ions' electron losses fall off sharply, as can be seen in Fig. 3-25. If ions must not be implanted in the substrate and the entire resist layer must be exposed, then the thickness of the resist layer and the maximal range of the projected ions must be identical. Such a requirement (for a given beam energy) limits the resist layer to being only a few tenths of a micron thick, which may not be enough to ensure that the resist layer has masking properties and will be free of defects. In these situations we must resort to a scheme from multilayer resist systems—two-layer[121] or three-layer.[124] The easiest way to increase the thickness of an exposed layer is, of course, to increase the beam energy to the point where the exposed layer becomes about 1 μm thick. Whether or not this is possible depends on how much ion energy can be obtained from the equipment being used.

3. Ion beam exposure etches a polymer resist and reduces the layer thickness. The etching may be a surface effect such as the physical sputtering of atoms; the theory of physical sputtering was discussed in Section 3-28. The efficiency of physical sputtering is low: maximal values of the sputtering coefficient S_a reach as high as ten atoms per fast ion. Therefore, at exposure doses of $\sim 10^{13}$ ions/cm^2 the thickness of sputtered polymer layer will not be more than a monolayer. Significantly more efficient is the volumetric process that includes the production of volatile, low-molecular-weight polymer fragments in the layer being exposed, their diffusion to the surface and removal into the vacuum. This process, which is inherent only to polymer systems, may be quite efficient: $S_a = 10^3–10^4$. A phenomenological theory of ion etching was developed in [125]. Polymer etching may proceed via a similar mechanism when any type of radiation capable of decomposing the molecules into volatile, low-molecular-weight fragments is used ion, electron, and photon in the ultraviolet and x-ray regions.[125–127] We can construct a unique formulation of a phenomenological theory of the etching processes for

each one of these cases.[125] The reduction in polymer resist layer thickness caused by the volumetric process described may be substantial and must be taken into consideration when developing a practical lithography process. We can easily estimate that when $S_a = 10^4$, exposure by a flux of 10^{13} ions/cm^2 removes approximately 100 atomic layers, i.e., a layer 30 nm thick.

The S_a coefficient for various polymer materials may vary between wide limits and, in addition, be reduced with increasing the dose. This behavior in S_a is easy to understand when we use mass spectroscopy methods to study the composition of the fragments lost by the polymer.[128,129] Such a study shows that it is basically the low-molecular-weight fragments at the ends of the polymer chain that are lost. A consequence of this is that polymers which have a large number of low-molecular-weight molecules on their ends will have a high bulk etching rate. The bulk diffusive etching rate is lowered for all types of hard radiation when the end radicals are lost from the polymer chains.

The etching of polymers by a focused beam of Ar^+ ions having 50 keV of energy at a current of 30 μA was described in [130]. Films from PMMA, PBS (polybutene-1-sulfone) and nitrocellulose were used as resists. The thickness of the PMMA and nitrocellulose layers was reduced by $h = R_0 - a \ln D$ for doses ranging from 10^{13} to 10^{16} ions/cm^2; an unetchable residue was formed in the PBS. At an exposure dose of 10^{14} ions/cm^2 a 520 nm thick nitrocellulose layer was completely removed and self developing occurred.

4. The change that takes place in polymer resists exposed to doses of from 10^{13} to 10^{16} ions/cm^2 is meaningful. When this happens, the entire complex of polymer material properties is altered: density, optical properties, vibrational spectra, surface topography, dissolution and plasma etching rates, and chemical composition. A composite study of the properties of PLCDA (3,4,9,10-perylene carboxyl dianhydride) and phthalocyanine of nickel irradiated with Ar ions having 2 MeV of energy was made in [131]. Ultraviolet, visible, infrared, and Raman spectroscopy, ESCA (electron spectroscopy for chemical analysis), RBS (Rutherford backscattering). ESR (electron spin resonance), and mass spectroscopy were all used on the low-molecular-weight fragments released during the radiation process. The different elements in the composition of the irradiated material were lost at a higher rate as the possibility of their forming volatile fragments was higher. Mass spectroscopy detected the release of H_2, CH_4, CO, CO_2, and C molecules from the irradiated material. As the composition of the irradiated material changed the rate at which volatile fragments were produced decreased according to $\log n = A - B \log D$. In most cases, the material became carbon enriched. The structure of the C enriched material is nearly amorphous. Free radicals (ruptured chemical bonds) are created during irradiation and can be detected by ESR methods. The maximum concentration of free radicals occurred at doses of 10^{14} ions/cm^2. It is interesting to note that the oxygen concentration increased after irradiation, probably because of a reaction between the free radicals and oxygen or water molecules entering the film from the surrounding atmosphere. Similar changes were observed in the resist materials. It was established in [132] via IR spectroscopy that when PMMA was irradiated with Ar^+ ions in doses ranging from 3×10^{13} to 3×10^{15} ions/cm^2 the number of C=O, C–O, and C–H chemical bonds in the polymer chains decreased, and the density of the material and its index of refraction (in the visible and infrared regions) increased. The thickness of an altered layer is roughly equal to the average projected range of the ions. The plasmachemical etching rate of altered PMMA films in a sulfur hexafluoride plasma, which has the highest level of activity to electron resists, decreases according to $v_e = A - B \log D$, which indicates

that the plasma stability of the resist films increases while they are being altered. Mass spectroscopy of the fragments being released detected the loss of molecules.[128,129]

The reaction between the free radicals (ruptured chemical bonds) and oxygen or water molecules entering from the surrounding atmosphere was mentioned earlier. It is obvious that reactions with the ruptured chemical bonds can be controlled by changing the composition of the gas surrounding the irradiated film. Irradiated PMMA films were subjected to acrylic acid $CH_2{=}CHCO_2H$ vapors for an hour in [133]. When this happened it is likely that the formation of block polymers of acrylic acid (AA) and MMA i.e., polymer chains whose chemical composition can be described by the formula

$$(\mathrm{AA})_x\text{–}(\mathrm{MMA})_y\text{–}(\mathrm{AA})_x\text{–}$$

was observed. Here the (AA) is a monomer of the acrylic acid CH_2–CH–CO_2H, and the MMA is a monomer of the methacrylate $CH_3CH_2CO_2CH_3C$–. The AA–PMA block polymer is insoluble in solvents used with PMMA and can possibly be used to produce a negative resist mask (the nonirradiated areas of the resist film are removed). It is important to note that the AA–MMA reaction described occurs, first, for all types of actions that create free radicals in the film and, second, at doses that are 1 to 2 orders of magnitude lower than those normally used in lithography to expose PMMA: for electron irradiation (25 keV) at doses of 10^{-6}–10^{-7} ions/cm^2; for ion irradiation (protons having 65 keV of energy) at doses of 10^{10}– 10^{12} ions/cm^2, and for ultraviolet radiation (5.3 eV) at doses of 10^{-3}–10^{-2} J/cm^2. This result suggests that the sensitivity of the films to irradiation can be increases by as much as 10 to 100 times.

3-36. ION-BEAM LITHOGRAPHY IN INORGANIC SOLIDS

Etching selectivity occurs in areas of inorganic solids irradiated by ions[5,87,91] and is caused by the creation of defects when the ion beam interacts with the material. As the density of these defects increases with increasing radiation dose, the dissolution (etching) rate of the material increases. The property that materials have to change their dissolution rate after being irradiated with ions makes it possible in a number of cases to produce dielectric, semiconductor, and metal masks without using a special organic resist.

We will discuss the results obtained from experiments in etching ion-irradiated silicon dioxide films. Following ion bombardment, the etching rate of SiO_2 films in hydrofluoric solutions increases. Depending on the thickness of the etched layer, two types of relationships can be distinguished as a function of the etching time. In the first type the increased etching rate is constant throughout the thickness of the film over the length of the ion's average projected range (Fig. 3-40a). In the second type the increased etching rate changes over the thickness of the film. The maximal etching rate is seen at a depth that is close to the average projected range R_p of the ion (Fig. 3-40b).

A characteristic that is important to the practice of microlithography is a film's etching selectivity, equal to the ratio of the etching rates in the irradiated and nonirradiated material. The change in the silicon dioxide etching rate with increasing dose has a threshold nature: the change begins at a given dose for a given type of ion and selectivity, and then increases with increasing dose. Finally, at a certain dose, the etching rate stops increasing ("saturation" sets in) as the radiation density increases. An example of this kind of relationship is shown in Fig. 3-41a. However, works have been published in which it is not "saturation" in etching selectivity that is noted, but a broad maximum,

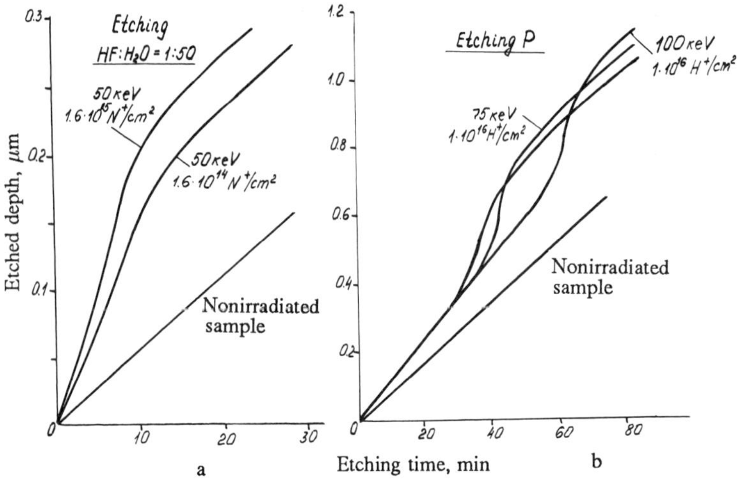

FIGURE 3-40. Thickness of an etched silicon dioxide layer as a function of etching time.

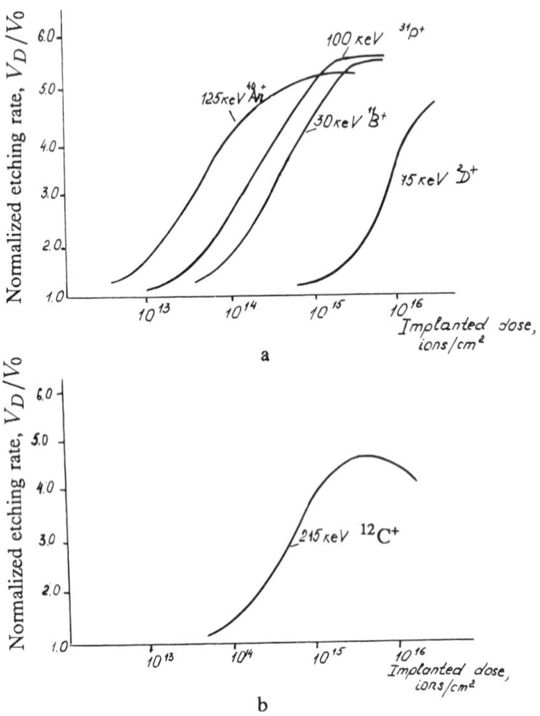

FIGURE 3-41. Etching selectivity in silicon dioxide films as a function of radiation dose.

after which etching selectivity decreases with increasing radiation dose (Fig. 3-41b). Still another relationship is evident in Fig. 3-41a the magnitude of the dose at which etching selectivity begins to increase as a function of the mass of the impinging ion.

The etching selectivity of SiO_2 films changes after they have been annealed.[88] Figure 3-42 shows the magnitudes of etching selectivity in silicon dioxide films that were subjected to P^+ ion bombardment and subsequently annealed in a nitrogen atmosphere for

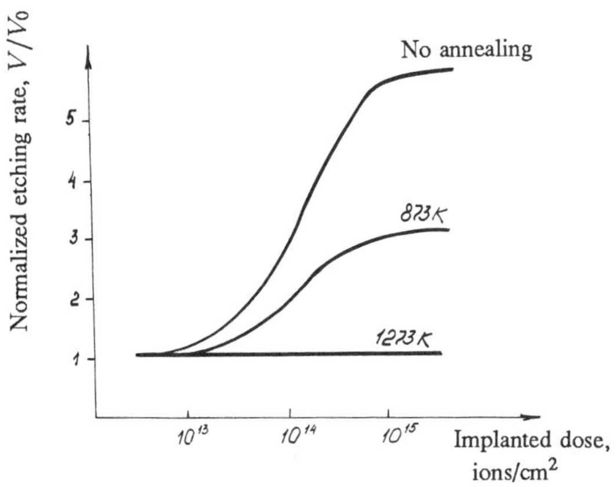

FIGURE 3-42. Etching selectivity in silicon dioxide films as a function of radiation dose at various annealing temperatures.

20 minutes as functions of the radiation doses. We can plainly see that etching selectivity decreases as the annealing temperature increases, and after annealing at a temperature of 1000° the etching rate in irradiated films becomes equal to that in a nonirradiated material.

Etching selectivity is absent from SiO_2 after it has been irradiated by electrons having up to 2 MeV of energy at a dose of up to 10^{18} C/cm^2.[90] These data suggest that electron collisions have no effect on the etching rate in silicon dioxide films.

In closing we will enumerate the fundamental results on whose basis we can suggest a physical model to describe how defects are created in silicon dioxide films during ion bombardment:

1. Ion irradiation introduces defects caused by nuclear collisions that change the properties of the material in a layer to a depth of as much as the average projected range of the ion, which increases the etching rate of this layer in hydrofluoric acid solutions. Fixed and varying etching rates were seen throughout the depth of the irradiated films.

2. The change in etching selectivity in silicon dioxide films depends on the radiation dose and the mass of the impinging ions. Two thresholds can be distinguished on the graphs of selectivity as a function of the dose of a particular type of ion: a "minimal" dose that corresponds to where etching selectivity begins to increase and a "maximal" dose that corresponds to reaching the maximal etching selectivity. At doses above the "maximal" two kinds of change in the degree of selectivity are seen: the maximal selectivity is constant ("saturation"), and selectivity decreases.

3. The maximum etching selectivity in silicon dioxide films irradiated by different ions at different energies fluctuates between 4.5 and 5.5.

4. Defects introduced into the subsurface layer of silicon dioxide during ion bombardment are annealed at fairly high temperatures (above 600°), which indicates that they will be exceptionally stable at room temperature.

If we assume that the increased etching rate in ion-irradiated silicon dioxide is a function of the density of the defects introduced by bombardment, then our first result points to different distributions of defect density over the depth of the film, or to the existence of different defects in the various layers of the film. The second result indicates that there is a maximum equilibrium defects density when irradiating by various ions. The third result shows that the maximum equilibrium defects density in irradiated films is roughly the same for different ions. Finally, the fourth result points to the possibility

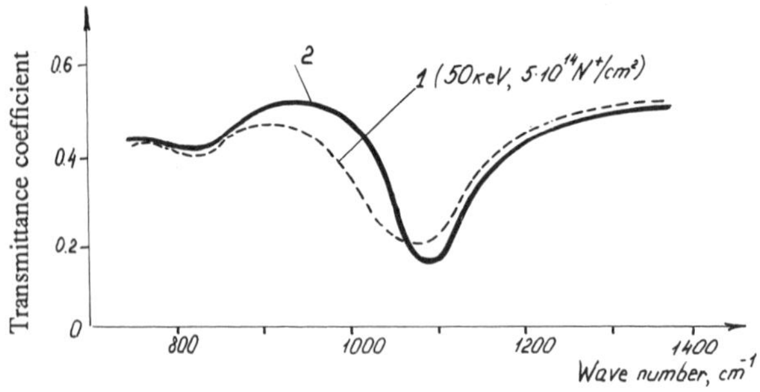

FIGURE 3-43. Infrared absorption spectrum of ion-irradiated (1) and nonirradiated (2) silicon dioxide films on silicon (the film thicknesses are 0.5 μm).

of annihilating these defects at elevated temperatures and thereby restoring the properties of a nonirradiated material.

We may suggest the following model for the process of defect creation in amorphous silicon dioxide films being bombarded by particles having medium energy. When ions are implanted into a film they lose about equal amounts of energy to electron and nuclear deceleration. Since electron collisions have little effect on etching selectivity, we can examine the effect that nuclear (elastic) collisions will have on the structure of an amorphous silicon dioxide film. This structure is characterized by SiO_4 tetrahedrons bound to one another by the Si–O–Si bridge bond.

The avalanche of nuclear collisions that occurs when one ion is implanted dislodges atoms (vacancies) and places atoms in the interstitial sites. This disruption of the structure distorts the bonds between the atoms.

The deformation of interatomic bonds can be detected by changing the vibrational spectra. Figure 3-43 shows such a change in the infrared absorption spectrum of SiO_2 films irradiated with nitrogen ions. Compared to nonirradiated films, substantial changes are made in the spectra of irradiated films in the 1000–1200 cm^{-1} range of wave numbers where the absorption bands that characterize the deformed vibrations of the Si–O–Si bond lie. This result agrees with the data from electron spin resonance studies of the defects created in ion-irradiated silicon dioxide films. These studies established that a major defect is the creation of a vacancy (dislodging the oxygen atom) in the Si–O–Si bridge bond.[134]

The distribution of "vacancy–interstitial atom" radiation defects over depth generally looks like a smooth curve having a broad maximum at a depth that is somewhat less than the projected range of the impinging ions.[55] This phenomenon is seen in experiments when the etching selectivity of silicon dioxide films is changed by irradiation by nitrogen ions at a dose not greater than the maximum dose.[93] Figure 3-44 shows etching selectivity as a function of the thickness of the etched layer for three values of the impinging ions' initial energy, 50, 120, and 200 keV. The respective average projected ranges are 110, 250 and 400 nm. The ratio of implantation depth for maximal selectivity to the average projected range of the ions was about 0.9 in all three cases, which coincides with the ratio of the depth of the maximal defects density to the average projected range of the ions calculated within the framework of the theory of avalanche collisions.[55]

The etching rates are roughly constant over depth when the films are irradiated by ions whose initial energies are low enough so that the average projected range is not greater than 100 nm. Since the distribution of defects over depth is described by a curve

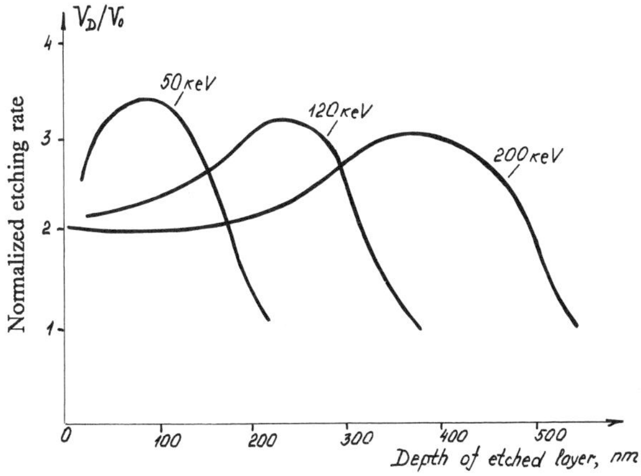

FIGURE 3-44. Etching selectivity in silicon dioxide films versus etched layer thickness.

having a broad maximum the defects density will change very little over depth when the range is short.

The etching rates will also be constant over depth when a maximum in the defects density is reached at every depth.

The reduced etching selectivity in silicon dioxide films seen after passing through its maximal value at radiation doses $D > 10^{15}$ ions/cm^2 may have two causes. One is the increase in current density at large doses. In this case the target may be heated to temperatures that anneal the defects and their density in the irradiated region will decrease. The second, which in our opinion is the more likely, is that the density of implanted particles per unit volume in the irradiated region increases to a value comparable to the density of point structural defects associated with recoil atoms. When this happens the size of the recombination region may change due to the interactions between two different types of defects.

The maximal etching selectivity in silicon dioxide films in hydrofluoric acid solutions ($v_D/v_0 = 5.0 \pm 0.5$) is independent of the mass and the energy of the ions, which shows that the size of the recombination region is independent of these parameters. This is explained by the fact that defects that accelerate the etching are created by avalanche collisions between the atoms in the film material (recoil atoms). It was established by electron spin resonance that three types of paramagnetic centers appear in SiO_2 films irradiated with neon ions (E_0 = 50 keV): P_a, P_b, and P_c. These paramagnetic centers are located at different depths in the film and are characterized by different g-factors. The P_a centers have the highest density and are most dense at a depth equal to 0.9 of the average projected range of the implanted ion. The authors(134) associated P_a centers with an oxygen vacancy in the Si–O–Si bridge bond. After annealing for 10 minutes in a nitrogen atmosphere at about 400°C the P_a centers disappeared completely (were annihilated). According to [88], etching selectivity is maintained up to annealing temperatures of 600°C, which also suggests the presence of several types of defects in irradiated silicon dioxide films that are responsible for the change in etching rate. Bands from radiation defects in SiO_2, called B_2 (245 nm), E_1 (215 nm), and E (160 nm) centers, were detected in the ultraviolet absorption spectra.

An important question for microlithography is how the defects in an irradiated layer are distributed when the particles have been implanted through a mask. Calculations done within the framework of elastic avalanche collisions allow us to construct a spatial distri-

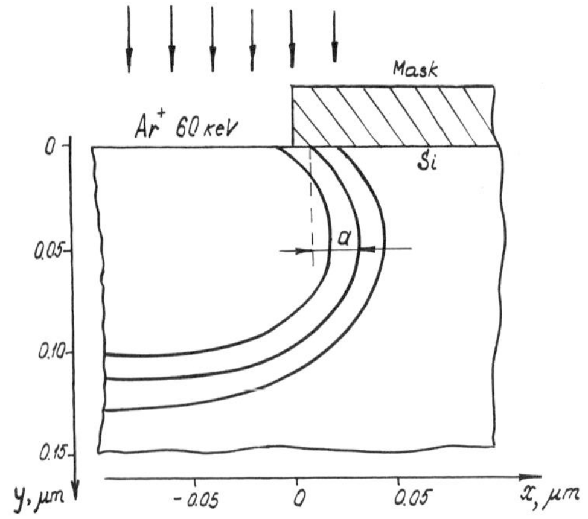

FIGURE 3-45. Distribution of the threshold concentration of defects in silicon irradiated by Ar^+ atoms near an impenetrable mask.

bution of the radiation defects density in the region bordering the topological mask.(136) Figure 3-45 shows an example of calculated lines for equal densities of defects in silicon bombarded by argon ions having 60 keV an initial energy for three different doses of radiation. All three curves correspond to the same threshold concentration, below which the irradiated material will not be etched in an aqueous solution of hydrofluoric acid, i.e., the calculated lines describe the etching profile of silicon for the three specific doses of radiation.

Silicon was irradiated through a mask under the conditions described above, etched in an aqueous solution of hydrofluoric acid, and the etched profile near the mask edge was measured by means of a scanning electron microscope over the chip. Good agreement between the calculated and experimental etching profiles was established. The average value of the a parameter (see Fig. 3-45), which corresponds to the maximal drift of the constant defects density line from the vertical is 20 nm in the example being considered. This quantity a is, in fact, a characteristic of the resolution of the ion lithography process presented here.

Etching selectivity after irradiation with medium energy ions was observed for a number of different materials: gallium arsenide, silicon nitride, garnets, nickel, and molybdenum.

The ion lithography process which obtains a negative image and uses a thin film of silver and germanium selenides as the negative resist is well known.(4,6) This system was used with in conjunction with organic resists that became the mask after ion implantation and subsequent developing. This mask was then used for selective reactive ion etching of the organic resist beneath it.

3-37. A THEORY OF WET ETCHING IN ION-IRRADIATED INORGANIC RESISTS

Experimental data concerning wet etching of ion-irradiated SiO_2 films (using aqueous solutions of HF) were described in Section 3-36. These facts can be explained from a simple, phenomenological point of view.(137,138)

We will say that the body of physical and chemical processes that make up dissolution (chemical reactions, diffusion, adsorption) obey the Arrhenius equation that relates etching rate to temperature:

$$v_e = A\exp(-U/kT) \tag{3-217}$$

The activation energy U for a given etchant is determined by the atomic interactions in the surface layer of the film. We may readily assume that U depends on the microstructure of the film, including the density of radiation defects N_i: $U = U(N_i)$. We will expand the function $U(N_i)$ over small atomic percentages of radiation defects:

$$U = U_0 + \sum_i B_i(N_i/N) + \frac{1}{2}\sum_{i,j} B_{i,y}(N_iN_j/N^2) + \ldots \tag{3-218}$$

where N_i/N is the atomic percentage of i type defects and N is the number of atoms (elementary cells) per unit volume. Hereafter, we will confine ourselves to the linear terms in the expansion, Eq. (3-218), which means that we will neglect interactions of the defects among themselves when ascertaining the activation energy. The coefficients in the expansion for U correspond to nonirradiated material: $B_i \equiv N(\partial U/\partial N_i)_0$.

The rate at which i type defects are created during exposure is

$$\frac{dN_i}{dt} = \sigma_i\Phi(N_{ci} - m_iN_i) \tag{2-219}$$

where σ_i is the effective i defect creation cross section, Φ is the flux of the bombarding particles (ions/cm^2 · sec), and N_{ci} is the total number (per cm) of centers that form i type defects (for Frenkel defects it is the number of lattice sites and for defects of the deformed bond type it is the number of bonds, etc.). The m_i parameter defines the maximum atomic percentage of i type defects: $N_{i\,\max} = N_{ci}/m_i$. Obviously, m_i can be related to the recombination region Ω_i by: $m_i = N_{ci}\Omega_i$. A solution of Eq. (3-219) has the form

$$N_i(t) = N_{ci}[1 - \exp(-G_it)] \tag{3-220}$$

where $G_i = m_i\sigma_i\Phi$.

We find the ratio of the etching rates $S = v_D/v_0$ (etching selectivity) from Eqs. (3-217) and (3-218):

$$\ln S = \ln\left(\frac{A_D}{A_0}\right) - (kT)^{-1}\sum B_i(N_i/N) \tag{3-221}$$

where A_D and A_0 are the pre-exponential factors in Eq. (3-217) for nonirradiated and irradiated targets. We will say that $A_0 = A_D$, since a stronger dependence of the etching rate on the density of defects relates to a change in the activation energy for etching. Substituting the solution of Eq. (3-220) into Eq. (3-221) we obtain

$$\ln S(D) = -(kT)^{-1}\sum_i B_i(N_{ci}/N)[1 - \exp(-\sigma_im_iD)] \tag{3-222}$$

where, according to the relationship $D = \Phi t$, the time dependence has been replaced by a dose dependence. At small doses when $\sigma_im_iD \ll 1$, ln $S(D)$ increases in proportion

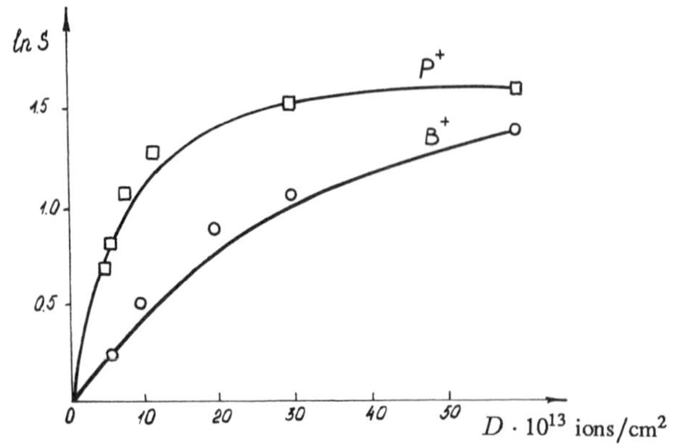

FIGURE 3-46. The ln S function for SiO_2 as a function of radiation dose. The solid lines are theoretical values; the dashed lines are experimental values.

to the dose:

$$\ln S(D) = -\frac{D}{kT} \sum_i B_i(N_{ci}/N)\sigma_i m_i \tag{3-223}$$

at large doses $\sigma_i m_i D \gg 1$, ln $S(D)$ reaches its maximal value (saturation):

$$\ln S_{\max} = -(kT)^{-1} \sum_i B_i(N_{ci}/N) \tag{3-224}$$

For defects of a type (in SiO_2 distortions of the Si–O–Si bonds are considered to be responsible for a change in the etching rate) we can write

$$U(N_d) = \frac{N - N_d}{N} U_0 + \frac{N_d}{N} U_d \tag{3-225}$$

where U_d is the activation energy of etching in SiO_2 when all of the Si–O–Si bonds have been distorted. Using Eqs. (3-223), (3-224), and (3-225) we can easily derive the relations

$$B = U_d - U_0$$

$$U_{\min} = U_0 + (U_d - U_0)(N_{d\max}/N)$$

$$\ln S(D) = -\frac{D}{kT}(U_d - U_0)\sigma_d m_d$$

$$\ln S_{\max} = -\frac{U_d - U_0}{kT}$$

We can construct the graphs of ln $S(D)$ obtained experimentally after irradiating SiO_2 with B^+ ions 50 keV and P^+ (100 keV), using Eq. (3-222) (Fig. 3-46) for the following parameter values[137]: $U_0 = 0.33$ eV, $U_d = 0.28$ eV; $\sigma_d = 2.7 \times 10^{-15}$ cm^2 (for ions B^+ 50 keV) and $\sigma_d = 10.5 \times 10^{-15}$ cm^2 (for ions P^+ 100 keV), $m = 1.25$.

REFERENCES

1. A. J. Muray and J. J. Muray, *Vac.*, **35**, Nos. 10, 11, 467–477 (1985).
2. J. Gulai, Proc. Summer School on Surface Physics, Pergamon Press (1980).
3. G. Stengl, R. Kaitna, H. Loscher, *et al.*, *J. Vac. Sci. Technol.*, **16**, No. 6, 1883–1885 (1979).
4. A. Wagner, D. Barr, T. Venkatesan, *et al.*, *J. Vac. Sci. Technol.*, **19**, No. 4, 1363–1373 (1981).
5. K. A. Valiev, V. A. Danilov, K. A. Drakin, *et al.*, *Mikroelektronika*, **11**, No. 4, 323–327 (1982).
6. L. Karapiperis, C. A. Lee, L. Ruoff, *et al.*, *J. Vac. Sci. Technol.*, **19**, No. 1, 18–22 (1981).
7. A. R. Neureuther, "Microlithography with soft x-rays" in: *Investigations by Synchrotron Radiation* (H. Winick and S. Doniach, editors) Plenum Press, New York (1980), pp. 223–275.
8. M. B. Heritage, *J. Vac. Sci. Technol.*, **12**, No. 6, 1135–1140 (1975).
9. D. B. Rensch, R. L. Seliger, G. Csonsky, *et al.*, *J. Vac. Sci. Technol.*, **16m**, No. 6, 1897–1900 (1979).
10. K. A. Valiev, L. V. Velikov, V. D. Verner, and A. V. Rakov, *Elektronnaya Promyshlennost'*, No. 1, 36–38 (1983).
11. M. A. Kumakhov and F. F. Komarov, *Energy Losses and Ion Ranges in Solids* [in Russian], Izdatelstvo Belorusskogo Universiteta, Minsk (1979).
12. M. A. Kumakhov and G. Shirmer, *Atom Collisions in Crystals* [in Russian], Atomizdat, Moscow (1980).
13. E. E. Müller and T. T. Tsong, *Field-Ion Microscopy: Principles and Applications* [Russian translation], Metallurgiya, Moscow (1972).
14. E. E. Müller and T. T. Tsong, *Field-Ion Microscopy, Field Ionization, and Field Evaporation* [Russian translation], Nauka, Moscow (1980).
15. T. Venkatesan, A. Wagner, and D. Barr, *Appl. Phys. Lett.*, **38**, No. 11, 943–945 (1981).
16. T. Venkatesan, A. Wagner, D. L. Mash, *et al.*, *Appl. Phys. Lett.*, **39**, No. 11, 9–11 (1981).
17. P. D. Prewett and D. K. Jefferies, *J. Appl. Phys.*, **13**, 1747–1755 (1980).
18. G. S. Gibson and H. C. Eaton, *J. Appl. Phys.*, **51**, No. 10, 5537–5539 (1980).
19. J. W. Ward, *J. Vac. Sci. Technol.*, **3**, No. 1, 207–213 (1985).
20. R. Gomer, *Appl. Phys.*, **19**, 365 (1979).
21. T. Ishitani and H. Tamura, *Appl. Phys.*, **23**, 193 (1980).
22. D. R. Kingham and L. W. Swanson, *Appl. Phys.*, **A41**, 157–169 (1986).
23. A. L. Pregenzer, *J. Appl. Phys.*, **58**, 4509 (1985).
24. L. Arian, A. L. Pregenzer, M. Barry, *et al.*, *J. Appl. Phys.*, **60**, No. 11, 3821–3824 (1986).
25. G. Taylor, *Proc. Roy. Soc.*, **A280**, 383–397 (1964).
26. M. D. Gabovich, *Usp. Fiz. Nauk*, **140**, No. 1, 137–151 (1983).
27. L. D. Landau and E. M. Lifshits, *Quantum Mechanics* [in Russian], Fizmatgiz, Moscow (1963).
28. P. D. Prewett, G. L. R. Mair, and S. P. Thompson, *J. Phys.*, **D15**, 1339 (1982).
29. L. W. Swanson, G. A. Schwind, and A. E. Bell, *J. Appl. Phys.*, **51**, 3453 (1980).
30. T. Ishitani, K. Umemura, and H. Tamura, *Jap. J. Appl. Phys.*, **24**, No. 6, 451–454 (1985).
31. G. L. R. Mair and T. Malvey, *J. Microscopy*, **142**, Pt. 2, 191–200 (1986).
32. G. L. R. Mair, *J. Phys. D.*, **17**, 2323 (1984).
33. A. E. Bell, G. A. Schwind, and L. W. Swanson, *J. Appl. Phys.*, **53**, No. 7, 4602–4665 (1982).
34. V. E. Krohn and G. R. Ringo, *Appl. Phys. Lett.*, **27**, No. 9, 479–481 (1975).
35. J. Orloff and L. W. Swanson, *J. Vac. Sci. Technol.*, **15**, No. 3, 845–848 (1978).
36. H. Liebl, *Vacuum*, **33**, No. 9, 525–531 (1983).
37. W. Lochte-Holtgreven, *Plasma Diagnostics*, North Holland, Amsterdam (1968).
38. K. Tokiguchi, N. Sakudo, and H. Koike, *J. Vac. Sci. Technol.*, **A2**, No. 1, 29–34 (1984).
39. G. Sidenius, The Niels Bohr Institute, Inst. Phys. Conf. Ser., No. 38, 1–11, (1978).
40. J. Asmussen and J. Root, *Appl. Phys. Lett.*, **44**, No. 4, 396–398 (1984).
41. J. Ishikawa and T. Takagi, *Jap. J. Appl. Phys.*, **22**, No. 3, 534–540 (1983).
42. I. S. Bitenskii, M. N. Murakhmetov, and E. S. Parilis, *Zh. Tekh. Fiz.*, **49**, 1044–1077 (1979).
43. K. Leiman, *Interactions Between Radiation and Solids and the Formation of Elementary Defects* [in Russian], Atomizdat, Moscow (1979).
44. L. D. Landau and E. M. Lifshits, *Mechanics* [in Russian], Fizmatgiz, Moscow (1958).
45. D. B. Firsov, *Zh. Eksp. Teor. Fiz.*, **32**, 1464–1469 (1957); **33**, 696–699 (1957).
46. E. J. Lindhard, M. Scharff, and H. Schiott, *Mat. Fys. Medd. Dan. Vid. Selsk.*, **33**, No. 14, 1–42 (1963).
47. D. R. Bates (editor), *Atomic and Molecular Processes*, Academic Press, New York–London (1962).
48. I. S. Gradshtein and I. M. Ryzhik, *Tables of Integrals, Sums, Series, and Products* [in Russian], Fizmatgiz, Moscow (1962).

49. O. B. Firsov, "A qualitative treatment of the average energy of electron excitation in atomic collisions," *Zh. Eksp. Teor. Fiz.*, **36**, 1517–1523 (1959).
50. J. Lindhard and M. Scharff, *Phys. Rev.*, **124**, No. 1, 128–130 (1961).
51. P. Sigmund, *Phys. Rev.*, **184**, 383–416 (1969).
52. "Ion beam exposure profiles in PMMA—a computer simulation," *J. Vac. Sci. Technol.*, **19**, No. 4, 1259–1263 (1981).
53. E. Bogh, P. Hogild, and I. Stensgaard, *Rad. Eff.*, **7**, 115–121 (1971).
54. N. F. Mott and E. A. Davis, *Electronic Processes in Noncrystalline Materials* [Russian translation], Mir, Moscow (1974).
55. P. Sigmund, *Rev. Roum. Phys.*, **17**, No. 7, 823–870 (1972).
56. R. Benedek, *J. Appl. Phys.*, **52**, No. 9, 5557–5565 (1981).
57. D. Pramanic and D. N. Seidman, *Appl. Phys. Lett.*, **43**, No. 7, 639–641 (1983); *J. Appl. Phys.*, No. 1, 137–150 (1986).
58. L. Meyer, *Phys. Stat. Sol.*, **44**, No. 2, 253–268 (1971).
59. G. Moliere, *Z. Naturforsch.*, **2a**, No. 3, 133–145 (1947); **3a**, No. 2, 78–97 (1948).
60. Yu. V. Gott, "Multiple scattering of slow particles," *Zh. Eksp. Teor. Fiz.*, **60**, 1291–1294 (1971).
61. H. W. Lewis, *Phys. Rev.*, **78**, 526–529 (1950).
62. A. M. Kol'chuzhkin and V. V. Uchaikin, *Introduction to the Transmission of Particles Through Matter* [in Russian], Atomizdat, Moscow (1978).
63. A. Marwick and P. Sigmund, *Nucl. Inst. Meth.*, **126**, No. 3, 317–323 (1975).
64. M. Komuro, N. Atoda, and H. Kawakatsu, *J. Electrochemical Soc.: Solid State Sci. Technol.*, **126**, No. 3, 483–490 (1979).
65. B. S. Danilin and Yu. V. Kireev, *Ion Etching in Microstructures* [in Russian], Sov. Radio, Moscow (1979).
66. I. A. MacHugh, "Secondary ion mass spectroscopy" in: *Surface Analysis Methods* [Russian translation], Mir, Moscow (1979), pp. 276–341.
67. G. Valyi, V. Schiller, J. Guimesi, and J. Gyulai, *J. Thin Solid Films*, **76**, No. 3, 215–219 (1981).
68. N. V. Pleshivtsev, *Cathodic Sputtering* [in Russian], Atomizdat, Moscow (1968).
69. P. Sigmund, *Rev. Roum. Phys.*, **17**, No. 8, 969–1000; No. 9, 1079–1106 (1972).
70. P. C. Zalm, *J. Appl. Phys.*, **54**, No. 5, 2660–2666 (1983).
71. W. D. Wilson, L. G. Haggmark, and J. P. Biersack, *Phys. Rev.*, **B15**, 2458 (1977).
72. N. Matsunami, Y. Yamamura, Y. Itikawa, N. Itoh, Y. Kazumata, S. Miyagawa, K. Morita, and R. Shimizu, *Rad. Eff. Lett.*, **57**, 15 (1980).
73. J. Bohdansky, J. Roth, and H. L. Bay, *J. Appl. Phys.*, **51**, 2861 (1980).
74. Y. Yamamura and J. Bohdansky, *Vacuum*, **35**, No. 12, 561–571 (1985).
75. H. Y. Kang, E. Kawato, and R. Shimizu, *Jap. J. Appl. Phys.*, **24**, No. 11, 1409–1416 (1985).
76. M. T. Robinson, *J. Appl. Phys.*, **54**, No. 5, 2650–2659 (1985).
77. K. H. Krebs, *Vacuum*, **33**, No. 9, 555–563 (1983).
78. A. Koyama, T. Shikata, and H. Sakaire, *Jap. J. Appl. Phys.*, **20**, No. 1, 65–70 (1981).
79. R. A. Baragiola, E. V. Alonso, and A. Oliva Florio, *Phys. Rev.*, **B19**, 121 (1979).
80. E. S. Mashkova and V. A. Molchanov, *Scattering Ions of Medium Energy on the Surfaces of Solids* [in Russian], Atomizdat, Moscow (1980).
81. V. I. Veksler, *Secondary Ion Emission from Metals* [in Russian], Nauka, Moscow (1978).
82. H. C. Pfeiffer, *J. Vac. Sci. Technol.*, **12**, No. 6, 1170–1173 (1975).
83. H. C. Pfeiffer, *J. Vac. Sci. Technol.*, **15**, No. 3, 887–890 (1978).
84. Yu. V. Vorob'ev and V. A. Zhukov, *Mikroelektronika*, **11**, 311–315 (1982).
85. V. A. Zhukov and K. A. Kalabushev, *Mikroelektronika*, **11**, 316–321 (1982).
86. P. D. Prewett, *Vacuum*, **34**, No. 10/11, 931–939 (1984).
87. "Plasma-developed ion-implanted resists with submicron resolution," *J. Vac. Sci. Technol.*, **19**, No. 4, 1379–1383 (1981).
88. A. Monfret and J. Bernard in: *Ion Implantation in Semiconductors*, Berlin (1971), pp. 389–396.
89. I. P. Akimchenko, V. V. Galkin, and V. V. Krasnopevtsev, *Mikroelektronika*, **2**, 166–172 (1973).
90. A. P. Webb, A. J. Houghton, and P. D. Townsend, *Rad. Eff.*, **30**, 177–182 (1976).
91. J. Mittenbacher and K. Gaztner, *Rad. Eff.*, **49**, No. 2, 113–118 (1980).
92. T. Kato, H. Morimoto, K. Saitoh, and H. Nakata, *J. Vac. Sci. Technol.*, **B3**, No. 1, 50–53 (1985).
93. K. A. Valiev, V. A. Danilov, S. V. Peshekhonov, A. V. Rakov, and A. G. Shchuchkin, *The Physics, Chemistry, and Mechanics of Surfaces*, **3**, 71–73 (1983).
94. M. Komuro, *Thin Solid Films*, **92**, 155–164 (1982).
95. H. Morimoto, Y. Sasaki, Y. Watakaba, and T. Kato, *J. Appl. Phys.*, 159–160 (1985).

96. T. Shiokawa, I. Miyamoto, P. N. Kim, Y. Ochiai, A. Masuyama, K. Toyada, and S. Namba, *Jap. J. Appl. Phys.*, **24**, No. 11, 87 (1985).
97. K. Gamo, N. Takakura, N. Samoto, R. Shimizu, and S. Namba, *Jap. J. Appl. Phys.*, **23**, No. 5, 293–295 (1984).
98. R. L. Kubena, J. Y. M. Lee, R. A. Jullens, R. G. Brault, R. L. Middleton, and E. H. Stevens, *IEEE Trans. Electron. Dev.*, **ED-31**, No. 9, 1186–1189 (1984).
99. S. Matsui, K. Mori, T. Shiokawa, K. Toyoda, and S. Namba, *Jap. J. Appl. Phys.*, **23**, No. 3, 172–174 (1984).
100. Y. Ochiai, K. Shihoyama, T. Shiokawa, A. Masuyama, K. Gamo, and S. Namba, *Jap. J. Appl. Phys.*, **25**, No. 7, 526–529 (1986).
101. H. Iguchi, Y. Hirayama, and H. Okamoto, *Jap. J. Appl. Phys.*, **25**, No. 7, 560–563 (1986).
102. S. Shukuri, Y. Wada, H. Masuda, T. Ishitani, and M. Tamura, *Jap. J. Appl. Phys.*, **23**, No. 8, 543–545 (1984).
103. R. L. Kubena, R. L. Seliger, and E. H. Stevens, *Thin Solid Films*, **92**, 165–169 (1982).
104. K. Gamo, K. Moriizumi, Y. Ochiai, M. Takai, S. Namba, T. Shiokawa, and T. Minamisono, *Jap. J. Appl. Phys.*, **23**, No. 8, 642–645 (1984).
105. D. H. Narum and R. F. W. Pease, *J. Vac. Sci. Technol.*, **B4**, No. 1, 154–158 (1986).
106. K. Shimizu, *Jap. J. Appl. Phys.*, **22**, No. 10, 1623–1626 (1983).
107. T. Morita, E. Miyauchi, H. Arimoto, A. Takamori, Y. Bamba, and J. Hashimoto, *J. Vac. Sci. Technol.*, **B4**, No. 4, 829–832 (1986).
108. G. Stengl, R. Kaitna, and H. Loschner, *Solid State Technol.*, **25**, No. 8, 104–109 (1982).
109. G. Stengl, R. Kaitna, and H. Loschner, *J. Vac. Sci. Technol.*, **19**, No. 4, 1358–1362 (1981).
110. G. Stengl, H. Loschner, and J. J. Muray, *Solid State Technol.*, Feb., 119–126 (1986).
111. G. Stengl, H. Loschner, W. Maurer, and P. Wolf, *J. Vac. Sci. Technol.*, **B4**, No. 1, 194–200 (1986).
112. J. L. Bartelt, *Solid State Technol.*, May, 215–220 (1986).
113. J. N. Randall, D. C. Flanders, and N. P. Economou, *J. Vac. Sci. Technol.*, **B1**, No. 4, 1152 (1983).
114. J. N. Randall, E. I. Bromley, and N. P. Economou, *J. Vac. Sci. Technol.*, **B4**, No. 1, 10–14 (1986).
115. J. N. Randall, D. C. Flanders, N. P. Economou, and J. P. Donnelly, *J. Vac. Sci. Technol.*, **B3**, No. 1, 58–61 (1985).
116. J. N. Randall, L. A. Stern, and J. P. Donnelly, *J. Vac. Sci. Technol.*, **B4**, No. 1, 201–204 (1986).
117. H. Ryssel, K. Haberger, and J. Kranz, *J. Vac. Sci. Technol.*, **19**, No. 4, 1358–1362 (1981).
118. T. M. Hall, A. Wagner, and L. F. Thompson, *J. Vac. Sci. Technol.*, **16**, No. 6, 1889–1892 (1979).
119. G. K. Selivanov, D. D. Mozzhukhin, and B. G. Gribov, *Mikroelektronika*, **9**, No. 6, 517–539 (1980).
120. L. Karapiperis, I. Adesida, C. A. Lee, and E. D. Wolf, *J. Vac. Sci. Technol.*, **19**, No. 4, 1259–1263 (1981).
121. S. Matsui, K. Mori, K. Saigo, T. Shiokawa, and S. Namba, *J. Vac. Sci. Technol.*, **B4**, No. 4, 845–849 (1986).
122. T. Shiokawa, Y. Aoyagi, P. H. Kim, K. Toyoda, and S. Namba, *Jap. J. Appl. Phys.*, **23**, No. 4, 232–233 (1984).
123. A. F. Burenkov, F. F. Komarov, M. A. Kumakhov, and M. M. Temkin, *Tables of Ion Implantation Spatial Distributions*. Translated from the Russian by S. I. Amoretty, Gordon and Breach (1986).
124. Y. Todokoro, *Electronics Lett.*, **18**, No. 13, 543–545 (1982).
125. K. A. Valiev, A. V. Rakov, and A. G. Shchuchkin, *Mikroelektronika*, **13**, No. 5, 387–391 (1984).
126. K. A. Valiev, T. M. Makhviladze, and M. E. Sarychev, *Mikroelektronika*, **14**, No. 4, 291–300 (1985).
127. K. A. Valiev, and T. M. Makhviladze, *The Physics, Chemistry, and Mechanics of Surfaces*, No. 3, 57–67 (1985).
128. L. V. Velikov, Yu. I. Dorofeev, A. S. Kramarenko, V. E. Skurat, and V. L. Tal'roze, *The Physics, Chemistry, and Mechanics of Surfaces*, No. 6, 86–95 (1985).
129. A. Ya. Vainer, L. V. Velikov, Yu. I. Dorofeev, K. M. Dyumaev, V. F. Limanova, V. E. Skurat, and V. L. Tal'roze, *Khim. Vys. Energ.*, **20**, No. 3, 252–257 (1986).
130. K. Harakawa and Y. Yasuoka, *J. Vac. Sci. Technol.*, **B4**, No. 1, 355–357 (1987).
131. T. Venkatesan, S. R. Forrest, M. L. Kaplan, P. H. Schmidt, C. A. Murray, W. L. Brown. B. J. Wilkens, R. F. Roberts, L. Rupp, Jr., and H. Schonhorn, *J. Appl. Phys.*, **56**, No. 10, 2778–2787 (1984).
132. K. A. Valiev, V. A. Galperin, D. V. Lomonosov, S. V. Peshekhonov, A. V. Rakov, and A. G. Shchuchkin, *Mikroelektronika*, **14**, No. 5, 420–423 (1985).
133. S. Y. Kim, J. Choi, D. Pulver, J. A. Moore, J. C. Corelli, and A. J. Steckl, *J. Vac. Sci. Technol.*, **B4**, No. 1, 403–408 (1986).
134. T. Izumi and T. Matsumori, in: *Ion Implantation in Semiconductors*, Osaka (1974), pp. 539–545.

135. N. N. Gerasimenko, G. M. Tseitlin, and S. V. Vasilev, *Phys. Stat. Sol.*, **(a)62**, k169–k174 (1980).
136. K. Moriwaki, L. Aritome, and S. Namba, *Jap. J. Appl. Phys.*, **20m**, No. 7, 1305–1309 (1981).
137. K. A. Valiev, T. M. Makhviladze, and A. V. Rakov, *Dokl. Akad. Nauk SSSR*, **289**, No. 4, 867–871 (1986).
138. K. A. Valiev, T. M. Makhviladze, and A. V. Rakov, *Mikroelektronika*, **15**, No. 5, 392–397 (1986).

4

The Physics of X-Ray Microlithography

4-1. THE PHYSICAL PHENOMENA USED IN X-RAY LITHOGRAPHY

We have repeatedly mentioned that the desire to use thin beams of ions or electrons in lithography is due to the impossibility of achieving submicron resolution with visible light (λ = 400–600 nm or light in the near ultraviolet region ($\lambda \geq 300$ nm). Electron or ion "optics" can compress corpuscular beams to submicron sizes and is the principal means of obtaining a "submicron" pattern by exposing a material via a scanning corpuscular ray (beam). At the same time, diffraction makes it difficult to obtain submicron beams of light or to transfer a submicron mask image by illuminating it with a plane light wave (at a wavelength in the visible spectrum).

The possibility of using shortwave electromagnetic radiation from the ultraviolet or x-ray region for photolithography does exist. Recall the definition of ultraviolet radiation. The visible spectrum is understood to be waves 380–780 nm long. There are three regions in the ultraviolet spectrum: near, 400–300 nm; far, 300–200 nm; and vacuum, 200– 4 nm.[1] Radiation having wavelengths less than λ = 0.1–4 nm is called soft x-rays. In this chapter we will examine the physical phenomena inherent to using wavelengths from 0.1 to 100 nm for photolithography. As we will see, this range includes soft x-rays and a significant portion of the vacuum ultraviolet region. A feature common to this range of wavelengths is that the materials lack transparency because the photon energy is $E_k > E_g$, where E_g is the width of the forbidden band in solid insulators. When $E_k > E_g$, strong photoelectron absorption of radiation quanta is observed (the so-called fundamental absorption in the range $E_k \geq E_g$). The result of all this is that it is impossible to design refractor lens optical systems for the $E_k > E_{g\,\max}$ range of wavelengths: the optical elements that are possible in the range $E_k > E_{g\,\max}$ are reflective elements and membranous elements whose thickness is $d < \mu^{-1}(\lambda)$, where μ (cm^{-1}) is a linear coefficient of absorption. For example, shadow masks (x-ray masks), optical windows, optical filters, and zone plates all relate to membranous elements. The current technology allows us to fabricate membranes $d \simeq 1$ μm thick that are uniform throughout their thickness and are sufficiently transparent to allow their use in laboratory and industrial conditions. Membranes $d \sim 1$ μm thick made from light-weight materials have relative transparency (transmittance $T > 50\%$) in the soft x-ray range at $\lambda = 0.1$–2 nm, which is the main reason for using this range in modern x-ray lithography systems. In the $\lambda = 2$–100 nm range, the absorption of radiation is so high, ($\mu \leq 10^5$ cm^{-1}), that membranes $d = 0.1$ μm thick are relatively transparent, which, along with a number of other factors, leave this range unused in lithography. However, the physical properties of materials and the phenomena that occur when they are exposed to radiation in the $\lambda = 0.1 - 100$ nm range have a great deal in common; therefore, the results discussed in this chapter can be related to the entire spectrum of wavelengths. The differences in the properties of materials for this range of wavelengths are quantitative, rather than qualitative. The range of wavelengths $\lambda > 100$ nm can be called optical because traditional lens (refractor) optical systems can be designed for use in this range. It is mainly this range of wavelengths for which the basic contemporary photolithography machines are designed: optical image generators, align and expose systems for proximity printing, and projection refractor and reflector systems (steppers). We will not consider the problems of lithography in the optical range of wavelengths in this chapter (see Chapter 5).

Diffraction limitations for radiation in the $\lambda = 0.1$–100 nm range lie in the submicron region and, in principle, x-ray and ultraviolet lithography methods can be developed which are comparable to corpuscular ion- and electron-beam methods in terms of resolution. We will examine the physics of x-ray lithography in this chapter. X-ray lithography has been investigated in a number of works in which the respective equipment and materials have also been developed.

In keeping with the approach employed in this book, we will relate to the physics of microlithography the physics of ultraviolet and x-ray sources and the physics of the interactions between the radiation and the materials. The physics of the interaction between the radiation and the material must explain several aspects of the lithography problem. In creating homogeneous sources of radiation, questions arise as to the choice of materials to be used as the windows in the radiation sources and the materials that define the spectrum and the spatial distribution of the radiation (filters, monochromators, reflectors, lenses, etc.). When fabricating photomasks we must choose the "transparent" substrate material and the "opaque" film material that is the image of the device on the photomask. The properties of transparency and opaqueness, equal in reflectance, are determined by the interactions between the photons of the incident radiation and the atoms of the material. These phenomena are determined by the process used to expose polymeric photosensitive materials called photoresists (x-ray resists when speaking of x-ray lithography. By analogy, the resists used for ultraviolet lithography may be called ultraphotoresists).

The fundamental phenomenon that occurs when ultraviolet or soft x-ray radiation passes through a material is the excitation or ionization of its atoms. When the photon energy is transferred to an atom's electron, the electron may make a transition to an unoccupied bound state (excitation), to a delocalized state (internal photoeffect), or into the vacuum (photoemission). Thus, we must know what the energy bands of the solid are in order to examine the processes by which energy quanta are absorbed. A quantum of energy can be transferred to the electrons in the inner shells, to electrons in the valence band or in the conduction band (in metals), and to the valence electrons in molecular solids. The transition to delocalized states in the conduction band is a phenomenon of photoemission.* It is also possible for electrons to make a transition to states in the

*It is usually assumed that forbidden bands no longer occur above the conduction band due to the filling of

forbidden band, which results in the creation of a bound electron–hole pair called an exciton.

A primary photoelectron created by absorbing a quantum of energy, may have a significant amount of energy $E = h\nu - I$, if $h\nu \gg I$. For example, the energy of a photoelectron created by absorbing x-ray quanta having an energy of $E = 1$ keV may be several hundreds of eV. Obviously, such an electron can, in the process of decelerating, produce a number of secondary excitations and ionizations in the outer shell electrons, e.g., electrons in the valence band, since the forbidden band is only a few eV wide. In a similar manner, the photoelectrons that appear in a molecular photoresist as a result of absorbing x-ray quanta of energy can induce secondary excitations and ionizations in the molecules which disrupt or knit the molecular chain.

Holes are created because of the photoeffect or the excitation of electrons in the inner shells or the valence band. The filling of these holes is accompanied by the radiation of a characteristic quantum (fluorescence) or passes as a two-electron Auger process in which one electron fills the hole and lowers its energy and the other electron carries off this excess energy, going into a delocalized state. The ratio of the probability of fluorescence to the probability of an Auger process increases rapidly as the atomic number Z increases (roughly as Z^4). Therefore, when materials composed of light-weight atoms are irradiated by x-rays, the filling of a hole is most often accompanied by the generation of Auger electrons which, just as with high-energy photoelectrons, can cause secondary excitations and ionizations that, e.g., disrupt the molecules in a resist.

The range of energies we are interested in for x-ray and ultraviolet lithography lie between 5 keV–10 eV. The fundamental process that is responsible for weakening radiation in this range of energies is photoelectron absorption. The familiar Compton scattering of x-ray photons by an atom's electrons, considered to be free, becomes a significant factor only at quantum energies of from a few tens of keV to several MeV. The quantum frequency shifts that occur in the Compton effect can be calculated directly from the laws of conservation of energy and momentum. In the range of energies from 10 eV to 5 keV the quantum frequency shift in Compton scattering is small and goes into elastic Rayleigh scattering. Because the Compton scattering cross section is much smaller than the photoelectron absorption cross section, we will not examine it (Compton scattering) in detail. We will likewise not examine combinational scattering because of its relative weakness. As far as the photogenesis of electron–positron pairs is concerned, its role becomes significant at photon energies in the tens of MeV.

Nonuniformities in the optical constants of a material make it possible in the ultraviolet range of quantum frequencies for radiation to be scattered. Such scattering has not however, been reported in the literature. Thus, we are faced with the need to discuss the physics of photoelectron absorption of shortwave radiation in solids and the physics of x-ray and ultraviolet sources.

We will examine the bremsstrahlung and characteristic radiation from x-ray and ultraviolet sources that occurs in x-ray tubes when a beam of accelerated electrons acts on a target (anode), the radiation from a high temperature plasma (e.g., a laser), and the radiation from relativistic electrons in an electron accelerator called a synchrotron (synchrotron radiation). The ultraviolet radiation obtained from the usual low pressure gas discharge lamps (the line spectra of inert gases He, Ar, H_2 at pressures of $p = 0.1$–0.2 mm Hg) or the radiation with the continuous spectrum of a gas discharge in He, Ar, Kr at

---

bands corresponding to states above those from which the conduction band is formed. The conduction band extends to the vacuum and beyond, forming a continuum of states.

higher p = 50–200 mm Hg pressures in the $\lambda < 100$ nm range have low levels of intensity and will not be considered in this chapter.

4-2. BREMSSTRAHLUNG WITH A CONTINUOUS X-RAY SPECTRUM

Electrons are accelerated in an x-ray tube to an energy of $E = eV$ and are decelerated at the anode (anticathode). The x-ray radiation that arises at the anode is made up of two components: a continuous x-ray spectrum called bremsstrahlung whose hardness is independent of the anode material, and a characteristic radiation whose line spectrum is determined by the transitions in the electron shells of the anode material's atoms. The hardest bremsstrahlung quanta are determined by the electrons' energy: $E = h\nu_{\text{max}}$.

X-ray bremsstrahlung with a continuous spectrum occurs because of the fast electrons in a beam colliding with the nuclei of the anode material's atoms. The fast electrons penetrate the electron shell of the target atoms and experience a Coulomb attraction from the nucleus as they pass near the unshielded nuclei. The acceleration that these electrons undergo due to this force stimulates radiation. According to classical electrodynamics, a charge e moving with an acceleration $\dot{v}$ radiates an amount of energy (power)

$$S = \frac{2}{3}\frac{e^2}{c^3}\dot{v}^2 \tag{4-1}$$

per unit time.[2,3] We will assume that the deceleration process is uniform so that $\dot{v} = v/\tau$, where τ is the duration of the deceleration process. Then the energy radiated by an electron as it decelerates is

$$E_r = \int_0^\tau S(\tau)\,d\tau = \frac{2}{3}\frac{e^2}{c^3}\frac{v^2}{\tau} \tag{4-2}$$

Sommerfeld[2] associated the deceleration time τ and the electron energy E by the uncertainty relation $E \cdot \tau = h$ such that

$$E_r = \frac{4}{3}\frac{e^2}{mc^3}\frac{E^2}{h} \tag{4-3}$$

Equation (4-3) explains the proportionality $E_r \sim V^2$ that is seen in experiments, where V is the anode voltage and the small amount of electron energy that goes into radiation:

$$\frac{E_r}{E} = \frac{2}{3}\frac{e^2}{hc}\frac{v^2}{c^2} \sim 10^{-4}$$

($v/c \approx 1/3$ when E = 30 keV).

The bremsstrahlung spectrum can be derived by expanding the time-dependent dipole moment of the electron–nucleus system in a Fourier series.[3] The instantaneous power of the dipole moment radiation $d(t)$ in the frequency band $d\omega$, integrated over all directions of radiation, is[3]

$$d\varepsilon_\omega = (8\pi/3c^3)\ddot{\mathbf{d}}_\omega^2\,d\omega \tag{4-4}$$

It follows from the definition of Fourier components that $\ddot{\mathbf{d}}_\omega = -\omega^2 \mathbf{d}_\omega$ and

$$d\varepsilon_\omega = (8\pi\omega^4/3c^2)|\mathbf{d}_\omega|^2\, d\omega \tag{4-5}$$

Because in an experiment we must deal with a flux of particles passing the nuclei and having different impact parameters ρ, the radiation given by Eq. (4-5) must be averaged over the impact parameters:

$$d\varkappa_\omega = \int_0^\infty d\varepsilon_\omega \cdot 2\pi\rho_0\, d\rho_0 = \frac{8\pi\omega^4\, d\omega}{3c^3}\int_0^\infty |\mathbf{d}_\omega|^2 2\pi\rho_0\, d\rho_0 \tag{4-6}$$

The dipole moment of a colliding electron and nucleus having the masses and charges m_1, e_1 and m_2, e_2, respectively, is

$$\mathbf{d} = e_1\mathbf{r}_1 + e_2\mathbf{r}_2 = \mathbf{r}(e_1 m_2 - e_2 m_1)/(m_1 + m_2) \tag{4-7}$$

where $\mathbf{r} = \mathbf{r}_1 - \mathbf{r}_2$ is the distance between them. Because of the Coulomb interaction between the electron and the nucleus, their relative motion is described by a hyperbolic trajectory[(4)]

$$r = a(1 - \varepsilon^2)/(1 - \varepsilon\cos\varphi) \tag{4-8}$$

having parameters

$$a = \frac{|e_1 e_2|}{2|\mathcal{E}|}, \qquad \varepsilon = \left[1 - \frac{2|\mathcal{E}|M^2}{\mu|e_1 e_2|^2}\right]^{1/2}, \qquad \mu = \frac{m_1 m_2}{m_1 + m_2}$$
$$\mathcal{E} = \mu\cdot\frac{v_0^2}{2}, \qquad M = \mu\cdot\rho_0\cdot v_0 \tag{4-9}$$

where v_0 is the velocity of the electron. The time dependence of the radius of the vector $r(t)$ or the coordinates $x(t)$ and $y(t)$ is specified by the parametric equations

$$r = a(\varepsilon\cosh\xi - 1), \qquad x = a(\cosh\xi - \varepsilon), \qquad y = a(\varepsilon^2 - 1)^{1/2}\sinh\xi \tag{4-10}$$

$$t = (\mu\cdot a^3/|e_1 e_2|)^{1/2}(\varepsilon\sinh\xi - \xi) \tag{4-11}$$

Calculating the Fourier components x_ω and y_ω given by

$$x_\omega = \frac{1}{2\pi}\int_{-\infty}^{\infty} x(t)e^{i\omega t}\, dt, \qquad y_\omega = \frac{1}{2\pi}\int_{-\infty}^{\infty} y(t)e^{i\omega t}\, dt \tag{4-12}$$

yields, according to [3], the result

$$x_\omega = -\frac{a}{2\omega} H^{(1)}_{i\omega/\omega_0}\left(\frac{i\omega\varepsilon}{\omega_0}\right), \qquad y_\omega = -\frac{a(\varepsilon^2 - 1)^{1/2}}{2\varepsilon\omega} H^{(1)}_{i\omega/\omega_0}\left(\frac{i\omega\varepsilon}{\omega_0}\right)$$
$$\omega_0 = \left(\frac{|e_1 e_2|}{\mu a^3}\right)^{1/3} \tag{4-13}$$

where $H_s^{(1)}$ is a first Hankel function of order s. Substituting Eq. (4-13) into Eqs. (4-5) and (4-6) yields[3]

$$\frac{d\varepsilon_\omega}{d\omega} = \frac{2\pi\omega^2 a^2}{3c^3}\left(\frac{e_1 m_2 - e_2 m_1}{m_1 + m_2}\right)^2 \left[\frac{1-\varepsilon^2}{\varepsilon^2}\left|H^{(1)}_{i\omega/\omega_0}\left(\frac{i\varepsilon\omega}{\omega_0}\right)\right|^2 + \left|H^{(1)\prime}_{i\omega/\omega_0}\left(\frac{i\varepsilon\omega}{\omega_0}\right)\right|^2\right] \tag{4-14}$$

$$\frac{d\varkappa_\omega}{d\omega} = i\frac{4\pi^2 a^2\omega}{3c^3}\left(\frac{ae_1e_2}{\mu}\right)^{1/2}\left(\frac{e_1 m_2 - e_2 m_1}{m_1 + m_2}\right)^2 H^{(1)}_{i\omega/\omega_0}\left(\frac{i\omega}{\omega_0}\right) H^{(1)\prime}_{i\omega/\omega_0}\left(\frac{i\omega}{\omega_0}\right) \tag{4-15}$$

These formulas can be simplified in asymptotic cases of low ($\omega \ll \omega_0$) and high ($\omega \gg \omega_0$) frequencies. According to the definition of a Hankel function as an integral

$$H^{(1)}_{i\omega/\omega_0}\left(\frac{i\omega}{\omega_0}\right) = -\frac{i}{\pi}\int_{-\infty}^{+\infty} \exp\left[\frac{i\omega}{\omega_0}(\xi - \sinh\xi)\right] d\xi \tag{4-16}$$

at low frequencies $\omega \ll \omega_0$ the range of integration over ξ to values $\xi - \sinh\xi = \omega_0/\omega \gg 1$ is significant when $\sinh\xi \gg \xi$. Then we can say that

$$H^{(1)}_{i\omega/\omega_0}\left(\frac{i\omega}{\omega_0}\right) \approx -\frac{i}{\pi}\int_{-\infty}^{\infty} \exp\left(-i\frac{\omega}{\omega_0}\sinh\xi\right) d\xi = H_0^{(1)}\left(\frac{i\omega}{\omega_0}\right) \tag{4-17}$$

An asymptotic representation of the function $H_0^{(1)}(ix)$ for small x is

$$H_0^{(1)}(ix) \approx \frac{2i}{\pi}\ln\frac{2}{\gamma x}, \qquad \gamma = 1.78107 \tag{4-18}$$

which gives

$$\frac{d\varkappa_\omega}{d\omega} = \frac{16|e_1e_2|^2}{3c^3 v_0^2}\left(\frac{e_1}{m_1} - \frac{e_2}{m_2}\right)^2 \ln\left[\frac{2m_1m_2v_0^3}{\gamma|e_1e_2|(m_1+m_2)\omega}\right] \tag{4-19}$$

instead of Eq. (4-15). When electrons are scattered at the nucleus we can say that $\mu = m_1 = m$, $e_1 = e$, $e_2 = Ze$ and rewrite Eq. (4-19) as

$$\frac{d\varkappa_\omega}{\hbar\, d\omega} = \frac{16Z^2e^6}{3m^2c^3\hbar v_0^2}\ln\left(\frac{2mv_0^3}{\gamma Ze^2\omega}\right) \tag{4-20}$$

If the logarithm of the electron's velocity v_0 is not considered the spectral density of the radiation energy in the low-frequency band $\omega \ll \omega_0$ falls off in inverse proportion to the square of the mass and velocity of the electron. The frequency dependence is logarithmic; as $\omega \to 0$ we observe a logarithmic divergence.

At high frequencies $\omega \gg \omega_0$ the exponent in the integral, Eq. (4-16), can be replaced by its value when ξ is small: $(i\omega/\omega_0)(\xi - \sinh\xi) \approx (i\omega/\omega_0)(\xi^3/6)$ and

$$H^{(1)}_{i\omega/\omega_0}\left(\frac{i\omega}{\omega_0}\right) \approx -\frac{i}{\pi}\int_{-\infty}^{\infty} \exp\left[-\frac{i\omega}{\omega_0}\frac{\xi^3}{6}\right] d\xi \tag{4-21}$$

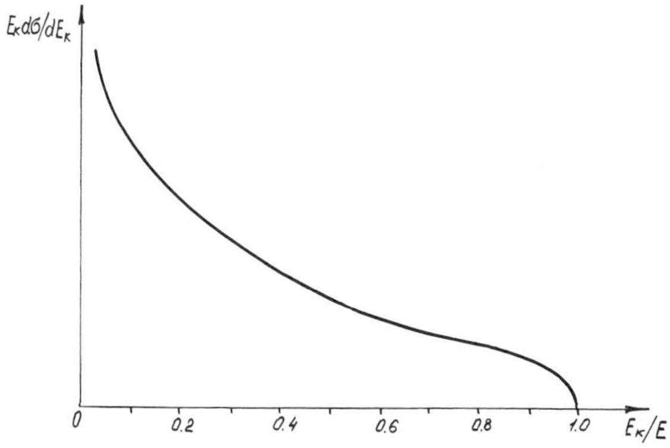

FIGURE 4-1. Spectral energy distribution of electron x-ray bremsstrahlung.

By making the change of variable $\eta = (i\omega/\omega_0)(\xi^3/6)$, the integral obtained reduces to a Γ function:

$$H^{(1)}_{i\omega/\omega_0}\left(\frac{i\omega}{\omega_0}\right) = -\frac{i}{\sqrt{3}\,\pi}\left(\frac{6\omega_0}{\omega}\right)^{1/3}\Gamma(1/3) \tag{4-22}$$

Similarly, we have

$$H^{(1)\prime}_{i\omega/\omega_0}\left(\frac{i\omega}{\omega_0}\right) = \frac{1}{3\sqrt{\pi}}\left(\frac{6\omega_0}{\omega}\right)^{2/3}\Gamma(2/3) \tag{4-23}$$

Substituting Eqs. (4-22) and (4-23) into (4-15) and considering the fact that $\Gamma(x)\Gamma(1-x) = \pi/\sin \pi x$, we find[3]

$$\frac{d\varkappa_\omega}{\hbar\, d\omega} = \frac{16\pi Z^2 e^6}{3\sqrt{3}\,\hbar m^2 c^3}\,\frac{1}{v_0^2} \tag{4-24}$$

Dividing Eqs. (4-20) and (4-24) by E_K and converting to a differentiation with respect to the quantum energy $E_K = \hbar\omega$, we obtain a scattering cross section that is the differential with respect to quantum energy. In addition, we make use of the definitions

$$a_0 = \frac{\hbar^2}{me^2}, \qquad \alpha = \frac{e^2}{\hbar c}, \qquad |n_1| = \left|-i\frac{Ze^2}{\hbar v_0}\right| \tag{4-25}$$

then, instead of Eqs. (4-20) and (4-24), respectively, we have

$$\frac{d\sigma}{dE_K} = \frac{16}{3}\alpha^3 a_0^2 |n_1|^2 \ln\left(\frac{4E_1}{\gamma|n_1|E_K}\right)\frac{1}{E_K} \qquad \text{when } E_K \ll \hbar\omega_0 \tag{4-26}$$

$$\frac{d\sigma}{dE_K} = \frac{16\pi}{3\sqrt{3}}\alpha^3 a_0^2 |n_1|^2 \frac{1}{E_K} \qquad \text{when } E_K \gg \hbar\omega_0 \tag{4-27}$$

The spectral energy distribution of the radiation is given by $I(E_K) = E_K\, d\sigma/dE_K$; Figure 4-1 shows the nature of this dependence.

Equations (4-26) and (4-27) are given in a form that is compatible with the results from quantum theory.[2,5] Quantum theory examines the collision between a nonrelativistic electron having $E = m\mathbf{v}_0^2/2 \ll mc^2$ of energy and momentum $\mathbf{p}_1 = m\mathbf{v}_0$, and a heavy nucleus having charge Ze and mass $M \gg m$. During collision a quantum of radiation having an energy of $\hbar\omega$ and momentum $\hbar\mathbf{k}$ is generated. In its final state the electron has an energy of $E_f = E - \hbar\omega$ and a momentum of $\mathbf{p}_2 = m\mathbf{v}_2$. In an approximation, we can say that energy is not transferred to the nucleus during a collision, but the momentum that is transferred must be finite. The energy of interaction that is responsible for this process is made up of two parts, the energy H_1 of the electrostatic interaction between the charge of the electrons and the nucleus and the energy of the interaction between the charge of the electron and the field of the photons. The scattering of an electron and subsequent generation of a photon arises in a second approximation from the quantum theory of perturbations. The calculation is introduced in a representation of secondary quantization. To conserve space, we will omit these calculations and produce only the most general form of the final formula from this theory[2,5]:

$$E_K \frac{d\sigma}{dE_K} = \frac{16\pi^2}{3} \alpha^3 a_0^2 |n_1|^2 f(n_1, n_2) \tag{4-28a}$$

where

$$f(n_1, n_2) = (e^{2\pi|n_1|} - 1)^{-1}(e^{2\pi|n_2|} - 1)^{-1} x_0 \frac{d}{dx_0} |F(x_0)|^2 \tag{4-28b}$$

$$F(x_0) = F(-n_1, -n_2, 1; x_0) \qquad \text{is a hypergeometric function} \tag{4-28c}$$

and

$$n_1 = -i\frac{Ze^2}{\hbar v_0} \qquad n_2 = -i\frac{Ze^2}{\hbar v_2}, \qquad x_0 = -\frac{4n_1 n_2}{(n_2 - n_1)^2} \tag{4-28d}$$

Near the high-frequency radiation limit $\hbar\omega \to E$ and $v_2 \to 0$ to which correspond parameter values of $|n_2| \gg 1$ and $|n_1| \ll 1$ at which we obtain Eq. (4-27) from Eq. (4-28) if we make use of the asymptotic form of $F(x_0)$. Similarly, in the low-frequency region $\hbar\omega \ll E/|n_1| \ln(h\omega|n_1|/E)$ and the asymptotic form of Eq. (4-28) agrees with Eq. (4-26). Equation (4-27) is called Kramer's relation and it describes the intensity of the fundamental component of the bremsstrahlung. The intensity of the radiation when $\omega_{\max} = E/\hbar$ has a finite value determined by Eq. (4-27). In keeping with the definition of the $|n_1|$ parameter the intensity of the electron bremsstrahlung is inversely proportional to the square of the velocity of the fast electrons: $E_K(d\sigma/dE_K) \sim v_0^{-2}$. This is physically because the duration of the collision $\Delta t = \rho/v_0$ during which the electron is subject to the acceleration $w \sim (Ze^2/\rho^2)(1/m) = F/m$ and radiates decreases as the velocity of the electron increases. However, as the energy of the electrons increases the range of frequencies of the radiated quanta increases, because it occupies the entire band of frequencies from $\omega = 0$ to $\omega_{\max} = E/\hbar$.

The parameters of an electron beam enter into the intensity of the bremsstrahlung spectrum as $(e^4/m^2v_0^2)$ or (e^4/mE), which means that for a specific particle energy E the intensity of the bremsstrahlung spectrum is inversely proportional to the mass m of the particle. If we deal with nuclear particles rather than electrons the bremsstrahlung will be a factor of (M/m) less (when the energy of the beam particles is equal). When the particle velocities are the same, the ratio of the intensity of the bremsstrahlung stimulated by electrons and ions is $(M/m)^2$. The intensity in the bremsstrahlung spectrum depends on the target material in which deceleration occurs as $(Ze)^2$. For this reason, heavy metals that have a high melting temperature must be used for the anodes of x-ray tubes.

It is worthwhile to show how the total energy E_r of the electron x-ray bremsstrahlung can be found from the spectral formulas, Eqs. (4-26) and (4-27). As we have seen, elementary considerations lead to a square law dependence of E_r on the electron energy E (see Eq. (4-3)). The spectral formula given by Eq. (4-27) is suitable for a very broad bremsstrahlung spectrum, and we will say that it can be used for the entire spectrum from $\omega = 0$ to $\omega = E/\hbar$. Then the average energy radiated during a single electron collision is

$$\delta E_r = \int_0^E I(E_K)\, dE_K = \frac{8\pi}{3\sqrt{3}}\, \alpha^3 a_0^2 \left(\frac{Ze}{\hbar}\right)^2 m \tag{4-29}$$

We find the total amount of energy radiated by an electron as it decelerates in the target material by multiplying Eq. (4-29) by the number of atoms per unit volume N and the total range R of an electron in the target:

$$E_r(E) = \delta E_r(E) N R(E) \tag{4-30}$$

The total range R of an electron has a roughly square law dependence on its energy E, according to Eq. (2-147). Substituting Eq. (2-147) into Eq. (4-30) yields

$$E_r(E) = \frac{2Ze^2}{3\sqrt{3}\ln\left(\frac{e}{2}\frac{E}{I}\right)} \frac{E^2}{\hbar m c^3} \tag{4-31}$$

which agrees with Eq. (4-3). Equation (4-31) also reveals the proportionality of the total energy of the x-ray electron bremsstrahlung to the atomic number Z of the target material.

The spectra given by Eqs. (4-26) and (4-27) can be observed directly when a thin target is bombarded by fast electrons. A target is said to be thin when its thickness is much less than the electron range for the given material, $d \ll R$, so that the velocity v_0 of an electron is essentially unchanged during deceleration in the thin layer. In addition, the target thickness must be compared with the x-ray radiation absorption depth $\mu^{-1}(\omega)$ in the target material, where $\mu(\omega)$ is the radiation absorption coefficient as a function of the frequency ω, cm^{-1}. The depth μ^{-1} is relatively shallow for soft x-rays, so that the low frequency boundary of the bremsstrahlung spectrum is more distorted due to absorption.

In this manner we can understand the basic factors that define the bremsstrahlung spectrum seen in thin targets. The average energy of an electron beam decreases with depth of penetration into a target, which shifts the shortwave boundary of the radiation toward the long-wave side. The deeper the decelerating layer is located in a target, the softer is the radiation spectrum that occurs in it. The soft spectrum is absorbed even more strongly, which causes the radiation that escapes from a given layer to lose the soft part of its spectrum to an even greater extent.

A quantitative spectral theory has been developed, specifically for local x-ray spectral analysis to reveal the composition of materials,[(6)] to explain the intensity of the radiation emitted from a target. This method places a thin electron probe having a Gaussian current distribution into the target being investigated. X-rays are emitted from the target area that have been stimulated by the electrons and the characteristic component of these x-rays contains information about the chemical composition of the material. The bremsstrahlung with a continuous spectrum that occurs when this is done is the background against which the intensity of the characteristic lines is measured. These lines must be related to the percent content of the various elements in the stimulated area of the target material.

The characteristic radiation is also used in x-ray lithography as the exposure medium. If x-ray filters that isolate a band of frequencies in the x-ray region are used, then characteristic radiation lines as well as a portion of the bremsstrahlung spectrum may be found in this band. For this reason, it is worth our while to now examine the question of the x-ray line spectra of matter.

4-3. THE FREQUENCIES AND THE TRANSITION PROBABILITIES OF CHARACTERISTIC X-RAY LINES

Ordinarily, the characteristic spectra of materials are stimulated by an electron collision, i.e., by a beam of accelerated electrons acting on a target. There are a number of factors that affect the intensity of the characteristic radiation emitted from the target: 1) a gradual reduction in the density of the electron flux due to elastic scattering and reflection of the electrons; 2) a change in the ionization cross section of the target atoms due to deceleration of the beam electrons and a reduction in their energy with depth as they travel into the target; 3) an increase in the path length ds travelled through a dx thick layer and caused by angular scattering such that dx more ionizations occur in a layer $R = ds/dx$ than when angular scattering is not a factor; 4) the existence of concurrent processes, which makes the probability that an ionized atom will emit a quantum $\hbar\omega$ of characteristic radiation a function of the fluorescence yield $W < 1$. We will find that these factors determine how the radiation that occurs is distributed over depth and that it is mapped as a curve having a maximum. Now we must consider: 5) that the radiation attenuates due to absorption as it emerges from a depth x in the target and, 6) the increase in intensity due to the fluorescent conversion of the x-rays, which have shorter wavelengths than the characteristic radiation being examined and is caused by the photoeffect and the radiation of characteristic quanta. The latter factor is taken into consideration in x-ray spectral analysis of the chemical composition of the target being irradiated. We will not examine this factor in detail because of the small amount of corrections introduced into the radiation intensity.

A special system of designations used in the theory of Auger electron spectra has been incorporated into the systematics of x-ray spectra. The electron shells that have principal quantum numbers $n = 1,2,3,4,5,6,\ldots$ are designated by the letters $K, L, M, N, O, P, \ldots$ and the succession of states that correspond to the values of the orbital number l are designated by Roman numerals[7]:

$1s$	$2s$	$2p$	$3s$	$3p$	$3d$	$4s$	$4p$	$4d$
K_{I}	L_{I}	$L_{\mathrm{II,III}}$	M_{I}	$M_{\mathrm{II,III}}$	$M_{\mathrm{IV,V}}$	N_{I}	$N_{\mathrm{II,III}}$	$N_{\mathrm{IV,V}}$
$4f$	$5s$	$5p$	$5d$	$5f$	$6s$	$6p$	$6d$	
$N_{\mathrm{VI,VII}}$	O_{I}	$O_{\mathrm{II,III}}$	$O_{\mathrm{IV,V}}$	$O_{\mathrm{VI,VII}}$	P_{I}	$P_{\mathrm{II,III}}$	$P_{\mathrm{IV,V}}$	

One shortwave K absorption edge, three L edges, five M edges, etc. are obtained in the photoelectron absorption spectra in the x-ray region.

The characteristic lines in the spectra of the atoms arise due to the dipole radiation transitions that match the selection rules for the quantum numbers[7] $\Delta l = \pm 1$ and

$\Delta J = 0, \pm 1$:

$$K\text{–}L_{\mathrm{II}},\ K\text{–}L_{\mathrm{III}},\ K\text{–}M_{\mathrm{II}},\ K\text{–}M_{\mathrm{III}},\ K\text{–}N_{\mathrm{II}},\ K\text{–}N_{\mathrm{III}},\ldots$$
$$L_{\mathrm{I}}\text{–}M_{\mathrm{II}},\ L_{\mathrm{I}}\text{–}M_{\mathrm{III}},\ L_{\mathrm{I}}\text{–}N_{\mathrm{II}},\ L_{\mathrm{I}}\text{–}N_{\mathrm{III}},\ldots$$

The aggregate of radiation spectral lines that occur when holes are filled in the K shell form a K series and form an L series when holes are filled in the L shell. The L series comprises the L_{I}, L_{II}, and L_{III} subseries, the M series comprises the M_{I}, M_{II}, M_{III}, M_{IV}, and M_{V} subseries, etc.

The frequency of a particular transition, e.g., K–L_{II}, monotonically increases with increasing atomic number Z. The frequency of the $n_2 \to n_1$ transition as a function of the nuclear charge Ze is well known from the theory of a one-electron atom:

$$\omega_{n_2 \to n_1} = 2\pi Ry Z^2(1/n_1^2 - 1/n_2^2) \tag{4-32}$$

Here, $Ry = (me^2/2\hbar^2)$ is the Rydberg constant (a rydberg), n_1 and n_2 are the principal quantum numbers of the states involved in the transition, and Ze is the charge on a one-electron (hydrogen-like) atom. For a multielectron atom having the same nuclear charge Eq. (4-32) is needed for a generalization that accounts for the shielding of the nucleus from an electron occupying the n and l state, and for all of the other electrons. The simplest way to account for shielding was suggested by Slater and we used this method in Chapter 2. According to Slater, an electron in the n, l states moves in the field of the effective (shielding) charge, $Z_{nl}^* e = (Z - \sigma_{nl})e$. The magnitude of the shielding charge $\sigma_{nl}e$ is calculated according to Slater's rules (see Chapter 2). Thus, with due regard for shielding, Eq. (4-32) can be written

$$\omega_{n_1 l_1 \to n_2 l_2} = 2\pi Ry \left(\frac{Z_{n_1 l_1}^{*2}}{n_1^{*2}} - \frac{Z_{n_2 l_2}^{*2}}{n_2^{*2}} \right) \tag{4-33}$$

where n_1^* and n_2^* are the effective quantum numbers according to Slater. Considering the inequalities $n_2 > n_1$, and $Z_{n_2 l_2}^* < Z_{n_1 l_1}^*$, we can say that

$$\omega_{n_1, l_1 \to n_2 l_2} = 2\pi Ry \frac{Z_{n_1 l_1}^{*2}}{n_1^{*2}} \left(1 - \frac{Z_{n_2 l_2}^{*2}}{Z_{n_1 l_1}^{*2}} \frac{n_1^{*2}}{n_2^{*2}} \right) \approx 2\pi Ry \frac{Z_{n_1 l_1}^{*2}}{n_1^{*2}} \tag{4-34}$$

From here we can easily find a relation for the frequency of the i band. This relation is called Moseley's law:

$$\sqrt{\omega_i / 2\pi Ry} = Z_i^* / n_i^* = (Z - \sigma_i)/n_i^* \tag{4-35}$$

Figure 4-2 shows the dependence on Z of the frequencies of the K, L, M, N, and O transitions,(7) where it is evident that Moseley's law is satisfied only approximately. Table 4-1 lists the frequencies of the characteristic lines that lie in the band used in x-ray lithography.(6)

When a hole is created as a result of ionization in the inner shell of an atom in the k state, the filling of the hole by an electron making a transition from a higher energy

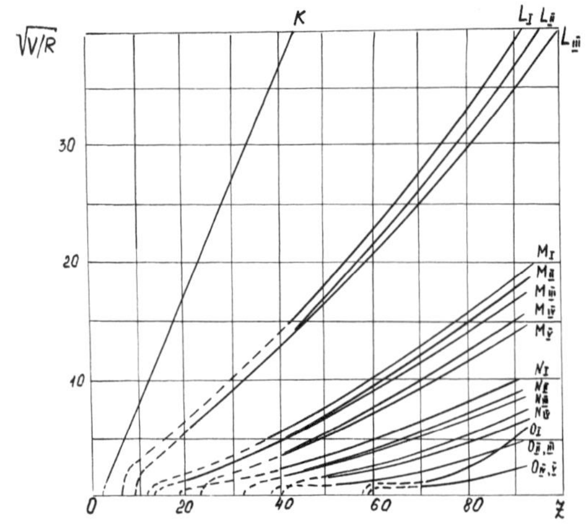

FIGURE 4-2. Frequency of an x-ray absorption edge as a function of the atomic number Z (Moseley diagram).

TABLE 4-1. K_α Lines of the Light-Weight Elements

Element	Z	λ, nm	Quantum energy, eV	Critical stimulation energy, eV
Be	4	11.40	109	112
B	5	6.76	183	192
C	6	4.47	277	284
N	7	3.16	392	400
O	8	2.36	525	532
F	9	1.83	677	687
Na	11	1.191	1041	1071
Mg	12	0.989	1253	1303
Al	13	0.834	1486	1560
Si	14	0.7126	1739	1840
P	15	0.6158	2013	2143

state i is accompanied by the radiation of a quantum $\hbar\omega_{ik}$; the probability of a transition is proportional to the square of the transition frequency ω_{ik}:

$$A_{ik} = \frac{2}{3}\frac{e^2\omega_{ik}^2}{mc^2} f_{ik} \tag{4-36}$$

Here, f_{ik} is the transition oscillator strength. The factor $\gamma = (2/3)(e^2\omega_{ik}^2/mc^2)$ in front of f_{ik} is the radiation attenuation $\gamma = 1/\tau$ of a classical electric oscillator having a frequency of ω_{ik}, when it is caused by the dipole radiation of an oscillating charge as in Eq. (4-1). Thus, the transition oscillator strength is equal to the ratio of the probability of a quantum transition W_{ik} to the attentuation γ of a classical electric dipole having the same frequency ω_{ik}. For electric dipole transitions the oscillator strength can be stated in terms of a matrix element of the dipole moment between the wave functions of the electron's initial and final states[7]:

$$f_{ik} = \frac{2m\omega_{ik}}{\hbar e^2} |\mathbf{p}_{ik}|^2 \tag{4-37}$$

With due regard for Moseley's law, Eq. (4-35),

$$A_{ik} = \frac{2}{3}\frac{e^2 f_{ik}}{mc^2}\frac{Z_k^{*4}}{n_k^{*4}}(2\pi Ry)^2 \tag{4-38}$$

i.e., the probability of radiating an x-ray characteristic quantum that in different atoms corresponds to the same transition, increases with Z as Z_k^{*4}. The probabilities of radiation transitions are especially high for heavy atoms, reaching values of $A = 10^{16}\ \text{sec}^{-1}$. From Eq. (4-38) we easily find[(7)]

$$A_{1n_2} = 2.6 \times 10^9\, Z_1^{*4} f_{1n_2}\ (\text{sec}^{-1}), \qquad A_{2n_2} = 1.6 \times 10^8\, Z_2^{*4} f_{2n_2}\ (\text{sec}^{-1})$$

$$A_{3n_2} = 3.2 \times 10^7\, Z_3^{*4} f_{3n_2}\ (\text{sec}^{-1})$$

for the lines in the K, L, and M series.

4-4. THE PROBABILITY OF AN AUGER TRANSITION; THE X-RAY FLUORESCENCE YIELD

The process by which a hole in a more interior atomic shell is filled is, on the strength of the law of conservation of energy, a two-particle process: in going from a high energy state to a lower energy state an electron must transfer the difference in energy to another particle, which during radiation turns out to be a quantum of radiation. However, it is quite natural to assume that another electron in the atom's shell may acquire this energy. Energy is transferred from one electron to another through the Coulomb interaction (repulsion) $V(r) = e^2/|\mathbf{r}_1 - \mathbf{r}_2|$ between them. In this process the electron that fills the hole drops into a more interior shell and loses $\hbar\omega_{ik}$ of energy; the other electron, by increasing its energy, is emitted (the atom is ionized). The two-electron process that we have just described is called the Auger process (Auger transition). The electron whose energy has been increased is called the Auger electron. If, prior to the Auger transition the energy of this electron was $E_j < 0$, it will be $E_f = E_j + \hbar\omega_{ik}$ afterward. Evidently, an Auger transition will only be "allowed" when the energy E_f of the Auger electron's final state refers to an allowed and unfilled state. The unfilled states of the discrete spectrum for the electrons in an atom (molecule, solid) and all states of the continuous spectrum of electron energies that lie above the discrete spectrum and describe the motion of the electrons that were liberated through ionization correspond to this condition. If an Auger electron has gone into a discrete spectrum state, the final state of the atom will be an excited state with holes in the i and j states and an electron in a outer (normally unoccupied) orbit. If this same electron has gone into a continuous spectrum state, we will have two particles in the final state—an ionization electron and an atom with holes in the i and j states. The kinetic energy of an Auger electron is E_f; in this case E_j is negative and is the ionization energy.

The symbols $K \rightarrow LL$ are used to denote Auger transitions, where the first letter denotes the state with the hole, the second letter denotes the state from which holes are filled, and the third letter denotes the state from which an electron is emitted due to the Auger process. Two new holes are created because of an Auger transition. These holes will also be filled via an Auger or radiation transition. Thus, we can observe cascades of Auger transitions by creating holes in a more interior electron layer (e.g., in the K

shell). Tables of the energy of electrons emitted as a result of Auger transitions and the probabilities of these transitions are available. These tables are used to find the chemical composition of materials from the energy spectra of Auger electrons (see, e.g., [8]).

Wentzel developed a theory for the Auger effect in 1927 which has subsequently been refined in many works.[8] We denote the initial one-electron wave functions of the two electrons that participate in an Auger transition as $\chi_i(\mathbf{r}_1)$ and $\psi_i(\mathbf{r}_2)$. The final state for one of these electrons will be $\chi_f(\mathbf{r}_2)$, which, prior to Auger transition, corresponds to a hole in an atom's shell. The final state of the other electron will be a positive energy state that corresponds to the emitted Auger electron: $\psi_f(\mathbf{r}_1)$. The function $\psi_f(\mathbf{r}_1)$ has been normalized for a single emitted electron over one second in a unit interval of energy.

The probability of an Auger transition is calculated in a first approximation from perturbation theory from

$$W_A = \frac{2\pi}{\hbar}\left|\iint \chi_f^*(\mathbf{r}_2)\psi_f^*(\mathbf{r}_1)\frac{e^2}{|\mathbf{r}_1 - \mathbf{r}_2|}\chi_i(\mathbf{r}_1)\psi_i(\mathbf{r}_2)\, d^3\mathbf{r}_1\, d^3\mathbf{r}_2\right|^2 \tag{4-39}$$

having a force when $W_A \ll \nu_{if}$, where ν_{if} is the quantum frequency corresponding to the transition.

It is interesting to note that an Auger transition can be thought of as a two-step process. A hole in the electron shell is filled in the first step. In the second step the quantum that was emitted as a result of the self-generated internal photoeffect, being unable to escape the atom, is absorbed by a second electron.[8]

According to the Pauli principle, the product of the wave functions from the initial and final states must be replaced by their antisymmetric combinations

$$\begin{aligned}\chi_i(\mathbf{r}_1)\psi_i(\mathbf{r}_2) &\to [\chi_i(\mathbf{r}_1)\psi_i(\mathbf{r}_2) - \chi_i(\mathbf{r}_2)\psi_i(\mathbf{r}_1)]/\sqrt{2}\\ \chi_f(\mathbf{r}_1)\psi_f(\mathbf{r}_2) &\to [\chi_f(\mathbf{r}_1)\psi_f(\mathbf{r}_2) - \chi_f(\mathbf{r}_2)\psi_f(\mathbf{r}_1)]/\sqrt{2}\end{aligned} \tag{4-40}$$

The probability of an Auger transition can then be written

$$W_A = (2\pi/\hbar)|f_1 - f_2|^2 \tag{4-41}$$

where f_1 coincides with the integral, Eq. (4-39), and the amplitude of f_2 describes the exchange transition and is

$$f_2 = \iint \chi_f^*(\mathbf{r}_2)\psi_f^*(\mathbf{r}_1)\frac{e^2}{|\mathbf{r}_1 - \mathbf{r}_2|}\chi_i(\mathbf{r}_1)\psi_i(\mathbf{r}_2)\, d^3\mathbf{r}_1\, d^3\mathbf{r}_2 \tag{4-42}$$

Integrals of the type given by Eqs. (4-39) and (4-42) are well-known in the theory of atomic spectra, where they are respectively called Coulomb and exchange integral and are distinguished from our integral by the fact that $\chi_f = \chi_i$ and $\psi_f = \psi_i$ in them. When these integrals are evaluated, a special procedure for expanding the $|\mathbf{r}_1 - \mathbf{r}_2|^{-1}$ factor in a series of Legendre polynomials

$$|\mathbf{r}_1 - \mathbf{r}_2|^{-1} = (r_1^2 + r_2^2 - 2r_1 r_2 \cos\omega)^{-1/2} = \begin{cases}\displaystyle\sum_{k=0}^{\infty}\frac{r_1^k}{r_2^{k+1}}P_k(\cos\omega), & r_1 < r_2\\ \displaystyle\sum_{k=0}^{\infty}\frac{r_2^k}{r_1^{k+1}}P_k(\cos\omega), & r_2 < r_1\end{cases} \tag{4-43}$$

is used. Here, ω is the angle between the $\mathbf{r}_1$ and $\mathbf{r}_2$ vectors. In turn, $P_k(\cos\omega)$ can be stated in terms of the spherical functions $Y_{lm}(\theta_1,\varphi_1)$ and $Y_{l'm'}(\theta_2,\varphi_2)$ of the angles that are defined by the directions of the vectors $\mathbf{r}_1$ and $\mathbf{r}_2$, according to the theorem concerning the addition of spherical functions:

$$P_k(\cos\omega) = \frac{4\pi}{2k+1}\sum_{q=-k}^{k} Y_{kq}(\theta_1,\varphi_1)Y^*_{kq}(\theta_2,\varphi_2) \tag{4-44}$$

As a result, we obtain

$$\frac{e^2}{|r_1 - r_2|} = e^2 \begin{cases} \displaystyle\sum_{k=0}^{\infty}\sum_{q=-k}^{k} \frac{r_1^k}{r_2^{k+1}}\frac{4\pi}{2k+1} Y_{kq}(\theta_1,\varphi_1)Y^*_{kq}(\theta_2,\varphi_2), & r_1 < r_2 \\ \displaystyle\sum_{k=0}^{\infty}\sum_{q=-k}^{k} \frac{r_2^k}{r_1^{k+1}}\frac{4\pi}{2k+1} Y_{kq}(\theta_1,\varphi_1)Y^*_{kq}(\theta_2,\varphi_2), & r_2 < r_1 \end{cases} \tag{4-45}$$

The wave functions $\chi_i(\mathbf{r}_1)$, $\psi_i(\mathbf{r}_2)$, and $\chi_f(\mathbf{r}_1)$ of the states of an atom's electron are given as the product of individually normed radial and angular components: $R_{nl}(r) \times Y_{lm}(\theta,\varphi)$. The wave function of a free-moving Auger electron can similarly be represented in a spherical coordinate system[9]:

$$\psi_{nlm}(r) = R_{nl}(r)Y_{lm}(\theta,\varphi)$$
$$R_{nl}(r) = (-1)^l \frac{1}{\sqrt{v}}\frac{r^l}{n^l}\left(\frac{1}{r}\frac{d}{dr}\right)^l \frac{e^{inr}}{r}, \qquad n = \frac{p}{\hbar} = \frac{\sqrt{2mE}}{\hbar} \tag{4-46}$$

It is fairly easy to verify the normalization of this wave function for the emission of a single electron in one second in a unit integral of energy

$$\int v|\psi_{nlm}(\mathbf{r})|^2\, df = 1, \qquad df = r^2\, dr \sin\theta\, d\theta\, d\varphi \tag{4-47}$$

By substituting Eq. (4-45) and the wave functions of the initial and final states, the integral of f_1 and f_2 can be written

$$f_1 = \sum_k a_k F_k, \qquad f_2 = \sum_k b_k G_k \tag{4-48}$$

where the quantities a_k and b_k are integral of the angular functions and define the selection rules during the Auger processes, and F_k and G_k are the respective radial integral.

The fundamental conclusion that can be drawn from these calculations is that the probabilities of Auger transitions are independent (or only weakly dependent) on the element's atomic number. The results from calculating the radiative, K–L, and Auger, K–LL, transitions are shown in Table 4-2 for several elements (Pincherle, 1935). As can be seen, the probability of a radiative transition W_r increases by a factor of 230 when Z increases from 10 (neon) to 36 (krypton), whereas the probability of an Auger

TABLE 4-2. Probabilities of Radiative and Auger Transitions as a Function of Z

Element	Z	W_A, 10^{14} sec^{-1}	W_r, 10^{14} sec^{-1}	$W_k = \dfrac{W_r}{W_A + W_r}$
Ne	10	7.40	0.063	0.0085
Mg	12	8.07	0.130	0.016
Ar	18	11.29	0.83	0.068
Ni	28	12.23	4.85	0.284
Kr	36	13.36	14.39	0.517

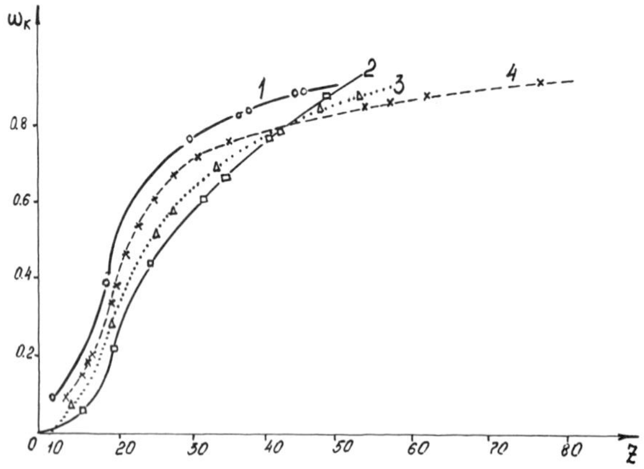

FIGURE 4-3. Fluorescence yield w_K when holes are filled in the K shell.[8]

transition increases by less than two. The small value of the ratio $w_K = W_r/(W_A+W_r)$ for light-weight atoms means that the holes in their shells are filled basically through Auger transitions, whereas they are filled by radiative transitions in the heavy materials. Figure 4-3 graphs the ratio $W_r/(W_r + W_A)$.[8]

As it applies to the intensity of the characteristic x-ray line, the ratio $w_K = W_r/(W_r + W_A)$ bears the name fluorescence yield. For the K shell, w_K can be found from an empirical equation[8]

$$\left(\frac{w_K}{1 - w_K}\right)^{1/4} = A + BZ + CZ^2 \tag{4-49}$$

where A, B, and C are constants determined from experimental data.

The fluorescence yield that corresponds to a specific transition, e.g., K–L_{II}, will be determined by the ratio of the probability of a radiative transition $W^{(r)}_{K\text{-}L_{\text{II}}}$ to the sum of the probabilities of all radiative transitions that end with the filling of a hole in the K shell and begin from states other than the L_{II} state, plus the probabilities of all Auger transitions that end with a hole in the K shell being filled:

$$W_{K\text{-}L_{\text{II}}} = W^{(r)}_{K\text{-}L_{\text{II}}} \Bigg/ \left[\sum_{i \neq L_{\text{II}}} W^{(r)}_{K-i} + \sum_{i,j} W^{(A)}_{K-ij}\right] \tag{4-50}$$

4-5. THE INTENSITY OF THE CHARACTERISTIC RADIATION LINES FROM A TARGET

We will now turn our attention to calculating the intensity of the characteristic x-rays emitted from a target subjected to electron bombardment, giving due regard to all of the factors that were enumerated at the beginning of Section 4-3. The number of ionizations dn that occur in the n and l shell of the atoms of an element of weight A on a path ds of the electron is

$$dn_{n,l} = \sigma_{nl}(N_A\rho/A)\,ds \tag{4-51}$$

where σ_{nl} is the total ionization cross section of the n and l shell of the atom, N_A is Avogadro's number, ρ is the density of the material, and A is the atomic weight of the element.

With Eq. (4-51) we can easily find the intensity of the radiation for thin films whose thickness Δx is much less than the range R_p of electrons having a given amount of energy in the material. Then we can ignore the dependence of the ionization cross section σ_{nl} on the energy of the electrons and we can also assume that the path of the electrons in the material is equal to the thickness of the film, and that reflection and beam attenuation are absent. Because the fluorescence yield w_{nl} is known, we can write the number of quanta $\hbar\omega_{inl}$ radiated when a single electron having an energy of E_0 passes through the film:

$$\Delta I_{inl} = w_{inl}\,\frac{N_A\rho}{A}\,\sigma_{nl}(E_0)\,\Delta x \tag{4-52}$$

For example, for copper when E_0 = 30 keV, $\sigma_{K_\alpha} = 3.55 \times 10^{-22}$ cm^2, $w_{K_\alpha} = 0.422$, $\rho = 8.93$ g/cm^3, $A = 63.54$, and $\Delta I_{K_\alpha} = 12.6\,\Delta x$; Δx is in cm. For a film thickness of 100 nm, $\Delta I_{K_\alpha} \approx 10^{-4}$ photon/electron.

According to Eq. (4-52), we can use experiments on thin films to establish how the total ionization cross section depends on the electron energy E_0. We now take a layer Δx thick at a depth x in a massive target. Generalizing Eq. (4-52) with due regard for the factors that change the intensity of the radiation (the number of quanta), we can write

$$\Delta I_{inl} = w_{inl}\,\frac{N_A\rho}{A}\,\sigma_{nl}[E(x)]n(\rho x)R(\rho x)\,\Delta x \tag{4-53}$$

Here, $n(\rho x)$ is the electron flux attenuation function, $\sigma_{nl}[E(x)]$ is the ionization cross section that corresponds to the energy $E(x)$ of electrons at a depth x, and $R(\rho x) = \Delta s/\Delta x$ is a function that indicates how much longer the true path Δs of an electron is in a layer Δx because of angular scattering. By taking the ratio of Eqs. (4-53) and (4-52) we introduce a function for the radiation distribution over the depth of a massive target[6]:

$$\varphi(\rho x) = \frac{\sigma_{nl}[E(x)]}{\sigma_{nl}(E_0)}\,n(\rho x)R(\rho x) \tag{4-54}$$

Different authors use different approximations for the functions that appear in Eq. (4-54). At this point we introduce functions used by Reiter in a theory of x-ray analysis.[10] Reiter used the linear function

$$n(\rho x) = 1–4 \times 10^4(\sqrt{Z}/E_0^{1.7})\rho x \tag{4-55}$$

for the electron beam attenuation in the initial interval and Lenard's rule

$$n(\rho x) = \exp(-\sigma \rho x), \qquad \sigma = 4.5 \times 10^5 \, E_0^{-1.65} \tag{4-56}$$

for the deeper layers wherein the transition from Eq. (4-55) to (4-56) occurs at the point where they are equal. Philibert[11] uses Lenard's law throughout the depth x.

The following considerations dictate the choice of $R(\rho x)$.[6] In Chapter 2 we showed that up to a depth equal to the diffusion depth x_D, beam spreading can be described by a Gaussian function having a variance that increases with depth; at depths $x > x_D$ the particle "forgets" what its initial direction of motion was and all directions become equiprobable and further movement in the target space is governed by the diffusion equation. A layer Δx that is at a depth greater than x_D is crossed by as many particles moving upward from below as those moving downward from above; also, the path Δs in a single crossing of the layer is longer than the layer is thick, by as much as a factor of two according to Philibert. Consequently, $R((\rho x)$ may be considered equal to $R_\infty = 4$ when $x > x_D$ (Philibert[11]). Reiter[10] took $R_\infty = 3$. The value of $R(\rho x)$ at the surface, $x = 0$, must be correlated with the probability of a particle being reflected. An empirical formula that yields this correlation is[12]

$$R(0) = R_0 = 1 + 2.8(1 - 0.9E_c/E)\eta \tag{4-57}$$

Here, η is the coefficient of electron reflection from the target. Having established the form and values of $R(\rho x)$ at the boundaries we can write[6]

$$R(\rho x) = R_\infty - (R_\infty - R_0)\exp(-K\rho x) \tag{4-58}$$

The K parameter is related to the diffusion depth x_D by[6]

$$\exp(-K\rho x_D) = 0.1 \tag{4-59}$$

according to Heinrich,[12] η can be found from

$$\eta = 0.016Z - 1.86 \times 10^{-4} \, Z^2 + 8.3 \times 10^{-7} \, Z^3 - 0.0254 \tag{4-60}$$

or[6]

$$\eta = (\ln Z)/6 - 1/4 \tag{4-61}$$

Tables of values of $\eta(Z)$ were obtained by the method of interpolation and are given in [6]. The coefficient of reflection for compound materials can be found from a formula that expresses the additive property of the η:

$$\eta = \sum_i C_i \eta_i \tag{4-62}$$

where the C_i are the mass percents of the elements in the material.

We will now turn our attention to the dependence of the ionization cross section σ_{nl} on the x coordinate of a fast electron in a target. This dependence is governed by the

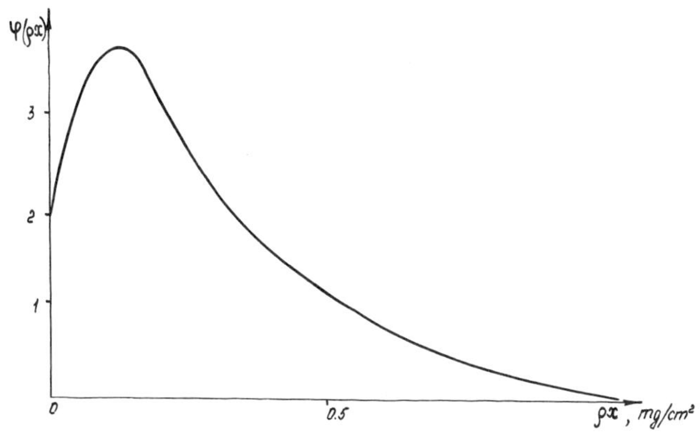

FIGURE 4-4. Shape of the x-ray radiation distribution function relative to target depth when stimulated by a beam of electrons.

fact that the cross section is a function of the energy E of the fast particle. In Bethe's theory, the ionization cross section can be written

$$\sigma_{nl}(E) = E_{nl}^{-2} a_{nl} \ln(b_{nl} E / E_{nl}) / (E / E_{nl}) \tag{4-63}$$

where a_{nl} and b_{nl} are numerical factors (see Section 2-10 for a more thorough discussion). The dependence of $E(x)$ is given in a first approximation by the Thomson–Whiddington law

$$E(x) = E_0(1 - x/R_B)^{1/2} \tag{4-64}$$

in which the range s of the electron is set equal to the depth x that the electron has reached. Bethe's law of deceleration must be used for more accurate calculations.

Figure 4-4 shows the function $\varphi(\rho x)$ found by computer experiment with the Monte Carlo method for electrons having an initial energy of E = 20 keV; the radiation from the K_α line of the carbon in Fe_3C was computed.[6] As is evident, scattering and reflection significantly increase the radiation in layers near the surface of the target. This increase is roughly a factor of two in the subsurface layer and reaches a maximum of almost four near the surface. Equation (4-54) for the function $\varphi(\rho x)$ yields a graph similar to what was calculated via the Monte Carlo method when the functions $n(\rho x)$, $R(\rho x)$, and $\sigma(\rho x)$ mentioned previously are substituted.

We will now find the intensity of the radiation emitted outward from a target into a specific solid angle. Let the distance l from the radiating layer at the depth x depend on the direction in which the solid angle dO was chosen: $l = x/\cos\psi$, where ψ is the angle between the normal to the surface outward and the axis of the solid angle dO. We will consider the mass coefficient of attenuation (absorption) for the radiation line being examined in the target material. Then the radiation being emitted from a layer at the depth x will be attenuated an amount $\exp(-\mu\rho x/\cos\psi)$ and the fraction of radiation emitted into the solid angle dO will be[6]

$$dI_\nu = \int_0^\infty \varphi_\nu(\rho x) \exp(-\mu_\nu \rho x/\cos\psi)\, d(\rho x)\, dO/(4\pi) \tag{4-65}$$

Going to an unnormalized function for the distribution of radiation over depth we find

$$dI_\nu = (1 - \eta_E) n_0 w_{inl} \frac{N_A \rho}{A} \sigma_{nl}(E_0) \int_0^\infty \varphi_\nu \, e^{-(\mu\rho x/\cos\psi)} \, d(\rho x) \frac{dO}{4\pi} \qquad (4\text{-}66)$$

for the total energy being radiated in a characteristic line from a target when that target is irradiated by a flux n_0 ($\text{cm}^{-2} \cdot \text{sec}^{-1}$) of electrons. The fact that the distribution of the characteristic radiation with respect to the angles is isotropic was used so that there is no preferential direction for an x-ray quantum to take as it leaves an atom. The factor $(1 - \eta_E)$ accounts for the fact that the fraction η_E of energy in an electron beam is carried off by reflected electrons.

We will introduce a function for the attentuation of x-rays (an absorption factor) by the relation

$$f(\chi) = \int_0^\infty \varphi(\rho x) \exp(-\chi\rho x) \, d(\rho x) \Big/ \int_0^\infty \varphi(\rho x) \, d(\rho x), \qquad \chi \equiv \mu/\cos\psi \qquad (4\text{-}67)$$

The function $f(\chi)$ monotonically decreases with increasing χ for a given level of electron energy. To see this, we will approximate $\varphi(\rho x)$ in Eq. (4-67) by the step function

$$\varphi(\rho, x) = \begin{cases} \varphi_0 & \text{when } x \le x_0 \\ 0 & \text{when } x > x_0 \end{cases} \qquad (4\text{-}68)$$

The quantity x_0 increases with increasing electron energy E_0 as R_p. Using Eq. (4-68) for $f(\chi)$ we find

$$f(\chi) = [1 - \exp(-\chi\rho x_0)]/\chi\rho x_0 \qquad (4\text{-}69)$$

The function $f(\chi)$ (the absorption function) is close to unity for low energy electrons (x_0 is small) and decreases with increasing E_0 (x_0 increases), which is physically understandable. When the bombarding electrons have low energy, radiation occurs mainly in the subsurface layer and is only weakly attenuated via absorption prior to emerging from the target. As beam energy increases, radiation arises in still deeper layers and is more strongly attenuated before it emerges from the target. Figure 4-5 shows values of $f(\chi)$ for radiation from copper. These values were found experimentally for three values of electron energy.[6] The total radiation energy in a characteristic line as E_0 increases passes through a maximum at some value of electron energy E_p. In fact, as E_0 increases the intensity of the x-ray radiation being stimulated increases, but at the same time the amount of this radiation that is absorbed within the target also increases. Figure 4-6 shows the intensity of the K_α line of radiation from the carbon in diamond as a function of the energy E_0 of the impinging electrons.[6] The electron energy at which the radiation energy in a specific line is a maximum depends on the absorption coefficient μ of the respective radiation line in the target[6]:

$$E_p = 2660(\cos\psi/\mu)^{0.6} \qquad (4\text{-}70)$$

Here, μ is the so-called mass absorption coefficient. From Eq. (4-70) we find for the K_α line for the carbon in diamond that $E_p \approx 16$ keV ($\cos\psi = 1$, and $\mu = 2280$ [6]),

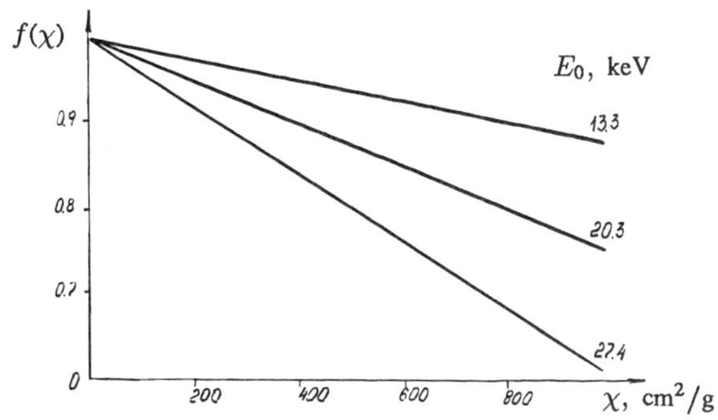

FIGURE 4-5. Graph of the absorption factor f_χ for copper for three values of energy of the stimulating electrons; Green's experimental data.[6]

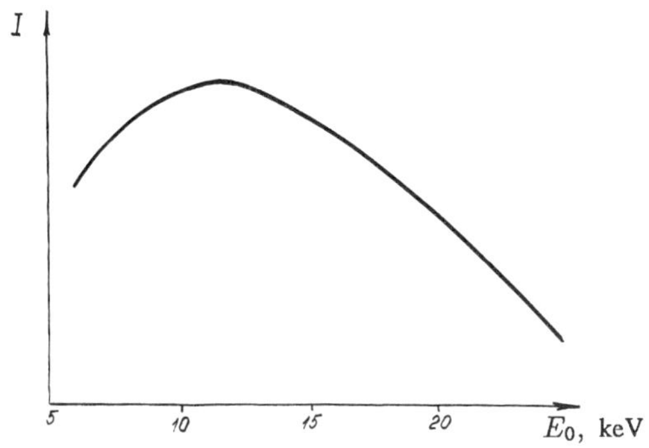

FIGURE 4-6. Intensity of the K_α radiation line in diamond as a function of the energy E_0 of the stimulating electrons.[6]

which agrees with the curve in Fig. 4-6 found experimentally. Similarly, for the K_α line for silicon in silicon we find the estimate $E_p \simeq 190$ keV, since $\mu = 328$ cm^2/g here.[6]

Equation (4-66) can be represented as a flux of quanta being radiated at a frequency ν for a given line of radiation into a unit solid angle:

$$dI_\nu/dO = \Phi_\nu f(\chi_\nu) \tag{4-71}$$

$$\Phi_\nu = \frac{1}{4\pi} n_0 \, \Delta S(1 - \eta_E) w_\nu \left(\frac{N_A \rho}{A}\right) \sigma_\nu(E_0) \int_0^\infty \varphi_\nu(\rho x) \, dx \tag{4-72}$$

Here, Φ_ν is the total number of quanta being radiated in a target into a unit solid angle, ΔS is the surface area of the target being irradiated by a flux of electrons (having a density of n_0 cm$^{-2} \cdot$ sec^{-1}), and $f(\chi_\nu)$ is the absorption factor. In order to convert to the radiation energy in a line, Eq. (4-71) must be multiplied by the quantum energy $h\nu$.

Using Eq. (4-72) we can estimate the number of x-ray quanta in a given line that are radiated by a thick target per unit time into a unit solid angle. Doing the calculations for a single impinging electron yields

$$\frac{\Phi_\nu f(\chi_\nu)}{n_0 \Delta S} = \frac{f(\chi_\nu)}{4\pi}(1 - \eta_E) w_\nu \left(\frac{N_A \rho}{A}\right) \sigma_\nu(E_0) \int_0^\infty \varphi_\nu(\rho x) \, dx \tag{4-73}$$

For the K_α line in copper, $w_{K_\alpha} = 0.422$, σ_{K_α} $(E_0 = 30$ keV$) = 3.55 \times 10^{-22}$ cm^2 and $(N_A\rho/A) = 8.5 \times 10^{22}$ cm^{-3}. We can estimate the quantity $1 - \eta_E$ from data contained in Table 13.2[6]: when $E_c/E_0 = 8.04$ keV$/30$ keV $= 0.266$, $1 - \eta_E = 0.8$. We will assume that $\int_0^\infty \varphi(\rho x)\,dx = 3 \times 10^{-4}$ cm, that we can find the absorption factor for $\mu = 54$ (the mass absorption coefficient for the K_α line of copper in copper), and that $\chi = \mu/\cos\psi = 100$ cm^2/g. According to Fig. 4-5, $f(100) \simeq 0.95$. Making the indicated substitutions gives us

$$\Phi_\nu f(\chi_\nu)/(n_0 \Delta S) = 2.4 \times 10^{-4} \quad \frac{\text{photon}}{\text{electron}}$$

The energy output from an x-ray line will be a factor of $h\nu/E_0 = 0.266$ less:

$$\frac{h\nu\Phi_\nu f(\chi_\nu)\ (\text{J}, K_\alpha)}{E_0 n_0\ \Delta S\ (\text{J, electron flux})} = 0.64 \times 10^{-4}$$

Empirical formulas are often used for the intensity of the characteristic lines and are derived. These formulas are derived by forcing the experimental data to fit in some range of the Z, E, and E_i parameters (E_i is the ionization energy of the i shell). By way of example, we present the empirical formulas for the intensity of the characteristic L_α lines from targets whose atomic numbers are in the range $Z = 50$–82 and the energy E_0 of the bombarding electrons ranges from threshold values to 30 keV[13]:

$$I_{L_\alpha} = 1.3 \times 10^{-7}\, Z^{1.3}(U_{L_3} - 1)^{1.46} \tag{4-74}$$

Here, $U_{L_3} = E_0/E_{L_3}$ is the so-called excess voltage ratio, equal to the ratio of the energy of the electrons in a beam E_0 to the ionization energy of the respective atomic shell (in this case, L_3). The radiation intensity has been defined as the number of characteristic photons per line into a unit of solid angle and calculated for a single electron in the beam.

4-6. ABSORPTION AND THE BREMSSTRAHLUNG INTENSITY FROM A TARGET

The intensity of the bremsstrahlung from an absorbing target can be calculated in much the same way that the intensity of the characteristic radiation is. Using the differential cross section given by Eq. (4-27) for the bremsstrahlung we can write

$$\frac{\delta E_r}{\delta E_K}\,\delta E_K = E_K\,\frac{d\sigma}{dE_K}\,dE_K N\,dx \tag{4-75}$$

for the energy of the bremsstrahlung photons in the interval $E_K, E_K + dE_K$ that arise in a layer $x, x + dx$ in the target. To find the intensity of the bremsstrahlung in the $E_K, E_K + dE_K$ range of quantum energies we multiply Eq. (4-75) by $(1 - \eta_E)$ to take electron reflection into account, by the function for the angular distribution of the bremsstrahlung $f(\psi)$, by the function $\exp(-\mu\rho x/\cos\psi)$ that describes the attenuation of the radiation as it emerges from the target in the ψ direction, by the function $R(\rho x) = ds/dx$ that indicates how many times the path ds of an electron in the target is longer than is the target thickness dx, and by the function $n(\rho x)$ for the electron flux attenuation. The radiation energy in an interval dE_K into a solid angle dO in the ψ direction is

$$\frac{dE_r}{dE_K}\,dE_K = \frac{(1-\eta_E)\,dO}{4\pi}\int_0^\infty E_K\,\frac{d\sigma}{dE_K}\,dE_K\,f(\psi)e^{-\mu\rho x/\cos\psi}\,n(\rho x)R(\rho x)\,dx \tag{4-76}$$

For the $R(\rho x)$ and $n(\rho x)$ we can use functions that we introduced when we calculated the intensity of the characteristic radiation. We can then state the dependence of $v_0^2(x)$ in $d\sigma/dE_K$ according to the Thomson–Whiddington law:

$$v_0^2(x) = v_0^2(0)(1 - x/R_B)^{1/2} \tag{4-77}$$

More complicated questions arise when we define the angular distribution function $f(\psi)$. The equations

$$d\varkappa_\omega = \frac{dO\,d\omega}{c^3}\left[A(\omega) + B(\omega)\,\frac{3\cos^2\psi - 1}{2}\right] \tag{4-78}$$

where

$$A(\omega) = \frac{2\omega^4}{3}\int_0^\infty d_\omega^2 2\pi\rho\,d\rho, \qquad B(\omega) = \frac{\omega^4}{3}\int_0^\infty (d_\omega^2 - d_{x\omega}^2)2\pi\rho\,d\rho \tag{4-79}$$

are the starting point for calculating this function.[3] Here the x axis is directed along the axis of the beam, being an isolated direction. A special investigation that we did showed that the anisotropy of the bremsstrahlung is not high and we can therefore assume that $f(\psi) = 1$. Then we have

$$\frac{dE_r}{dE_K}\,dE_K = \frac{dO\,dE_K}{4\pi}\,E_K\,\frac{d\sigma(v_0(0))}{dE_K}(1-\eta_E)J(E_K) \tag{4-80}$$

$$J(E_K) = \int_0^{R_B-\Delta(E_K)} \frac{[\exp(-\mu\rho x/\cos\psi - \sigma\rho x)][R_\infty - (R_\infty - R_0)e^{-K\rho x}}{(1 - x/R_B)^{1/2}}\,dx \tag{4-81}$$

instead of Eq. (4-76). The limit of integration over depth is limited here by the distance at which the electron energy decreases to a value of $E(R_B - \Delta) = E_K$, so that when $x > R_B - \Delta$, the bremsstrahlung quanta E_K cannot be generated. Using the Thomson–Whiddington law, we find

$$\Delta(E_K) = R_B(E_K/E(0))^2 \tag{4-82}$$

Evaluating the integral, Eq. (4-81), by numerical integration for specific values of the $E(0) = E_0$, E_K, $\mu(E_K)$, σ (see Eq. (4-56)), etc. parameters we can find the spectral intensity of the bremsstrahlung calculated for a single electron according to Eq. (4-80).

Empirical formulas are normally used to describe the spectral intensity of the bremsstrahlung from a solid target, giving due consideration to absorption in the target. The simplest in form is the so-called Kramers formula[14]

$$I(E_K) = KZ(E - E_K) \tag{4-83}$$

where Z is the atomic number of a one-element target and K is a constant.

A more attentive analysis shows that the experimental curves must be described by a more complex formula(15)

$$I(E_K) = K' Z^n (E - E_K)^{1.11} \tag{4-84}$$

where K' is a constant and the exponent n is a weak function of the E, E_K, and Z parameters and varies between 1.2 and 1.37. Apparently, one of these latter empirical formulas is suggested in [16]:

$$\begin{aligned} I(E_K) &= 10^5 \, e^B [Z(E/E_K - 1)]^M \\ B &\equiv 5.80 - 0.0322E \\ M &\equiv 1.05 + 0.00599E \end{aligned} \tag{4-85}$$

Here the intensity is given in terms of the number of photons per coulomb of charge delivered to the target by the current and for a photon energy range of 10 keV. The formula shown describes the bremsstrahlung for a range of electron-beam energies between 10 and 40 keV that are commonly used in lithography and for a large number of elements whose atomic numbers range from 4 to 92. Equations (4-83) to (4-85) can be used to calculate the background noise in resist layers by the x-ray bremsstrahlung from a solid target (see below).

4-7. THE THERMAL LIMITATION ON THE POWER IN AN X-RAY SOURCE

A stationary source of x-rays is desirable for x-ray lithography. The properties of the source must be stable over its lifetime. Obviously, the stability of an x-ray tube type of source will be defined by the operating stability of the electron gun and the stability of the anode being bombarded by an electron beam to produce characteristic x-rays. When an anode is bombarded by electrons in a beam its stability is governed to large extent by its thermal stability. The maximum elevation of anode temperature ΔT, due to being bombarded by electrons having an energy of V_0 (kV) and current I (A) from a beam of diameter d (cm) is(17)

$$\Delta T(K) = 4.8 \frac{V_0 I}{kd} \tag{4-86}$$

Here, k is the thermal conductivity of the anode material in $\mathrm{W \cdot cm^{-1} \cdot K^{-1}}$. We will assume, for estimating purposes, that a copper anode remains stable during operation in a vacuum when locally heated by an electron beam to a temperature of 300°C. The thermal conductivity of copper at T = 300°C is k = 3.76 $\mathrm{W \cdot cm^{-1} \cdot K^{-1}}$. We will assume that the beam diameter d is 0.1 cm. Then, the allowable power dissipation for the electron beam is, from Eq. (4-86), 22 kW. When the coefficient that describes the conversion of the electrical power in the beam to x-ray power is on the order of 10^{-4}, e.g., a K_α line of copper, we find that the radiation power in this line is about 0.17 W/steradian. By using rapidly spinning anodes to facilitate heat removal, the electrical power can be held to 25 kW. Nevertheless, the radiation power in a single characteristic line will only be about 0.1 mW/cm^2 when the distance between the anode and the wafer being irradiated (in x-ray lithography) is 40 cm.(18)

4-8. THE BREMSSTRAHLUNG IN A HOT MAXWELL PLASMA

It has been shown in numerous experiments that a hot ($T \simeq 10^6$ K) plasma is a source of intense x-ray and ultraviolet radiation. Because the relationship between the radiation spectrum and the temperature of an electron plasma is relatively simple, measurements of the x-rays are an important tool in diagnosing a plasma, e.g., in experiments in laser thermonuclear reaction. Here, as in [6], we will give a brief derivation of the formula for the intensity of the x-ray radiation emitted from a so-called Maxwell plasma. It is assumed for such a plasma that there is internal equilibrium in the system of electrons that are described by a Maxwell distribution relative to the velocities with a temperature T:

$$dW = 4\pi \left(\frac{m}{2\pi kT}\right)^{3/2} \exp\left(-\frac{mv^2}{2kT}\right) v^2\, dv \tag{4-87}$$

The radiation of quanta in a high temperature plasma occurs because of the scattering of fast thermal electrons by the nuclei of the ions, similar to the bremsstrahlung of monochromatic electrons in a beam being decelerated in a target. We will use Kramers' formula

$$E_K \frac{d\sigma}{dE_K} = \frac{16\pi}{3\sqrt{3}}\, \alpha^3 a_0^2 \left|\frac{Ze^2}{\hbar v}\right|^2 \tag{4-88}$$

as the cross section for the bremsstrahlung process. The number of collisions per 1 cm^3 between electrons having velocities of $v, v + dv$ and atomic nuclei, which lead to the radiation of quanta having an energy of $E_K, E_K + dE_K$ is

$$N_i N_e\, dW(v, v + dv) \frac{d\sigma(v)}{dE_K}\, dE_K v \tag{4-89}$$

where N_i and N_e are the ion and electron densities, respectively.

Multiplying Eq. (4-89) by the quantum energy E_K and integrating over the electron velocities from a value v^* that corresponds to the frequency of the quantum being radiated, $E_K = mv^{*2}/2$, gives us

$$\begin{aligned} dE_r(E_K, E_K + dE_K) &= \frac{16\pi}{3\sqrt{3}}\, \alpha^3 a_0^2 \left(\frac{Ze^2}{\hbar}\right)^2 N_i N_e \\ &\quad \times 4\pi \left(\frac{m}{2\pi kT}\right)^{3/2} dE_K \int_{\sqrt{2E_K/m}}^{\infty} \exp\left(-\frac{mv^2}{2kT}\right) v\, dv \end{aligned} \tag{4-90}$$

This integral is easily evaluated and we find that

$$\frac{dE_r}{dE_K}\, dE_K \equiv I(E_K)\, dE_K = \frac{32\pi}{3\sqrt{3}}\, \alpha^3 a_0^2 \left(\frac{Ze^2}{\hbar}\right)^2 \left(\frac{m}{2\pi kT}\right)^{1/2} N_e N_i e^{-E_K/kT}\, dE_K \tag{4-91}$$

This formula gives the spectral distribution of the radiation energy from 1 cm^3 of a plasma. For a volume of ΔV, we must multiply Eq. (4-91) by ΔV.

It is obvious that in order to stimulate the radiation of quanta having an energy of E_K the electron temperature of the plasma must be raised to approximately $T = E_K/k$, which, for a photon energy of $E_K = 1$ keV for x-ray lithography, corresponds to a temperature of $T = 10^6$–10^7 K. A dense, solid-state plasma having these kinds of temperatures occurs when the surface of a solid is acted upon by high power pulses of laser radiation and by pulses of high density electric current (the pinch effect). Experiments are being carried out on using these pulsed sources of x-ray radiation for lithography.[8] Another mechanism for the generation of radiation with a continuous spectrum in a plasma comprises recombination processes in which an ion captures a free electron at a discrete, unoccupied ion energy level that corresponds to a bound state for the electron in an atom. We can distinguish the processes of direct recombination

$$A^{+(Z_i)} + e^- \to A^{+(Z_i-1)} + \hbar\omega \tag{4-92}$$

and bi-electron recombination

$$A^{+(Z_i)} + e^- + e^- \to A^{+(Z_i-1)} + \hbar\omega + e^- \tag{4-93}$$

in which a portion of the energy and momentum of the electron being captured are transferred to a second electron. The electron capture cross section in the direct recombination process with a coulomb center having a charge of Ze (the bare nucleus having atomic number Z) at a level whose principal quantum number is n, is[19]

$$\sigma_n = \frac{8\pi}{3\sqrt{3}}\frac{\alpha^5}{n^3}\frac{Z^4}{E(E+E_n)}$$

where the cross section is given in the so-called atomic units $\hbar = c = m_e = 1$, and $\alpha = 1/137.037$, E is the electron energy, and $E_n = Z^2\alpha^2/2n^2$ is the binding energy of the captured electron. The recombination radiation, by adding to the bremsstrahlung, creates plasma x-rays having a continuous spectrum.

Naturally, the characteristic spectral lines of the atoms of a solid target accompany the bremsstrahlung having the continuous spectrum in a solid-state hot plasma. If the temperature T of the plasma is high, the plasma contains ions that have been highly ionized so that the equilibrium state in the plasma is characterized by an ions distribution with respect to the degree of ionization and the populations of the levels for these ions. The electron component of the plasma is capable of ionizing the inner shells of the atoms if we consider the high energy "tail" of the plasma electrons distribution with respect to energy. The intensity of a line that corresponds to an $i \to k$ transition between the levels of a r ionized atom whose concentration in the plasma is N^r is given by the expression given in [5, p. 361]:

$$I_{ik} = \hbar\omega_{ik}A_{ik}N^r\frac{g_i}{S_r}e^{-E_i/kT} \tag{4-94}$$

Here, the A_{ik} is the probability of a radiation $i \to k$ transition, g_i is the statistical weight (the degree of degeneration) of the i state, and $S_r = \sum_i g_i \exp(-E_i/kT)$ is a statistical sum. For a given temperature and atom density N in a plasma, we can calculate the concentrations N^r of the ions of multiplicity r if we assume that the plasma is in internal equilibrium. The intensity of a given line as a function of temperature is given by a Boltzmann factor that describes the probability of excitation in the E_i level.

If the plasma layer is much thicker than μ_{ik}^{-1}, where μ_{ik} is the absorption coefficient of the plasma at a frequency ω_{ik}, the intensity of the I_{ik} line must then be calculated with due regard for absorption. For more information on this subject, we refer the reader to [5].

The control parameters for producing a plasma by pulses of laser radiation on a plane, solid target are the power density (intensity) of the laser pulse Φ (W/cm^2), the size of the laser spot (μ) on the target and energy E, the duration τ of the laser pulse, and the wavelength λ of the laser radiation. Intensities of 10^{14}–10^{16} W/cm^2 can be obtained when laser pulses only picoseconds in duration are used and x-ray quanta up to 50 keV in energy are generated under these conditions. This kind of hard radiation is unsuitable for x-ray lithography.[20–22] The plasma generation parameters must be chosen such that the high energy boundary of the x-ray spectrum falls off when the quantum energy is 2–3 keV and exhibits intense radiation in the $E_K \approx 1$ keV range of energies. A laser plasma having the desired properties can be produced by nanosecond solid-state and excimer lasers having an intensity of $\Phi = 10^{11}$–10^{13} W/cm^2.

A systematic investigation of the dependence of the x-ray radiation from a laser-generated plasma on the intensity of the laser radiation Φ was conducted in [28] for intensities of $\Phi = 3 \times 10^{11}$–10^{13} W/cm^2 on targets having atomic numbers ranging from 6 to 73. A neodymium laser having a pulse duration of 5 nsec and pulse power of 1 GW was used in the experiments. The radiation was focused onto the target by a lens as a spot 100 μm in diameter. The intensities of the pulses were changed by varying the energy of the laser pulse. An aluminized polymer film 0.48 mg/cm^2 thick (75 μg/cm^2 oxygen, 0.30 mg/cm^2 carbon, 0.10 mg/cm^2 aluminum) was used as the filter. The intensity of the x-rays from a laser-generated plasma (with due regard for the transparency of the filter) can be described by a power function of Φ:

$$I_x = k\Phi^\alpha \tag{4-95}$$

in the range $3 \times 10^{11} \leq \Phi \leq 10^{13}$ W/cm^2 for targets having atomic numbers Z from 6 to 73. For the various targets α varies between 1.9 and 2.8. For a given value of Φ the value of I_x as a function of the target's atomic number Z is given by a curve having maxima and minima; the variation in I_x for the different elements is as much as an order of magnitude.

The experimental results can be interpreted on the basis of the models for heating a solid-state plasma by laser radiation and ionization equilibrium in the plasma. For intensities of $\Phi \leq 10^{13}$ W/cm^2 and laser pulse durations of $\tau > 1$ nsec, the plasma heating process can be thought of as a stationary self-regulating process. The time needed to establish a stationary state can be estimated as $t = R/v_s$, where R is the radius of the laser spot; v_s is the velocity of the ion wave; and $t < 0.5$ nsec, which is much shorter than the $\tau = 5$ nsec duration of the laser pulse, so that the state of the laser plasma can be considered stationary. For the stationary model the energy flux density Φ in a laser pulse determines the electron temperature T_e in the plasma according to

$$T_e = k_e\Phi^{4/9} \tag{4-96}$$

given in [28] for the range of values $3 \times 10^{11} \leq \Phi \leq 10^{13}$ W/cm^2 that correspond to electron temperatures of $T_e = 60$–250 eV. The electron temperature of the plasma makes it possible for us to calculate the intensity of the x-ray radiation from the plasma. To do the calculations we need a model that will allow us to estimate the relative concentrations of the different charge states of the ions in the plasma, e.g., the corona model is so called because it has been borrowed from the theory of stellar coronas.(29) Calculations of the intensities of the various components of plasma x-rays in the characteristic radiation lines I_L (bound–bound transitions), in recombination radiation I_{fb} (free–bound transitions), and in bremsstrahlung I_{ff} (free–free transitions) show that in the range T_e = 60–250 eV of plasma temperatures all of the components of x-rays increase rapidly with temperature T_e: for the total radiation being observed a dependence (with due regard for absorption in the filter)

$$I_x \sim T_e^{5.6} \tag{4-97}$$

has been predicted. Combining Eqs. (4-95) and (4-96) gives us $I_x \sim T_e^{2.25\alpha}$, which is nearly Eq. (4-97), since $1.8 \leq \alpha \leq 2.8$.

It is significant to note that the laser radiation efficiency in heating a plasma increases as the wavelength λ of the radiation decreases. For example, by converting the infrared portion of laser radiation λ = 1.05 μm to its third harmonic λ = 0.35 μm we lose 50% of the radiation energy; however, these losses are compensated by the increase in the efficiency of plasma heating at the wavelength λ = 0.35 μm. Also, shortwave laser radiation can be focused into a smaller spot, which reduces the size of the laser x-ray source.(26) When resists are exposed to pulsed x-ray radiation their sensitivity is seen to increase, possibly because of indirect heating in the resist layer.(26)

High concentrations of multiply ionized ions ($N_i > 10^{18}$ cm^{-3}) are generated in a gaseous plasma from a Z-pinch discharge. In the area in which the discharge radius is $r \simeq 0.4$ mm, the electron temperature increases to $T_e \simeq 150$ eV; therefore, the discharge can serve as a source of soft x-rays and far ultraviolet radiation.(33)

To obtain a Z-pinch discharge(30) a gas in the form of an annular flux r_{mean} = 1.25 cm, $r_2 - r_1$ = 0.3 cm) is spewed through an ultrasonic nozzle into the space between the anode and cathode and perpendicular to the plane of the electrodes (the electrodes are wire grids). A capacitor bank is capable of providing a discharge current of 1.45 MA through a pre-ionized gas. About 100 nsec after the current pulse begins, the magnetic field compresses and heats the ionized gas to such an extent that intense x-ray radiation, associated with transitions into the $1s$ state of the K shell (neon atoms that have been ionized nine and ten times) appears.(30) The energy of these transitions lies in the 915–1334 eV ($\lambda \simeq 1$ nm) range, corresponding to an optimum for x-ray lithography. A pulse of x-rays lasts for about 40 nsec and has an energy of 1–2.5 KJ, more than 90% of which is concentrated in the characteristic lines.(30) The radiation intensity is different at different points along the Z axis of the discharge. Measurements of the radiation from a Z-pinch discharge in argon were made in [31, 32] for wavelengths in the 4–30 nm range.

4-9. PROPERTIES OF SYNCHROTRON RADIATION

In the electron accelerators called synchrotrons because of the acceleration principles used (the electron's period of rotation in its orbit matches that of the high-frequency accelerating electric field), the electrons are accelerated to energies of 1 GeV = 10^9 eV. The ratio of this energy to the characteristic energy m_0c^2 describes the electron's degree of relativism. Because m_0c^2 = 0.5 MeV, for electrons having an energy E = 1 GeV the ratio is

$$\frac{E}{m_0c^2} \equiv \gamma = \frac{m_0c^2}{\sqrt{1 - v^2/c^2}} \frac{1}{m_0c^2} = (1 - \beta^2)^{1/2} \simeq 2 \times 10^3 \tag{4-98}$$

i.e., is much greater than unity. By convention, electrons having these large γ are called ultrarelativistic.

An electron's orbit in an accelerator is either circular or contains segments in the form of the arcs of a circle. In these sections of its trajectory an electron has acceleration and, like any charged particle being accelerated, must radiate an electromagnetic field. In this manner we see that the radiation from an electron in an accelerator (synchrotron radiation) and the bremsstrahlung examined previously are, in principle, the product of a single process: the radiation from an electron that is being accelerated. The difference is that electrons in an accelerator have been driven to ultrarelativistic energies, whereas earlier we examined the bremsstrahlung from nonrelativistic electrons having energies of 10^4 eV, when

$$\gamma = \frac{m_0v^2}{2} \Big/ (m_0c^2) \leq 1$$

I. M. Ternov *et al.* have discussed synchrotron radiation theory in detail in their books(34,35); other references are [36–38]. Our brief discussion will follow these sources.

An electron's orbit becomes circular through the action of the magnetic field H perpendicular to this orbit. The (Larmor) radius r of the orbit, the Larmor frequency ω_0, and the centripetal acceleration w are, in the relativistic cases,(35)

$$r = \frac{m_0cv}{eH\sqrt{1 - v^2/c^2}} = \gamma r^{(nr)} \gg r^{(nr)} \tag{4-99}$$

$$\omega_0 = \frac{v}{r} = \frac{eH}{m_0c}\sqrt{1 - v^2/c^2} = \frac{\omega_0^{(nr)}}{\gamma} \ll \omega_0^{(nr)} \tag{4-100}$$

$$|\mathbf{w}| = \frac{e}{m_0c\gamma}\,|[\mathbf{v} \times \mathbf{H}]| = v\omega_0 = \omega_0^2 r = \frac{v^2}{r} \tag{4-101}$$

Here, $r^{(nr)}$ and $\omega_0^{(nr)}$ denote the radius of the orbit and the Larmor frequency, respectively, of an electron in a field H. As we can see, a heavy ($m = \gamma m_0$) relativistic electron has a larger (by γ times) orbit radius and lower (by γ times) rotational frequency in this orbit.

We will choose a coordinate system and a point P for observing the radiations as shown in Fig. 4-7. The coordinates of an electron in the plane of the orbit are

$$x = R\cos\psi, \qquad y = R\sin\psi, \qquad \psi = \omega\tau \tag{4-102}$$

The observation point in a spherical coordinate system is given by the R', θ, and φ parameters. We will also designate $\chi = \psi - \varphi$ and $\alpha = \chi + \pi/2$. The power flow through an area $R'^2\,dO$ perpendicular to $\mathbf{R}'$ at the point P is

$$dI = \frac{c}{4\pi}|\mathbf{E}|^2 R'^2\,dO \tag{4-103}$$

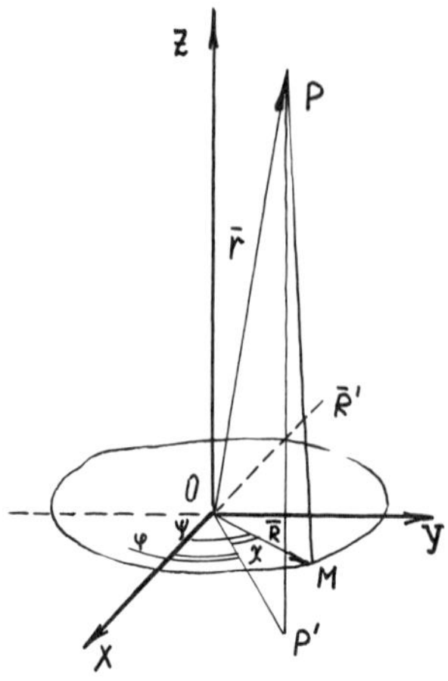

FIGURE 4-7. The coordinates of an electron in a circular orbit (the point M) and the point P at which its radiation is observed.

In order to find the **H** (or **E**) vector in the wave zone of radiation we calculate the vector-potential **A** ($\mathbf{R}'$) of the field at P and generated by an electron moving in a circular orbit. Because the velocity of the electron and the velocity of light c are comparable in magnitude, we must take the effects of delay into account.

A solution of d'Alembert's equation for the vector potential, with regard for delay is

$$\mathbf{A}(\mathbf{r}) = \frac{e}{c} \oint \frac{\mathbf{v}(\tau)}{R'} \delta\left(\tau - t + \frac{|\mathbf{R}'|}{c}\right) d\tau \tag{4-104}$$

τ is the instant of radiation in a laboratory coordinate system $Oxyz$, and $t = \tau + R'/c$ is the instant at which the radiation is observed at the point P. The integration is done over a period of time equal to the electron's period of revolution about its orbit, T. The quantity R' in the denominator of Eq. (4-104) can be replaced by r, whereas in the numerator we can use the approximation

$$R' = r\left(1 - \frac{2\mathbf{Rr}}{r^2} + \frac{R^2}{r^2}\right)^{1/2} \simeq r\left(1 - \frac{\mathbf{r}\cdot\mathbf{R}}{r^2}\right) = r - R\sin\theta\cos\chi \tag{4-105}$$

We will seek the components of the Fourier expansion of the vector potential as a series of frequencies that are multiples of the Larmor frequency ω_0:

$$\mathbf{A}(\mathbf{r}) = \sum_{\nu=-\infty}^{+\infty} \mathbf{A}(\nu) e^{-i\nu\Phi} \tag{4-106}$$

If in Eq. (4-104) we replace the δ-function with the series

$$\delta\left(\tau - t + \frac{r}{c} - \frac{R}{c}\sin\theta\cos\chi\right) = \frac{\omega_0}{2\pi} \sum_{\nu=-\infty}^{+\infty} \exp\left[i\nu\omega_0\left(\tau - t + \frac{r}{c} - \frac{R}{c}\sin\theta\cos\chi\right)\right] \tag{4-107}$$

we find

$$\mathbf{A}(\nu) = \frac{e}{2\pi c r} \int_{-\pi}^{\pi} \mathbf{v} \exp[i(\nu\alpha - \nu \sin\theta \sin\alpha)]\, d\alpha \tag{4-108}$$

$$\alpha = \chi + \pi/2, \qquad \Phi = \omega t - \omega\frac{r}{c} - \varphi + \frac{\pi}{2} \tag{4-109}$$

Taking the projection of the velocity vector onto the unit vectors $\mathbf{e}_\varphi$ and $\mathbf{e}_\theta$ in a spherical coordinate system

$$v_\varphi = v\cos\chi - v\sin\alpha \tag{4-110}$$

$$v_\theta = v\cos\theta\sin\chi = -v\cos\theta\cos\alpha \tag{4-111}$$

we write the projections of the Fourier components of the vector potential $\mathbf{A}(\nu)$ in these directions:

$$A_\varphi(\nu) = \frac{ev}{2\pi c r} \int_{-\pi}^{\pi} \sin\alpha \exp[i\nu(\alpha - \beta\sin\theta\sin\alpha)]\, d\alpha \tag{4-112}$$

$$A_\theta(\nu) = -\frac{ev\cos\theta}{2\pi c r} \int_{-\pi}^{\pi} \cos\alpha \exp[i\nu(\alpha - \beta\sin\theta\sin\alpha)]\, d\alpha \tag{4-113}$$

$$A_r = 0$$

Using the integral form of Bessel functions

$$J_\nu(x) = \frac{1}{2\pi} \int_{-\pi}^{\pi} \exp[i\nu(\alpha - \sin\alpha)]\, d\alpha \tag{4-114}$$

we find

$$A_\varphi(\nu) = i\frac{ev}{rc} J'_\nu(\nu\beta\sin\theta) \tag{4-115}$$

$$A_\theta(\nu) = -\frac{e}{r}\cot\theta \cdot J_\nu(\nu\beta\sin\theta) \tag{4-116}$$

instead of Eq. (4-112). Using the relationship between the vector potential and the magnetic field strength $\mathbf{H} = \operatorname{rot}\mathbf{A}$ we find the projection of the magnetic field in the directions of the unit vectors $\mathbf{e}_\varphi$ and $\mathbf{e}_\theta$:

$$H_\varphi = \frac{1}{r}\frac{\partial}{\partial r}(rA_\theta) = \frac{2e\beta\cot\theta}{Rr} \sum_{\nu=1}^{\infty} \nu J_\nu(\nu\beta\sin\theta)\sin\nu\Phi \tag{4-117}$$

$$H_\theta = -\frac{1}{r}\frac{\partial}{\partial r}(rA_\varphi) = -\frac{2e\beta^2}{Rr} \sum_{\nu=1}^{\infty} \nu J'_\nu(\nu\beta\sin\theta)\cos\nu\Phi \tag{4-118}$$

The flux of radiation energy into the solid angle dO is

$$dI = r^2(c/4\pi)(H_\varphi^2 + H_\theta^2)\, dO \tag{4-119}$$

Our interest lies in the average radiation energy period of motion, because there are no methods for finding the instantaneous values of electron radiation power. In the course of taking the average we use the orthonormality of the trigonometric functions:

$$\frac{1}{T}\int_0^T \cos\nu\Phi\cos\nu'\Phi\,dt = \frac{1}{T}\int_0^T \sin\nu\Phi\sin\nu'\Phi\,dt = \frac{1}{2}\,\delta_{\nu,\nu'}$$

$$\frac{1}{T}\int_0^T \cos\nu\Phi\sin\nu'\Phi\,dt = 0$$

As a result, we obtain a function for the angular distribution of the spectral component of the radiation in a ν-multiple frequency of the electron's orbital frequency ω_0:

$$dI(\nu,\theta) = \frac{ce^2\beta^2}{2\pi R^2}\,\nu^2[\cot^2\theta\cdot J_\nu^2(\nu\beta\sin\theta) + \beta^2 J_\nu'(\nu\beta\sin\theta)]\,dO \tag{4-120}$$

The formula obtained contains a remarkable amount of information. Integrating it over all angles we obtain the total radiation energy at a given frequency. By summing the result obtained over all frequencies (in ν from 1 to ∞), we find the total radiation energy. Asymptotic forms of Eq. (4-120) for high ($\nu \gg \gamma^3$) and low ($1 \ll \nu \ll \gamma^3$) frequencies reveal the nature of the synchrotron radiation spectrum in the shortwave (x-ray) region and as viewed from the long wave region.

Equation (4-120) must be augmented with information about how the synchrotron radiation is polarized. We can separate the electron's radiation into two components that are related to the geometry of the experiment: we call the component with the electric vector of the field in the plane of the orbit a σ-component, and the component with the electric vector along the external magnetic field (perpendicular to the plane of the orbit) a π-component. We write the electric vectors for the field of this components as

$$E_\sigma = c_\sigma\cos(\omega t + \varphi_\sigma), \qquad E_\pi = c_\pi\cos(\omega t + \varphi_\pi) \tag{4-121}$$

The polarization is uniquely defined if the ratio c_σ/c_π and phase difference $\delta = \varphi_\sigma - \varphi_\pi$ is given. Two mutually perpendicular linear oscillations are adequate to describe the two circular oscillations:

$$E_\sigma + iE_\pi = c_1 e^{i(\omega t+\varphi_1)} + c_{-1}e^{-i(\omega t+\varphi_{-1})} \tag{4-122}$$

The polarization components are given by[35]

$$dI_\sigma(\nu,\theta) = \frac{e^2c^2\beta^2}{2\pi R^2}\,\nu^2\beta^2 J_\nu'^2(\nu\beta\sin\theta)\,dO \tag{4-123}$$

$$dI_\pi(\nu,\theta) = \frac{e^2c^2\beta^2}{2\pi R^2}\,\nu^2 J_\nu^2(\nu\beta\sin\theta)\cot^2\theta\,dO \tag{4-124}$$

A breakdown into right-hand and left-hand components that is equivalent to Eq. (4-123) is

$$dI_{\pm 1} = \frac{e^2c^2\beta^2\nu^2}{4\pi R^2}\{\beta J_\nu'(\nu\beta\sin\theta) \pm \cot\theta\,J_\nu(\nu\beta\sin\theta)\}^2\,dO \tag{4-125}$$

The results obtained denote the fact that at a frequency $\omega = \nu\omega_0$, elliptically polarized radiation whose characteristics are defined by the ratio of the amplitudes and phase difference of two linearly polarized components propagates in the θ direction:

$$\frac{dI_\pi}{dI_\sigma} = \frac{\cot^2\theta}{\beta^2}\frac{J_\nu^2(\nu\beta\sin\theta)}{J_\nu'^2(\nu\beta\sin\theta)} \tag{4-126}$$

$$\delta = \arcsin\frac{dI_1 - dI_{-1}}{2\sqrt{dI_\pi\, dI_\sigma}} = \arcsin\frac{\cos\theta}{|\cos\theta|} = \pm\frac{\pi}{2} \tag{4-127}$$

In the plane of the orbit $\theta = \pi/2$ and $dI_\pi = 0$ and the radiation is completely linearly polarized. When $0 < \theta < \pi/2$ and $\delta = \pi/2$ light with right-hand circular polarization is radiated; in the lower half-space $0 < \theta < \pi/2$ and $\delta = -\pi/2$ light with left-hand circular polarization is radiated.

When x-rays act on polymer materials used as resists, the fundamental process that takes place is the ionization of the atoms as they absorb radiation quanta (the photoeffect). The direction in which a photoemitted electron moves is, as will be shown in Section 4-13, related to the direction of the electric vector in the field: photoelectrons are emitted mainly in the direction of the polarization field and may affect the resolution that can be attained. The reflection of x-rays is also dependent on polarization.

We now turn to an analysis of the physical relations given by Eqs. (4-123) and (4-124) that exist in synchrotron radiation.

4-10. THE ANGULAR DISTRIBUTION OF SYNCHROTRON RADIATION: THE PROJECTOR EFFECT

Synchrotron radiation is concentrated inside a cone having a small opening $\Delta\theta \simeq 1/\gamma$ with an axis along the tangent to the trajectory of an electron in an orbit. We can say that at any point on a circular trajectory an electron "lights" like a projector whose axis is directed along the tangent to the trajectory and this property is sometimes called the projector effect. Concentration into a narrow cone along the direction of motion is a feature common to other forms of radiation from ultrarelativistic particles (Cerenkov and Kumakhov radiation). In the final analysis, this feature is associated with the space–time relationships from the theory of relativity. Because of space limitations, we will not go into this side of the question. The projector property is inherent to each of the frequency components of synchrotron radiation. It follows from Eq. (4-120) that the width of the angular radiation distribution is a function of the radiation wavelength λ [39]:

$$\Psi_\lambda = \frac{2}{\gamma}\left(\frac{\lambda}{\lambda_c}\right)^{1/3} \qquad \text{when } \frac{\lambda}{\lambda_c} \gg 1$$

$$\Psi_\lambda = \frac{2}{\gamma}\left(\frac{3\lambda}{\lambda_c}\right)^{1/2} \qquad \text{when } \frac{\lambda}{\lambda_c} \ll 1,\ \lambda_c = \frac{4\pi R}{3\gamma^3} \tag{4-128}$$

The "harder" the radiation is, the narrower will be the cone in which it is concentrated. The presence of the projector effect can easily be seen if we sum Eq. (4-123) over all frequencies using relations from the theory of Bessel functions:

$$\sum_{\nu=1}^{\infty} \nu^2 J_\nu(\nu x) = \frac{x^2(4+x^2)}{16(1-x^2)^{7/2}}, \qquad \sum_{\nu=1}^{\infty} \nu^2 J_\nu'^2(\nu x) = \frac{4+3x^2}{16(1-x^2)^{5/2}} \tag{4-129}$$

We then find

$$dI_\sigma(\theta) = \sum_{\nu=1}^{\infty} dI_\sigma(\nu,\theta) = \frac{e^2 c\beta^4}{32\pi R^2} \frac{4+3\beta^2\sin^2\theta}{(1-\beta^2\sin^2\theta)^{5/2}}\, dO \tag{4-130}$$

$$dI_\pi(\theta) = \sum_{\nu=1}^{\infty} dI_\pi(\nu,\theta) = \frac{e^2 c\beta^4}{32\pi R^2} \frac{(4+\beta^2\sin^2\theta)\cos^2\theta}{(1-\beta^2\sin^2\theta)^{7/2}}\, dO \tag{4-131}$$

The total $dI = dI_\sigma + dI_\pi$ has a sharp maximum at $\theta = \pi/2$; a radiation minimum corresponds to the directions $\theta = 0, \pi$. If we introduce the variable $\xi = \gamma\cos\theta$, we can write the angular distribution function as

$$dI(\theta) = \frac{e^2 c\beta^4 f_0(\xi)\, dO}{32\pi R^2(1-\beta^2)^{5/2}} \tag{4-132}$$

$$f_0 = f_\pi + f_\sigma, \qquad f_\sigma = \frac{7}{(1+\xi^2)^{5/2}}, \qquad f_\pi = \frac{5\xi^2}{(1+\xi^2)^{7/2}} \tag{4-133}$$

We can see from a graph of the $f_0(\xi)$ function that the basic part of the radiation intensity is concentrated within the range $\Delta\xi = \pm 1$, i.e., the opening of the cone is roughly $\Delta\psi = 1/\gamma$, which, when $\gamma = 10^3$, corresponds to an angular divergence of roughly half a degree.

The component having σ-polarization has the form of an angular distribution with a single narrow maximum; the π-component has a minimum of zero at $\theta = \pi/2$ and two maxima at $\theta \simeq \pi/2 \pm \sqrt{2/5}\,(1/\gamma)$. In the ultrarelativistic case, 7/8 of the radiation is σ-polarized and the rest is π-polarized. All this means that, as a whole, synchrotron radiation is highly polarized.

4-11. THE FREQUENCY DISTRIBUTION OF SYNCHROTRON RADIATION

It is difficult to explain the dependence of synchrotron radiation intensity given by Eq. (4-120) (by Eq. (4-123) for the polarized components) because of the difficulties in obtaining asymptotic expressions for these functions for high and low frequencies. A curve of the frequency dependence, after integrating over the angles and making a number of conversions, reduces to an integral of a third-order Bessel function $K_{5/3}(x)$, for which asymptotic forms are known[35]:

$$dI(\nu) = \frac{3\sqrt{3}\, e^2 c\gamma^4}{4\pi R^2} \left[y \int_y^{\infty} K_{5/3}(x)\, dx \right] dy \tag{4-134}$$

$$y = \frac{\nu\omega_0}{\omega_c} \qquad \omega_c = \frac{3}{2}\gamma^3\omega_0$$

TABLE 4-3. The Parameters of Domestic Storage Rings and Synchrotrons

Sources	E, GeV	R, m	I, mA	λ_c, nm
Storage rings:				
VÉPP-2M (Novosibirsk)	0.67	1.22	100	2.3
VÉPP-3 (Novosibirsk)	2.2	6.15	80–500	0.3
VÉPP-4 (Novosibirsk)	7.0	33.0	100	0.054
N-100 (Khar'kov)	0.1	0.5	250	230
Synchrotrons:				
FIAN-S-60	0.68	2.0	10	3.5
FIAN-Pakhra	1.3	4.0	100	1.0
SIRIUS (Tomsk)	1.2	4.2	20	1.4
ARUS (Erevan)	6.0	24.65	20	0.074

Equation (4-134) has a maximum at a frequency of $\omega_{\max} \simeq \gamma^3\omega_0$; lower frequencies correspond to the inequality $y = \nu\omega_0/\omega_c \ll 1$, and higher frequencies to $y = \nu\omega_0/\omega_0 \gg 1$. At low frequencies we can say that

$$K_s(x)|_{x\to 0} = \frac{2^{s-1}\Gamma(s)}{x^s}, \qquad y\int_y^{\infty} K_{5/3}(x)\,dx \simeq 2^{2/3}\Gamma\left(\frac{2}{3}\right) y^{1/3} \tag{4-135}$$

which yields

$$dI(\omega \ll \omega_c) = \frac{3^{1/6}}{\pi}\Gamma\left(\frac{2}{3}\right)\frac{e^2 c}{R^2}\frac{\omega^{1/3}\,d\omega}{\omega_0^{4/3}} \tag{4-136}$$

for the spectral distribution of radiation power. Thus, starting from the frequency ω_0 the intensity of the radiation increases slowly, as $\omega^{1/3}$, with increasing frequency, reaching (for $E \approx 1$ GeV) a maximum at a frequency that corresponds to roughly the 10^9 harmonic of the fundamental, ω_0: $\omega_{\max} = \gamma^3\omega_0$.

There is an asymptotic expression for $K_s(x)$ when the values of its argument are large:

$$K_s(x)|_{x\to\infty} = \sqrt{\frac{\pi}{2x}}\,e^{-x} \tag{4-137}$$

which yields for the spectrum

$$dI(\omega \gg \omega_c) = \frac{1}{2\pi}\frac{e^2 c}{R^2}e^{-\omega/\omega_c}\frac{\omega^{1/2}\,d\omega}{\omega_0^3\sqrt{\gamma}} \tag{4-138}$$

Thus, the intensity of the radiation increases rapidly as it passes through its maximum value and then decays exponentially. The frequency $\omega_c = (3/2)\gamma^3\omega_0$ is called the critical frequency.

4-12. THE RADIATION FROM SYNCHROTRONS AND STORAGE RINGS

Table 4-3 shows the fundamental parameters of domestic synchrotrons and storage rings.[35]

The theory of synchrotron radiation that we have just discussed corresponds, generally speaking, to the radiation seen in storage rings in which electrons are kept in orbit for several hours while the orbit radius and the energy are held constant: radiation energy losses are compensated by accelerating the electrons via a high-frequency generator in the straight line segments of the trajectory; radiation takes place in the circular segments of the trajectory. The electrons in a storage ring are grouped into one or more clusters several centimeters long and the total current is equal to the product of the charge of all the electrons on the orbit and the frequency of revolution of a single electron: $I = en\omega_0/2\pi$. To reduce the loss of electrons to scattering by residual gas atoms the degree of evacuation is maintained at 10^{-9} mm Hg. Nevertheless, without replenishment from the source-injector, the current in a storage ring will gradually decrease.

Unlike storage rings, synchrotrons are pulsed machines: the electron injection–acceleration–discharge cycle, as a rule, is done at a frequency of 50 Hz. The electron energy is increased every cycle, depending on the device type, according to $E(t) = E_m \sin(\pi t/2T)$, or $E(t) = E_m \sin^2(\pi t/2T)$ (FIAN-Pakhra), or $E(t) = E_m(t/T)$ (FIAN-S-60). The spectral and angular distributions of the radiation that are seen in this case will be the result of time averaging the distributions obtained earlier for a monoenergetic electron.

It must be pointed out that even with monoenergetic electrons the spectra from actual sources will be the result of averaging the formulas presented over some distribution of orbit radii R and electron energies E because the aggregate of electrons in the clusters is described by a distribution with respect to cross section and to energy.

In terms of its spectral content, synchrotron radiation reminds us of the radiation from an absolutely black body (Planck distribution). Specifically, a maximum in the synchrotron radiation spectrum can be compared with the maximum in black body radiation:

$$\lambda_{\max}\ (\text{Å}) = 2.896 \times 10^7/T \qquad \text{(black body)} \tag{4-139}$$

$$\lambda_{\max}\ (\text{Å}) = 0.42\lambda_c = 2.54\,\frac{R\ (\text{M})}{E^3\ (\text{GeV})} \tag{4-140}$$

The location of the radiation maximum from an FIAN-S-60 synchrotron at an energy of $E = 0.68$ GeV reminds us of the radiation from a black body at a temperature of roughly 1.8 million degrees.

We will present still another radiation spectral energy distribution for a monoenergetic electron on a wavelength scale[39]:

$$\begin{gathered} I(\lambda) = \frac{3^{5/2} e^2 c}{16\pi^2 R^3}\,\gamma^7 G(y) \qquad \left(\frac{\text{ergs}}{\text{sec} \cdot \text{Å} \cdot \text{electron}}\right) \\ G(y) = y^3 \int_y^\infty K_{5/3}(x)\,dx, \qquad y = \frac{\lambda}{\lambda_c}, \qquad \lambda_c = \frac{4\pi R}{3\gamma^3} \end{gathered} \tag{4-141}$$

or in numbers (E in GeV, λ_c in Å):

$$I(\lambda) = 4.88 \times 10^5 E \cdot \lambda_c^{-2} G(y) \qquad \left(\frac{\text{ergs}}{\text{sec} \cdot \text{Å} \cdot \text{electron}}\right) \tag{4-142}$$

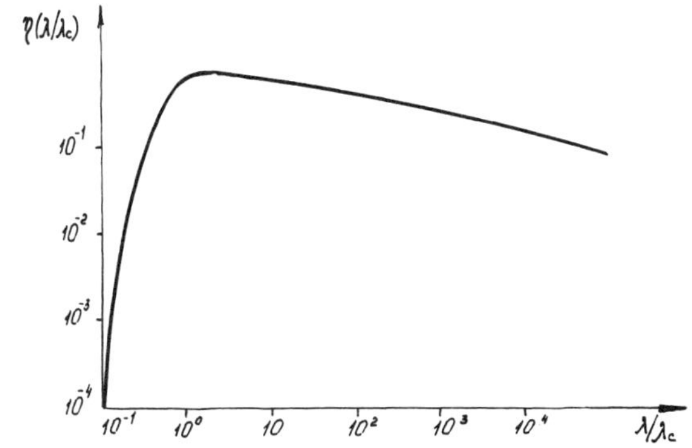

FIGURE 4-8. The universal spectral function $\eta(\lambda/\lambda_c)$ in logarithmic scale.(39)

The distribution for the number of photons is(39)

$$N(\lambda) = 2.46 \times 10^{13} \frac{E\ (\text{GeV})}{\lambda\ (\text{Å})} y^{-2} G(y) \qquad \left(\frac{\text{photons}}{\text{sec} \cdot \text{Å} \cdot \text{mrad} \cdot \text{mA}} \right) \tag{4-143}$$

The amount of energy that an electron loses during a single revolution,

$$\Delta E = \frac{4\pi}{3} \frac{e^2}{R} \gamma^4 = 88.5 \frac{E^4\ (\text{GeV})}{R\ (\text{m})} \qquad (\text{keV}) \tag{4-144}$$

is of interest and we obtain, for a FIAN-S-60 synchrotron at E = 0.68 GeV, ΔE = 9.46 keV and the quantum energy λ_m = 1.6 nm corresponding to λ_m is 0.7676 keV. We can say that an electron radiates 12 photons per revolution, each of which has an average energy of 0.7676 keV.

The spectral illuminance of a surface being exposed in a relative span of wavelengths is given by(38)

$$\varepsilon_\lambda \left(\frac{\text{W}}{\text{mm}^2} \right) = 59 \frac{E^2 I}{\lambda L^2} \eta \left(\frac{\lambda}{\lambda_c} \right) \nu^{-1} \left(\frac{\lambda}{\lambda_c} \right) \frac{\Delta\lambda}{\lambda} \tag{4-145}$$

where λ is in Å, E is in GeV, I is in A, and L is in m; $\eta(\lambda/\lambda_c)$ and $\nu(\lambda/\lambda_c)$ are, respectively, a universal spectral function and a universal angular function(38) and L is the distance from the point at which radiation occurs to the point at which it is observed.

The function $\nu(\lambda/\lambda_c)$ is defined as $\nu = \psi_\lambda/\psi_{\lambda_c}$, where ψ_λ and ψ_{λ_c} are the half-widths of the angular distribution over ψ, calculated from the plane of the orbit. Figures 4-8 and 4-9 are graphs of the $\eta(\lambda/\lambda_c)$ and $\nu(\lambda/\lambda_c)$ functions.(38)

All of these formulas are useful in setting up experiments in exposing materials to a field of synchrotron radiation by segregating one spectral range or another through filters or monochromators.

For a synchrotron the time averaged distribution of the radiation with respect to wavelength, according to

$$\bar{I}(\lambda) = \frac{1}{T} \int_0^T I(\lambda)\, d\lambda \tag{4-146}$$

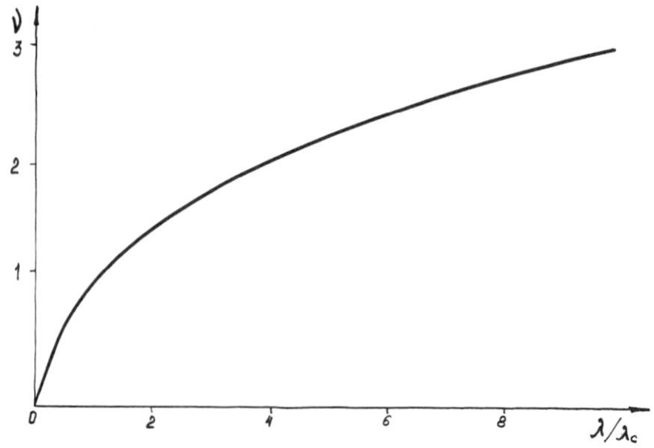

FIGURE 4-9. The universal angular function $\nu(\lambda/\lambda_c) = \psi_\lambda/\psi_{\lambda_c}$, ψ_{λ_c} (mrad) = $0.82/E$ (GeV).(39)

when $E(t) = E_m \sin^2(\pi t/2t)$ is

$$\bar{I}(\lambda) = \frac{3^{3/2}}{8\pi^3}\frac{e^2 c}{R^3}\gamma^7 y^4 L(y), \qquad y = \frac{\lambda}{\lambda_c} \tag{4-147}$$

$$L(y) = \int_1^\infty (\zeta^{1/3} - 1)^{1/2}(\zeta^{1/2} + 2)K_{5/3}(y\zeta)\, d\zeta, \qquad \zeta = \left(\frac{E_m}{E(t)}\right)^3 \tag{4-148}$$

and when $E(T) = E_m \sin(\pi t/2T)$, $0 \le t \le T$, is

$$I(\lambda) = \frac{3^{5/2}}{8\pi^3}\frac{e^2 c}{R^3}\left(\frac{E_m}{m_0 c^2}\right)^7 y^4 G(y) \tag{4-149}$$

$$G(y) = \int_1^\infty (\tau^{2/3} - 1)^{1/2} K_{5/3}(y\tau)\, d\tau$$

$$y = \frac{\lambda_{\max}}{\lambda} = \frac{4\pi R}{3\lambda}\left(\frac{m_0 c^2}{E_m}\right)^3 \tag{4-150}$$

Figure 4-10 is a graph of the universal function $y^4 G(y)$ used in Eq. (4-149). The $y^4 L(y)$ function in Eq. (4-147) has been tabulated in [40].

Until recently experiments in using synchrotron radiation for x-ray lithography were carried out in accelerators and storage rings designed and built for basic research work. Several years ago work began on producing storage rings specially designed for x-ray lithography.(41–43) The first problem to be solved was to reduce the size of a storage ring to that of a desk-sized device: the outer diameter of such a compact synchrotron source (COSI) was about 2 m. No strict requirements were imposed on focusing the electron beam; the cross section of the beam is artificially increased by stimulating transverse oscillations. The parameters of the COSI are shown in Table 4-4.(42)

The use of only the I, R, and E parameters, whose values can be very precisely determined, in synchrotron radiation theory allows us to use synchrotron radiation as a reference source of electromagnetic radiation over a wide range of wavelengths.(44)

Radiation that is similar to that obtained from a synchrotron is generated by moving a relativistic electron in a periodic field, the so-called undulatory radiation.(45) The radiation

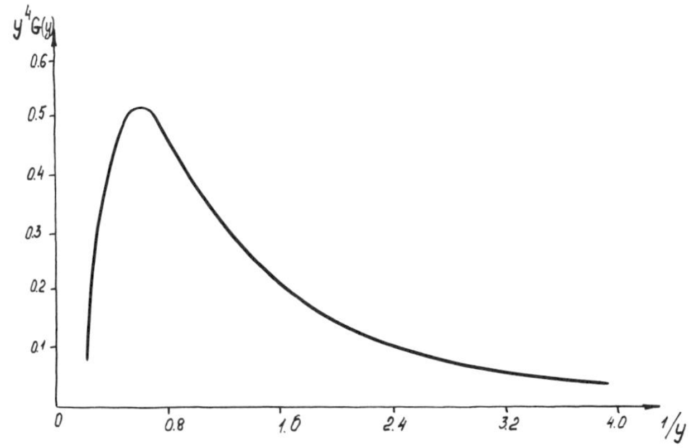

FIGURE 4-10. The universal curve for the spectral distribution of electron radiation when the energy is increased sinusoidally during the first quarter cycle of acceleration.

TABLE 4-4. The Parameters of a Compact Storage Ring for X-Ray Lithography[41,42]

Parameter	Design value
Critical wavelength, nm $\lambda_c = 4\pi R/(3\gamma^3)$	2
Electron current in the orbit, mA	300
Angular aperture of the radiation, mrad	5
Number of x-ray channels,	5–10
External dimensions (including shielding), m	2
Weight, tons	10
Power required, kW	50–100
Cost (excluding the cost of devices in channels), Deutsche marks	5 million

frequency ω at an angle θ to the direction of electron motion is[46]

$$\omega = \frac{c}{\lambda_0(1 - \beta \cos \theta)}, \qquad \omega_m = \frac{2c}{\lambda_0}\gamma^2 \tag{4-151}$$

where ω_m is the maximal radiation frequency in the direction of motion ($\theta = 0$), and λ_0 is the lattice period of the field. At macroscopic values of the field period, $\lambda_0 \simeq 10$ cm and $\gamma = 10^3$, the frequency ω_m falls into the optical region. In order that the frequency ω_m fall into the soft x-ray region ($\omega_m \simeq 2 \times 10^{18}$ sec^{-1}), when $\gamma = 10^3$, the period of the field must be $\lambda_0 = 0.3$ mm, and when $\gamma = 10^2$, it must be $\lambda_0 = 3$ μm. It was suggested in [47] that a superlattice having an appropriate superlattice period be used as a solid-state undulator. When relativistic electrons and positrons are channeled in crystals an intercrystal electric field having a period $\lambda \simeq 10^{-8}$ cm causes the so-called Kumakhov radiation of relativistic electrons in the hard γ radiation range: $\omega_m \simeq 6 \times 10^{24}$ sec^{-1}.[48] We also recall experiments in generating soft x-rays when passing a 90 MeV beam of electrons through a packet of thin films (30 Al films 1 μm thick, 25 mylar films 1.5 μm thick, and 23 Be films 1 μm thick)[49]: the peaks of the photon spectra were at E_K = 1100, 1000, and 800 eV, respectively. In physical nature this radiation is the so-called transition radiation.[50]

4-13. THE THEORY OF SOFT X-RAY ABSORPTION IN SOLIDS

The absorption of soft x-rays by matter is due almost entirely to the photoeffect. Attenuation of the radiation is possible because of nonuniformities in the density (the coefficient of refraction). However, these nonuniformities are small in the x-ray region (as is the difference $n - 1$) and scattering is routinely ignored. Therefore, the theory of soft x-ray absorption reduces, in essence, to the theory of the atomic and the interband photoeffect. We will now examine this theory.

1. The interrelationship between classical and quantum methods for describing an electromagnetic field. The classical method is based on Maxwell's equations and uses the concepts of a vector potential $\mathbf{A}(\mathbf{r}, t)$, an electric field $\mathbf{E}(t)$, and a magnetic field $\mathbf{H}(t)$ that are related by

$$\mathbf{E} = -\frac{1}{c}\frac{\partial A}{\partial t}, \qquad \mathbf{H} = \operatorname{curl}\mathbf{A} \tag{4-152}$$

The energy density in the field is

$$\rho = \frac{1}{4\pi}|\mathbf{E}|^2 = \frac{1}{4\pi}|\mathbf{H}|^2 \tag{4-153}$$

and the energy flux into the solid angle dO is

$$ds = \frac{c}{8\pi}(|\mathbf{E}|^2 + |\mathbf{H}|^2)\, dO \tag{4-154}$$

An electromagnetic field can be represented by standing plane waves, by a propagating plane or spherical waves, etc. depending on the symmetry of the problem to be solved. In addition, there is a static (Coulomb) electric field by means of which charged particles interact and which is contained in the Hamiltonian of the particles system.

As it interacts with atoms, which are essentially quantum objects, a field displays quantum properties: when particles absorb or radiate energy, the energy of the field changes discretely by an amount equal to a quantum of the field energy $\hbar\omega$. It is this phenomenon that gives rise to the quantum description of a field: an electromagnetic field is treated as a group of quantum oscillators having energies of $\varepsilon_n = \hbar\omega(n + 1/2)$, $n = 0, 1, 2, \ldots$. The physical world in quantum electrodynamics is a space populated by field quantum oscillators on the one hand, and by particles on the other. When particles radiate or absorb energy the number of field quanta n changes by ± 1.

The correspondence between the classical and the quantum description of a field is established by equating the density of the field energy in the same method.[51] In the quantum description: let there be in 1 cm^3 one quantum of a given frequency ω; then the energy density is $\rho_\omega = \hbar\omega$. In the classical description: let the amplitude of the vector potential $\mathbf{A}$ be a; then

$$\mathbf{A} = a\mathbf{e}\cos(\omega t - \mathbf{kx}) \tag{4-155}$$

$$E = -\frac{1}{c}\frac{\partial \mathbf{A}}{\partial t} = \frac{\omega a}{c}\mathbf{e}\sin(\omega t - \mathbf{kx}), \qquad |\mathbf{H}| = |\mathbf{E}| \tag{4-156}$$

$$\rho = \frac{\omega^2 a^2}{4\pi c^2}\overline{\sin^2(\omega t - \mathbf{kx})} = \frac{\omega^2 a^2}{8\pi c^2} \tag{4-157}$$

Here, $\mathbf{e}$ is a unit vector in the $\mathbf{E}$ direction (a polarization vector).

Equating the field energy densities given in classical and quantum terms we find the amplitude of the vector potential, which turns out to be dependent on the frequency by an amount

$$a = (8\pi\hbar c^2/\omega)^{1/2} \tag{4-158}$$

(because a quantum of energy depends on the oscillator frequency!)

$$\mathbf{A}(\mathbf{x}, t) = \left(\frac{8\pi\hbar c^2}{\omega}\right)^{1/2} \mathbf{e}[e^{-i(\omega t - \mathbf{k}\cdot\mathbf{x})} + e^{i(\omega t - \mathbf{k}\cdot\mathbf{x})}] \tag{4-159}$$

When calculating the probability of quantum absorption we usually take the term $\exp[-i(\omega t - \mathbf{k}\cdot\mathbf{x})]$ in Eq. (4-159) and take the term $\exp[i(\omega t - \mathbf{kx})]$ when calculating the probability of radiation.

The field oscillators density per unit volume and in the range $\mathbf{k}, \mathbf{k} + d\mathbf{k}$ of wave numbers is

$$dn\ (1\ \text{cm}^3, d^3\mathbf{k}) = d^3\mathbf{k}/(2\pi)^3 \tag{4-160}$$

The quantum momentum is $\hbar\omega/c$ and the wave number is $k = 2\pi/\lambda = \omega/c$.

2. The probability of a transition between states in an atom when a photon is absorbed. The probability that an atomic particle will absorb a photon is equal to the probability of a transition between states in the atomic system, the transition being accomplished under the action of a classical electromagnetic field that corresponds to a photon.[(51)] The energy of one of the atom's electrons, including the interaction with the field $\mathbf{A}(\mathbf{x}, t)$ of the photon and the electrostatic field $V(\mathbf{x})$ of the other atomic particles, is[(51)]

$$\begin{aligned}\hat{H} &= \frac{1}{2m}\left(\hat{\mathbf{p}} - \frac{e}{c}\mathbf{A}\right)^2 - \frac{e\hbar}{2mc}\boldsymbol{\sigma}\cdot\nabla\mathbf{A} + eV \\ &= \frac{\hat{\mathbf{p}}^2}{2m} + eV - \frac{e}{2mc}(\hat{\mathbf{p}}\cdot\mathbf{A} + \mathbf{A}\cdot\hat{\mathbf{p}}) - \frac{e\hbar}{2mc}\boldsymbol{\sigma}\nabla\mathbf{A} + \frac{e^2}{2mc^2}\mathbf{A}\cdot\mathbf{A}\end{aligned} \tag{4-161}$$

The $(\hat{\mathbf{p}}^2/2m) + eV$ terms are included in the Hamiltonian of an isolated atom and define the state of the electron in the atom. We can ignore the energy of interaction between the magnetic moment of an electron $(e\hbar/2mc)\boldsymbol{\sigma}$ and the magnetic field $\nabla\mathbf{A}$ of a photon because it is much less than $(e/2mc)(\hat{\mathbf{p}}\cdot\mathbf{A})$. Finally, we can also ignore the $\mathbf{A}$ squared term, since we are looking at a (one quantum) process that is linear with respect to the field. Thus, the energy of interaction between an atom's electron and the field of a photon is

$$\hat{H}' = -\frac{e}{2mc}(\hat{\mathbf{p}}\cdot\mathbf{A} + \mathbf{A}\cdot\hat{\mathbf{p}}) \tag{4-162}$$

The probability that an electron will make a transition between the i and f states when acted upon by the field of a photon is

$$W_{i\to f} = \frac{2\pi}{\hbar}|H'_{if}|^2\rho_E \tag{4-163}$$

In the process we are examining (the photoeffect) the final state of an electron f corresponds to free motion having energy and momentum that are determined by the conservation laws:

$$\hbar\omega = p^2/2m + I, \qquad \hbar\mathbf{k} = \mathbf{p} + \hbar\mathbf{q} \tag{4-164}$$

Here, I is the energy needed to ionize an electron out of the i state, $p^2/2m$ and $\mathbf{p}$ are the energy and momentum of the electron after ionization (in the final state), and $\hbar\mathbf{q}$ is the momentum transferred to the atom. Because the momentum of a photon $\hbar\mathbf{k}$ is very small, $\hbar\mathbf{q} \approx -\mathbf{p}$, i.e., the momentum of the recoil atom is roughly the same (in absolute value) as that of the electron.

The quantity ρ_E in Eq. (4-163) is the density of the final electron states in the continuous spectrum for a single band of energies in the solid angle dO, defined by the direction in which the photoemitted electron is moving and per unit volume ($V_0 = 1$ cm):

$$\rho_E = \frac{p^2\,dp}{(2\pi\hbar)^3}\frac{dO\,V_0}{dE} = \frac{mp\,dO}{(2\pi\hbar)^3} \tag{4-165}$$

The matrix element between the i and f states of an electron that are described by the wave functions $\Psi_i = \varphi_i(\mathbf{x})\exp(iE_it/\hbar)$ and $\Psi_f = \varphi_f(\mathbf{x})\exp(-E_ft/\hbar)$ is

$$U_{if} = u_{if}\exp[it(E_f - E_i - \hbar\omega)/\hbar] \tag{4-166}$$

$$u_{if} = \int \varphi_f^*(\mathbf{x})[-\frac{a\mathbf{e}\mathbf{e}}{2mc}(\hat{\mathbf{p}}e^{i\mathbf{kx}} + e^{i\mathbf{kx}}\hat{\mathbf{p}})]\varphi_i(\mathbf{x})\,d^3\mathbf{x} \tag{4-167}$$

In keeping with the adopted rule, the term $\exp[-i(\omega t - \mathbf{kx})]$ from Eq. (4-159) is chosen. The exponential in U_{if} defines the law of conservation of energy during the transition: $\hbar\omega = E_f - E_i$.

We will now examine the $\hat{\mathbf{p}}e^{i\mathbf{kx}}\varphi_i(\mathbf{x})$ function inside the integral of Eq. (4-167). The electron momentum operator $\hat{\mathbf{p}} = -i\hbar\nabla$ acts on both the function $\varphi(\mathbf{x})$ of the electron's $\mathbf{x}$ coordinate and on the field function $e^{i\mathbf{kx}}$, since the field we must examine is at the point where the electron is found; consequently,

$$\hat{\mathbf{p}}e^{i\mathbf{k}\cdot\mathbf{x}}\varphi_i(\mathbf{x}) = e^{i\mathbf{k}\cdot\mathbf{x}}(\hat{\mathbf{p}} + \hbar\mathbf{k})\varphi_i(\mathbf{x}) \tag{4-168}$$

$$\mathbf{e}(\hat{\mathbf{p}}e^{i\mathbf{k}\cdot\mathbf{x}} + e^{i\mathbf{k}\cdot\mathbf{x}}\hat{\mathbf{p}})\varphi_i(\mathbf{x}) = 2e^{i\mathbf{k}\cdot\mathbf{x}}\mathbf{e}\hat{\mathbf{p}}\varphi_i(\mathbf{x}) \tag{4-169}$$

$$u_{if} = -\frac{ea}{mc}\int \varphi_f^* e^{i\mathbf{k}\cdot\mathbf{x}}\mathbf{e}\hat{\mathbf{p}}\varphi_i(\mathbf{x})\,d^3\mathbf{x} \tag{4-170}$$

Hereafter, we will designate the radius vector $\mathbf{x}$ of an electron as $\mathbf{r} \equiv \mathbf{x}$.

3. The photoeffect from the K shell of an atom. In order to further calculate the transition matrix element, Eq. (4-170), we need an explicit definition for the wave functions for the initial and final electron states, $\varphi_i(\mathbf{r})$ and $\varphi_f(\mathbf{r})$. We will assume that an atom's electron is in the $1s$ state, i.e., belongs to the innermost shell that contains two $1s$ electrons. For an atom whose atomic number is Z the wave function for an electron in the $1s$ state is[52,53]

$$\Psi_i(\mathbf{r}) = \Psi_{1s}(r) = (\pi a^3)^{-1/2}\exp\left(-\frac{r}{a}\right), \qquad a = \frac{\hbar^2}{me^2}\frac{1}{Z} = \frac{a_B}{Z} \tag{4-171}$$

where a_B is a Bohr radius.

The wave function for the final state is a plane wave having the wave number $\varkappa = \mathbf{p}/\hbar$: $\Psi_f(\mathbf{r}) = \exp(i\varkappa\mathbf{r})$ (normalization is per cm^3, in accord with the choice of ρ_E). Substituting the Ψ_i and Ψ_f into the integral in Eq. (4-170) gives us

$$u_{if} = \frac{ea^*}{mc}(\pi a^3)^{-1/2}(i\hbar)\mathbf{e} \int e^{i(\mathbf{k}-\varkappa)\cdot\mathbf{r}} \nabla e^{-r/a}\, d^3\mathbf{r} \tag{4-172}$$

The amplitude of the photon wave is marked by an asterisk to distinguish it from the radius of the electron orbit a. The subintegral expression in Eq. (4-172) is $u\nabla v$, where $u = \exp[i(\mathbf{k} - \varkappa) \cdot \mathbf{r}]$ and $v = \exp(-r/a)$. Using the identity $\nabla(uv) = u\nabla v + v\nabla u$ allows us to write

$$\int u\nabla v\, d^3\mathbf{r} = \int \nabla(uv)\, d^3\mathbf{r} - \int v\nabla u\, d^3\mathbf{r} \tag{4-173}$$

According to Stokes' theorem, a volume integral from $\nabla(uv)$ can be converted to a surface integral:

$$\int \nabla(uv)\, d^3\mathbf{r} = \int_s uv\, ds \tag{4-174}$$

We can choose as s a large sphere in which the function $v = \exp(-r/a)$ is exponentially small; this shows that $\int uv\, ds = 0$ and therefore,

$$\begin{aligned} \int e^{i(\mathbf{k}-\varkappa)\mathbf{r}} \nabla e^{-r/a}\, d^3\mathbf{r} &= -\int e^{-r/a} \nabla e^{i(\mathbf{k}-\varkappa)\mathbf{r}}\, d^3\mathbf{r} \\ &= -i(\mathbf{k} - \varkappa) \int \exp\left[-\frac{r}{a} + i(\mathbf{k} - \varkappa) \cdot \mathbf{r}\right] d^3\mathbf{r} \end{aligned} \tag{4-175}$$

We take the direction $\mathbf{q} = \mathbf{k} - \varkappa$ to be the axis of the spherical coordinate system and use α to denote the angle between $\mathbf{q}$ and $\mathbf{r}$. Then

$$\begin{aligned} I &= \int \exp\left[-\frac{r}{a} + i(\mathbf{k} - \varkappa) \cdot \mathbf{r}\right] d^3\mathbf{r} \\ &= \int_0^\infty r^2\, dr e^{-r/a} \int_0^{2\pi} d\varphi \int_0^\pi \sin\alpha \cdot d\alpha e^{iqr\cos\alpha} \\ &= 2\pi \int_0^\infty \frac{2q\sin qr}{qr} r^2\, dr \end{aligned} \tag{4-176}$$

The integral obtained corresponds to the tabulated integral

$$\int_0^\infty xe^{-bx} \sin ax\, dx = 2ab(a^2 + b^2)^{-2}$$

which yields

$$I = 8\pi a^3/(1 + q^2a^2)^2 \tag{4-177}$$

Thus, we have found

$$u_{if} = 8\pi \frac{a^* e\hbar}{mc} \frac{\mathbf{e} \cdot \mathbf{q}}{\sqrt{\pi a^3}} \frac{a^3}{(1 + q^2 a^2)} \tag{4-178}$$

for the transition matrix element. Substituting Eq. (4-178) and $a^* = (\hbar c^2/\omega)^{1/2}$ into Eq. (4-163) gives us the probability of a transition that results in a photoelectron being emitted into the solid angle dO:

$$dP = \frac{32e^2 a^3}{\hbar\omega} \frac{p}{m} \frac{(\mathbf{e} \cdot \mathbf{q})^2 \, dO}{(1 + q^2 a^2)^4} \tag{4-179}$$

The amount of energy absorbed from the field in 1 sec calculated for one atom in 1 cm^3 will be $d\varepsilon = \hbar\omega \, dP$. Dividing the amount of energy absorbed by an atom by the magnitude of the field energy flux $S = \hbar\omega c$ gives us the atomic photon absorption coefficient (the differential absorption cross section) in the photoeffect:

$$d\mu_{\text{at}} = \frac{\hbar\omega \, dP}{\hbar\omega c} = \frac{32e^2 a^2}{\hbar\omega} \frac{(\mathbf{e} \cdot \mathbf{q})^2}{(1 + q^2 a^2)^4} \frac{p}{mc} \, dO \tag{4-180}$$

The total photoeffect cross section can be found by integrating over all angles dO.

We will explain the angular dependence of the photoeffect cross section contained in Eq. (4-180). We will introduce a spherical coordinate system whose z axis lies along the wave vector $\mathbf{k}$ for the impinging photons; the x axis lies along $\mathbf{e} \perp \mathbf{k}$. We specify the direction of the photoelectron's momentum $\mathbf{p} = \hbar\boldsymbol{\varkappa}$ by the spherical angles θ and Φ. Because $\mathbf{q} = \mathbf{k} - \boldsymbol{\varkappa}$,

$$q^2 = \varkappa^2 + k^2 - 2k\varkappa \cos\theta \tag{4-181}$$

$$\mathbf{eq} = \mathbf{ek} - \mathbf{e}\boldsymbol{\varkappa} = -\mathbf{e}\boldsymbol{\varkappa} = \varkappa \sin\theta \cos\Phi \tag{4-182}$$

from which

$$d\mu_{\text{at}}(\theta, \Phi) = \frac{32e^2 a^3 v}{\hbar\omega c} \frac{\varkappa^2 \sin^2\theta \cos^2\Phi \, dO}{[1 + a^2(\varkappa^2 - 2k\varkappa \cos\theta)]^4} \tag{4-183}$$

The main part of the angular dependence of the photoeffect cross section is given by the $\sin^2\theta \cos^2\Phi$ factor. This factor assumes its maximal value of unity when $\boldsymbol{\varkappa} \| \mathbf{e}$, i.e., when the photoelectron escapes along the polarization vector of the photon's electric field. When unpolarized light impinges the photoelectrons are distributed according to $\sin^2\theta$, because in unpolarized light all directions of the angle Φ are equiprobable. At a given frequency of light the photoelectrons from a given shell have the same energy, regardless of the direction in which they escape: $\varepsilon = \hbar\omega - I_K$. Stating p, $\varkappa$, and k in terms of ω by means of the equations

$$p = \sqrt{2m\varepsilon} = \sqrt{2m(\hbar\omega - I_K)}, \qquad \varkappa = \frac{p}{\hbar}, \qquad k = \frac{\omega}{c} \tag{4-184}$$

we can write $d\mu$ as a function of the frequency of the light:

$$d\mu_{\text{at}}(\omega, \theta, \Phi) = \frac{32e^2a^3(2m)^{3/2}}{m\hbar^3\omega} \times \frac{(\hbar\omega - I_K)^{3/2}\cos^2\Phi\sin^2\theta\, dO}{\left\{1 + a^2\left[\left(\frac{2m}{\hbar^2}\right)(\hbar\omega - I_K) + \left(\frac{\omega}{c}\right)^2 - 2\frac{\omega}{c}\left(\frac{2m}{\hbar^2}\right)^{1/2}\cos\theta(\hbar\omega - I_K)^{1/2}\right]\right\}} \tag{4-185}$$

This formula can be simplified if the quantum energy $h\omega$ is much greater than the ionization energy: $\hbar\omega \gg I_K$ and $p^2/2m \approx \hbar\omega$. Then,

$$\frac{k}{\varkappa} = \frac{\hbar\omega}{pc} = \frac{mv^2/2}{mvc} = \frac{1}{2}\frac{v}{c} = \frac{1}{2}\beta \ll 1 \tag{4-186}$$

$$a\varkappa = \frac{\hbar^2}{me^2Z}\frac{mv}{\hbar} = \left[\frac{mv^2/2}{Z^2e^4m/2\hbar^2}\right]^{1/2} \approx \left(\frac{h\omega}{I_K}\right)^{1/2} \gg 1 \tag{4-187}$$

$$q^2 = \varkappa^2(1 + \beta^2/4 - \beta\cos\theta) \approx \varkappa^2(1 - \beta\cos\theta) \tag{4-188}$$

$$1 + q^2a^2 = 1 + a^2\varkappa^2(1 - \beta\cos\theta) \simeq a^2\varkappa^2(1 - \beta\cos\theta) \tag{4-189}$$

$$d\mu_{\text{at}}(\omega, \theta, \Phi) = \frac{e^2}{\hbar c}\frac{4\sqrt{2}\,Z^5(\hbar/m)^{7/2}}{a_{\text{B}}^5\omega^{7/2}}\frac{\sin^2\theta\cos^2\Phi}{(1 - \beta\cos\theta)^4}\,dO \tag{4-190}$$

In the photoeffect in x-ray beams (in x-ray lithography) $h\omega \approx 1$ keV, $\beta \ll 1$, and we can assume that $1 - \beta\cos\theta \approx 1$ in the denominator of Eq. (4-190). The photoeffect cross section as a function depends on an atom's parameters as Z^5, and on a photon's parameters as $\omega^{-7/2}$. At a given frequency ω ($\hbar\omega \gg I_K$) the photoeffect cross section from the K shell increases rapidly from element to element in Mendeleev's periodic table. The heavier atoms absorb x-ray and gamma radiation more strongly, which is the principle upon which high-contrast x-ray pictures are obtained through "exposure" to hard x-rays and the use of blocks containing heavy elements for radiation protection. On the other hand, the transparency of a material when Z is fixed increases rapidly, as $\omega^{-7/2}$, with the frequency of the radiation quanta. The high penetrating power of hard γ-radiation that occurs during a nuclear reaction or the deceleration of ultrarelativistic particles is related to this phenomenon.

Equation (4-190) can be put into an especially simple form if we introduce a frequency ν that corresponds to the ionization energy of the $1s$ shell[(52)]:

$$\nu_1 = Z^2Ry, \qquad Ry = \frac{1}{4\pi}\frac{me^4}{\hbar^3} = \frac{1}{4\pi}4.1314 \times 10^{16}\ \text{sec}^{-1} \tag{4-191}$$

then

$$d\mu_{\text{at}\,1s}(\nu, \theta, \Phi) = \frac{2^4}{\pi}\frac{e^2}{mc}\frac{\nu_1^{5/2}}{\nu^{7/2}}\sin^2\theta\cos^2\Phi\, dO \tag{4-192}$$

The cross section of photon absorption by an electron in the ns state in an atom having a single electron is given by[52]

$$d\mu_{\text{at}\, ns} = \frac{2^4}{\pi}\,\frac{e^2}{mc}\,\frac{\nu_n^{5/2}}{n^3\nu^{7/2}}\sin^2\theta\cos^2\Phi\, dO \tag{4-193}$$

which shows that the absorption cross section decreases as n^{-3} when the principal quantum number increases; also, in an atom having many electrons the reduction in the $\nu_n^{5/2}$ factor due to shielding must be considered.

If we integrate Eq. (4-193) over the angles we will find the total quantum absorption cross section, which is often written[54]

$$\mu_{\text{at}\, 1s} = 2^7\,\frac{e^2}{mc}\,\frac{\nu_1^{5/2}}{\nu^{7/2}} \equiv 4\sqrt{2}\,\alpha^4\sigma_0 Z^5\left(\frac{mc^2}{\hbar\omega}\right)^{7/2} \tag{4-194}$$

Here $\alpha = e^2/\hbar c = 1/137$ (the constant for a thin structure) and $\sigma_0 = (8\pi/3)(e^2/mc^2)^2 = (8\pi/3)r_0^2 = 6.653 \times 10^{-25}$ cm^2 is the so-called Thomson scattering cross section for a photon by an electron.

The formulas obtained for the photoelectric absorption cross section of a photon are extremely interesting, although the conditions that $\hbar\omega \gg I_K$ and $mv^2/2 \approx \hbar\omega \gg I_K$ are rarely satisfied during the exposure process in x-ray and ultraviolet lithography. In these processes the photon frequency is close to a photoelectric absorption edge: $\hbar\omega \approx I_K$ and $mv^2/2 = h\omega - I_K \ll I_K$. Under these conditions we can simplify the general expression for the photoelectron absorption cross section, Eq. (4-185). However, in deriving Eq. (4-185) a plane wave $\Psi_f = \exp(i\boldsymbol{\varkappa}\mathbf{r})$ was chosen for the wave function of a photoelectron, which is permissible in only the so-called Born approximation; i.e., when the electron's kinetic energy $mv^2/2$ is much greater than its potential energy in the atom's field, $mv^2/2 \gg I_K$. Formulas that are correct when $mv^2/2 \leq I_K$ can be obtained if functions that correspond to the motion of an electron having positive energy in the field of the atom are used as the wave functions for the photoelectrons final state. In a first approximation the field of an atom is understood to be the field of the atom's nucleus; more accurate calculations take the shielding of the nuclear field by the electrons into consideration. Since the potential of an atom is spherically symmetric, the wave functions of an atom's electrons in the continuous spectrum and having positive energy $\varepsilon = \hbar^2\varkappa^2/2m$ will be[2]

$$\Psi_f(r) = R_n(\rho)Y_{lm}(\theta,\varphi), \qquad n = Z/i\varkappa a_{\text{B}}, \qquad \rho = 2i\varkappa r \tag{4-195}$$

i.e., are described by a principal n, an orbital l, and an azimuthal m quantum numbers. However, when the principal quantum number and the argument of ρ are imaginary the radial portion of the wave function is given not by a polynomial as it is in the wave function in the discrete spectrum when n is real, but by an infinite series. (Integral forms are known for these series). A study of these solutions can be found in handbooks on the theory of atomic spectra and we will not reproduce them here. Replacing the plane wave $\Psi_f = \exp(i\boldsymbol{\varkappa}\mathbf{r})$ with a wave function of the type described, we can iterate the calculations of the photoelectron absorption cross section for a photon, the results of which will be[2,52]

$$\mu_{\text{at}\,1s} = \frac{2^8\pi}{3}\frac{e^2}{mc}\frac{\nu_1^3}{\nu^4}f(n') \tag{4-196}$$

$$f(n') = \frac{\exp[-4n' \operatorname{arccot} n']}{1-\exp(-2\pi n')} \tag{4-197}$$

$$n' = \left(\frac{I_K}{\varepsilon}\right)^{1/2}, \qquad \nu_1 = Z^2 Ry$$

Expanding $f(n')$ for small ε (large n') we find

$$f(n' \gg 1) = (e_n)^{-4}\left(1+\frac{4}{3}\frac{\varepsilon}{I_K}\right) \approx e_n^{-4}\left(1+\frac{\varepsilon}{I_K}\right)^{4/3} = e_n^{-4}\left(\frac{\nu}{\nu_1}\right)^{4/3} \tag{4-198}$$

(here e_n is the base of natural logarithms) and

$$\mu_{\text{at}\,1s}(\nu \geq \nu_1) = \frac{2^8\pi}{3}\frac{e^2}{mc}\frac{1}{e_n^4}\frac{\nu_1^{5/3}}{\nu^{8/3}} \tag{4-199}$$

Near $\varepsilon \geq I_K$ we can say that, approximately, $f(n') = e_n^{-4}(4/3)(\nu/\nu_1)$, and

$$\mu_{\text{at}\,2s}(\nu - \nu_1 \approx I_K) = \frac{2^{10}\pi}{9e_n^4}\frac{e^2}{mc}\frac{\nu_1^2}{\nu^3} \tag{4-200}$$

Finally, when $\varepsilon \gg I_K$ we obtain the previous formula, Eq. (4-194).

We will also present the cross sections obtained in these approximations for the photoelectron absorption of a photon by the $2s$ and $2p$ electrons of an atom. For the two $2s$ electrons[52]

$$\mu_{\text{at}\,2s} \equiv \mu_{L_{\text{I}}} = \frac{2^{11}\pi}{3}\frac{e^2}{mc}\frac{\nu_2^3}{\nu^4}\left(1+3\frac{\nu_2}{\nu}\right)\frac{\exp\left[-4n_2' \operatorname{arccot}\left(\frac{1}{2}n_2'\right)\right]}{1-\exp(-2\pi n_2')} \tag{4-201}$$

for the six $2p$ electrons

$$\mu_{\text{at}\,L_{\text{II}}} + \mu_{\text{at}\,L_{\text{III}}} = \frac{2^{11}\pi}{3}\frac{e^2}{mc}\frac{\nu_2^4}{\nu^5}\left(3+8\frac{\nu_2}{\nu}\right)\frac{\exp\left[-4n_2' \operatorname{arccot}\left(\frac{1}{2}n_2'\right)\right]}{1-\exp(-2\pi n_2')} \tag{4-202}$$

$$\nu_2 = \frac{1}{4}(Z-s_2)^2 Ry, \qquad s_2 = 4.15, \qquad n_2' = 2\left(\frac{\nu_2}{\nu-\nu_2}\right)^{1/2} \tag{4-203}$$

s_2 is a shielding parameter for the electrons in the shell, $n = 2$.

We will formulate the physical results predicted by theory for the photoelectron absorption cross section.

For a specific frequency of incident radiation ν the coefficient of photoelectron absorption is equal to the sum of the electron absorption coefficients of every atomic shell:

$$\mu_{\text{at}}(\nu) = \mu_{\text{at}\,K}(\nu) + \mu_{\text{at}\,L}(\nu) + \mu_{\text{at}\,M}(\nu) + \dots \tag{4-204}$$

We obtain a linear absorption coefficient $\tau(\nu)$ cm^{-1} by multiplying the atomic coefficient $\mu_{\text{at}}(\nu)$ by the number N of the material's atoms per cm^3:

$$\tau(\nu) = N\mu_{\text{at}}(\nu) = \frac{N_A \rho}{A}\,\mu_{\text{at}}(\nu) \tag{4-205}$$

and

$$\tau(\nu) = N_A \rho \sum_i (C_i/A_i)\mu_{\text{at}\,i}(\nu) \tag{4-206}$$

for a multielement material; the C_i are weighted fractions of the elements in the material.

If the quantum energy $h\nu$ is higher than the ionization energy in every shell ($\nu_i/\nu \ll 1$), absorption is relatively low, and the radiation is hard with respect to the given material and has a great deal of penetrating power. As the photon energy is reduced (the radiation wavelength increases), in the area of $\nu \geq \nu_K$ the photoelectron absorption by $1s$ electrons reaches a maximum, which is so great that the contribution from the other electrons, although there are $(Z/2 - 1)$ times more of them, is relatively small. There is a discontinuity in the K absorption edge; the magnitude of which is

$$\delta_K = (\mu_K + \mu_L + \mu_M + \dots)/(\mu_L + \mu_M + \dots) \tag{4-207}$$

From estimates made according to the formulas given earlier, $\delta_K = 9.2$, 7.4, 6.5 for Fe, Ag, and W, respectively.[52]

At frequencies $\nu < \nu_K$ only $2s$, $2p$ and other electrons having higher n ($n = 3; 4$) participate in absorption. As the frequency decreases the L edge is approached and absorption increases, reaching values that are much higher than in absorption at the K maximum. This means that reducing the cross section by a factor of $2^3 = 8$ in the calculation for one electron with $n = 2$ is compensated by $2n^2 = 8$ electrons at the $n = 2$ shell and the increase in absorption is caused by the increase in the frequency factor ν_2^4/ν^5 near the absorption edge. The absorption coefficient in the area of the L edge is much greater than it is near the K edge. Figure 4-11 shows the relations described, where the atomic absorption coefficient for an atom of tin is shown as a function of the radiation wavelength.

If we study the absorption that takes place over a broad range of frequencies from $\nu \gg \nu_K$ to the range $\nu = \nu_{\text{val}}$, where ν_{val} is the edge for photoelectron absorption by valence electrons, we can say that the absorption cross sections in areas near the edge increase gradually as ν decreases: $\mu(\nu_K) < \mu(\nu_L) < \mu(\nu_M), \dots$; their envelope, as can be seen from Eqs. (4-200), (4-201), or (4-192) and (4-193), behave as $1/\nu$. Figure 4-12 shows the general scheme of the frequency dependence of $\mu_{\text{at}}(\nu)$.

Absorption by a given group of electrons, e.g., by $1s$ electrons, depends somewhat differently on Z in different parts of the frequency range. Indeed, near the absorption edge itself $\mu_{1s} \sim \nu_1^{5/3}$, which means that $\mu_{1s} \sim Z^{10/3}$, because $\nu_1 \sim Z^2$. As we move away from the edge by about I_K ($\varepsilon \geq I_K$), $\mu_{1s} \sim \nu_1^2 \sim Z^4$. Finally, at a large distance from this absorption edge ($\hbar\omega \gg I_K$), $\mu_{1s} \sim \nu_1^{5/2} \sim Z^5$. Thus, at an

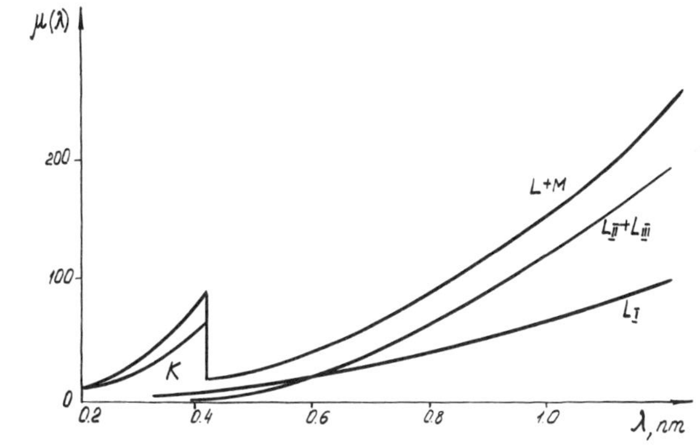

FIGURE 4-11. Atomic coefficient of photoelectron absorption of x-rays in tin as a function of wavelength (in units of 10^{-22} $cm^2/atom$).

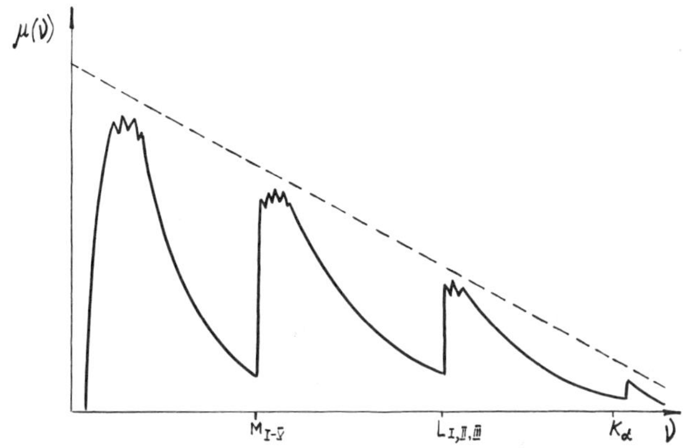

FIGURE 4-12. General relations for the frequency dependence of the photoelectron absorption of atoms.

absorption maximum the dependence $\mu_{1s}(Z)$ is roughly cubic, and at greater distances is $-Z^5$. The same results will be valid for absorption by electrons in the other shells.

As the photon frequency decreases we gradually make the transition from soft 1–10 nm x-rays to the 10–100 nm ultraviolet region, to 100–200 nm vacuum ultraviolet, and to 200–450 nm near-ultraviolet radiation. In these regions the fundamental mechanism for the absorption of photons is also the photoeffect being produced form the more exterior shells. Only the outermost, the so-called valence electrons, experience the photoeffect in the optical and near-ultraviolet regions. The question arises as to whether or not the formulas obtained earlier can be used to calculate the photon absorption cross section for the photoeffect in the outer electrons. The answer to this question depends on what the wave functions are for the initial and final states of the electron experiencing the photoeffect.

These wave functions were previously chosen to be hydrogen-like (with due regard for shielding, e.g., Slater's). The wave functions of the outer valence electrons in molecules and solids are far from being "hydrogen-like." Calculation of the wave functions for the valence electrons in molecules is a problem in quantum chemistry; these functions have the form of molecular orbits that embrace one group of atoms or another and sometimes all of the atoms in a molecule. Such a molecular orbit will be the wave function of the initial state of a valence electron during the photoeffect. If the electron has a great deal of kinetic energy its final state can be represented by a wave function in the form of a plane wave. More interesting situations however, arise near the absorption edge when absorption and the photoeffect cross section are maximal; in this case the state of the electron in the continuous spectrum must be found with regard for the potential in the molecular field. These calculations present a great deal of difficulty and because of this, experimental data on molecular absorption in the optical and ultraviolet regions are so valuable.

The valence electrons in crystalline solids fill the states of an energy band called the valence band. The valence band is almost completely filled by valence electrons in insulators and semiconductors and partially filled in metals. In insulators the next band of allowed states and energies is separated from the valence band by a forbidden band E_g. Thus, the states of electrons in the valence and the conduction bands look like Bloch functions plane waves having an amplitude in the form of an atomic wave function. The wave functions for electrons in bands and in other approximations are constructed along similar lines. Thus, for the internal photoeffect for valence electrons in insulators the process includes taking the valence electrons out of the valence band and transferring them into the conduction band; the initial and final wave functions will be Bloch-like functions. The theory of interband transitions (the photoeffect) in semiconductors can be found in books on semiconductor physics.(55)

Figure 4-13 shows the absorption coefficient measured for LiF.(56) The absorption edge lies at 12.7 eV, which corresponds to 105 nm. This material has a transmission band that is shifted more deeply into the vacuum ultraviolet region and can be used as a transparent material in the fabrication of lenses, windows, and substrates for photomasks in the ultraviolet region.

4-14. THE PERMITTIVITY OF MATERIALS TO ULTRASOFT X-RAYS

The spectral range (0.10–10.0) nm for ultrasoft x-rays is the most favorable for x-ray lithography. In this spectral range we can use powerful synchrotron radiation, choose resist materials that are sensitive to these wavelengths, and weakly absorbent substrate materials in the x-ray masks. One of the requirements on x-ray lithography is the control of the radiation, e.g., isolating a narrow band in the spectrum and deflecting the beam to a specific angle. This part of the chapter (Sections 4-14 to 4-17) is devoted to studying the fundamental optical phenomena that occur with ultrasoft x-rays and which can be seen in the optical components of x-ray lithography machines and in special devices for studying materials in this region of the spectrum. There are several monographs available that discuss the optics of x-ray beams that we can recommend to the reader interested in a more detailed analysis of this subject.(57,86)

We will present a brief discussion of the electronic theory of dispersion. Under the influence of an external electromagnetic field the electrons in an atom are displaced a distance r, converting the atom into a dipole with a moment of $p = re$ (e is the electronic charge). The electron density n_e is

$$n_e = \sum_q n_q = \frac{ZN_A\rho}{A} \tag{4-208}$$

where Z is the atomic number of the element, N_A is Avogadro's number, ρ is the density of the material, A is the atomic weight, and n_q is the number of electrons at the

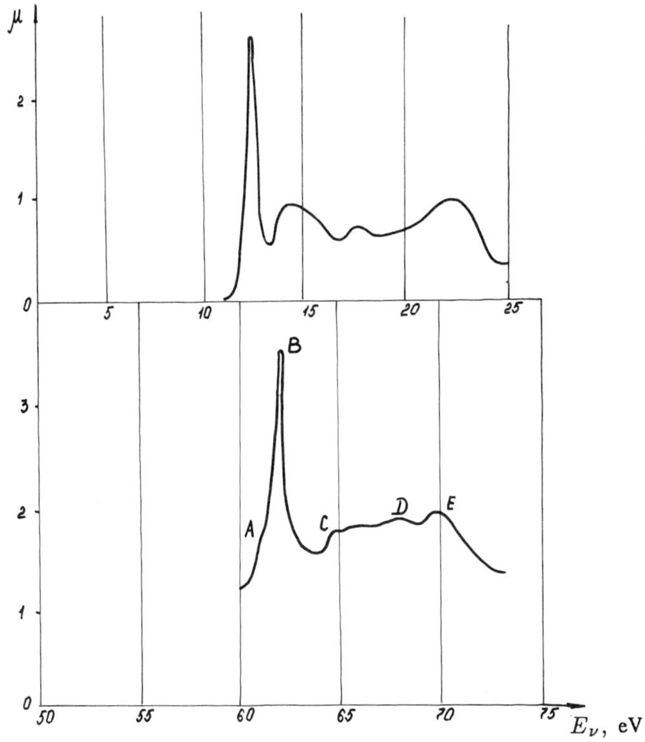

FIGURE 4-13. Photoelectron absorption in LiF.

qth level. The polarization of the medium will then be

$$P = n_e \cdot er \tag{4-209}$$

The electrical induction D of the medium is given by

$$D = \varepsilon E = E + 4\pi P \tag{4-210}$$

where E is the electric field strength. Using Eq. (4-210) gives us

$$\varepsilon = 1 + \frac{4\pi}{E} n_e er = 1 + \frac{4\pi e}{E} \sum_q n_q \cdot r_q \tag{4-211}$$

We will now find out the extent r_q to which an electron is displaced by the action of an external, periodic field. We will think of an atom as an aggregate of harmonic oscillators with respective frequencies ω_q (the vibrational frequency of the q electron relative to its equilibrium position). The equation for the motion of an electron being acted upon by an external field is

$$\ddot{r}_q + \omega_q r_q = \frac{e}{m} E \tag{4-212}$$

where m is the electron mass and the field is a wave $E = E_0 \sin \omega t$. A solution of this equation is

$$r_q(t) = eE(t)/m(\omega_q^2 - \omega^2) \tag{4-213}$$

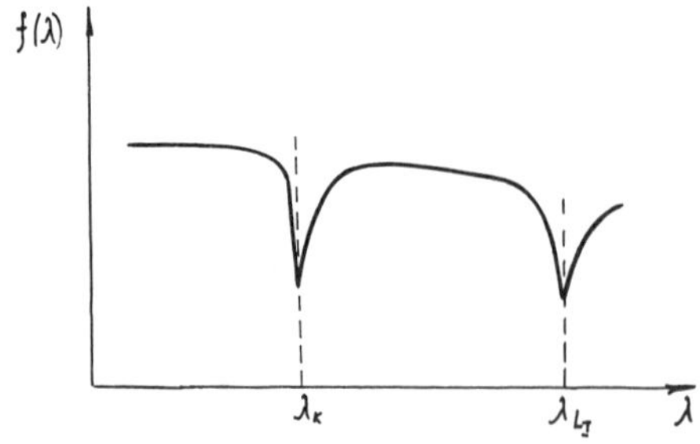

FIGURE 4-14. Spectral dependence of the ratio δ/λ^2.

Substituting Eq. (4-213) into (4-211) gives us

$$\varepsilon = 1 + \frac{4\pi e^2}{m} \sum_q \frac{n_q}{(\omega_q^2 - \omega^2)} \tag{4-214}$$

According to this formula, ε is a function of the frequency ω of the external field, i.e., the formula obtained gives the dispersion effect.

In the x-ray region, $\omega > \omega_q$ for most ω_q, i.e.,

$$\sum_q \frac{n_q}{(\omega_q^2 - \omega^2)} < 0$$

Therefore, Eq. (4-214) can be put into a more convenient form

$$\varepsilon = 1 - \delta = 1 + \frac{4\pi e^2}{m} \sum_q \frac{n_q}{(\omega_q^2 - \omega^2)} < 1 \tag{4-215}$$

for the x-ray region. We can estimate the magnitude of δ, assuming that $\omega_q \ll \omega$ (wavelength $\lambda < 0.01$ nm). Then, taking Eq. (4-208) into account, we will have

$$\delta \approx \frac{e^2\lambda^2}{\pi mc^2} n_e = \frac{e^2 N_A Z\rho}{\pi mc^2 A} \lambda^2 \tag{4-216}$$

i.e., δ/λ^2 = const for any given chemical element. For example, we find for aluminum ($Z = 13$ $A = 27$, and $\rho = 2.7$ g/cm^3) when $\lambda = 0.708 \times 10^{-8}$ cm that $\delta = 1.74 \times 10^{-6}$. Since $\delta \ll 1$, then $\varepsilon = n^2 = 1 - \delta$, and $n = 1 - \delta/2$. The quantity $0 < \delta \ll 1$ (for the x-ray region) is called a unit decrement in the index of refraction.[57]

The strict quantum theory of dispersion shows that the constancy of δ/λ^2 is approximately satisfied far from an absorption edge. For wavelengths near absorption edges δ/λ^2 is a function of the wavelength: $\delta/\lambda^2 = f(\lambda)$. Figure 4-14 shows schematically how this function changes when the wavelength λ changes (λ_K and λ_{L_I} are the wavelengths corresponding to the K and L_I absorption edges).

An absorbing medium is described by a permittivity and index of refraction that are complex numbers:

$$\varepsilon = 1 - \delta + i\beta, \qquad n^* = n - i\varkappa \tag{4-217}$$

TABLE 4-5. Spectral Dependence of the Dimensionless Index of Absorption $\varkappa$

	Dimensionless index of absorption		
λ, nm	$\mu = 1.0\ \mu m$	$\mu = 10\ \mu m^{-1}$	$\mu = 50\ \mu m^{-1}$
0.5	3.9×10^{-5}	3.9×10^{-4}	1.35×10^{-3}
1.0	7.9×10^{-5}	7.9×10^{-4}	3.95×10^{-3}
2.0	1.6×10^{-4}	1.6×10^{-3}	8.0×10^{-3}
5.0	4.0×10^{-4}	4.0×10^{-3}	2.0×10^{-2}
10.0	–	7.9×10^{-3}	3.95×10^{-2}

The relationship between these parameters, $\varepsilon = (n^*)^2$, is maintained from which

$$1 - \delta = n^2 - \varkappa^2, \qquad \beta = 2n\varkappa \tag{4-218}$$

follows. The quantity $n = 1 - \delta/2$ is Eqs. (4-217) and (4-218) is an absolute index of refraction for the medium and $\varkappa$ is a dimensionless index of absorption that is related to the linear index of absorption μ by

$$\mu = 4\pi\varkappa/\lambda \tag{4-219}$$

where λ is the wavelength of the incident radiation. For elements whose $Z \leq 50$ the quantity μ in the spectral range $0.1 \leq \lambda \leq 10.0$ nm never exceeds 50 μm^{-1} on the average. To illustrate the order of magnitude of the dimensionless index of absorption $\varkappa$ various wavelengths, $\mu = 1.0,\ 10$, and 50 μm^{-1}.

As can be seen from the table, the magnitude of $\varkappa$ lies in the range $4 \times 10^{-5} < \varkappa < 4 \times 10^{-2}$ for the wavelengths indicated, i.e., $\varkappa \ll 1$. Thus, in the x-ray spectrum when interacting with a material the latter is described by the complex index of refraction $n^* = n - i\varkappa$. The real part is $n < 1$, $n = 1 - (1/2)\delta$, and $\delta \ll 1$. The imaginary part is $|n^* - n| \ll 1$. The inequalities $0 < \delta \ll 1$ and $\varkappa \ll 1$, characteristic for the indicated spectrum, significantly complicate the design of optical systems for the deflection and refraction of x-rays.

4-15. THE REFLECTION OF X-RAYS THAT IMPINGE FROM THE NORMAL

We will examine the reflection and refraction of x-rays at the interface between two media. Let φ be the angle of incidence, φ' be the angle of reflection, and Ψ be the angle of refraction for a ray (Fig. 4-15). In the space $Z > 0$ the index of refraction is n_1 and in the space $Z < 0$ is n_2. The laws of reflection and refraction can then be written as

$$\varphi = \varphi', \qquad n_1 \sin\varphi = n_2 \sin\Psi \tag{4-220}$$

The amplitude coefficients of reflection (r) and refraction (t) are given by the Fresnel

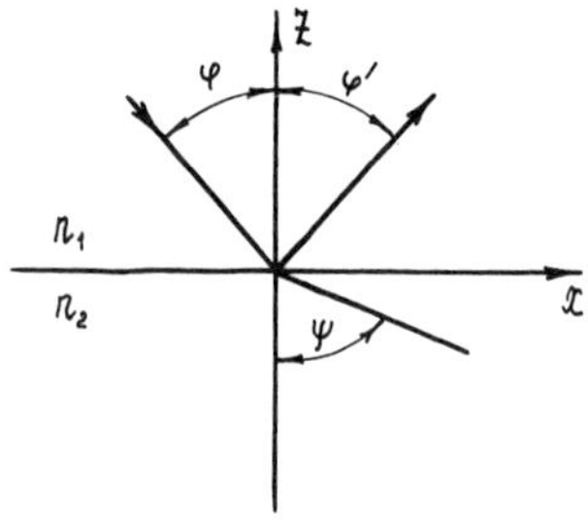

FIGURE 4-15. Reflection and refraction at the interface between two media.

equations[59]

$$r_{\perp} = -\frac{\sin(\varphi - \Psi)}{\sin(\varphi + \Psi)}, \qquad t_{\perp} = \frac{2 \sin \Psi \cos \varphi}{\sin(\varphi + \Psi)}$$
$$r_{\parallel} = -\frac{\tan(\varphi - \Psi)}{\tan(\varphi + \Psi)}, \qquad t_{\parallel} = \frac{2 \sin \Psi \cos \varphi}{\sin(\varphi + \Psi) \cos(\varphi - \Psi)} \tag{4-221}$$

(where the subscripts $\perp$ and $\parallel$ denote the components corresponding to the E-plane, perpendicular ($\perp$) and parallel ($\parallel$) to the plane of incidence (which corresponds to the plane of the drawing). Because the material is absorbent, n_2 is a complex quantity: $n_2^* = n_2 - i\varkappa_2$, and the angle of refraction will also be a complex quantity:

$$\sin(\Psi *) = \frac{n_1 \sin \varphi}{n_2^*} = \frac{n_1 \sin \varphi}{n_2 - i\varkappa_2} \tag{4-222}$$

We will examine the case in which the rays impinge at an angle that is close to the normal: φ and Ψ are small. In this case the law of refraction is

$$\Psi^* = (\Psi_1 - i\Psi_2) = \frac{n_1 \varphi}{n_2 - i\varkappa_2} = \frac{n_1 n_2 \varphi}{n_2^2 + \varkappa_2^2} + i\frac{n_1 \varkappa_2 \varphi}{n_2^2 + \varkappa_2^2} \tag{4-223}$$

The magnitude of the complex angle Ψ^* does not have the simple sense of an angle of refraction and points to the fact that an electromagnetic wave in an absorbent medium is a nonhomogeneous wave. In metal optics it was shown that for such a nonhomogeneous wave the planes of constant amplitude do not coincide with the planes of constant real phase. The angle θ_2 between the normal to the plane of the constant real phase and the plane of the interface between two media is thought of as the angle of refraction. Then the law of refraction can be put into its usual form:

$$n_1(\varphi_1) = (n_2)_{\text{eff}} \cdot \theta_2 \tag{4-224}$$

where $n_{2\,\text{eff}}$ is a function of n_2, $\varkappa_2$, and φ:

$$(n_2)_{\text{eff}} = f(n_2, \varkappa_2, \varphi)$$

We will now derive an expression for the coefficient of reflection of an electromagnetic wave from an absorbent medium when the radiation is incident from an angle that is close to the normal, i.e., the angle of incidence is small. It follows from Eq. (4-221) that

$$r_\perp = r_\parallel = -\frac{\varphi - \Psi^*}{\varphi + \Psi^*} \tag{4-225}$$

Replacing Ψ^* in Eq. (4-225) by $n_1\varphi/(n_2 - i\varkappa_2)$, we find

$$r_\perp = r_\parallel = -\frac{(n_2 - n_1) - i\varkappa_2}{(n_2 + n_1) - i\varkappa_2} \tag{4-226}$$

The coefficient of reflection will then be

$$R = |r_\perp|^2 = |r_\parallel|^2 = \frac{(n_2 - n_1)^2 + \varkappa_2^2}{(n_2 + n_1)^2 + \varkappa_2^2} \tag{4-227}$$

It was shown earlier that for x-rays in the $0.1 \leq \lambda \leq 100$ nm range $n_2 = I - (1/2)\delta$ for all media except a vacuum, where δ is small and the index of absorption $\varkappa_2$ is also small. If the first medium is a vacuum, $n_1 = 1$. Then, according to Eq. (4-227) the coefficient of reflection is

$$R = \frac{\delta^2 + \varkappa_2^2}{(2 + \delta)^2 + \varkappa^2} \simeq \frac{\delta^2 + \varkappa_2^2}{4} \ll 1 \tag{4-228}$$

Thus, when x-rays impinge at the interface of a homogeneous medium at an angle that is close to the normal, reflection is almost nonexistent. This is also true for metals in which $R \simeq 1$ in the optical region. Later on we will see that multilayer structures that are highly reflective to x-rays impinging from the normal (and close to the normal) can be created. We will now examine reflection when the x-rays have grazing incidence at the interface of a homogeneous medium.

4-16. REFLECTION AND REFRACTION OF X-RAYS THAT GRAZE THE INTERFACE BETWEEN HOMOGENEOUS MEDIA

It follows immediately from Eq. (4-221) that $R = 1$ when an electromagnetic wave grazes the interface between two media, because when $\varphi = (\pi/2)$, $|r_\perp| = |r_\parallel| = 1$. In this regard the reflection and refraction of an x-ray when the ray grazes the interface between the media are of interest, i.e., when $\varphi_0 = (\pi/2 - \varphi)$ is small (φ_0 is the grazing angle). We make the designation $\Psi_0 = (\pi/2 - \Psi)$, and then obtain an expression for refraction in another form from Eq. (4-220) (neglecting absorption):

$$\cos\varphi_0 = n_2 \cos\Psi_0 \qquad (n_1 = 1) \tag{4-229}$$

or

$$n_2 = 1 - \frac{1}{2}\delta = \cos\varphi_0 / \cos\Psi_0 \tag{4-230}$$

Because δ is small the difference $(\varphi_0 - \Psi_0) = \Delta\varphi_0 > 0$ is also small, i.e., the refracted ray is only slightly deflected from its initial direction (see Fig. 4-16). We will ascertain

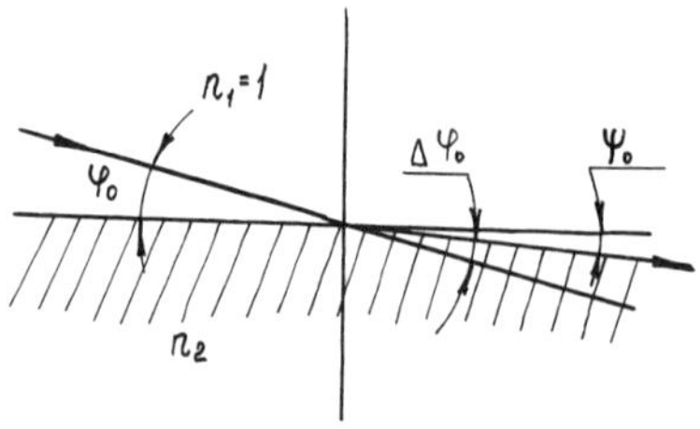

FIGURE 4-16. X-ray refraction at the interface between a medium and a vacuum.

how $\Delta\varphi_0$ depends on the angle φ_0. We rewrite Eq. (4-230) as

$$\cos\varphi_0 - \cos\Psi_0 = -\frac{1}{2}\,\delta \cdot \cos\varphi_0 \tag{4-231}$$

Since $\Psi_0 \approx \varphi_0 - \Delta\varphi_0$, we have

$$\cos\varphi_0 - \cos\Psi_0 = -\Delta\varphi_0 \cdot \sin\varphi_0 \tag{4-232}$$

It follows from Eqs. (4-231) and (4-232) that

$$\varphi_0 - \Psi_0 \simeq \Delta\varphi = \frac{1}{2}\,\delta \cdot \cot\varphi_0 \tag{4-233}$$

Thus, $\Delta\varphi_0$ increases as φ_0 decreases.

We will estimate the magnitude of $\Delta\varphi_0$. In the λ = 1.0–2.0 nm range the unit decrement in the index of refraction reaches values of $(1/2)\delta \approx 1 \times 10^{-4}$. Let the angle be $\varphi_0 = 0.1$ rad (5.7°). Then we find from Eq. (4-233) that $\Delta\varphi_0 \simeq 10^{-3} \simeq 21''$.

The possibility of total external reflection follows from Eq. (4-230). We will find the critical angle φ_{0_c} at which total external reflection occurs. Making the substitution $\cos\Psi_0 = 1$ in Eq. (4-230) gives us

$$1 - \frac{1}{2}\,\delta = \cos\varphi_{0_c} \tag{4-234}$$

Disregarding the second-order infinitesimals in Eq. (4-234), we find that

$$\varphi_{0_c} \simeq \sqrt{\delta} \tag{4-235}$$

The formula for the angle of refraction is valid for angles $\varphi_0 \geq \varphi_{0_c}$. When a ray impinges at an angle $\varphi_0 < \varphi_{0_c}$ it does not penetrate the medium, but is reflected from the surface. This phenomenon is called total external reflection.

We will now ascertain how the coefficient of reflection depends on the grazing angle φ_0 when absorption of the x-rays by the medium is considered (the index of refraction is a complex quantity: $n^* = n - i\varkappa$).

We rewrite the Fresnel equations for the amplitude coefficients of reflection, Eq. (4-221), going from φ and Ψ to φ_0 and Ψ_0 via $\varphi = (\pi/2) - \varphi_0$ and $\Psi = (\pi/2) - \Psi_0$:

$$r_\perp = \frac{\sin(\varphi_0 - \Psi_0)}{\sin(\varphi_0 + \Psi_0)}, \qquad r_\parallel = -\frac{\tan(\varphi_0 - \Psi_0)}{\tan(\varphi_0 + \Psi_0)} = -\frac{\sin 2\varphi_0 - \sin 2\Psi_0}{\sin 2\varphi_0 + \sin 2\Psi_0} \tag{4-236}$$

After straightforward trigonometric conversions and making the substitution $\cos\varphi_0 = n^*\cos\Psi_0$, we have

$$r_\perp = \frac{\sin\varphi_0 - n^*\sin\Psi_0}{\sin\varphi_0 + n^*\sin\Psi_0}, \qquad r_\parallel = \frac{-n^*\sin\varphi_0 - \sin\Psi_0}{n^*\sin\varphi_0 + \sin\Psi_0} \tag{4-237}$$

We now give Ψ_0 in terms of φ_0:

$$\sin\Psi_0 = \sqrt{1-\cos^2\Psi_0} = \frac{1}{n^*}\sqrt{(n^*)^2 - 1 + \sin^2\varphi_0} \tag{4-238}$$

Since $n^* = n - i\varkappa = (1-\delta) - i\varkappa$, and $\varkappa \ll 1$, we have

$$(n^*)^2 = (1-2\delta) - i2\varkappa \tag{4-239}$$

Substituting the values of $\sin\Psi_0$ and n^* in Eq. (4-237) according to Eqs. (4-238) and saying that $\sin\varphi_0 \approx \varphi_0$, we find

$$\left.\begin{aligned} r_\perp &= \frac{\varphi_0 - \sqrt{(\varphi_0^2 - 2\delta) - i2\varkappa}}{\varphi_0 + \sqrt{(\varphi_0^2 - 2\delta) - i2\varkappa}} \\ r_\parallel &= \frac{(1-2\delta - i2\varkappa)\varphi_0 - \sqrt{(\varphi_0^2 - 2\delta) - i2\varkappa}}{(1-2\delta - i2\varkappa)\varphi_0 + \sqrt{(\varphi_0^2 - 2\delta) - i2\varkappa}} \end{aligned}\right\} \tag{4-240}$$

The quantities a and b, introduced in

$$\sqrt{(\varphi_0^2 - 2\delta) - i2\varkappa} = a + ib \tag{2-241}$$

$$\left.\begin{aligned} a^2 &= \frac{1}{2}\delta\left[\sqrt{\left(\frac{\varphi_0}{\delta} - 1\right)^2 + \frac{\varkappa^2}{\delta^2}} + \left(\frac{\varphi_0^2}{\delta} - 1\right)\right] \\ b^2 &= \frac{1}{2}\delta\left[\sqrt{\left(\frac{\varphi_0}{\delta} - 1\right)^2 + \frac{\varkappa^2}{\delta^2}} - \left(\frac{\varphi_0^2}{\delta} - 1\right)\right] \end{aligned}\right\} \tag{2-242}$$

are often used. With the help of Eq. (4-241) we easily find

$$\left.\begin{aligned} |r_\perp|^2 &= \frac{(\varphi_0 - a)^2 + b^2}{(\varphi_0 + a)^2 + b^2} \\ |r_\parallel|^2 &= \frac{(\varphi_0 - \delta\varphi_0 - a)^2 + (b - \varkappa\varphi_0)^2}{(\varphi_0 - \delta\varphi_0 + a)^2 + (b - \varkappa\varphi_0)^2} \end{aligned}\right\} \tag{4-243}$$

Because $\delta \ll 1$ and φ_0 are equal to $\varphi_0 \leq \sqrt{\delta}$ in order of magnitude, the product of $\delta\varphi_0$ and $\varkappa\varphi_0$ will be second-order infinitesimals and can be ignored, in which case

$$|r_\perp|^2 = |r_\parallel|^2 = \frac{(\varphi_0 - a)^2 + b^2}{(\varphi_0 + a)^2 + b^2}$$

TABLE 4-6. The Coefficient of Reflection R as a Function of φ_0/φ_{0_c}

φ_0/φ_{0_c}	Coefficient of reflection						
	$\varkappa/\delta = 0.01$	$\varkappa/\delta = 0.02$	$\varkappa/\delta = 0.03$	$\varkappa/\delta = 0.05$	$\varkappa/\delta = 0.1$	$\varkappa/\delta = 1.0$	$\varkappa/\delta = 10$
0.5	0.985	0.98	0.96	0.95	0.89	0.50	0.66
0.8	0.98	0.95	0.92	0.875	0.82	0.29	0.50
0.9	0.96	0.92	0.88	0.81	0.675	0.20	0.45
1.0	0.76	0.67	0.61	0.53	0.42	0.16	0.42
1.2	0.08	0.08	0.08	0.08	0.08	0.09	0.35
1.4	0.03	0.03	0.03	0.03	0.03	0.05	0.29

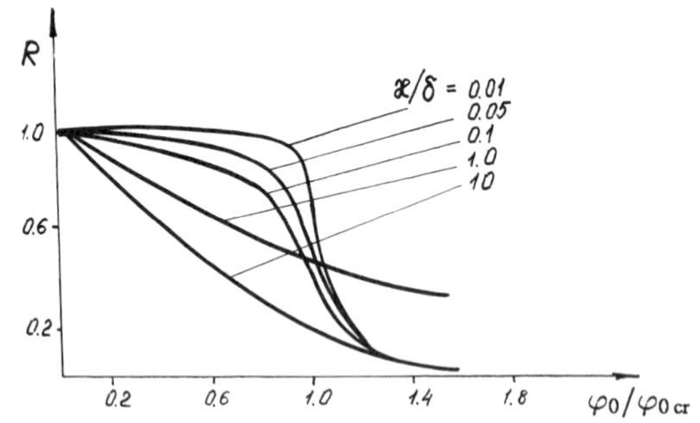

FIGURE 4-17. Coefficient of reflection R as a function of the ratio $\varphi_0/\varphi_{0\,\text{cr}}$.

and the coefficient of reflection $R = (1/2)[|r_\perp|^2 + |r_\parallel|^2]$ will be

$$R \cong \frac{(\varphi_0 - a)^2 + b^2}{(\varphi_0 + a)^2 + b^2} \tag{2-244}$$

Since $\varphi_{0_c}^2 = \delta$, it follows from Eq. (4-241) that $a = b = \sqrt{\varkappa/2}$ when $\varphi_0 = \varphi_{0_c}$. Substituting this quantity for a and b into Eq. (4-244) gives us

$$R_{\varphi_{0_c}} = \frac{\varphi_{0_c}^2 + \varkappa - 2\varphi_{0_c}\sqrt{\varkappa/2}}{\varphi_{0_c}^2 + \varkappa + 2\varphi_{0_c}\sqrt{\varkappa/2}} < 1 \tag{4-245}$$

It follows from Eq. (4-244) that $R_{\varphi_0=0} = 1$ when $\varphi_0 = 0$.

Thus, as the grazing angle φ_0 increases from 0 to φ_{0_c}, the coefficient of reflection from an absorbing medium decreases.

We will pause to examine the dependence $R = R(\varphi_0)$ more thoroughly. Figure 4-17 shows graphs of $R(\varphi_0)$ for different ratios $\varkappa/\delta$. For a more accurate perception, numerical values are given in Table 4-6 for some characteristic points on the abscissa when $\delta = 7.0 \times 10^{-4}$.

As we can see from the figure and the table the coefficient of reflection decreases as φ_0 increases. A sharp decrease in R is seen near φ_{0_c} for slightly absorbent materials ($\varkappa/\delta \ll 1$). Increasing absorption (more accurately, the ratio $\varkappa/\delta$) leads to distinct spreading in the threshold effect of total external reflection and the concept of a critical angle becomes meaningless. The behavior of the function $R(\varphi_0)$ as $\varkappa/\delta$ increases is of particular practical importance. Initially as $\varkappa/\delta$ increases from 0 to 1 when $\varphi_0 < \varphi_{0_c}$, the magnitude of R decreases and as R is increased further ($\varkappa/\delta > 1$) once again increases, whereby the decrease in $R(\varphi_0)$ when $\varphi_0 > \varphi_{0_c}$ slows down and we may again have $R \simeq 0.2$ at angles $\varphi_0 = 2\varphi_{0_c}$.

TABLE 4-7. The Nature of the Decrease in the Coefficient of Reflection R at Large Values φ_0/φ_{0_c} for $\varkappa/\delta = 10$, $\delta = 3.5 \times 10^{-4}$

$\frac{\varphi_0}{\varphi_{0_c}}$	0.5	0.8	0.9	1.0	1.2	1.4	1.6	2.0	5.0	10.0
R	0.66	0.50	0.45	0.42	0.35	0.29	0.24	0.16	0.07	0.0006

The presence of a minimum in $R(\varkappa/\delta)$ at any fixed angle φ_0 follows from the general expression, Eq. (4-244). Finding this minimum is much easier when $\varphi_0 = \varphi_{0_c}$ because Eq. (4-245) can be greatly simplified. In fact, making the substitution $\varphi_{0_c}^2 = \delta$ in Eq. (4-245) gives us

$$R = \frac{1 + x - \sqrt{2x}}{1 + x + \sqrt{2x}}, \qquad x = \varkappa/\delta \tag{4-246}$$

Equation (4-246) has an extremum (minimum) when $x = 1$. To evaluate the nature of the decrease in $R(\varphi_0)$ when $\varphi_0 > \varphi_{0_c}$ we find an asymptotic value for the quantities a and b when $\varphi_0 \gg \varphi_{0_c}$. We rewrite Eq. (4-242), keeping in mind that $\delta = (\varphi_{0_c})^2$:

$$a^2 = \frac{1}{2}\varphi_0^2 \left\{ \sqrt{\left[1 - \frac{\varphi_{0_c}^2}{\varphi_0^2}\right]^2 + \frac{\varkappa^2}{\varphi_0^4}} + \left[1 - \frac{\varphi_{0_c}^2}{\varphi_0^2}\right] \right\}$$

$$b^2 = \frac{1}{2}\varphi_0^2 \left\{ \sqrt{\left[1 - \frac{\varphi_{0_c}}{\varphi_0^2}\right] + \frac{\varkappa^2}{\varphi_0^4}} - \left[1 - \frac{\varphi_{0_c}^2}{\varphi_0^2}\right] \right\}$$

As we can see, when $\varphi_0 \gg \varphi_{0_c}$, $a^2 \approx \varphi_0^2$ and $b^2 = 0$. We will demonstrate this conclusion with a specific example. Let $\varkappa/\delta = 10$ and $\delta = 7 \times 10^{-4}$. Then, using Eqs. (4-241) and (4-244) we find the values given in Table 4-7.

The characteristics of x-ray reflection discussed above for the case when the grazing angles φ_0 are less than or equal to the angle φ_{0_c} corresponding to total external reflection are used in fabricating reflectors. A. P. Lukirskii and his coworkers studied the spectral and angular dependencies of the coefficients of reflection for several materials over a wide range of wavelengths from 0.7 to 20.0 nm (the range up to 30 nm was studied for some materials). These dependencies are shown as an example in Fig. 4-18 for polystyrene. The spectral behavior of the coefficient of reflection for polystyrene has, as can be seen from Fig. 4-18, a monotonic nature in the long wave ($\lambda > 4.5$ nm) region because the material does not have an absorption edge at these wavelengths. At small grazing angles (from $\varphi_0 = 3°$ to $\varphi_0 = 9°$) a polystyrene reflector can be used as a filter for shortwave radiation. A reflector made from nickel, for example, is fairly effective for wavelengths ranging from 1.5 to 4.0 nm (the location of the $L_{\mathrm{II,III}}$ absorption edges in nickel is $\lambda = 1.45$ nm).

4-17. THE SCATTERING AND REFLECTION OF SOFT X-RAYS FROM A ROUGH SURFACE

Surface roughness, the product of processing defects or contamination, causes the incident radiation to be scattered and the reflected ray to be attenuated. Measurements

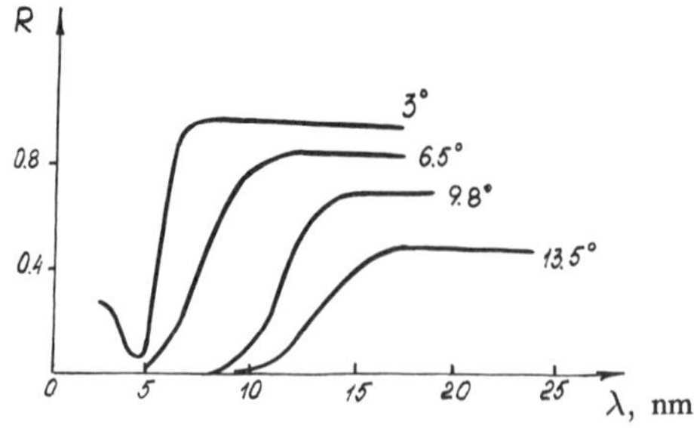

FIGURE 4-18. Spectral dependencies of the coefficient of reflection R of a polystyrene reflector at different grazing angles φ_0.

of the total integrated scattering are widely used to describe the processing quality of optical surfaces.(61) Total integrated scattering is defined as the ratio of the total intensity of the scattered light within a hemisphere to the sum of the total integrated scattering plus the intensity of the secularly reflected beam. The amount of total integrated scattering depends only on the ratio of the average height of the projections δ to the wavelength λ of the incident (laser) radiation(61):

$$S = \left(\frac{4\pi\delta}{\lambda}\right)^2 \tag{4-247}$$

The degree of roughness $\delta = (\lambda/4\pi)S^{1/2}$, found from measurements of S, can be used to create quality standards for surface processing.(61)

If the details of the angular radiation scattering distribution must be known, a more detailed description of the surface of the interface between the two media at which the radiation is scattered is needed. Solving this problem inspired the works [62, 63]; important results from recent research are contained in [64–66], and our discussion will follow [67, 68]. We will use a geometric function $\zeta(\boldsymbol{\rho})$ to describe the unevenness of the interface surface, where $\boldsymbol{\rho}(x, y, 0)$ is the radius vector of the distances in the plane of the interface, and ζ is the height of the unevenness. We will say that $\overline{\zeta(\boldsymbol{\rho})} = 0$ and $\overline{\zeta(\bar{\rho})\zeta(\bar{\rho}')} = \xi^2 \cdot \chi(\boldsymbol{\rho} - \boldsymbol{\rho}')$,

$$\chi(\boldsymbol{\rho} - \boldsymbol{\rho}') = \exp(-|\boldsymbol{\rho} - \boldsymbol{\rho}'|^2/a^2) \tag{4-248}$$

where a is the correlation radius. Thus, statistical characteristics of roughness are the mean-squared height of the projections ξ^2, the correlation radius a, as well as the type of correlation function—a uniform Gaussian correlation function. Using $\zeta(\rho)$, the dielectric properties of the interface medium having permittivities of $\varepsilon_1(z < 0)$ and $\varepsilon_2(z > 0)$ are specified:

$$\varepsilon(\mathbf{r}) = \varepsilon_0(z) + \Delta\varepsilon(\boldsymbol{\rho}) \tag{4-249}$$

$$\varepsilon_0(\mathbf{r}) = \varepsilon_1 + \theta(z)(\varepsilon_2 - \varepsilon_1) \tag{4-250}$$

$$\Delta\varepsilon(\boldsymbol{\rho}) = [\theta(z - \xi(\boldsymbol{\rho})) - \theta(z)](\varepsilon_2 - \varepsilon_1) \tag{4-251}$$

Here, $\theta(z)$ is the Student's function $O(z < 0) = 0$, $\theta(Z > 0) = 1$. We easily find the statistical properties of the random part of the permittivity $\Delta\varepsilon(\rho)$ from Eq. (4-251):

$$\overline{\Delta\varepsilon} = (\delta\varepsilon)\bar{\xi}^2\delta'(z) \tag{4-252}$$

$$\overline{\Delta\varepsilon(\mathbf{r})\Delta\varepsilon(\mathbf{r}')} = (\delta\varepsilon)^2\delta(z)\delta(z')\xi^2\cdot\chi(\rho-\rho') \tag{4-253}$$

We use the statistical description of the dielectric properties of an interface just given to solve the problem of the scattering of a scalar field at the interface between media.[67] The unknown scalar field $\psi(\mathbf{r})$ must satisfy

$$[\Delta + k^2\varepsilon_0(z)]\psi(\mathbf{r}) = -k^2\Delta\varepsilon(\mathbf{r})\psi(\mathbf{r}) \tag{4-254}$$

or, in integral form,

$$\psi(\mathbf{r}) = \psi_0(\mathbf{r}) - k^2\int\Delta\varepsilon(\mathbf{r}')G(\mathbf{r},\mathbf{r}')\psi(\mathbf{r}')\,d^3\mathbf{r}' \tag{4-255}$$

Green's function $G(\mathbf{r},\mathbf{r}')$ for Eq. (4-254) is so complicated that we must refer the reader to Vinogradov *et al.*[67] for the details of the calculations, where it is shown that the energy flux through an area σ perpendicular to the z axis,

$$Q = \frac{1}{k}\int_\sigma \frac{\partial\psi(\mathbf{r})}{\partial z}\,\mathrm{Im}\,\psi^*(\mathbf{r})\,d^2\rho \tag{4-256}$$

contains terms corresponding to specular reflection from an ideal interface between media and to diffusive scattering at the nonuniformities. In a first approximation of the theory of perturbations in the small parameter $\Delta\varepsilon$ for the range of small grazing angles $\alpha \ll 1$, $\alpha^2 < 2[1-(\varepsilon_2/\varepsilon_1)]^{1/2}$, a small difference $(\varepsilon_1 - \varepsilon_2) \ll 1$, and the conditions that $ka(\varepsilon_1-\varepsilon_2) \ll 1$, which are satisfied for x-ray wavelengths, we obtain a result for the reflectance of a rough surface[67]:

$$|R'|^2 = |R|^2 - \delta R \tag{4-257}$$

$$\delta R = C|R|^2(k\xi)^2(ka)^{3/2}\frac{\delta\varepsilon^2}{\varepsilon_1^{1/4}}\alpha \qquad \text{if } \left(\frac{5\delta\varepsilon}{4\varepsilon_1}\right)^{1/2} < \frac{ka\delta\varepsilon}{4\varepsilon_1^{1/2}} \ll 1 \tag{4-258}$$

$$\delta R = C|R|^2(k\xi)^2(ka)^{-1/2}\frac{\delta\varepsilon}{\varepsilon_1^{1/4}}\alpha \qquad \text{if } \frac{ka\delta\varepsilon}{4\varepsilon_1^{1/2}} < \left(\frac{5\delta\varepsilon}{4\varepsilon_1}\right)^{1/2} \ll 1 \tag{4-259}$$

$$\delta R = 4|R|^2(k\xi\sin\alpha)^2\varepsilon_1^{1/2}, \qquad \text{if } \mu = \frac{(ka)\varepsilon_1^{1/2}}{4}\alpha^2 > 1 \tag{4-260}$$

Here, $C = 5\Gamma(3/4)/\sqrt{\pi}$. Equation (4-260) describes grazing reflection from a rough surface when the grazing angles are relatively large, ($\mu > 1$); in this case scattering is defined by the apparent height of the roughness $\xi\sin\alpha$ to the radiation wavelength λ: $k\xi\sin\alpha = 2\pi(\xi\sin\alpha)/\lambda$. At angles α so small that $\mu < 1$, scattering is proportional to α and depends on the ratios ξ/λ and a/λ. Let, e.g., $\xi = 10$ Å, $a = 100\ \mu$m, $\lambda = 10$ Å, $\xi/\lambda \simeq 1$, $a/\lambda \approx 10^5$, $\delta\varepsilon \simeq 10^{-3}$, $\alpha = 10^{-2}$, and $\delta R/|R|^2 \approx 3\%$. When $\alpha = 10^{-1}$, the reduction in the coefficient of reflection due to scattering by the roughness reaches 30%.

From an analysis of Eqs. (2-258)–(4-260) we can say that the amount of roughness allowed depends on the grazing angle of the ray. According to Eq. (4-260) this criterion has the form

$$\xi \ll \frac{\lambda}{2\varepsilon_1 \alpha} \tag{4-261}$$

when the grazing angles are fairly large ($\alpha > (\lambda/a)^{1/2}$). For small angles ($\alpha < (\lambda/a)^{1/2}$) the condition that scattering due to roughness be small will, according to Eq. (4-258), be

$$\xi \ll \frac{(\lambda^3 a)^{1/4}}{\alpha^{1/2}} \tag{4-262}$$

For the sake of clarity, factors on the order of unity have been omitted in Eqs. (4-261) and (4-262). According to Eq. (4-261), when $\lambda = 1$ nm, $\varepsilon_1 \simeq 1$, and $2\alpha = 10^{-2}$, we have the condition that $\xi \ll 100$ nm. We can say that a roughness of $\xi = 10$ nm is tolerable for reflectors of grazing incidence in the soft x-ray spectrum. The current level of polishing technology makes it possible to achieve the indicated quality for a number of materials.

4-18. MULTILAYER REFLECTORS FOR SOFT X-RAYS

Multilayer interference filters and reflectors for the optical range of wavelengths are well known and widely used. It is quite natural to extend the concepts of producing multilayer interference optical components (reflectors, filters) to soft x-rays. At least three factors complicate this task: 1) the permittivities of the materials are close to unity: $1 - \varepsilon \simeq 10^{-1}$–$10^{-3}$; 2) the thickness d of the layers in a multilayer structure must be on the order of a wavelength, i.e., $d \simeq 1$ nm in the range under consideration; and 3) in the soft x-ray spectrum, all materials are highly absorbent: the linear coefficients of absorption are 10^3–10^5 cm^{-1}.

The technology of microelectronics can be thought of as making it possible to obtain multilayer structures of satisfactory quality that are 1 nm thick. It is possible that new methods such as molecular epitaxy will yield multilayer devices of still better quality. Devices with Ni–C, Cr–C, CoC, MoSi, and W–C, etc. layers have been studied; the layers are created by electron beam and laser vaporization and by magnetron sputtering.[69–72]

According to their principle of operation, x-ray reflectors provide the very familiar Bragg reflection when

$$\lambda = 2d \cos\theta \tag{4-263}$$

where d is the period of the structure and θ is the angle of incidence. When incidence is from the normal the period is $d = \lambda/2$, i.e., the thickness of each layer is roughly $1/4\lambda$.

We can represent a multilayer periodic structure (MPS) as a succession of layers having permittivities ε_1 and ε_2. Let the material with ε_1 be highly absorbent and the material with ε_2 be weakly absorbent, i.e., $\text{Im}\,\varepsilon_1 > \text{Im}\,\varepsilon_2$. The material with ε_1 occupies a fraction β of the period d so that the average permittivity of the structure is $\mu =$

$\beta\varepsilon_1 + (1-\beta)\varepsilon_2$. According to [73], the reflection of radiation having a wavelength of λ near the first Bragg peak is described by the coefficient

$$R^{(s,p)} = \left| \frac{1}{2} B\sigma \bigg/ \left[b - i\sqrt{\left(\frac{B\sigma}{2}\right)^2 - b^2} \right] \right|^2 \tag{4-264}$$

$$b = \sin^2\theta + (\lambda/2d)^2 - \mu, \qquad B = 2(\varepsilon_1 - \varepsilon_2)\frac{\sin\pi\beta}{\pi} \tag{4-265}$$

the indexes s and p describe the reflection of the radiation s and p the polarization, whereby $\sigma(s) = 1$ and $\sigma(p) = 1 - 2(\lambda/2d)^2$, θ is the angle of incidence of the radiation on the MPS.

Values of $R^{(s,p)}$, calculated from Eq. (4-264), can be optimized by choosing those d and β^* for which R is a maximum. It turns out that such d and β^* are related by[74–76]

$$\left(\frac{\lambda_0}{2d}\right)^2 = \cos^2\theta_0 - \mathrm{Re}(1-\varepsilon_2) + \mathrm{Re}(\varepsilon_1 - \varepsilon_2)\left(\beta^* - \frac{\sin 2\pi\beta^*}{2\pi}\sigma^2\right) \tag{4-266}$$

$$\tan(\pi\beta^*) = \pi\left[\beta^* + \frac{\mathrm{Im}\,\varepsilon_2}{\mathrm{Im}(\varepsilon_1 - \varepsilon_2)}\right] \tag{4-267}$$

and the values of $R^{(s,p)}$ at a maximum are

$$R^{(s,p)}_{\max} = \frac{1 - u^{(s,p)}}{1 + u^{(s,p)}}, \qquad u^{(s,p)} = \left(\frac{1 - \nu^{(s,p)2}\cos^2\pi\beta^*}{1 + \nu^{(s,p)2} f^2 \cos^2\pi\beta^*}\right)^{1/2} \tag{4-268a}$$

$$f = \frac{\mathrm{Re}(\varepsilon_1 - \varepsilon_2)}{\mathrm{Im}(\varepsilon_1 - \varepsilon_2)}, \qquad g = \frac{\mathrm{Im}\,\varepsilon_2}{\mathrm{Im}(\varepsilon_1 - \varepsilon_2)}; \qquad \beta^* = \beta^*(g) \tag{4-268b}$$

$$\nu^{(s)} = 1, \qquad \nu^{(p)} = \cos 2\theta_0 \tag{4-268c}$$

Values of λ_0 and θ_0 that satisfy Eq. (4-266) are called resonant. Equation (4-266) is itself the Bragg condition for an MPS.

It is important to note that $R^{(s,p)}_{\max}$ for an MPS is defined in terms of the f and g parameters, which are *ratios* (for the soft x-ray spectrum) of small quantities such that f and g can vary between wide limits of from $f,\ g \gg 1$ to $f,\ g \ll 1$ and, consequently, even $R_{\max}$ itself may acquire values $\sim$1. This situation must be contrasted with reflection from a single interface where, according to the Fresnel equations (see Section 1-15), $R \sim |\varepsilon_1 - \varepsilon_2|^2$ and R is then a second-order infinitesimal in $|\varepsilon_1 - \varepsilon_2|$. Figure 4-19 shows curves of $R_{\max}(f, g)$ = const for radiation impinging at the normal. Obviously, the larger values of $R_{\max}$ are seen when the f are large and g are small. How an MPS operates to produce a high degree of soft x-ray reflection can be explained by comparing the depth to which a ray $L(f, g)$ penetrates when the penetration depth L_{abs} is due only to absorption (without regard for interference): $L_{\mathrm{abs}} = (\cos\theta/\mathrm{Im}\,\mu)(\lambda/\pi)$. Referring to Vinogradov *et al.*[74] because of its similarity to our analysis, we will say that the case $f\cos\pi\beta^* \gg 1$ corresponds to a low degree of absorption in the structure. When $|\mathrm{Re}(\varepsilon_1 - \varepsilon_2)| \gg \mathrm{Im}(\varepsilon_1\varepsilon_2)$, $L(f, g) < L_{\mathrm{abs}}$, which means that interference repulsion is present in the field from the MPS, which results in reduced absorption and a higher

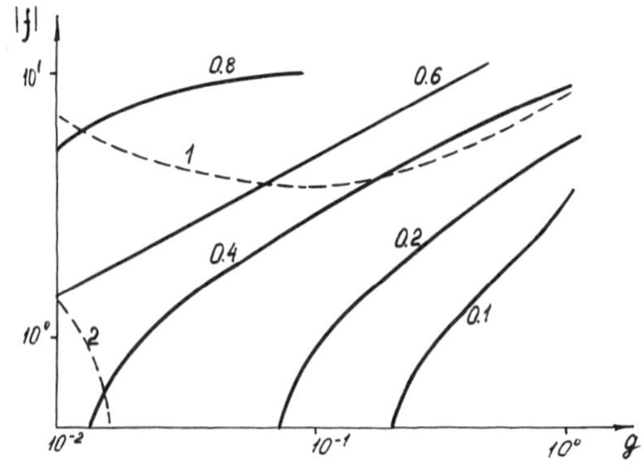

FIGURE 4-19. Curves of the constant values of the maximum coefficient of reflection R_{max} = const in a plane with parameters f and g when the radiation impinges from the normal. The interference reflectors range ($L < 0.5L_{abs}$) is located above the dotted curve 1 and the Borrmann reflector's range ($L > 2L_{abs}$) is below dotted curve 2.

coefficient of reflection. The reflection of soft x-rays from an MPS will in this situation be primarily due to interference. We can conditionally say that if $L(f,g) < (1/2)L_{abs}$, the reflector is an interference type of reflector.

Now let the MPS comprise very thin layers of highly absorbent material separated by "thick" layers of weakly absorbent material, such that $\beta \ll 1$, $\text{Im}\,\varepsilon_1 \gg \text{Im}\,\varepsilon_2$, and $\text{Re}(\varepsilon_1 - \varepsilon_2)$ is small. Then a standing wave whose nodes are located in the highly absorbent layers of the MPS arises and energy does not flow into the depths of the MPS. These conditions correspond to the so-called Borrmann effect in the x-ray physics of crystals. When this happens, the depth to which the standing wave penetrates into the medium (the MPS) is

$$L(f,g) = \lambda_0/(\beta^* \,\text{Im}\,\mu) = L_{abs}/\beta^* \gg L_{abs} \qquad (4\text{-}269)$$

In the Borrmann effect, the high degree of reflection from an MPS is due to the large number of periods in the structure that participate in reflection. When $L > 2L_{abs}$ the reflector can be called a Borrmann reflector.[74]

The analysis of the reflection of soft x-rays from an MPS was done for the ideal case of smooth, sharp interfaces between the layers. When roughness is a factor, we assume that

$$R' = R\exp[-(2\pi\xi/d)^2] \qquad (4\text{-}270)$$

where ξ is the mean-squared height of the roughness, d is the period of the structure, and R is the coefficient of reflection from a structure with smooth interfaces. The effect of roughness on reflection is slight when $\xi < (1/20)d$.

Values of the ε_1 and ε_2 parameters in the layers of an MPS may depend in large degree on the methods used to deposit the layers because these methods determine the density ρ of the layers. Thus, the density of carbon films can vary between 0.9–2.2 g/cm^3,[76] and the density of silicon layers can vary between 1.5–2.2 g/cm^3.[77] Reducing the density of weakly absorbent materials may lead to an increase in the coefficient of reflection R in an MPS. We mention the work [78] in which increasing the density (and thickness) of the W layers was studied in W–C structures during the heat treating that followed their growth. The effect that heat treating has on the properties of an MPS was also studied in [71].

X-ray diffraction in an MPS can be used for the spectral selection of soft x-rays. Defining the line width $\Delta\lambda$ for Bragg reflection by the condition that $R(\lambda_0 + \Delta\lambda) = (1/2)R(\lambda_0)$ (at a fixed angle of incidence θ_0), it is not hard to find the resolution of an MPS, $\lambda_0/\Delta\lambda$. There is a connection between the magnitude of $\lambda_0/\Delta\lambda$ and the depth to which a wave penetrates into an MPS, because $\lambda_0/\Delta\lambda \sim N_{\text{eff}}$, where $N_{\text{eff}} = L(g, f)/d$ is the number of periods that participate in reflecting the wave. When absorption in the MPS is slight

$$L \sim \frac{\lambda_0}{|R(\varepsilon_1 - \varepsilon_2)|} \to \frac{\lambda_0}{\Delta\lambda} = \frac{2\sqrt{2}}{3} \frac{\pi}{\sin \pi\beta} \frac{\cos^2\theta_0}{|\nu||\text{Re}(\varepsilon_1 - \varepsilon_2)|} \tag{4-271}$$

Similarly, when absorption is high (extreme)

$$\begin{gathered} L \sim \frac{\lambda_0}{\beta^* \operatorname{Im}\mu} \to \frac{\lambda_0}{\Delta\lambda} = \frac{\cos^2\theta_0}{\operatorname{Im}\mu} F^{-3/4}(f, g, \beta), \\ F = \left(1 - \frac{\nu^2}{y^2}\right)\left(1 + \frac{f^2\nu^2}{y^2}\right), \qquad y = \frac{\pi(\beta + g)}{\sin \pi\beta} \end{gathered} \tag{4-272}$$

The quantity $\lambda_0/\Delta\lambda$ reaches values of $\sim 10^2$ and decreases as absorption in the structure increases.

Multilayer periodic structures are successfully used as the dispersing element of a spectral goniometer.[79] The angular width of the Bragg reflection peak lines with λ = 13.3 Å(Cu L_α), 44.8 Å(C K_α), and 67.7 Å(B K_α) from an MPS (21 periods, WRe–C, d = 106 Å) are roughly twice those from a pseudocrystal of lead stearate; however, the intensity of reflection is about 5–15 rad greater from the MPS than from the pseudocrystal.

Besides being highly reflective, multilayer periodic structures may have a high degree of transmittance, which allows them to be used as semitransparent reflectors.[75] Figure 4-20 graphs the absorption Q, the reflectance R, and the transmittance T as a function of the number of layers in a multilayer periodic structure, calculated for radiation having a wavelength λ = 6.63 nm and impinging a Ru–B device from the normal. As we can see, when the number of layers is $\sim$ 50 both reflectance R and transmittance T are high.

In concluding our discussion of the properties of an MPS used as a reflector of soft x-rays impinging from the normal, we note that although research into multilayer periodic devices has begun only recently and there are comparatively few results, the prospects for using an MPS in soft x-ray and vacuum ultraviolet lithography are obvious. As the period in an MPS is reduced to $\sim$ 5, the MPS as man-made crystals continually evolve into natural multicomponent (layered) crystals. Apparently, manufacturing capabilities limit the period of an MPS to maximum values of $d \geq d_{\text{min}}$ which, in turn, puts a lower limit on the wavelength for which an MPS reflector of soft x-rays can be produced: $\lambda_{\text{min}} = 2d_{\text{min}}$. For example, when λ = 10 Å, d_{min} = 5 Å, and in this situation every period in the MPS must comprise two atomic monolayers. It is possible that molecular epitaxy will make structures with alternating monolayers possible in the future. At present however, estimates of $d_{\text{min}} \geq 1$ nm and λ_{min} = 2 nm are more realistic and define the maximum wavelength for which we can design x-ray lithography systems with optics for reflecting radiation that impinges from the normal.

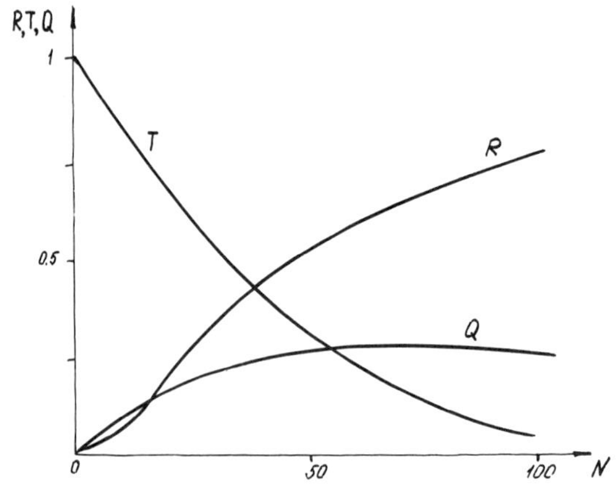

FIGURE 4-20. Coefficients of reflection R, transmission T, and absorption $Q = 1 - R - T$ as a function of the number of pairs N in an MPS comprising alternating layers of boron and ruthenium.

4-19. X-RAY OPTICAL COMPONENTS

The physical effects of absorption, total external reflection, diffraction in crystals, multilayer periodic structures, and zone plates makes it possible to produce a number of x-ray optical components.

1. In the x-ray spectrum windows and filters are the simplest optical components and they perform their function when radiation is propagated. The absorption of soft x-rays is due to the photoemission of electrons and for this reason transmittance $T(\lambda)$ is described by a linear coefficient of absorption $\mu(\lambda)$: $T(\lambda) = \exp(-\mu(\lambda)h)$, where h is the layer thickness. The function $\mu(\lambda)$ has a minimum between absorption edges; a window can serve as a band pass filter, stopping radiation in the range of maxima and transmitting radiation that lies between the maxima. In essence, substrate-membranes for x-ray lithography masks can serve as windows (but not filters). Windows are made as films or foils only a few microns thick. Obviously, the main optical requirement imposed on these is that they be uniform in thickness and density. If we introduce a mass absorption coefficient $\alpha(\lambda) = \mu(\lambda)/\rho$ $(1/\text{g} \cdot \text{cm}^2)$ and a mass thickness $H = \rho h$ (g/cm^2) for the material, $T = \exp(-\alpha \cdot H)$. At micron thicknesses the nonuniformities in the film may become substantial: $\Delta H/H \sim 1$–3. For example, by measuring the losses in the electron beam, changes in H of as much as four times in aluminum and one to three times in beryllium films produced by rolling were detected. Films obtained through vacuum deposition have a much higher degree of uniformity.

2. Plane mirrors for total external reflection are widely used in x-ray optics and their properties have been thoroughly discussed in Section 4-16. At this point we will add information regarding multiple total external reflection and devices that use multiple reflection. Because of multiple reflection, a ray that is reflected within a cylindrical surface at a grazing angle is returned at an angle of $\alpha = 2n\theta$. The Fresnel coefficient of reflection from a medium $\varepsilon = 1 - \delta + i\beta$, $\delta, \beta \ll 1$ when $\theta < \theta_c = \sqrt{\delta}$ is[80]

$$r_F(\theta) = -1 + \frac{2\theta}{\sqrt{\varepsilon - 1}} \qquad (4\text{-}273)$$

and after n-tuple reflection is

$$R_n(\alpha,\theta) = |r_F(\theta)|^{2n} = \left(1 - \frac{2\theta}{\sqrt{\varepsilon - 1}}\right)^{\alpha/2\theta} \simeq \left|\exp\left(-\frac{\alpha}{\sqrt{\varepsilon - 1}}\right)\right|^2 \qquad (4\text{-}274)$$

If $\beta/\delta \ll 1$ (a weakly absorbing medium) we find

$$R(\alpha, \theta \to 0) \simeq \exp(-\alpha\beta/\delta^{3/2}) \qquad (4\text{-}275)$$

Equations (4-273)–(4-275) are valid when $\theta < \theta_c$. At fixed α (e.g., $\alpha = \pi/2$) and θ, we can write $R(\alpha, \theta)$ as

$$R(\alpha,\theta) = |r_F(\theta)|^{2ent(\alpha/2\theta)+2} \qquad (4\text{-}276)$$

Here, $entx$ is the integer portion of x and $r_F(\theta)$ is found from Eq. (4-273). The function $R(\alpha, \theta)$ increases with increasing θ (Fig. 5b, [80]).

Estimates show that when $\theta < 4°$, a rotation to $\alpha = \pi/2$ because of multiple grazing reflection can be achieved with a value of $R(\pi/2, \theta) \simeq, 0.3.$(80) Estimates were made for $\lambda = 6.02$ nm and the carbon surface of a mirror. The possibility of using concave cylindrical mirrors to rotate x-ray beams to large angles and segregate them and as laser resonators in the x-ray spectrum was suggested in [80]. In these resonators the generating modes must produce grazing propagation along an interior cylindrical surface. Because $R(\theta)$ decreases more rapidly in the shortwave part of the spectrum, filtration of shortwaves can be seen when beams are reflected. Nothing has yet been published regarding the creation of such systems.

3. Multiple external reflection from the surface of a solid can be used to create x-ray waveguides.(82,83) Wave propagation in a plane waveguide proceeds by way of multiple reflection from two plane surfaces and in cylindrical waveguides by reflecting from the interior walls of a pipe. The angle between modes in cylindrical waveguides is $\Delta\theta \approx (\lambda/r_0)$, where r_0 is the inside radius of the waveguide. Geometric optics can be used if $\Delta\theta = (\lambda/r_0) \ll \theta_F$, i.e., if a large number of modes lie within the Fresnel angle $\theta_F = \sqrt{\beta}$. In an approximation from geometric optics, the transmittance of a circular waveguide was calculated in [82]. Let an isotropic point source of soft x-rays with a power of W_0 be placed on the axis of a cylinder of radius r_0 and L long. The radiation energy passing unreflected through the waveguide is

$$W_1 = W_0 \left(\frac{r_0}{2L}\right)^2 \qquad (4\text{-}277)$$

and the reflected portion is given by the integral

$$W_2 = W_0 \int_{\theta_{\min}}^{\theta_{\max}} [R(\theta)]^{N(\theta)} \sin\theta \, d\theta \qquad (4\text{-}278)$$

where $R(\theta)$ is the coefficient of reflection for grazing incidence at an angle θ, and $N(\theta)$ is the number of reflections that occur per unit length L of the waveguide; $N(\theta)$ is the integer part of $(L\theta/2r_0) + 1$, and $\theta_{\min}$, and $\theta_{\max}$ can easily be found from the geometry of the problem: $\theta_{\min} = r_0/L$, $\theta_{\max} = \pi/2$. Denoting the permittivity of the material by $\varepsilon = 1 - \delta + i\beta$, just like we did earlier, we can write $R(\theta)$ in terms of the Fresnel equation ($\theta \ll 1$):

$$R(\theta) = \frac{(\theta - \theta_1)^2 + \theta_2^2}{(\theta + \theta_1)^2 + \theta_2^2} \tag{4-279a}$$

$$\theta_{1,2} = \frac{1}{\sqrt{2}} \left[\sqrt{(\theta^2 - \delta)^2 + \beta^2} \pm \theta^2 \mp \alpha \right]^{1/2} \tag{4-279b}$$

Using Eqs. (4-278) and (4-279), we can find the transmittance η of a waveguide:

$$\eta = \frac{W_2 + W_1}{W_0} \approx \left(\frac{r_0}{4L}\right) A(\delta, \beta) - \left(\frac{r_0^2}{24L^2}\right) B(\delta, \beta) \tag{4-280a}$$

$$A^2(\delta, \beta) = \frac{\delta}{2} \frac{1 + (\beta/\delta)^2}{\sqrt{1 + \beta^2/\delta^2} - 1}, \qquad B(\delta, \beta) = \frac{\sqrt{1 + \beta^2/\delta^2} + 2}{\sqrt{1 + \beta^2/\delta^2} - 1} \tag{4-280b}$$

A remarkable feature of Eq. (4-280a) is the presence of a term $2r_0/L$ that is linear in W_2/W_0 and which decreases much more slowly as L increases than does $W_1/W_0 \sim (2r_0/L)^2$. We can easily see from Eq. (4-280b) that when $(\beta/\delta) \ll 1$, $A = (\delta^{3/2}/\beta)$ and

$$\eta = W/W_0 \simeq \frac{r_0}{4L} \frac{\delta^{3/2}}{\beta} \tag{4-281}$$

The dependence of $\eta(\lambda)$ on the wavelength of the radiation that is contained in Eq. (4-281) by $\delta^{3/2}(\lambda)/\beta(\lambda)$ shows that η increases as λ decreases in the $\lambda < 2$ nm range. This is related to the increase in the ratio δ/β in the $\lambda < \lambda_K$ range, where λ_K is the edge of a K absorption line. The condition that $\lambda < \lambda_K$ occurs first for light-weight elements and then for the heavier elements. As we can see, the wave properties of cylindrical pipes are manifested in the $\lambda = 0.5$–2 nm range, and are more interesting from the viewpoint of x-ray lithography. The preferred materials for waveguide pipes in the $\lambda < 2$ nm range are Be, B, and C.[82] Experiments show that a waveguide (a glass pipe) with parameters $L = 167$ and $r_0 = 0.4$ cm can transmit 32 times more radiation power in the $\lambda = 1$–2 nm range from a laser plasma than can be transmitted without a waveguide.[84]

4. Multilayer interference mirrors that reflect soft x-rays with a coefficient of $R \sim 0.3$–0.8 when the radiation impinges from the normal can be used in the manufacture of the reflective optical components used in the so-called spherical and aspherical large aperture mirrors. A scheme was suggested in [85] for a projection x-ray lithography system using spherical mirrors (Fig. 4-21). The system comprises an isotropically radiating source, condensing and focusing mirrors with multiple coatings, and an x-ray mask. A reduced image of the mask is focused on the surface of a resist. Methods are available for illuminating the entire surface of the x-ray mask, as well as for scanning the surface of the mask with a beam. The energy efficiency of the system is governed by the angular dimensions α of the focusing mirror, which is visible from a point on the mask (for a given coefficient of reflection). Even the simplest mirror projection systems can collect 10–30% of the source energy onto the resist. More sophisticated (multicomponent) optical systems make it possible to increase the fraction of usable source energy.

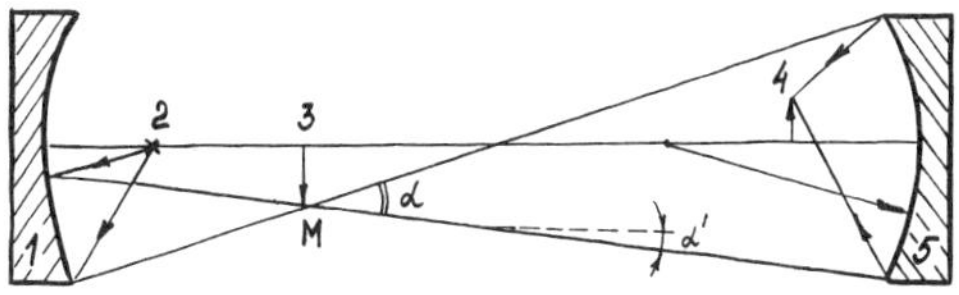

FIGURE 4-21. Simplified diagram of projection x-ray lithography: 1) multilayer spherical mirror; 2) source of soft x-rays; 3) mask; 4) x-ray resist; 5) multilayer spherical focusing mirror.

It must be emphasized that it is the period of the multilayer structure that defines the wavelength of the radiation reflected from a mirror; this limitation shifts the wavelength of the radiation that can be used to expose a resist into the λ = 3–4 nm range. The technology of multilayer structures must be developed even further if radiation in the $\lambda = 1$ nm range is to be used.

5. Zone plates for wavelengths in the soft x-ray range are familiar optical components in x-ray microscopy.[86–90] The methods of scanning and projection electron lithography, as well as interference (holographic) methods are used to fabricate them. They are circular amplitude or phase gratings that have an increasing number of annular lines, called zones, along the radius. The radius r_n of the edge of the n zone, the focal length of the zone plate f, and the wavelength λ are related through

$$r_n^2 = nf\lambda + \left(\frac{n\lambda}{2}\right)^2 \tag{4-282}$$

When a zone plate has $n \geq 100$ zones, its focusing properties are those of a thin refractive lens.

We can use Eq. (4-282) to give the focal length f in terms of the radius r_1 of the first zone: $f_1 = r_1^2/\lambda$; for diffractive m-order focusing:

$$f_m = \frac{r_1^2}{m\lambda} \equiv f_1/m \tag{4-283}$$

Thus, when radiation having a continuous spectrum impinges on a zone plate the plate disperses the radiation at any order of diffraction. An aperture d_A in diameter placed in the focal region of a zone plate D in diameter extracts quasimonochromatic radiation having a degree of monochromaticity of

$$\lambda/\Delta\lambda = D/2d_A \to \Delta\lambda/\lambda = 2d_A/D \tag{4-284}$$

from the continuous spectrum.[88] On the other hand, the monochromaticity of a zone plate's focused radiation must not be worse than $\lambda/\Delta\lambda \simeq nm/2$. The width of the least n external zone is $dr_n = r_n/(2n)$. An aberration-free zone plate has a resolution (according to Rayleigh's criterion) of

$$\delta_m = 1.22 dr_n/m \tag{4-285}$$

Finally, note that the diffraction efficiency of an amplitude zone plate is low; it focuses 10% of the incident energy in the first order of diffraction; 2.6% in the second order, and 1.2% in the third order. At the same time, phase zone plates are capable of focusing up to 40% of the incident beam energy in the first order of diffraction and up to 60% for radiation incident from other than the normal.[(86)]

The formulas presented above make it easy for us to find the limitations in the parameters of zone plates as equivalents of optical lenses. We will specify the resolution Δx (μm) of a process for manufacturing zone plates. Since it is obvious that $\Delta x \leq dr_n = r_n/(2n)$, the resolution of the process determines the maximum diameter $D = 2r_n = n\Delta x$ of a zone plate: for $\Delta x = 0.1$ μm and $n = 10^3$, $D = 400 \mu$m.

Condenser zone plates and projection (focusing) zone plates are distinct from one another; the former are designed to produce the radiation that illuminates a specimen in a microscope and they must be as large in diameter as possible to collect the energy from the point source. The diameter of condenser zone plates is as much as 10 mm; there are 10^4 zones, the radius of the first being 25–50 μm, and the focal length is 30–50 cm. The resolution of a condenser lens is limited by aberrations and is 1–10 μm (it is difficult to avoid aberrations when there are 10^4 zones). We wish to avoid aberrations in a focusing lens and obtain the highest diffraction-limited resolution; therefore, there are no more than a few hundred zones less than 1 mm in diameter. Of course, lenses such as these can only be used in scanning systems. Notice also that the use of zone plates in the infeed and outfeed apertures of x-ray waveguides is also a possibility (see Paragraph 3).

We will supplement our discussion with information on the relationship between a material's $\varepsilon = 1 - \delta + i\beta$ parameters and the properties of a zone pair. When a ray passes through a zone plate at a point r where the thickness is $h(r)$, the phase of the wave changes by an amount $2\pi h(r)(\delta - i\beta)$. The intensity of the wave at the first order focal point will be, according to scalar theory,[(91)]

$$I^{(1)} = A(n) \left| \int_0^{r_1} \frac{2\pi r\, dr}{r_0} \exp[2\pi i(r_0 - h(r)[\delta - \beta i])] \right|^2 \tag{4-286}$$

Here the contributions from all zones are considered equal; the coefficient $A(n)$ depends on the number of zones. Assuming, e.g., that $h(r) = h_0$ when $0 < r < d_1$, and $h = 0$ when $d_1 \leq r \leq r_1$, we find[(91)] that

$$(I^{(1)}/I_{\text{IN}})_{\max} = \frac{1}{\pi^2} \sin^2(2\pi h_0 \delta)(1 + (\delta/\beta)^2) \tag{4-287}$$

where the values $h = h_0$ and d_1 that maximize $I^{(1)}/I_{\text{IN}}$ are found from

$$\exp[-2\pi h\kappa] = \cos(2\pi h\delta - \gamma)/\cos\gamma \tag{4-288}$$

$$\cos\gamma = \beta/(\beta^2 + \delta^2)^{1/2} \tag{4-289}$$

$$d_1 = a_1 - \sqrt{a_1^2 - L \cdot \lambda} \tag{4-290}$$

Thus, the material parameters δ and β determine the diffraction efficiency of a zone plate. If $\beta \ll \delta$, radiation is weakly absorbed and the zone plate acts for the most part like a phase plate and has a high degree of diffraction efficiency. If we know what the functions $\delta(\lambda)$ and $\beta(\lambda)$ are for the different materials we can choose a material which, at a specified wavelength λ^*, satisfies the condition that $\delta(\lambda^*) \gg \beta(\lambda^*)$, and a zone plate made from this material will have the properties of a phase zone plate. The creation of a phase zone plate from a silver film 0.55 μm thick and having a diffraction efficiency of $I^{(1)}/I_{\text{IN}} \simeq 0.26$ was reported in [89].

A small gap (radius R) in an opaque screen is the distinguishing feature of a zone plate. This gap can be thought of as the first Fresnel zone of a zone plate. It can be shown that the sharpest image of an object created by the gap on a screen will occur when[92]

$$1/a + 1/b = 0.81 \frac{1}{(R^2/\lambda)} \equiv 0.81 \frac{1}{f_1} \tag{4-291}$$

Here a and b are the distances from an object and its image to the screen, and $f_1 = R^2/\lambda$ is the focal length of the zone plate, defined in terms of the radius of the central zone. Equation (4-291) corresponds to optimal image sharpness when the radii of the diffraction ($R_D \simeq \lambda/R \cdot b$) and geometric ($Rg \simeq R$ when $a \gg b$) blurring circles of the image are identical: $R = \lambda b/R \rightarrow b = f_1 = R^2/\lambda$. Gaps in screens continue to be used in research into the images of small x-ray sources. In order that $f_1 = 10^2$ cm when $\lambda = 1$ nm, it is necessary that $R \simeq 30$ μm.

4-20. X-RAY LITHOGRAPHY SYSTEMS FOR SHADOW PRINTING WITH A GAP

To date, many firms have reported the creation of experimental shadow x-ray lithography machines,[93–98] some of which, the Micronix MX-1600[95] for example, are recommended for use in production.

There is no need to present detailed diagrams of the respective machines. A simplified diagram of the process of shadow printing with a gap using a source of finite size makes it possible to evaluate the limitations that arise in this process (Fig. 4-22). The width of the half-shadow is $\delta = S(2r/L)$; the displacement of the shadow relative to a detail on the mask is $b = S(R + r)/L$; the change in displacement because of an inaccuracy ΔS in the gap is $\Delta b = \Delta S(R + r)/L$. All of these errors decrease as L increases; however, when this happens, the fraction of source energy that goes into illuminating the mask $E_{\text{M}}/E_{\text{tot}} \sim 1/L^2$, also decreases and the exposure time for the source increases as $t_{\text{ex}} \sim L^2$. Reducing the size of the source to that of a point eliminates the half-shadow, but changes the displacement $b + \Delta b$ slightly ($R \gg r$). Synchrotron radiation, which is fairly uniform in intensity when its cross section is about $\Delta y \Delta x \approx 3 \times 30$ mm^2 and has an angular divergence of $\Delta\varphi \sim 1$ mrad is an alternative. Scanning the mask + wafer system with a ray from a synchrotron source uniformly exposes a wafer $\sim$30 $\times$ 30 mm^2 in area and in this case a half-shadow that is only $\delta = S \cdot \Delta\varphi \simeq 0.01\mu$m wide is maintained.

The analysis of x-ray lithography systems for maximal productivity (minimal exposure time) is of great interest. Wafer exposure time is governed by many factors, chief among which are: the power and spectral content of the radiation source, $dP/d\lambda$; absorption in the semitransparent window of the source, $\exp(-\mu_W(\lambda) \cdot h_W)$; absorption in a window of the mask, $\exp(-\mu_M(\lambda) \cdot h_M)$; absorption in the resist, $\exp(-\mu_M(\lambda) h_M)$; and the sensitivity of the resist, (in J/cm^2 or J/cm^3). In the manner of [99] we will examine the process of x-ray exposure in a synchrotron source. The radiation power

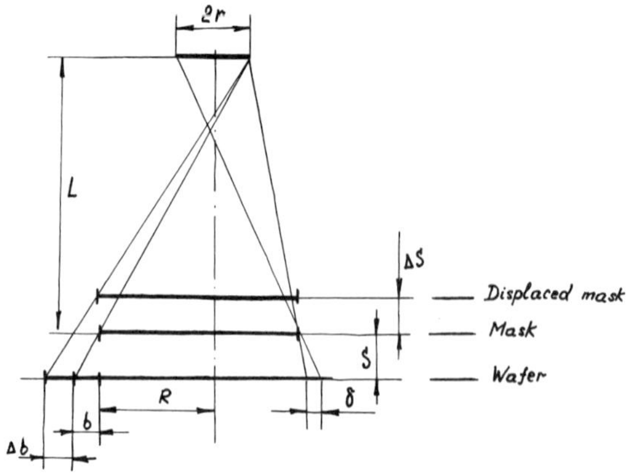

FIGURE 4-22. Diagram of the scheme for transferring the mask image onto a wafer in an x-ray lithography machine: r is the radius of the radiation source; S is the gap between the mask and the wafer; ΔS is the gap error in the machine (due to warpage in the mask or the wafer); b is the displacement of the edge of the shadow of the mask; Δb is the positioning error with displacement, due to ΔS; and δ is the width of the half-shadow.

absorbed in a layer Δd_R

$$F = \int_{\lambda_{\min}}^{\lambda_{\max}} \left(\frac{dP}{d\lambda}\right) \exp[-\mu_W(\lambda)d_W - \mu_M(\lambda)d_M - \mu_R(\lambda)d_R](\mu_R(\lambda)\Delta d_R \, d\lambda \tag{4-292}$$

The spectral power in a vertical band W wide for a synchrotron source is

$$\frac{dP}{d\lambda} = 15788I \frac{E^7}{R^2} \frac{W}{L} G\left(\frac{\lambda}{\lambda_c}\right) \quad \left(\frac{\mathrm{W}}{\mathrm{nm}} \frac{\mathrm{m}^2}{\mathrm{A} \cdot (\mathrm{GeV})^7}\right) \tag{4-293}$$

$$\lambda_c = 0.559R/E^3 \quad \left(\mathrm{nm} \frac{(\mathrm{GeV})^3}{\mathrm{m}}\right) \tag{4-294}$$

$$E_{\max} = 0.299R \cdot B_{\max} \quad \frac{\mathrm{GeV}}{\mathrm{m} \cdot \mathrm{T}} \tag{4-295}$$

The L in Eq. (4-293) is the distance from the point on the electron orbit at which radiation occurs to the point at which it is observed; I is the beam current; R is the radius of curvature of the orbit; and E is the energy of the electrons on the orbit. Equation (4-295) gives the maximal energy of the electrons in the accelerator, for which the maximal value of the magnetic field is $B_{\max}$. The range of radiation wavelengths exposing the resist is governed by the values of $\lambda_1 = 0.1$ and $\lambda_2 = 2$ nm. Using experimental values of $\mu_W(\lambda)$, $\mu_M(\lambda)$, and $\mu_R(\lambda)$ chosen for the windows, membranes, and resists we can easily calculate the density F of the power absorbed by the resist and the energy density $E_R = F \cdot \tau_e$, where τ_e is the exposure time. The value of E_R, which is $E_R = 500$ J/cm^3 for PMMA, determines the sensitivity of the resist. In this manner we can find the exposure time $\tau_e = E_R/F$ as a function of the E parameters of a synchrotron source. In all combinations of window thicknesses d_W (Be—200, 100, 50, 30, 10 μm; Kapton, 30, 10 μm), mask membrane d_M (Si—5, 2, 0.5 μm), and resist d_R (PMMA—0.5 μm) τ_e falls off as the energy E of the electrons in the orbit increases: rather rapidly in the E = 0.6–0.9 GeV range and more slowly when $E > 1$ GeV. On the other hand, the contrast of a mask (the illuminance of the resist) calculated according to

$$C = \int_{\lambda_1}^{\lambda_2} \left(\frac{dP}{d\lambda}\right)_R d\lambda \Big/ \int_{\lambda_1}^{\lambda_2} \left(\frac{dP}{d\lambda}\right)_R \exp(-\mu_{\text{abs}}(\lambda) d_{\text{abs}})\, d\lambda \tag{4-296}$$

[$(dP/d\lambda)_R$ is the amount of power absorbed in the exposed part of the resist and is given by the subintegral function in Eq. (4-296)] decreases as the energy E of the electrons increases. In fact, $\lambda \to 0$, $\exp(-\mu_{\text{abs}} d_{\text{abs}}) \to 1$ and $C \to 1$ in the limit. It follows from this that the maximal energy of the electrons in an orbit is governed by the condition that $C(E_{\max}) = C_{\min}$, where $C_{\min}$ is the minimal mask contrast that can be tolerated. If we say that $C_{\min} = 3$ for PMMA, then $E_{\max} = 0.9$ GeV (when $d_{\text{abs}} = 0.3$ μm Au, and a Be window 30 μm). The values of contrast depend to large degree on the window material and its thickness; contrast is lower when the window is thicker (a thick window attenuates the long wave range of the spectrum more strongly than the shortwave range). Contrast is, of course, also highly dependent on the thickness d_{abs} of the absorbent layer in the mask; although d_{abs} is limited from above by the ability to fabricate a submicron pattern in this layer.

Exposure of the lowest resist layer by secondary photoelectrons and Auger electrons generated in the substrate has a significant impact on the reproducibility of the mask pattern (resolution). In this circumstance a small change in the wavelength of the exposing radiation may cause a significant change in the effect. For example, for a silicon substrate $\lambda = 0.6$ nm $< \lambda_K$, $\lambda = 0.7 > \lambda_K$, where λ_K is the K band absorption edge for silicon. The result from absorbing $\lambda < \lambda_K$ radiation is that Auger electrons having energy $E_{AK} = 1.65$ keV are generated, whereas when $\lambda > \lambda_K$ the energy of Auger L-electrons is $E_{AL} = 0.15$ keV.[99]

Equations (4-292) and (4-296) can be used to find the minimal exposure time in systems that use a point source of x-rays when the power spectrum $(dP/d\lambda)$ of the source is known. For x-ray tubes,

$$\frac{dP}{d\lambda} = \sum_{\lambda_i} I_{\lambda_i} + \left(\frac{dP}{d\lambda}\right)_{\text{cont}} \tag{4-297}$$

Recall that both the intensities I_{λ_i} of the lines and the intensity in the continuous radiation spectrum are functions of the energy E of the electron beam bombarding the anode of the tube. It follows from this that, according to Eq. (4-292), the density F of the energy absorbed by a resist will be a function of the beam energy: $F = F(E)$, and at some value $E^* F(E)$ will be a maximum and the exposure time will be a minimum. The magnitude of E^* will be governed by the entire complex of anode, window, mask membrane, and resist parameters. Computer calculations done in [100] show that the optimal (in terms of minimal exposure time) wavelengths lie in the 0.6–1.3 nm range, depending on the thickness of the window and membrane material. A contrast level of $C > 10$ can be provided when the thickness of the absorbent layer on a Au mask is $d_{\text{abs}} > 0.6$ μm.

Synchrotron sources can provide a radiation power density of F = 50–100 mW/cm^2 on the surface of a resist, whereas x-ray tubes cannot provide more than 1 mW/cm^2. The power densities available from plasma sources (laser plasma, Z-pinch discharge and plasma focus) have an intermediate value of $F \leq 10$ mW/cm^2. In evaluating the prospects for using the various sources in x-ray lithography systems it is necessary that an important parameter—their useful life—bc indicated: synchrotron sources, apparently, having once been evacuated, can be used for a long time (in the thousands of hours); x-ray tubes typically have lifetimes measured in the hundreds of hours, whereas plasma systems require that the electrodes be replaced after 10^2–10^3 pulses. It is quite possible that this limitation on pulsed sources will exclude them from being used in production equipment for some time to come.

4-21. MASK AND WAFER ALIGNMENT SYSTEMS IN X-RAY LITHOGRAPHY MACHINES

Ordinarily, x-ray lithography systems are thought of as having a resolution of 0.1 μm, which is several times better than what can be obtained from optical systems, including ultraviolet, (0.25–0.5 μm). For this reason, alignment in these systems takes on special significance. It is desirable that alignment accuracy be close to 0.01 μm. The alignment systems that have been suggested and produced to date are capable of achieving this level of accuracy.

Peculiar to x-ray steppers is a scheme for shadow printing with a small ~20 μm gap between mask and wafer. No sensors can be placed in this gap; they must be located above the mask. Alignment marks are made on the mask and on the wafer. Two possible alignment procedures have been studied: 1) simultaneous detection of the marks in the planes of the mask and the wafer and alignment of these marks[(101,102)] and 2) alignment according to the intensity of the light beams diffracted at the mark-gratings.[(103–108)] Both methods are optical and therefore require that the mask substrates be optically transparent (at least for thicknesses of 1–2 μm). Diffraction alignment methods require that the marks be larger than in systems that sense the marks. Figure 4-23 shows a simplified diagram for alignment with a diffraction grating as marks on the wafer.[(107)] Laser light through a rectangular window in the mask impinges at the normal to the grating and the detector senses the intensity of the reflected first-order diffraction maximum. As the diffracted light passes through the window in the mask above the grating on the wafer its intensity is given by a curve having a maximum and this curve is used to align the mask with the wafer. Another possibility is to detect the intensity of a first-order diffraction beam as it passes through the grating in the mask above the grating in the wafer, in which case the intensity of the first-order diffraction beam varies periodically. An alignment system with three gratings was suggested in [108]. The two gratings G_2 and G_2' send the reflected first-order beams $u(+1)$ and $u(-1)$ to the grating G_1 in the mask; as interference and diffraction of the beams $u(+1)$ and $u(-1)$ takes place at grating G_1, beams $u(-1,1)+u(1,-2)$ and $u(-1,2)+u(1;-1)$ are generated and are then sensed by detectors D_1 and D_2 (Fig. 4-23b). The sensitivity to intensity is such that mutual alignment between the mask and wafer of 0.0025–0.005 μm is easily detected when the gap is 80–90 μm, which attests to the fact that the gap can be reduced to 10 μm. Piezoelectric sensors can be used for alignment when the movements are $\sim 10^{-3}$ μm. It is important to note that the intensity of the signal is dependent upon only the mutual movements between the mask and the wafer and does not change as the distance between them changes by small amounts.[(108)]

We will not discuss alignment schemes in which the marks on the mask and the wafer are detected simultaneously (see, e.g., [101]). We will merely say that in this scheme the marks on the mask and the wafer can be detected at different points in time by aligning them transversely with a 10×-reduced image of the bench mark figures

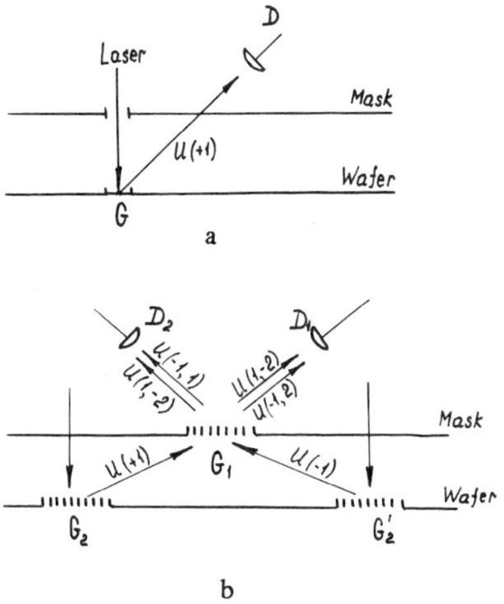

FIGURE 4-23. Alignment scheme in systems for shadow printing with a gap; a) alignment scheme with a diffraction grating on the wafer; b) alignment scheme with three gratings.

(rectangles) on an auxiliary mask.[102] Mutual alignment of 0.04 μm between the mask and the wafer can be detected with this scheme.

4-22. CONTRAST IN THE SHADOW MASKS USED IN X-RAY LITHOGRAPHY

A shadow mask pattern for x-ray lithography is fabricated in a 1× scale on a large Z metal film (Au) using electron-beam lithography.[109,110] A gold film $d > 0.5$ μm thick is deposited onto a thin membrane made from materials having a low Z. In the early days, films made from organic materials (polyamides, etc.[111–113]) were often used as the membranes for x-ray masks; however, the large coefficient of thermal expansion in these films and their sensitivity to mechanical and other influences led to their abandonment. At present, silicon, SiO_2, Si_3N_4, Al_2O_3, and SiC are the popular inorganic materials for x-ray mask membranes.[114–117]

We will pause to discuss contrast in x-ray masks and the deformations caused in them by mechanical stresses. As usual, we will use

$$C = (I_{max} - I_{min})/(I_{max} + I_{in}) \tag{4-298}$$

to define the contrast in the illumination transmitted by a mask. If we consider the attenuation of the uniform illuminating radiation only in the mask membrane and the absorbing layer, then

$$I_{max} = I_0 \exp(-\mu_1 t_1), \qquad I_{min} = I_0 \exp(-\mu_1 t_1 - \mu_2 t_2) \tag{4-299}$$

$$C_{RS}^{(0)} = (1 - e^{-\mu_2 t_2})/(1 + e^{-\mu_2 t_2}) \tag{4-300}$$

where μ_1, t_1 and μ_2, t_2 are a linear absorption coefficient and the thickness of the membrane and the pattern-bearing film, respectively. The subscript RS denotes the fact that contrast relates to the illumination at the surface of the resist. However, the secondary radiation that arises when x-rays are absorbed in the mask membrane and in the absorbing film must be considered. This secondary radiation is a flux composed of photoelectrons and Auger electrons, as well as the characteristic x-ray radiation, the bremsstrahlung, and the ultraviolet radiation from the secondary electrons. The angular distribution of the photoelectrons is, as we have seen, anisotropic. When there is a gap h between the mask and the wafer such that $h \gg (a, b)$, where a and b are the light and dark intervals in the mask, we can assume that secondary radiation in the mask is uniformly distributed on the surface of the resist; therefore, the contrast in illumination at the surface of the resist will be

$$C_{RS} = (1 - e^{-\mu_2 t_2})/(1 + e^{-\mu_2 t_2} + 2I_f/I_0 e^{-\mu_1 t_1}) \tag{4-301}$$

where I_f is the total secondary radiation background intensity. Since $2I_f/I_0 e^{-\mu_1 t_1} \ll 1$, $C_{RS} \simeq C_{RS}^{(0)}$. However, the background plays a much greater role in determining the contrast of the energy absorbed in the resist and this property is unique to x-ray exposure through a mask. The components of the background are photoelectrons and Auger electrons whose range in the resist is $R \leq 0.2$ μm, and soft x-rays and ultraviolet radiation. We can therefore say that the background radiation is completely absorbed by the resist layer, giving an exposure that decreases exponentially with depth in the resist layer. At the same time, the exposing x-rays are only partially absorbed by the resist, the amount absorbed being equal to $\mu_R t_R$, and $\mu_R t_R \ll 1$. Consequently, we find

$$C_{Rv} = \frac{(1 - e^{-\mu_2 t_2})}{[1 + e^{-\mu_2 t_2} + 2I_f/(I_0 e^{-\mu_1 t_1} \mu_R t_R)} \tag{4-302}$$

for the contrast in terms of the amount of energy absorbed in the resist. The subscripts Rv denote the fact that the contrast is defined in terms of the amount of energy absorbed in the volume of the resist. Although the background intensity is low, e.g., $2I_f/I_0 e^{-\mu_2 t_1} \approx 0.1$ the quantity $\mu_R t_R$ is also small, e.g., $\mu_R t_R \leq 0.1$. Therefore, the contrast in terms of absorbed energy C_{Rv} according to Eq. (4-302) may be 2–3 times less than the contrast in terms of illumination C_{RS}.

For stencil x-ray masks in which the transparent windows are physical openings[118] we must assume that $t_1 = 0$, and the contrast for these masks will be

$$C_{Rv}^{(\text{Stencil})} = \frac{1 - \exp(-\mu_2 t_2)}{1 + \exp(-\mu_2 t_2) + 2I_f/(I_0 \mu_R t_R)} \tag{4-303}$$

An obvious advantage to using stencil masks is the ability to choose an x-ray wavelength that has been optimized with respect to absorption in the resist, so that $\mu_R t_R \approx 1$. In this case, the contrast with respect to the amount of energy absorbed increases to that with respect to the illumination of the surface: $C_{RV}^{(\text{Stencil})} \to C_{RS}$. On the other hand, the thickness of the absorbing areas of a stencil mask must be chosen so that $\mu_2 t_2 \geq 1$, which, obviously, gives the same result $I_f/I_0 \to 0$ since secondary radiation at the mask surface does not occur because the primary radiation has been attenuated in the mask layer. Under these conditions the contrast defined according to Eq. (4-298) tends to its maximum possible value of $C_{RV} = C_{RS} = 1$. In truth, experiments have pointed to the possibility that the image of stencil masks may be degraded due to total external x-ray reflection from the vertical walls of the openings[118]; increasing the roughness of the walls rapidly decreases reflectance.

Secondary radiation from the Au film that carries the pattern makes the largest contribution to I_f (this layer has a large Z and is turned toward the resist). A coating of light-weight material $h \geq R$ thick on the pattern lowers I_f and increases the observed contrast with respect to the amount of absorbed energy, where R is the range of the secondary electrons in the coating material.[119]

4-23. DEFORMATIONS IN SHADOW MASKS USED IN X-RAY LITHOGRAPHY

From a mechanical viewpoint, an x-ray mask is an elastic shell that has been deformed by stresses in the membrane and in the metal that bears the mask pattern. Because a pattern may have extremely diverse configurations, it is hard to give the stresses in the metal layer as a universal function of the deformations in the membrane. Compression or extension of the membrane shifts the pattern elements by an amount Δ in the plane of the membrane. We define the relative shift (distortion) as $\epsilon = \Delta/l$, where l is a dimension of the mask. If the mask dimension coincides with the chip dimension we can say that $l = c\sqrt{K}$, where K is the degree of integration on the chip (Mbits). It is evident that the relative distortion that can be tolerated is a function of the dimensions of the chip: $\epsilon = \Delta/c\sqrt{K} \leq \epsilon_{\max}$. For a 16 Mbit IC memory chip, $\epsilon_{\max} = 6 \times 10^{-6}$, according to [120].

The stress σ in a gold film obtained by electrodeposition is $\sigma = 3 \times 10^7$ N/m^2; annealing at different temperatures may give different values to σ. When $\sigma = 10^7$ N/m^2 the relative deformation of a free Au film is $\epsilon_p = \sigma/E_A = 1.2 \times 10^{-4}$ ($E_A = 8 \times 10^{11}$ N/m^2). The deformation of a film on a membrane with parameters E_c and t_c (E_c is the modulus of elasticity and t_c is the thickness of the membrane) will be $\epsilon = \epsilon_p/(1 + f(q))$, $q = E_c t_c/E_A t_A$. The form of the $f(q)$ function depends on how the film-absorbent is configured; for a system having horizontal and vertical bands, $f(q) = 8q(1 + q)$. For a mask having a (2 μm + 1 μm)-thick silicon and gold structure, $q = 2.5$ and $f(q) = 70$. The deformation of such a mask is $\epsilon = 1.7 \times 10^{-6}$, which allows it to be used to fabricate 16 Mbit ICs.[120] Thus, the technology of mask fabrication must ensure that the stresses σ that arise in a gold film will not cause greater shifts in the mask pattern than can be tolerated. If the mask membrane has also been stretched (some tension in the membrane is needed to avoid flexing) mask deformation depends on the stresses in each layer of the structure. Some layers in the structure are made especially for the purpose of reducing the stresses in other layers.[121]

The amount of warpage h in the mask membrane is governed by the limitations on the size of the half-shadow δ around the shaded elements of the mask. These shaded elements are caused by the transverse dimension of the source d and the displacement of the shadow edge Δ due to the rays impinging on the edge of the mask from an angle (Fig. 4-24)[119]:

$$\delta = \frac{d(S + H)}{L} < \delta_{\max} \tag{4-304}$$

$$\Delta_{\max} = \frac{S + h}{L}\frac{1}{2}l_{\max} \tag{4-305}$$

Here, L is the distance between the source and the mask; $S + h$ is the distance between the mask and the substrate, including the warpage h; and $(1/2)l$ is half the diagonal of

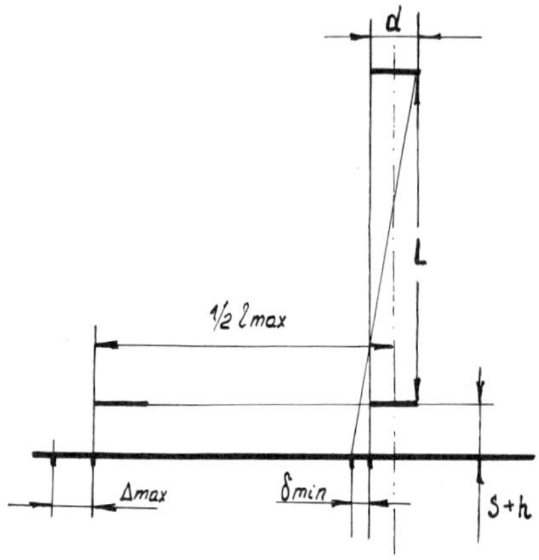

FIGURE 4-24. The displacements Δ and δ of the image of the elements in an x-ray mask for shadow printing with a gap: d is a dimension of the x-ray source; L is the distance between the source and the mask; $S + h$ is the gap between the mask and the resist; h is mask warpage; l_{max} is the largest dimension of the mask (along the diagonal of the rectangle); δ is the width of the half-shadow at the edge of a mask element and is due to the limited size d of the source; and Δ_{max} is the shift in the edge of an element at the most remote point of the mask.

the rectangular mask. We will use realistic parameter values: d = 2.5 mm (the x-ray source at the anode), L = 200 mm, S = 15 μm, and l_{max} = 20 mm; then, it follows from the requirement that δ_{max} = 0.2 μm and Δ_{max} = 0.1 μm that the warpage h be not greater than 1 μm. Actual warpages in a SiN_x membrane reach magnitudes of 5–15 μm. Optimizing the processes for manufacturing membrane layers and an absorbing layer will hold warpage to within acceptable limits.

We will not pause to discuss the obvious limitation on changes in mask temperature prior to and during exposure. We find the value of ΔT_{max} from the condition that

$$\varepsilon_{T_{max}} = \frac{\Delta_{T_{max}}}{l} = \alpha \Delta T_{max} \tag{4-306}$$

When $\Delta_{T_{max}}$ = 0.1 μm, l = 2 cm, $\alpha = 10^{-6}$, we find that $\Delta T_{max} = 5°$. Pulsed exposure methods can lead to significant overheating of the islands on the absorbing film of the mask and even to destruction of the mask, e.g., because of peeling in the absorbing film. Peeling sets in when the force of compression on an island on the film exceeds the force of adhesion between the film and the mask membrane. When x-ray radiation D_0 = 1.5 mJ/cm^2 is pulsed and the coefficient of absorption in the gold is $\mu = 4 \times 10^4$ cm^{-1} the density of the absorbed energy is E = 60 J/cm^3 and the heating of an island on the gold film is ΔT = 24°C; the compressive stress with this kind of heating is $\sigma = E\alpha\Delta T \approx 2 \times 10^8$ dynes/cm^2, which is somewhat less than the force of adhesion. When pulses of D_0 = 100 mJ/cm^2 are used, peeling of the masking elements is unavoidable.[122] The stability of masks over long periods of use also depends on the stresses in the mask layers.[123]

4-24. X-RAY RESISTS

The close kinship between electron resists and x-ray resists is frequently emphasized. Indeed, electron resists can, as a rule, be used as x-ray resists. Moreover, within a group of kindred materials there may be a proportionality relationship between the sensitivity to electron beams (D_e, C/cm^2) and to x-rays (D_x, J/cm^2),

$$D_e = K D_x, \qquad K = \text{const} \tag{4-307}$$

e.g., for chloromethylated polystyrene having different molecular weights, $K = 3.2 \times 10^{-8}$ C/mJ.[124] However, there are significant differences between x-ray and electron resists, which compels us to segregate x-ray resists into an independent group of materials. First of all, we notice the essential difference in the mechanisms by which the resists absorb energy. Inelastic scattering of the beam electrons (with excitation or ionization of the target atoms and molecules) transmits a large part (up to 3/4) of the electrons' energy in small amounts (less than 5 eV) to the secondary electrons. X-ray quanta having $\sim$2 keV of energy are completely absorbed and photoelectrons and Auger electrons whose energy is only slightly less than that of the quanta: $E_{fe} = E_k - I \simeq 1.5$ keV. The cross sections of photoelectron quanta absorption $E_k\sigma(E_k)$ and inelastic electron scattering $\sigma_{ne}(E)$ are also vastly different: $\sigma_{ne}(E) \gg \sigma(E_k)$. The cross sections weighed by absorbed energy are close: $\Delta\tilde{E}\sigma_{ne}(E) \simeq E_k\sigma(E_k)$, because $\Delta\tilde{E} \sim 5$ eV and $E_k \sim 2$ keV. In the final analysis, this is reflected in the fact that the attenuation coefficients for the energy carried by an electron beam, $\mu_E = (1/E)(dE/dx)$, and by x-rays, $\mu_x = (1/I_0)(dI/dx)$, are comparable in magnitude: $|dE/dx| \simeq 27$ MeV/cm for a beam having 10 keV of energy and $\mu_E \simeq 2.7 \times 10^3$ cm^{-1} = 0.27 μm^{-1}. Values of $\mu_x(\lambda)$ for PMMA are 2–4 μm^{-1} for wavelengths of 5–30 nm and 0.1–1 μm^{-1} for wavelengths of λ = 0.2–2 nm. Thus, electron-beam and x-ray exposure processes are comparable in terms of their efficiency in transmitting energy to a resist, despite the fact that the mechanisms by which this energy is transmitted are quite different. (The approximate equality $\mu_E \sim \mu_x$ must be considered as happenstance; it exists only for a few quantum E_k and electron E energy ranges.)

The difference in the mechanisms by which energy is transmitted into a resist from electron beams and from x-rays dictates a different strategy in choosing materials for the purpose of increasing the μ_E and μ_x factors. The magnitude of the electron beam energy attenuation coefficient

$$\mu_E = \frac{1}{E}\left|\frac{dE}{dx}\right| = \frac{2\pi e^4 n_e}{E^2}\ln(1.166E/I) \tag{4-308}$$

shows a $\mu_E \sim \bar{Z}/E^2$ dependencies. A logarithmic dependence on the ratio of energy to average ionization energy of unity has no practical significance. Incorporating heavier elements into the composition of a resist increases $\mu_E \sim \bar{Z}$ somewhat. However, because of other requirements imposed on the properties of the films hydrocarbon polymers are chosen as the basis for the resist material and an increase in μ_E is not attained as heavier elements are incorporated. The possibilities for lowering the beam energy E are also limited by the requirements that the proximity effect be small and that the energy transmitted to a $h \simeq 1$ μm layer be uniform throughout its thickness, which requires that the design range of the electrons be $R_p \gg h$. Because of these requirements, energies of $E \geq 10$ keV are chosen. Thus, the physics of the process by which energy is transmitted from electron beams to a resist does not make it possible for us to increase μ_E significantly. There is a way to chemically increase the sensitivity of electron resists: the chemical composition of the resist must be such that at a given level of energy transmission into the resist the largest possible number of chemical reactions directed toward changing the solubility of the resist film occur. A very thorough review of the chemical reactions that have been studied is found in [125].

The energy from soft x-rays that is absorbed by resists is a more complex matter. For a multicomponent material

$$\mu_x(\nu) = N_A\rho \sum_i \frac{C_i}{A_i}\mu_{\text{at}}^{(i)}(\nu)$$
$$\mu_{\text{at}}^{(i)} = \mu_{\text{at}\,K}^{(i)}(\nu) + \mu_{\text{at}\,L}^{(i)}(\nu) + \mu_{\text{at}\,M}^{(i)}(\nu) + \dots \tag{4-309}$$

A characteristic trait of $\mu_{\text{at}}^{(i)}(\nu)$ is 1) the rapid increase (by roughly a factor of 10) in absorption as ν increases to $\nu = I_K,\ I_L^{(i)},\ I_M^{(i)} \dots$, and 2) the very strong dependence of μ_{at} on atomic number $\mu_{\text{at}} \sim Z^{10/3}$–$Z^5$. Thus, the physics of transmitting energy from x-rays prompts us to increase μ_{at} (at a given frequency): atoms of the heavier elements must be incorporated into the resist, whereby the band of frequencies chosen for x-ray radiation, $\nu + \Delta\nu$, will fall into the absorption band of the principal element in the compound, i.e., $\nu \pm \Delta\nu \geq \nu_j$, $j = K$, or L, or $M \dots$ (the principal element in a compound $(C_i/A_i)\mu^{(i)}(\nu + \Delta\nu)$ is that which makes the greatest contribution to absorption). It is not hard, of course, to find a maximum for $\mu_x(\nu)$ from Eq. (4-309).

The means for incorporating atoms of the heavier elements into the composition of the resist molecules for the purpose of increasing sensitivity has been studied many times: chlorine was used in [126] and copper was used in [127]. Figure 4-25 is an example of the characteristic curves for positive resists of CuMA–MA, a copolymer of methacrylic acid and a copper salt of methacrylic acid. Incorporating 5.2% of copper by weight into the copolymer increases the sensitivity of the resist to x-rays in the λ = 0.7–2.3 nm range by a factor of eight.[127] Ordinarily, sensitivity is described by the flow of x-ray energy through the surface of a polymer, J/cm^2. This parameter is, naturally, a function of the radiation wavelength and varies between 10 and 1000 mJ/cm^2 for the L line of molybdenum ($\lambda = 0.541$) nm and the K line of aluminum ($\lambda = 0.843$ nm).[18] These quantities can be multiplied by μ_ν (cm^{-1}) to give the density of the absorbed energy. The sensitivity of PMMA when exposed by the K line of Al is 1000 mJ/cm^2; assuming that for PMMA $\mu_\nu = 0.1\ \mu\text{m}^{-1}$ at this wavelength, we find that the density of the absorbed energy is D = 100 J/cm^3. This is in rough agreement with the density of the energy absorbed by PMMA when exposed by an electron beam (see Chapter 2).

It must be noted that a knowledge of the processes by which energy is taken from a beam is inadequate for following the exposure of a polymer resist to its conclusion, i.e., to the disruption (positive resists) or knitting (negative resists) of the molecular chains. For lack of a microscopic theory of molecular chain knitting and disruption we will use empirical relationships, chief among which are the average energy ΔE per act of molecular chain disruption (knitting) and, e.g., for positive resists, an empirical formula for the rate of dissolution $R(0)$ as a function of the molecular weight $M(0)$ of the nonirradiated polymer

$$R(0) = R_0 + BM(0)^\alpha \tag{4-310}$$

Here, R_0 and B are constants. Applying this equation to a polymer irradiated by a dose D we can write

$$R(D) = R_0 + BM(D)^\alpha \tag{4-311}$$

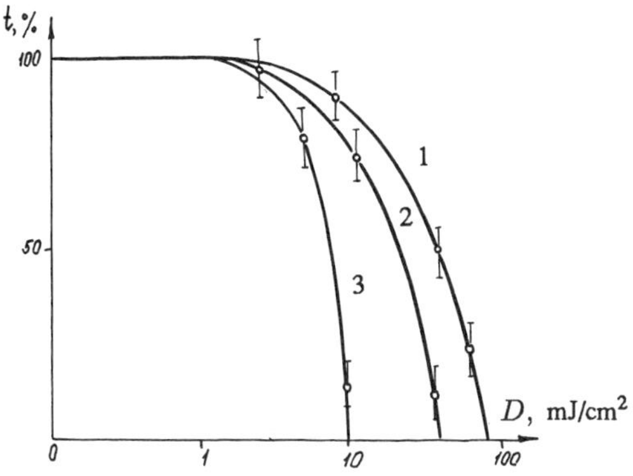

FIGURE 4-25. Characteristic curves for a Cu MA–MA x-ray resist having different amounts of copper: 1) 0% copper by weight; 2) 2.6% copper; 3) 5.2% copper.

It is obvious that each disruption of a molecular chain creates a new molecule so that the number of molecules per 1 cm^3 after absorbing the energy from D J/cm^3 is

$$N(D) = N(0) + D/\Delta E \qquad (4\text{-}312)$$

$$N(0) = N_A \cdot \rho/M(0) \qquad (4\text{-}313)$$

Here, $N(0)$ and $M(0)$ are the initial (pre-radiation) number of molecules per unit volume and average molecular weight of the polymer, N_A is Avogadro's number, and ρ is the density of the polymer. For a polymer that received a dose of D J/cm^3 we have

$$N(D) = N_A \cdot \rho/M(D) \qquad (4\text{-}314)$$

from which we easily find that

$$M(D) = M(0)/(1 + D/D_0), \qquad D_0 = \Delta E \frac{\rho N_A}{M(0)} \qquad (4\text{-}315)$$

At a dose of $D = D_0$ the molecular weight of a polymer decreases by half. As we can see, D_0 is a function of the initial molecular weight $M(0)$. For PMMA it is assumed that there are $G = 1.3$ disruptions of a molecular chain per 100 eV of energy absorbed, or $\Delta E = 80$ eV $= 1.28 \times 10^{-17}$ J per disruption of a molecular chain. If $M(0) = 10^5$ g, we find that $D_0 = 92$ J/cm^3. Values of D_0 obtained through experiment for different resists are 200–300 J/cm^3.

Making the substitution $M(D)$ according to Eq. (4-315) in Eq. (4-311) we find

$$R(D) = R_0 + BM(0)^\alpha(1 + D/D_0)^\alpha \qquad (4\text{-}316)$$

which, for practical purposes, agrees with the empirical equation[98]

$$R(D) = R_1(C_m + D/D_0)^\alpha \qquad (4\text{-}317)$$

where R_1 (nm/sec), D_0 (J/cm^3), C_m, and a are constants that are typical for a specific resist–solvent system under given conditions (temperature). Molecular knitting is the dominant process in negative resists; the dose at which a polymer is insoluble corresponds to the onset of gelation. Negative resists are more sensitive than positive resists; x-ray systems with point sources may have an acceptable throughput only when working with negative resists.

4-25. RESOLUTION IN X-RAY RESISTS

A knowledge of the physical relations of the atomic photoeffect and the radiation or Auger process that follows it makes it possible to explain the factors that govern resolution in resists during exposure. We can formulate a theory of latent image formation in x-ray resists in much the same way as when we analyzed the proximity effects in electron- and ion-beam exposure.

Let a plane $0xy$ coincide with the surface of a resist film illuminated by photons distributed according to $I_\nu(x, y)$ photons/cm^2 · sec · Hz. Although a photon is not a particle that can be localized and followed along its trajectory, it "localizes" itself at the moment the photoeffect occurs, because we can indicate the location $r_f(x_f, y_f, z_f)$ of the atom experiencing photoemission of an electron from one shell or another (the shell is identified by the energy of the photoelectron, $\epsilon_f = E_K - I_i$). Ignoring diffraction, we can say that a photon moves in a straight line (parallel to the z axis) right up to the point r_f at which it is absorbed by an atom, causing photoionization. Therefore, the function $I_\nu(x, y)$ for the spatial distribution of the spectral density of the radiation can be thought of as a distribution function for the number of photons impinging on an area of the resist surface. (When x-ray diffraction is taken into account, an angular distribution for the photons' directions of motion according to the Fourier expansion of the diffraction field with respect to plane waves must be introduced, adopting the wave vectors of the plane waves as the direction in which the photons move, and the normalized coefficients for the expansion $C(\bar{K})$ will be the distribution functions.)

We will introduce an x-ray proximity function $I_\nu(r)$, similar to the proximity function in electron-beam exposure, as a function that has been normalized for a single photon in the spatial distribution of the energy of absorbed photons entering the resist layer at the normal to its surface at the coordinates origin:

$$I_\nu(x, y) = I_\nu \, \delta(x) \, \delta(y) \tag{4-318}$$

(In order to leave diffraction out of our analysis, we will say that this is a δ-shaped stream of photon, mentally segregated from a wide beam of photons). The energy distribution over the volume of the resist for the photons absorbed in the resist at the point $r_f(0, 0, z_f)$ can be written as the product

$$I_\nu(r, r_f) = I_\nu^{\|}(z_f) I_\nu^{\perp}(x^2 + y^2 + (z - z_f)^2) \tag{4-319}$$

We easily find the $I_\nu^{\|}$ function, considering the linear coefficient of photoelectron absorption $\mu_\nu(\nu)$ to be known. The probability that photon absorption will occur in a layer $z, z + dz$ is equal to the product of the probability $\exp(-\mu_\nu z)$ that a photon will not be absorbed in the layer from $z = 0$ to z and the probability $dW = dz/\mu_\nu^{-1}$ that the photon will be absorbed in the $z, z + dz$ layer:

$$dP_\nu(z, z + dz) = \exp(-\mu_\nu z)\mu_\nu \, dz \tag{4-320}$$

From here we find the density of the energy absorbed, normalized for a single photon, to be

$$I_\nu^{\|}(z) = E_\nu \mu_\nu \exp(-\mu_\nu z) \tag{4-321}$$

where E_ν is the photon's energy. Much greater difficulties are encountered in finding the transverse proximity function $I_\nu^{\perp}(x^2 + y^2 + (z - z_f)^2)$. As a result of a photoeffect in an atom at $r_f(0, 0, z_f)$ the photon energy $E_\nu = h\nu$ is distributed between the photoelectron—this part is equal to $\varepsilon = E_\nu - I$—and the atom, which in an ionized state has an excess of energy equal to the ionization energy $I \equiv I_{nl}$. The energy I is, with probability $W_R(W_R + W_A)^{-1}$, radiated as a quantum of characteristic radiation and is, with probability $W_A/(W_R + W_A)$, transmitted to the emitted Auger electrons. For this reason, the function $I_\nu^{\perp}$ is represented as a linear combination of the functions $I_{\nu\,\mathrm{FE}}$, $I_{\nu\,\mathrm{AE}}$, $I_{\nu\,\mathrm{CX}}$ with the respective probabilities and energy weighing functions:

$$I_\nu^{\perp}(r') = \frac{E_\nu - I}{E_\nu} I_{\nu\,\mathrm{FE}}(r') + \frac{I}{E_\nu}\frac{W_A}{W_R + W_A} I_{\nu\,\mathrm{AE}}(r') + \frac{I}{E_\nu}\frac{W_R}{W_R + W_A} I_{\mathrm{CX}}(r')$$
$$r'^2 = x^2 + y^2 + (z_f - z)^2 \tag{4-322}$$

An obvious simplification of the phenomena was made when Eq. (4-322) was written: it was assumed that the atom is relieved of its excess energy I by radiating either one characteristic photon $E_{\mathrm{CX}} = I$, or one Auger electron having energy $E_{\mathrm{AE}} = I$ (as a matter of fact, cascades of radiation and Auger emissions are possible in complex atoms).

The contribution to absorbed energy at points $r_f(x, y, z_f)$ on a plane $z = z_f$ comes from the absorption of quanta that occurs in atoms in this plane and from the absorption in adjacent planes $z'_f \gtrless z_f$ by virtue of the fact that the quanta of energy absorbed in them is distributed by secondary particles (electrons, quanta) in a layer whose thickness is $\Delta z \approx 2R_z$, where R_z is the range of the secondary particles projected onto the z axis. The need to sum the contributions from the various layers points to the existence of a self-generated, longitudinal, proximity effect which, evidently, will occur every time, if we take the finiteness of the secondary particle ranges into account. Starting from statements that have already been made, we find the unknown proximity function by integrating a type of convolution over the vertical:

$$I_\nu(r_\perp, r_f) = \int I_\nu^{\|}(z'_f) I_\nu^{\perp}(r_\perp^2 + (z'_f - z_f)^2)\, dz'_f \tag{4-323}$$

Equation (4-323) is a definition of the proximity function during x-ray exposure of the resists. If $I^{\|}(z_f)$ changes slightly in the interval $(z - z_f) \sim 2R_z$, i.e., we can take $\mu^{-1} \gg R_z$, $I^{\|}(z'_f)$ outside the integral in Eq. (4-323), replacing it with its value when $z'_f = z_f$. Then, approximately,

$$I_\nu(r_\perp, z_f) \approx I_\nu^{\|}(z_f) \int I_\nu^{\perp}(r_1^2 + (z'_f - z_f)^2)\, dz'_f \tag{4-324}$$

The functions given by Eqs. (4-323) and (4-324) are defined in three-dimensional space: they describe the fuzziness in the image of a point source of x-ray quanta in the $z = z_f$ plane. We can simplify Eqs. (4-323) and (4-324) by averaging them over the thickness h_R of the resist layer, $0 \leq z_f \leq h_R$. This averaging makes sense when

TABLE 4-8. Maximal Energies of Photoelectrons and Auger Electrons and Their Range When Exposing a Resist by Rh L and Al K Radiation

Parameter	Exposing radiation: Rh L, $\lambda = 0.46$ nm, $E_\nu = 2.7$ keV, $\mu_\nu = 0.05\ \mu$m			Al K, $\lambda = .083$ nm, $E_\nu = 1.49$ keV, $\mu_\nu = 0.2\ \mu$m			Shell being ionized
Maximal energy, keV	2.17	0.52	0.525	0.98	0.52	0.525	$1s^2$ O
	2.42	0.28	0.27	1.21	0.28	0.277	$1s^2$ C
Range, μm	0.15	0.0126	1	0.04	0.012	1.0	$1s^2$ O
	0.18	0.0001	0.2	0.055	0.0001	0.2	$1s^2$ C
	Photo-electrons	Auger electrons	K_α Line photons	Photo-electrons	Auger electrons	K_α Line photons	

$h_R \ll \mu_R^{-1}(\nu)$ and the functions given by Eqs. (4-323) and (4-324) change slightly within the thickness of the resist layer $0 \le z_f \le h_R$:

$$\tilde{I}_\nu(r_\perp) = \frac{1}{h_R} \int_0^{h_R} I_\nu(r_\perp, z_f)\, dz_f \tag{4-325}$$

The function $\tilde{I}_\nu(r_\perp)$ is two-dimensional and is similar to the two-dimensional function in the theory of the proximity effect that occurs in electron-beam lithography.

Table 4-8 shows the energies of the photoelectrons and Auger electrons emitted by O and C atoms when exposing resists by Rh L and Al K lines. Using values of $E_{\text{AE}} =$ 0.52 keV for the energy of the Auger electrons (or a characteristic quantum) for O atoms being irradiated by Rh L lines (E_ν = 2.7 keV) and the ratio $W_R/(W_A + W_R) \simeq 0.1$ according to the data in Section 4 we find that

$$\frac{E_{\text{cx}}}{E_\nu} \frac{W_R}{W_A + W_R} = \frac{0.52}{2.7} \cdot 0.1 \simeq 0.02$$

This means that the fraction of energy being fluoresced in a characteristic line of radiation is small. Since the range of these photons is about 1 μm the energy of the characteristic photons can be thought of as a uniform background in the submicron pattern.

The amount of energy being released by Auger electrons is somewhat higher:

$$\frac{E_{\text{AE}}}{E_\nu} \frac{W_A}{W_A + W_R} = \frac{0.52}{2.7} \cdot 0.9 \simeq 0.17$$

Thus, photoelectrons are the main factor in dissipating the energy of the primary photons. The direction in which the photoelectrons escape is governed by the distribution $\sin^2\theta$ (see Eq. (4-192)). The electron loses E_{EF} of energy on a path Rg of its range in the resist (Grun range) which for an electron having an energy E_{FE} is $Rg = 0.46\rho^{-1} E_{\text{FE}}^{1.75}$ (ρ is the density, g/cm^3; E is in keV, and Rg is in μm). The rate at which a photoelectron loses energy is

$$\frac{dE}{dS} = \frac{E_{\text{FE}}}{Rg} f(x), \qquad x \equiv S/Rg, \qquad 0 \le x \le 1 \tag{4-326}$$

$$f(x) = 0.74 + 4.7x - 8.9x^2 + 3.5x^3, \qquad 5 < x < 12,$$

$$f(x) = 0.6 + 6.21x - 12.4x^2 + 5.69x^3, \qquad 10 < x < 15$$

The function $f(x)$ has been normalized to unity and has a weak maximum at $x \simeq 0.33$ and varies between 0.6 and 1.3 in the interval $0 \le x \le 0.8$. The function $f(x)$ rapidly goes to zero in the interval $0.8 \le x \le 1$. On this basis we can replace the function $f(x)$ with the simple step function

$$f(x) = \begin{cases} 1, & 0 \le x \le 1 \\ 0, & x > 1 \end{cases} \tag{4-327}$$

when estimating the size of the exposure area. It follows from this that when exposing a resist by an infinitesimally narrow beam of photons $I_\nu(x, y) = \delta(x)$, the width of the exposed region is $\Delta x_{\text{min}} = 2Rg(E_{\text{FE}})$, which can be taken to be the physical limit on resolution in resists during x-ray exposure. Using the expression for Rg that was presented earlier we can write

$$\Delta x_{\text{min}} = 0.092\rho^{-1}(E_\nu - I)^{1.75} \tag{4-328}$$

From here it follows that incorporating heavy elements into the composition of a resist to increase sensitivity also increases resolution, because $\Delta x_{\text{min}} \sim 1/\rho$. If photoelectron quanta are provided mainly by atoms of the heavy elements in the resist when this is done, we must then take care to choose the frequency of the exposing radiation so that the energy of the photoelectrons $E_{\text{FE}} = E_\nu - 1$ will be as low as possible.

Nothing has been published in which the proximity functions for x-ray exposure $I_\nu(r')$ have been modelled according to the Monte Carlo method. An obvious obstacle to doing this is the need for extensive knowledge of many parameters. Nevertheless, such calculations would be extremely valuable because they would make it possible for us to choose a simplified form of the empirical functions that describe the proximity function. It is possible, e.g., that the function $I_\nu^{\perp}(\dot{r}')$ in Eq. (4-322) could be described as the superposition of three Gaussian curves with their respective weights and variances.

Obviously, additional exposure of the lowest film layer, connected with the substrate effect, occurs in x-ray lithography and is similar to the exposure of a resist by backscattered electrons that occurs in electron-beam lithography. Indeed, intense photoelectron absorption of the radiation passing through the resist takes place in substrates having a large Z, and when this happens an external photoelectron arises, in this case in the resist layer from below. Thus, the resist layer turns out to be exposed by an additional dose of photoelectrons from the substrate. Figure 4-26 shows that this effect is present, where the thickness of the dissolved resist layer as a function of dissolution time[128] is shown. In these experiments a thin metal layer was deposited on the surface of the resist layer to cause additional exposure of the upper layer of the resist film by photoelectrons. It is clearly visible that the upper layer, attached to the metal film, dissolves much more rapidly than does the fundamental part of the film. The thickness of the layer exposed

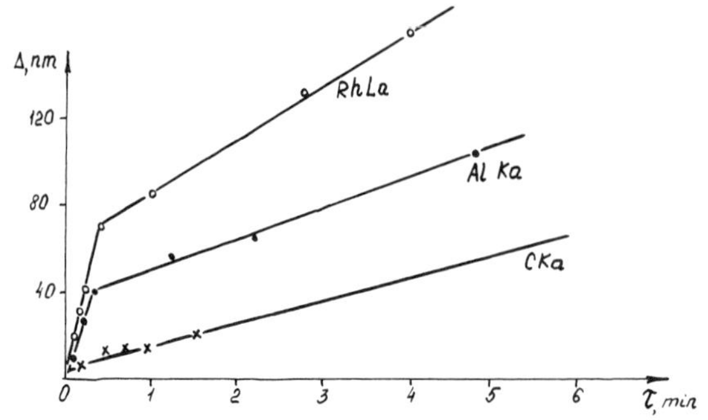

FIGURE 4-26. Thickness of an exposed layer of x-ray resist as a function of dissolution time.[128]

by photoelectrons from the metal film is roughly equal to the projected range of the photoelectrons. This effect goes almost unnoticed when soft K_α C radiation is used.

It is not hard to estimate the intensity of the radiation in the resist due to photoelectrons from the substrate. The intensity of the primary x-rays entering the substrate is $I_{\nu S} = I_{\nu_0} \exp(-\mu_R h_R)$. Photoelectrons can escape from a layer $\Delta z_S = R_S$ thick into the resist, in which case $\mu_S R_{\text{FES}}$ photoelectrons are generated (μ_S is the coefficient of photoelectron absorption in the substrate and R_S is the length of the photoelectron range in the substrate). Considering the fact that, on the average, a photoelectron carries $(1/2)E_{\text{FES}}$ of energy, and that 1/6 of them escape, we can write a function for the photoelectron exposure of a resist from the substrate:

$$D_{\text{FES}} \approx \frac{1}{6}\frac{1}{2} E_{\text{FES}} I_{\nu 0} e^{-\mu_R h_R} \mu_S R_S I_{\text{FES}}(r'') \tag{4-329}$$

$$r^2 = x^2 + y^2 + (z - h_R)^2$$

Since the dose of x-ray exposure in the lowest layer of a resist film is $D(h_R) = E_\nu I_{\nu 0} \exp(-\mu_R h_R)\mu_R$, the ratio $D_{\text{FES}}/D(h_R)$ is

$$\frac{D_{\text{FES}}(h_R)}{D(h_R)} \simeq \frac{1}{12}\frac{E_{\text{FES}}}{E_\nu}\frac{\mu_S R_S}{\mu_R R_R}$$

since $I_{\text{FES}} \approx 1/R_R$. If $\mu_S R_S \gg \mu_R R_R$, this ratio may be almost unity. We will add that, considering photoelectron exposure of the resist from the substrate, a term $I_{\text{FES}}(r)$ with respective weight appears in the proximity function, Eq. (4-322).

We will now turn to writing the formulas that describe a latent image in an x-ray resist. An x-ray mask illuminated by a uniform beam of x-rays creates an illuminance of $I_{\nu_0}(x_f, y_f)$ (photons/cm^2 · sec) on the surface of a resist layer; the band of radiation frequencies is $\Delta\nu$. The dose of energy absorbed at a point $r_\perp(x, y)$ on the resist layer is given by taking the convolution of the illuminance function $I_{\nu_0}(x_f, y_f)$ and the proximity function $\tilde{I}_\nu(|r_{\perp f} - r_\perp|)$ and by integrating over the frequency distribution of the exposing radiation:

$$D(r_\perp) = t \int_{\Delta\nu} E_\nu \mu_R(\nu)\, d_\nu \int I_{\nu_0}(\mathbf{r}_{\perp f})\tilde{I}_\nu(|\mathbf{r}_\perp - \mathbf{r}_{\perp f}|)\, d^2\mathbf{r}'_{\perp f} \tag{4-330}$$

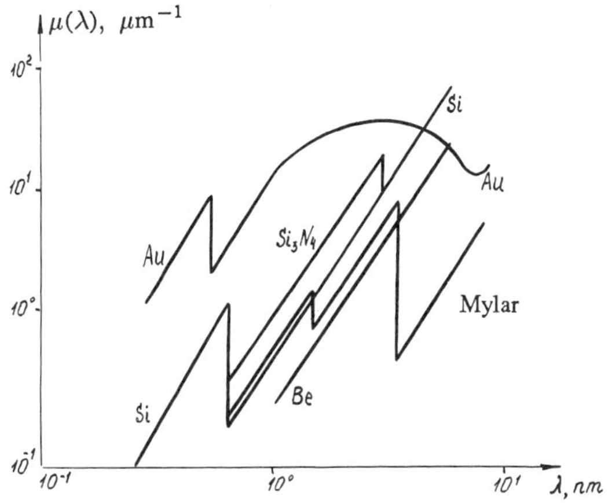

FIGURE 4-27. Absorption coefficients $\mu(\lambda)$ for some materials in the soft x-rays range.[18]

where t is the exposure time. If we ignore energy dissipation near the point $r_{\perp f}$ and assume that $\tilde{I}_\nu(|\mathbf{r}_\perp - r_{\perp f}|) = \delta(|\mathbf{r}_\perp - \mathbf{r}_{\perp f}|)$, then

$$D(r_\perp) = \left[t \int_{\Delta\nu} E_\nu \mu_R(\nu)\, d\nu\right] I_{\nu_0}(r_{\perp f}) = \kappa I_{\nu_0}(r_{\perp f}) \tag{4-331}$$

The latent image in this approximation is a copy of the illumination.

If a filter segregates a narrow spectral region $\Delta\nu$ in which $\mu_R(\nu)$ = const, $I_{\nu_0} = I_0$ = const, then

$$D(r_\perp) = I_0 \Delta\nu E_\nu \mu_R(\nu) \tag{4-332}$$

The last formula is exact when the exposure is done by the characteristic radiation line of a material and $I_\nu \Delta\nu$ cm^{-2} · sec^{-1} is then simply the flux of characteristic photons from radiation at the wavelength of this line. The quantity μ_ν for resists is strongly dependent on ν and this dependence is close to the $\mu(\nu)$ for mylar shown in Fig. 4-27. As we can see, in the λ = 0.7–5 nm range of wavelengths μ_ν increases by more than an order of magnitude from 0.1 to 5 μm^{-1}. This means that by changing the wavelength of the radiation in a relatively narrow band of wavelengths, we can very markedly change the exposure conditions. It is obvious that stable conditions can more easily be achieved by using sources of monochromatic x-rays within the limits of one characteristic line. In order to expose a resist layer more or less uniformly throughout its depth, the layer thickness d_R must satisfy the condition that $\mu_\nu d_R < 1$. As we can see in Fig. 4-27, this condition is not satisfied for a layer d_R = 1 μm when the wavelengths are $\lambda > 2$ nm.

We will compare the process of exposing PMMA at two wavelengths: Rh L, λ = 0.46 nm and Al K, λ = 0.83 nm (see Table 4-8).

From the viewpoint of resolution, the Al K line is the more appropriate characteristic line for exposing PMMA: it ensures the absorption of a significant amount $(1 - \exp(-\mu_\nu d)) = 0.2$ of radiation energy in a 1 μm thick layer and resolution better than 0.1 μm, owing to the short range of the photoelectrons. At the same time, the sensitivity of the resist (in J/cm^2) for a Rh L line is lower because only $\mu_\nu d_R$ = 0.05 of the energy from the incident radiation is absorbed and the resolution is roughly 0.2 μm, owing to the longer range of the photoelectrons.

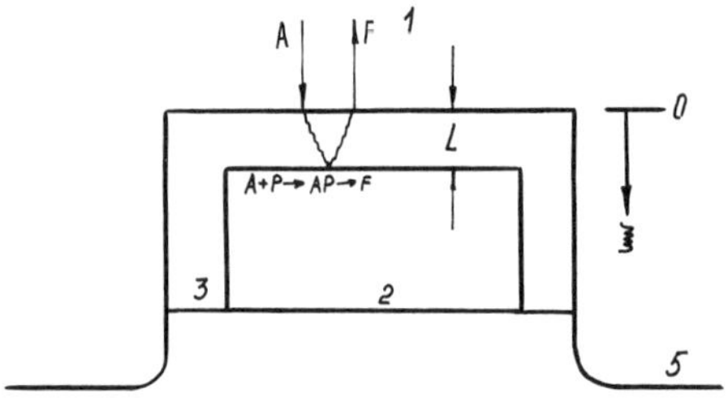

FIGURE 4-28. Diagram of the model for plasmachemical etching of a resist film: 1) plasma; 2) polymer mask on the substrate; 3, 4) the K layer of a polymer film modified by plasma action; 5) the plasmachemical etching region of a structural layer (substrate). The active particles A diffuse through the K layer to the polymer molecules; the fragments F generated in the reaction A + P go into the plasma region.

From what has been said, we can make a practical recommendation for choosing the radiation wavelength to be used in x-ray lithography: the wavelength must be so close to the absorption edge of the resist atoms (usually C and O) that it will satisfy the requirement for uniform exposure throughout the depth of a resist layer of given thickness d_R: $\mu_\nu d_R < 1$.

4-26. THE STABILITY OF RESIST FILMS IN CHEMICALLY ACTIVE PLASMAS

An indispensable requirement on x-ray resists is that they be plasma stable because the resist masks must be preserved when the lower structural layers are plasma etched. It is hardly possible to use liquid etchants with patterns having these dimensions. A model for the plasma chemical etching of polymer films that makes it possible to isolate modes having the lowest etching rate was first constructed in [129] in which etching in a reactor with perforated screens, which precludes high-energy ion bombardment and physical sputtering, was examined. For example, in a standard fluorine-oxygen plasma the surface of a resist is subjected to the action of neutral atoms (A) and molecules (M) in their ground (F, O) and excited (F*, O*) states, of excited O_2^* molecules and CF_x radicals, $x = 1, 2, 3$. Due to the action of A, A* and M, M* particles, a modified layer is formed on the surface of the film L thick (the K layer is a skin layer) and is more stable to plasma etching than is an unmodified layer. (This is quite natural, since the possible reactions between the A, A* and M, M* particles have already taken place). Therefore, for reactions between the A, A* and M, M* particles and the polymer molecules to proceed it is necessary that the active particles from the plasma as well as the products of reaction (molecule fragments) F (in the opposite direction) diffuse through the K layer. The model described (Fig. 4-28) enables us to construct a phenomenological theory of plasmachemical etching in polymers which makes it possible to interpret the facts observed in experiment, in a very natural fashion.

In a coordinate system that moves with velocity u (nm/sec), the polymer etching rate in a plasma, the diffusion equations for the active A particles (from the plasma through the K layer to the polymer) and for the volatile fragments F from the $A + P \to AP \to F$

reaction are[129]

$$\left(\frac{\partial}{\partial t} - u\frac{\partial}{\partial \xi}\right) N_A = D_A \frac{\partial^2 N_A}{\partial \xi^2} \tag{4-333a}$$

$$\left(\frac{\partial}{\partial t} - u\frac{\partial}{\partial \xi}\right) N_F = D_F \frac{\partial^2 N_F}{\partial \xi^2} \tag{4-333b}$$

The boundary conditions for Eqs. (4-333a) and (4-333b) are

$$N_A|_{\xi=0} = N_A(0), \qquad -D_A \left.\frac{\partial N_A}{\partial \xi}\right|_{\xi=L} = \kappa_1(1-\eta)N_P N_A|_{\xi=L} \tag{4-334a}$$

$$N_F|_{\xi=L} = g\eta N_P, \qquad D_F \left.\frac{\partial N_F}{\partial \xi}\right|_{\xi=0} = v_0 \cdot N_F|_{\xi=0} \tag{4-334b}$$

Here, for the sake of clarity, it is assumed that only one type of active particle A and one type of volatile fragment F exist, where N_A, N_F, D_A, and D_F are their concentrations and diffusion coefficients through the K layer; $N_A(0)$ is the known concentration of A particles beneath the surface of the layer K and is proportional to their concentration in the plasma; k_1 is the reaction constant for the generation of the F fragments at the boundary between the K layer and the polymer; η is the fraction of the surface between the K layer and the resist occupied by AP complexes that decompose into fragments; g is the number of fragments produced by the destruction of a single polymer molecule; and v_0 is the velocity at which the F particles leave the surface of the K layer and enter the plasma. Under stationary etching conditions $\partial/\partial t = 0$, u = const, and L = const, which makes it possible for us to define the η parameter from the requirement that the concentration of F particles at the $\xi = L$ boundary be constant:

$$D_F \left.\frac{\partial N_F}{\partial \xi}\right|_{\xi=L} - g(1-\eta)\kappa_1 N_P N_A = 0 \tag{4-335}$$

By the definition of $N_F(\xi)$ from Eqs (4-333) and (4-334) the etching rate u can be found from one of the equations

$$u = \frac{m}{\rho} N_F|_{\xi=0} v_0 = \frac{m}{\rho} D_F \left.\frac{\partial N_F}{\partial \xi}\right|_{\xi=0} \tag{4-336}$$

We can write a simplified version of the expression for u, using the inequalities $u \ll \min(D_A/L, D_F/L)$, $D_F/L \ll v_0$ for the simplifications (see [129]), which are purposely satisfied in experiment, since $u \approx 1$ nm/sec, and $D_A > D_F \simeq 10^{-14}$ m^2/sec, where $L \ll h$, $h \simeq 1$ μm and v_0 is the order of thermal velocities

$$u = \frac{m}{2\rho\beta}\{\alpha + \gamma(\beta + D_A) - [(\alpha + \beta\gamma + \gamma D_A)^2 - 4\alpha\beta\gamma]^{1/2}\} \tag{4-337a}$$

$$\alpha = g\kappa_1 D_A N_P N_A(0), \qquad \beta = \kappa_1 N_P L, \qquad \gamma = g D_F N_P/L \tag{4-337b}$$

We will investigate the solution of Eq. (4-337) for different conditions. Experiment shows that the dependence of u on temperature is Arrhenian in nature and that the activation energy E_a varies from E_a = 3–6.5 kcal/mole at $T < T_g$ to E_a = 10–27 kcal/mole at $T > T_g$, where T_g is the polymer vitrification temperature.[130] We will attempt to explain these phenomena. It is well known that in multistage processes one of them creates a "bottleneck" that determines the rate of the entire process. We will assume that when $T < T_g$ this bottleneck is the diffusion of active particles through the K layer

$$D_{\rm A} \ll \min(k_1 N_{\rm P} L, D_{\rm F} N_{\rm P}/N_{\rm A}(0)) \tag{4-338}$$

then we find from Eq. (4-337) that

$$u = \frac{mg}{\rho L} N_{\rm A}(0) D_{\rm A}(T) \sim N_{\rm A}(0) D_{\rm A}^0 \exp(-E_{\rm DA}/RT) \tag{4-339}$$

E_{DA} is the activation energy for the diffusion of A particles into the K layer material. Experiments on the diffusion of light-weight gas molecules through a polymer have found that $E_D \simeq$ 5–6 kcal/mole, which agrees with the activation energy of the etching process ($u(T) \sim \exp(-E_{\rm etch}/RT)$ found from experiment.[130] It is interesting to note that the activation energy depends on the power of the discharge in the reactor, W: $E_{\rm etch} = E_0 - \delta W$.[130] This result has a natural explanation in the fact that the extent to which the K layer is modified depends on W (the number of modified molecules $N_K/N_{\rm P}$): $N_K \sim N_{\rm M}$, whereby $N_{\rm M}$ is the concentration of the modifying active particles in the plasma, and $N_{\rm M} \sim W$, then, $N_K \sim W$

$$E_{\rm DA}(N_K/N_{\rm P}) = E_{\rm DA}(0) + \left(\frac{\partial E_{\rm DA}}{\partial(N_K/N_{\rm P})}\right)\frac{N_K}{N_{\rm P}} \equiv E_{\rm DA}(0) + \delta W \tag{4-340}$$

We will now turn our attention to the fact that the concentration of active particles A is also proportional to W and then, according to Eq. (4-339), $\log u = c_1 + \ln W + c_2 W$. This dependence is also in close correspondence with the observed facts.[130] The manners in which $E_{\rm DA} = E_a$ and the etching rate u depend on the composition of the $CF_4 + O_2$ gaseous mixture are also explained in a natural manner and we refer the reader to [129] for a thorough discussion.

We will now consider the range of temperatures $T > T_g$. We will assume that, because the diffusion coefficient increases with increasing temperature, the destruction of polymer molecules $\rm A + P \to AP \to F$ has become the "bottleneck":

$$\kappa_1 \ll \min\{D_{\rm A}/LN_{\rm P}, D_{\rm F}/LN_{\rm A}(0)\} \tag{4-341}$$

Then, according to Eq. (4-339) we have

$$u = \frac{m}{\rho} g N_{\rm P} Na(0) \kappa_1 \sim W \exp(-E_1/RT)$$

First of all, notice that u, according to Eq. (4-342) does not have any of the K layer parameters and it is possible that a K layer will not form if etching proceeds rapidly (we can easily verify that $u(T > T_g) \gg u(T < T_g)$). Since E_1 the activation energy of the chemical reaction is independent of W, $\log u = C + \log W$; it is namely this weak dependence that is seen in experiment.[130] Finally, the activation energy for the chemical reactions is $E_1 \simeq$ 10–30 kcal/mole, which is in good agreement with the activation energy of the etching process when $T > T_g$: E_a = 10–27 kcal/mole.

The etching rate was seen to change over time in [131] and was much lower at the end of the process than it was at the beginning. The authors[131] assumed that the etching rate at the end of the process was due to etching the K layer, which is more resistant to etching. The K layer model can possibly be applied to the plasmachemical etching of structural layers (Si, SiO_2, Me) if the formation of polymer films (e.g., $(CF_x)_n$ when etching in a $CF_4 + O_2$ plasma) is possible.

We will examine the question from the viewpoint of increasing the plasma stability of polymer masks when etching the structural layers IC. It follows from our analysis of plasmachemical etching that the formation of a K layer on the surface of a polymer film stabilizes the process at a low etching rate. Therefore, the composition of the gas mixture in the plasma and the resist must be chosen so that the formation of a K layer will be enhanced. At the same time, a K layer on the structural layer is undesirable. With these goals in mind, we must optimize the multifactor etching process with due regard for the composition of the gas, the polymer film, and the structural layer, as well as for those parameters such as the reactor power W, the substrate temperature T, and the gas pressure P.

REFERENCES

1. L. R. Koller, *Ultraviolet Radiation*, Wiley, New York (1965).
2. A. Sommerfel'd, *The Structure of the Atom and Its Spectrum* [in Russian], Gostekhizdat, Moscow (1956).
3. L. D. Landau and E. M. Lifshits, *Fields Theory* [in Russian], Gostekhizdat, Moscow–Leningrad (1948).
4. L. D. Landau and E. M. Lifshits, *Mechanics* [in Russian], Gostekhizdat, Moscow (1958).
5. I. I. Sobel'man, *Introduction to the Theory of Atomic Spectra* [in Russian], Fizmatgiz, Moscow (1963).
6. S. Reed, *Electron Probe Microanalysis* [Russian translation], Mir, Moscow (1979).
7. M. A. El'yashevits, *Atomic and Molecular Spectroscopy* [in Russian], Fizmatgiz, Moscow (1962).
8. É. S. Parilis, *The Auger Effect* [in Russian], Izd. Akad. Nauk UzbSSR, Tashkent (1969).
9. L. D. Landau and E. M. Lifshits, *Quantum Mechanics* [in Russian], Fizmatgiz, Moscow (1963).
10. G. Shinoda, K. Kohra, and T. Ishinokawa (editors), 6th Int. Conf. on X-ray Optics and Microanalysis, Tokyo University Press, Tokyo (1972).
11. H. H. Pattee, V. E. Cosslett, and A. Engstrom (editors), *X-Ray Optics and X-Ray Microanalysis*, Academic Press, New York (1963).
12. R. Castaing, P. Deschamps, and J. Philibert (editors), *Optique des Rayons X et Microanalyse*, Herrman, Paris (1966).
13. K. Shima, M. Okuda, E. Suzuki, T. Tsubota, and T. Mikumo, *J. Appl. Phys.*, **54**, No. 3, 1202–1208 (1983).
14. H. A. Kramers, *Philos. Mag.*, **46**, 836 (1923).
15. T. S. Rap-Sahib and D. B. Wittry in: 6th Int. Conf. on X-Ray Optics and Microanalysis, G. Shinhoda, K. Kohra, and T. Ishinokawa (editors), Tokyo University Press, Tokyo (1972).
16. J. A. Small, S. D. Leigh, D. E. Newbury, and R. L. Myklebust, *J. Appl. Phys.*, **61**, No. 2, 459–469 (1987).
17. S. T. Lins and H. S. Kukuk, "A resonance frequency shift thin film thickness monitor" in: Trans. 7th Nat. Symp. on Vac. Sci. Technol., Pergamon Press, New York (1960), pp. 333–338.
18. A. R. Neureuther, "Microlithography with soft x-rays" in: *Investigations in Synchrotron Radiation*, H. Winick and S. Doniach (editors), Plenum Press, New York–London (1980), pp. 223–275.
19. Y. S. Kim and R. H. Pratt, *Phys. Rev.*, **27**, No. 6, 2913–2924 (1983).
20. D. L. Matthews, E. M. Campbell, N. M. Ceglio, G. Hermes, R. Kauffman, L. Koppel, R. Lee, K. Manes, V. Rupert, V. W. Slivinsky, R. Turner, and F. Ze, *J. Appl. Phys.*, **54**, No. 8, 4260–4268 (1983).
21. D. G. Colombant and W. M. Manheimer, *J. Appl. Phys.*, **51**, No. 12, 6121–6123 (1980).
22. S. Goldsmith, J. F. Seely, U. Feldman, W. E. Behring, and L. Cohen, *J. Appl. Phys.*, **58**, No. 11, 4011–4014 (1985).
23. P. D. Gupta, P. A. Naik, and H. C. Pant, *J. Appl. Phys.*, **56**, No. 5, 1371–1373 (1984).
24. H. C. Pant, S. Sharma, and D. D. Bhawalkar, *J. Appl. Phys.*, **51**, No. 11, 6030–6032 (1980).

25. H. G. Gerritsen, H. van Brug, F. Bijkerk, and M. J. van der Wiel, *J. Appl. Phys.*, **59**, No. 7, 2337–2344 (1986).
26. B. Yaakobi, H. Kim, J. M. Soures, H. W. Deckman, and J. Dunsmuir, *J. Appl. Phys. Lett.*, **43**, No. 7, 686–688 (1983).
27. F. O'Neill, M. C. Gower, I. C. E. Tureu, and Y. Owadano, *Appl. Optics*, **25**, No. 4, 464–465 (1986).
28. H. Puell, *Z. Naturforsch.*, **A25**, 1807 (1970).
29. D. Mosher, *Phys. Rev.*, **A10**, 1807 (1974).
30. G. Mellman, P. G. Burkhalter, S. J. Stephanakis, F. C. Young, and D. J. Nagel, *J. Appl. Phys.*, **60**, No. 10, 3427–3432 (1986).
31. R. E. Stewart, D. D. Dietrich, P. O. Egan, R. J. Fortner, and R. Dukart, *J. Appl. Phys.*, **61**, No. 1, 126–130 (1987).
32. R. E. Marrs, D. D. Dietrich, R. J. Fortner, M. A. Levine, D. F. Price, R. E. Stewart, and B. K. F. Young, *Appl. Phys. Lett.*, **42**, No. 11, 946–948 (1983).
33. R. B. Speilman, D. L. Hayson, M. A. Palmer, *et al.*, *J. Appl. Phys.*, **57**, No. 3, 830–833 (1985).
34. V. A. Veretennikov, Yu. S. Leonov, V. I. Mishachev, and O. G. Semenov, *Pis'ma Zh. Tekh. Fiz.*, **11**, No. 19, 1200–1203 (1985).
35. I. M. Ternov, V. V. Mikhailin, and V. R. Khalilov, *Synchrotron Radiation and Its Applications* [in Russian], Izd. MGU, Moscow (1980).
36. A. A. Sokolov and I. M. Ternov, *A Relativistic Electron* [in Russian], Nauka, Moscow (1983).
37. D. D. Ivanchenko and A. A. Sokolov, *Classical Fields Theory* [in Russian], Gostekhizdat, Moscow–Leningrad (1949).
38. *Synchrotron Radiation in the Study of Solids: A Collection of Articles* [Russian translation], Mir, Moscow (1970).
39. G. N. Kulipanov and A. N. Skrinskii, *Usp. Fiz. Nauk*, **122**, No. 3, 369–418 (1977).
40. D. H. Tombulian and D. E. Bedo, *J. Appl. Phys.*, **29**, 804–809 (1958).
41. A. Heuberger, "Photolithography and x-ray lithography" in: *Methods and Materials in Microelectronics Technology*, J. Bargon (editor), Plenum Press, New York–London (1984), pp. 99–126.
42. U. Trinks, F. Nolden, and A. Jahnke, *Nucl. Instr. and Methods*, **200**, 475 (1982).
43. E. Gronid *et al.*, Microcircuit Engineering '81, Swiss Federal Institute Technol., Lausanne, Switzerland (1981), p. 121; J. Le Duff *et al.*, ibid.
44. 1. H. Suzuki and N. Saito, *Jap. J. Appl. Phys.*, **25**, No. 1, 130–136 (1986).
45. M. N. Yakimenko, *Usp. Fiz. Nauk*, **114**, No. 1, 55–66 (1974).
46. V. L. Ginzburg, *Dokl. Akad. Nauk SSSR*, **56**, 145 (1947); *Izv. Akad. Nauk SSSR, Ser. Fiz.*, **11**, 165 (1947).
47. S. A. Bogacz and J. B. Ketterson, *J. Appl. Phys.*, **60**, No. 1, 177–178 (1986).
48. M. A. Kumakhov, *Phys. Lett.*, **57**, 17 (1976); *Phys. Stat. Sol. (b)*, **84**, 41 (1977).
49. A. N. Chu, M. A. Piestrog, R. H. Pantell, and F. R. Buskirk, *J. Appl. Phys.*, **52**, No. 1, 22–24 (1981).
50. V. Ginzburg and I. M. Frank, *Zh. Eksp. Teor. Fiz.*, **16**, No. 15 (1949).
51. R. Feynman, *Quantum Electrodynamics* [Russian translation], Mir, Moscow (1964).
52. H. Bethe and E. Salpeter, *Quantum Mechanics of Atoms with One and Two Electrons* [in Russian], Fizmatgiz, Moscow (1960).
53. P. Gombash, *The Many Particle Problem in Quantum Mechanics* [Russian translation], Inostrannaya Literatura, Moscow (1953).
54. W. Heitler, *The Quantum Theory of Radiation* [Russian translation], Inostrannaya Literatura, Moscow (1956).
55. A. I. Ansel'm, *Introduction to Semiconductor Theory* [in Russian], Nauka, Moscow (1978).
56. R. Hensel, *Research into the Optical Properties of Alkaline-Halide Compounds in Solid Inert Gases: A Collection of Articles* [Russian translation], Mir, Moscow (1970).
57. T. M. Zimkina and V. A. Fomichev, *Ultrasoft X-Ray Spectroscopy* [in Russian], Izd. LGU, Leningrad (1971).
58. G. S. Landsberg, *Optics* [in Russian], Nauka, Moscow (1976).
59. A. P. Lukirskii, E. P. Savinov, O. A. Ershov, and Yu. F. Shepelev, *Opt. Spektrosk.*, **26**, 310–319 (1964).
60. A. P. Lukirskii, E. P. Savinov, O. A. Ershov, *et al.*, *Opt. Spektrosk.*, **29**, 425–433 (1965).
61. J. M. Bennett, *Thin Solid Films*, **123**, 27–44 (1985).
62. L. I. Mandelshtam, *Ann. Physik*, **41**, 609 (1913).
63. A. A. Andronov and M. A. Leontovich, *Z. Phys.*, **38**, 485 (1926).
64. G. Hasinger, Preprint MPI-PAE/Extratter, No. 163 (1980).
65. B. Aschenbach, H. Brauninger, G. Hasenger, and J. Trumper, *Proc. SPIE*, **257**, 223 (1980).

66. M. V. Zombeck, H. Brauninger, A. Ondrusch, and P. Predehl, *Proc. SPIE*, **316**, (1981).
67. A. V. Vinogradov, N. N. Zorev, I. V. Kozhevnikov, and I. G. Yakushkin, *Zh. Eksp. Teor. Fiz.*, **89**, No. 6(12), 2124–2132 (1985).
68. A. V. Vinogradov, N. N. Zorev, I. V. Kozhevnikov, S. I. Sagitov, and A. G. Tur'yanskii, Preprint FIAN, Moscow, No. 319, 1–23, (1986).
69. S. V. Gaponov, S. A. Gusev, Yu. Ya. Platonov, and N. N. Salashchenko, *Zh. Tekh. Fiz.*, **54**, 747 (1984); *Pis'ma Zh. Tekh. Fiz.*, **9**, 1140 (1983).
70. P. Dhez, in: X-Ray Microscopy, Proc. Int. Symp., Gottingen, FRG, 1983, G. Schmal and D. Rudolph (editors), Springer-Verlag, Berlin–Heidelberg–New York–Tokyo (1984), pp. 139–143.
71. T. W. Barbee, Jr., *ibid.*, pp. 144–162.
72. W. K. Warburton, Z. U. Rek, and T. W. Barbee, Jr., *ibid.*, pp. 163–171.
73. A. V. Vinogradov and B. Ya. Zeldovich, *Appl. Opt.*, **16**, 89 (1977).
74. A. V. Vinogradov and I. V. Kozhevnikov, Preprint FIAN, Nos. 102, 103, Moscow (1986).
75. A. V. Vinogradov and I. V. Kozhevnikov, *Opt. Spektrosk.*, **58**, No. 4. 895–899 (1985).
76. A. V. Vinogradov, V. M. Elinson, N. N. Zorev, G. F. Ivanovskii, I. V. Kozhevnikov, S. I. Sagitov, and V. V. Sleptsov, *Opt. Spektrosk.*, **59**, No. 3, 703–705 (1985).
77. A. G. Revesz, *Phys. Stat. Sol. (a)*, **60**, 619 (1980).
78. Y. Lepetre, E. Ziegler, and I. K. Shuller, *J. Appl. Phys.*, **60**, No. 7, 2301–2303 (1986).
79. M. Arbaoui, J. M. Andre, P. Coullaux, and R. Barchewitz, *Rev. Sci. Instrum.*, **56**, No. 11, 2055–2058 (1985).
80. A. V. Vinogradov, V. F. Kovalev, I. V. Kozhevnikov, and V. V. Pustovalov, *Zh. Tekh. Fiz.*, **55**, No. 2, 244–250 (1985).
81. A. V. Vinogradov, V. F. Kovalev, I. V. Kozhevnikov, and V. V. Pustovalov, *Zh. Tekh. Fiz.*, **55**, No. 3, 567–574 (1985).
82. A. V. Vinogradov and I. V. Kozhevnikov, *Zh. Tekh. Fiz.*, **54**, No. 9, 1755–176Z (1984).
83. Yu. M. Aleksandrov, K. A. Valiev, L. V. Velikov, S. D. Dushenkov, *et al.*, *Pis'ma Zh. Tekh. Fiz.*, **13**, No. 5, 257–268 (1987).
84. P. J. Mallozzi, H. M. Epstein, R. G. Jung, *et al.*, *J. Appl. Phys.*, **45**, No. 4, 1891–1895 (1974).
85. A. V. Vinogradov and N. N. Zorev, Preprint FIAN, Moscow (1987).
86. G. Schmal, D. Rudolph, P. Guttman, and O. Christ in: X-Ray Microscopy, Proc. Int. Symp., Göttingen, FRG, 1983, G. Schmal and D. Rudolph (editors), Springer-Verlag, Berlin–Heidelberg–New York–Tokyo (1984), pp. 63–74.
87. P. Guttman, *ibid.*, pp. 75–90.
88. J. Thieme, *ibid.*, pp. 91–96.
89. N. M. Ceglio, *ibid.*, pp. 97–108.
90. A. G. Michette, M. T. Browne, *et al.*, *ibid.*, pp. 109–118.
91. R. O. Tatchyn, *ibid.*, pp. 40–50.
92. D. V. Sivukhin, *Optics* [in Russian], Nauka, Moscow (1980).
93. A. Heuberger, *Sol. State Technol*. 93–101 (1986).
94. A. Heuberger, *Vac. Sci. Technol.*, **B6**, No. 1, 107–121 (1988).
95. A. Tobey, *Microelectronic Manufacturing and Testing*, July, 1986.
96. E. N. Fuls, *Proc. SPIE*, March, 1982, pp. 113–117.
97. R. T. Gallagher, *Electronics*, No. 3, 81–82 (1981).
98. K. Suzuki *et al.*, *IEEE El. Dev. Lett.*, **EDL-6** (1985).
99. H. Betz, F. K. Fey, A. Heuberger, and P. Tischer, *IEEE Trans. on Electron Devices*, **ED-26**, 693–697 (1979).
100. P. A. Sullivan and H. John, *IEEE Trans. on Electron Devices*, **ED-23**, No. 4, 412–418 (1976).
101. I. Brodie and J. J. Muray, *The Physics of Microfabrication*, Plenum Press (1982).
102. N. Bobroff, R. Tibbets, J. Wilczynski, and A. Wilson, *J. Vac. Sci. Technol*, **B4**, No. 1, 285–289 (1986).
103. D. C. Flanders, H. I. Smith, and S. Austin, *Appl. Phys. Lett.*, **31**, 426 (1977).
104. B. Fay, J. Trotel, and A. Fzichet, *J. Vac. Sci. Technol.*, **16**, 1954 (1980).
105. P. Nelson, Jr., V. DiMilia, and J. Warlaumont, *J. Vac. Sci. Technol.*, **19**, 1219 (1981).
106. M. Feldman, A. P. White, and D. L. White, *J. Vac. Sci. Technol.*, **19**, 1224 (1981).
107. E. Cullman in: Brookhaven Conf. on Advances in Soft X-Ray Science and Technology, Upton, NY (1983).
108. J. Itoh and T. Kanayama, *Jap. J. Appl. Phys.*, **25**, No. 6, 487–489 (1986).
109. D. Maydan, G. A. Coquin, J. R. Maldonado, *et al.*, *IEEE Trans. Electron. Dev.*, **ED-22**, 425–439 (1975).
110. T. Funayama, K. Yanagida, N. Nakayama, *et al.*, *J. Vac. Sci. Technol.*, **15**, No. 3, 998–1000 (1978).

111. D. Flanders and H. I. Smith, *J. Vac. Sci. Technol.*, **15**, No. 1, 99–109 (1978).
112. B. M. Gong and Y. D. Ye, *J. Vac. Sci. Technol.*, **19**, No. 4, 1204–1207 (1981).
113. T. Wada, S. Sakurai, and K. Kawabuchi, *J. Vac. Sci. Technol.*, **19**, No. 4, 1208–1210 (1981).
114. O. I. Spears and H. I. Smith, *Solid State Technol.*, **15**, No. 1, 21–26 (1972).
115. K. Suzuki, J. Matsui, T. Kadata, and T. Ono, *Jap. J. Appl. Phys.*, **17**, 1447–1448 (1978).
116. T. Funayama, Y. Takayama, T. Inagaki, and M. Nakamura, *J. Vac. Sci. Technol.*, **12**, No. 4, 1324 (1975).
117. R. K. Wats, K. E. Bean, and T. L. Brewer, "X-ray lithography with aluminum radiation and SiC mask" in: Proc. of the Eighth Intern. Conf. on Electron and Ion Beam Science and Technology, New Jersey, 1978, pp. 453–457.
118. Yu. M. Aleksandrov, K. A. Valiev, L. V. Velikov, S. D. Dushenkov, R. Kh. Makhmutov, and M. I. Yakimenko, *Mikroelektronika*, **15**, No. 1, 66–69 (1986).
119. K. Suzuki and J. Matsui, *J. Vac. Sci. Technol.*, **B4**(1), 221–225 (1986).
120. K. H. Muller, P. Tisher, and W. Windbracke, *J. Vac. Sci. Technol.*, **B(4)**, 230–234 (1986).
121. T. Ebata, M. Sekimoto, T. Ono, K. Suzuki, J. Matsui, and S. Nakayama, *Jap. J. Appl. Phys.*, **21**, No. 5, 762–767 (1982).
122. A. Zacharias, *Solid State Technol.*, August, 57–59 (1981).
123. M. Karnezos, *J. Vac. Sci. Technol.*, **B(4)**, 226–229 (1981).
124. S. Imamura and S. Suguwara, *Jap. J. Appl. Phys.*, **21**, No. 5, 776–782 (1982).
125. J. Bargon, "Lithographic materials" in: *Methods and Materials in Microelectronic Technology*, J. Bargon (editor), Plenum Press, New York–London (1984).
126. G. N. Taylor, G. A. Coquin, and S. Somekh, *Polymer Engineering and Science*, **17**, No. 6, 420–429 (1977).
127. Yu. M. Aleksandrov, K. A. Valiev, L. V. Velikov, *et al.*, *Mikroelektronika*, **12**, No. 1, 3–10 (1983).
128. R. Feder, E. Spiller, and J. Topalian, *Polymer Engineering and Science*, **17**, No. 6, 385–389 (1977).
129. K. A. Valiev, T. M. Makhviladze, and M. E. Sarychev, Preprint IOFAN No. 45, 1–30 (1985); *Dokl. Akad. Nauk SSSR*, **283**, No. 2, 366–369 (1985); *Zh. Tekh. Fiz.*, **56**, No. 7, 1329–1335 (1986).
130. K. Harada, *J. Appl. Polym. Sci.*, **26**, No. 6, 1961–1973 (1981); No. 10, 3395–3408.
131. K. A. Valiev, K. Ya. Mokrousov, and A. A. Orlikovskii, *Poverkhnost'*, No. 1, 53–58 (1987).

5

Optical Lithography

5-1. INTRODUCTION

The opinion that optical lithography has exhausted all of its possibilities for improving resolution and is incapable of encroaching into the submicron region—that only electron-beam, ion-beam, and x-ray lithography can be used in this region—has been expressed several times in the last ten years. Recently however, this opinion has been revised and experts now believe that optical lithography via shadow and projection systems with stepwise align and expose (steppers) can provide the resolution and reproduce elements having minimal sizes of 0.25–0.5 μm.[1–3] Thus, optical lithography can be classified with microlithography methods, and, when this is done, x-ray lithography becomes the successor to optical lithography for printing elements having sizes less than 0.2 μm.[4,5]

Optical methods are used to reproduce the image of a mask on a photosensitive film via contact and proximity printing (Fig. 5-1) and projection printing (Fig. 5-2). The properties of the light source in optical systems are especially significant in creating the image of a mask-stencil. Figure 5-1a shows the diagram of a proximity printing device when the dimensions of the source are so small that it can be considered a point source. In this case, a condenser lens, at whose focal point the source is located, forms a highly collimated beam of coherent light. When a mask is illuminated by coherent radiation, diffraction maxima in the light distribution in the plane of the wafer being exposed are seen in the image of a transparent window (Fig. 5-1b). The degree of coherence in the light on the mask can be reduced by increasing the size of the glow source, which is equivalent to increasing the angular spectrum of the radiation that illuminates the stencil (Fig. 5-1c). It follows from the theory of coherence in radiation that the width of the angular radiation spectrum and the extent of its coherence are mutually induced: the larger the angular spectrum, the lower the degree of coherence in the light. As the coherence of the light is reduced, diffraction maxima (fluctuations) in the light in the image of a slit in the mask are reduced, since the various angular components of the light field produce fluctuations at different points; the fluctuations are smoothed when the intensities are added. Degradation in the edge of the slit image due to the formation of a half-shadow occurs when the angular spectrum is expanded. The width of the half-shadow can easily be estimated by the quantity $\Delta x \simeq \theta h$, where h is the distance between the mask and the wafer being exposed (Fig. 5-1d). When competing factors in the degradation of a mask image are present, one of which increases with increasing source size and the other increases with decreasing source size, an optimal source size (or optimal width θ_{opt} of the angular spectrum) must exist for which the quality of the mask edge image is least degraded. According to [6], the optimal value of the angular light spectrum can be found from the empirical formula

$$\theta_{opt} = 0.2\sqrt{\lambda/h}(1 + 2h\lambda/a_{min}^2) \tag{5-1}$$

where λ is the wavelength of the light, and a_{min} is the width of the smallest element in the mask pattern (in the shadow method of image transfer for a mask and image ratio of 1:1). The value of θ_{opt} is related to the source size by $l_{opt} = L \cdot \theta_{opt}$, where L is the distance between the source and the condenser lens L_c. We see from Eq. (5-1) that when $h = 10\ \mu$m and $a_{min} = 2\ \mu$m, $\theta_{opt} = 5°$. This means that a glow source located $L = 200$ mm from the condenser lens must have a dimension of $l_{opt} = 17$ mm.

Light having various degrees of coherence can also be used in projection systems[7] to illuminate a mask. Figure 5-2 is a diagram for a projection system that uses a point radiation system. The point source is projected to the center of the objective lens L_p by the condenser lens. The mask M is located near the condenser lens. All of the mask elements are exposed to coherent light and diffraction coherent beams are formed at each of these elements. Low-order $(0, \pm 1)$ diffraction beams enter the aperture of the objective lens. By blending together in the image plane of the beams diffracted at the point P form an image P' of this point. When the subject is exposed to coherent light the amplitudes of the beams add together (interference) and when the light is incoherent, the intensities add.

Increasing the size of an incoherent source to a finite value of $2l$ lowers the degree of coherence in the light on the mask elements. The angular spectrum of the light wave is proportional to the linear dimension of the source. The measure of the coherence of the radiation from a source (Fig. 5-2b) is understood to be the amount of filling σ of the input aperture of the objective lens (the object), which is equal to the ratio of the diameter of the source image D_s to the diameter of the input aperture of the objective lens D_p: $\sigma = D_s/D_p$. It is apparent that as the size of the source $2l$ increases, the amount of filling increases such that larger values σ correspond to less coherent light. For sources that are perfectly coherent, $\sigma = 0$. Indeed, the image of a point source also has a point dimension (in a paraxial approximation of geometric optics). At some finite source dimension $2l_{max}$ the entrance aperture of the source is completely filled and further increasing the size of the source has no effect on image quality, since the side rays emitted by points on the source that are beyond $2l_{max}$ do not impinge on the entrance aperture of the objective lens. Nevertheless, it is assumed that $\sigma = \infty$ for perfectly incoherent sources.

Figure 5-2c shows the beam of rays that are diffracted at a point P and the image P' that they form in the plane of the mask image. The different points of an incoherent source transmit rays that are incoherent among themselves to the point P. The incoherent rays diffract independently of one another and form an image at point P'. The intensity of the image is equal to the sum of the intensities of each diffracted ray. This situation is realized in the extreme case of perfect incoherence, when the filling of the aperture is $\sigma = \infty$. For finite values σ the light (the source) can be thought of as being partially coherent.

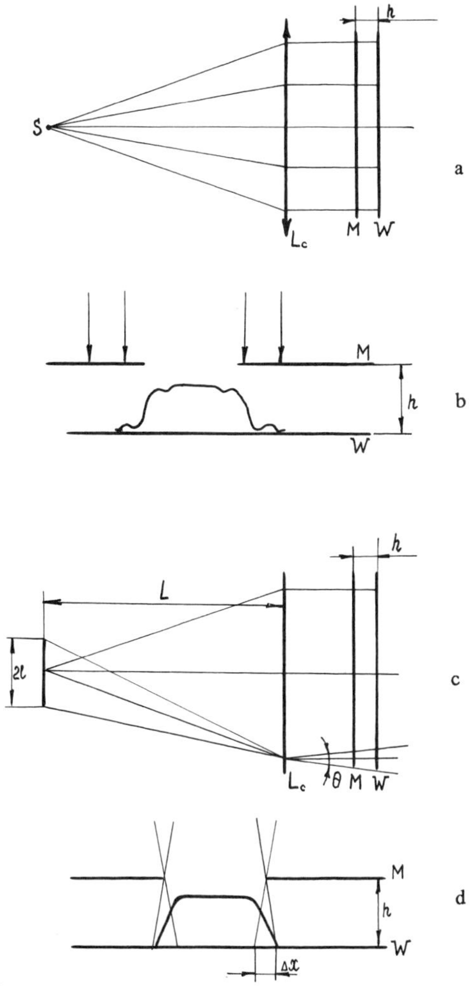

FIGURE 5-1. A proximity printing scheme: a) the source is point size; c) the source is finite size; L_c is the condenser lens, M is the mask, W is the wafer being exposed, and h is the distance between the wafer and the mask. The distribution of light in the image of a slit in the mask is shown in b); the creation of a half-shadow in a system using a finite-size source is shown in d).

Laser sources are characterized by the fact that their radiation is highly directed (the angular divergence is less than one degree), and by their brightness and the power of their radiation. Because laser radiation is so highly directed it can be focused into a spot having a diameter approximately equal to the wavelength of the radiation. Because such a spot can be thought of as a virtual source, a laser can be considered equivalent to a point source and we can repeat what was said earlier about the coherence of the light on the mask from point sources in shadow and projection lithography systems with respect to lasers. It is however, impossible to completely identify laser sources with point sources. The fact is that the radiation energy distribution in lasers is highly nonuniform over the cross section of the beam. This distribution is a function of the resonator geometry and the stimulation mode and, consequently, of the mode content of the radiation. For example, the intensity distribution in a TEM_{00} mode is Gaussian over the cross section; in multimode radiation the cross-sectional energy distribution may

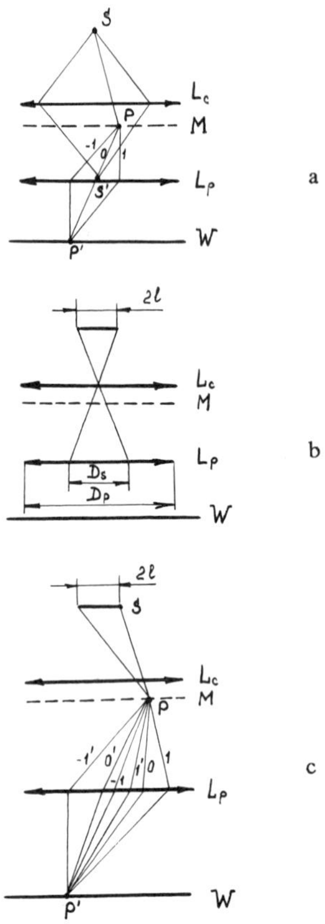

FIGURE 5-2. An optical projection exposure system. L_c is the condenser lens, L_p is the projection lens, M is the mask, and W is the wafer being exposed. Detailed explanations can be found in the text.

be extremely nonuniform and diverse. The nonuniform distribution of energy over the beam's cross section is expressed, obviously, in the nonuniform angular distribution of the radiation from a point source, which is equivalent to a laser. Topics in the photometry of partially coherent (laser) sources have been thoroughly examined in [8–10].

The nonuniform distribution of the energy in laser radiation over the cross section of the beam requires that special illuminating systems be used. These systems are designed to equalize (homogenize) the radiation intensities and reduce the degree of spatial (transverse) and temporal (lengthwise) coherence in the radiation. In these systems the image from a virtual source created in the input aperture of the objective lens must guarantee a certain (optimal) amount of filling in the entrance pupil of the objective lens. Experience has shown that an amount of entrance pupil filling $\sigma = 0.7$ may be considered optimal.(7) Thus, we see that illuminating systems must create a standard virtual source of light, regardless of the difference in the sources at the entrance of the optical system. Standard filling of the aperture of an objective lens can be done by scanning with a laser beam,(7) using raster of lenses,(11) using the so-called Brewster tunnel,(1) etc.

We will now turn to the question of what parameters can be used to describe a mask and its image. It is obvious that the relationship between the parameters of a mask and its image is governed by the optical system (including the light source). Thus, the need to also include parameters that describe the optical system arises.

Ordinarily, amplitude masks (alternative phase masks) are studied. These masks are characterized by a two-dimensional intensity transmission function $T(x, y)$ such that the intensity of radiation propagating through the mask is $I(x, y) = I_0 T(x, y)$, where I_0 is the intensity of the light on the mask and is said to be uniform over the entire mask. Masks designed for IC fabrication usually have a number of transparent rectangular windows placed on an opaque background. The boundaries between the transparent and opaque regions and the angular elements are the simplest elements in this kind of configuration. If we assume that the sides of the rectangles are parallel to the x and y axes of a rectangular coordinate system (this condition is not a substantial limitation), we can expand the function $I(x, y)$ in a two-dimensional Fourier spectrum:

$$I(x, y) = \iint J(f_x, f_y) \exp[-2\pi i(f_x x + f_y y)]\, df_x\, df_y \tag{5-2}$$

$$J(f_x, f_y) = (2\pi)^{-2} \iint I(x, y) \exp[2\pi i(f_x x + f_y y)]\, dx\, dy \tag{5-3}$$

If the boundaries between the transparent and opaque regions are sharp (in mathematical terms, a step function), there will be components $J(f_x, f_y)$ in the spectrum of the mask with spatial frequencies f_x, f_y that are as large as we like.

Although the functions $I(x, y)$ and $J(f_x, f_y)$ completely describe a mask, they unfortunately relate to only a specific mask. It is therefore worthwhile to introduce more general characteristics (parameters) for a mask. The first of these, contrast, is the ratio of the intensities of the radiation that passes through the transparent and opaque regions of the mask, respectively:

$$C = I_{\max}/I_{\min}, \qquad 1 < C < \infty \tag{5-4}$$

The ratio

$$M = \frac{I_{\max} - I_{\min}}{I_{\max} + I_{\min}}, \qquad 0 < M \leq 1 \tag{5-5}$$

is called mask modulation. There is a relationship between constrast C and modulation M:

$$C = \frac{1 + M}{1 - M} \tag{5-6}$$

In a completely analogous fashion, we can define (and measure) the contrast and modulation in the image of a mask, created by an optical system on the surface of a photosensitive layer. Then the ratio of the modulations of a mask and its image,

$$\mathrm{MTF} = M_{\mathrm{image}}/M_{\mathrm{mask}} \tag{5-7}$$

describes the optical system. This ratio is called the modulation transfer function (MTF).

All of the definitions given by Eqs. (5-4)–(5-7) can be repeated in terms of the amplitudes, rather than the intensities of the light field; however, it is the intensities of the light field that are measured by sensors.

We will now assume that sinusoidal reticles having different spatial frequencies $f_x = 0\text{–}f_{\max}$ have been chosen as the masks. The MTF of Eq. (5-7) can be found by measurements at every frequency f_x and, consequently, the frequency dependence

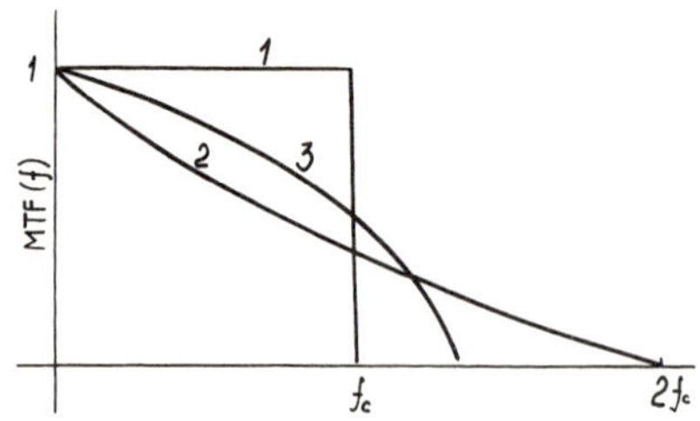

FIGURE 5-3. The frequency dependence of the modulation transfer function for a projection optical system: the mask is illuminated by 1) coherent light; 2) incoherent light; 3) partially coherent light.

MTF(f_x) can be found. How the x axis is oriented in a plane perpendicular to the axis of the system is inconsequential in axially symmetric optical systems.

As we can see, the MTF(f) is the weighting function with which the various spatial intensity components of a field passing through a mask are propagated further by an optical system. When the light is incoherent the phases of the different spatial frequency components are uncorrelated and the intensity in the image is found by adding MTF(f)-weighted intensities transmitted through the mask. Thus, when a mask is irradiated by incoherent light, knowing the MTF(f) function of the optical system makes it possible to reconstruct an image, provided the Fourier expansion of the mask transmission function is also known. When a mask is irradiated by coherent light the phases of the different spatial components are correlated and the phases of the spatial and frequency components, as well as function MTF(f) must be known in order to reconstruct the image. The phases of the Fourier components of the field can be found from theoretical calculations, starting from the phase relations in the source and the differences in the optical paths of the optical system. Aberrations in the optical system contribute to the phase of a wave propagating through the system. For this reason, image quality in systems using coherent light is extremely sensitive to the magnitude of the aberrations in the system. Optical noise caused by interference between the waves that create an image in the image plane and waves scattered from specks of dust, nonuniformities in the optical system, on lens surfaces and mountings, etc., becomes significant in a coherent system.

The modulation transfer function for aberration-free, coherent and incoherent optical systems is shown schematically in Fig. (5-3). For coherent light MTF(f) is constant and equal to one, right up to the limiting frequency $f_c = 1/\lambda F$, where λ is the wavelength of the light and F is the effective f number, equal to the distance between the plane of the objective lens and the image divided by the size of its aperture. As the frequency in an incoherent optical system increases, the MTF(f) function decays monotonically and becomes zero at a frequency that is twice the cutoff frequency of a coherent system. The MTF(f) function for a partially coherent system is somewhere between these two extremes. The MTF(f) in systems having large aberrations is lower than the MTF(f) of an incoherent system and is often nonmonotonic, depending on what kind of aberrations are dominant.

Methods for finding the MTF(f) for optical systems by measuring the parameters of the resist masks that reproduce the image of test masks have been developed.[12] Judging by the nature of the MTF(f) function, the proximity printing machines studied in [12] can be classified as coherent light systems. For projection systems the MTF(f) function corresponds to a system with partial coherence in one case and to a system with significant aberrations (exceeding the Rayleigh limit on the difference in optical paths, equal to a quarter wavelength, by a factor of two[12]) in another case.

Projection photolithography systems (steppers) are characterized by three parameters that directly depend on the properties of its optical system[1]: 1) the minimal linewidth l that can be reproduced in the mask image:

$$l = k_1 \lambda/(\mathrm{NA}) \tag{5-8}$$

2) the allowable changes in focusing for which the minimal line width in the image can be maintained:

$$\pm d = \pm k_2 \lambda/(\mathrm{NA})^2 \tag{5-9}$$

(the amount of defocusing that is allowed is often called the depth of focus), and 3) the size of the image field, η, within which the minimal linewidth l can be reproduced:

$$\eta = \frac{H}{k_3(\mathrm{NA})} \tag{5-10}$$

In these formulas λ is the wavelength of the radiation; NA = $N' \sin \mu'$ is the numerical aperture of the objective lens that reproduces an image of the mask on the substrate; N' is the coefficient of refraction of the medium in the image space, which for air is $N' = 1$; and μ' is the angular aperture of the rays that converge at an image point (on the substrate). The H factor is a constant for a Helmholtz–Lagrange optical system. The coefficients k_1, k_2, k_3 must be thought of as empirical parameters that are chosen for the purpose of reproducing a minimal line width l in the image.

The numerical factor k_1 in Eq. (5-8) is an implicit function of several auxiliary system parameters, e.g., the magnitude of the aberrations in the objective lens, the degree of coherence in the light, etc. It is essential that k_1 depend also on the properties of the photosensitive layer of the resist in which the image of a line is to be reproduced and on the resist layer processing modes. k_1 should be as small as possible. Reproducing a line having a width approximately equal to the radiation wavelength λ is possible if $k_1 \leq$ NA. An optimal choice of resists and their processing modes, the coherence of the light, the focusing, and the size of the image field allows us to reproduce a line that is as wide as the exposing radiation wavelength. Increasing the tolerances on the technological modes for exposing and processing the resist layers,the types of resists, etc. increases the minimal linewidth l that can be reproduced in an image and thereby increases the value of the k_1 coefficient in Eq. (5-8). According to [1], for a Lentar objective lens having a 5$\times$ reduction, the numerical aperture is NA = 0.35 when working at a wavelength of $\lambda = 404.7$ nm with both positive and negative resists, with 10% accuracy in reproducing lines whose width is given by Eq. (5-8) for a value of $k_1 = 0.75$, i.e., linewidths of $l \simeq 2\lambda$. As we can see, extending the range of exposure conditions and resist materials properties increases k_1 and the minimal linewidth l. Modelling of the exposure conditions shows that an aperture filling coefficient of $\sigma = 0.7$ is optimal for many cases. The use of three-layer resists or contrast enhancement systems makes a photolithography process having a resolution of $l \simeq \lambda$ possible.[1]

The k_2 coefficient in Eq. (5-9) for the depth of focus is also chosen on the basis of criteria on the magnitude of the distortions that can be allowed in an image. The amount of defocusing found when $k_2 = 0.25$ causes barely noticeable distortions in the image. This value k_2 may be used to estimate the amount of defocusing that can be allowed according to Eq. (5-9); and corresponds to distortions in a wavefront that produces an image not larger than $\lambda/8$.[1] However, because the depth of focus has square-law dependence on the numerical aperture, distortions of the wavefront increase rapidly during defocussing so that k_2 cannot be chosen much larger than the value of 0.25 presented earlier.

Equation (5-10) for the size of the image field is in fact a form of the Helmholz–Lagrange constant H. The Helmholtz–Lagrange theorem states that in conjugate planes of the constant, the magnitude is $H = l \cdot N' \cdot \sin\alpha$, where $2l$ is a linear dimension of an object or its image, α is the angular aperture for the convergence of rays into a point on these planes, and N' is the coefficient of refraction of the medium in the region in which a conjugate plane is found. If the final image of the mask is in air and $N' = 1$, the Helmholtz constant is $H = \eta_i \sin\alpha'_p$; here $\eta = \eta_i$ is the size of the mask image field M, and $\alpha'_p = D_p/2L$ is the angular aperture of the rays converging at a point in the image plane. With partially coherent light the aperture of the objective lens is only partially filled. The filling diameter is $D'_p = \sigma D_p$, where σ is a filling coefficient that gives the degree of coherence in the light. Obviously, the diameter of the filled portion of the objective lens diameter emerges as the effective aperture diameter and the Helmholtz constant is

$$H_M = M_i \sin\left(\frac{\sigma D}{2L}\right) \simeq M_i \sigma \sin\left(\frac{D_p}{2L}\right) = M_i \sigma(\mathrm{NA}) \tag{5-11a}$$

$$M_i \equiv \eta = H_M/\sigma \cdot \mathrm{NA} \tag{5-11b}$$

Thus, the k_3 in Eq. (5-10) has the sense of a filling coefficient σ: $k_3 \equiv \sigma$.

The analysis presented here of the problems of projection photolithography indicates the critical paths in the development of this method. The resolution that can be obtained with this method can be increased by using shorter wavelengths in the exposing radiation. The best sources in the ultraviolet region are excimer lasers. For this reason, the effect that coherent light has on image quality in projection photolithography systems is a genuine problem.

Information from the most diverse areas of geometric and physical optics is used in the study of photolithography systems and major difficulties arise when choosing and discussing theoretical material. Attempts to give a logical discussion of the problems invariably make it necessary to repeat material that can be found in a large number of monographs devoted to physical optics.[13–16] For this reason we can only outline a demonstrative discussion, referring the reader to the well-known monographs for the details.

It is a well-known fact that a successive theory of optical systems must be built upon Maxwell's equations. However, in using Maxwell's equations to solve optical problems we encounter such immense mathematical difficulties that we can obtain practically useful results for only the simplest of cases, such as, e.g., the diffraction of waves on conducting screens having the proper geometric shape (a half-plane, a circle). A simplified version of wave theory, the so-called theory of scalar (unpolarized) waves is the most productive and therefore, the most widely used in optical systems theory. The diffraction theory of images and the spectral analysis of optical systems, objects, and their images that is based on it are its most important accomplishments in the area we are interested in. We will briefly discuss the results from the scalar theory in this chapter, paying particular attention to the role of coherence in the light to creating an image of amplitude objects.

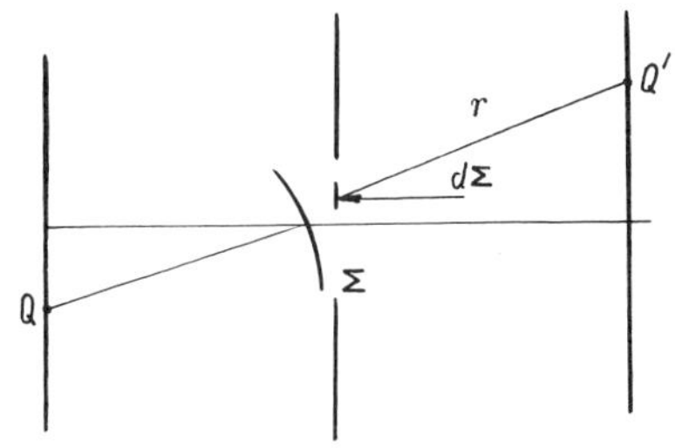

FIGURE 5-4. Image formation by a point source due to diffraction of a wave irradiated by the source at the aperture of the diaphragm: Q is a point source of waves; Q' is a point on the image plane whose illumination is defined by Huygens' principle.

The third area of optics—beam geometry optics—is also widely used in analyzing optical systems. In the most general form its results are contained in a matrix, ABCD-representation of optical systems.

5-2. FUNDAMENTALS OF THE DIFFRACTION THEORY OF IMAGE FORMATION

In the scalar theory a propagating wave, $\varphi(r,t) = u(r)\exp(i\omega t)$, is a solution of the wave equation

$$\nabla^2\varphi = \frac{1}{C^2}\frac{\partial^2\varphi}{\partial t^2} \tag{5-12}$$

and its amplitude $u(r)$ is a solution of the Helmholtz equation

$$\nabla^2 u + k^2 u = 0, \qquad k = 2\pi/\lambda \tag{5-13}$$

for some boundary conditions.

Figure 5-4 shows a model for an image forming system. An opaque screen with a diaphragm Σ separates a point source of scalar waves at the point Q and the point Q' at which we find the amplitude of a wave transmitted by a source at Q. The integral

$$u(Q') = -\frac{ik}{2\pi}\iint_\Sigma u\frac{e^{ikr}}{r}\,d\Sigma \tag{5-14}$$

is a solution of this problem[14] and is the mathematical form of Huygens' principle. According to Eq. (5-14), we need only know the amplitude of the field u at the diaphragm Σ in order to find the field at any point Q'. This is valid even if the field u at the diaphragm is produced by a system of points (a distributed source) since the wave equation, Eq. (5-13), is linear.

We denote the coordinates of the point Q' in the image plane as $(x, y, 0)$ and the coordinates of a differential area $d\Sigma$ in the diaphragm Σ as (u, v, z) (Fig. 5-5). Let P' be an ideal image of the point P. If the optical system were ideal, it would convert a spherical diverging front propagating from the point P into an image of the spherical front that converges at P'. In this situation, the amplitude of the wave converging at P' would be $\exp(-ikR)/R$. The optical system does, however, change the phase slightly so that the surface of the constant phase will differ from a spherical surface by a small

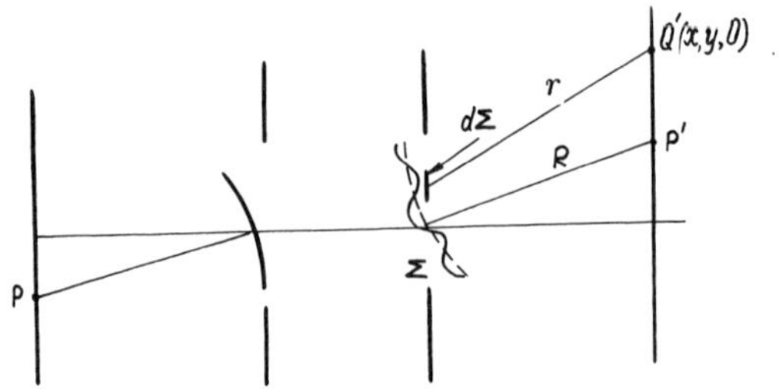

FIGURE 5-5. Image formation by a point source in an optical system with aberrations. The dotted line shows an ideal spherical front converging at P'; the solid line shows a wave front distorted by the aberrations in the optical system.

amount $L(u, v, z)$. Let the function $|F(u, v)|$ describe the amplitude of a wave at the diaphragm. Then,

$$u_\Sigma = |F(u, v)|e^{-ik(R+L)}/R = F(u, v)\frac{e^{ikR}}{R} \tag{5-15}$$

Substituting Eq. (5-15) into (5-14) gives us

$$u(Q') = \frac{-ik\exp(-kR)}{2\pi R}\iint_\Sigma F(u, v)\frac{e^{ikr}}{r}\,d\Sigma \tag{5-16}$$

$$r^2 = z^2 + (x - u)^2 + (y - v)^2 \tag{5-17}$$

When evaluating the integral over the aperture in Σ, the quantity $1/r$ will not change markedly if the aperture diameter D is much less than the distance between the planes of the aperture and the image: $D \ll z$. This allows us to say that $1/r \simeq 1/z$ and take the quantity $1/z$ outside the integral. Conversely, the phase factor $\exp(ikr)$ can, because k is so large ($k \simeq 10^5$ in the optical range), change significantly even when r changes by only a small amount. We expand r in a series

$$r = z\left[1 + \frac{x^2 + y^2}{2z} + \frac{u^2 + v^2}{2z} - \frac{ux + vy}{z} + \ldots\right] \tag{5-18}$$

Using Eq. (5-18), we find for the amplitude of $u(Q')$ that

$$u(Q') = \frac{-ik\exp(-ikR + ikz)}{2\pi Rz}\exp\left(ik\frac{x^2 + y^2}{2z}\right)$$
$$\times \iint F(u, v)\exp\left(ik\frac{u^2 + v^2}{2z}\right)\exp\left(-ik\frac{ux + vy}{z}\right)du\,dv \tag{5-19}$$

Equation (5-19) gives the amplitude of a wave at the point Q' for Fresnel diffraction at the diaphragm Σ.

If the distance z is large enough that the phase factor $\exp[ik(u^2 + v^2)/2z]$ is roughly unity for all u and v on the diaphragm, i.e., $k(u^2 + v^2)/2z \ll 1$, we have Fraunhofer diffraction at the diaphragm. Then

$$u(Q') = A \iint_{-\infty}^{+\infty} F(u,v) \exp\left(-ik\,\frac{ux+vy}{z}\right) du\,dv \tag{5-20}$$

Here, the function $F(u,v)$ is defined on the entire u,v plane in such a way that $F(u,v) \equiv 0$ outside the aperture on the Σ diaphragm.

We now introduce the spatial frequencies $\beta = uk/z$, $\gamma = kv/z$ as variables in the u, v plane; then

$$u(x,y) = A' \iint_{-\infty}^{+\infty} F(\beta,\gamma) \exp(-i(\beta x + \gamma y))\, d\beta\, d\gamma \tag{5-21}$$

Thus, the amplitude of a field produced by a point source after Fraunhofer diffraction at the aperture of a diaphragm is a two-dimensional Fourier expansion of the field amplitude at the aperture. If the aperture is a circle of radius $l/2$ we find that

$$u(r) = \exp(ikz) \exp\left(i\,\frac{kr^2}{2z}\right) \frac{kl^2}{8iz} \left[2\,\frac{J_1(klr/2z)}{(klr/2z)}\right] \tag{5-22}$$

in polar coordinates.[15] The intensity distribution in the image of the point is given by an Airy function

$$I(r) = |u(r)|^2 = \left[2\,\frac{J_1(klr/2z)}{klr/2z}\right]^2 \tag{5-23}$$

that describes the alternation of the light and dark rings; the radius of the first dark ring that borders a light circle at the center (the amount of diffraction blurring in the image of the point) is

$$\Delta r_1 = 1.22\,\frac{\lambda \cdot z}{l} \tag{5-24}$$

When the diameter l is small and the distances z are large, the image's diffraction radius may be as large as we like.

5-3. THE DIFFRACTION THEORY OF IMAGES PRODUCED BY A THIN LENS

It is not hard to show that a plane monochromatic wave, in passing through a thin lens parallel to its axis, undergoes phase conversion[15]:

$$U_l'(x,y) = \exp[ikn\Delta_0] \exp\left[-i\,\frac{k}{2f}(x^2+y^2)\right]$$
$$f^{-1} = (n-1)(1/R_1 - 1/R_2) \tag{5-25}$$

where R_1, R_2 are the radii of curvature of the lens surfaces, n is the coefficient of refraction, Δ_0 is its thickness, f is the focal length of the lens, x, y are the coordinates of a point in the plane of the lens through which a ray passes, and the coordinates origin is at the center of the lens. We will examine the process by which a wave produces an image of a subject through a lens, according to Fig. 5-6. We denote the distance from the lens to the object and to its image, respectively, as d_0 and d_i. The variables x, y are the coordinates of a point in the plane of the lens and are limited by the area of the lens aperture in Σ:

$$P(x,y) = \begin{cases} 1, & x,y \in \Sigma \\ 0, & x,y \notin \Sigma \end{cases} \tag{5-26}$$

We will formulate the problem of finding the amplitude of a wave in the image of a point source located at a point x_0, y_0 in the object plane. A spherical wave whose amplitude in a paraxial approximation is

$$U_l(x,y) = \frac{1}{i\lambda d_0} \exp\left\{ ik \frac{(x-x_0)^2 + (y-y_0)^2}{2d_0} \right\} \tag{5-27}$$

impinges on the lens from this point. A wave passing through a lens has an amplitude of

$$U_l'(x,y) = U_l(x,y)P(x,y)\exp\left[-i\frac{k}{2f}(x^2+y^2)\right] \tag{5-28}$$

The wave of Eq. (5-28) undergoes Fresnel diffraction at the lens aperture and as a result, produces the total amplitude

$$h(x_i y_i, x_0, y_0) = \frac{1}{i\lambda d_i} \int U_l'(x,y) \exp\left\{ ik \frac{(x-x_i)^2 + (y-y_i)^2}{2d_i} \right\} dx\, dy \tag{5-29}$$

at this point x_i, y_i. Combining the formulas obtained, we find the distribution function for the amplitude in the image of a point source:

$$\begin{aligned} h(x_i, y_i, x_0, y_0) = {} & \frac{\exp(i\Phi)}{\lambda^2 d_i d_0} \iint_{-\infty}^{+\infty} P(x,y) \exp\left[\frac{ik}{2}\left(\frac{1}{d_0} + \frac{1}{d_i} - \frac{1}{f}\right)(x^2+y^2)\right] \\ & \times \exp\left\{ -ik\left[\left(\frac{x_0}{d_0} + \frac{x_i}{d_i}\right)x + \left(\frac{y_0}{d_0} + \frac{y_i}{d_i}\right)y\right]\right\} dx\, dy \end{aligned} \tag{5-30}$$

$$\Phi = ik\left[\frac{x_i^2 + y_i^2}{2d_i} + \frac{x_0^2 + y_0^2}{2d_0}\right]$$

We will say that the planes of the subject and the image are conjugate, i.e., related by the lens equation $d_0^{-1} + d_i^{-1} - f^{-1} = 0$; also, we will introduce the image magnification factor $M = d_i/d_0$. Then

$$\begin{aligned} h(x_i, y_i, x_0, y_0) = {} & \frac{\exp(i\Phi)}{\lambda^2 d_i d_0} \iint_{-\infty}^{+\infty} P(x,y) \\ & \times \exp\left\{ -\frac{ik}{d_i}[(x_i + Mx_0)x + (y_i + My_0)y] \right\} dx\, dy \end{aligned} \tag{5-31}$$

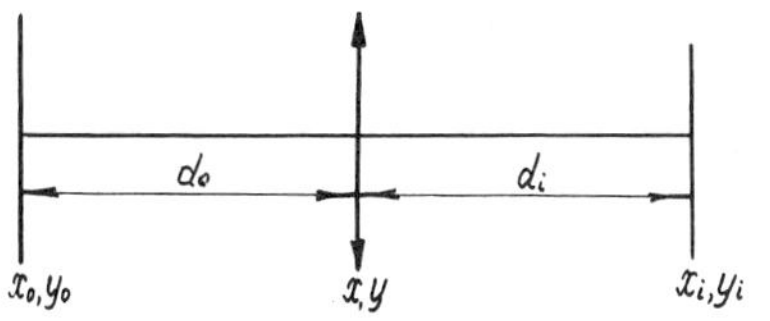

FIGURE 5-6. Designation of the coordinates of a point in the object plane (x_0, y_0), the image plane (x_i, y_i), and the lens aperture (x, y).

As might be expected from Huygens' principle, the distribution of amplitudes in the image of a point is due to Fraunhofer diffraction at the lens aperture. The influence of a lens' refracting surfaces is given by the magnification M of the coordinates of the central point of the image: $x_i = -Mx_0$, $y_i = -My_0$. It is interesting to convert to geometric optics by going to the limit $\lambda \to 0$. When $\lambda \to 0$, the phase factor in Eq. (5-31) oscillates very rapidly and the fundamental contribution to the integral comes from those values x and y that satisfy the conditions $(k/d_i)(x_i + Mx_0)x \ll 1$ and $(k/d_i)(y_i + My_0)y \ll 1$; therefore, the function $P(x, y)$ can be assumed equal to unity in the entire plane. Consequently, as an image of a δ-source (a point) we again obtain a δ-source with a $1/M$ magnification of the amplitude[(15)]:

$$h_g(x_i, y_i, x_0, y_0) = \frac{1}{M}\,\delta\left(\frac{x_i}{M} + x_0, \frac{y_i}{M} + y_0\right) \tag{5-32}$$

Accordingly, an object represented by a set of δ-sources distributed according to a $U_0(x_0, y_0)$ law is imaged, in an approximation of geometric optics, by the amplitude distribution

$$U_g(x_i, y_i) = \frac{1}{M}\,U_0\left(-\frac{x_i}{M}, -\frac{y_i}{M}\right) \tag{5-33}$$

which follows immediately from the superposition integral

$$U_i(x_i, y_i) = \iint_{-\infty}^{+\infty} h(x_i, y_i, x_0, y_0)U_0(x_0, y_0)\,dx_0\,dy_0 \tag{5-34}$$

when Eq. (5-32) is substituted.

Making the substitution of variables $\tilde{x} = x/\lambda d_i$, $\tilde{y} = y/\lambda d_i$ in Eq. (5-31) and designating $\tilde{x}_0 = -Mx_0$, $\tilde{y}_0 = -My_0$, we find

$$\begin{aligned} h(x_i, y_i, x_0, y_0) = \exp(i\Phi)M \iint_{-\infty}^{+\infty} & P(\lambda d_i\tilde{x}, \lambda d_i\tilde{y}) \\ & \times \exp\{-i2\pi[(x_i - \tilde{x}_0)\tilde{x} + (y_i - \tilde{y}_0)\tilde{y}]\}\,d\tilde{x}\,d\tilde{y} \equiv \tilde{h}M \end{aligned}$$

In this form, the scattering function h for a point is a spatially invariant quantity that depends only of the differences of the coordinates $x_i - \tilde{x}_0$, $y_i - \tilde{y}_0$. Using this function, the amplitude in the image of a subject $U_i(x_i, y_i)$ will be the convolution of the pulse response and the ideal (geometric) image of the subject[15]:

$$\begin{aligned} U_i(x_i, y_i) &= \iint_{-\infty}^{+\infty} h(x_i - \tilde{x}_0, y_i - \tilde{y}_0)U_0\left(-\frac{\tilde{x}_0}{M}, -\frac{\tilde{y}_0}{M}\right)\frac{1}{M}\,d\tilde{x}_0\,d\tilde{y}_0 \\ &\equiv \tilde{h}(x_i - \tilde{x}_0, y_i - \tilde{y}_0) * U_g(\tilde{x}_0, \tilde{y}_0) \end{aligned} \tag{5-35}$$

5-4. INTENSITY IN AN OPTICAL IMAGE: COHERENT AND INCOHERENT LIGHT

Because the response of photosensitive materials (detectors) during exposure is a function of light intensity, it is mainly intensity that is the measurable quantity. Having restored the temporal dependence of the wave amplitude in the image plane we can write

$$I_i(x_i, y_i) = \langle U_i(x_i, y_i, t) U_i^*(x_i, y_i, t)\rangle \tag{5-36}$$

for the light intensity in the image plane, where the angle brackets indicate time averaging. Substituting Eq. (5-35) into the formula for the light intensity in an image gives us

$$\begin{aligned} I_i(x_i, y_i) = \iint_{-\infty}^{+\infty} d\tilde{x}_0\, d\tilde{y}_0 \iint_{-\infty}^{+\infty} d\tilde{\tilde{x}}_0\, d\tilde{\tilde{y}}_0\, \tilde{h}(x_i - \tilde{x}_0, y_i - \tilde{y}_0) \\ \times \tilde{h}(x_i - \tilde{\tilde{x}}_0, y_i - \tilde{\tilde{y}}_0)\langle U_g(\tilde{x}_0, \tilde{y}_0, t) U_g^*(\tilde{\tilde{x}}_0, \tilde{\tilde{y}}_0, t)\rangle \end{aligned} \tag{5-37}$$

Obviously, the averaging operation can be done inside the integral. Now we have the mean value of the product of the amplitudes at two different points in the object plane. This value describes the correlation of the light at different points of the object taken at the same instant in time. When correlation exists in optics, we speak of coherence between the signals; in this case, spatial coherence. We can easily find final results from Eq. (5-37) for two extreme cases: perfectly coherent and perfectly incoherent light on the object. Without going to a limit, the general formula given by Eq. (5-37) corresponds to the case of partially coherent light on the subject.

For perfectly incoherent light we say that spatial correlation of the amplitudes corresponds to the so-called δ-correlation:

$$\langle U_g(\tilde{x}_0, \tilde{y}_0, t) U_g^*(\tilde{\tilde{x}}_0, \tilde{\tilde{y}}_0, t)\rangle = I_g(\tilde{x}_0, \tilde{y}_0)\delta(\tilde{x}_0 - \tilde{\tilde{x}}_0, \tilde{y}_0 - \tilde{\tilde{y}}_0) \tag{5-38}$$

Integrating Eq. (5-37) with the δ function of Eq. (5-38) gives us

$$I_i^{nc}(x_i, y_i) = \iint_{-\infty}^{+\infty} |\tilde{h}(x_i - \tilde{x}_0, y_i - \tilde{y}_0)|^2 I_g(\tilde{x}_0, \tilde{y}_0)\, d\tilde{x}_0\, d\tilde{y}_0 \tag{5-39}$$

Physically, Eq. (5-39) means the following: when a subject is exposed to incoherent light the optical system linearly transforms the intensity of the object when it produces the image of the subject; the square of the absolute value of the pulse response,

$$s(x_i - \tilde{x}_0, y_i - \tilde{y}_0) \equiv |\tilde{h}(x_i - \tilde{x}_0, y_i - \tilde{y}_0)|^2 \tag{5-40}$$

serves as the weighting function.

When an object is exposed to perfectly coherent light, we say that the correlation between the amplitudes at two points is such that a constant difference in the phases of the amplitudes is maintained for as long as we wish to observe them, so that

$$\langle U_g(\tilde{x}_0, \tilde{y}_0, t) U_g^*(\tilde{\tilde{x}}_0, \tilde{\tilde{y}}_0, t)\rangle = U_g(\tilde{x}_0, \tilde{y}_0) U_g^*(\tilde{\tilde{x}}_0, \tilde{\tilde{y}}_0) \tag{5-41}$$

Here, $U_g(\tilde{x}_0, \tilde{y}_0)$ and $U_g(\tilde{\tilde{x}}_0, \tilde{\tilde{y}}_0)$ are complex variables of the amplitudes at the respective points of an object being illuminated and are time independent. Substituting the result obtained into Eq. (5-37) for the intensity gives us

$$I_i^c(x_i, y_i) = \left| \iint_{-\infty}^{+\infty} \tilde{h}(x_i - \tilde{x}_0, y_i - \tilde{y}_0) U_g(\tilde{x}_0, \tilde{y}_0)\, d\tilde{x}_0\, d\tilde{y}_0 \right|^2 \tag{5-42}$$

Thus, with coherent light the intensity in an image is described by the square of the absolute value of the linear transformation of the signal amplitude given by Eq. (5-38).

In the diffraction theory of images produced by optical systems, cases of perfectly coherent and perfectly incoherent light on an object are given as extremes of the more usual case of partially coherent light, since in the latter case the transformation turns out to be not linear, but the square of the signal amplitude. We will discuss the theory of images when the light is partially coherent later; for now we will confine out attention to the spatial theory of images in the two extreme cases of incoherent and coherent light.

5-5. THE SPECTRAL REPRESENTATION OF LINEAR OPTICAL CONVERSIONS FOR COHERENT AND INCOHERENT LIGHT

The linearity of the integral conversions relative to amplitude (coherent light) or intensity (incoherent light) allows us to find linear algebraic relationships between the spectral components of the respective quantities and to thereby introduce an extremely elegant means of characterizing optical systems.

We will introduce spectral components for the amplitudes that take part in a linear integral conversion given by Eq. (5-35):

$$G_i(\mathbf{f}) = \iint_{-\infty}^{+\infty} U_i(\mathbf{r}_i) \exp[-i2\pi(\mathbf{r}_i\mathbf{f})]\, d^2\mathbf{r}_i \tag{5-43}$$

$$G_g(\mathbf{f}) = \iint_{-\infty}^{+\infty} U_g(\tilde{\mathbf{r}}_0) \exp[-i2\pi(\tilde{\mathbf{r}}_0\mathbf{f})]\, d^2\tilde{\mathbf{r}}_0 \tag{5-44}$$

$$H(\mathbf{f}) = \iint_{-\infty}^{+\infty} \tilde{h}(\mathbf{r}_i) \exp[-i2\pi(\mathbf{r}_i\mathbf{f})]\, d^2\mathbf{r}_i \tag{5-45}$$

Here and hereafter, $\mathbf{f}$, $\mathbf{r}_i$, and $\tilde{\mathbf{r}}_0$ stand for $\mathbf{f}(f_x, f_y)$, $\mathbf{r}_i(x_i, y_i)$, $\tilde{\mathbf{r}}_0(\tilde{x}_0, \tilde{y}_0)$.

According to the theorem of the Fourier transform of a convolution, there is a linear algebraic relationship between the spectral components:

$$G(\mathbf{f}) = H(\mathbf{f}) G_g(\mathbf{f}) \tag{5-46}$$

The ratio $H(f_x, f_y) = G_i(f_x, f_y)/G_g(f_x, f_y)$ may be called a coherent transfer function that indicates the changes in amplitude and phase (distortions) that every spatial-frequency component of light amplitude to be transmitted by the optical system experiences. In this fashion, $H(\mathbf{f})$ emerges as the exhaustive characteristic of an optical system for producing an image in coherent light. Because the pulse response of an aberration-free, diffraction-limited systems is the Fourier transform of the pupil function $P(\lambda d_i \tilde{x}, \lambda d_i \tilde{y})$, we find by the theorem of double sequential Fourier transforms that[15]

$$H(\mathbf{f}) = \mathcal{F}\{\mathcal{F}\{P(\lambda d_i \tilde{r})\}\} = P(-\lambda d_i \mathbf{f}) \tag{5-47}$$

The function for a circular pupil of radius $l/2$ is equal to unity in a circle of radius $l/2$, i.e., in the range of frequencies given by

$$(\lambda d_i)^2 f^2 \leq (l/2)^2$$

It follows from this that the coherent transfer function $H(f)$ is equal to unity for all frequencies below the cutoff frequency f_c:

$$f_c = \frac{l}{2\lambda d_i} \equiv \frac{\mathrm{NA}}{\lambda} \tag{5-48}$$

if we introduce the numerical aperture of the entrance pupil of the system as NA = $(l/2d_i)$. For a square aperture $l/2$ on a side, the transfer function is equal to unity in a square bounded by the lines $f_x = \pm l/2\lambda d_i$, and $f_y = \pm l/2\lambda d_i$.[15]

We now compute the normalized spectral components of those quantities that participate in a linear integral conversion of the intensity, Eq. (5-39), when creating the image of an object illuminated by incoherent light:

$$S_i(\mathbf{f}) = \frac{\iint_{-\infty}^{+\infty} I_i(\mathbf{r}_i) \exp[-2\pi i(\mathbf{f}\mathbf{r}_i)]\, d^2\mathbf{r}_i}{\iint_{-\infty}^{+\infty} I_i(\mathbf{r}_i)\, d^2\mathbf{r}_i} \tag{5-49}$$

$$S_g(\mathbf{f}) = \frac{\iint_{-\infty}^{+\infty} I_g(\tilde{\mathbf{r}}_0) \exp[-2\pi i(\mathbf{f}\tilde{\mathbf{r}}_0)]\, d^2\tilde{\mathbf{r}}_0}{\iint_{-\infty}^{+\infty} I_g(\tilde{\mathbf{r}}_0)\, d^2\tilde{\mathbf{r}}_0} \tag{5-50}$$

$$S(\mathbf{f}) = \frac{\iint_{-\infty}^{+\infty} |\tilde{h}(\mathbf{r}_i)|^2 \exp[-2\pi i(\mathbf{f}\mathbf{r}_i)]\, d^2\mathbf{r}_i}{\iint_{-\infty}^{+\infty} |\tilde{h}(\mathbf{r}_i)|^2\, d^2\mathbf{r}_i} \tag{5-51}$$

According to a theorem regarding the Fourier transform of a convolution between the spatial-frequency components in an object and its image, we have a linear algebraic relationship

$$S_i(\mathbf{f}) = S(\mathbf{f}) \cdot S_g(\mathbf{f}) \tag{5-52}$$

The function

$$S(\mathbf{f}) = S_i(\mathbf{f})/S_g(\mathbf{f}) \tag{5-53}$$

is called the optical transfer function; obviously, it has the sense of an incoherent transfer function for an optical system. The absolute value of this transfer function is called the modulation transfer function (MTF), which indicates how much lower the transmittance of the system is at the spatial frequencies $\mathbf{f}(f_x, f_y)$ in comparison with its transmittance at the frequency $\mathbf{f} = 0$.

We can show that the incoherent transfer function $S(\mathbf{f})$ can be stated in terms of the coherent transfer function $H(\mathbf{f})$. If we write Eq. (5-51) as

$$S(\mathbf{f}) = \mathcal{F}(|h|^2)_{\mathbf{f}} / \mathcal{F}(|h|^2)_{\mathbf{f}=0} \tag{5-54}$$

and employ a theorem on the representation of the Fourier transform of the square of the function $g(\mathbf{r})$ in terms of the convolution of its Fourier components $G(\boldsymbol{\rho})$:

$$\mathcal{F}\{|g(\mathbf{r})|^2\} = \iint_{-\infty}^{+\infty} G(\boldsymbol{\rho})G^*(\boldsymbol{\rho}+\mathbf{f})\, d^2\boldsymbol{\rho}$$

we find from Eq. (5-54) that[15]

$$S(\mathbf{f}) = \frac{\iint_{-\infty}^{+\infty} H(\boldsymbol{\rho}-\mathbf{f}/2)H^*(\boldsymbol{\rho}+\mathbf{f}/2)\, d^2\boldsymbol{\rho}}{\iint_{-\infty}^{+\infty} |H(\boldsymbol{\rho})|^2\, d^2\boldsymbol{\rho}} \tag{5-55}$$

Using the frequency function for the pupil, $H(\boldsymbol{\rho}) = P(\lambda d_i \boldsymbol{\rho})$, as the coherent transfer function (the last equation is valid for an aberration-free, diffraction-limited system) we can write

$$S(f) = \frac{\iint_{-\infty}^{+\infty} P(\lambda d_i \boldsymbol{\rho} - \lambda d_i \mathbf{f}/2)P(\lambda d_i \boldsymbol{\rho} + \lambda d_i \mathbf{f}/2)\, d^2\boldsymbol{\rho}}{\iint_{-\infty}^{+\infty} |P(\lambda d_i \boldsymbol{\rho})|^2\, d^2\boldsymbol{\rho}} \tag{5-56}$$

In geometric terms, the value of the ratio in Eq. (5-56) is equal to the overlap area of two pupil functions displaced by vectors $\pm\lambda d_i \mathbf{f}/2$ relative to the center of the frequency plane to the total area of the pupil. A straightforward calculation yields the following result for a circular pupil[15]:

$$S(f) = \frac{2}{\pi}\left[\arccos\left(\frac{f}{2f_c}\right) - \frac{f}{2f_c}\sqrt{1-(f/2f_c)^2}\right], \qquad f \le 2f_c \tag{5-57}$$

It is apparent that the optical (incoherent) transfer function goes to zero at twice the frequency of a coherent transfer function, although it is less than unity for all spatial-frequency components in the range $0 < f \le 2f_c$. A similar result has been obtained for a square pupil: its optical transfer function is a pyramid having an altitude of unity and a square base $2f_c = l/\lambda d_i$ on a side.[15]

Since radiation detectors react to intensity, coherent and incoherent light systems for producing an image must be compared relative to the spatial frequency spectrum of the intensity in the image. The Fourier transforms of the convolutions that give the intensity in an image with incoherent and coherent light, $I_i^{nc} = |\tilde{h}|^2 * |U_g|^2$ and $I_i^c = |\tilde{h} * U_g|^2$, can be written

$$\mathcal{F}\{I_i^{nc}\} = [H * H][G_g * G_g] \tag{5-58}$$

$$\mathcal{F}\{I_i^{c}\} = [H * G_g][H * G_g] \tag{5-59}$$

where the asterisk is used to denote calculation of the autocorrelation function:

$$H * H \equiv \int H(\boldsymbol{\rho})H^*(\boldsymbol{\rho}+\mathbf{f})\, d^2\boldsymbol{\rho}$$

Equations (5-58) and (5-59) are used in comparing coherent and incoherent light systems.

In masks designed for integrated circuit fabrication the most typical element is the edge of the plane (the boundaries of the larger figures emerge as these elements).

Regarding light or dark bands, as well as rectangular windows in the opaque region, we must speak of those cases when their linear dimensions are close to the value of the smallest linear dimension l, Eq. (5-8), and the edges of these "complex" elements affect one another by changing the distribution of the light. The concept of "image transfer quality" of the elements must include indexes such as the intensity gradient at a boundary, the coincidence of a boundary defined at some intensity level $\varkappa I_{i\,\max}$, $\varkappa < 1$, with the boundary of an ideal image, the change in these indexes with defocusing, etc. Noise (speckles) in the intensity distribution that are caused by interference between the amplitudes of a coherent wave in an image with waves scattered in the system are highly significant. The linear dimensions of interference spots (speckles) are roughly equal to the maximum resolution in the system, $l_{\min} = \lambda/\mathrm{NA}$,[17] so that the images of the smallest figures deteriorate first because of optical noise in systems with coherent light. Reduction in the degree of coherence in the light is accompanied by a reduction in the optical noise, so that light with a given degree of coherence (partially coherent light) turns out to be optimal.

5-6. THE EFFECT OF ABERRATIONS IN THE OPTICAL SYSTEM ON IMAGE QUALITY

Design and the fabrication of objective lenses for projection photolithography has been developed to the extent that these objective lenses are considered to be diffraction-limited (aberration-free). Because it is impossible to obtain a system that is completely free of aberrations, systems having a low level of aberrations relative to some criterion are called aberration-free systems. For an optical system with aberrations, the wavefront that emerges from the system pupil at the point x, y of the pupil has a phase error of $kL(x, y)$, where $L(x, y)$ is the effective error in the optical path length of a ray that passes through that point. Taking aberration phase errors into account, the pupil function looks like

$$P_a(x, y) = P(x, y) \exp[ikL(x, y)] \tag{5-60}$$

According to Marechal's criterion, optical systems for which the mean squared value of L is not greater than $\lambda/14$ are diffraction-limited (aberration-free) systems. Strehl's criterion is based on a comparison of the maximum value in the intensity distribution in the image of a point source: if $I_{\max} \geq 0.8 I_{\max}^{(na)}$ ($I_{\max}^{(na)}$ is the intensity maximum for a perfectly aberration-free system), the objective lens can be classified as aberration-free. Finally, Hopkins' criterion considers the effect of aberrations on the optical transfer function: if at a frequency $f = (1/2) f_{\max}$ the reduction in the optical transfer function due to aberrations is not more than 20% the system can be considered aberration-free.[13,16]

By applying, according to Eq. (5-47), two successive Fourier transforms to Eq. (5-60) we find a coherent transfer function when aberrations are present:

$$H_a(\mathbf{f}) = P(\lambda d_i \mathbf{f}) \exp(ikL(\lambda d_i \mathbf{f})), \qquad \mathbf{f} = f_x, f_y \tag{5-61}$$

Thus, aberrations cause phase shifts in the transfer function. The effect of aberrations on image quality can be taken into account according to Eqs. (5-58) and (5-59) if a specific function $L(\mathbf{f})$ and spectrum $G_g(\mathbf{f})$ can be found for the object.

In incoherent light systems, an optical transfer function (OTF) that takes aberrations into account can be calculated according to Eq. (5-56), but with the function given by Eq. (5-61):

$$s(\mathbf{f}) = \frac{\iint_{\Sigma(f)} \exp\left\{ik\left[L\left(\lambda d_i\boldsymbol{\rho} - \tfrac{1}{2}\lambda d_i\mathbf{f}\right) - L\left(\lambda d_i\boldsymbol{\rho} + \tfrac{1}{2}\lambda d_i\mathbf{f}\right)\right]\right\} d^2\rho}{\iint_{\Sigma(0)} d^2\rho} \tag{5-62}$$

The ranges in which the pupil functions overlap for their indicated shift vectors in the frequency plane are denoted through $\Sigma(\mathbf{f})$, $\Sigma(0)$. Because the range of integration in the numerator of Eq. (5-62) remains constant (in comparison with the aberration-free case), the appearance of a phase function inside the integral denotes a reduced value of the OTF at the respective frequency. It is not hard to calculate the OTF for the special case when

$$L(x, y) = \frac{1}{2}\,\varepsilon(x^2 + y^2) \tag{5-63}$$

The function given by Eq. (5-63) arises when image defocusing occurs. In fact, if the observation plane is displaced relative to the conjugate image plane by an amount Δd_i, then, instead of the lens equation $1/d_0 + 1/d_i - 1/f = 0$, we will have

$$1/d_0 + 1/d_i - 1/f = \frac{\Delta d_i}{d_i^2} \equiv \varepsilon \tag{5-64}$$

When defocusing, described by the quantity $\varepsilon = \Delta d_i/d_i^2$, is present the optical difference in beam paths will, according to Eq. (5-30), be

$$L(x, y) = \frac{\varepsilon}{2}(x^2 + y^2) = \frac{\Delta d_i}{2d_i^2}(x^2 + y^2) \tag{5-65}$$

We will assume that the lens aperture is a square l on a side. The maximum difference in path length on the x and y axes is

$$L_{\max} = \frac{\Delta d_i}{d_i^2}\,\frac{l^2}{8} \tag{5-66}$$

We chose the condition for smallness in defocusing to be $L_{\max} \leq \lambda/8$. The condition that the displacement of the image observation plane be small with respect to the conjugate plane follows from this:

$$\Delta d_i \leq \frac{1}{4}\,\frac{\lambda}{(\mathrm{NA})^2} \tag{5-67}$$

because $((l/2)/d_i) = (\mathrm{NA})$. We have arrived, as we will see, at the result given by Eq. (5-9) cited at the beginning of this chapter for the depth of focus.

An explicit form of the OTF in the situation being discussed is[15]

$$S(f_x, f_y) = \Lambda(f_x/2f_0)\,\mathrm{sinc}\left[\frac{8L_m}{\lambda}\frac{f_x}{2f_0}\left(1-\frac{|f_x|}{2f_0}\right)\right]$$
$$\times\,\Lambda(f_y/2f_0)\,\mathrm{sinc}\left[\frac{8L_m}{\lambda}\frac{f_y}{2f_0}\left(1-\frac{|f_y|}{2f_0}\right)\right] \qquad (5\text{-}68)$$
$$\Lambda(x) = \begin{cases} 1-|x| & |x|\le 1 \\ 0 & |x|>1; \end{cases} \quad \mathrm{sinc}\,x = \frac{\sin\pi x}{\pi x}; \quad f_0 = (l/2)/d_i$$

Graphs of Eq. (5-68) for different $L_{\max}$ are shown in the figures in [15]. When $L_{\max} = \lambda/4$, the reduction in the transmission of the spatial frequency components of an object that is caused by aberrations is still relatively small. However, when $L_{\max} \geq \lambda/2$ the band of frequencies being transmitted is sharply reduced (by a factor of several times); in addition, contrast inversion takes place (minima are observed where maxima are expected and conversely), which is confirmed by the change of sign in the OTF. It must be noted that contrast inversion will actually be seen when regular structures having a period that corresponds to frequencies having a negative OTF are present. In regions without regular structures, the effect of contrast inversion for a band of spatial frequencies is manifested by the strong deterioration of contrast in the image of the object.

For optical systems with large aberrations $L_{\max} \geq \lambda/2$ the aberration blurring of the image of a point source becomes greater than the diffraction blurring $\Delta r_i \simeq 1.22\lambda/(\mathrm{NA})$ given by Eq. (5-24). In this case all of the system parameters can be found from geometric optics, i.e., it is unnecessary to apply the wave theory of images to systems with large aberrations. Specifically, the OTF of a system will be the Fourier transform of the intensity distribution in an aberration spot that images a point source.

It is interesting to study the spatial light distribution near the focal point of an optical system more thoroughly. In the focal plane of a system with a circular aperture the intensity in the image of a point is a system of concentric rings that can be described by Airy's function; the central light spot has a radius of $r_\perp = 0.61\lambda/\alpha$. If the observation plane is moved from the focal plane closer to or further away from the lens, the intensity distribution maintains the form of concentric rings. At a significant distance from the focal plane the intensity in the central spot gradually decays and becomes zero at a distance $r_\| = \lambda/(1-\cos\alpha) \simeq 2\lambda/\alpha^2$. Thus, a cigar-shaped solid of revolution extracted along the optical axis surrounds the focal point. The light intensity in this solid of revolution decays from a maximum at the center to zero at the periphery; the size of this solid is roughly

$$2r_\| \cdot r_\perp^2 \simeq 1.44\,\frac{\lambda^3}{\alpha^4}$$

The depth of focus, evaluated according to Eq. (5-67) comprises $1/8$ of the distance from the focal point to the apex of the central illuminated solid. We can easily find equations for the light intensity by departing from the general formula that expresses Huygens' principle. Let the optical system form a spherical front bounded by a circular aperture whose angular radius is α = NA that converges at the focal point O (Fig. 5-7). We select an observation point P a distance $\mathbf{r}$ from the focal point. The amplitude of a

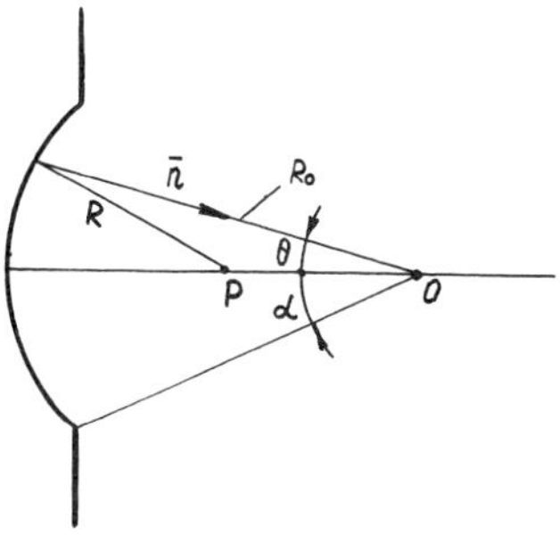

FIGURE 5-7. Diagram for calculating the light at points on the optical axis near the focal point of a system: O is the focal point, P is an arbitrary point on the axis, $OP = r$, and $\alpha = \mathrm{NA}$ is the numerical aperture of the system.

wave at the observation point P can be described, according to Huygens' principle, as

$$U_p = c \int_{\Sigma} (e^{ik(R_0 - R)}/R)\, d\Sigma \tag{5-69}$$

We can say that, approximately, $R - R_0 \simeq \mathbf{n} \cdot \mathbf{r} = -r\cos\theta$. Then

$$\begin{aligned} U_p &= 2\pi c \int_0^{\alpha} e^{-ikr\cos\theta} \sin\theta\, d\theta = \frac{2\pi c}{-ikr}(e^{-ikr} - e^{-ikr\cos\alpha}) \\ \mathrm{Re}\, U_p &\simeq \frac{1}{r} \sin\left[\frac{kr(1-\cos\alpha)}{2}\right] \cos\left[\frac{kr(1+\cos\alpha)}{2}\right] \end{aligned} \tag{5-70}$$

As we can see from Eq. (5-70), the amplitude of a wave goes to zero at the point $r_{\|}^{(m)} = m\lambda/(1-\cos\alpha) \approx 2m\lambda/\alpha^2$ on the axis; when $m = 1$, $r_{\|}^{(1)} = 2\lambda/\alpha^2$.

5-7. THE IMAGE OF A BOUNDARY BETWEEN A LIGHT AND A DARK FIELD IN COHERENT AND INCOHERENT LIGHT[16]

We will now examine the question of how the images of typical elements in integrated circuit masks are transferred. We will begin with the simplest element—the edge of a light field.

We assume that the light field has unity intensity for all $\tilde{y}_0 > 0$ and zero elsewhere. When a mask is illuminated with incoherent light according to Eq. (5-39) and (5-40), we can write

$$I_i^{nc}(x_i, y_i) = \int_0^{\infty} \int_{-\infty}^{+\infty} s(x_i - \tilde{x}_0, y_i - \tilde{y}_0)\, d\tilde{x}_0\, d\tilde{y}_0 \equiv \int_0^{\infty} C(y_i - \tilde{y}_0)\, d\tilde{y}_0 \tag{5-71}$$

$$C(y_i - \tilde{y}_0) = \int_{-\infty}^{+\infty} s(x_i - \tilde{x}_0, y_i - \tilde{y}_0)\, d\tilde{x}_0 \tag{5-72}$$

for the intensity in the image plane. The function $C(y_i - \tilde{y}_0)$ introduced here is numerically equal to the area of the cross section of the scattering function for a point

$s(x_i - \tilde{x}_0, y_i - \tilde{y}_0)$ by the plane x_i = const. Writing $s(x_i - \tilde{x}_0, y_i - \tilde{y}_0)$ in terms of the field amplitude in the image plane we have

$$C(y_i - \tilde{y}_0) = \int |h(x_i - \tilde{x}_0, y_i - \tilde{y}_0)|^2 \, d\tilde{x}_0 \tag{5-73}$$

We can give the amplitude from a point source as an integral over the exit aperture of an optical system:

$$h(x_i - \tilde{x}_0, y_i - \tilde{y}_0) = \frac{iR}{\lambda} \iint F(\beta, \gamma) e^{-ik(\beta x_i + \gamma y_i)} \, d\beta \, d\gamma \tag{5-74}$$

The conversion to angular variables has been made here (see Eq. (5-21)). To make further computations easier, it is convenient to introduce a function that is one-dimensional in linear or angular variables:

$$f(y_i, \beta) = \int F(\beta, \gamma) e^{-ik\gamma y_i} \, d\gamma \tag{5-75}$$

This function, the one-dimensional intensity function (transverse to the light field boundary), can be written

$$C(y_i) = \frac{R^2}{\lambda} \int |f(y_i, \beta)|^2 \, d\beta \tag{5-76}$$

The field of a point source at the exit aperture has a constant amplitude $F(\beta, \gamma) = E_0 =$ const. The angular variables at the aperture are related by the equation $\beta^2 + \gamma^2 \leq \alpha^2$, where α is the angular dimension of the aperture. Considering what has been said, Eq. (5-75) can be written as

$$f(y_i, \beta) = E_0 \int_{-\sqrt{\alpha^2 - \beta^2}}^{\sqrt{\alpha^2 - \beta^2}} \exp(-ik\gamma y_i) \, d\gamma = \frac{2E_0}{ky_i} \sin\left[ky_i\sqrt{\alpha^2 - \beta^2}\right] \tag{5-77}$$

For a one-dimensional intensity function we have, according to Eq. (5-76),

$$C(y_i) = \frac{\lambda R^2 E_0^2}{\pi^2 y_i^2} \int_{-\alpha}^{+\alpha} \sin^2\left[ky_i\sqrt{\alpha^2 - \beta^2}\right] \, d\beta \tag{5-78}$$

Making the substitution $\beta = \alpha \sin\theta$, $C(y_i)$ is stated in terms of a first-order Struve function $H_1(2z)$:

$$\begin{aligned} C(y_i) &= \frac{\lambda R^2 E_0^2 \alpha}{\pi^2 y_i^2} \int_0^{\alpha} (1 - \cos[2ky_i\alpha \sin\theta]) \cos\theta \, d\theta \\ &= kR^2 E_0^2 \alpha^3 H_1(2ky_i\alpha)/(ky_i\alpha)^2 \end{aligned} \tag{5-79}$$

Replacing the variable of integration with $y_i - \tilde{y}_0 = Y$ and designating $Z = ky_i\alpha$, we can write

$$I^{nc}(y_i) = \int_{-\infty}^{y_i} C(Y) \, dY = R^2 E_0^2 \alpha^2 \int_{-\infty}^{Z} H_1(2Z) Z^{-2} \, dZ \tag{5-80}$$

for the intensity $I_i^{nc}(y_i)$.

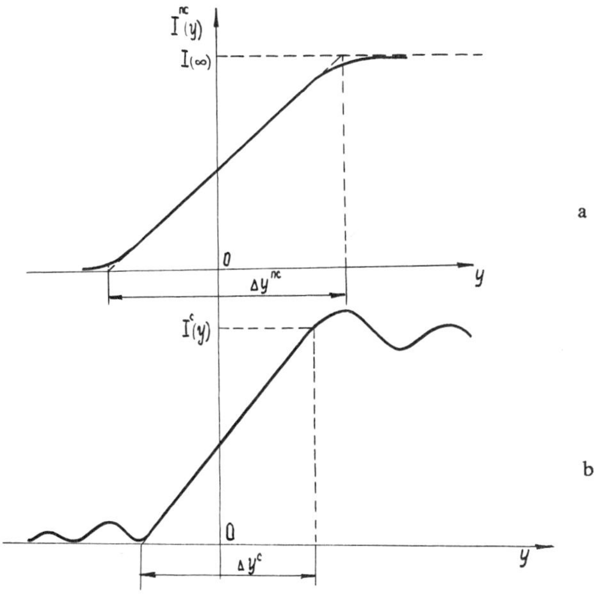

FIGURE 5-8. Intensity changes in the projected image of a boundary for a dark and a light field when illuminated by: a) incoherent light, b) coherent light, and Δy is the transition zone width.

It follows from Eq. (5-80) that the intensity gradient along the normal to the boundary of a light field dI^{nc}/dy_i is equal to the function $C(y_i)$. The course of the function $I^{nc}(y_i)$ is shown in Fig. 5-8. If we draw a straight line having a slope $C(0)$ from the point where it intersects the maximal intensity $I^{nc}(\infty)$ to where it intersects the zero intensity level $I^{nc} = 0$, we can introduce the transition zone width Δy_i^{nc}:

$$\Delta y_i^{nc} = \frac{I^{nc}(\infty)}{C(0)} \tag{5-81}$$

From the expansion of the Struve function,[18]

$$H_\nu(Z) = \sum_{m=0}^{\infty}(-1)^m \frac{(Z/2)^{2m+\nu+1}}{\Gamma(m+3/2)\Gamma(\nu+m+3/2)}$$

the asymptotic form $H_1(2Z) \simeq (8/3\pi)Z^2$ follows when $Z \ll 1$ so that

$$C(0) = \frac{16R^2E_0^2}{3\lambda}\,\alpha^3 \tag{5-82}$$

The maximum intensity $I^{nc}(\infty) = I_0$ can be found by computing the illumination in the image of a uniformly lighted plane:

$$\begin{aligned} I_0 &= \iint s(y_i - \tilde{y}_0, x_i - \tilde{x}_0)\, d\tilde{x}_0\, d\tilde{y}_0 = R^2 \iint |F(\beta,\gamma)|^2\, d\beta\, d\gamma \\ &= R^2E_0^2\omega = \pi\alpha^2R^2E_0^2 \end{aligned} \tag{5-83}$$

Substituting the result obtained into Eq. (5-81) gives us the transition zone width for the image of the edge of an incoherent light field:

$$\Delta y_i^{nc} = I_0/C(0) = \frac{3\pi}{16}\frac{\lambda}{\alpha} \tag{5-84}$$

As we can see, the width of the edge decreases when light with a shorter wavelength is used; an inversely proportional dependence on the numerical aperture of the objective lens occurs.

We will now examine the intensity distribution in the edge of a light field for coherent light. Let the amplitude in the object plane be $\Omega(x_0, y_0) = 1$ for $y_0 > 0$ and $\Omega(x_0, y_0) = 0$ for $y_0 < 0$. The amplitude in the image plane is given, according to Eq. (5-34), by the integral of the coherent transfer function $h(x_i - \tilde{x}_0, y_i - \tilde{y}_0)$ over the lighted half-plane of the object:

$$U(y_i) = \int_0^{\infty} d\tilde{y}_0 \int_{-\infty}^{+\infty} d\tilde{x}_0 \tilde{h}(x_i - \tilde{x}_0, y_i - \tilde{y}_0) = \int_0^{\infty} e(y_i - \tilde{y}_0)\, d\tilde{y}_0 \tag{5-85a}$$

$$e(y_i) \equiv \int_{-\infty}^{+\infty} \tilde{h}(y_i, x_i)\, dx_i \tag{5-85b}$$

We write the equations for the inverse Fourier expansion of the functions $\tilde{h}(x_i, y_i)$ and $F(\beta, \gamma)$:

$$\tilde{h}(x_i, y_i) = (2\pi)^{-2}\left[\frac{ie^{ikR}}{\lambda R}\right]^{-1} \iint F(\beta,\gamma)\exp(-ik(\beta x_i + \gamma y_i))\, d\beta\, d\gamma \tag{5-86a}$$

$$F(\beta, \gamma) = -\frac{ie^{ikR}}{\lambda R}\iint \tilde{h}(x_i, y_i)\exp(ik(\beta x_i + \gamma y_i))\, dx_i\, dy_i \tag{5-86b}$$

We introduce the one-dimensional function $F(\gamma)$, assuming that $\beta = 0$ in $F(\beta, \gamma)$:

$$F(\gamma) = (iR)^{-1}\int \frac{dy_i}{\lambda} e^{ik\gamma y_i} \int \tilde{h}(x_i, y_i)\, dx_i = (iR^{-1})\int e(y_i)e^{ik\gamma y_i}\frac{dy_i}{\lambda} \tag{5-87}$$

As we will see, the function $F(\gamma)$ is the Fourier expansion of the function $e(y_i)$. Using the inverse Fourier transform of Eq. (5-87) gives us

$$e(y_i) = (iR)\int F(\gamma, 0)e^{-ik\gamma y_i}\, d\gamma \tag{5-88}$$

For a circular (angular dimension α) aperture we find

$$e(y_i) = (iR)\int_{-\alpha}^{+\alpha} E_0 e^{-iky_i\gamma}\, d\gamma = 2iRE_0\alpha\frac{\sin(ky_i\alpha)}{ky_i\alpha} \tag{5-89}$$

By making the substitution $y_i - \tilde{y}_0 = Y'$, we find for the amplitude $U(y_i)$ that

$$U(y_i) = \int_{-\infty}^{y_i} e(Y')\, dY' \tag{5-90}$$

Substituting for Eq. (5-89), we calculate

$$U(y_i) = \frac{2iRE_0}{k}\left[\frac{\pi}{2} + Si(k\alpha y_i)\right]$$
$$Si(Y) = \int_0^Y \frac{\sin t}{t}\,dt \qquad Si(\infty) = \frac{\pi}{2} \tag{5-91}$$

The presence of an integral sine function in the amplitude equation predicts oscillations in the amplitude when the observation point is displaced perpendicular to the boundary of a light field. For small $Y \ll 1$, $Si(Y) = Y$, and

$$U(y_i) = \frac{2iRE_0}{k}\left[\frac{\pi}{2} + k\alpha y_i\right]$$

Consequently, a linear change in amplitude is observed near the image of a light field boundary and oscillations are absent. For large values $Y \gg 1$, $Si(Y) \approx \pi/2 - \cos Y/Y$. Thus, amplitude oscillations (resonance effects) arise at a sufficient distance from a boundary line in the image. Analyzing the behavior of the amplitude is speculative, since the amplitudes are not observable quantities. The magnitude of the quantity that can be observed—intensity—is given by

$$I^c(y_i) = |U(y_i)|^2 = \frac{4R^2E_0^2}{k^2}\left[\frac{\pi}{2} + Si(Y)\right]^2 \tag{5-92}$$

The intensity gradient along the normal to the boundary of a coherent light field is

$$\frac{dI^c}{dy_i} = \frac{8R^2E_0^2\alpha}{k}\left[\frac{\pi}{2} + Si(Y)\right]\frac{\sin Y}{Y}, \qquad Y = k\alpha y_i \tag{5-93}$$

It is also easy to find the extent of uniform illumination for a light field as $Z \to \infty$:

$$I^c(\infty) = |U(\infty)|^2 = \frac{4R^2E_0^2}{k^2}\pi^2 \tag{5-94}$$

because $Si(\infty) = \pi/2$. We find the transition zone width along a straight line having a slope of $dI^c/dy_i|_{y_i=0}$ from the ratio

$$\Delta y_i^c = \frac{I^c(\infty)}{(dI^c/dy_i)_{y_i=0}} = \frac{\lambda}{2\alpha} \tag{5-95}$$

We can also find the transition zone width that corresponds to the maximal slope dI^c/dy_i = maximum[16]:

$$\Delta y_i^c = \frac{I^c(\infty)}{(dI^c/dy_i)_{\max}} = 0.37\frac{\lambda}{2\alpha} \tag{5-96}$$

It is apparent that the transition zone is much narrower for coherent light than it is for incoherent light. Self-compensation for this is the appearance of intensity oscillations in both the shadow and in the light regions. Narrower transition zones mean, of course, larger values of the intensity gradient dI^c/dy_i with coherent light.

The behavior of the function $I^c(y_i)$ in Eq. (5-92) is shown in Fig. 5-8b.

5-8. THE IMAGE OF LIGHT AND DARK BANDS IN COHERENT AND INCOHERENT LIGHT

Studying the intensity distribution in the image of a light field boundary allows us to define the criteria for the width of the light or dark band in the image, when the bands cease to be two independent boundaries in the light field. In fact, the intensity distribution in the image of a light band will be different from the sum of the intensities of two parallel boundaries of a light field if the width l of the band is less than the transition zone width: $l < \Delta y_i$. In the inverse inequality there will be a flat portion that attests to the independence of the two distributions at the boundaries of the light field. With coherent light, the condition that $l < \Delta y_i^c$ does not take intensity fluctuations far from the center of the transition zone into consideration. Nevertheless, because of the low degree to which the intensity is modulated by these oscillations the criterion that $l < \Delta y_i^c$ can also be applied to coherent light. Thus, the intensity distribution in the light and dark bands must be computed separately if the band is

$$l < \Delta y_i^{nc} = \left(\frac{3\pi}{16}\right)\frac{\lambda}{\alpha}$$

wide for incoherent light, and

$$l < \Delta y_i^c = \frac{\lambda}{\alpha}$$

wide for coherent light.

Similar to what we did in Eq. (5-71) we can write the intensity distribution in the image of a light band with incoherent light, saying that the band at $-l/2 \leq \tilde{y}_0 \leq l/2$ has been illuminated:

$$I_l^{nc}(y_i) = \int_{-l/2}^{+l/2} C(y_i - \tilde{y}_0)\, d\tilde{y}_0$$

After substituting the variable of integration $y_i - \tilde{y}_0 = Y$ and considering the form of the $C(y_i)$ function given by Eq. (5-79) we find

$$I_l^{nc}(y_i) = R^2 E_0^2 \alpha^2 \int_{k\alpha(y_i+l/2)}^{k\alpha(y_i-l/2)} H_1(2t)t^{-2}\, dt \tag{5-97}$$

Graphs of the intensity distribution given by Eq. (5-97) can only be obtained by numerical integration. In the extreme case of narrow bands, $l \ll \Delta y_i$, when within a band l the function $H_1(2k\alpha y_i)/(k\alpha y_i)^2$ changes only slightly we can say that, approximately,[16]

$$I^{nc}(y_i) = kR^2 E_0^2 \alpha^3 l H_1(2k\alpha y_i)(k\alpha y_i)^{-2} \tag{5-98}$$

An intensity maximum in the image of a narrow, incoherent light band is $I_{\max}^{nc} = (8/3\pi)kR^2E_0^2\alpha^3 l$. As we can see from Eq. (5-98), the light in the image of a slit is nonzero, however narrow the slit in the mask, $l \to 0$. This result must be made more precise in an exact theory of diffraction founded on Maxwell's equations; specifically, transmission by a narrow slit must be a function of the polarization of the wave.

When there is a dark band in a plane illuminated by incoherent light the intensity distribution in an image is written

$$I^{nc}(y_i) = \int_{-\infty}^{+\infty} C(y_i - \tilde{y}_0)\, d\tilde{y}_0 - \int_{-l/2}^{+l/2} C(y_i - \tilde{y}_0)\, d\tilde{y}_0 = I_0 - \int_{-l/2}^{+l/2} C(y_i - \tilde{y}_0)\, d\tilde{y}_0 \tag{5-99}$$

The part I_l^{nc} subtracted from a uniform background I_0 coincides with Eq. (5-97) or (5-98) and its maximum, as was mentioned earlier, is $I_{l_1\,\max}^{nc} = I_0(16/3\pi)(\alpha l/\lambda)$. We find for the intensity modulation in the image of a dark band that

$$M = \frac{I_{\max} - I_{\min}}{I_{\max} + I_{\min}} = \frac{\frac{16}{3\pi}\frac{\alpha l}{\lambda}}{2 - 2\frac{16}{3\pi}\frac{\alpha l}{\lambda}} = \frac{(8/3\pi)\frac{\alpha l}{\lambda}}{1 - (16/3\pi)\frac{\alpha l}{\lambda}} \tag{5-100}$$

because $I_{\max} = I_0$. The modulation decays rapidly as the width l of the band decreases.

We will examine the question of the image of a band illuminated by coherent light (or a dark band in a plane illuminated by coherent light). We can, according to Eq. (5-85), write

$$U^c(y_i) = \int_{-l/2}^{+l/2} e(y_i - \tilde{y}_0)\, d\tilde{y}_0, \qquad e(y) = \frac{2iRE_0\alpha}{ky\alpha}\sin(ky\alpha) \tag{5-101}$$

for the amplitude of a wave in the image of the illuminated band $-l/2 \le \tilde{y}_0 \le l/2$. After some straightforward conversions, the unknown quantities are written in terms of the integral sine function $Si(x) = \int_0^x (\sin t/t)dt$:

$$U_l^c(y_i) = \frac{2iRE_0}{k}\left[Si(k\alpha(y_i - l/2)) - Si(k\alpha(y_i + l/2))\right] \tag{5-102}$$

$$I_l^c(y_i) = \frac{4R^2E^2}{k^2}\left|Si(k\alpha(y_i - l/2)) - Si(k\alpha(y_i + l/2))\right|^2 \tag{5-103}$$

For small l we easily find

$$I_l^c(y_i) \simeq 4R^2E_0^2l^2\alpha^2\,\mathrm{sinc}^2(k\alpha y_i) \tag{5-104}$$

It is interesting to estimate the width of the transition zone for a narrow band l. Differentiating Eq. (5-104) and expanding the result obtained from differentiation when $k\alpha y \ll 1$, we find

$$\frac{dI^c}{dy_i} = 8R^2E_0^2\alpha^2l^2\,\mathrm{sinc}(ky\alpha)\left(-\frac{1}{3}\,ky\alpha\right)(k\alpha)$$

The transition zone width turns out to be

$$|\Delta y_i^c| = \frac{I_l^c(0)}{dI_l^c/dy_i|_{l/2}} = \frac{3}{k^2\alpha^2 l\,\mathrm{sinc}(k\alpha l/2)} = \frac{3\lambda^2}{\alpha^2 l(2\pi)^2}$$

We can see that as the width l of the band decreases, the width of the transition increases as l^{-1}, which confirms the accelerating deterioration in the image of the band.

For a dark band $-l/2 \leq y_i \leq l/2$ we find the intensity distribution in a plane illuminated by coherent light to be

$$I_b^c(y_i) = I_0 - I_l^c(y_i)$$

where $I_l^c(y_i)$ is defined by Eq. (5-103). The modulation in the image of a narrow dark band will be ($I_{max} \approx I_0$)

$$M = \frac{I_{max} - I_{min}}{I_{max} + I_{min}} = \frac{4l\alpha/\lambda}{2 - 8l\alpha/\lambda}$$

Modulation falls off as l decreases.

The intensity distribution for right angles can be shown from the results of modelling the image of light squares in a projection system on a computer.[19,20] A highly irregular picture, which is symmetric with respect to the axes and diagonals of the square, is seen during illumination with coherent light. These irregularities are not present when incoherent light is used. An edge is steeper in coherent light than in incoherent light; also, the steepness decreases to a lesser extent during defocusing if the light is coherent. In actual systems these properties provide a significantly higher image quality in coherent light. It is interesting to note that the steepness of an edge dI/dx in coherent light fluctuates as the size of the square changes. At the same time, the light in the center of the square fluctuates.

The irregularity of the light in the image of a square reveals an "optical proximity effect." We will understand this term to mean the interference between beams diffracting at closely spaced boundaries on the same or adjacent figures. When the image of a square is formed, the wave diffracts at the four sides of the square; the diffracting beams interfere in the image plane, forming an image. Under rigid, invariant conditions, as the size of the square increases the difference in the path of the interfering beams changes, causing oscillations in the steepness of the edge or the intensity at the center of the square. These intensity oscillations become insignificant ($\lesssim$ 10%) when the sides of the square are $a \geq 5l_{min}$; the distance $a = 5l_{min}$ can conditionally be taken as a measure of the range at which the mutual effect of the elements on their image in coherent light disappears (l_{min} is the Rayleigh resolution). In Section 5-9 we will show that the transverse coherence length of an incoherent light source is $r_k \approx \lambda$; at these distances proximity effects appear when an image is produced in incoherent light. Optical proximity effects appear even when a shadow image of the masks is formed in incoherent light (see Section 5-16).

5-9. THE PHOTOMETRY OF SOURCES OF COHERENT (LASER) LIGHT

A number of measurable quantities that relate to a specific light frequency ω are incorporated in the traditional photometry of stationary, incoherent, planar sources.[8]

1. The brightness (radiance) $B_\omega(\bar{r}, \bar{s})$ at a point $\bar{r}$ of a source having an overall area σ in the $\bar{s}$ direction is defined by

$$d\Phi_\omega = B_\omega(\bar{r}, \bar{s}) \cos\theta \, d\Omega \, d\sigma \tag{5-105}$$

Here, $d\Phi_\omega$ is the power radiated into a solid angle $d\Omega$ by a source of area $d\sigma$; the element of the solid angle $d\Omega$ was chosen along the unit vector $\bar{s}$, and θ is the angle between the normal to $d\sigma$ and the vector $\bar{s}$.

2. The emissivity $E_\omega(\bar{r})$ is equal to

$$E_\omega(\bar{r}) = \int_{(2\pi)} B_\omega(\bar{r}, \bar{s}) \cos\theta \, d\Omega \tag{5-106}$$

In Eq. (5-106) the radiation energy is summed over all directions $\bar{s}$ in the half-space $z > 0$. Thus, $E_\omega(\bar{r})$ is the total energy radiated into the half-space $z > 0$ from a unit area on the source.

3. The radiation intensity of a source in the direction $\bar{s}$ (radiant intensity) is defined by

$$J_\omega(\bar{s}) = \int_\sigma B_\omega(\bar{r}, \bar{s}) \cos\theta \, d\sigma \tag{5-107}$$

The total power radiated from the source at a frequency ω (total power) is

$$\Phi_\omega = \int_{2\pi} J_\omega(\bar{s}) \, d\Omega \tag{5-108}$$

When the photometric quantities given by Eqs. (5-105) through (5-108) were introduced, it was assumed that the contributions from the different areas of the source are additive and diffraction was not considered.

If we incorporate photometric quantities based on the wave theory of light, we will have an opportunity to consider diffraction and interference in the radiation process. We will say that the radiation is described by its field amplitude $U(\bar{r}, t)$, and that $U(\bar{r}, t)$ is a scalar, stationary in time, ergodic (the time average is equal to the ensemble average) quantity. The values of $U(\bar{r}, t)$ at different points $\bar{r}$ and times t fluctuate, i.e., $U(\bar{r}, t)$ is a random process. This allows us to use tools from the theory of the correlation between random processes on $U(\bar{r}, t)$, with the exception that correlation between the $U(r, t)$ at different points in space is "spatial coherence" and correlation at different times is "temporal coherence." It is sufficient for our purposes to consider the quadratic correlations, which can be expressed as the mutual correlation function (mutual coherence function) $\Gamma(\bar{r}_1, \bar{r}_2, \tau)$:

$$\Gamma(\bar{r}_1, \bar{r}_2, \tau) = \langle U(\bar{r}_1, t) U^*(\bar{r}_2, t + \tau) \rangle \tag{5-109}$$

The angle brackets denote time or ensemble averaging (because the random process is assumed to be ergodic, these averages are equivalent); the result of averaging is independent of t (because the process is stationary). We find the spectral density of $\Gamma(\bar{r}_1, \bar{r}_2, \tau)$ as its Fourier component:

$$W(\bar{r}_1, \bar{r}_2, \omega) = \frac{1}{2\pi} \int_{-\infty}^{+\infty} \Gamma(\bar{r}_1, \bar{r}_2, \tau) e^{i\omega\tau} \, d\tau \equiv I(\bar{r}, \omega), \qquad \text{if } r_1 = r_2 = r \tag{5-110}$$

and with Eq. (5-110) define the value of complex spatial coherence in the radiation:

$$\mu(\bar{r}_1, \bar{r}_2, \omega) = W(\bar{r}_1, \bar{r}_2, \omega) / [I(\bar{r}_1, \omega) I(\bar{r}_2, \omega)]^{1/2} \tag{5-111}$$

It was shown in [9] that the photometric quantities can be stated in terms of the spectral density of the mutual coherence function; specifically, the total radiation from a source in the $\bar{s}$ direction is

$$J_\omega(\bar{s}) = \lambda^{-2}\cos^2\theta \iint_{-\infty}^{+\infty} W^{(0)}(\bar{r}_1, \bar{r}_2, \omega)e^{-iks_\perp(\bar{r}_1-\bar{r}_2)}\, d\bar{r}_1^2\, d\bar{r}_2^2 \tag{5-112}$$

The superscript (0) on the function $W^{(0)}$ denotes the fact that the function is evaluated in the $z = 0$ plane where the source is located; $\mathbf{s}_\perp$ is the projection of the vector $\mathbf{s}$ onto the plane of the source. If the integral in Eq. (5-112) is given as a Fourier expansion of $W^{(0)}(\bar{r}_1, \bar{r}_2, \omega)$ according to

$$\tilde{W}^{(0)}(\mathbf{f}_1, \mathbf{f}_2, \omega) = (2\pi)^{-4} \iint_{-\infty}^{+\infty} W^{(0)}(\bar{r}_1, \bar{r}_2, \omega)e^{-i(\bar{f}_1\bar{r}_1+\bar{f}_2\bar{r}_2)}\, d^2\bar{r}_1\, d^2\bar{r}_2$$

then Eq. (5-112) can be written in abbreviated form as

$$J_\omega(\bar{s}) = (2\pi k)^2\cos^2\theta\tilde{W}^{(0)}(ks_\perp, -ks_\perp; \omega) \tag{5-113}$$

Equation (5-112) allows us to establish the extent to which the directionality of radiation depends on the coherence of the source. In the extreme case of an incoherent (δ-correlated) source,

$$W^{(0)}(\bar{r}_1, \bar{r}_2, \omega) = I^0(\bar{r}_1, \omega)\delta(\bar{r}_1 - \bar{r}_2) \tag{5-114}$$

and

$$J_\omega^{nc}(\bar{s}) = \lambda^{-2}\cos^2\theta \int_\sigma I^{(0)}(\bar{r}_1, \omega)\, d\sigma \equiv J_{\omega_0}\cos^2\theta \tag{5-115}$$

It is apparent that the directionality of radiation from a δ-correlated source obeys a $\cos^2\theta$ law rather than the $\cos\theta$ law that a Lambert source, which is ordinarily assumed to be incoherent, obeys. We may assume that a Lambert source is partially coherent. In fact, it was shown in [8] that for black-body radiation, spatial coherence is described by the equation

$$\mu^{(0)}(r, \omega) = \frac{\sin kr}{kr} \tag{5-116}$$

i.e., the radiation at a source element $\Delta\sigma \approx \lambda^2$ in area is coherent. The function given by Eq. (5-116) corresponds to the special case of sources for which $\mu^{(0)}(\bar{r}, \omega)$ is given in terms of spherical Bessel functions[8]:

$$\mu^{(0)}(r, \omega) = \frac{2}{\sqrt{\pi}}\Gamma(\nu + 3/2)J_\nu(k|r|)/\left(\frac{1}{2}k|r|\right)^\nu \tag{5-117}$$

For these sources[8]

$$J_\omega(\bar{s}) = J_{\omega 0}(\cos\theta)^{2\nu+1} \tag{5-118}$$

$$J_{\omega 0} = \frac{\nu + 1/2}{\pi}\int I^{(0)}(\bar{r}, \omega)\, d^2\bar{r} \tag{5-119}$$

In the special case $\nu = 0$, we obtain the result that describes a Lambert source:

$$J_\omega(\bar{s}) = J_{\omega 0} \cdot \cos\theta \tag{5-120}$$

For a source having perfect spatial coherence, we can assume that

$$W^{(0)}(\bar{r}_1, \bar{r}_2, \omega) = U^{(0)}(\bar{r}_1, \omega)U^{(0)*}(\bar{r}_2, \omega) \tag{5-121}$$

$$J_\omega(\bar{s}) = (2\pi k)^2 \cos^2\theta |\tilde{U}^{(0)}(ks_\perp, \omega|^2 \tag{5-122}$$

$$\tilde{U}^{(0)}(\mathbf{f}) = (2\pi)^{-2} \int_{-\infty}^{+\infty} U^{(0)}(\bar{r}, \omega)e^{-i\bar{f}\cdot\bar{r}}\, d^2\bar{r} \tag{5-123}$$

If the coherent source is infinite and uniform, i.e., $U^{(0)}(\bar{r}, \omega) = U^{(0)} = \text{const}$, we obtain

$$J_\omega(\bar{s}) = (2\pi k)^2 \cos^2\theta |U^{(0)}|^2 \delta^2(ks_\perp) \sim \delta(\theta) \tag{5-124}$$

the source radiates a plane wave in the direction normal to the surface of the source. In this case the radiation from the source is described by maximum directionality.

We will now examine a Gaussian intensity distribution from a laser (perfectly coherent) source[8]:

$$U^{(0)}(\bar{r}, \omega) = \sqrt{I_0}\exp(-\bar{r}^2/4\sigma_I^2) \tag{5-125}$$

Then

$$J_\omega(\bar{s}) = J_{\omega 0} \cdot \cos^2\theta \cdot \exp(-2(k\sigma_I)^2 \sin^2\theta) \tag{5-126a}$$

$$J_{\omega 0} = (2k\sigma_I^2)^2 I_0 \tag{5-126b}$$

When $(k\sigma_I) \to \infty$, we obtain a δ-directed source from Eqs. (5-126); when $(k\sigma_I) \to 0$, $J_\omega(s) \sim \cos^2\theta$.

Partially coherent sources are modelled by specifying their specific intensity distributions $I^{(0)}(\bar{r}, \omega)$ and their normalized coherence function $\mu^{(0)}(\bar{r}_1, \bar{r}_2, \omega)$. Planar sources are related to quasiuniform sources if we can say for such sources that

$$W^{(0)}(\bar{r}_1, \bar{r}_2, \omega) = I^{(0)}\left(\frac{1}{2}(\bar{r}_1 + \bar{r}_2), \omega\right)\mu^{(0)}(\bar{r}_1 - \bar{r}_2, \omega) \tag{5-127}$$

Equation (5-127) is a qualitative description of finite size real sources, if: 1) the correlation length r_k, i.e., the range of values $|\bar{r}_1 - \bar{r}_2|$, where $\mu^{(0)}(\bar{r}_1 - \bar{r}_2, \omega)$ is notably greater than zero, much smaller than the size of the source; 2) the intensity of the source $I^{(0)}(\bar{r})$ is a slowly varying function such that the average geometric value $[I^{(0)}(\bar{r}_1, \omega)I^{(0)}(\bar{r}_2, \omega)]^{1/2}$ in Eq. (5-111) can be replaced by the value $I^{(0)}(\bar{r})$ at the center point $\bar{r} = (\bar{r}_1 + \bar{r}_2)/2$. In addition, we will assume that the size of the source is much larger than the wavelength of the source, which is always true. Using Eq. (5-127) for quasiuniform sources, we obtain[8]

$$J_\omega(\mathbf{s}) = k^2 \cos^2\theta \tilde{I}^{(0)}(0, \omega) \int_{-\infty}^{+\infty} \mu^{(0)}(\mathbf{r}', \omega)\exp[-ik\mathbf{s}_\perp \cdot \mathbf{r}']\, d^2\mathbf{r}' \tag{5-128}$$

$$\tilde{I}^{(0)}(\mathbf{f}, \omega) = (2\pi)^{-2} \int_{-\infty}^{+\infty} \tilde{I}^{(0)}(\bar{r}, \omega)\exp(-i\bar{f}\bar{r})\, d^2\bar{r} \tag{5-129}$$

If the geometric mean value of the intensity is not replaced by the intensity at the central point, we have Schell's model for a source[8,21]:

$$W^{(0)}(\bar{r}_1, \bar{r}_2, \omega) = [I^{(0)}(\bar{r}_1, \omega) I^{(0)}(\bar{r}_2, \omega)]^{1/2} \mu^{(0)}(\bar{r}_1 - \bar{r}_2, \omega) \tag{5-130}$$

The results of calculating the radiation parameters for three types of normalized coherence functions—Gaussian, exponential, and Bessel (see Eq. (5-117))—are shown in Table 5-1. We can see from the table that the properties of Gaussian, quasiuniform, and Shell sources are similar.

A remarkable feature of partially coherent sources is that they may have an angular intensity distribution $J_\omega(\bar{s})$ that is identical to the angular distribution of a perfectly coherent laser source. This possibility is explained by the fact that the angular intensity distribution is defined as both a coherence function, $\mu^{(0)}(\bar{r}_1, \bar{r}_2, \omega)$, of the source, as well as an intensity distribution function of the source, $I^{(0)}(\bar{r}, \omega)$. For example,[8] a laser source ($\mu_L \equiv 1$) having a Gaussian intensity distribution

$$I_L^{(0)}(\bar{r}, \omega) = I_L \exp(-\bar{r}^2/2\sigma_L^2)$$

and a partially coherent Shell source for which the parameters I_0, σ_μ, σ_I satisfy the conditions

$$I_0 = (\sigma_L/\sigma_I)^2 I_L$$

$$\sigma_\mu^{-2} + (2\sigma_I)^{-2} = (2\sigma_L)^{-2}$$

have the same angular radiation intensity distribution:

$$J_\omega(\bar{s}) = (2k\sigma_L^2)^2 I_L \cos^2\theta \exp[-2(k\sigma_L)^2 \sin^2\theta]$$

See [8] for a discussion of the brightness and emissivity of sources as a function of coherence.

Obviously, we would like to be able to control the properties of partially coherent radiation, i.e., the properties of the $I(\bar{r}, \omega)$ and $\mu(\bar{r}_1, \bar{r}_2, \omega)$ functions. There are two ways to do this: 1) control the laser (the laser design, its generating modes), and 2) optically convert the laser radiation. We cannot even begin to touch upon the means for controlling a laser. Optical methods are extremely interesting to us and it turns out that we can find wave-type equations for the mutual correlation function for the radiation propagating in free air so that we can speak of propagation in the space of the correlation function that describes the radiation. The mutual correlation function and the normalized spatial coherence function are converted when the radiation passes through optical systems. We will now turn our attention to a study of these topics.

5-10. CHANGING THE COHERENCE OF A LIGHT FIELD WHILE THE RADIATION IS PROPAGATING

The so-called method of slowly varying amplitudes or the parabolic equation method first developed by Leontovich[22] is used to solve optics problems based on wave equations. (A detailed discussion can be found in [23].)

TABLE 5-1

Model for the source	Quasiuniform source			Shell source
Intensity distribution	Irrelevant			Gaussian
Correlation function	Gaussian $\exp(-r^2/2\sigma_\mu^2)$	Exponential $\exp(-\lvert r\rvert/D)$	Bessel $\frac{2}{\sqrt{\pi}}\Gamma\left(\nu+\frac{3}{2}\right)\frac{j_\nu(k\lvert\bar{r}\rvert)}{(\frac{1}{2}k\lvert\bar{r}\rvert)^\nu}$	Gaussian $\exp(-r^2/2\Delta^2)$
Angular distribution	$J_{\omega_0}\cos^2\theta e^{-(1/2)(k\sigma_\mu)^2\sin^2\theta}$ $J_{\omega_0}=\frac{k\sigma_\mu^2}{2\pi}\int I^{(0)}(\bar{r},\omega)\,d^2\bar{r}$	$J_{\omega_0}\cos^2\theta(1+k^2D^2\sin^2\theta)^{-3/2}$ $J_{\omega_0}=\frac{(kD)^2}{2\pi}\int I^{(0)}(\bar{r},\omega)\,d^2\bar{r}$	$J_{\omega_0}\cos^{2\nu+1}\theta$ $J_{\omega_0}=\frac{1}{\pi}\left(\nu+\frac{1}{2}\right)\int I^{(0)}(\bar{r},\omega)\,d^2\bar{r}$	$J_{\omega_0}\cos^2\theta_1\exp\left[-\frac{1}{2}(k\Delta)^2\sin^2\theta\right]$ $J_{\omega_0}=(k\Delta\sigma_I)^2I_0$

We will examine propagation in a medium of quasiplanar waves having a narrow angular spectrum. The integro-differential equation for the strength of the electric field **E** follows from Maxwell's equations[24]:

$$c^2 \operatorname{curl} \operatorname{curl} \mathbf{E} + \frac{\partial^2 \mathbf{E}}{\partial t^2} + 4\pi \frac{\partial^2 \mathbf{P}}{\partial t^2} = 0 \tag{5-131a}$$

$$\mathbf{P}(r,t) = \int_0^{\infty} H(t')\mathbf{E}(t-t',z)\,dt' \tag{5-131b}$$

$$\mathbf{E}(r,t) = \mathbf{e}U(r,t)\exp(i\omega_0 t - ik_0 z) \tag{5-131c}$$

where $H(t)$ is the susceptibility of the linear medium and $\mathbf{P}(r,t)$ is its polarization. Since the rapid changes in a wave along its direction of propagation z are accounted for by the exponential term in Eq. (5-131c), we may consider the amplitude $U(\bar{r},t)$ to be a slowly varying function of $\bar{r},t$. We will assume that $U(\bar{r},t)$ changes only slightly at distances of about a wavelength λ and during the period of the light waves T so that

$$\left|\frac{\partial U}{\partial \bar{r}}\right|\lambda \ll |U|, \qquad \left|\frac{\partial U}{\partial t}\right| T \ll |U|, \qquad \left|\frac{\partial^2 U}{\partial z^2}\right| \ll k_0\left|\frac{\partial U}{\partial z}\right| \tag{5-132}$$

When Eq. (5-132) is satisfied, we obtain a parabolic equation for the field amplitude $U(\bar{r}_1,z)$ from Maxwell's equations[24]:

$$\left(\frac{\partial}{\partial z} + \frac{\Delta_\perp}{2k}\right)U(\bar{r},z) = 0, \qquad \Delta_\perp = \frac{\partial^2}{\partial x^2} + \frac{\partial^2}{\partial y^2}, \qquad \bar{r} = x, y \tag{5-133}$$

A solution of Eq. (5-133) in terms of Green's function $H_1(\bar{r}_2 - \bar{r}_1)$ is[24]

$$U(\bar{r}_2,z) = \int_{-\infty}^{+\infty} U(\bar{r}_1)H_1(\bar{r}_2-\bar{r}_1)\,d^2\bar{r}_1 \tag{5-134a}$$

$$H_1(\bar{r}_2-\bar{r}_1) = \frac{ik}{2\pi z}\exp\left(-\frac{ik}{2z}(\bar{r}_2-\bar{r}_1)^2\right) \tag{5-134b}$$

The solution of Eq. (5-134a,b) corresponds to Fresnel diffraction.

We will now write the equations for the amplitude of two points $\bar{r}_1$ and $\bar{r}_2$ in the plane z:

$$\left(\frac{\partial}{\partial z} + i\frac{\Delta_{\perp 1}}{2k}\right)U(\bar{r}_1,z) = 0$$

$$\left(\frac{\partial}{\partial z} - i\frac{\Delta_{\perp 2}}{2k}\right)U^*(\bar{r}_2,z) = 0$$

we multiply the first equation by $U^*(\bar{r}_2,z)$ and the second by $U(\bar{r}_1,z)$ and then add the two together. After averaging over the set we obtain an equation for the transverse mutual correlation function[24] $W_\perp(\bar{r}_1,\bar{r}_2;z)$:

$$\left[\frac{\partial}{\partial z} + \frac{i}{2k}(\Delta_{\perp 1} - \Delta_{\perp 2})\right]W_\perp(\bar{r}_1,\bar{r}_2,z) = 0 \tag{5-135}$$

Equation (5-135) describes the "propagation" in space of the mutual correlation function of the field amplitude.

In the variables $\bar{r} = \bar{r}_1 - \bar{r}_2$ and $R = (r_1 + r_2)/2$, we obtain the equation

$$\left(\frac{\partial}{\partial z} + \frac{i}{k}\frac{\partial^2}{\partial \bar{r}\,\partial \bar{R}}\right) W_\perp(\bar{r}, \bar{R}; z) = 0 \tag{5-136}$$

instead of Eq. (5-135). A solution of Eq. (5-136) has the form[24]

$$W_\perp(\bar{r}, \bar{R}; z) = \frac{1}{(\lambda z)^2} \int_{-\infty}^{+\infty} W_\perp^{(0)}(\bar{r}', \bar{R}') \exp\left[-i\frac{k}{z}(\bar{r} - \bar{r}')(\bar{R} - \bar{R}')\right] d^2\bar{r}'\, d^2\bar{R}' \tag{5-137}$$

The solution of Eq. (5-137) defines the correlation function in a plane $z \neq 0$ in terms of its values in the plane $z = 0$.

We will investigate two special cases: 1) the field is incoherent in the plane $z = 0$, and 2) the field is partially coherent in the plane $z = 0$.

1) In the first case the δ-correlated radiation intensity

$$W_\perp^{(0)}(\bar{r}', \bar{R}') = h^2 I(\bar{R}')\delta(\bar{r}') \tag{5-138}$$

$$\int_{-\infty}^{+\infty} W_\perp^{(0)}(\bar{r}', \bar{R}')\, d^2\bar{r}' = h^2 I(\bar{R}') \tag{5-139}$$

will be an explicit expression for the amplitude correlation function in the plane $z = 0$. The function $I(\bar{R}')$ describes how the average intensity of an incoherent field is distributed in the plane of the source, $z = 0$. Performing the integration with a δ-function in Eq. (5-137) yields

$$W_\perp(\bar{r}, \bar{R}; z) = \left(\frac{hk}{2\pi z}\right)^2 \exp\left(-i\frac{k}{z}\bar{r}\bar{R}\right) \int_{-\infty}^{+\infty} I(\bar{R}') \exp\left(i\frac{k}{z}\bar{r}\bar{R}'\right) d^2\bar{R}' \tag{5-140}$$

Just as it was before, h is a normalizing factor.

Thus, the correlation function (coherence) of a field in an arbitrary plane z is (to within a factor) a Fourier component of the intensity distribution $I(\bar{R}')$ in the original plane $z = 0$ at the spatial frequency $kr/z = |\mathbf{f}|$. The shape of the correlation function is governed by the properties of the function $I(\bar{R}')$ and $\exp(i(k/z)\bar{r}\bar{R}')$. As $k \to \infty$ (the approximation of geometric optics) or $z \to 0$ the term $\exp(i(k/z)\bar{r}\bar{R}')$ oscillates very rapidly. The range of integration and the parameter values defined by the phase interval $(k/z)\bar{r} \cdot \bar{R}' \leq 1$ make the fundamental contribution to the integral. The maximum value is $R' = a$, where a is the radius of the beam in the plane $z = 0$. From here we find that the correlation radius for the wave amplitude in the plane z is

$$r_k(z) \approx \frac{z}{ka} = \frac{1}{k\theta_0} \tag{5-141}$$

$\theta_0 = a/z$ is the angular radius of the source in the plane $z = 0$, which can be seen from a plane $z \gg a$.

If a δ-correlated light field is uniform in a circle of radius a,

$$I(\bar{R}') = \begin{cases} I_0 & R' \leq a \\ 0 & R' > a \end{cases} \tag{5-142}$$

Eq. (5-137) is easily evaluated and we obtain (assuming that $\bar{r}_1 = 0$ and $\bar{r}_2 = \bar{s}$)

$$W_\perp(\bar{s}; z) = \frac{I_0 ka}{2\pi zs} e^{-iks^2/z} J_1(kas/z) \tag{5-143}$$

Using a Bessel functions expansion $J_p(x) \simeq x^p/p!\,2^p$, $p > 0$, $x \ll 1$, we easily find that $W_\perp(0, z) = I_0 ka/4\pi z^2$ and write the normalized correlation function as

$$\gamma(s, z) = \frac{W_\perp(s; z)}{W_\perp(0, z)} = 2\left|\frac{J_1(kas/z)}{(kas/z)}\right| e^{-iks^2/z} \tag{5-144}$$

The correlation radius depends on how it is defined. If for r_k we consider the first zero to be $J_1(s)$, then $r_k = 0.61\lambda z/a$. If we adopt a stricter condition of coherence $\gamma(r_k) = 0.88$, then $r_k = 0.16\lambda z/a$. The radiation from a point source is totally coherent, since when $a = 0$, $r_k = \infty$ and $\gamma(s, z) = 1$. The ratio a/z is equal to the angular radius of the source, θ_0, and this angle is equal to the half-width of the angular spectrum of the radiation in the plane z. From here it follows that the radius of coherence and the width of the angular spectrum $\Delta\theta$ of the radiation in a given plane are related by an inverse proportion $r_k \sim \lambda/\Delta\theta$, $\Delta\theta = 2\theta_0$.

We use a relationship that exists between the diffraction angle at a circular diaphragm of radius a and the angular dimension of an incoherent source θ_0:

$$\theta_d = \frac{0.61\lambda}{a} = \frac{0.61\lambda}{z}\frac{1}{\theta_0}$$

which gives us the relationship $r_k(z) = 0.16\lambda z/a = 0.262z \cdot \theta_d$. These results are valid only for the radiation from a δ-correlated source in the plane $z = 0$.

A major conclusion to be drawn from our examination is that a perfectly incoherent (δ-correlated) source illuminates an object with light having a high degree of coherence if the ratio $a/z = \theta_0$ is small; i.e., the distance z between the source and the lighted object is many times greater than the source dimension a. In actual lithography machines, the condition $a/z \ll 1$ cannot be met because of energy considerations; the light on an object decreases as z^{-2} when z is increased. Therefore, the coherence that appears in the light in lithography machines with incoherent sources may be neglected.

2) We will now find the radius of radiation coherence in a plane z when the radiation in the original plane $z = 0$ has a finite coherence radius r_0. Since there are no optical components between the planes $z = 0$ and $z > 0$ the beam can experience only diffraction spreading in this interval. We can easily find an analytical solution for this problem if the transverse intensity distribution and the correlation function are Gaussian.[24] We will first describe a Gaussian source in the plane $z = 0$:

$$U(\bar{r}) = U_0 \exp(-\bar{r}^2/a^2 + i\alpha_0\bar{r}^2/2) \tag{5-145}$$

$$W_\perp^{(0)}(\bar{r}_1, \bar{r}_2) = I_0 \exp\left\{-\frac{\bar{r}_1^2 + \bar{r}_2^2}{a^2} - \frac{(\bar{r}_1 - \bar{r}_2)^2}{r_0^2} - \frac{i\alpha_0}{2}(\bar{r}_2^2 - \bar{r}_1^2)\right\} \tag{5-146a}$$

or in the variables $\bar{r} = \bar{r}_1 - \bar{r}_2$, $\bar{R} = (\bar{r}_1 + \bar{r}_2)/2$:

$$W_{\perp}^{(0)}(\bar{r}_1, \bar{R}) = I_0 \exp\left\{-\frac{2R^2}{a^2} - \frac{r^2}{r_e^2} - i\alpha_0 \bar{R}\bar{r}\right\} \tag{5-146b}$$

$$r_e^{-2} = r_0^{-2} + \frac{1}{2}a^{-2}$$

Here, a is the beam "radius" and r_0 is the initial correlation radius; the phase coefficient α_0 (cm^{-2}) describes the initial divergence of the beam (a value $\alpha_0 < 0$ corresponds to a converging beam).

The correlation function $W(\bar{R}, \bar{r}; z)$ conforms to Eq. (5-137). We will seek a solution for this equation in the form

$$W_{\perp}(\bar{R}, \bar{r}; z) = I_0 \varphi_1(z) \exp\left\{-\varphi_2(z)\frac{2R^2}{a^2} - \varphi_3 \frac{r^2}{r_e^2} - i\varphi_4(z)\alpha_0 \bar{R}\cdot\bar{r}\right\} \tag{5-147}$$

Substituting Eq. (5-147) into Eq. (5-137) and setting the coefficients when $R^k r^\lambda$ to zero, we obtain equations for the functions $\varphi_1(z), \ldots, \varphi_4(z)$. The final calculated result will be[24]

$$W_{\perp}(\bar{R}_1, \bar{r}_2; z) = I_0 u^{-2}(z) \exp\left\{-\frac{1}{u^2(z)}\left[\frac{R}{a^2} + \frac{r^2}{r_0^2} - ik\frac{(u^2)}{4}\bar{R}\bar{r}\right]\right\} \tag{5-148a}$$

$$u^2(z) = 1 + 2\alpha_0 z/k + z^2/l_d^2 + z^2\alpha_0^2/k^2, \qquad l_d = kar_e/2\sqrt{2} \tag{5-148b}$$

As we can see, the beam radius $a(z)$ and the radiation correlation radius $r_k(z)$ are changed in the same manner:

$$r_k(z) = r_0 u(z), \qquad a(z) = au(z) \tag{5-149}$$

so that there exists a constant (a global coherence function[8])

$$C = r_k(z)/a(z) = r_0/a \tag{5-150}$$

that describes the fields in all planes z simultaneously.

According to Eqs. (5-148a) and (5-149), the radii $r_k(z)$ and $a(z)$ for a converging beam have minima for some $z_0 = -(\alpha_0/k)[l_d^{-2} + (\alpha_0/k)^2]^{-1}$, whereas when $\alpha_0 > 0$ (a diverging beam) the radii $r_k(z)$ and $a(z)$ are monotonically increasing with the distance z.

With a transverse correlation radius r_0 is associated a lengthwise correlation length $l_{\parallel} = kr_0^2/2$.[24] The field amplitudes in planes separated a distance $z > l_{\parallel}$ are uncorrelated.

We will estimate the effective beam radius $a(z)$, the radius of coherence $z_k(z)$, and the diffraction length l_d for an excimer laser. According to our measurements,[25] the radius of transverse coherence for an XeCl laser is anisotropic: $r_{0x} \approx 0.725$ mm, and $r_{0y} = 0.65$ mm. For estimating purposes we will say that $r_0 = 0.1$ cm and $r_e \approx r_0 = 0.1$ cm. Then $l_d = 5 \times 10^2$ cm. We can isolate two distances between the planes, $z \ll l_d$ and $z \gg l_d$, which can be called the near and far zones. In the near zone (assuming that $\alpha_0 = 0$) $u \simeq 1$ and the radii $r_k(z)$ and $a(z)$ of a beam retain their original values. In the far zone, when $r_0 \ll a$ (which occurs for excimer lasers), and $\alpha_0 = 0$ the correlation radius is determined by the initial radius of the beam: $r_k = \sqrt{2}\,\lambda z/\pi a$, whereas the beam radius is determined by the initial correlation radius of the radiation r_0: $a(z) = \sqrt{2}\,\lambda z/\pi r_0$.

5-11. THE PROPAGATION OF THE MUTUAL COHERENCE FUNCTION WHEN DIFFRACTION OF THE FIELD AT THE MASK (OBJECT) IS CONSIDERED

Let a mask be illuminated in transmitted light by a field $U_s(\bar{r})$ so that the field amplitude immediately beyond the mask having transmittance $t(\bar{r}) = m(\bar{r})\exp(i\phi(\bar{r}))$ is equal to $U(\bar{r}_1) = U_s(\bar{r}_1)t(\bar{r}_1)$. The field is now diffracted at not only its intrinsic aperture, but at the mask described by the transmittance $t(\bar{r})$. We will show that, under certain conditions that limit diffraction at the mask, the field in an arbitrary plane beyond the mask can be written as a solution of the parabolic equation given by Eq. (5-133):

$$U_1(\bar{r}_2, z) = \iint_{-\infty}^{+\infty} U(\bar{r}_1)H_1(\bar{r}_2 - \bar{r}_1)\, d^2\bar{r}_1 \tag{5-134a}$$

$$H_1(\bar{r}_2 - \bar{r}_1) = \frac{ik}{2\pi z}\exp\left(-i\frac{k}{2z}(\bar{r}_2 - \bar{r}_1)^2\right) \tag{5-134b}$$

For the proof, we use Rayleigh's method.

We expand with respect to plane waves in the z plane beyond the mask a field which contains the result from additional diffraction at nonuniformities in the mask:

$$U(\bar{R}) \equiv \iint_{-\infty}^{+\infty} \varepsilon(\bar{k})e^{-i\bar{k}\bar{R}}\, d^3\bar{k} \tag{5-151}$$

We introduce the two-dimensional vectors $\bar{\varkappa}$ and $\bar{r}$ according to

$$\bar{k}\bar{R} = k_z z + \bar{\varkappa}\bar{r}, \qquad k^2 = k_z^2 + \varkappa^2$$

Then, instead of Eq. (5-151), we can write

$$U(\bar{R}) \equiv U(\bar{r}, z) = \int_{-\infty}^{+\infty} \varepsilon(\bar{\varkappa})e^{-i(\bar{\varkappa}\bar{r} + \sqrt{k^2 - \varkappa^2}\, z)}\, d^2\bar{\varkappa} \tag{5-152}$$

Waves with $\varkappa^2 < k^2$ are travelling waves along the z axis. Values $\varkappa^2 > k^2$ correspond to attenuating (so-called nonuniform) waves.

The Fourier components of the expansion with respect to plane waves $\varepsilon(\varkappa)$ in Rayleigh's method are found from the boundary condition at $z = 0$:

$$U(\bar{r}_1, z = 0) \equiv U_s(\bar{r}_1)t(\bar{r}_1) = \int_{-\infty}^{+\infty} \varepsilon(\bar{\varkappa})e^{-i\bar{\varkappa}\bar{r}_1}\, d^2\bar{\varkappa} \tag{5-153}$$

from which,

$$\varepsilon(\bar{\varkappa}) = (2\pi)^{-2} \int_{-\infty}^{+\infty} U_s(\bar{r}_1)t(\bar{r}_1)e^{i\bar{\varkappa}\bar{r}_1}\, d^2\bar{r}_1 \tag{5-154}$$

Substituting Eq. (5-154) into (5-152) gives us

$$U(\bar{r}_2, z) = (2\pi)^{-2} \iint_{-\infty}^{+\infty} U_s(\bar{r}_1)t(\bar{r}_1)e^{-i[\boldsymbol{\varkappa}(\bar{r}_2-\bar{r}_1)+\sqrt{k^2-\varkappa^2}\,z]}\, d\bar{r}_1^2\, d^2\boldsymbol{\varkappa} \tag{5-155}$$

which is a general solution of the problem and is valid for any relationship between the sizes of the nonuniformities in a mask (object) and wavelength, which means for any diffraction angle at the mask. We will assume however, that typical dimensions r_0 of the transparent and opaque regions of the mask are large in comparison with the wavelength: $(\lambda/\pi r_0) \ll 1$. This is equivalent to saying that the diffraction angles (first-order) at mask elements $\theta_{\text{dif}} \approx \lambda/r_0$ are relatively small. From here it also follows that the values $|\boldsymbol{\varkappa}|$ for which $\varepsilon(\boldsymbol{\varkappa})$ in Eq. (5-154) are significant, are limited by the inequality $\varkappa r_0 < 1$, $\varkappa/k \ll 1$. We can then say, in an approximation, that

$$(k^2 - \varkappa^2)^{1/2} \approx k - \varkappa^2/2k \tag{5-156}$$

Substituting Eq. (5-156) into (5-155) and integrating over $\varkappa$ gives us

$$U(\bar{r}_2, z) = \frac{ik}{2\pi z}e^{-ikz} \int_{-\infty}^{+\infty} U_s(\bar{r}_1)t(\bar{r}_1)e^{-(ik/2z)(\bar{r}_2-\bar{r}_1)^2}\, d^2\bar{r}_1 \tag{5-157}$$

which agrees with Eq. (5-134). Thus, a solution for the parabolic equation in the form given by Eq. (5-134) can also be used when a diffracting element (mask) is placed in the $\bar{r}_1(z = 0)$ plane, provided that first-order diffraction angles at the mask are small. In projection systems the inequality $\theta_d \leq \alpha_{\text{inp}}$, where α_{inp} is the entrance angular aperture of the projection lens and θ_d is the (first-order) diffraction angle at the mask, must be satisfied in order to produce an image which, of course, puts an upper limit on θ_d so that the condition $\theta_d \sim \lambda/r_0 \leq \alpha \leq 1$ may be thought of as being always satisfied.

It is interesting to estimate the angular spectrum of the radiation that passes through a mask. The mask is illuminated with partially coherent light having an amplitude of $U_s(\bar{r}_1)$. We find the transverse correlation function for the amplitude of a light wave in the plane of the mask $z = 0$ and write its Fourier expansion:

$$W_\perp^{(s)}(\bar{r}_2 - \bar{r}_1) = \langle U_s(\bar{r}_1)U_s^*(\bar{r}_2)\rangle = \iint_{-\infty}^{+\infty} G(\bar{\varkappa})e^{i\bar{\varkappa}(\bar{r}_2-\bar{r}_1)}\, d^2\bar{\varkappa}$$

the function $G(\bar{\varkappa})$ can be thought of as the spectrum of a statistically uniform field having a correlation function $W_\perp^{(s)}(\bar{r}_2 - \bar{r}_1)$:

$$G^{(s)}(\bar{\varkappa}) = (2\pi)^{-4} \int_{-\infty}^{+\infty} W_\perp^{(s)}(\bar{r}_2 - \bar{r}_1)e^{-i\bar{\varkappa}(\bar{r}_2-\bar{r}_1)}\, d^2\bar{r}_1\, d^2\bar{r}_2 \tag{5-158}$$

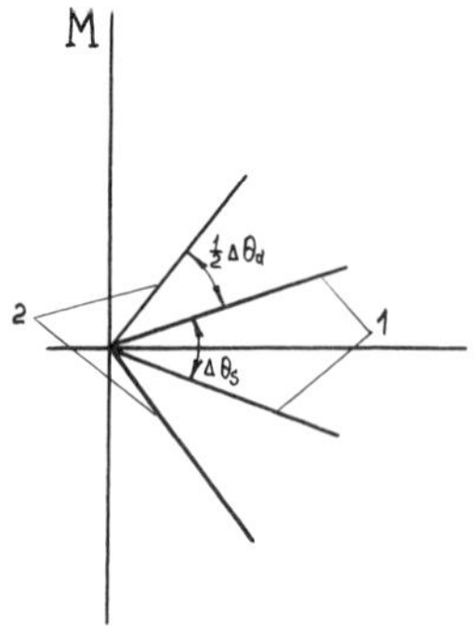

FIGURE 5-9. Diffraction pattern for partially coherent light on a mask: 1) direction of propagation for boundary plane waves in the wave packet of a partially coherent wave; 2) direction of propagation for boundary waves diffracted at the mask in the wave packet of the beams.

In a similar fashion, we can write

$$G^{(M)}(\bar{\varkappa}) = (2\pi)^{-4} \iint_{-\infty}^{+\infty} W_{\perp}^{(M)}(\bar{r}_2 - \bar{r}_1)e^{-i\bar{\varkappa}(\bar{r}_2-\bar{r}_1)}\, d^2\bar{r}_1\, d^2\bar{r}_2 \qquad \text{(5-159a)}$$

$$W_{\perp}^{(M)}(\bar{r}_2 - \bar{r}_1) = \langle U_s(\bar{r}_1)t(\bar{r}_1)U_s^*(\bar{r}_2)t(\bar{r}_2)\rangle \qquad \text{(5-159b)}$$

for the spectrum of a wave passing through a mask. It is obvious that the width of the angular spectrum $\Delta\theta_M$ for a field $U_s(\bar{r})t(\bar{r})$ is greater than the angular spectrum $\Delta\theta_s$ for the field $U_s(\bar{r})$. For estimating purposes we can assume that $\Delta\theta_M \simeq \Delta\theta_s + \Delta\theta_d$, where $\Delta\theta_s$ is the width of the angular spectrum of the illuminating field $U_s(\bar{r})$, $\Delta\theta_d$ is the (first-order) diffraction angle of the field $U_s(\bar{r})$ at the mask; $\Delta\theta_M$ is the width of the angular spectrum for the process $U_s(\bar{r})t(\bar{r})$. The "additivity rule" for the widths of the angular spectra is explained by the diagram shown in Fig. 5-9.

We will write Eq. (5-159b) for the case when the amplitude $U_s(\bar{r})$ in the plane of the subject is produced by a primary source having an area σ and whose radiation is altered and directed onto the mask by an illuminator. We denote the contribution from a point $\bar{s}$ on the primary source to the amplitude $U_s(\bar{r})$ as $U(\bar{s},\bar{r})$. Then, the mutual intensity of the radiation that lights the subject (mask) is[13]

$$W_s(\bar{r}_1,\bar{r}_2) = \langle U_s(\bar{r}_1)U_s^*(\bar{r}_2)\rangle = \int_\sigma U(\bar{s},\bar{r}_1)U^*(\bar{s},\bar{r}_2)\, d^2s \qquad \text{(5-160)}$$

The contribution from a point $\bar{s}$ to the amplitude of a wave passing through a mask at the point $\bar{r}$ is $U_M(\bar{s},\bar{r}) = U(\bar{s},\bar{r})t(\bar{r})$, and the mutual intensity of a wave passing through the mask is

$$U_M(\bar{r}_1,\bar{r}_2) = \int_\sigma U(\bar{s},\bar{r}_1)t(\bar{r}_1)U^*(\bar{s},\bar{r}_2)t^*(\bar{r}_2)\, d^2\bar{s} = W(\bar{r}_1,\bar{r}_2)t(\bar{r}_1)t^*(\bar{r}_2) \qquad \text{(5-161)}$$

The result obtained, Eq. (5-161), is needed in a theory of how a mask image is formed when the mask is illuminated by partially coherent light.

5-12. THE SPECTRAL REPRESENTATION OF A MASK IMAGE IN PARTIALLY COHERENT LIGHT

We use Eq. (5-37) for the intensity at a point $r_i(x_i, y_i)$ in the image plane, replacing the mutual intensity function for a wave passing through the mask with Eq. (5-161):

$$I_i(r_i) = \int d^2\bar{r}_{01} \int d^2\bar{r}_{02} h(\bar{r}_i - \bar{r}_{01}) h^*(\bar{r}_i - \bar{r}_{02}) W_s(\bar{r}_{01}, \bar{r}_{02}) t(\bar{r}_{01}) t^*(\bar{r}_{02}) \quad (5\text{-}163)$$

All the factors that affect an image are shown in the integral in Eq. (5-163); the role of the light (its coherence and uniformity) is described by the mutual intensity function for a wave $W_s(\bar{r}_{01}, \bar{r}_{02})$ that illuminates the mask; the mask is described by a complex transmittance $t(\bar{r}_0)$; and the optical system that produces the image is described by the amplitude scattering function (the pulse response) $h(\bar{r}_i - \bar{r}_0)$. Ordinarily, the mutual intensity function $W_s(\bar{r}_{01}, \bar{r}_{02})$ is assumed to depend on only the difference $\bar{r}_{01} - r_{02}$: $W_s(\bar{r}_{01}, \bar{r}_{02}) = W_s(\bar{r}_{01} - \bar{r}_{02})$. We replace the functions $W_s(\bar{r}_{01} - \bar{r}_{02})$, $t(\bar{r}_0)$, and $h(\bar{r})$ by their Fourier expansions:

$$W_s(\bar{r}) = \int d^2\bar{f} \cdot \omega_s(\bar{f}) \exp[-2\pi i \bar{f} \cdot \bar{r}] \quad (5\text{-}164a)$$

$$t(\bar{r}) = \int d^2\bar{f} \tau(\bar{f}) \exp[-2\pi i \bar{f} \cdot \bar{r}] \quad (5\text{-}164b)$$

$$h(\bar{r}) = \int d^2\bar{f} H(\bar{f}) \exp[-2\pi i \bar{f} \cdot \bar{r}] \quad (5\text{-}164c)$$

Substituting Eqs. (5-164a,b,c) into (5-161) we convert the expression for $I_i(\bar{r}_i)$ to the form[13]

$$I_i(\bar{r}_i) = \int \tau(\bar{f}')\tau^*(\bar{f}'') J(\bar{f}', \bar{f}'') \exp[-2\pi i(\bar{f}' - \bar{f}'') \cdot \bar{r}_i]\, d^2\bar{f}'\, d^2\bar{f}'' \quad (5\text{-}165)$$

where the mutual transmittance function for the optical system $J(\bar{f}', \bar{f}'')$ is

$$J(\bar{f}', f'') = \int \omega_s(\bar{f}) H(\bar{f} + \bar{f}') H^*(\bar{f} + \bar{f}'')\, d^2\bar{f} \quad (5\text{-}166)$$

As we can see from Eq. (5-165), the intensity at a point $\bar{r}_i$ on the image emerges from the quadratic form $\tau(\bar{f}')\tau^*(\bar{f}'')$ of the Fourier component $\tau(\bar{f})$ in the amplitude transmittance of the mask and the matrix elements of the system mutual transmittance $J(\bar{f}', \bar{f}'')$ serve as the weighting coefficients for the superposition. We will assume that an "ideal" optical system exists which transmits all spatial frequencies in the spectrum of the mask transmittance equally, i.e., $J(\bar{f}', \bar{f}") = J_0 = \text{const}$, and in this case we have, instead of Eq. (5-165),

$$\begin{aligned} I_i^{id}(\bar{r}_i) &= J_0 \int \tau(\bar{f}')\tau^*(\bar{f}'') \exp[-2\pi i(\bar{f}' - \bar{f}'') \cdot \bar{r}_i]\, d^2\bar{f}'\, d^2\bar{f}'' \\ &= J_0 |t(\bar{r}_i|^2 \end{aligned} \quad (5\text{-}167)$$

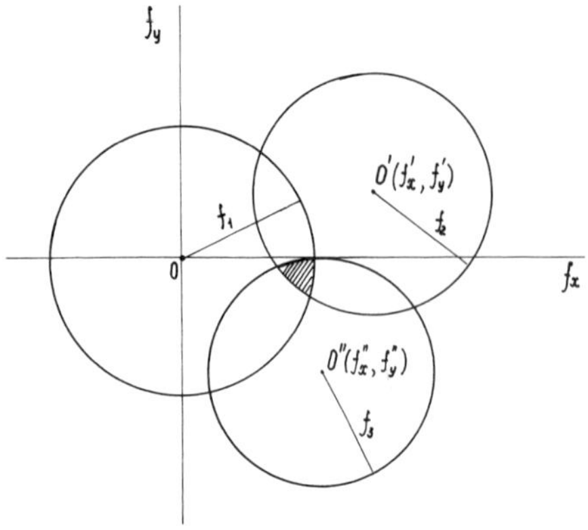

FIGURE 5-10. The area of the overlap region for three circles is equal to the value of the matrix element $J(\bar{f}', \bar{f}'')$ of the mutual transmittance function given by Eq. (5-166). The radius of circle 0 is $f_1 = m(\text{NA})_0/\lambda$. The radii of circles 0′ and 0″ are $f_2 = f_3 = \text{NA}/\lambda = (l/2d_i\lambda)$. The coordinates of circles 0′ and 0″ are related by the arguments of $J(\bar{f}', \bar{f}'')$.

Thus, in the ideal case an image is an exact replica of the original (with a change of scale). Using a theorem regarding autocorrelation we can write the Fourier spectrum of an ideal image $J_0|t(\bar{r}_i)|^2$ as

$$I_i^{id}(\bar{f}) = J_0 \int \tau^*(\bar{f})\tau(\bar{f}' + \bar{f})\, d^2\bar{f}' \tag{5-168}$$

A "real" optical system, by weighting the terms of the quadratic form of $\tau(\bar{f}')\tau^*(\bar{f}'')$ by the weighted factors $J(\bar{f}', \bar{f}'')$, introducing distortions into the image. The geometrical treatment of the element values of $J(\bar{f}', \bar{f}'')$ as the overlap areas of the functions $\omega_s(\bar{f})$, $H(\bar{f} + \bar{f}')$, and $H^*(\bar{f} + \bar{f}'')$ in the frequency plane $\bar{f}(f_x, f_y)$ follows from Eq. (5-166). As we saw in Section 5 the functions $H(\bar{f} + \bar{f}')$, $H^*(\bar{f} + \bar{f}'')$ are equal to unity in circles having centers at $-\bar{f}'$, $-\bar{f}''$ and radii $f = (l/2\lambda d_i) = (\text{NA})_0/\lambda$, where $l/2$ is the radius of the lens aperture and d_i is the distance from the lens to the image plane (Fig. 5-10). It is not hard to find the Fourier components $\omega_s(\bar{f})$ of the mutual intensity function $W_s(\bar{r}_1 - \bar{r}_2)$ if an explicit form of $W_s(\bar{r}_1 - \bar{r}_2)$ is known. For a Kohler illumination system or critical illumination[13]

$$W_s(\bar{r}_1 - \bar{r}_2) = I_{0s} \cdot \left(\frac{2J_1(mv)}{mv}\right) \tag{5-169a}$$

$$v = k|\bar{r}_1 - \bar{r}_2| \cdot (\text{NA})_0 \tag{5-169b}$$

where I_{0s} is the intensity of the incident light and $m = (\text{NA})_c/(\text{NA})_0$ is the ratio of the numerical apertures of the condenser and objective lenses. Equation (5-169a) is the Fourier form of the function for a circle:

$$\omega_s(f) = I_{s0}\left(\frac{\lambda}{m(\text{NA})_0}\right)^2 \frac{1}{\pi} = \text{const} \qquad \text{inside the circle } |\bar{f}| \le \frac{m(\text{NA})_0}{\lambda} \tag{5-170}$$

$$\omega_s(f) = 0 \qquad \text{outside the circle}$$

The region in which the subintegral functions in Eq. (5-166) overlap is shown by dotted lines in Fig. 5-10.

It is interesting to trace the transition to the conditions for incoherent light. In this case the ratio of the numerical apertures, $m = (\mathrm{NA})_c/(\mathrm{NA})_0$, must be great and $2J_1(mv)/mv \to 0$. For the normalization conditions, $W_s(\bar{r}_1 - \bar{r}_2) = I_{0s}\delta(\bar{r}_1 - \bar{r}_2)$; therefore, $\omega_s(\bar{f}) = \text{const}$ for any $\bar{f}$. This circumstance is reflected in Eq. (5-170) by the fact that the radius of the circle C is $f_c = m(\mathrm{NA})_0/\lambda \to \infty$ as $m \to \infty$; therefore, the circles C' and C'' are inside C and the value of $J(\bar{f}', \bar{f}'')$ is found from the total overlap area of the circles C' and C'', as occurred in the incoherent case.

In order to find the relationship between the Fourier components in the image and the Fourier components in the object of transmission, we multiply Eq. (5-165) by $\exp[2\pi i\bar{f} \cdot \bar{r}_i]$ and integrate with respect to $d^2\bar{r}_i$; the result will be[13]

$$J_i(\bar{f}) = \int d^2\bar{f}'\, J(\bar{f} + \bar{f}', \bar{f}'')\tau(\bar{f} + \bar{f}')\tau^*(\bar{f}') \tag{5-171}$$

Substituting Eq. (5-166) for $J(\bar{f}', \bar{f}'')$ into (5-171) gives us

$$J_i(\bar{f}) = \int d^2\bar{f}'\tau(\bar{f}+\bar{f}')\tau^*(\bar{f}') \int d^2\bar{f}'''\omega_s(\bar{f}''')H(\bar{f}'''+\bar{f}'+\bar{f})H^*(\bar{f}'''+\bar{f}') \tag{5-172}$$

We can compare the coordinate representations of a mask image produced by an ideal (Eq. (5-167) and a real (Eq. (5-163)) optical system; the spectral representations of these quantities are given by Eqs. (5-168) and (5-172). Equations (5-163) and (5-172) are "complicated" by additional factors that affect image quality: the light whose role is expressed by the functions $W_s(\bar{r}_{10} - \bar{r}_{20})$ or $W_s(\bar{f})$; the optical system can be described by the transfer function $h(\bar{r}_i - \bar{r}_0)$ or its Fourier transform $H(\bar{f})$.

We will compare the spectral representation of the image intensity for partially coherent light (Eq. (5-172)) with the respective equations for coherent and incoherent light (Eqs. (5-46) and (5-52)). In the extreme cases of coherent and incoherent light, the spectral components (amplitudes or intensities) in an image are related by a linear (proportional) dependence with the spectral components (amplitudes or intensities) in the object. With partially coherent light, the spectral component of intensity $I_i(\bar{f})$ is related by a nonlinear integral dependence to the spectral components of the amplitude transmittance of the object.

5-13. THE COHERENCE OF LIGHT IN THE IMAGE PLANE

We will now approach the interesting problem of coherent light in the image plane of a mask, the image being produced by a lens having a focal length f. Aberrations in the lens will not be considered so that the phase function of the lens has the form

$$M_p(\bar{r}) = \exp\left(i\frac{k}{2f}\bar{r}^2\right)$$

Figure 5-11 shows the designations for the distances from the subject l_1 and from the image l_2 to the lens and also their sizes y_1 and y_2. Designating $l_1 = (1 + \psi)f$, we can easily find relationships for the system magnification $M = y_2/y_1 = 1/\psi = l_2/l_1$

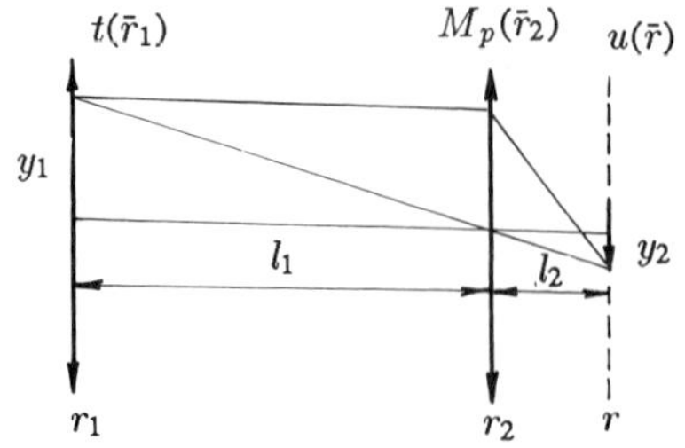

FIGURE 5-11. Diagram of a projection system that produces a mask image: $t(r_i)$ is the amplitude transmittance of a mask located in the plane r_i; $M_p(r_2)$ is the phase function for a projection lens in the plane r_2; $u(r)$ is the field amplitude in the image plane r. The system magnification is $V = y_2/y_1 = l_2/l_1$.

with their help. In projection lithography systems $M = 0.1$ or 0.2 (a ten-fold or five-fold reduction of the mask). Thus, reductions occur, respectively, at $l_1 = 11f$ and $6f$; $l_2 = 1.1f$, $1.2f$.

We will say that a mask is illuminated at a distance by a field $U_s(\bar{r})$ so that the amplitude of the field immediately beyond the mask having transmittance $t(\bar{r})$ is $U_M(\bar{r}) = U_s(\bar{r})t_M(\bar{r})$.

According to Eq. (5-134), we can write the field amplitude directly ahead of the lens as

$$U_1(\bar{r}_2) = \iint_{-\infty}^{+\infty} U_M(\bar{r}_1)H_1(\bar{r}_2 - \bar{r}_1)\,d^2\bar{r}_1$$

where $U_M(\bar{r}_1)$ is the field diffracted at the mask. Multiplying by the phase function of the lens gives us the field immediately beyond the lens: $U_2(\bar{r}_2) = U_1(\bar{r}_2)M_p(\bar{r}_2)$. Finally, the field in a plane l_2 behind the lens is written

$$U_3(\bar{r}) = \iint_{-\infty}^{+\infty} U_2(\bar{r}_2)H_2(\bar{r} - \bar{r}_2)\,d^2\bar{r}_2$$

Combining these formulas yields an expression for the field amplitude $U_3(\bar{r})$ in the image plane given in terms of the field amplitude in the mask plane $U_M(\bar{r}_1)$:

$$U_3(\bar{r}) = \iint_{-\infty}^{+\infty} U_M(\bar{r}_1)H(\bar{r}, \bar{r}_1)\,d^2\bar{r}_1 \qquad \text{(5-173a)}$$

$$H(\bar{r}, \bar{r}_1) = \iint_{-\infty}^{+\infty} H_1(\bar{r}_2 - \bar{r}_1)H_2(\bar{r} - \bar{r}_2)M_p(\bar{r}_2)\,d^2\bar{r}_2 \qquad \text{(5-173b)}$$

Here, $H(\bar{r}, \bar{r}_1)$ is the total propagation function from the mask plane to the image plane. By designating $q = l_1^{-1} + l_2^{-1} - f^{-1}$, and $\bar{h} = \bar{r}_1/l_1 + \bar{r}_2/l_2$ we can easily evaluate the integral in Eq. (5-173b):

$$H(\bar{r}, \bar{r}_1) = (\pi l_1 l_2 q^2)^{-1} \exp\left[-i\frac{k}{2}\left(\frac{\bar{r}^2}{l_2} + \frac{\bar{r}_1^2}{l_1}\right) - i\frac{k}{q}\bar{h}^2\right] \qquad \text{(5-174)}$$

For the conjugate planes of the subject and its image, $q = 0$. We make use of the representation of a two-dimensional δ-function in the form

$$\delta(\bar{h}) = \lim_{q \to 0} \frac{1}{\pi q^2} \exp\left(-i\frac{k}{q}\bar{h}^2\right)$$

and instead of Eq. (5-174), we have

$$H_{\text{conj}}(\bar{r}, \bar{r}_1) = \left(-\frac{1}{l_1 l_2}\right) \exp\left[-\frac{ik}{2}\left(\frac{\bar{r}^2}{l_2} + \frac{r_1^2}{l_1}\right)\right] \delta(\bar{h}) \tag{5-175}$$

The presence of a δ-function inside the integral allows us to easily integrate Eq. (5-173). The result will be[24]

$$U_3(\bar{r}) = \left(-\frac{l_1}{l_2}\right) U_M \left(-\frac{l_1}{l_2}\,\bar{r}\right) \exp\left\{-ik\bar{r}^2 \frac{l_1 + l_2}{2l_2^2}\right\} \tag{5-176}$$

As we can see, the distribution of the field amplitude in the image plane is similar to the distribution of the amplitude in the object plane (with a scaling factor of $-l_1/l_2 = -1/M$, where M is the magnification), but the amplitude is a factor of l_1/l_2 greater. The point $\bar{r}$ in the image plane corresponds to a point $\bar{r}_M = -(l_1/l_2)\bar{r}$ in the mask plane, or $\bar{r} = -M \cdot \bar{r}_M$. With Eq. (5-176) we can easily obtain

$$W_\perp^{(i)}(\bar{r}_1, \bar{r}_2) = \left(\frac{l_1}{l_2}\right)^2 W_\perp^{(0)}\left(-\frac{l_1}{l_2}\,\bar{r}_1, -\frac{l_1}{l_2}\,\bar{r}_2\right) \exp\left\{\frac{ik(l_1 + l_2)}{2l_2^2}(\bar{r}_2^2 - \bar{r}_1^2)\right\} \tag{5-177}$$

for the coherent correlation function.[24] For the normalized coherence function $\gamma^{(i)}(s)$ and the correlation radius $r_k^{(i)}$ we obtain

$$\gamma^i(s) = \gamma^{(0)}\left(\frac{l_1}{l_2}\,s\right) = \gamma^{(0)}(s/M) \tag{5-178}$$

$$r_k^{(i)} = r_k^{(0)} \frac{l_2}{l_1} = M \cdot r_k^{(0)} \tag{5-179}$$

Thus, when a projection system produces an M-fold reduction in the image of the mask, the correlation radius of the light is reduced by as many times in the image plane. Obviously, the ratio of the correlation radius r_k of the beam to the size of the object image (generally, the size of the object is the beam radius a) is a system constant:

$$C = \frac{r_k(z)}{a(z)} = \text{const} \tag{5-180}$$

This constant may be called the (global) coherence function of the beam, because it has the same value for all sections in the beam. A rigid proof of this relation for Gaussian beams having Gaussian correlation functions was done in [26,8]; it was reproduced in briefer form in Section 5-10.

5-14. THE LENS AS A SPECTRAL DEVICE

We can use the equations in Section 5-12 to derive the spectral properties of a lens, which is the fact that the intensity distribution in the focal plane of the lens corresponds to the angular spectrum of the radiation that impinges on the lens. We will say that $l_1 = 0$ so that $U_0(\bar{r}_1)$ is the field amplitude in front of the lens, $l_2 = f$, and write

$$U(\bar{r}, z = f) = \iint_{-\infty}^{+\infty} U_0(\bar{r}_1) e^{i(k/2f)\bar{r}_1^2} H_1(\bar{r}_1 - \bar{r})\, d^2\bar{r}_1 \tag{5-181a}$$

$$H_1(\bar{r}_1 - \bar{r}) = \frac{ik}{2\pi f} \exp\left[-\frac{ik(\bar{r} - \bar{r}_1)^2}{2f}\right] \tag{5-181b}$$

for the amplitude in the focal plane. After some straightforward conversions of Eq. (5-181a) we find that

$$U(\bar{r}, z = f) = \frac{ik}{2\pi f} e^{-ik\bar{r}^2/2f} \iint_{-\infty}^{+\infty} U_0(\bar{r}_1) e^{-ik\bar{r}\cdot\bar{r}_1/f}\, d^2\bar{r}_1 \tag{5-182a}$$

$$I(\bar{r}) = |U(\bar{r})|^2 = \left(\frac{k}{2\pi f}\right)^2 \iint_{-\infty}^{+\infty} W_\perp^0(\bar{r}_1, \bar{r}_2) e^{(ik/f)\bar{r}\cdot(\bar{r}_1 - \bar{r}_2)}\, d^2\bar{r}_1\, d^2\bar{r}_2 \tag{5-182b}$$

As we can see, to within a constant the intensity $I(\bar{r})$ in the focal plane of the lens is the Fourier expansion of the transverse correlation function for the field amplitude in the plane of the lens; i.e., the angular spectrum of this field. The intensity (light) at a point $\bar{r}(x, y)$ in the focal plane corresponds to the intensity of a radiation component having a spatial frequency of $f_x = kx/f$, $f_y = ky/f$. As we can see, the light from the low-frequency components of the field is concentrated at the center of the focal plane and light from the high-frequency components is concentrated at the periphery. The distance from the center of the focal plane of the highest frequency components f_{max} is $r_{max} = f f_{max}/k$; when $k \simeq 10^5$ cm^{-1}, and $f_{max} \approx 10^4$ cm^{-1}, and $r_{max} \approx 0.1f$. Good spatial separation in the frequency components of the radiation is achieved when the lens focal distance is large. Placing an amplitude filter having the necessary transmission distribution $t(\bar{r})$ in the focal plane of the objective lens allows us to improve the image quality of the mask. These methods of improving image quality are well-known in photography,[(15)] although there are no reports of their being used in projection photolithography.

5-15. OPTICAL NOISE IN THE IMAGE OF A MASK ILLUMINATED BY COHERENT LIGHT

When an object is illuminated with coherent light, coherent noise, manifested as randomly placed light and dark spots (speckles), is seen in its image. Topical literature[(17)] has been devoted to the optics of speckles and has found many useful applications. Under the standard conditions for observing coherent noise (speckles) laser radiation is passed through a diffusive object (a focusing screen) called a diffuser; a lens produces the image of the diffuser D on the surface of a screen S (Fig. 5-12).

A plane wave of laser radiation passing through a diffuser acquires random phase increases at different points on the diffuser and remains coherent. Consequently, the waves that emerge from the various points on the diffuser interfere at each point on the screen, producing light or dark spots at random. The diameter of the smallest of these

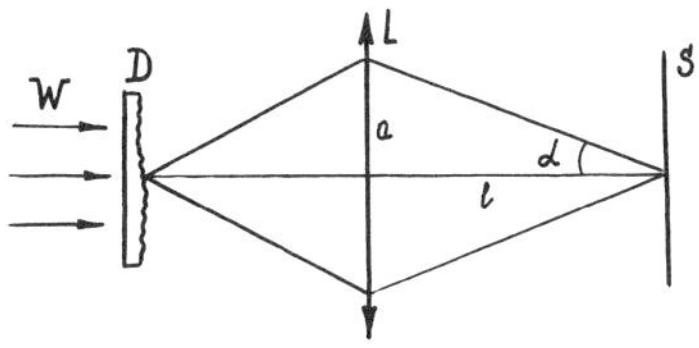

FIGURE 5-12. Scheme for observing optical noise (speckles) in coherent light: W is a plane wave of coherent light; D is a diffuser that introduces random phase distortions into a plane wavefront; L is the focusing lens; and S is the observation plane for the speckles.

spots is roughly equal to the diameter of the diffraction blurring in the image of a point produced by the lens:

$$d_{\min} \simeq \frac{\lambda}{\alpha} = \frac{\lambda}{\mathrm{NA}} \tag{5-183}$$

where α is the angular aperture of the lens and is visible from the center of the screen. In a scheme with a diffuser the modulation $M = (I_{\max} - I_{\min})/(I_{\max} + I_{\min})$ of the speckles reaches a maximum value of $M = 1$ ($I_{\min} = 0$).

Speckle noise occurs when images of masks are produced in coherent laser light, although a diffuser in the light track is, naturally, absent. The fact is that in passing through an optical system some portion of the coherent light is scattered (by reflection, optical nonuniformities, dust, including mask defects). The scattered light interferes with the image-producing light in the image plane, which creates optical noise in the form of speckles on the image. It can be shown that interference between a scattered wave and a signal-bearing wave (the image) causes the signal to be significantly modulated by noise for relatively low intensities of scattered light (less than 5%).[11]

In general, a signal (image)-bearing wave having amplitude U_s is partially coherent and is described in the image plane of a system by coherence function $\mu_{12}^{(s)}(\bar{r})$. The scattered (noise) wave U_N in the same plane can be described by a noise coherence function $\mu_{12}^{(N)}(\bar{r})$ as well as by a signal-to-noise coherence function:

$$\langle (U_{s1} + U_{N1})(U_{s2}^* + U_{N2}^*) \rangle = I_s \mu^{(s)}(\bar{r}_1, \bar{r}_2) + I_N \mu^{(N)}(\bar{r}_1, \bar{r}_2) + 2\sqrt{I_s I_N}\, \mu^{(SN)}(\bar{r}_1, \bar{r}_2) \tag{5-184}$$

When points 1 and 2 blend together, $\mu^{(s)}(\bar{r}_1, \bar{r}_2) = \mu^{(N)}(\bar{r}_1, \bar{r}_2) = 1$, and instead of Eq. (5-184) we have

$$I(\bar{r}_1) = I_s + I_N + 2\sqrt{I_s I_N}\, |\mu^{(SN)}(0)| \cos(\alpha_{SN}(\bar{r}_1)) \tag{5-185}$$

where $\alpha_{SN}(\bar{r})$ is the stationary phase difference between the signal amplitude and noise at a point $\bar{r}$ in the image plane. Using Eq. (5-185) we can easily find the depth M of random modulation of an image $I_s(\bar{r})$ by the optical noise from scattered light:

$$M = \frac{I_{\max} - I_{\min}}{I_{\max} + I_{\min}} = \frac{2\sqrt{I_s I_N}\, |\mu^{(SN)}(0)|}{I_s + I_N} \tag{5-186}$$

We rewrite Eq. (5-186), denoting the fraction of scattered light intensity by $I_N/I_s = \rho$:

$$M = \frac{2\sqrt{\rho}\, |\mu^{(SN)}(0)|}{1 + \rho} \tag{5-187}$$

It immediately follows from Eq. (5-187) that optical noise is absent from incoherent light: $M = 0$ when $\mu^{(SN)} = 0$. Equation (5-187) reveals a surprisingly large value of noise modulation in the image of a light field for extremely low intensity levels ρ in the scattered light. Thus, when $\rho = 0.04$, and assuming that $|\mu^{(SN)}| = 1$, we find that $M \simeq 0.4 = 40\%$. The maximum value of the depth of modulation, $M_{\max} = |\mu^{(SN)}|$, occurs when $\rho = I_N/I_s = 1$, which we can easily see from Eq. (5-187). The degree of mutual coherence $\mu^{(SN)}$ between the signal and noise can be found from Eq. (5-187) if we measure the depth of noise modulation M in the image experimentally and find $\rho = I_N/I_s$ by measuring I_N and I_s independently.

It is interesting to note that deep noise modulation in an image only occurs in a light field where $I_s \gg I_N$. Conversely, in a dark image field, where $I_s \ll I_N$, $\rho \gg 1$ the depth of modulation is $M \simeq 2|\mu^{(SN)}|/\sqrt{\rho} \ll 1$. When there is no signal, $I_s = 0$, the intensity modulation is governed by modulating the intensity of the intrinsic noise.

Since it is very difficult to reduce the intensity of the scattered light to values of $\rho = 10^{-2}$–10^{-3}, reducing optical noise to an acceptable level requires that coherence in the light on the mask be provided by space and time averaging methods. Various scanning, diffraction, and raster systems for illuminations that ensure that the light is equalized and that spatial coherence is reduced have been described in [11].

5-16. THE DIFFRACTION THEORY OF SHADOW EXPOSURE IN OPTICAL LITHOGRAPHY

5-16-1. *A General Examination of the Problem of the Shadow Method for Image Transfer*

In discussing the theory of shadow exposure in optical lithography we will follow [6] in which the topic is discussed in a rather systematic manner. We will assume that a plane wave of monochromatic (coherent) light impinges on a mask having an amplitude transmittance of $t(\bar{r}') = |t(\bar{r}')| \exp(i\varphi)$, $\varphi = \varphi(\bar{r}')$, where $\bar{r}'$ is a point in the plane of the mask. The wavefront of light passing through the mask will be distorted because of absorption, phase changes, and diffraction. According to Kirchoff's diffraction theory, the amplitude of a wave at a point $r(x, y)$ on the surface of the wafer being exposed will be[6]

$$U(\bar{r}) = A \int_{\sigma} \frac{\exp(iks)}{s}(1 + \cos\chi)t(\bar{r}')\, d^2\bar{r}' \qquad \text{(5-188a)}$$

$$s^2 = h^2 + (\bar{r}' - \bar{r})^2, \qquad \cos\chi = h/s = h/[h^2 + (\bar{r}' - \bar{r})^2]^{1/2} \qquad \text{(5-188b)}$$

where h is the gap between the mask and the wafer; σ is the area of the mask; χ is the angle between the normal to the wavefront at a point $\bar{r}'$ on the surface of a mask and the direction from a point $\bar{r}'$ to the point $\bar{r}$; $k = 2\pi/\lambda$ is the wave number; and A is a constant.

As $|\bar{r}' - \bar{r}|$ increases, the $\exp(iks)$ factor fluctuates rapidly if $|\bar{r}' - \bar{r}| \geq h$; a notable contribution to the integral comes from the range of values $|\bar{r}' - \bar{r}| \leq h$. Therefore, we can use the expansion $h(1 + |\bar{r} - \bar{r}'|^2/2h)$ in $\exp(iks)$ and in the denominator of $\cos\chi$ make the substitution $s = h$. Then we can write

$$I(\bar{r}) = |U(\bar{r})|^2 = \left| A' \iint_\sigma h^{-1} \exp\left\{ i\pi \frac{|\bar{r} - \bar{r}'|^2}{W^2} \right\} t(\bar{r}')\, d^2\bar{r}' \right|^2 \tag{5-189}$$

for the light at the point $\bar{r}$. Here we introduce the length $W = \sqrt{\lambda h}$ which is equal to the Fresnel zone width. We usually assume that the transmittance $t(\bar{r})$ changes only slightly over the length $\lambda/2\pi$. If the gap between the mask and the wafer decreases ($h \rightarrow 0$) then the major contribution to the integral comes from only those points on the mask for which $|\bar{r}' - \bar{r}| \rightarrow 0$ i.e.,

$$I(\bar{r}) \sim |t(\bar{r}|^2 \tag{5-190}$$

This means that when the mask contacts the wafer the transmittance of the pattern on the mask is accurately reproduced on the wafer.* For limited gaps a mask pattern, distorted by diffraction effects, will appear on the mask.

Two diffraction situations, Fresnel and Fraunhofer, are usually distinguished and are associated with the minimal dimension l of the elements in a mask pattern. We will assume that a light band l wide on a shaded background is such an element: $t(\bar{r}') = 1$ when $0 \leq x \leq l$. Then the $|\bar{r} - \bar{r}'|$ are also close in value to l: $|\bar{r} - \bar{r}'| \approx l$. If $W = \sqrt{\lambda h} \ll l \approx |\bar{r} - \bar{r}'|$, we must evaluate the diffraction integral given by Eq. (5-189) and these conditions correspond to Fresnel diffraction. Conversely, if $W \gg l$ we can say in an approximation that

$$\frac{|\bar{r} - \bar{r}'|^2}{W^2} = \frac{\bar{r}^2 + \bar{r}^{12}}{W^2} - 2\frac{\bar{r} \cdot \bar{r}'}{W^2} \simeq \frac{2\bar{r}^2}{W^2} - 2\frac{\bar{r} \cdot \bar{r}'}{W^2}$$

and we then have

$$I(\bar{r}) = \left| \frac{2A}{h} \exp\left(2i\pi \frac{\bar{r}^2}{W^2}\right) \int \exp\left(-2i\pi \frac{\bar{r} \cdot \bar{r}'}{W^2}\right) t(\bar{r}')\, d^2\bar{r}' \right|^2 \tag{5-191}$$

in the Fraunhofer diffraction conditions rather than Eq. (5-189). With Fraunhofer diffraction the light in the image plane is the square of the absolute value of the Fourier transform of the amplitude coefficient of mask transmittance, whereas with Fresnel diffraction a more complicated transformation of the amplitude coefficient of transmittance is done according to Eq. (5-189). We will evaluate the size of the gap at which Fraunhofer diffraction is accomplished. We will say that $\lambda = 0.365$ nm and $l = 1\ \mu$m. Then, from the condition $W^* = \sqrt{\lambda h^*} = l$ we find the "boundary" value of the gap h^*: $h^* = 2.75\ \mu$m. Consequently, Fraunhofer diffraction is seen with gaps h = 5–10 μm, whereas Fresnel diffraction occurs with gaps of $h < 2.75\ \mu$m. If $l = 2\ \mu$m (which is more realistic in practical situations), then $h^* \simeq 13\ \mu$m. In align and expose machines attempts are made to hold the gap to h = 5–10 μm to avoid a great deal of diffraction-induced degradation in an image; as we can see, in masks having $l_{\min} = 2\ \mu$m, Fresnel

*In practice, the gap h cannot be zero because the thickness of the photosensitive layer on the wafer must be considered equal to h when contact is made.

diffraction takes place with this kind of gap. Notice that in shadow x-ray lithography Fresnel diffraction is seen with gaps of as much as $h \sim 10^3$ μm. From this point on, we will consider only Fresnel diffraction.

The scaling law for a diffractive field is preserved in Eq. (5-189). We will increase the size of the mask elements and their image on a wafer by m times: $t(\bar{r}') \to t(m\bar{r}')$, $r \to mr$ and the size of the gap h by m^2 times: $h \to m^2 h$. Obviously, the change $h \to m^2 h$ is equivalent to $W \to mW$ when λ = const. Then the diffraction integral is written

$$I^*(\bar{r}) = \int t(m\bar{r}') \exp\left[i\pi \frac{m^2|\bar{r}' - \bar{r}|^2}{2m^2\lambda h}\right] d^2(m^2\bar{r}'^2) = I(m\bar{r})$$

The quantity $W = \sqrt{\lambda h}$ is a natural measure of length in the problems being examined; therefore, we will hereafter use the dimensionless coordinates $\xi = \sqrt{2}\,x/W$ and $\eta = \sqrt{2}\,y/W$.

Before we turn to coherent problems, we will rewrite Eq. (5-191) in a more highly specialized form. The mask area comprises transparent areas having transmittance $t_l = \varepsilon_l \exp(i\varphi_l)$ and dark (absorbent) areas having transmittances $t_b = \varepsilon_b \exp(i\varphi_b)$. The contrast in the mask is $C = (\varepsilon_l/\varepsilon_b)^2$. Consequently, the integral in Eq. (5-189) can be separated into two integrals:

$$I(\bar{r}) = \varepsilon_l^2 \left| \frac{A}{h} \frac{\varepsilon_b}{\varepsilon_l} \exp\left[i(\varphi_b - \varphi_l)\right] \cdot \iint_{\sigma_b} \exp\left[i\pi \frac{|\bar{r}' - \bar{r}|^2}{2W^2}\right] d^2\bar{r}' + \frac{A}{h} \iint_{\sigma_l} \exp\left[i\pi \frac{|\bar{r}' - \bar{r}|^2}{2W^2}\right] d^2\bar{r}' \right|^2 \qquad (5\text{-}192)$$

where σ_b and σ_l are the areas of the dark and light (transparent) regions on the mask. We will use the designations $\varepsilon = \varepsilon_b/\varepsilon_l$ and $\varphi = \varphi_b - \varphi_l$ from now on.

5-16-2. The Image of the Edge of a Dark Half-Plane

In the scenario for Fresnel diffraction at the edge of a semitransparent half-plane the expression for the light is, according to Eq. (5-192)[6],

$$I(x) = \left| \frac{A}{h} \varepsilon \exp(i\varphi) \int_{-\infty}^{0} \exp\left(i\pi \frac{(x' - x)^2}{W^2}\right) dx' + \frac{A}{h} \int_{0}^{\infty} \exp\left(i\pi \frac{(x' - x)^2}{W^2}\right) dx' \right|^2 \qquad (5\text{-}193)$$

Designating $\xi = \sqrt{2}\,x/W$ we easily find from Eq. (5-193) that

$$I(\xi) = \frac{I_0}{2}\left\{\left[\frac{1}{2} + \varepsilon\cos\varphi\left(\frac{1}{2} - C(\xi)\right) - \varepsilon\sin\varphi\left(\frac{1}{2} - S(\xi)\right) + C(\xi)\right]^2 + \left[\frac{1}{2} + \varepsilon\cos\varphi\left(\frac{1}{2} - S(\xi)\right) + \varepsilon\sin\varphi\left(\frac{1}{2} - C(\xi)\right) + S(\xi)\right]^2\right\} \tag{5-194a}$$

$$\begin{pmatrix} C(\xi) \\ S(\xi) \end{pmatrix} = \int_0^{\xi} \begin{pmatrix} \cos \\ \sin \end{pmatrix} \frac{\pi t^2}{2}\, dt \tag{5-194b}$$

We can easily verify that the light on the surface of a wafer being exposed is I_0 in a light field: $I(\xi \to \infty) = I_0$, and $\varepsilon^2 I_0$ in a dark field: $I(\xi \to -\infty) = \varepsilon^2 I_0$.

Figure 5-13 shows how $I(\xi)$ depends on the edge of the half-plane for three situations: 1) $\varepsilon = 0$; 2) $\varepsilon = 0.3$ and $\varphi = 0$; and 3) $\varepsilon = 0.3$ and $\varphi = \pi$. As we can see, the partial transmittance in a dark region and zero phase difference in the waves passing through two half-planes ($\varphi = 0$) leads to the most rapid decrease in intensity when they pass into the dark half-plane. Conversely, phase rotation ($\varphi = \pi$) almost doubles the width of the attenuating region and reduces the gradient of the intensity by the same amount. A second consequence of partial transmittance in a dark half-plane is fluctuations in intensity in the dark region; these fluctuations do not occur if $\varepsilon = 0$ (the solid line). Fluctuations are present in the light region for all cases. The depth of light modulation is significant. The width of the transition zone when $\varepsilon = 0$ or $\varepsilon = 0.3$, $\varphi = 0$ can be estimated in dimensionless units with the interval $\Delta\xi \simeq 1$ or, in units of length, with

$$\Delta x \simeq \frac{W}{\sqrt{2}}\Delta\xi \simeq \sqrt{\lambda h}/\sqrt{2}$$

It is apparent that the width of the transition zone increases with increasing h according to $\Delta x \sim h^{1/2}$. The result obtained for the transition zone width is the basis for the statement that the minimal diffraction-limited linewidth that can be reproduced in proximity printing systems is

$$l_{\min} = \sqrt{\lambda h}$$

Curves of the reduction in intensity are shown in Fig. 5-14 for a few gap distances h ($\varepsilon = 0$) and we can see that the edge of the image deteriorates rapidly as the gap increases.

5-16-3. The Diffraction Theory for the Shadow Image of a Band

$$-a \le x \le a, \qquad l = 2a$$

It follows from Eq. (5-192) that the diffracted intensity distribution in a shadow image of a dark band (having a relative transmittance of $t = t_b/t_l = \varepsilon\exp(i\varphi)$) is given

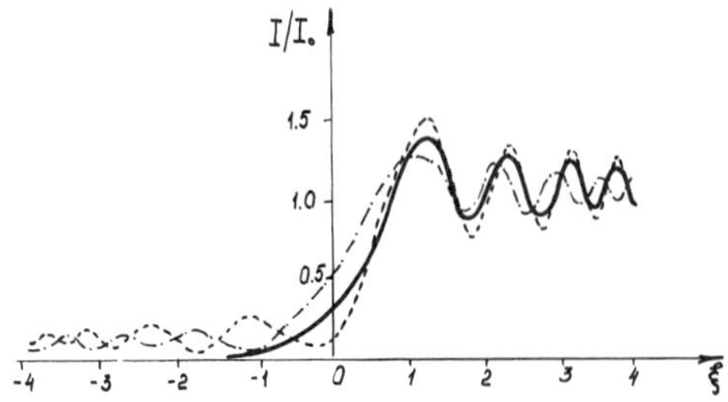

FIGURE 5-13. $I(\xi)$ as a function of the edge of the half-plane for three cases: 1) $\varepsilon = 0$ (solid line); 2) $\varepsilon = 0.3$ and $\varphi = 0$ (dotted line); 3) $\varepsilon = 0.3$ and $\varphi = \pi$.

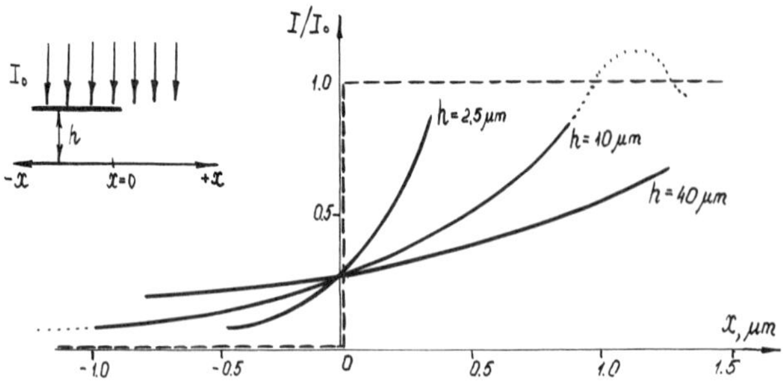

FIGURE 5-14. Deterioration in the image of a light half-plane as the gap h between mask and wafer is increased. As h increases, the intensity gradient decreases rapidly.

by[6]

$$I(x) = B\left|\varepsilon \exp(i\varphi)\int_{-a}^{a} \exp\left[i\pi\left(\frac{x-x'}{W}\right)^2\right]dx' + \int_{-\infty}^{-a} \exp\left[i\pi\left(\frac{x-x'}{W}\right)^2\right]dx' + \int_{a}^{\infty} \exp\left[i\pi\left(\frac{x-x'}{W}\right)^2\right]dx'\right|^2 \tag{5-195}$$

If we introduce a dimensionless linewidth $2\alpha = 2a\sqrt{2}/\sqrt{h\lambda}$ and a dimensionless variable $\xi = \sqrt{2}\,x/\sqrt{h\lambda}$, we can easily convert the expression given for Eq. (5-195) to

$$I(\xi;\alpha) = \frac{I_0}{2}\{[1 - (1 - \varepsilon\cos\varphi)C(\xi,\alpha) - (\varepsilon\sin\varphi)\cdot S(\xi,\alpha)]^2 + [1 - (1 - \varepsilon\cos\varphi)S(\xi,\alpha) + (\varepsilon\sin\varphi)\cdot C(\xi,\alpha)]^2\} \tag{5-196a}$$

$$\begin{pmatrix} C(\xi,\alpha) \\ S(\xi,\alpha) \end{pmatrix} = \int_{\xi-\alpha}^{\xi+\alpha} \begin{pmatrix} \cos \\ \sin \end{pmatrix} \frac{\pi t^2}{2}\,dt \tag{5-196b}$$

The light band $-a \leq x \leq a$ on a dark background having a relative transmittance for the dark background of $t = t_b/t_l = \varepsilon \exp(\varphi)$ yields an intensity distribution of

$$I(x) = B\left|\int_{-a}^{+a} +\varepsilon \exp(i\varphi)\left[\int_{-\infty}^{-a} + \int_{a}^{\infty}\right]\right|^2$$
$$= B\varepsilon^2\left|\frac{\exp(-i\varphi)}{\varepsilon}\int_{-a}^{+a} + \int_{-\infty}^{-a} + \int_{a}^{\infty}\right|^2 \tag{5-197}$$

As we can see, $I(\xi, \alpha)$ will in this case have the form given by Eq. (5-196) when the substitutions $I_0 \to I_0\varepsilon^2$, $\varepsilon \to 1/\varepsilon$, and $\sin\varphi \to -\sin\varphi$ are made:

$$I(\xi,\alpha) = \frac{I_0\varepsilon^2}{2}\left\{\left[1 - \left(\frac{1}{\varepsilon}\cos\varphi\right) C(\xi,\alpha) + \frac{1}{\varepsilon}\sin\varphi \cdot S(\xi,\alpha)\right]^2\right.$$
$$\left. + \left[1 - \left(1 - \frac{1}{\varepsilon}\cos\varphi\right) \cdot S(\xi,\alpha) - \frac{\sin\varphi}{\varepsilon} \cdot C(\xi,\alpha)\right]^2\right\} \tag{5-198}$$

We can easily verify that Eq. (5-198) has limits: $I(\xi, \alpha \to 0) = I_0\varepsilon^2$ and $I(\xi, \alpha \to \infty) = I_0$, that correspond to the shadow "image" of a completely dark or a completely light plane.

Figure 5-15 shows graphs of the intensity distribution at the edge of a dark band having a reduced width $2\alpha = 10$ for two situations: $\varepsilon = 0.3$, $\varphi = 0$ and $\varepsilon = 0.3$, $\varphi = \pi$. A mask that does not introduce a phase change ($\varphi = 0$, the solid line) ensures a high rate of change in the intensity when converting from a shadow region to a light region; however, a greater rate of change in the curve is accompanied by stronger fluctuations in the intensity in both shadow and light regions. Fluctuations can produce specific defects in a resist mask: when a positive resist is overexposed a band may appear parallel to the edge of the resist near the maximum in the shadow region and when underexposed resist band may remain in the lighted area along the region of a minimum (near the point B). Notice that the magnitude of the "shadow" maximum at the point A is roughly equal to εI_0, and not $\varepsilon^2 I_0$ (this can easily be verified by isolating the terms in Eq. (5-196) that are linear in the small quantity ε). Thus, for a mask transmittance in the shadow region $\varepsilon^2 = 0.1$, the intensity in the maximum at point A will be $\varepsilon I_0 = 0.3I_0$. The linear dependence of intensity on contrast with amplitude is a consequence of the coherence in the mask illumination.

5-16-4. The Diffraction Intensity Distribution for Shadow Printing of a Right Angle

A right angle is a detail that is often encountered in integrated circuit masks. Let a dark angle occupy the third quadrant in a rectangular coordinate system, i.e., bounded by the semiaxes $x < 0$ and $y < 0$. The relative transmittance of the dark region is

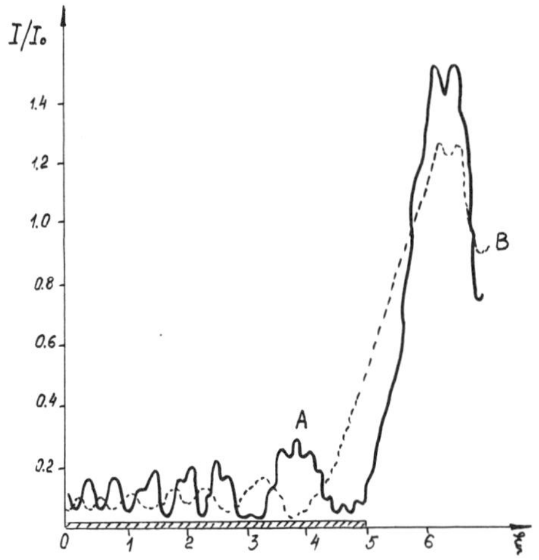

FIGURE 5-15. Graphs of the intensity distribution $I(\xi)/I_0$ at the edge of a dark band having a reduced width of $2\alpha = 10$ for two cases: $\varepsilon = 0.3$, $\varphi = 0$ and $\varepsilon = 0.3$, $\varphi = \pi$.

$t = t_b/t_l = \varepsilon \exp(i\varphi)$. Then we can write

$$I(r) = B\left|\varepsilon \exp(i\varphi)\int_{-\infty}^{0} dx' \int_{-\infty}^{0} dy' \exp\left[i\pi \frac{|\bar{r}-\bar{r}'|^2}{W^2}\right] + \int_{-\infty}^{+\infty} dx' \int_{0}^{\infty} dy' \exp\left[i\pi \frac{|\bar{r}-\bar{r}'|^2}{W^2}\right] + \int_{0}^{\infty} dx' \int_{-\infty}^{0} dy' \exp\left[i\pi \frac{|\bar{r}-\bar{r}'|^2}{W^2}\right]\right|^2 \tag{5-199}$$

for the intensity at a point $r(x, y)$ on a plane and a distance h from the mask according to Eq. (5-192). The integrals given by Eq. (5-199) can easily be reduced to a combination of the Fresnel integrals of Eq. (5-194b)[(6)]:

$$\begin{aligned} I(\xi,\eta) = I_0 \Bigg\{ &\frac{3}{4} + \frac{\varepsilon}{4}\cos\varphi + \frac{1}{2}(\varepsilon\cos\varphi - 1)\Big[C_\xi S_\eta + C_\eta S_\xi \\ &- \frac{1}{2}(C_\xi + C_\eta + S_\xi + S_\eta)\Big] - \frac{\varepsilon}{2}\sin\varphi\Big[S_\eta S_\xi - C_\eta C_\xi \\ &+ \frac{1}{2}(C_\xi + C_\eta - S_\xi - S_\eta)\Big]\Bigg\}^2 \\ &+ \Bigg\{\frac{\varepsilon}{4}\sin\varphi + \frac{\varepsilon}{2}\sin\varphi\Big[C_\xi S_\eta + C_\eta S_\xi - \frac{1}{2}(C_\xi + C_\eta + S_\xi + S_\eta)\Big] \\ &+ \frac{1}{2}(\varepsilon\cos\varphi - 1)\Big[S_\xi S_\eta - C_\xi C_\eta + \frac{1}{2}(C_\xi + C_\eta - S_\xi - S_\eta)\Big]\Bigg\}^2 \end{aligned} \tag{5-200}$$

If the right angle of a light figure is placed in the third quadrant the intensity distribution for this case will be given by Eq. (5-200) when the substitutions $I_0 \to I_0\varepsilon^2$, $\varepsilon \to 1/\varepsilon$, and $\sin\varphi \to -\sin\varphi$ are made, just as in the image of a light band. Isoluxes of the intensity distributions are shown in Fig. 5-16a,b for diffraction at light and dark angles. The lines downward show the linear scale of the pattern for different gaps h. Typical defects in the resist masks that represent the angles of the figures depend on what kind of isolux the resist manifests itself as. Thus, the accurate transfer of a dimension of a dark figure according to Fig. 5-16a is ensured when developing at an isolux of 25% for which an angle on the resist mask will have been cut off. The dimensions of a light figure are also transferred at an isolux of 25%, for which an angle of the light figure will have been covered by the resist (Fig. 5-16b). If the angles of the figures are projected as rounded (reduced radius of curvature $R = r\sqrt{2}/\sqrt{h\lambda}$) then we can simultaneously reproduce both the dimensions of the figures as well as the contours of the arcs at the vertexes of the angles when R = 2–3.[6] Additional types of defects appear in the dark region for finite values of transmittance and phase.[6]

5-16-5. Proximity Effects in the Shadow Printing of Mask Images

As we can see from the preceding, interference between beams that have been diffracted at the boundaries of the figures is a fundamental reason for image deterioration in shadow printing. If two boundaries are fairly close to one another, interference between beams formed by diffracting at the boundaries changes the image of every boundary and, consequently, the image of the figure that these boundaries form. By analogy with the terminology used in electron-beam lithography, the effects due to mutual (diffraction-interference) effects from different parts of the figure's boundaries may be called proximity effects (the authors of [6] call them collective diffraction effects). In essence, the terms due to mutual boundary effects are contained in Eqs. (5-196)–(5-198) and (5-200). For example, a change in intensity caused by interference between beams diffracted at the boundaries $\Delta I(\xi)$ and $\xi = -\alpha, \alpha$ of a dark band can be distinguished by the following means. We write the intensity distribution for a dark half-plane $\xi \leq +\alpha$. This distribution will be described by the function $I(\xi)$ given by Eq. (5-194) when the substitution $\xi \rightarrow \xi - \alpha$: $I(\xi - \alpha)$ is made. A similar means can be used to find the intensity distribution for a dark half-plane $\xi \geq -\alpha$: $I(-\xi - \alpha)$, $\xi < -\alpha$. Then the difference

$$\Delta I_{\text{p.e.}} = I(\xi; \alpha) - I(\xi - \alpha) - I(-\xi - \alpha) \tag{5-201}$$

gives the change in intensity in the image of a band that is due to optical proximity effects. It is clear from the symmetry of the problem that beams diffracted at the edges come together in phase at the center of the shadow in a dark band and yield an intensity maximum. Basically, the function $\Delta I_{\text{p.e.}}$ given by Eq. (5-201) describes this maximum.

5-16-6. The Diffraction Theory of Shadow Image Formation in Partially Coherent Light

All of the results obtained above from the theory of producing a shadow image of the masks relate to a plane wave of monochromatic light impinging on the mask, i.e., perfectly coherent light. However, partially coherent light is used in actual machines. It was shown (Section 5-10) that the angular spectrum of the light can be used as a characteristic of the degree of coherence in the light: the wider the angular spectrum, the less coherent is the light. Thus, partially coherent light can be represented as a packet of plane (coherent) waves that are incoherent among themselves. If we find the intensity distribution $I(x, \gamma)$ in an image, produced by a plane wave having a wave vector in

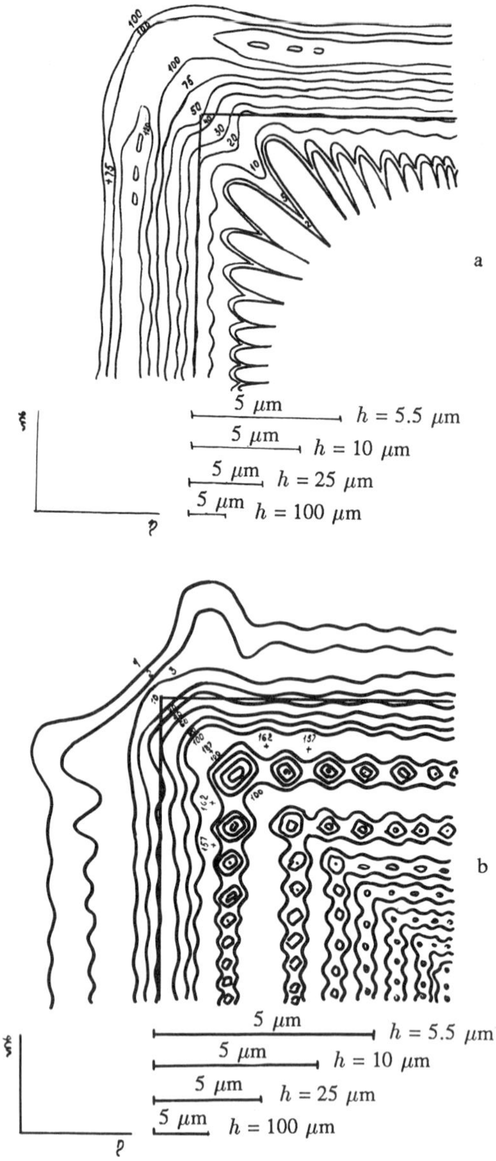

FIGURE 5-16. The diffraction intensity distribution for light (a) and dark (b) angles of a figure for $\varepsilon = 0$.

the γ direction, then the observed average intensity distribution can be calculated by averaging the intensity over the angular spectrum:

$$\tilde{I}(x) = \int I(x, \gamma) g(\gamma)\, d\gamma \tag{5-202}$$

where $g(\gamma)$ is the angular radiation spectrum (with respect to intensity). Obviously, the method for averaging Eq. (5-202) is quite universal and can be used with different means of image formation. A particularly simple method of averaging can be used in the shadow method for producing an image if the angular radiation spectrum has been concentrated in a narrow band around the normal to the plane of the mask. In this case

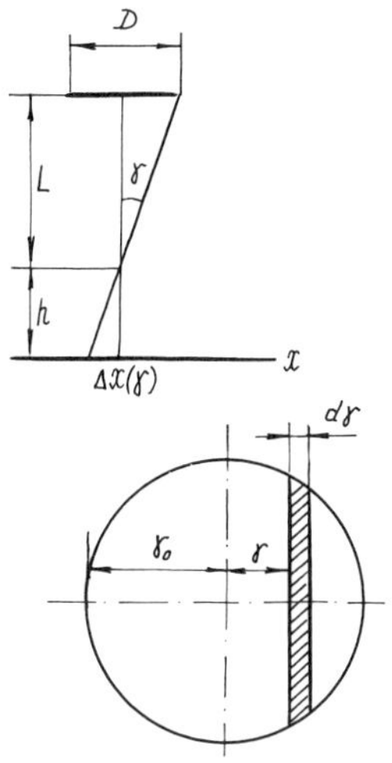

FIGURE 5-17. Displacement of the image of the edge of a figure produced by sources that emit waves at an angle γ to the normal: $\Delta x(\gamma) \approx h\gamma$ (a). The area of the illuminating disk, visible at an angle γ is $ds = \pi(D/2)^2\sqrt{1-(\gamma/\gamma_0)^2}\,d\gamma$, where $\gamma_0 = D/2L$ is the angular radius of the disk.

the function $I(x,\gamma) = I(x - \Delta x(\gamma))$, i.e., an image produced by a plane wave in the γ direction, is identical to an image produced by a wave impinging along the normal ($\gamma = 0$), but displaced an amount $\Delta x(\gamma) = h\tan\gamma \simeq h\gamma$ (Fig. 5-17). In this case the averaging given by Eq. (5-202) becomes

$$\tilde{I}(x) = \int I(x - h\gamma)g(\gamma)\,d\gamma \tag{5-203}$$

We will state the problem of choosing an angular spectrum that will ensure that oscillations in the area of the image are decreased. The distance between fluctuating maxima is roughly equal to $W = \sqrt{\lambda h}$. The angle at which $\sqrt{\lambda h} = \Delta x = h\gamma^*$, is $\gamma^* = \sqrt{\lambda/h}$. If, e.g., $\lambda = 406$ nm, $h = 10$ μm, and $\gamma^* = 0.2$ radians $= 11.4°$. If we choose $g(\gamma) = (1/2)(\delta(\gamma - (1/2)\gamma^*) + \delta(\gamma + (1/2)\gamma^*))$ then this kind of distribution from two plane waves will ensure fairly good smoothing of the intensity oscillations. It is roughly this kind of angular distribution that is used to suppress intensity fluctuations in industrial exposure machines having a controllable gap.[27] When this happens, the attenuation gradient of intensity at the boundary of a figure is somewhat reduced.

Let the degree of coherence in the light on a mask be defined by the finite diameter D of an incoherent circular source located a distance L from the mask (Fig. 5-17b). In the range of angles γ, $\gamma + d\gamma$ point sources whose area is $ds = (\pi D^2/4)\sqrt{1-(\gamma/\gamma_0)^2}\,d\gamma$ emit rays, and γ_0 is the angular radius of the disk. From here the equation for the angular radiation distribution from the source follows:

$$g(\gamma) = ds/(\pi D^2/4) = \sqrt{1 - \gamma^2/\gamma_0^2}$$

The intensity distribution in the image of the edge of a half-plane, averaged over the

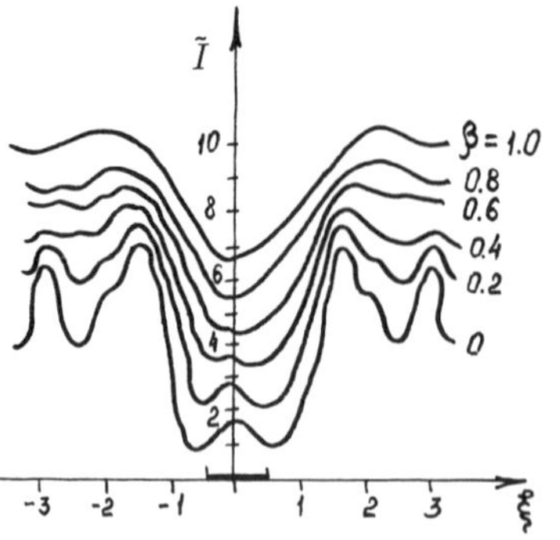

FIGURE 5-18. The intensity distribution in a shadow image of a dark band (transmittance $\varepsilon = 0$, width $\alpha = 0.5$) for different β.

angular spectrum of a disk-shaped source, has the form[(6)]

$$\tilde{I}(\xi) = \int_{-1}^{1} I(\xi - \beta u)\sqrt{1-u^2}\,du \Big/ \int_{-1}^{1} du \tag{5-204}$$

Here, we have said that $\beta = \gamma_0\sqrt{2h/\lambda} = \sqrt{2}\,\gamma_0/\gamma^*$.

Graphs of the intensity distribution in a shadow image of a dark band are shown in Fig. 5-18 (transmittance $\varepsilon = 0$, width $\alpha = 0.5$) for values of $\beta = \sqrt{2}\,\gamma_0/\gamma^* = 0, 0.2, 0.4, 0.6, 0.8, 1$.[(6)] The disappearance of intensity fluctuations as the degree of coherence in the light on the mask is reduced is distinctly visible. According to the criterion that the dimension of a band be transferred with 10% accuracy, a parameter value of $\beta_{opt} = 0.8$–0.9 may be considered optimal. Since $\beta = (D/L)\sqrt{h/\lambda}\,\sqrt{2}/2$, at specific values of h and λ we can speak of an optimal D/L:

$$\left(\frac{D}{L}\right)_{opt} = \sqrt{2}\sqrt{\frac{\lambda'}{h}}\,\beta_{opt} \approx 1.1\sqrt{\frac{\lambda}{h}}$$

However, the empirical formula for $(D/L)_{opt}$ given in [6] corresponds to a lower value of the β_{opt} parameter:

$$\left(\frac{D}{L}\right)_{opt} = 0.2\sqrt{\frac{\lambda}{h}}\left(1 + \frac{2\lambda h}{l_{min}^2}\right)$$

A factor that considers the effect of the minimal size l_{min} in the mask pattern on choosing $(D/L)_{opt}$ has also been introduced here. When $\lambda = 400$ nm and $h = 10\ \mu$m, $l_{min} = \sqrt{2\lambda h}$, $(D/L)_{opt} = 0.08$ and if $D = 20$ mm, $L \simeq 25$ cm.

The possibility that optical noise (speckles) will appear when a mask image is produced by the shadow method in coherent light must also be noted. For contact (with a gap) printing the multiple reflection and scattering of light from contacting surfaces that are highly reflective is typical (see Section 5-15). Choosing an optimal degree of coherence in the light to be used for shadow exposure must be done in conjunction with the necessary extent of optical noise reduction.

TABLE 5-2

Thermoprocessing temperature (1 h)	E_1	E_2	E_3
70°C	5.63	7.43	–12.6
100°C	4.39	5.69	–9.0

5-17. EXPOSING PHOTORESISTS

Up to now we have examined topics in how the light is distributed in the image of masks on the surface of wafers coated with a photosensitive polymer–photoresist layer. Photoresists are multicomponent materials, one component being an inhibitor that performs two very important functions. First, the inhibitor impedes the developer from etching the resist compound; second, because it has a high absorption index μ_λ at the wavelength of the exposing (actinic) radiation, it enters into photochemical reactions with the other components in the resist. An increase in the rate of dissolution in an exposed resist results from the loss of inhibitor due to photochemical reaction. We will not go into an analysis of the chemical composition of the resists or the chemical processes that take place in them; we will instead study only the changes that occur in the physical parameters of the resists, which are governed by the resist exposure and dissolution processes.

How the dissolution rate depends on the concentration of the inhibitor is a function that is unique to each resist, and this dependence changes with the layer fabrication conditions (prebake). As an example, we will present an empirical formula for an AZ-1350J resist etched at t = 21°C [28]:

$$r_{\text{dev}}(M) = \exp[E_1 + E_2 M + E_3 M^2] \tag{5-205}$$

The values of the E_1, E_2, and E_3 parameters depend on the thermoprocessing temperature (see Table 5-2).

As is evident from Eq. (5-205), we must know what the concentration of the light sensitive component $M(z,\xi)$ in the exposed resist is in order to find the etching rate. Assuming that changes in the concentration M are caused by one-quantum processes we can write

$$\frac{\partial M(r,t)}{\partial Q(r,t)} = -C \cdot M(r,t) \tag{5-206}$$

where $dQ = I(t)\,dt$ is the increase in exposure, and C is the quantum yield of the photosensitive component (PSC) reaction rate, $\text{J}^{-1} \cdot \text{cm}^2$. Integrating Eq. (5-206), we find that

$$M(r,t) = \exp(-CQ(r,t)) \tag{5-207a}$$

$$Q(r,t) = \int_0^t I(r,t')\,dt' \tag{5-207b}$$

It is not hard to see that a problem in defining the function $I(r,t)$ for the intensity distribution of the light in the exposed area of the resist has occurred. Hereafter, we will write $I(r,z,t)$, understanding the r points on a plane z = const. We denote the portion of exposing radiation $\rho_1^2 I_l(r,0,t)$ that penetrates a resist as $I(\bar{r},0)$ and the respective field amplitude as $E(\bar{r},0)$. (The coefficient of refraction from the resist layer is $\rho_1^2 = (n-1)^2/(n+1)^2 \simeq 6\%$, $n \simeq 1.5$). A wave $E(\bar{r},0,t)$, in being absorbed passes through the resist layer and is reflected from the substrate. The reflected wave is coherent with respect to the incident wave; standing waves are generated when the direct and reflected waves interfere; the intensity of the exposing wave is modulated on the z axis.

We will formulate these processes in mathematical terms.[6] At a depth z the amplitude of a wave passing through is

$$E_1(z,\bar{r},t) = E(\bar{r},0)\exp\left(iknz - \int_0^z \mu(z',t)\,dz'\right) \tag{5-208}$$

The amplitude coefficient of absorption μ depends on the exposure time t (because of the reaction rate of the PSC); the coefficient of refraction can be assumed to be constant. The amplitude of the wave reflected from the substrate is (the complex coefficient of reflection $\rho_2 = |\rho_2|\exp(i\varphi)$):

$$E_2(z,\bar{r},t) = E(\bar{r},0)\exp\left[ikn(2d-z) - \int_0^d \mu(z',t)\,dz'\right]\rho_2\exp\left(-\int_z^d \mu(z',t)\,dz'\right) \tag{5-209}$$

Considering the fact that waves formed during successive reflections can be neglected, we can write

$$E(\bar{r},t,z) = E_1(\bar{r},z,t) + E_2(\bar{r},z,t) \tag{5-210}$$

The coefficient of refraction in the resist is a function of the concentration $M(\bar{r},z,t)$:

$$\mu(z,\bar{r},t) = \mu_\infty + \psi M(\bar{r},z,t) = \mu_\infty + \psi\exp(-CQ(\bar{r},z,t)) \tag{5-211}$$

Substituting Eq. (5-210) into (5-208) and (5-209) gives us

$$\begin{aligned} E(\bar{r},z,t) = E(\bar{r},0,t)\Bigg\{&\exp\left[iknz - \int_0^z (\mu_\infty + \psi e^{-CQ(z',\bar{r},t)})\,dz'\right] \\ &+ \rho_2\exp\left[ikn(2d-z) - \left(\int_0^d + \int_z^d\right)(\mu_\infty + \psi e^{-CQ(z',\bar{r},t)})\,dz'\right]\Bigg\} \end{aligned} \tag{5-212}$$

Equation (5-212) cannot be solved for $E(\bar{r},z,t)$, because $Q(\bar{r},z,t)$ is an integral of $E(\bar{r},z,t)$. Equation (5-212) is convenient for solving on a computer by an iteration method. We will write Eq. (5-212) for the initial moment $t = 0$:

$$\begin{aligned} E(\bar{r},z,0) = E(\bar{r},0,0)\{&\exp[iknz - (\mu_\infty + \psi)z] \\ &+ \rho_2\exp[ikn(2d-z) - (\mu_\infty+\psi)z - 2(\mu_\infty+\psi)d]\} \end{aligned} \tag{5-213}$$

Using Eq. (5-213) to define $Q(\bar{r},z,\Delta t)$ at time $t = \Delta t$ (Δt is small),

$$Q(\bar{r},z,\Delta t) = |E(\bar{r},z,0)|^2\Delta t$$

we substitute the numbers found for the $Q(\bar{r},z,\Delta t)$ into the right-hand side of Eq. (5-212) and find $E(\bar{r},z,\Delta t)$. Repeating the procedure n times we find values of $E(\bar{r},z,t)$

for any moment $t = n\Delta t$ in time. The radiation intensity will be found as $I(\bar{r}, z, t) = |E(\bar{r}, z, t)|^2$. At the same time, the exposure $Q(\bar{r}, z, t)$ and the function $r_{\text{dev}}(Q)$ will be found (if the empirical function $r_{\text{dev}}(Q)$ is known).

As we can see from Eq. (5-212), we must know the values of many physical parameters of the resist in order to be able to calculate $r_{\text{dev}}(Q)$: n_λ, $\mu_{\infty\lambda}$, ψ_λ, d, C_λ, the wavelength λ, the intensity (or amplitude) $I_\lambda(\bar{r}, 0, t)$, the angular spectrum (degree of coherence) of the light, the exposure time t_l and developing time t_{dev}, the function $r_{\text{dev}}(Q)$, and the complex coefficient of reflection from the substrate $\rho_{2\lambda} = |\rho| \exp(i\varphi_\lambda)$. (The subscript λ means that the parameter relates to the respective wavelength of exposing light). For most of the familiar positive resists the spectral sensitivity lies in the 340–450 nm range (for some it is 260–450 nm); the coefficients of refraction are $n = 1.4$–1.7; thicknesses are $d = 0.5$–2 μm and absorption coefficients of absorption are $\mu(M) = \mu_\infty + \psi M = 0.7$–$0.9$ μm^{-1}. For example, the experimental curves of the $\mu \sim M$ and $M \sim Q$ dependence for the AZ-1350J resists are described by the equations $\mu = \mu_\infty + \psi M$ and $M(Q) = \exp(-C \cdot Q)$ for parameter values of

TABLE 5-3

Wavelength, nm	μ_∞, μm^{-1}	ψ, μm^{-1}	C, cm^2/mJ	n
435.8	0.06	0.54	0.010	1.68
404.7	0.07	0.86	0.016	1.70
365.0	0.20	0.74	0.012	1.72

The coefficient of reflection at the resist–substrate interface, $\rho_2 = |\rho_2| \exp(i\varphi)$, has a significant effect on how exposure is distributed in the resist. Standing waves are generated when the incident and reflected waves interfere with one another. Phase rotation occurs when the waves are reflected from a metal film: $\varphi = \pi$, and for this reason an amplitude minimum occurs at the boundary between the resist and the substrate, leaving the resist unexposed and poorly etched. When a wave is reflected from a metal and phase reversal occurs, the maxima and minima can be found in the planes

$$z_{\text{min}} = \frac{m\lambda}{2n}, \qquad m = 0, 1, 2, \ldots, N$$

$$z_{\text{max}} = \frac{m\lambda}{4n}, \qquad m = 1, 3, \ldots, (2N + 1)$$

where N is the total number of periods in the standing wave. The distance between adjacent layers is $z_{\text{max}} - z_{\text{min}} = \Delta l = \lambda/4n = 0.07$ μm when $\lambda = 400$ nm, $n = 1.5$.

Standing waves produce a number of undesirable effects: the edge of the resist mask acquires a corrugated shape (along the z coordinate), and the weakly exposed area is incompletely etched. Standing waves are very pronounced in metal substrates; the reflection can be weakened by depositing antireflective dielectric coatings on the metal layers. In contact printing systems the effect from standing waves can be reduced by increasing the spectral width $\Delta\lambda$ of the exposing radiation. This effect can also be reduced by using multilayer resists. One such system is the two-layer contrast enhancement lithography (CEL) system.[29] The upper layer in the CEL system contains a component that bleaches at the exposure wavelength so that it becomes less absorbent with exposure according to the principle $\alpha = AM(\bar{r}, z, t) + B$. The lower layer is usually a positive resist. The time integral of the intensity contrast in the image that passes through the bleaching layer proves to be much higher than is the contrast in the light on the surface of a two-layer system: in the areas that receive less light during the exposure process the upper layer is not "bleached" and in these areas the lower layer is almost completely unexposed. After being exposed the upper layer is removed (e.g., by dissolving in water).

The intensity of the radiation that exposes the lower layer is equal to the intensity of the radiation passing through the upper layer of the CEL layer (reflections at the interface can be ignored):

$$I_{\text{trans}}(\bar{r}, d_{\text{CEL}}, t) = I_0 \exp\left\{ -\int_0^{d_{\text{CEL}}} [A_{\text{CEL}} \exp(-C_{\text{CEL}} Q(\bar{r}, z', t)) + B_{\text{CEL}}]\, dz' \right\} \tag{5-214}$$

The $I_{\text{trans}}(\bar{r}, d_{\text{CEL}}, t)$ can be found for any moment in time $t = m\Delta t$ from Eq. (5-214) by an iteration method. At the initial moment $t = 0$ we find

$$I(\bar{r}, z, 0) = I_0 \exp\{-(A + B)_{\text{CEL}} z\} \tag{5-215}$$

We use the result from Eq. (5-215) to find $Q(\bar{r}, z, \Delta t) = I(\bar{r}, z, 0)\Delta t$; the values of $Q(\bar{r}, z, \Delta t)$ found are put into the right side of Eq. (5-214) to find $I_{\text{trans}}(\bar{r}, d_{\text{CEL}}, \Delta t)$, etc. Finally, using $I_{\text{trans}}(\bar{r}, d_{\text{CEL}}, t)$ as the radiation intensity in the lower layer we can model the process by which the lower light-sensitive layer is exposed. The modelling that was done in [30] showed that in order to enhance the image contrast in the lower layer the A and C parameters in the upper (contrast enhancing) layer must be as large as possible, whereas the B parameter must be small. The C parameter can only be changed if the chemistry of the bleachant is changed. At the same time, the A parameter can be increased by increasing the initial concentration of the staining chemical. A diazoquinone compound for which $A = 3.8\ \mu\text{m}^{-1}$, $B = 0.22\ \mu\text{m}^{-1}$, and $C = 0.04\ \text{cm}^2/\text{mJ}$ was suggested in [30] for use as the staining agent. Modelling and natural experiments showed that the use of a contrast enhancement system allows wider tolerances on developing time, but exposure time tolerances are not increased. In two-layer systems with contrast enhancement masks having dimensions smaller than 0.5 μm can be produced by photolithography methods.[30] The shortcoming of the method is that the exposure time is increased and is now the combination of the bleaching time in the upper layer t_{CEL} and the exposure time in the lower layer t_{ph}: $t_e = t_{\text{CEL}} + t_{\text{ph}}$.

5-18. THE INVERSE PROBLEM OF PHOTOLITHOGRAPHY: FINDING A MASK FOR A SPECIFIC IMAGE OR RESIST MASK

In the traditional approach to photolithography problems, the required shape of the resist mask was reproduced (sometimes with minor corrections) in the mask (photomask) and the optical system was expected to reproduce the configuration of the mask as accurately as possible on the wafer. The inverse of this problem was first stated in [31]: for a given resist mask and complex optical system transfer function, define the complex transmittance of the mask. If the inverse problem of photolithography can be solved with a given resist mask terrain, then the parameters of the resist exposure and latent image development processes are taken as known factors. Such a statement of the inverse problem can be called "broad." In a narrow statement of the inverse problem, the intensity distribution function in the mask image is specified; the complex transfer function for the optical system is also known and the complex transmittance of the mask must be found.

Obviously, the inverse problem can be solved as it applies to all lithography methods (optical, x-ray, electron-, and ion-beam) and to image transfer if the system transfer function that produces the image is known. In Chapter 2 we studied the system transfer function for shadow printing in wide electron beams; the corresponding transfer function can be used to solve the inverse problem in electron-beam lithography.

It is obvious that the inverse problem of lithography is a method for modelling optical masks. The pattern in masks such as these may have nothing in common with the geometry of the required resist mask; therefore, optimal masks may be quite different from traditional masks (it may turn out that a new technology will have to be developed for these masks).

Inverse problems of lithography are obviously related to the so-called incorrect inverse problems for which methods of solution were developed by A. N. Tikhonov[(32)]; these problems can be solved on a computer using nonlinear programming methods.[(33)]

We will say that we are required to create a given light pattern $I(\rho) = |\mathbf{E}(\rho)|^2$ on a wafer, where $\mathbf{E}(\rho)$ is the field amplitude in the image plane. The amplitude $\mathbf{E}(\rho)$ is the result of a (linear or nonlinear) transformation of the complex amplitude transmittance coefficient of the mask $T(\rho')$:

$$\mathbf{E}(\rho) = \hat{D}[T(\rho')] \tag{5-216}$$

where ρ' is a point in the plane of the mask. In a shadow printing system with a microgap the Fresnel transformation (see Section 5-16),

$$E(\rho) = \frac{1}{\sqrt{2}} \iint T(\rho') \exp\left[\frac{i\pi}{2}(\rho - \rho')^2\right] d^2\rho' \tag{5-217}$$

emerges as the transformation $\hat{D}$, where $\rho = (\xi, \eta)$, $\xi = x/\sqrt{\lambda h/2}$, and $\eta = y/\sqrt{\lambda h/2}$, λ is the wavelength of the radiation, and h is the gap between the mask and the image plane. The transformation given by Eq. (5-217) is linear; however, for an intensity $I(\rho) = |\mathbf{E}(\rho)|^2$ the relationship between $I(\rho)$ and $T(\rho')$ will become nonlinear. A similar situation (from a mathematical point of view) arises in a projection system with coherent light on the mask, where the transformation $\hat{D}$ has the form (see Eq. (5-34))

$$E(\rho) = \iint h(\rho - \rho')T(\rho')\, d^2\rho' \tag{5-218}$$

where $h(\rho - \rho')$ is the coherent transfer function for the optical system. It is assumed in Eqs. (5-217) and (5-218) that the mask is irradiated by a plane wave having unit amplitude (coherent light).

A transformation that is nonlinear in $T(\bar{\rho}')$ occurs in systems with incoherent light (see Eq. (5-39)):

$$I(\bar{\rho}) = \iint s(\bar{\rho} - \bar{\rho}')|T(\bar{\rho}')|^2\, d^2\bar{\rho}' \tag{5-219}$$

where it is assumed that the intensity of the unit amplitude wave passing through the mask is $I_g(\bar{\rho}') = |T(\bar{\rho}')|^2$. Finally, for the sake of completeness, we will reproduce the transformation $\hat{D}$ for projection systems with partially coherent light (see Section 5-12, Eq. (5-163)):

$$I(\bar{\rho}) = \int d^2\bar{\rho}_1' \int d^2\bar{\rho}_2' h(\bar{\rho} - \bar{\rho}_2')h^*(\bar{\rho} - \bar{\rho}_1')W_s(\bar{\rho}_1', \bar{\rho}_2')T(\bar{\rho}_1')T(\bar{\rho}_2') \tag{5-220}$$

As we can see, a nonlinear transformation of the type $I(\bar{\rho}) = \hat{L}[T(\bar{\rho}')]$ occurs in every case.

We will break down the area occupied by the image in the plane $\bar{\rho}$ into the light area L_i, the shadow area S_i, and the boundary areas G_i between the light and shadow areas (these areas correspond to the intervals of points x_i in the one-dimensional case). Restrictions (the intensity distribution in the image plane) that define the required image quality can be imposed on the values of $I(\bar{\rho})$ in these areas, e.g.:

$$\min\{I(\bar{\rho})\}\bar{\rho} \in L > A \max\{I(\bar{\rho})\}_{\bar{\rho}\in s}, \qquad A > 1 \tag{5-221a}$$

$$|\partial I(\bar{\rho})/\partial x|_{x\in g}, \qquad |\partial I(\bar{\rho})/\partial y|_{y\in g} > B \tag{5-221b}$$

$$\max\{|I(\bar{x}_i) - I(\bar{x}_j)|\}_{x_i\in G, x_j\in G} < C \tag{5-221c}$$

where $\bar{x}_k$ is a median point in the interval G_k. The problem of finding an optimal mask is that the unknown transparence function $T(\bar{\rho}')$ must provide an extremum in the objective function (functional) of $T(\bar{\rho}')$ for all or some of the constraints given by Eqs. (5-221). For example, we can require that the total illumination in the light areas be a maximum:

$$\int_L I(\bar{\rho})\, d^2\bar{\rho} \to \max \tag{5-222}$$

or a minimum in a shadow area:

$$\int_S I(\bar{\rho})\, d^2\bar{\rho} \to \min \tag{5-223}$$

The image contrast may serve as the objective function:

$$\int_L I(\bar{\rho})\, d^2\bar{\rho} \bigg/ \int_S I(\bar{\rho})\, d^2\bar{\rho} \to \max \tag{5-224}$$

The boundary gradients may also be thought of as a functional and we require that

$$|\partial I(\bar{\rho})/\partial\bar{\rho}|_{\bar{\rho}\in G} \to \max \tag{5-225}$$

(when this is done, Eq. (5-221b) must not be one of the conditions). We will point out that the function $T(\bar{\rho}')$ must also belong to those solutions that correspond to the actual physical and technological constraints.

In solving the problem given by Eq. (5-219)–(5-221) on a computer, the chosen objective function becomes its own discrete analogs. Thus, the inverse problem of lithography reduces to a problem in nonlinear programming.

As an example, we will discuss a solution of the inverse problem of photolithography for a system that has a controllable microgap when the transformation $\hat{D}$ is described by Eq. (5-215) and the desired image is one-dimensional. Then

$$E(\xi) = \frac{1}{\sqrt{2}} \int_{-\infty}^{+\infty} T(\xi') \exp\left[\frac{i\pi}{2}(\xi - \xi')^2\right] d\xi', \qquad \xi = x/\sqrt{\lambda h/2} \tag{5-226}$$

Let the discrete analog $T(\xi')$ be the even, piecewise-constant function

$$\begin{gathered} T(-\xi') = T(\xi'), \qquad T(\xi_j^l \le \xi' \le \xi_j^r) = T_j \\ \xi_j^l = (j-1)h_1, \qquad \xi_j^r = jh_1, \qquad j = 1, 2, \ldots N \end{gathered} \tag{5-227}$$

Thus, the desired mask is a set of infinitely long bands h_1 wide having amplitude transmittance coefficients T_j that are constant within the bands. The width of the mask is $L = 2Nh_1$. We specify the discrete points in the image as $\xi_k = (k-1)h_2$; $h_2 = mh_1$ and k, m are integers. When the discrete function $T(\xi')$ given by Eq. (5-227) is substituted into Eq. (5-226) we easily find that

$$E_k \equiv E(\xi_k) = \sum_{j=1}^{N} \phi_{kj} T_j \tag{5-228a}$$

$$\phi_{kj} = \frac{1}{\sqrt{2}}(\varphi_{jm-k+1} - \varphi_{jm-m+k-1} + \varphi_{jm+k-1} - \varphi_{jm-m-k+1}) \tag{5-228b}$$

$$\varphi_n = C(nh_2) + iS(nh_2) \tag{5-228c}$$

where C and S are standard Fresnel integrals. As we can see from Eq. (5-228a) the discrete transformation $E(\xi) = \hat{D}[T(\xi')]$ is linear; the mask image has been specified as a discrete function $E(\xi_k)$, i.e., a band h_2 wide having a constant amplitude $E(\xi_k \le \xi \le \xi_{k+1}) = E(\xi_k)$.

Putting the complex quantities of Eq. (5-228a) into the form

$$\phi_{kj} = a_{kj} \exp(i\alpha_{kj}), \qquad T_j = t_j \exp(i\beta_j), \qquad E_k = e_k \exp(i\gamma_k)$$

we can find the light intensity at a point ξ_k in the image:

$$I_k \equiv I(\xi_k) = \left(\sum_{j=1}^{N} a_{kj} t_j \cos(\alpha_{kj} + \beta_j)\right)^2 + \left(\sum_{j=1}^{N} a_{kj} t_j \sin(\alpha_{kj} + \beta_j)\right)^2 \tag{5-229}$$

which, along with the discrete analogs of the objective function and Eqs. (5-228a-c), are used to optimize the masks. Conditions may be imposed on the variables t_j and β_j that describe the amplitude and phase properties of the masks. These conditions arise, e.g., due to fabrication technology and may be constraints for the optimization problem. It is quite natural to single out, e.g., amplitude, ($\beta_j = 0$), and phase, ($t_j = 1$), masks. We will use the functional

$$F(\{T_j\}) = \min\{I_j\}_{j \in L} - \max\{I_j\}_{j \in s} \tag{5-230}$$

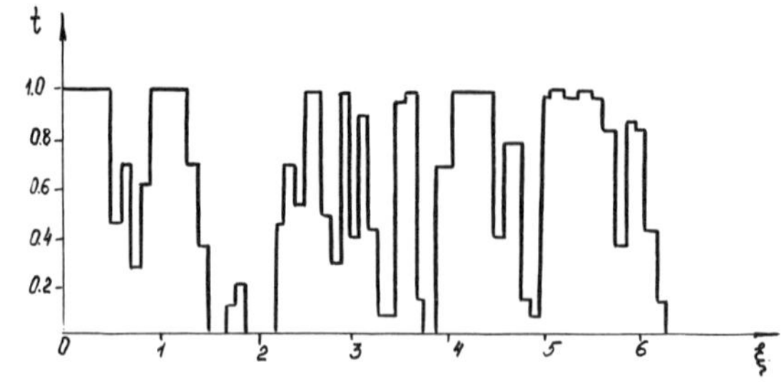

a

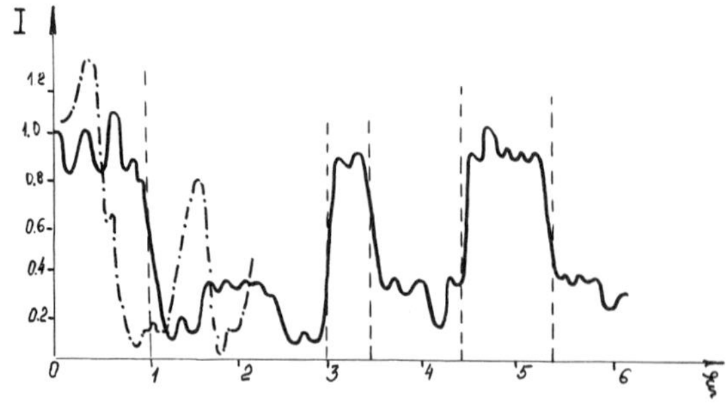

b

FIGURE 5-19. a) Mask transmittance calculated for a given intensity distribution on a wafer (the inverse problem of photolithography); b) the solid line is the intensity distribution in the image of the mask; the dashed and dotted line is the image of a mask that consists of light and dark bands; the vertical dashed lines are the boundaries between the light and dark bands.

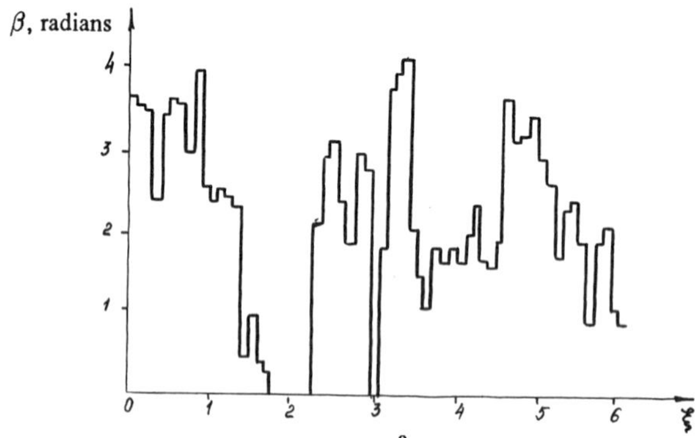

a

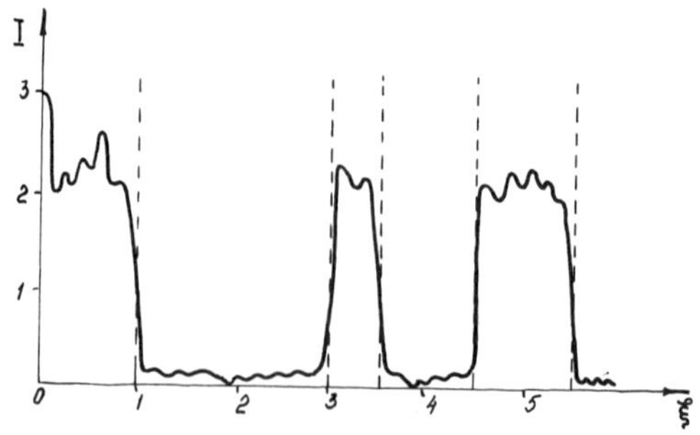

b

FIGURE 5-20. a) Phase shifts in a mask calculated for a given intensity distribution on a wafer; b) the vertical dashed lines are the specific boundaries between the light and dark bands in the image of the mask; the solid line is the light in the image of a phase mask.

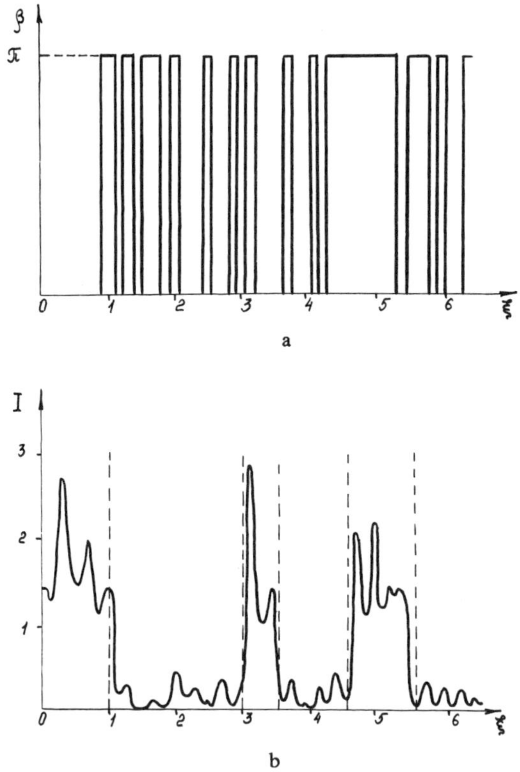

FIGURE 5-21. a) Distribution of phase shifts in a phase "binary" mask; b) the solid line is the intensity distribution in the shadow image of a binary phase mask; the vertical dashed lines are the specific boundaries between the light and dark bands.

which is the difference between the minimal illumination in a light area and maximal light in a shadow area, as the objective function; it is quite natural to seek a $F(\{t_j\}) \rightarrow$ max. The problem is solved by the gradient method for amplitude and phase masks. Equations (5-219)–(5-221) are not used, but even then, the masks that were modelled produce an intensity distribution in the image which is very close to what is desired.

An example of finding optimal masks is shown in Figs. 5-19 and 5-20 for phase and amplitude masks. The desired image comprises three light bands having dimensionless widths of 1, 0.5, and 1; the dark bands are 2 and 1 units wide, and the transition from a light to a dark area is a step function. Values $N = 63$ and $h_1 = h_2 = 0.1$ were used in the calculations. This value of width for the dimensionless band corresponds to a band $\Delta x = \Delta\xi\sqrt{\lambda h/2} = 0.1$ μm when $\lambda = 0.3$ μm and $h = 7$ μm. It is apparent from the patterns that a phase mask produces a high quality image (high contrast, sharp boundaries between the light and shadow areas), whereas the light produced in the image plane by an optimized amplitude mask is markedly different from what was required. Phase masks also have the advantage in reducing exposure time; they transmit all of the light flux, merely redistributing it in the image plane.

We now return to the subject of fabricating optimized masks. We will assume that a phase difference $\Delta\beta = (2\pi/\lambda)(n-1)\Delta d$ is created by depositing a thin coating onto a uniform, perfectly transparent, mask substrate. If we assume, e.g., that the step change in the phase difference is $\Delta\beta/2\pi = 0.1$, then the step change in the coating thickness is $\Delta d = (\Delta\beta/2\pi)(\lambda/(n-1)) = 0.1\lambda/(n-1)$. For a given wavelength λ the quantity

Δd will be larger if a material having low optical density is used for the coating. When $n - 1 = 0.5$, $\lambda = 0.3$ μm and $\Delta d = 0.6$ μm the step is $\Delta d = 0.6$ μm. Ten percent accuracy in thickness control requires that $\delta(\Delta d) \leq 60$ nm. Controlling the deposition of thin films can produce the indicated parameter values of $\Delta d \pm \delta(\Delta d)$; however, significant obstacles to mask fabrication are encountered in a lithography process that has high resolution at each thickness step Δd. When $2\pi/\Delta\beta = 10$, nine such processes must be performed which, by the number of lithographies is equivalent to fabricating a medium scale integrated circuit. A unique way to reduce the number of lithographies is to increase the size of the discrete step in the phase change; when $\Delta\beta = \pi/2$ the number of lithographies is $n_L = 3$.

It is not hard to evaluate the parameters of absorbent coatings for fabricating amplitude masks. From the transmittance function $t(d) = |T(d)|^2 = \exp(-\mu d)$ we easily find $\Delta d = (1/\mu)(\Delta t/t)$; if $\Delta t/t = 0.1$ and $\mu = 1$ μm^{-1}, then $\Delta d \pm \delta(\Delta d) = 0.1 \pm 0.01$ μm. In terms of the number of lithography processes, everything that was said earlier remains valid.

Calculations done for the one-dimensional case show that a significant improvement in image quality is seen when binary masks are used. As an example, the intensity distribution (Fig. 5-21a) is shown for a binary phase mask (Fig. 5-21b). Fabricating a binary mask requires only one film layer to be deposited and only one lithography, just as in fabricating ordinary masks. However, the number of details on the mask is much greater than in the image; consequently, the resolution of the mask fabrication process must be higher.

REFERENCES

1. J. S. Wilczynski, *J. Vac. Sci. Technol.*, **5**, No. 1, 288–292 (1987).
2. A. N. Broers, *Methods and Materials in Microelectronic Technology*, J. Barton (editor), Plenum Press, New York (1984), pp. 147–179.
3. B. J. Lin, *Solid State Technol.*, 63–68 (1986).
4. A. Heuberger, *Methods and Materials in Microelectronic Technology*, J. Barton (editor), Plenum Press, New York (1984), pp. 99–126.
5. A. Heuberger, *Microelectronic Engineering*, **5**, No. 1–4, 3–38 (1986).
6. G. N. Berezin, A. V. Nikitin, and R. A. Suris, *Optical Fundamentals of Contact Photolithography* [in Russian], Radio i Svyaz', Moscow (1982).
7. M. Lacombat, G. M. Dubroeucq, J. Massin, and M. Brevignon, *Solid State Technol.*, 115–121 (1980).
8. A. T. Friberg, SPIE, 194, Appl. of Optical Coherence, 55–70, 71–83 (1979).
9. E. W. Marchand and E. J. Wolf, J. *Opt. Soc. Am.*, **64**, No. 9, 1219–1226 (1974).
10. E. Wolf, *J. Opt. Soc. Am.*, **68**, No. 11, 1597–1605 (1978).
11. B. N. Kotletsov, *Microimages. Optical Methods for Obtaining and Controlling Them* [in Russian], Mashinostroenie, Leningrad (1985).
12. M. C. King and M. R. Goldrick, *Sol. State Technol.*, **20**, No. 2, 37–40 (1977).
13. M. Born and E. Wolf, *Principles of Optics*, Pergamon Press, New York (1964).
14. E. L. O'Neill, *Introduction to Statistical Optics*, Addison-Wesley, Palo Alto (1963).
15. J. W. Goodman, *Introduction to Fourier Optics*, McGraw-Hill, New York (1968).
16. A. Marechal and M. Franson, *Diffraction Structure des Images*, Paris (1960).
17. M. Franson, *La Granularite Laser (Spekle) et ses Applications en Optique*, Paris–New York (1978).
18. I. S. Gradshtein and I. M. Ryzhik, *Tables of Integrals, Sums, and Series* [in Russian], Moscow (1962).
19. J. Cuthbert, *Solid State Technol.*, No. 8, 59–69 (1977).
20. R. Hershel, *Proc. SPIE*, **135**, 24–29 (1978).

21. A. C. Schell, *IEEE Trans.*, **AP-15**, No. 1, 187–188 (1967).
22. M. A. Leontovich, *Izv. Akad. Nauk SSSR, Fiz.*, **8**, 16 (1944).
23. S. A. Akhmanov and R. V. Khokhlov, *Problems in Nonlinear Optics* [in Russian], Moscow (1964).
24. S. A. Akhmanov, Yu. E. D'yakov, and A. S. Chirkin, *Introduction to Statistical Radiophysics and Optics* [in Russian], Nauka, Moscow (1981).
25. K. A. Valiev, L. V. Velikov, G. S. Volkov, and D. Yu. Zaroslov, *Kvant. Elektron.*, **14**, No. 6, 1266–1268 (1987).
26. H. Kogelnik, *Appl. Optics*, **4**, No. 12, 1562–1569 (1965).
27. E. Cullman, Semiconductor International, May, 1985.
28. F. H. Dill, W. P. Hornberger, P. S. Hauge, and J. M. Shaw, *IEEE Trans. Electron Devices*, **ED-22**, No. 7, 445–452 (1975).
29. B. F. Griffing and P. R. West, *Polym. Eng. Sci.*, **23**, 947 (1982); *IEEE Electron Device Lett.*, **EDL-4**, 14 (1983).
30. M. Nakase, H. Niki, T. Satoh, and A. Kumagae, *Jpn. J. Appl. Phys.*, **26**, No. 2, 293–297 (1987).
31. K. A. Valiev, V. P. Kudrya, and T. M. Makhviladze, *Mikrolektronika*, **15**, No. 5 (1986).
32. A. N. Tikhonov and V. Ya. Arsenin, *Methods for Solving Incorrect Problems* [in Russian], Moscow (1979).
33. F. Gill, U. Murray, and M. Wright, *Practical Optimization* [Russian translation], Mir, Moscow (1985).

6

Procedures for Processing Exposed Resist Films and Resist Mask Topography

6-1. THE RESIST TRANSFER FUNCTION

The ultimate goal of lithography is to produce a resist mask that will serve for the local processing of a lower structural layer. Figure 6-1 shows the cross section of part of a resist mask. The length of the base, W_2, is taken as the mask width, since the mask covers the surface of the structural layer in this space, preserving the W_2-wide portion of the surface when processing through the resist mask. The width $W_2 < 1\ \mu$m of a band on a mask must be evaluated from two different perspectives. First, we can compare W_2 with the width W_M of a dark band on the mask-photostencil pattern (in a 1:1 scale or an n-fold reduction). The algebraic difference $\Delta W = W_2 - W_M$ can be called distortion. The amount of distortion and the other mask parameters depend to large degree on the properties of the optical system that transfers the photomask image onto the surface of the resist film as a two-dimensional intensity distribution $I(x, y)$. Second, the size of W_2 is a function of the resist material parameters and the resist layer processing parameters. Uncontrolled variations in the resist film parameters and its processing modes will produce variations in W_2 even when the light $I(x, y)$ is invariant. Ordinarily, we assume that the relative error in linewidth reproduction in a resist mask cannot exceed 10%. As a result, we must face the problem of revealing those factors that cause deviations in W_2 and of defining the necessary accuracy with which these factors must be controlled when resist processing procedures, excluding exposure, are performed.

The steepness of the slope of the side walls on a resist mask, $\tan\theta = h/(1/2)(W_2 - W_1)$, is the second major feature of a resist mask and it determines the width of the transition zone $\Delta x = (1/2)(W_2 - W_1) = h/\tan\theta$ in which the structural layer is protected by a thinner mask. We say that the quantity Δx defines minimal linewidth in a resist mask: $l_{\min} = 2\Delta x$. In turn, $l_{\min}$ defines the minimal lattice period d that can be reproduced in the resist layer: $d = 2l_{\min}$. The reciprocal of the minimal period

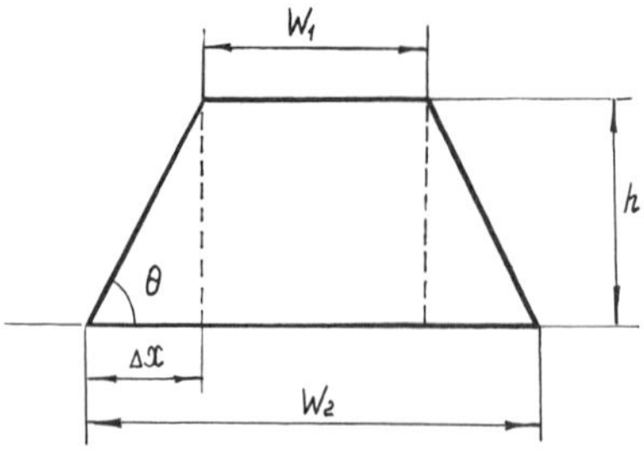

FIGURE 6-1. Cross section of a band in a resist mask represented by a vertical plane: h is the thickness, W_2 is the mask width, $\Delta x = (1/2)(W_2 - W_1)$ is the edge width, and $\tan\theta = 2h/(W_2 - W_1)$ is the slope of a side wall in the mask.

R is the resolution of the lithographic process: $R = 1/d$ (lines/mm). In most cases, a resist mask whose side walls are very nearly vertical is preferred. In order for lift-off lithography to be successfully employed the slope of the side walls must be negative, ($W_2 < W_1$). Thus, resist mask production methods must control both the width of the resist mask, W_2 , and the steepness of the vertical walls, $\tan\theta$, to within a certain accuracy. Consequently, we must clarify the material and processing parameters upon which the indicated resist mask characteristics depend.

The topography of a resist mask $h(x,y)$ is related to the topography of the radiation dose $I(x,y)$ by a correspondence principle. For positive resists this correspondence is inverted in the sense that a maximum radiation dose $I(x,y)$ corresponds to minimum resist layer thickness $h(x,y)$, and conversely for negative resists. We will further examine the case of a one-dimensional dose distribution $I(x,y)$. It is obvious that a resist mask topography defined by a thickness function $h(x)$ is implicitly related to the dose function $I(x)$: $h(x) = h(I(x))$. Consequently, we can write[1]

$$\frac{\partial h}{\partial x} = \frac{\partial h}{\partial I}\frac{\partial I}{\partial x} \tag{6-1a}$$

For positive resists the regions in which the dose falls off correspond to regions in which the resist mask thickness increases, which means that $\partial h/\partial I < 0$. Physically, Eq. (6-1) means that the steepness of an edge on a resist mask $\partial h/\partial x$ is determined by two independent (in a first approximation) groups of factors: the dose gradient depends only on the parameters of the optical system that exposes the resist, and the energy gradient of $\partial h/\partial I$ depends on the properties of the resist layer and the procedures used to process it. (It has been suggested[1] that the function $\partial h/\partial I$ be called the resist transfer function; however, this nomenclature has not been widely accepted.) If we rewrite Eq. (6-1) in the form

$$\varphi(I) = \frac{\partial h}{\partial I} = \frac{\partial h/\partial x}{\partial I/\partial x} \tag{6-1b}$$

it follows that the resist transfer function is the ratio of the derivatives with respect to the coordinates of the functions that describe the resist mask profile and the respective dose profile. The resist transfer function $\varphi(I) = \partial h/\partial I$ allows us to find the mask profile in terms of the dose profile and vice-versa:

$$h(x) = \int \varphi(I(x))(\partial I/\partial x)\,dx \tag{6-2a}$$

$$I(x) = \int \varphi^{-1}(I(x))(\partial h/\partial x)\,dx \tag{6-2b}$$

With the introduction of the resist transfer function we must make some additional clarifications. As we saw in previous chapters, the dose distribution function (the electron or ion current density in electron or ion lithography) along with some integral conversion generates an absorbed energy density distribution function: $E(x, z) = D\{I(x)\}$. The integral conversions contain proximity effects in corpuscular lithography methods and diffraction effects, reflections, and interference in optical lithography. The resist mask profile problem is solved in two steps: from the function $I(x, y)$ we find the absorbed energy density $E(x, y, z) = D(I(x, y))$ and then, using an empirical relationship $v_{\text{etch}} = f(E(x, y, z))$, we find the profile of the resist mask. In striving toward an analytical formulation of the problem, we will simplify it, adopting a "multiplicative" rule for the conversion of $E = D\{I\}$:

$$E(x, z) = D\{I\} = \varkappa(z) \cdot I(x) \tag{6-3}$$

This is tantamount to ignoring the effects from side scattering and diffraction in the resist film. The function $\varkappa(z)$ describes how the light is attenuated due to absorption or the effects due to increasing energy losses when a beam of fast electrons penetrates a target:

$$\varkappa_{\text{ph}}(z) = \alpha \exp(-\alpha z) \tag{6-4}$$

$$\varkappa_{\text{el}}(z) = \alpha(1 + \alpha_2 z) \tag{6-5}$$

Equation (6-4) describes how the light intensity inside a resist layer with an absorption coefficient of α attenuates. Absorption modulation due to standing waves is understood here to be absent. The linear function in Eq. (6-5) gives the absorbed energy density distribution during exposure to an electron beam; a linear approximation for this function is valid when $R_p \gg h_0$, where R_p is the range of electrons in the resist layer and the electrons are only slightly reflected from the substrate.[2]

6-2. THE KINETIC CURVES FOR RESISTS

Experimental results from etching exposed resist are given in two different forms. The first form gives the normalized residual thickness $h(t)/h_0$ and etching rate $r(t)$ (measured in nm/sec) as a function of time t; these are called kinetic curves. The *kinetic* curves define how the profiles of the resist masks will evolve with time as the latent image is developed. The kinetic curves for nonlinear resists are shown schematically in Fig. 6-2. The presence of a so-called induction (delay) time in the etching process is typical for these curves. This induction time is connected with the initial stage in which the resist and etchant interact with one another and no etching takes place. The presence of an induction time for etching is prominent in both the curves of $h(t)/h_0$ and especially in the curves of $r(t)$. The presence of induction time causes anomalies in curves of the

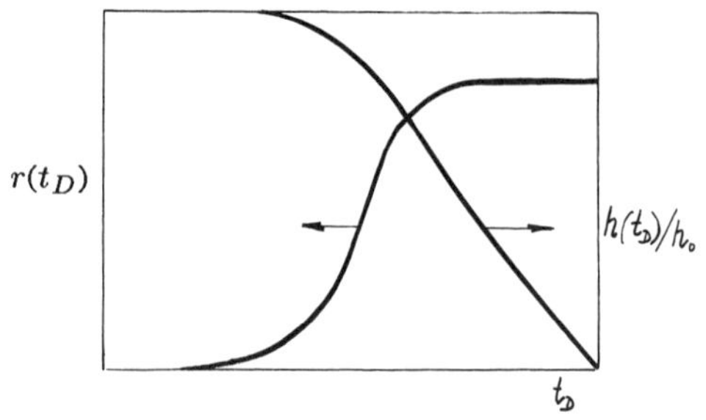

FIGURE 6-2. Kinetic curves for nonlinear resists.

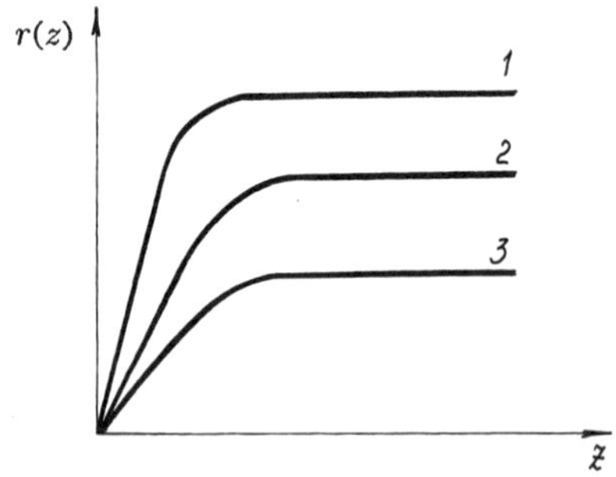

FIGURE 6-3. The dependence $r(z)$ inside a nonlinear resist layer. Curves 1, 2, and 3 correspond to radiation doses of $I_1 > I_2 > I_3$.

function $r(z)$ for the etching rate as a function of depth in the layer z: the upper layers are etched at a very slow rate (Fig. 6-3). Induction time is absent in linear resists; the function $r(z)$ is due entirely to the absorbed energy density distribution $E(z)$ inside the resist layer. We can assume that the etching rate as a function of the z coordinate in nonlinear resists can be broken down into two parts: the "energy" part $H(E(z))$ and the part due to the nonlinearity of the resist, $G(z)$: $r(z) = G(z)H(E(z))$.[2] For linear resists, $G(z) \equiv 1$.

The etching rate as a function of the absorbed energy density can be written as a power function[3]:

$$H(E) = kE^{\beta} \tag{6-6}$$

The $G(z)$ function in [2] was chosen in its "Bose–Einstein" form

$$G(z) = [(1 + F_1/z)^{F_2} + (1 + F_1/z)^{-F_3}]^{-1} \tag{6-7a}$$

Since the function $E(z)$ is known (see Eqs. (6-4) or (6-5)), we can write

$$r(z) = \frac{k[E(z)]^{\beta}}{[(1 + F_1/z)^{F_2} + (1 + F_1/z)^{-F_3}]} \tag{6-7b}$$

Tuning parameters k, β, F_1, F_2, and F_3 can be found from experimental kinetic curves of $h(t)/h_0$ and $r(t)$ by the method of least squares. To do this, we (graphically) invert the equation for $h(t)$ and find $t = \varphi(h)$. We substitute $t = \varphi(h)$ and $h = h_0 - z$ into the function $r(t)$; the result will be the experimental curve for $r(z)$. Fitting the curve for $r(t)$ under the experimental curve gives us the parameters for the function in

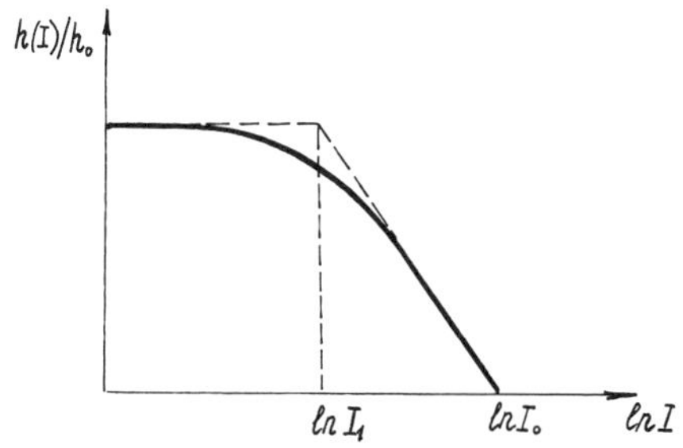

FIGURE 6-4. Characteristic curve for a positive resist: I_1 is the exposure dose at which noticeable etching in the resist begins (under specified conditions); I_0 is the dose at which the resist layer is completely etched.

Eq. (6-7b). Substituting Eq. (6-3) into (6-7), we find the correspondence between the radiation dose distribution $I(x, y)$ and the etching rate function $r(x, y, z)$.

6-3. CHARACTERISTIC CURVES FOR RESISTS

We will now describe procedures for etching exposed resists, making use of the so-called characteristic curves. The latter describe the normalized residual thickness $h(I)/h_0$ as a function of the radiation dose $I(x, y)$ (C/cm^2, J/cm^2) under specific etching conditions (Fig. 6-4). (The relationship with the etching time t_D is developed in the characteristic curves if we construct their families, obtained for different values of t_D.)

We will examine the definition of etching rate as the differential equation

$$-\frac{dh}{dt} = r \tag{6-8}$$

If the function $r(t)$ is known from experimental data the relationship between the kinetic curves for $r(t)$ and $h(t)$ is

$$h(t) = h_0 - \int_0^{t_D} r(t')\,dt'$$

We know that the functions $G(z)$ and $H(E)$ have been found from the kinetic curves. In addition, the exposure dose $I(x, y, z)$ (J/cm^2) and the absorbed energy density $E(z)$ are related by

$$E(z) = E_s \cdot \varepsilon(z), \qquad E_s = I \cdot \alpha \tag{6-9}$$

where E_s is the density of the energy absorbed at the surface of the resist layer, α is the coefficient of absorption for the resist, and $\varepsilon(z)$ is a function that describes how $E(z)$ changes inside the layer; $\varepsilon(0) = 1$. We can then find a function $r(z) = G(z)H(E(z))$ that contains the exposure dose I as a parameter. Integrating Eq. (6-8) over z gives us

$$\int_0^{h_0 - h} \frac{dz'}{r(z'; I)} = t_D \tag{6-10}$$

Solving Eq. (6-10) for h, we can find an equation for the characteristic curve of a resist $h(I)/h_0$ that depends on t_D as a parameter.

Let the resist have been uniformly exposed[1] over its area. We will assume also that $r(E(z), z)$ is not an explicit function of z (a linear and uniform resist). Then, according to Eq. (6-6), $r(E) = kE_s^\beta \varepsilon^\beta(z)$, and

$$\frac{1}{kE_s^\beta} \int_0^{h_0 - h(t_D)} \varepsilon^{-\beta}(z')\, dz' = t_D \tag{6-11}$$

We can easily evaluate the integral in Eq. (6-11) for optical lithography when $\varepsilon(z) = \exp(-\alpha z)$. For a given developing time t_D we find from Eq. (6-11) that

$$h = h_0 - \frac{1}{\alpha\beta} \ln(1 + kE_s^\beta \alpha\beta t_D) \tag{6-12}$$

Using the equality $E_s = I \cdot \alpha$ we find an equation for the characteristic curve of $h(I)/h_0$:

$$\frac{h(I)}{h_0} = 1 - \frac{1}{\alpha\beta h_0} \ln(1 + k\alpha^{\beta+1} \beta t_D I^\beta) \tag{6-13}$$

Equations (6-12) and (6-13) were obtained by integrating the differential equation for the etching process; they allow us to find the time for complete developing of a resist at a point (x, y) having a surface density of absorbed energy $E_s(x, y)$. We assume that $h = 0$ in Eq. (6-12) and define the time t_D for complete etching at a point with a dose of $E_s(x, y)$:

$$t_D(E_s) = \frac{1}{\alpha\beta} \frac{1}{kE_s^\beta(x, y)} (\exp(\alpha\beta h_0) - 1) \tag{6-14}$$

For a thin (weakly absorbing) resist film ($\alpha\beta h_0 \ll 1$) we find from Eq. (6-14) the obvious equation

$$t_D^* = h_0 / kE_s^\beta = h_0 / r_s \tag{6-15}$$

for a strongly absorbing ("thick") resist film, $\alpha\beta h_0 \gg 1$, the developing time increases exponentially:

$$t_D^*(\alpha\beta h_0 \gg 1) = \frac{h_0}{r_s} \frac{e^t - 1}{t}, \qquad t = \alpha\beta h_0 \tag{6-16}$$

Using the relationship $E_s(x, y) = \alpha I(x, y)$ in Eqs. (6-14)–(6-16), we can easily convert from the absorbed energy density E_s to the dose I.

The equation for the characteristic curve, Eq. (6-12), allows us to find the exposure dose E_0 for which a resist will be completely developed (under specific conditions for the developing process). The dose E_0 is used in part as a measure of the sensitivity of the resist: the lower the E_0, the higher the sensitivity. Assuming that $h = 0$, we can solve Eq. (6-12) for $E_s (= E_0)$; the result is

$$E_0 = \left[\frac{\exp(\alpha\beta h_0) - 1}{k\alpha\beta t_D} \right]^{1/\beta} \tag{6-17}$$

For a thin layer ($\alpha\beta h_0 \gg 1$), $E_0 = [h_0 / kt_D]^{1/\beta}$; for a thick layer ($\alpha\beta h_0 \gg 1$), the sensitivity is seen to decrease exponentially as thickness increases:

$$E_0(\alpha\beta h_0 \gg 1) \simeq \frac{\exp(\alpha h_0)}{(k\alpha\beta t_D)^{1/\beta}} \tag{6-18}$$

The reason for this is that the etching rate of an exposed resist falls off inside the layer.

We will define "contrast" (the γ parameter) of a resist (see Fig. 6-4) according to

$$\gamma = \left.\frac{\partial h}{h_0 \partial \ln(I_0/I)}\right|_{I_0} = \left.\frac{1}{h_0}\frac{I\,\partial h}{\partial I}\right|_{I_0} \equiv \left.\frac{1}{h_0}\frac{E_s\,\partial h}{\partial E_s}\right|_{E_0} \tag{6-19}$$

The last identity in Eq. (6-19) follows from the equality $E_s = I \cdot \alpha$. Differentiating the equation for the characteristic curve, Eq. (6-12), we find

$$\gamma(E_s) = \frac{E_s}{h_0}\frac{\partial h}{\partial E_s} = -\beta\frac{r_s t_D}{h_0(1 + r_s t_D \alpha\beta)} < 0 \tag{6-20}$$

By defining contrast in a resist as the slope at the point E_0 that corresponds to $h = 0$, we find from Eq. (6-12) that

$$\gamma = \gamma(E_0) = \frac{1}{\alpha h_0}[\exp(-\alpha\beta h_0) - 1] < 0 \tag{6-21}$$

For a thin, weakly absorbing resist layer ($\alpha\beta h_0 \ll 1$) we have $\gamma = -\beta$, and in this case the etching process defines the contrast of the resist film. The values of γ and β found from experiments are actually close to one another in value; e.g., for an AZ-1370 resist, $\beta = 1.63$ and $\gamma = 1.85$.[(4)] For strongly absorbing resist films ($\alpha\beta h_0 \gg 1$) we find from Eq. (6-21) that $\gamma = -1/h_0\alpha$; in this case the value of contrast is defined by the absorbed energy distribution throughout the resist thickness. The contrast as a function of the parameter $t = \alpha\beta h_0$ is given by the relation

$$\gamma(\alpha\beta h_0 = t) = -\beta\,\frac{1 - e^{-t}}{t} \tag{6-22}$$

The function $\varphi(t) = (1 - e^{-t})/t$ decays rather slowly from a value of $\varphi(0) = 1$ when $t = 0$ to $\varphi(1) = 0.6321$. Apparently, the situation $\alpha\beta h_0 \gg 1$ never occurs in practice; in practically important situations $\alpha\beta h_0 \leq 1$ the contrast $|\gamma|$ is close in value to β. It is interesting to compare Eq. (6-13) for a characteristic curve with its empirical form, widely known in the literature of lithography:

$$h(I)/h_0 = \gamma \ln(I_0/I), \qquad \gamma = 1/\ln(I_0/I_1) \tag{6-23}$$

where $h(I)$ is the equation for the tangent to the characteristic curve at the point I_0 that corresponds to the "boundary condition" $h(I_0) = 0$. The dose I_1 is defined by where $H(I)$ intersects with the horizontal line $h = h_0$, i.e., from the condition that $h(I_1) = h_0$. The contrast is $\gamma = -\tan\theta$ and θ is the angle between the tangent to the characteristic curve at the point I_0 and the axis $\ln I$. Equation (6-23) describes only that part of the characteristic curve that coincides with the tangent, i.e., its straight portion at the points h and lnI. Conversely, Eq. (6-13), which we derived, describes a characteristic curve throughout its extent, satisfying the natural boundary conditions $h(I_0) = 0$ and $h(0) = h_0$ and having at the point I_1 a "deficit" $\Delta h = h_0 - h$ equal to

$$\Delta h = \frac{1}{\alpha\beta}\ln(1 + kI_1^{\beta}\alpha\beta t_D) \tag{6-24}$$

From the empirical Eq. (6-23) we can derive a number of new results if we convert from the dose I to the absorbed energy density. In optical lithography, when a resist is described by the coefficient of absorption α, we have

$$E = E_s \exp[-\alpha(h_0 - h)], \qquad E_s = \alpha I \tag{6-25}$$

From Eq. (6-25) we can find energy densities I_0 and I that correspond to the residual thicknesses $h = 0$ and h after developing:

$$I_0 = E_0 \alpha^{-1} \exp[+\alpha h_0], \qquad I = E\alpha^{-1} \exp[\alpha(h_0 - h)]$$

and substitute these values into Eq. (6-23). The result will be

$$h = h_0 \gamma \left[\ln \left(\frac{E_0}{E} \right) - \alpha h \right] \tag{6-26}$$

Solving Eq. (6-26) for h we find[5] that

$$h(E) = h_0(\gamma^{-1} + \alpha h_0)^{-1} \ln(E_0/E) \tag{6-27a}$$

If we consider the absorbed energy density to be a function of the coordinates x and y in the plane of the resist mask, then Eq. (6-27a) is a function that describes a resist mask profile obtained by developing after exposure:

$$h(E(x,y)) = h_0(\gamma^{-1} + \alpha h_0)^{-1} \ln(E_0/E(x,y)) \tag{6-27b}$$

For weakly absorbing resists with low contrast γ we will assume that the condition $\gamma^{-1} \gg \alpha h_0 \ll 1$ is satisfied; the resist mask profile is then defined by the value of the contrast γ and the energy release profile $E(x, y)$:

$$h(x,y)/h_0 = \gamma \ln(E_0/E(x,y)) \tag{6-27c}$$

As we can see, when exposure is uniform throughout the thickness ($\alpha h_0 \ll 1$) of a resist the value of α does not affect the resist mask profile.

In the opposite case of a strongly absorbing resist ($\alpha h_0 \lesssim 1$) and a high contrast resist ($\gamma^{-1} \ll \alpha h_0$) we have

$$h(x,y)/h_0 = \frac{1}{\alpha h_0} \ln(E_0/E(x,y)) \tag{6-27d}$$

Under these conditions the profile of a resist mask obtained after developing is defined by the $\alpha h_0 = h_0/\alpha^{-1} = \tilde{h}_0$ parameter, equal to the original dimensionless thickness of the resist mask.

It is evident that Eq. (6-27b) can satisfy the boundary condition $h(E) = 0$ at any point x, y on the surface, i.e., the thickness of the resist can be made zero at any point on the exposed surface if the corresponding developing time is chosen. However, in the same Eq. (6-27b) the rate and time of developing are not present as parameters of the characteristic curve; therefore, the empirical Eq. (6-23) gives no indication of the amount of time t_D needed for complete developing.

We will now find an equation for the characteristic curve for the case when $\varepsilon(z) = (1 + \alpha_2 z)$ (an approximation of the $E(z)$ function from electron-beam lithography, see Eq. (6-5)). Integrating Eq. (6-11) with $\varepsilon(z) = (1 + \alpha_2 z)$ yields

$$h = h_0 - \frac{1}{\alpha_2}\left\{[1 + \alpha_2(1 - \beta)r_s t_D]^{1/(1-\beta)} - 1\right\}, \qquad r_s = kE_s^\beta \tag{6-28}$$

The time needed to completely develop a resist, t_D, is

$$t_D = \frac{(1 + \alpha_2 h_0)^{1-\beta} - 1}{r_s \alpha_2 (1 - \beta)} \tag{6-29}$$

For the case of a thin layer ($\alpha_2 h_0 \ll 1$), we find the developing time for a uniformly exposed resist from Eq. (6-29): $t_D^* = h_0/r_s$. For a nonuniformly exposed resist we can write

$$t_D^* = \frac{h_0}{r_s}\varphi(\alpha_2 h_0), \qquad \varphi(\xi) = \frac{(1 + \xi)^{1-\beta} - 1}{\xi(1 - \beta)} \tag{6-30}$$

The function $\varphi(\xi)$ when $\xi \gg 1$ behaves like $\varphi(\xi) \simeq 1/\xi(\beta - 1)$; this means that the developing time decreases as $1/\alpha_2 h_0$, since the lower lying areas are etched more quickly because of the larger amount of absorbed energy.

The slope of the tangent to the characteristic curve $h \sim \ln(E_0/E)$ at the point $E = E_s$ on the E axis is

$$\gamma(E_s) = \frac{E_s}{h_0}\frac{\partial h}{\partial E_s} = -\frac{\beta r_s t_D}{h_0}[1 + \alpha_2(1 - \beta)t_D r_s]^{\beta/(\beta-1)} \tag{6-31}$$

Substituting the value of t_D from Eq. (6-29) that corresponds to complete developing, ($h = 0$), into Eq. (6-31) gives us the contrast in a resist during electron-beam exposure:

$$\gamma(E_0) = -\frac{\beta}{h_0}\frac{(1 + \alpha_2 h_0)^{1-\beta}}{\alpha_2(1 - \beta)}(1 + \alpha_2 h_0)^\beta \tag{6-32}$$

For a thin layer ($\alpha h_0 \ll 1$), we again find from Eq. (6-32) that $\gamma \approx \beta$.

We will now find the sensitivity of a resist during electron-beam exposure. Setting $h = 0$ in Eq. (6-28) we will solve the equation for $E_s (\equiv E_0)$:

$$E_0 = \left[\frac{1 - (1 + \alpha_2 h_0)^{1-\beta}}{k\alpha_2 t_D(\beta - 1)}\right]^{1/\beta} \tag{6-33a}$$

for a thin layer ($\alpha_2 h_0 \ll 1$), $E_0 = [h_0/kt_D]^{1/\beta}$. For the condition $\alpha_2 h_0 \gg 1$ we have

$$E_0 = [k\alpha_2 t_D(\beta - 1)]^{-1/\beta} \tag{6-33b}$$

The formulas obtained establish how the parameters of an electron resist (contrast, sensitivity) depend on the parameters of the etching processes (t_D, β).

6-4. THE STEEPNESS OF THE SIDE WALLS IN A RESIST MASK

If the characteristic equation is known for a resist, we can easily find the slope of its side walls after developing. We will say that we are dealing with a positive resist: $\partial h/\partial I < 0$. We will assume that the slope of the mask profile has been produced in a region in which the light falls off: $\partial I/\partial x < 0$. Then, $\partial h/\partial x > 0$, i.e., the slope $\partial h/\partial x$ describes the support wall of the resist mask.

Equation (6-1) can be rewritten as

$$\left.\frac{\partial h}{\partial x}\right|_{E_s} = h_0\left(-\frac{E_s}{h_0}\frac{\partial h}{\partial E_s}\right)\left(-\frac{1}{E_s}\frac{\partial E_s}{\partial x}\right) = -h_0\gamma(E_s)\left(-\frac{1}{E_s}\frac{\partial E_s}{\partial x}\right) \tag{6-34}$$

In optical lithography, when light inside the layer decreases according to Bouguer's law, using the result for γ from Eq. (6-21) we obtain

$$\frac{\partial h}{\partial x} = -\frac{1}{\alpha}[\exp(-\alpha\beta h_0) - 1]\left(-\frac{1}{E_s}\frac{\partial E_s}{\partial x}\right) \tag{6-35a}$$

for the slope. When $\alpha\beta h_0 \ll 1$, we obtain

$$\frac{\partial h}{\partial x} = \beta h_0\left(-\frac{1}{E_s}\frac{\partial E_s}{\partial x}\right) \tag{6-35b}$$

When $\alpha\beta h_0 \gg 1$,

$$\frac{\partial h}{\partial x} = \frac{1}{\alpha}\left(-\frac{1}{E_s}\frac{\partial E_s}{\partial x}\right) \tag{6-35c}$$

During electron-beam exposure when the change in exposure inside a layer can be described by the function $E(x, z) = E_s(x)(1 + \alpha_2(h_0 - h))$ we find, with the help of Eq. (6-31), that

$$\frac{\partial h}{\partial x} = -\beta\,\frac{(1 + \alpha_2 h_0)^{1-\beta} - 1}{\alpha_2(1 - \beta)}(1 + \alpha_2 h_0)^{\beta}\left(-\frac{1}{E}\frac{\partial E}{\partial x}\right) \tag{6-36}$$

Using the empirical characteristic equation for a curve, Eq. (6-23), gives us

$$\frac{\partial h}{\partial x} = \left[\frac{h_0\gamma}{1 + h_0\alpha\gamma}\right]\left(-\frac{1}{E(x)}\frac{\partial E(x)}{\partial x}\right) \tag{6-37}$$

It follows from Eqs. (6-34)–(6-36) that the slope of a side wall in a resist mask is greater when: 1) the contrast in the resist, γ, or the etching rate parameter β of an exposed resist is higher; 2) the coefficient of absorption in the resist, α, is smaller; and 3) the rate of attenuation in the light $((1/E)(\partial E/\partial x))$ is greater.

To make our discussion complete, we will recall the rate of light attenuation as a function of the optical system parameters and the light. For proximity printing,

$$\left(-\frac{1}{E}\frac{dE}{dx}\right) = \frac{k_1}{\sqrt{\lambda h/2}} \tag{6-38}$$

and for projection printing,

$$\left(-\frac{1}{E}\frac{dE}{dx}\right) = \frac{k_2(\mathrm{NA})}{\lambda} \tag{6-39}$$

where λ is the wavelength of the exposing light; h is the gap between mask and wafer; NA is the numerical aperture of the projection lens; and k_1 and k_2 are numerical constants ~ 1. We can also find $(1/E)(dE/dx)$ as a function of the beam parameters for electron, ion, and x-ray lithography systems.

6-5. LINEWIDTH IN A RESIST MASK

Linewidth in a resist mask was defined in Section 6-1 as the width W_2 of the base of the cross section of a "line" in a vertical plane (Fig. 6-1). In the practice of lithography, we must control W_2 to within 10% of a specified nominal value and the variance in W_2 to within 10% of its mean value. In microlithography this means that ΔW_2 cannot deviate from its nominal value by more than 0.1 μm. This kind of accuracy in controlling linewidth in resist masks can only be achieved if all of the factors that affect ΔW_2 are revealed and carefully controlled.

An analytical model for the process of developing a latent image in a positive resist was constructed in [4] and the basic factors that regulate linewidth were established. Figure 6-5a shows the light distribution on the surface of a resist layer that forms a "light" line W_M wide in the photomask. After the latent image is developed a window $W_2(\approx W_M)$ wide will be obtained in the resist mask. The window is developed in two steps. In the first step, all points on the surface of the substrate in the area $2x_s$ wide that corresponds to the flat portion of the light distribution $I(x)$ are developed. If the width of the flat portion of $I(x)$ is zero (the function $I(x)$ looks like a curve with a maximum), then at the end of the first step the central point 0 is revealed. In the second step etching the side walls of a slit in the resist mask produces a gap that extends to the points $\pm x_0$; when this happens the width of the light band, $2x_0$, must be equal to the nominal width of the light band; $W_M = 2x_0$.

The complete developing time for a band $2x_0$ wide is[4]

$$t_D = \int_0^{h_0} \frac{dz'}{r(x < x_s, z')} + \int_{x_s}^{x_0} \frac{dx'}{r(z = h_0, x')} \tag{6-40}$$

here, $r(z; x < x_s)$ is the "vertical etching rate" of the resist in the flat area $-x_s \leq x \leq x_s$ of the light $I(x)$, and $r(x', z = h)$ is the "horizontal etching rate" along the interface between the resist and the substrate in the area $x_s \leq |x| \leq x_0$.

We will say that the resist etching rate as a function of the absorbed energy density obeys a power law: $r = (1/k)E^{\beta}$. For the interval $-x_s \leq x \leq x_s$ we can write[4]

$$E(z) = I \cdot f_s \cdot \varepsilon(z) \tag{6-41}$$

where I is the integral intensity (dose) of radiation on a large area, $f_s \leq 1$ is a coefficient that distinguishes the intensity in a finitely wide band from the intensity in the large area,

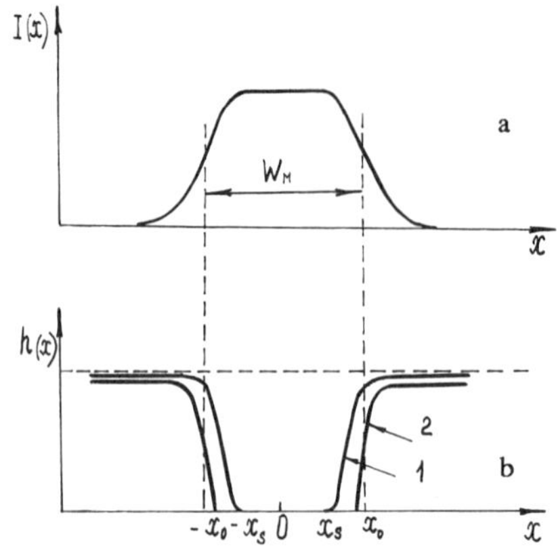

FIGURE 6-5. a) Profile of the intensity distribution on the surface of a resist layer produced by an optical system: W_M is the width of the transparent window in the mask; b) curve 1 is the profile of the developed latent image (resist mask) at the moment when the surface of the substrate is revealed in the area $I(x) = I_{\max} = \text{const}$; curve 2 is the profile of the resist mask at the moment when the width of the developed area is $2x_0 = W_M$.

and $\varepsilon(z)$ is the fraction of radiation intensity $I \cdot f_s$ that was absorbed at the depth z. Similarly, in the areas of decreasing intensity $x_s \leq |x| \leq x_0$ we can assume that

$$E(z_0 = h_0, x) = I \cdot f_s \cdot \varepsilon(h_0) \cdot f(x) \tag{6-42}$$

for the E at the interface between the resist and the substrate, where $f(x)$ is a function that describes the decrease in intensity. Substituting Eqs. (6-41) and (6-42) into (6-40) gives us

$$t_D = k(If_s)^{-\beta} \left\{ \int_0^{h_0} \varepsilon^{-\beta}(z')\, dz' + \varepsilon^{-\beta}(h_0) \int_{x_s}^{x_0} f^{-\beta}(x')\, dx' \right\} \tag{6-43}$$

If we neglect the function $\varepsilon(z)$, assuming that $\varepsilon(I) = \text{const}$, then

$$t_D = k(I \cdot f_s \cdot \varepsilon(I))^{-\beta} \left\{ h_0 + \int_{x_s}^{x_0} f^{-\beta}(x)\, dx \right\} \tag{6-44}$$

We now use Eq. (6-44) to study how the various factors in the lithography process will affect linewidth $W = 2x_0$: the exposure dose I, time t_B, and thermoprocessing temperature t_B that takes place prior to exposure; the time t_D and developing process temperature T_D, and the resist sensitivity.

The "sensitivity" of linewidth to developing time t_D is given by the relative derivative

$$\frac{1}{t_D}\frac{\delta t_D}{\delta x_0} = f^{-\beta}(x_0) \Big/ \left[h_0 + \int_{x_s}^{x_0} f^{-\beta}(x)\, dx \right] \equiv L \tag{6-45a}$$

$$\delta x_0 = \frac{1}{L}\frac{\delta t_D}{t_D} \tag{6-45b}$$

As we can see, the sensitivity $\delta x_0 \sim \delta t_D$ depends only on the properties of the resist (in terms of the β parameter) and on the properties of the optical system (in terms of the intensity decay function $f(x)$).

We can find the sensitivity of the linewidth x_0 to changes in the radiation dose I by differentiating Eq. (6-44) with respect to x_0 and I (in an approximation, $\varepsilon(I)$ = const)):

$$\frac{1}{I}\frac{\delta I}{\delta x_0} = L/\beta \rightarrow \delta x_0 = \beta\frac{1}{L}\frac{\delta I}{I} \tag{6-46}$$

As we can see, the sensitivity of the linewidth to changes in radiation dose is β times higher than to changes in the developing time t_D. This means that a radiation dose I must be controlled at least β times more stringently than the developing time is. According to [4], for the AZ-1370 resist, $\beta = 1.6$. During exposure in a projection system with a numerical aperture of NA = 0.28 by light at a wavelength of $\lambda = 436$ nm and coherence $\sigma = 0.7$ (these parameters define the light decay function $f(x)$, see Section 6-4) the value calculated from Eq. (6-45a) for a linewidth of $2x_0 = 1$ μm is $L = 2.5$. We find the accuracy of control on the developing time from Eq. (6-45b):

$$\frac{\delta t_D}{t_D} = L\frac{\delta x_0}{2x_0}(2x_0)$$

for 10% accuracy in linewidth control, $\delta x_0/x_0 = \pm 0.05$ (the total error on both sides of the line is $2\delta x_0 = 0.1$ μm), developing time must be controlled to within 12.5% accuracy; the radiation dose must be controlled β more stringently; 7.8%. Experiments have shown that constant $1/k$ in the formula for etching rate $r(E) = (1/k)E^{\beta}$ of an irradiated AZ-1370 resist depends on the developing temperature T_D according to Arrhenius' law:

$$\frac{1}{k} = A\exp(-H_D/RT_D) \tag{6-47}$$

Taking the logarithm of Eq. (6-44) with (6-47) for t_D and T_D, we easily find[4] that

$$\frac{1}{T_D}\frac{\delta T_D}{\delta x_0} = \frac{R}{H_D}L \rightarrow \delta x_0 = \frac{H_D}{R}\frac{1}{L}\frac{\delta T_D}{T_D} \tag{6-48}$$

A deviation of $\delta x_0 = \pm 0.05$ μm in a line 1 μm wide occurs when the developing temperature is controlled to within ± 2.1°C (for the AZ-1370 resist, $H_D = 10$ kcal/mole [4]).

Similarly, it has been observed in experiments that the constant k in the relation $r = (1/k)E^{\beta}$ is proportional to the duration of the thermal processing t_B prior to exposure (prebake) and, according to Arrhenius, depends on the thermoprocessing temperature T_B [4]:

$$k = At_B\exp(-H_B/RT_B) \tag{6-49}$$

With Eq. (6-49) we easily find that

$$\frac{1}{t_B}\frac{\delta t_B}{\delta x_0} = -L \rightarrow \delta x_0 = -\frac{1}{L}\frac{\delta t_B}{t_B} \tag{6-50}$$

$$\frac{1}{T_B}\frac{\delta T_B}{\delta x_0} = -\frac{R}{H_B}L \rightarrow \delta x_0 = -\frac{H_B}{L}\frac{\delta T_B}{T_B} \tag{6-51}$$

For 10% tolerance in the width of a line one micron wide ($2x_0 = 1$ μm) the tolerance on the thermoprocessing time for the resist is $\pm 12.5\%$ and on the thermoprocessing temperature is $\pm 6.5\%$ ($H_B = 4.8$ kcal/mole for the AZ-1370 resist; $T_B = 80°$C).[(4)]

In closing, we will consider how linewidth W_2 is affected by changes in resist sensitivity, defined as the dose I_s at which large areas of irradiated resist are completely removed. Under these conditions, we must assume in Eq. (6-44) that $f_s = 1$ and $I = I_s$:

$$t_D = k(I_s\varepsilon(I_s))^{-\beta}\left\{h_0 + \int_{x_s}^{x_0} f^{-f}(x)\,dx\right\} \tag{6-52}$$

Differentiating Eq. (6-52) with respect to I_s and x_0 gives us (assuming that $\varepsilon(I_s) = \text{const}$)

$$\frac{1}{I_s}\frac{\delta I_s}{\delta x_0} = -\frac{1}{\beta}L = \frac{1}{I}\frac{\delta I}{\delta x_0} \rightarrow \delta x_0 = -\frac{\beta}{L}\frac{\delta I_s}{I_s} \tag{6-53}$$

In practical situations the total deviation in linewidth is the sum of all the independent deviations:

$$(\Delta W_{\text{total}})^2 = 4\left[\left(\frac{\delta x_0}{\delta t_D}\Delta t_D\right)^2 + \left(\frac{\delta x_0}{\delta t_B}\Delta t_b\right)^2 + \left(\frac{\delta x_0}{\delta I}\Delta I\right)^2 + \left(\frac{\delta x_0}{\delta I_s}\Delta I_s\right)^2 + \left(\frac{\delta x_0}{\delta T_B}\Delta T_B\right)^2 + \left(\frac{\delta x_0}{\delta T_D}\Delta T_D\right)^2\right] + R^2 \tag{6-54}$$

where R^2 is the possible deviation due to additional factors that were not considered. It is apparent from what has already been said that a numerical estimate of linewidth variation requires a great deal of a priori information. Foremost, information about the etching rate in the exposed resist as a function of the absorbed energy E is needed, the thermoprocessing parameters T_B and t_B, the developing parameters T_D and t_D, the resist sensitivity I_s, and the radiation dose I along with the optical system parameters that define the intensity distribution in the mask image (NA, σ, the amount of defocussing, aberrations). It must be pointed out that for many resists, the amount of information that is actually available is far from being enough. This discussion of linewidth control in lithography processes here clearly demonstrates the nature of the information that is required and the direction for developing this subject further.

6-6. CHARACTERISTIC CURVES FOR A PHOTORESIST SUBJECTED TO PULSED EXPOSURE

A number of photoresists, sensibilized by diazonaphthoquinones (AZ-2400, AZ-2415, FP-617) display higher values of contrast γ and sensitivity when exposed by short ($\tau = 20$ nsec, $E_i \simeq 30$–50 mJ/cm^2) pulses of laser radiation than when exposed to continuous light (Fig. 6-6 [6]). The power density $P_i = E_i/\tau$ in laser radiation pulses is as much as $\sim 5 \times 10^6$ W/cm^2, which is 10^6–10^7 times higher than the P_c obtained by exposure to continuous light. According to [6], structural changes as well as chemical

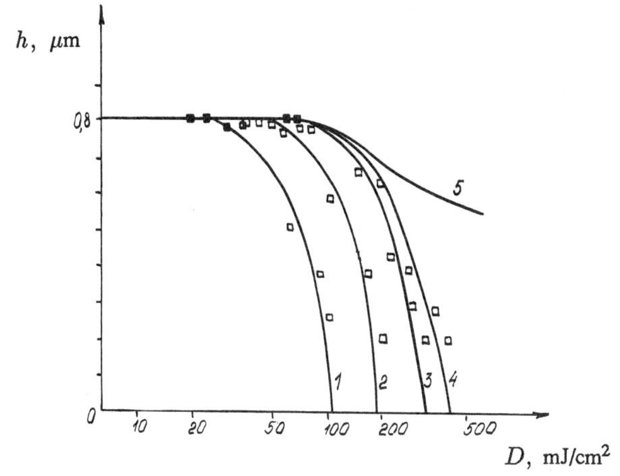

FIGURE 6-6. Experimental characteristic curves for pulsed laser exposure of an AZ-2400 resist. The doses D for curves 1, 2, 3, and 4 were obtained by exposing a resist 1, 2, 3, and 4 times, respectively. Curve 5 was obtained by exposing a resist to continuous light. The wavelength of the exposing radiation was $\lambda = 308$ nm (an excimer laser).

changes occur during pulse exposure. Those structural changes affect the etching process that occurs when the latent image in an exposed resist is developed.

The photochemical reactions that occur in the resists being studied are accompanied by the release of gaseous nitrogen. The typical time required for the gas molecules to diffuse from a resist film $h_0 = 1$ μm thick is $t_D = h_0^2/D \simeq 0.1$ sec, since $D \simeq 10^{-7}$ cm^2/sec. It is evident from this that a film exposed to continuous light for a period of time $t_e \gg t_D$ has no gaseous molecules or structural changes that can affect the etching process. The effect on etching, described by the equation $r(E) = kE^{\beta}$, is due only to chemical reactions. During pulse (laser) exposure gas molecules are formed suddenly, which causes an internal gas kinetic pressure. At a power density of $P_i \geq 10^7$ W/cm^2 material scattering (ablation) occurs. At preablation power densities of $P_i \leq 5 \times 10^6$ W/cm^2, the pressure of the gas being released is insufficient to cause material ablation (scattering); however, it is sufficient to cause preablation "agitation" and to create structural disruptions (micropores).(6) These micropores, at a certain concentration, form a solid percolation grid capable of penetrating an etchant in the body of the resist; the process of etching from a surface turns into a bulk process and the etching rate increases rapidly. Using a mathematical foundation to describe the phenomena discussed earlier,(6) we will find an equation for the characteristic curve of resists during pulse exposure.

We can assume that the gas kinetic pressure at some point in the body of the resist at a depth z is proportional to the absorbed energy density $E(z)$:

$$p(E(z)) = kE(z) = k\alpha I_s \exp(-\alpha z) \tag{6-55}$$

Here, I_s is the power density (W/cm^2) produced on the surface of the resist by a single pulse or pulse train during time $t_e \ll t_D$, and k is a constant proportional to the reaction yield of gas formation.

We will say that there is a threshold gas pressure p_c at which structural disruptions in a polymer are created, so that the concentration of structural disruptions (micropores) $\eta(E, z)$ is related to $p(E(z))$ by a relation of the type

$$\eta(E, z) = \varkappa(p - p_c)^q, \qquad p > p_c \tag{6-56}$$

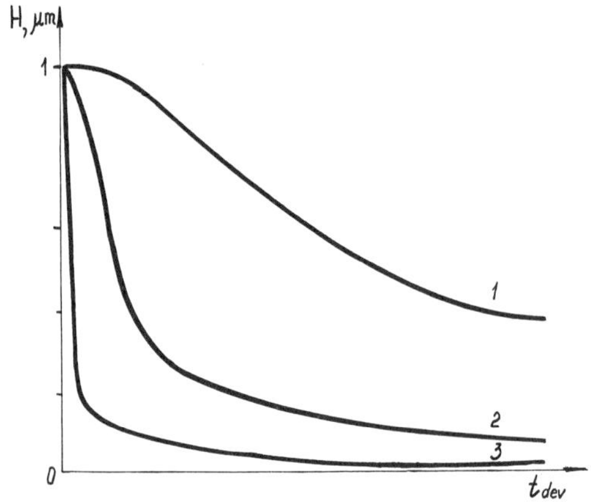

FIGURE 6-7. Kinetic curves obtained by modelling the etching in polymer resists exposed by pulsed radiation. Curves 1, 2, and 3 correspond to pulse energies of $E_1 < E_2 < E_3$. For curves 2 and 3, the upper layer of the resist, in which micropores were created, was etched more rapidly.

where $\varkappa$ and q are constants of the pore creation process. We may also assume that the pore concentration increases additively with the number i of laser pulses (in certain ranges of energy E_i and number of pulses i), since the process of pore creation is irreversible (the pores that are produced do not annihilate):

$$\eta(E_i, z, i) = i\eta(E_i, z, 1) = i\varkappa(p - p_c)^q \tag{6-57}$$

There is obviously a second threshold, namely a threshold pores concentration η_c such that when $\eta \gg \eta_c$ the pores are associated with an infinite percolation grid. If $\eta \geq \eta_c$ in the upper area of the resist layer and $\eta < \eta_c$ in the lower, thanks to the exponential decay of the light, then the upper area containing the connected pores grid will be etched more rapidly than will the lower part of the layer. Two-dimensional modelling of the process by which a solvent penetrates into a resist through the pores and the etching of the resist confirm these expectations (Fig. 6-7 [7]). Kinetic curves of $h(t)$ are shown in Fig. 6-7, where t is the developing (etching) time. A rapidly etching upper resist layer and a more slowly etching lower layer is seen at power densities that are close to the ablation threshold (curve 2). When the power density in a pulse is increased the thickness of the rapidly etching layer increases (curve 3).

If we define the characteristic curve of a resist, specifying a relatively short developing time t_d, such that during the time t_d we can etch only the upper, rapidly etching layer, then we can find the coordinates z^* of the boundary of the rapidly etching layer from the equation

$$\eta(E, z^*, i) = i\varkappa(p - p_c)^q = \eta_c \tag{6-58}$$

The characteristic equation for a resist under the indicated etching conditions follows from Eq. (6-58). Substituting the $p(E(z))$ from Eq. (6-55) and $z^* = h_0 - h$ into Eq. (6-58) yields

$$i\varkappa(k\alpha I_s e^{-\alpha(h_0 - h)} - p_c) = \eta_c \tag{6-59}$$

It was assumed for convenience's sake that $q = 1$. Solving Eq. (6-59) for h/h_0 we find an equation for the characteristic curve of a resist exposed in the preablation mode:

$$h/h_0 = 1 + \frac{1}{\alpha h_0} \ln \left[\frac{\eta_c + i\varkappa p_c}{k\alpha\varkappa i I_s} \right] \tag{6-60}$$

If we denote the complete radiation dose by i pulses as $D = iI_s$, and $D_{1i} = (\eta_c + i\varkappa p_c)/k\alpha\varkappa$, Eq. (6-60) becomes the familiar

$$h = h_0 - \frac{1}{\alpha} \ln(D/D_{1i}) \tag{6-61}$$

and the threshold dose D_{1i} when etching begins depends on the number i of pulses that constitute the dose D. Assuming that $h = 0$, we find from Eq. (6-61) that the sensitivity of a resist exposed by laser pulses in the preablation mode is

$$D_0 = (iI_s)|_{h=0} = \frac{\exp(\alpha h_0)}{k\alpha} \left(\frac{\eta_c}{\varkappa} + ip_c \right) \tag{6-62}$$

As we can see, resist sensitivity decreases when exposed by a large number of pulses (D_0 increases). Finally, from Eq. (6-61) we find the contrast in a resist during pulsed exposure:

$$\gamma = \left. \frac{\partial h}{h_0 \, \partial \ln D} \right|_{D_0} = -\frac{1}{\alpha h_0} \tag{6-63}$$

We can see that contrast is independent of the parameters that describe the etching rate; it is determined by the reduced thickness $\tilde{h} = \alpha h_0$ of the resist.

6-7. PHOTOETCHING OF POLYMER RESISTS BY VACUUM ULTRAVIOLET LIGHT

6-7-1. Experiments in Photoetching Polymer Resists

Experiments in observing photoetching in polymer resists exposed to vacuum ultraviolet light began in 1981 and it was suggested that this process be used to produce submicron resist masks.[8,9] Two factors associated with the wavelength (quantum energy E_ν) of the light promoted the process of photoetching by vacuum ultraviolet light: 1) the quantum energy $E_\nu \simeq 10$ eV is sufficient to disrupt almost all types of chemical bonds that can exist in polymers, and 2) the coefficient of absorption at these frequencies is so large ($\alpha \sim 10^5$ cm^{-1}) that the radiation is absorbed by a polymer surface layer only $\delta = \alpha^{-1} \approx 0.1$ μm thick. Photolysis of the molecules in a polymer layer δ thick causes low-molecular-weight fragments to diffuse to the surface of the polymer film and escape into the surrounding medium (gas or vacuum).

We will discuss the fundamental experimental relations in the process of photoetching polymethyl methacrylate (PMMA).[10] A fundamental dependence exists on the atmosphere in which the etching surface is located: in an oxygen atmosphere (at pressures of $p \geq 10$ kPa) the etching rate is constant, whereas in a vacuum or an inert atmosphere the etching rate decreases over time according to $u(t) = u_1/(1 + t/\tau)$, where τ is a

TABLE 6-1. Initial Quantum Yields for the Products of Photolysis of PMMA in a Vacuum by Light Having Wavelengths of 147 and 123.6 nm

Products	Quantum yield	
	147 nm	123.6 nm
CO	0.1 ± 0.02	0.09 ± 0.02
CO_2	0.06 ± 0.01	0.055 ± 0.01
CH_4	–	0.025 ± 0.004
H_2	0.02 ± 0.002	–
Methyl formate	0.09 ± 0.02	0.055 ± 0.01
Formaldehyde	0.17 ± 0.04	0.125 ± 0.03
Methanol	0.04 ± 0.01	0.035 ± 0.001
Methyl methacrylate	< 0.0001	< 0.0001
Double bonds (C=C type)	0.3 ± 0.06	0.4 ± 0.08

constant. The change in resist layer thickness when etching in an oxygen atmosphere obeys the linear equation

$$h(t) = h_0 - u_1 t \tag{6-64}$$

and when etching in a vacuum (an inert atmosphere) obeys the logarithmic equation

$$h(t) = h_0 - u_1 \tau \ln(1 + t/\tau) \tag{6-65}$$

The decrease in the etching rate $u(t)$ when etching in a vacuum (an inert atmosphere) attests to the chemical change in the surface layer in which the radiation-chemistry reactions occur. Etching relations that are close to those given by Eq. (6-65) are seen when PMMA is etched with soft x-rays and with electron and ion beams in a vacuum.[11–13] We might expect that the change in the surface layer of a resist is in every case caused by roughly the same factors: depletion of the polymer molecules by the lateral low-molecular-weight fragments, accumulation of medium-molecular-weight bonds, binary C=C bonds, etc. This changes the density, average molecular mass, and optical properties of the polymer, and, what is important, the radiation yield of the products of molecule destruction. A theory was constructed in [14] for the radiation change in a polymer that leads to the etching equation, Eq. (6-65). This theory is based on equations in polymers. The constants u_1 and τ in Eq. (6-65) are given in terms of the reaction constants (the cross sections) that change the polymer and, because of the diffusion mechanism of etching, in terms of the coefficient of diffusion D for low-molecular-weight fragments and the film thickness h.

Obviously, the logarithmic relation given by Eq. (6-65) for etching in a vacuum (an inert atmosphere) makes these processes inefficient for etching through a resist layer and creating a resist mask at resist thicknesses of $h \approx 1$ μm. At the same time, vacuum alteration processes are extremely useful if the alteration is done after the resist mask is formed: altered resist masks show more stability when structural layers are plasma etched through the resist mask.[13,15] Table 6-1 shows the quantum yields for the products of PMMA photolysis in a vacuum by light at a wavelength of λ = 147 nm (light from a xenon resonance lamp) and 123.6 nm (light from a krypton resonance lamp). The data in Table 6-1 relate to the initial stage in etching when the layer is unaltered.

A linear etching relation, Eq. (6-64), is seen when photoetching resists with vacuum ultraviolet light at wavelengths of λ = 110–220 nm in an oxygen atmosphere at pressures

$p \gtrsim 50$ torr. The photoetching rate is proportional to the radiation intensity u_1 and depends on the radiation wavelength. (When a wide range of the spectrum is used, the magnitude must be understood to be $u_1 = \sum_\lambda u_\lambda$). The oxygen atmosphere may play a dual role in photoetching. First, the products from the photolysis of gaseous oxygen can diffuse into the irradiated layer of the polymer and react with the polymer radicals that were formed in the photodestruction reactions, preventing the knitting processes, etc. that alter the polymer.[14] This process prevents the polymer from being altered during photoetching under stationary conditions. Second, chemical etching from the surface by the chemically active particles that were formed during vacuum ultraviolet photolysis of oxygen (O, O_2 in excited states, and ozone O_3) is possible. We might expect that the surface (chemical) etching mechanism will dominate at elevated oxygen pressures and when the distance H between the mask and the surface of a polymer impregnated with oxygen is large, when most of the light $k_{O_2} = 1 - \exp[-\sigma \cdot H \cdot [O_2]]$ is absorbed by the oxygen. Here, σ is the radiation absorption cross section for O_2, $[O_2]$ is the concentration of O_2 molecules per unit volume, and H is the thickness of the oxygen layer. Conversely, at low pressures and small gaps, when $\sigma[O_2]H \ll 1$, the bulk mechanism will dominate and most $\exp[-\sigma[O_2]H]$ of the light passes through the oxygen layer and is absorbed in the bulk of the polymer.

The etching rate equation $r_v = r_s$ defines the boundary between two ranges of parameters P_{O_2} and H in which the bulk and surface polymer etching mechanisms, respectively, dominate.

Figure 6-8 shows the experimental curves of $h(t)$ for PMMA. Etching was observed when the PMMA was exposed to light from a VMF-25 hydrogen lamp. The radiation intensity of the lamp in the range $\lambda = 110\text{–}220$ nm is $\sim 10^{15}$ quanta/cm$^2 \cdot$ sec and the amount of energy distributed in the 110–135, 137–170, and 170–220 nm subgroups are 23%, 46%, and 9%, respectively. We will define the sensitivity of a resist as the dose D_0 needed to etch the entire resist layer:

$$D_0 = I\tilde{E}_\nu \cdot t_e$$

Assuming that $I = 10^{15}$ quanta/cm$^2\cdot$sec, $t_e = 3 \times 10^3$ sec, and $\tilde{E}_\nu = 10^{-18}$ J, we find that $D_0 = 3$ J/cm^2. Thus, we see by comparing the sensitivity of photoresists in processes that use subsequent liquid developing that the sensitivity in photoetching processes is 10–100 times less. Approximately the same relation holds true for the sensitivity in x-ray lithography (for the purpose of etching during the exposure process) and in electron- and ion-beam lithography.

Table 6-2 shows the quantum yields of the products of PMMA photolysis when O_2 is present ($P_{O_2} = 6$ kPa, layer thickness is $H = 0.4$ mm) at wavelengths of 147 and 123.6 nm.[10] The basic products of PMMA photolysis are CO_2 and formaldehyde, H_2CO. The quantum yield increases as the wavelength of the light decreases and is equal to 0.77 at $\lambda = 123.6$ nm.

6-7-2. *The Theory of Bulk (Diffusive) Photoetching in Polymers*

We designate the initial layer thickness as h_0. For simplicity, we assume that photoetching is associated with the production, diffusion, and conversion into the gaseous phase of only a single type of volatile fragments and that the alteration process is absent (etching in an O_2 atmosphere).

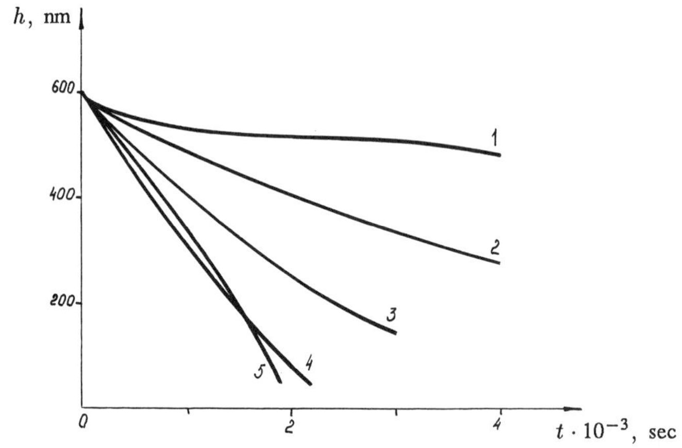

FIGURE 6-8. Etching kinetics of PMMA exposed to light from a VMF-25 hydrogen lamp: in a vacuum (1) and in oxygen at pressures of: 2) 0.4 kPa; 3) 1.3 kPa; 4) 4 kPa; and 5) 13 kPa.

TABLE 6-2. Quantum Yields of the Products of PMMA Photolysis by Light at Wavelengths of 147 and 123.6 nm in the Presence of Oxygen (oxygen pressure, 6 kPa; photolysis time, 2 h; distance between lamp window and polymer film, 0.4 mm)

Products	Quantum yield	
	147 nm	123.6 nm
CO_2	0.22 ± 0.02	0.45 ± 0.04
H_2CO	0.12 ± 0.02	0.2 ± 0.04
Methyl formate	0.03 ± 0.006	0.05 ± 0.001
CH_3OH	0.012 ± 0.002	0.03 ± 0.04
H_2O	0.02	0.04
CO	?	?

We write the kinetic equation for the concentration $n(x,t)$ of the volatile fragments:

$$\frac{\partial n(x,t)}{\partial t} = D\frac{\partial^2 n(x,t)}{\partial x^2} + G_1(x,t) - G_2 \cdot n(x,t) \tag{6-66}$$

where $G_1(x,t) = g \cdot \alpha \cdot I(x,t)$ is the "birth" rate of the volatile fragments, α is the coefficient of radiation absorption, $I(x,t)$ is the radiation intensity at the point x at time t, g is the quantum yield of fragments per unit of absorbed energy, G_2 is the recombination (mortality) coefficient for the volatile fragments with the macromolecules, and D is the coefficient of fragments diffusion. The initial condition for Eq. (6-66) is $n(x,t) = 0$. The boundary condition at the interface between the resist and the gaseous medium follows from the continuity of fragment flow from the resist into the gaseous medium:

$$D\left.\frac{\partial n(x,t)}{\partial x}\right|_{x=h(t)} = v_0 \cdot n(x,t)|_{x=h(t)} \tag{6-67}$$

where v_0 is the average rate at which the fragments escape the film surface. Fragments do not flow through the resist–substrate interface:

$$D \left. \frac{\partial n(x,t)}{\partial x} \right|_{x=0} = 0 \tag{6-68}$$

The rate at which the etching surface moves is governed by the rate at which the polymer mass decreases due to volatile fragments going into the gaseous medium:

$$u(t) = -\frac{dh(t)}{dt} = \left(\frac{m}{\rho}\right) v_0 n(x,t) \bigg|_{x=h(t)} \tag{6-69}$$

where m is the fragment mass and ρ is the polymer density. Solving Eq. (6-66) with the initial and boundary conditions defines $n(x,t)$, and, according to Eq. (6-69), the etching rate $u(t)$ in a polymer film.

A nonstationary solution of Eq. (6-66) was given in [16]. We will examine only a quasistationary solution here; $\partial/\partial t = 0$ and u_{st} = const correspond to the established etching mode during time $t_D \ll t \ll h_0/u_{st}$, where $t_D = h_0^2/D$ is the time needed to establish a quasistationary distribution $n(x,t)$ in the film. Quasistationary solutions of Eq. (6-66) can be found(11,17,18) in an approximation of a step function of radiation intensity in a film:

$$I(x,t) = \begin{cases} I_0 & h(t) - \delta \le x \le h(t) \\ 0 & x < h(t) - \delta \end{cases} \tag{6-70}$$

where $\delta = \alpha^{-1}$, and α is the coefficient of light absorption in the polymer.

We will present the final formula for the rate of a stationary polymer etching process. This formula was obtained by solving Eq. (6-66) according to Eq. (6-69) (under the constraint that $u_{st} \ll \sqrt{D \cdot G_2}$):

$$u_{st} = u_0 \frac{2\sinh(\delta/2l_D)\cosh[(h_0 - \delta/2)/l_D]}{\cosh(h_0/l_D) + (\sqrt{DG_2}/v_0)\sinh(h_0/l_D)} \frac{l_D}{\delta} \tag{6-71a}$$

$$u_0 = \frac{m}{\rho} G_{10}\delta \tag{6-71b}$$

Here, $l_D = \sqrt{D/G_2}$ is the diffusion length, $G_{10} = g\alpha I_0$, and m is the average mass of the volatile fragments. It is obvious that $mG_{10}\delta = mg\alpha I_0\delta = m_F$ in Eq. (6-71b) is the total mass of all the volatile fragments produced per unit time in a layer $\delta = \alpha^{-1}$ thick and u_0 is the maximal etching rate under the condition that all of the fragments produced escape into the gaseous medium. The multiplier on u_0 in Eq. (6-71a) is the fraction of the fragments that escape into the gaseous medium, less the number that recombine during the diffusion process. For a semiinfinite layer $h_0/l_D \to \infty$, we find from Eq. (6-71a) that

$$u_{st}(h_0/l_D \to \infty) = u_0 \frac{l_D}{\delta} \frac{v_0}{v_0 + \sqrt{D \cdot G_2}} \cdot [1 - \exp(-\delta/l_D)] \tag{6-72}$$

In actual situations we can consider the inequality $v_0 \gg \sqrt{D \cdot G_2}$ to be satisfied, since v_0 is on the same order of magnitude as the thermal velocity of the fragments in the gaseous medium. The correction in Eq. (6-72) is then defined by the relationship that exists between the diffusion length l_D and the radiation absorption length δ. When $l_D \gg \delta$, all of the fragments produced escape from the polymer and $u_{st} = u_0$. When $l_D \ll \delta$, only the fragments produced in a layer l_D thick escape from the polymer and

$$u_{st} = u_0 l_D/\delta = (m/\rho)G_{10}l_D \tag{6-73}$$

The situation in which $l_D \gg \delta$ or $l_D \ll \delta$ can be identified by the temperature dependence of the etching rate. If we assume that the quantum yield of fragments g (which means G_{10}) is not strongly temperature dependent, then when $l_D \gg \delta$ the etching rate need not depend on T; when $l_D \ll \delta$ we can expect an exponential dependence on $u_{st} \sim \exp(-E_D/2kT)$, where E_D is the energy barrier for the diffusion of fragments in the polymer.

Notice that when $l_D \gg \delta$, the polymer etching rate is independent of the coefficient of light absorption α:

$$u_{st} = u_0 = (m/\rho)gI_0 \tag{6-74}$$

We will estimate the etching rate according to Eq. (6-74) and assume that $m = 44$ amu (CO_2), and $\rho = 1.2$ g/cm^3, $g = 0.5$, $I_0 = 10^{15}$ quanta/cm$^2 \cdot$ sec (a hydrogen lamp in the experiments in [10]). Then, $u_0 = 0.3$ nm/sec. The measured value of u_{st} for PMMA is 0.25 nm/sec.(10) Apparently, when etching polymers with vacuum ultraviolet light, when $\delta = \alpha^{-1} < 0.1\ \mu$m, the inequality $l_D \gg \delta$ is satisfied. When etching polymers with soft x-rays, when $\delta > 1\ \mu$m, the condition $l_D < \delta$ may be achieved.

6-7-3. Photoinduced Chemical Etching in Polymers in an Oxygen Atmosphere

When studying the process of chemical etching in a polymer film by the products of photolysis of gaseous oxygen, we must first establish the concentration of active particles in the gaseous phase. We will examine the photolysis of gaseous oxygen in a slot H wide that separates two glass wafers. The chemical equations for the concentration n_i of the products of photolysis are(19)

$$\frac{\partial n_i}{\partial t} = -\text{div}(n_i\mathbf{u} - D_i\nabla n_i) + Q_i(\{n\}, I) \tag{6-75}$$

Here, D_i is the coefficient of diffusion of i-type products in the gas, $\mathbf{u}$ is the gas dynamic flow rate of the oxygen (the scavenging rate), I is the radiation density, and Q_i is a term that takes particle generation and annihilation into account, including the heterogeneous reactions that occur on the reaction vessel walls.

Computer methods were used in [19] to solve Eq. (6-75) and find the concentrations n_i and flows $j_i = n_i v_i$ of the active particles that form during the photolysis of oxygen; the chemical etching rate U_{ch} in a polymer due to these particles can be written

$$U_{ch} = \sum_i (m_i/\rho)\alpha_i j_i \tag{6-76}$$

where m_i is the average mass of the volatile fragments generated by i-type particles that attack the polymer, and α_i is the total probability of the channels of interaction between the particles and the surface that will produce volatile products. We will assume that α_i is a function of the radiation intensity on the surface being etched, i.e., there are

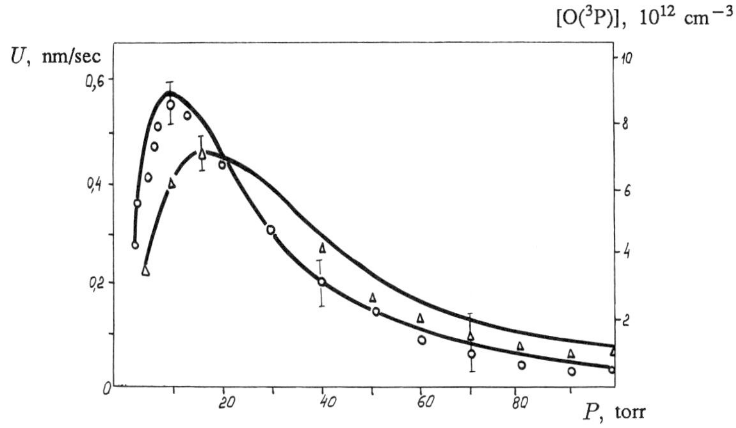

FIGURE 6-9. Experimental dependencies of the rate of etching of PMMA irradiated by a KSR-2 lamp on the oxygen pressure at H = 0.9 mm (triangles) and H = 1.8 mm (circles) and calculated dependencies of the concentration of $O(^3P)$ (solid curves).

radiation-induced channels for the production of volatile fragments; this is taken into consideration by the second term in the following formula:

$$\alpha_i = \alpha_{i0}[1 + a_i I_0 \exp(-\sigma H[O_2])] \tag{6-77}$$

Here, $\exp(-\sigma H[O_2])$ is the amount of radiation that passes through a gaseous layer H thick, $[O_2]$ is the oxygen content in the gaseous layer, and σ is the cross section of light absorption.

Taking Eq. (6-77) into account, the radiation-induced "chemical" etching in a polymer can be written

$$U_{ch} = \sum_i (m_i/\rho)\alpha_{i0}[1 + a_i I_0 \exp(-\sigma H[O_2])] v_i n_{is} \tag{6-78}$$

where n_{is} are the concentrations of particles near the surface being etched and are a function of the $[O_2]$ and H parameters.

The experimentally observed functions $U_{ch}([O_2], H)$ are best explained by Eq. (6-78) if for n_{is} we use the concentration of atomic oxygen in the ^3P state (Fig. 6-9 [19]). Curves 1 and 2 are graphs of the concentrations of $[O(^3P)]$ atomic oxygen in the ^3P state as a function of the oxygen pressure P_{O_2} and the points indicate the measured polymer etching rates for two values of gas layer thickness (H = 0.9 and 1.8 mm); a xenon lamp provided the radiation. (The $[O(^3P)](P_{O_2})$ and $U_{ch}(P_{O_2})$ curves were adjusted by equating their magnitudes at their maxima on the curves.) From what has been said, we can conclude that for oxygen pressures of P = 20–50 torr and a gas layer thickness $H \geq 1$ mm, that atomic oxygen in the ^3P state plays the dominant role in etching PMMA polymers. Equation (6-78) explains still more properties of photoetching polymers, the shift in the location of the maximum in the $U_{ch} \sim P_{O_2}$ curve when the layer thickness H is changed, etc.(19)

In conclusion, we present a photograph of a resist mask profile that we obtained in our laboratory by etching PMMA in an oxygen atmosphere (Fig. 6-10).

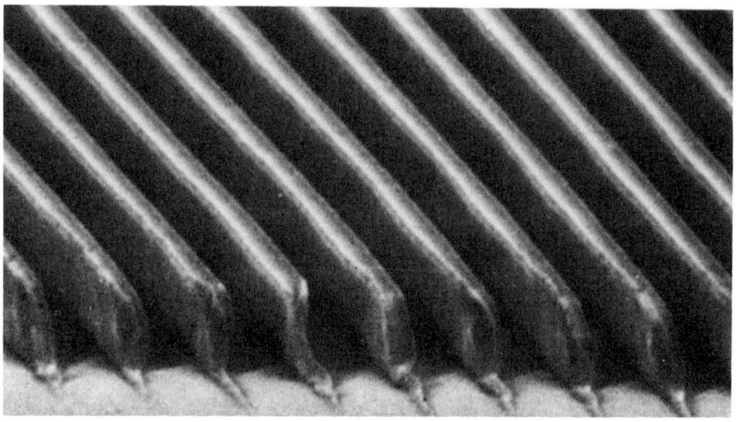

FIGURE 6-10. Profile of a resist mask obtained by photoetching PMMA in an oxygen atmosphere.

6-8. PATTERN FORMATION IN POLYMER LAYERS BY PULSED LASER ABLATION

The first studies of the removal (ablation) of polymer films by pulses of excimer laser radiation were done in 1982[20,21] and had the goal of producing polymer masks with submicron dimensions. Laser-pulsed ablation of metallic films had previously been used in fabricating metallized masks and electronic components.[22] The use of pulsed radiation on polymer films to produce optical disks for one-time recording has attracted a great deal of attention.[23,24] Various materials have been used in experiments with laser photoablation: PMMA,[25–28] polyamide,[29,31] polyacetylene,[32] poly(α-methyl styrene),[26] polystyrene,[28] poly(ethylene terephthalate),[33,34] silicon organic polymers,[35] and biological fabrics.[36]

Pulsed photoablation of a polymer film is the result of heating in the film due to radiation absorption, thermal and photodecomposition of the polymer, and evaporation of the decomposed material. A simplified model for the photoablation of polymer films has been developed in [33,37,38] and a theory that describes how a profile is created in a polymer film following photoablation from a Gaussian laser pulse (the process of one-time recording on an optical disk) was given in [24].

Photoablation of polymer films is seen at wavelengths from 193 to 351 nm; energy density thresholds ϕ in a laser pulse range from a few tens to hundreds of mJ/cm^2; the absorbed energy density is $E = \alpha \cdot \phi = 3 \times 10^2–5 \times 10^4$ J/cm^3,[27,30] and α is the coefficient of light absorption in a polymer material. Photodecomposition occurs because of absorbed light energy and because of heating and thermal decomposition of the polymer molecules; the gas kinetic pressure in the body in which absorption occurs grows to magnitudes of 100 atmospheres (10^7 Pa), which expels the products of destruction from the polymer film at supersonic velocities.[39]

Mass spectroscopy studies of the products of destruction show that the volatile products of ablation are not simple fragments of the molecules in the original polymer; during ablation, high-speed reactions between the products of destruction are possible.[9]

Following in the manner of [14], we will characterize the thermophysical properties of a polymer by the threshold energy necessary to sublimate a unit volume of the polymer:

$$E_{th} = H + \int_{T_0}^{T_c} \rho C_p \, dT \tag{6-79}$$

where H is the heat of conversion into the gaseous state per unit volume, calculated by taking the difference between the energy needed to form chemical bonds in the polymer molecules and energy of the gaseous products that form during photoablation, T_0 is the initial temperature, T_c is the critical temperature at which thermal decomposition in the layer due to light absorption occurs, and ρ and C_p are, respectively, the density and specific heat of the polymer.

If the radiation loss function $d\phi/dx = E(x)$ is known, where $E(x)$ is the absorbed energy density at a depth x, then the thickness Δh_1 of the expelled portion of the polymer layer is found from the condition that $E(x_1) = E_{th}$, and

$$\Delta h_1 = x_1 = E^{-1}(E_{th}) \tag{6-80}$$

If the luminous flux density decreases according to $\phi(x) = (1 - R)\alpha\phi_0 \exp(-\alpha x)$, where R is the coefficient of reflection, we find that

$$\Delta h_1 = \frac{1}{\alpha} \ln\left(\frac{E_s}{E_{th}}\right) \tag{6-81}$$

where $E_s = (1 - R)\alpha\phi_0$ is the density of the energy absorbed at the surface of the irradiated film. Equation (6-81) is satisfied, e.g., when photoablating polyethylene terephthalate with XeCl radiation from an excimer laser (λ = 308 nm).[33] Photoablation by a train of pulses when thermal carryover from the adjacent (in time) pulses occurs was examined in [37]; the heat generated by a pulse is retained in the polymer when the next pulse begins to act.

A more thorough examination of the polymer film profile that remains after local ablation showed that the shape of the profile is governed by the viscous flow of material under the influence of reactive pressure caused by ablation expulsion.[24] The forces of surface tension play a subordinate role. The shape of the profile of polymer resist masks produced by laser photoablation have as yet been studied only superficially. The potential of this method for producing submicron resist masks remains to be clarified.

REFERENCES

1. P. D. Plais, *Solid State Technol.*, No. 8, 76–85 (1977).
2. N. K. Eib, *J. Vac. Sci. Technol.*, **B3**, No. 8, 425–428 (1985).
3. D. Kyser and R. Pyle, *IBM J. Res. Develop.*, **24**, 426 (1980).
4. M. P. C. Watts, *J. Vac. Sci. Technol.*, **B3**, No. 1, 434–440 (1985).
5. W. Scot Ruska, *Microelectronic Processing*, McGraw-Hill, New York (1987).
6. A. L. Bogdanov, K. A. Valiev, L. V. Velikov, and D. Yu. Zaroslov, Proceedings of the 30th Intern. Symp. on Electron, Ion, and Photon Beams, New York (1987), pp. 391–395.
7. A. L. Bogdanov, A. A. Polyakov, K. A. Valiev, and L. V. Zaroslov, *Proc. SPIE*, **771**, 167–172 (1987).
8. K. A. Valiev, L. V. Velikov, S. D. Dushenkov, A. V. Mitrofanov, and A. M. Prokhorov, *Pis'ma Zh. Tekh. Fiz.*, **8**, No. 1, 33–36; *op. cit.* 48–53 (1982).
9. N. Ueno, S. Nonishi, K. Tanimoto, and K. Sugita, *Jpn. J. Appl. Phys.*, **20**, No. 10, L709–L712 (1981).

10. K. A. Valiev, L. V. Velikov, Yu. I. Dorofeev, A. S. Kramarenko, V. E. Skurat, and V. L. Tal'roze, *Poverkhnost'*, No. 6, 86–95 (1985).
11. K. A. Valiev, L. V. Velikov, T. M. Makhviladze, and A. M. Prokhorov, in: *Problems in Solid State Physics*, A. M. Prokhorov and A. S. Prokhorov (editors) [in Russian], Mir, Moscow (1984), pp. 291–322.
12. K. A. Valiev, L. V. Velikov, and S. D. Dushenkov, *Poverkhnost'*, No. 5, 73–79 (1985).
13. K. A. Valiev, V. A. Danilov, A. V. Rakov, and A. G. Shchuchkin, *Mikrolektronika*, **13**, No. 2, 167–169; *op. cit.* **12**, No. 3, 195–199 (1984).
14. K. A. Valiev, T. M. Makhviladze, and M. E. Sarychev, Preprint IOFAN, No. 291, Moscow (1986).
15. A. V. Rakov and A. G. Shchuchkin, *Trudy IOFAN*, **8**, 87–100 (1987).
16. T. M. Makhviladze, E. G. Panteleev, M. E. Sarychev, and A. B.Khorev, *Trudy IOFAN*, **8**, 40–60 (1987).
17. K. A. Valiev, L. V. Velikov, and T. M. Makhviladze, *Poverkhnost'*, No. 12, 99–112 (1983).
18. K. A. Valiev and T. M. Makhviladze, *Poverkhnost'*, No. 3, 57–67 (1985).
19. K. A. Valiev, T. M. Makhviladze, and S. B. Pekarchuk, Preprint IOFAN, No. 360, Moscow (1986).
20. Y. Kawamura, K. Toyoda, and S. Namba, *Appl. Phys. Lett.*, **40**, 374 (1982); *J. Appl. Phys.*, **53**, 6489 (1982).
21. R. Srinivasan and V. Maine-Banton, *Appl. Phys. Lett.*, **41**, 576 (1982).
22. M. I. Cohen, "Materials processing" in: *Laser Handbook*, F. Arecchi and E. Schutz-Dubois (editors), North Holland, Amsterdam (1972).
23. G. Bouwhuis, J. Braat, A. Huijser, J. Pasman, G. Van Rosmalen, and Immink K. Schouhamer, *Principles of Optical Disk Systems*, Hilger, Bristol (1985).
24. D. A. Hill and D. S. Soohg, *J. Appl. Phys.*, **61**, No. 6, 2132–2139 (1987).
25. R. Srinivasan, *J. Vac. Sci. Technol.*, **B1**, No. 4, 923–926 (1983).
26. M. C. Burrell, Y. S. Liu, and H. S. Cole, *J. Vac. Sci. Technol.*, **A4**, No. 6, 2459–2462 (1986).
27. G. M. Davis and M. C. Gover, *J. Appl. Phys.*, **61**, No. 5, 2090–2092 (1987).
28. R. Larciprete and N. Stuke, *Appl. Phys.*, **B42**, 181–184 (1987).
29. G. Koren and J. J. C. Yeh, *Appl. Phys. Lett.*, **44**, No. 12, 1112–1114 (1984).
30. R. Srinivasan, B. Braren, and R. W. Dreyfus, *J. Appl. Phys.*, **61**, No. 1, 372–376 (1987).
31. G. Koren, *Appl. Phys. Lett.*, **50**, No. 16, 1030–1032 (1987).
32. M. Golombok, M. C. Gower, S. J. Kirbi, and P. T. Rumsby, *J. Appl. Phys.*, **61**, No. 3, 1222–1224 (1987).
33. J. E. Andrew, P. E. Dyer, D. Forster, and P. H. Key, *Appl. Phys. Lett.*, **43**, No. 8, 717–719 (1983).
34. S. Lazare, J. C. Sculignac, and P. Fragnaud, *Appl. Phys. Lett.*, **50**, No. 10, 624–625 (1987).
35. E. E. Marinero and R. D. Miller, *Appl. Phys. Lett.*, **50**, No. 16, 1041–1043 (1987).
36. F. W. Cross, R. K. Al-Dharir, P. E. Dyer, and A. J. MacRobert, *Appl. Phys. Lett.*, **50**, No. 15, 1019–1021 (1987).
37. K. A. Valiev and T. M. Makhviladze, *Mikrolektronika*, **14**, No. 4, 301–307 (1985).
38. E. Sutcliffe and R. Srinivasan, *J. Appl. Phys.*, **60**, No. 9, 3315–3322 (1986).
39. P. E. Dyer and R. Srinivasan, *Appl. Phys. Lett.*, **48**, 1986 (1986).

Index